Dynamics and Characterization of Marine Organic Matter

Ocean Sciences Research (OSR) Volume 2

The titles in this series are listed at the end of this volume.

Cover figures: (Left) deployment of a sediment trap; (right) red tide of blue-green algae in the Coral Sea.

Dynamics and Characterization of Marine Organic Matter

Edited by

Nobuhiko HANDA
Faculty of Information Science and Technology, Aichi Prefectural University, Aichi-gun, Japan

Eiichiro TANOUE
Institute for Hydrospheric-Atmospheric Sciences, Nagoya University, Nagoya, Japan

Takeo HAMA
Institute of Biological Sciences, Tsukuba University, Tsukuba, Japan

Terra Scientific Publishing Company, Tokyo

Kluwer Academic Publishers, Dordrecht, London, Boston

C.I.P. Catalogue record for this book is available from the Library of Congress.

ISBN 0-7923-6293-4 (Kluwer)

Published by Terra Scientific Publishing Company, 2003 Sansei Jiyugaoka Haimu, 27-19 Okusawa 5-chome, Setagaya-ku, Tokyo 158-0083, Japan / Kluwer Academic Publishers, P.O. Box 17, 3300 AA Dordrecht, The Netherlands.

Kluwer Academic Publishers incorporates
the publishing programmes of
D. Reidel, Martinus Nijhoff, Dr W. Junk and MTP Press.

Sold and Distributed in the U.S.A. and Canada
by Kluwer Academic Publishers,
101 Philip Drive, Assinippi Park, Norwell, MA 02061, U.S.A.
in Japan by Terra Scientific Publishing Company (TERRAPUB),
2003 Sansei Jiyugaoka Haimu, 27-19 Okusawa 5-chome, Setagaya-ku,
Tokyo 158-0083, Japan.

In all other countries, sold and distributed
by Kluwer Academic Publishers,
P.O. Box 322, 3300 AH Dordrecht, The Netherlands.

(This book is published by Grant-in-Aid publication of Scientific Research Result of the Ministry of Education, Science and Culture of Japan.)

Printed in Japan

Preface

Over the past decade the scientific activities of the Joint Global Ocean Flux Study (JGOFS), which focuses on the role of the oceans in controlling climate change via the transport and storage of greenhouse gases and organic matter, have led to an increased interest in the study of the biogeochemistry of organic matter. There is also a growing interest in global climate fluctuations. This, and the need for a precise assessment of the dynamics of carbon and other bio-elements, has led to a demand for an improved understanding of biogeochemical processes and the chemical characteristics of both particulate and dissolved organic matter in the ocean.

A large amount of proxy data has been published, describing the changes of the oceanic environment, but qualitative and quantitative estimates of the vertical flux of (proxy) organic compounds have not been well documented. There is thus an urgent need to pursue this line of study and, to this end, the present book starts with several papers dealing with the primary production of organic matter in the upper ocean. Thereafter, the book goes on to follow the flux and characterization of particulate organic matter, discussed in relation to the primary production in the euphotic zone and resuspension in the deep waters, including the vertical flux of proxy organic compounds. It goes on to explain the decomposition and transformation of organic matter in the ocean environment due to photochemical and biological agents, and the reactivity of bulk and specific organic compounds, including the air–sea interaction of biogenic gases.

The 22 papers in the book reflect the interests of JGOFS and will thus serve as a valuable reference source for future biogeochemical investigations of both bio-elements and organic matter in seawater, clarifying the role of the ocean in global climate change.

All papers deal with the authors' current work in progress and all have been refereed and revised in accord with the comments received. We would like to express our thanks to all of the contributors for their patience in waiting for this publication. We also thank the reviewers who kindly gave the authors critical and helpful suggestions and comments. Our thanks are also due to the Ministry of Education, Science, Sport and Culture for the financial support they gave to this publication.

Finally, the book was primarily planned as volume to commemorate the retirement of one of the editors (N.H.) from Nagoya University after 35 years of service to education and research in the biogeochemistry of organic matter.

September 18, 1999

N. Handa, E. Tanoue and T. Hama

Contents

Chapter 5. Air-Sea Interactions of Biogenic Gases and Organic Matter

Dynamics and Characterization of Marine Organic Matter, Eds., N. Handa, E. Tanoue and T. Hama, pp. 1–38.

Production and Turnover of Organic Compounds through Phytoplankton Photosynthesis

Takeo HAMA*

Institute for Hydrospheric-Atmospheric Sciences, Nagoya University, Chikusa-ku, Nagoya 464-8601, Japan

(Synthesis, Received 15 November 1996; accepted 31 January 1997)

INTRODUCTION

Primary production of phytoplankton in the surface layer of the ocean down to 200 m depth harvests the sunlight and converts light energy to chemical energy as chemical bonds of organic compounds. The life of marine heterotrophs including zooplankton, bacteria, fish and mammals depends on the organic compounds produced by phytoplankton. Through biological activity, organic compounds are partly decomposed, transformed and transported to the deep layer. The scale of primary productivity in the euphotic layer accordingly regulates the amount of biomass and accompanying material flux throughout the water column including the sediment.

Although the number of chemical species of inorganic carbon dissolved in sea water such as carbon dioxide, carbonic acid, bicarbonate and carbonate is limited, the number of organic compounds produced by phytoplankton photosynthesis are incomparably greater. The difference in the constituent elements, type of bonds between the elements, and molecular weight results in an infinite variety of chemical and biological characteristics of organic compounds.

Among parameters inherent in the biogeochemical cycle in aquatic environments, the lability of organic compounds is one of the most important. A refractory compound may not be easily decomposed through the food web in the surface layer and may survive and be transported to deeper layers. A labile one, on the contrary, may be rapidly decomposed into inorganic compounds within the surface layer. Thus, the composition of organic matter produced by phytoplankton can have a great effect on the biogeochemical cycle in aquatic environments because organic matter produced by phytoplankton is the starting point for the flow of organic matter in the ocean.

The composition of cellular organic matter of phytoplankton and particulate matter was first studied on an elemental composition level, such as carbon and nitrogen. Major components constituting the organic matter of phytoplankton

* Present address: Institute of Biological Sciences, University of Tsukuba, Tsukuba, Ibaraki 305-8572, Japan.

and particulate organic matter (POM), including proteins/amino acids, carbohydrates and lipids, were also determined by colorimetric and fluorometric methods; subsequently, the detailed composition of monomers such as amino acids, monosaccharides and fatty acids were estimated by gas and/or liquid chromatography. The analysis of the photosynthetic products has been developed from elemental level to monomer level as well as the composition of cellular organic matter and POM.

Meanwhile, determination of the chemical structure of an organic compound seems to be insufficient in itself to estimate its lability. Concerning the lability of an organic compound in natural waters, an evaluation through its parameters measurable in units of time such as the decomposition rate, the turnover rate (time) and age, is sometimes much more informative than analysis of its chemical structure. Thus, to evaluate dynamics of organic matter in the ocean, it can be valid to characterize organic compounds both by chemical analysis and by temporal parameters.

Concerning the production processes of organic matter by phytoplankton photosynthesis, when both the production rate and the concentration of each organic compound are available, the turnover time (TT) of organic compounds can be estimated through the production process. The TT of organic compounds through phytoplankton photosynthesis is likely to be specific for individual compounds, reflecting the lability of each compound in the biogeochemical cycle.

Although phytoplankton photosynthesis is the most important source of organic compounds existing in aquatic environments, it is not a sole source. Thus, it is noteworthy that the TT estimated from the production rate trough phytoplankton photosynthesis is likely biased from the "true TT", which by definition, is a function of all sources and sinks.

In this report, I will first compile the information on primary productivity and the sizes of organic carbon reservoirs such as particulate organic carbon (POC) and dissolved organic carbon (DOC) on a global ocean scale, and then estimate the TT of these components through primary productivity.

Secondly, a brief review of the organic composition of cellular organic matter and photosynthetic products will be presented, followed by the experimental results of the variability of the composition of photosynthetic products as measured by the ^{13}C tracer and gas chromatography/mass spectrometry (GC/MS) method.

Finally, I will show the spectra of the TT of organic constituents both in particulate and dissolved form through phytoplankton photosynthesis, as obtained from both lacustrine and oceanic waters, and will discuss the relationship between TT and the lability of organic compounds.

GLOBAL ESTIMATES OF PRIMARY PRODUCTIVITY AND TURNOVER TIME OF ORGANIC MATTER IN THE OCEAN

Global ocean primary productivity

Since Steeman Nielsen (1952) introduced radioactive carbon (^{14}C) as a tool

Table 1. Estimates of the global ocean primary productivity.

REFERENCE	Productivity (× 10^{15}gC y^{-1})
Fleming (1957)	20
Koblenz-Mishke *et al.* (1970)	23
Berger *et al.* (1987)	27
Martin *et al.* (1987)	51
Knauer (1993)	51
Longhurst *et al.* (1995)	45–50

for the measurement of the primary productivity of a natural phytoplankton population, the ^{14}C method has been widely used. The first quantitative global map of ocean productivity was presented by Fleming (1957), partly using the results of the ^{14}C method by Steeman Nielsen and coworkers. The first estimation of global primary productivity was also attempted, and a value of 20×10^{15} gC y^{-1} was reported (Table 1). The most famous map of world productivity was presented by Koblenz-Mishke *et al.* (1970). They divided the waters into five categories using the more than 7,000 stations cited in the literature up to 1968. They proposed a slightly higher global ocean productivity (23×10^{15} gC y^{-1}) than Fleming. This value together with the global map of ocean productivity, has been widely used until recently.

In the latter half of the 1980s, Berger *et al.* (1987) compiled about 8,000 measurements of primary productivity in the world's oceans reported between 1944 and 1985. They used vertically integrated measurements only, though Koblenz-Mishke *et al.* (1970) sometimes estimated the water column productivity from measurements for surface water. However, the map obtained by Berger *et al.* (1987) was generally comparable with Koblenz-Mishke *et al.* (1970) and their estimated global primary productivity, 27×10^{15}gC y^{-1}, was also similar to the values already reported.

The possibility of underestimating primary productivity by the ^{14}C method has been pointed out from the late 1970s (Gieskes *et al.*, 1979; Leftley *et al.*, 1983) by comparison with other estimates such as the consumption rate of dissolved oxygen in the aphotic zone (Jenkins, 1982; Jenkins and Goldman, 1985). Some problems with the ^{14}C method have been pointed out (Gieskes *et al.*, 1979; Fitzwater *et al.*, 1982; Leftley *et al.*, 1983), and improvements in protocols which were mainly designed for protection against the contamination of inhibiting metal elements during the incubation experiments have been applied (Laws *et al.*, 1987; Marra and Heinemann, 1987). Reported productivity using the clean technique was generally 2 wo 4 times higher than former estimates: values of 100–200 μgC m^{-2} d^{-1} had generally been reported for primary productivity in the central North Pacific Gyre (Heyward *et al.*, 1983), but values of 300–600 μgC m^{-2} d^{-1} were obtained recently in almost the same area (Laws *et al.*, 1989; Letelier *et al.*, 1996). Although other factors such as the increase in depth for the measurements of productivity (Laws *et al.*, 1987) are considered as one of the reasons why water column productivity is greater in recent determinations, it has generally been accepted that the phytoplankton population is protected from inhibiting elements during incubation by the application of the "clean technique". Extrapolating the results obtained from the "clean technique" Martin *et al.* (1987) and Knauer (1993) re-estimated the global ocean primary productivity, and their values increased to 51×10^{15}gC y^{-1}. These recent estimates are about 2 times higher than previous ones and nearly comparable with terrestrial primary productivity (Table 1).

Using both the global distribution of near-surface chlorophyll a (Chl *a*) obtained by the Nimbus 7 CZCS and the photosynthesis-light relationship, Longhurst *et al.* (1995) presented a global map of primary productivity in the ocean which they calculated as $45–50 \times 10^{15}$ gC y^{-1}. It is notable that although their estimation of water column productivity is " independent of the historical data of the direct measurement at sea " (Longhurst *et al.*, 1995), the calculated global ocean productivity is quite similar to the most recent estimates. A value of around 50×10^{15} gC y^{-1} seems to be adequate for global ocean productivity at the present time.

It has been pointed out that phytoplankton excrete photosynthetic products as dissolved organic matter (DOM) (Fogg, 1966; Watt, 1966; Hellebust, 1974; Harris, 1980). The proportion of excretion to total photosynthetic production has been a matter of debate. Early studies reported that a higher amount of photosynthetic products (up to 75% of the total production) was excreted as DOM (Table 2). Sharp (1977) examined the experimental protocols and asked whether excretion was not a normal property of phytoplankton, concluding that the reported evidence of excretion was mainly due to artifacts. Since then, however, a critical examination of the protocols was reported (e.g., Mague *et al.*, 1980), with results

Table 2. Percent of excretion to total primary productivity in oceanic environments. The average value over the euphotic zone and/or the time interval is presented when available.

Study area	% of excretion	average	REFERENCES
neritic North Atlantic	4–16		Hellebust (1965)
inshore Atlantic	1–38		Horne *et al.* (1969)
coastal North Pacific	7–17		Anerson and Zeutschel (1970)
inshore Indian Ocean	10–19	14	Samuel *et al.* (1971)
coastal North Atlantic	0–27		Thomas (1971)
Strait of Georgia	9–26		Wallen and Geen (1971)
coastal Atlantic	0–23	6.9	Williams and Yentsch (1976)
off North West Africa	9		Smith *et al.* (1977)
Baltic Sea	2–21		Iturriaga and Hoppe (1977)
Gulf of Maine	5–30	10	Mague *et al.* (1980)
Baltic Sea	12–16		Lasson and Hagstrom (1982)
Baltic Sea	5–10	7.1	Lingell (1990)

that at the excretion of at least some DOM is an essential property of phytoplankton photosynthesis.

Estimations of the global ocean primary productivity listed in Table 1 do not take into account the release of photosynthetic products in dissolved form. As summarized in Table 2, it is rather difficult to fix the proportion of excretion of the total primary productivity when estimating the global productivity of DOC through phytoplankton excretion. However, the value of percentage excretion averaged throughout the year or integrated through the water column generally ranged from 5 to 15%, though individual determinations varied within a somewhat wider range. Here we assume that the rate of excretion is one-tenth of the total (particulate and dissolved) production, as also did Murray (1992). When using the value of 51×10^{15} gC y^{-1} (Martin *et al.*, 1987; Knauer, 1993) as the global primary productivity of POC, the total (POC and DOC) primary productivity and the dissolved productivity are calculated as 57×10^{15}gC y^{-1} and 5.7×10^{15}gC y^{-1}, respectively.

Organic reservoirs

The major reservoir of marine organic carbon is in dissolved form (Table 3), defined as material passing through 0.2 or 0.4μm filter paper, and accounts for 92, 98, and 97% of thetotal organic carbon in the surface (0–300 m), deep (300–3,800 m) and whole ocean, respectively (Williams, 1971). The contribution of POC, on the contrary, accounted for only 8, 2, and 3% of the total organic carbon in the surface, deep and whole ocean, respectively. Moreover, the living fraction of POC is remarkably low, even in the surface ocean, accounting for about 5% of the

Table 3. The standing stock, the primary productivity and the turnover time though photosynthetic productivity of various ocean organic carbon reservoirs and land plants. Turnover time through photosynthetic productivity of the oceanic POC and DOC was calculated by the comparison of the POC and DOC productivity and their standing stocks in the surface layer (0 - 300 m).

Organic carbon reservoir	Standing Stock ($\times 10^{15}$C)	Productivity ($\times 10^{15}$gC y^{-1})	Turnover time (y)
POC(total)	24[a)]		
POC(0–300 m)	11[a)]	51[d)]	0.22
POC(300–3,800 m)	13[a)]		
DOC(total)	740[a)]		
DOC(0–300 m)	110[a)]	5.7[e)]	19
DOC(300–3,800 m)	630[a)]		
Phytoplankton	0.55[b)]	51[d)]	0.011
Land plant	950[c)]	60[c)]	16

*[a)] Willams (1971), [b)] Muray (1992), [c)] Olson *et al.* (1985), [d)] Martin *et al.* (1987), [e)] Martin *et al.* (1987) and 10% of excretion rate to total productivity (see Table 2 and the text).

total POC. Non-living particles, referred to as detritus, are the major component of POC except in eutrophic waters (Cauwet, 1981). Thus, dead carbon constitutes the major part of organic carbon in the ocean.

Turnover time of organic matter

As the primary productivity is the major source of organic matter, it is worth estimating the turnover time (TT) of organic carbon originating from photosynthesis from the reservoir size and production rate, and assuming that the reservoir size is at steady state. The TT of phytoplankton carbon is thus 0.011 y (4 d) which clearly shows the dynamic status of phytoplankton compared with the long TT (16 y) of land plants (Table 3). The specific growth rate of phytoplankton living in an oceanic environment reportedly varies from 0.04–5 d^{-1} (Goldman *et al.*, 1979), thus the TT of phytoplankton carbon estimated here seems to be reasonable. An extremely short TT of phytoplankton carbon implies that changes in physical, chemical and biological factors affecting phytoplankton photosynthesis easily result in both quantitative and qualitative variations in the organic compounds constituting phytoplankton cells. On the other hand, the TT calculated from primary productivity and the standing stock of bulk POC in the surface layer is considerably longer (0.22 y, 80 d) than that of phytoplankton carbon. When phytoplankton photosynthesis is assumed to be the main source of non-living detrital organic carbon, the longer TT of POC than phytoplankton organic carbon implies the more refractory nature of detrital organic compounds.

When the standing stock of DOC in the surface layer is divided by the production rate of DOC by phytoplankton excretion in the global ocean, 19 y is obtained as the TT of DOC in the surface layer. This value is 86 times longer than the TT of POC, reflecting the high standing stock of DOC and its low production rate. Williams and Druffel (1987) estimated the "apparent age" of DOC in the surface waters of the oligotrophic Central North Pacific Gyre from the radiocarbon measurements, and reported a value of 1,310 y BP, which is about 70 times longer than the TT estimated here. This estimate of TT of DOC fails to account for the sources of DOC other than phytoplankton excretion; e.g., sloppy feeding (Jumars, 1989), riverine discharge (Meybeck, 1982) and eolian input (Romankevich, 1984). Nevertheless, DOC from phytoplankton excretion is, by itself, more than enough to account for all DOC produced in the surface layer, where DOC input to the DOC reservoir is calculated to be only about 0.08×10^{15} gC y^{-1} (based on the "apparent age" of DOC in the surface layer)". Such a comparison of TT and the "apparent age" of surface DOC assumes that the major part of DOC is not easily "turned over", and that the TT of the labile fraction is considerably shorter.

The TT estimated here for several organic carbon reservoirs in the ocean represent only "average" values. The striking difference in the TT of DOC and its "apparent age" clearly demonstrates that there is an exceedingly broad range in the biological and chemical stability of organic compounds existing in oceanic environments.

VARIABILITY OF PHOTOSYNTHETIC PRODUCTS IN OCEANIC ENVIRONMENTS

The chemical composition of phytoplankton cells has generally been expressed as $C_{106}N_{16}P_1$ in atomic ratio (Redfied *et al.*, 1963). The production and decomposition processes of organic compounds have accordingly been viewed as taking place based on this fixed elemental proportion. However, studies using cultured phytoplankton demonstrated that elemental ratios were extremely variable, suggesting that the composition of source materials of the organic carbon flow in the ocean can change with space and time.

This section begins by examining the basic features of the composition of phytoplankton organic matter and photosynthetic products. Next, experimental results on the variability of the composition of photosynthetic products in upwelling areas are shown, followed by a discussion of the implications of the composition of photosynthetic products on the biogeochemical cycle.

Basic features of the composition of phytoplankton cellular organic material and photosynthetic products

Cellular organic material

Environmental factors such as nutrient availability, water temperature and light conditions affect the elemental composition of phytoplankton cells (Sakshaug and Holm-Hansen, 1977; Goldman *et al.*, 1979; Goldman, 1980). Nutrient availability, especially, has been well studied. Sakshaug and Holm-Hansen (1977) observed the variation in the C/N ratio of cellular organic compounds of Skeletonema costatum as a function of nitrate-, phosphate-, and iron-limited growth. The C/N ratio (atomic) for the exponentially growing cell subject to an abundance of nutrients ranged from 5 to 7.5, which was almost comparable with the "Redfield ratio". However, values increased up to 35 for nitrogen-deficiency and to 13 for phosphate-deficiency. Goldman and his coworkers (Goldman *et al.*, 1979; Goldman, 1980) examined the relationship between the C/N ratio and the specific growth rates for several phytoplankton species, and observed that the C/N ratio decreased with an increase in a specific growth rate. They concluded that an elemental ratio around the "Redfield ratio" implies that the nutrient status of phytoplankton was near complete saturation.

The elemental composition of phytoplankton cells reflects the composition of major organic compounds such as proteins/amino acids, carbohydrates and lipids. Parsons *et al.* (1961) measured the organic composition of various species of marine phytoplankton in their exponential growth phase. Proteins/amino acids are the most abundant compounds in almost all phytoplankton species, accounting for 17–57% of the total organic carbon. On the other hand, the contribution of carbohydrates ranged from 6.6 to 37% and lipids from 2.9 to 18%. Proteins, all of which contain nitrogen, are the major components. Thus, the C/N ratio discussed previously mainly reflects the compositional ratio between proteins and carbohydrates/lipids. Considering that the C/N ratio changes with the degree of nutrient limitation, organic composition must vary with nutrient status.

Myklestad (1974) observed changes in the protein and carbohydrate contents of several species of cultured diatoms. Carbohydrate/protein ratios were low during the exponential growth and increased with the degree of nutrient limitation regardless of species. Further fractionation of carbohydrates into glucan and other carbohydrates revealed that the increase in the carbohydrate/protein ratio under nutrient-limited conditions was exclusively due to the increase in glucan content. This phenomenon has been repeatedly observed in other cultured phytoplankton, though the degree of accumulation was different for each species (Myklestad and Haug, 1972; Myklestad, 1974, 1977; Moal *et al.*, 1987). An accumulation of glucan suggests that reserved carbohydrates are the major components of carbohydrates under nutrient-limited conditions, and that structural carbohydrates constituting the cell wall are not actively produced under such condition.

Photosynthetic products

Since ^{14}C is generally used as a tracer for photosynthetic metabolism, determination of the composition of photosynthetic products requires fractionation of the products into organic compound groups prior to the measurement of radioactivity. Solvent extraction can be used to fractionate organic compounds into chloroform (lipids), methanol (low molecular weight (LMW) compounds), TCA (carbohydrates and nucleic acids) and residual (proteins) fractions (Morris and Skea, 1978). Although problems exist with this fractionation procedure (Knopka and Schnur, 1980; Hitchcock, 1983; Hama and Handa, 1987a), the composition of photosynthetic products of a natural phytoplankton population has been elucidated in both lacustrine (Olive and Morrison, 1967; Olive *et al.*, 1969; Hama and Handa, 1987a; Hawes, 1990) and oceanic (Morris and Skea, 1978; Hitchcock, 1978; Li *et al.*, 1980; Smith and Morris, 1980; Barlow, 1982a, 1982b, Rivkin, 1989; Fernandez *et al.*, 1992) waters.

Studies concerning the relationship of photosynthetic products and nutrient availability have been carried out. Barlow (1982a) measured the composition of photosynthetic products in the southern Benguela Current, and observed that the polysaccharide/protein ratio was low in the developing stage of a phytoplankton bloom but was high during and after the peak of the bloom; the results are almost comparable with those obtained from compositional studies. On the other hand, increased polysaccharide production was sometimes measured at the developmental stage of the phytoplankton bloom (Barlow, 1982b; Priscu and Priscu, 1984; Fernandez *et al.*, 1992), a finding which seems to disagree with the previous results. However, Fernandez *et al.* (1992) suggested that the enhanced polysaccharide production during the developmental stage of Phaeocystis sp. was mainly due to the formation of mucilaginous colonial matrix. Colonial matrix is considered to be a "structural" material, and the fractionation of newly incorporated carbon by solvent extraction is unable to distinguish the difference in the structure and function of polysaccharides.

Structure and function of carbohydrates can be estimated from their monosaccharide composition because the composition of structural and reserved carbohydrates are different (Handa, 1969). The degree of lability is also likely to differ between reserve and structural polysaccharides. Methodological improvements for obtaining further information on the structure and function of carbohydrates and other photosynthetic products would be desirable.

Composition of photosynthetic products and environmental factors

^{13}C/GC/MS method

Stable carbon isotope (^{13}C) has been applied as a tracer to measure the primary productivity in addition to the ^{14}C method (Slawyk *et al.*, 1977, 1979; Hama *et al.*, 1983), since stable isotopes pose no hazardous radioactivity problems. To determine the total uptake of carbon, a seawater sample with added ^{13}C-enriched CO_2, POC filtered from the incubated sample is then combusted, and the CO_2 produced is used for the measurement of the ^{13}C atom %. Primary productivity can then be estimated from the ^{13}C atom % and the concentration of POC (Hama *et al.*, 1983). When it is possible to estimate the ^{13}C atom % and the concentration of individual organic compounds instead of bulk POC, the production rate of each organic compound can be estimated.

The combination of gas chromatography (GC), widely used to separate organic molecules, and mass spectrometry (MS) in the stable isotope tracer method was first applied in biomedical studies by Gordon and Frigerio (1972) and Summonds *et al.* (1974). The estimation of the ^{13}C content of organic molecules by GC/MS is based on the fact that the weight of a ^{13}C enriched molecule increased over that of a non-enriched molecule and its increase is in proportion to their ^{13}C content. The ^{13}C atom % of an organic molecule can be calculated from the difference in the relative intensities of the isotopic peaks of mass spectra between non-enriched and enriched organic molecules (Van den Heuvel *et al.*, 1970; Kouchi, 1982).

Hama and Handa (1986) and Hama *et al.* (1987) applied the combined method of ^{13}C tracer and gas chromatography/mass spectrometry (GC/MS) for the determination of production rates of organic constituents. They determined the composition of the photosynthetic products of a natural phytoplankton population on a molecular level (i.e., monosaccharides, amino acids and fatty acids). This method affords new and significant information not only on the relation between photosynthetic products and phytoplankton growth, but also on the significance of photosynthetic products in the biogeochemical cycle (Hama *et al.*, 1993).

Photosynthetic products and nutrient availability

Nutrient availability affects the composition of photosynthetic products as discussed previously. The ^{13}C/GC/MS method was applied to an upwelling area to study the relation between nutrient availability and phytoplankton metabolism; the developmental stage of a phytoplankton bloom can be followed from the relation between the ambient concentration of inorganic nutrients and Chl *a* (Takahashi *et al.*, 1986). Hama (1988) determined the monosaccharide and amino acid composition of photosynthetic products in "maturing" and "aged" upwelling waters around the Izu Islands, Japan. Below, the results will be summarized and the implications for the geochemical cycle discussed.

Estimation of the aging status of upwelling waters

The relation between the concentration of Chl *a* and nitrate obtained at 10 m depth around the Izu Islands is shown in Fig. 1. Newly upwelling water contains a high concentration of nitrate but low Chl *a* concentration since the response of phytoplankton to the addition of nutrients takes a few days. The concentration of Chl *a* increases with a decrease in the concentration of nutrients until the nutrients are used up. After the nutrients are exhausted, the concentration of Chl *a* decreases due to grazing or sinking losses.

Incubation experiments with ^{13}C were carried out at Stns. 49 and I-1 (Table 4). Judging from the relation between the concentration of Chl *a* and nitrate, Stn. 49 was a "maturing" upwelling site where the uptake of nitrate by the phytoplankton population is likely to be active, indicating rapid phytoplankton growth. On the other hand, nitrate was almost depleted at Stn. I-1, indicating that Stn. I-1 was an "aged" upwelling station. The higher concentration of ammonium at Stn I-1 than at Stn. 49, probably reflects the high decompositional activity of the post bloom condition and suggests that regenerated production might be the major production process at Stn. I-1. Primary productivity was higher at Stn.. 49 than

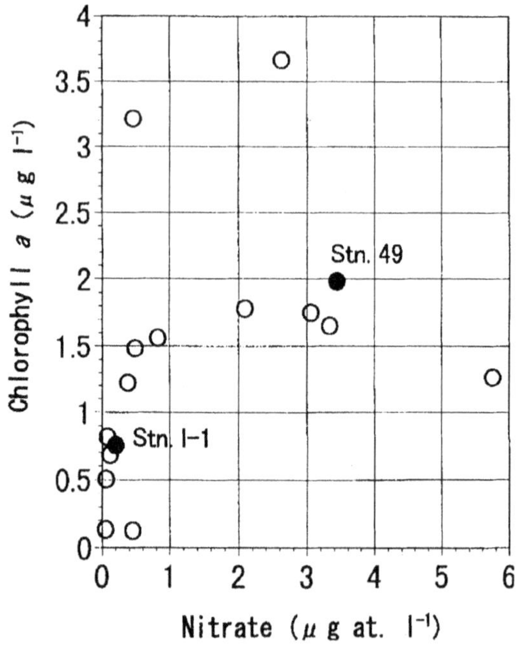

Fig. 1. The relationship between the concentration of nitrate and Chl. *a* at 10 m depth around the Izu Islands, Japan.

at Stn. I-1, but it is noteworthy that the Chl *a* specific production rate was lower at the "maturing" than at the "aged" upwelling station (Table 4). Higher Chl *a* specific productivity has been considered to be an indicator of a healthy state of phytoplankton in nutrient-abundant waters (Curl and Small, 1965; Thomas, 1970; Malone, 1971). Therefore, the lower Chl *a* specific productivity at the nitrate-abundant "maturing" rather than at the nitrate-depleted "aged" upwelling seems to be inconsistent with the results reported so far.

GC/MS analysis of photosynthetic products

^{13}C atom % values of eight monosaccharides and 12 amino acids, which were obtained after acid hydrolysis of particulate samples from Stns. 49 and I-1 were estimated by GC/MS analysis. Figure 2 shows the chemical ionization mass spectra of mannose and glucose which are the major constituents of carbohydrates. A base peak of m/z (mass to charge ratio) 452 was the quasi-molecular ion of acetyl derivatives of mannose and glucose ($C_{18}H_{26}O_{12}$ plus NH_4^+; reagent gas). In the non-incubated sample, the relative intensity of the isotope peak (m/z: 453) to the base peak was 0.220 in the mannose. The relative intensity of the isotope peak (m/z: 453) of the mannose from the incubated sample at Stn. 49 showed a distinct increase to 0.582, indicating the incorporation of ^{13}C as the constituents of the carbon skeleton. At "aged" upwelling Stn. I-1, the isotope peak also showed an increase, but the degree of the increase in the relative intensity was lower (0.372) than at Stn. 49.

In the case of glucose, the relative intensity of the isotope peak of the incubated sample at Stn. 49 was almost comparable with mannose, suggesting that the fraction of newly incorporated carbon to the standing stock of these monosaccharides was almost the same between mannose and glucose at this station. However, at Stn. I-1, the relative intensity of the isotope peak in glucose (0.612) was much higher than the intensity in mannose. A comparison of the mass spectra of mannose and glucose clearly showed that the production pattern of carbohydrates was remarkably different between the two stations as will be discussed later.

The ^{13}C atom % of 12 amino acids was determined by GC/MS, and the mass spectra of *N*-trifluoroacetic *n*-butyl esters of aspartic and glutamic acid are illustrated as examples in Fig. 3. The increase in the relative intensity of isotopic peaks was higher at Stn. 49 than at Stn. I-1 both in aspartic and glutamic acid.

Table 4. Phisicocheical conditions and primary productivity at 10 m of Stns. 49 and I-1. *AIU were calculated after Takahashi *et al.* (1986).

Station	Water temperature (°C)	NO_3^-	NO_2^- (μ g-at. l^{-1})	NH_4^+	Chlorophyll *a* ($\mu g\ l^{-1}$)	AIU*	Primary productivity ($\mu gC\ l^{-1}$ daytime^{-1})	Chlorophyll specific productivity ($\mu gC\ \mu g$Chl. a^{-1} daytime^{-1})
49	21.2	3.44	0.05	0.10	1.98	0.285	150	76
I-1	21.7	0.20	0.50	0.79	0.75	0.677	100	130

These results indicated that the incorporation of carbon to amino acids was higher at the nitrate-rich "maturing" Stn. 49 than at the nitrate-depleted "aged" Stn. I-1.

Production rates of monosaccharides and amino acids, and nutrient availability

Production rates of eight monosaccharides and 12 amino acids at Stns. 49 and I-1 were estimated from the ^{13}C atom % and the concentration of each compound (Fig. 4). The most striking difference in the composition of photosynthetic products was the production rate of glucose. Glucose accounted for 50% of

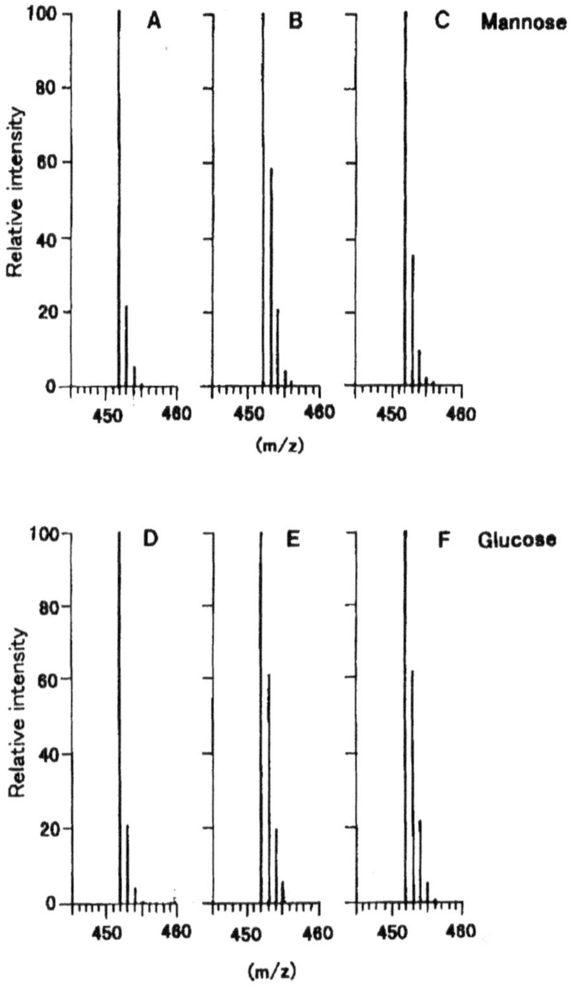

Fig. 2 Chemical ionization mass spectra of acetyl derivatives of mannose (mannitol acetate) and glucose (glucitol acetate) of non–incubated and incubated samples for 10 m at Stns. 49 and I–1. A and D :Non–incubaed sample from Stn. 49; B and E: Incubated sample from Stn. 49; C and F:incubated sample from Stn. I–1. (modified from Hama (1988)).

the total carbohydrate production rate at Stn. 49, which was calculated by adding the rates of eight monosaccharides. Other monosaccharides such as galactose, mannose and ribose, also showed relatively high production rates, each accounting for more than 9% of the total carbohydrate production. At the "aged" upwelling Stn. I-1, however, an extremely high production of glucose was noticed, accounting for 90 and 36% of total carbohydrates and total POC production, respectively. At this station, the production rates of the seven other monosaccharides were quite low, each forming only 0.5 to 2.5% of the total

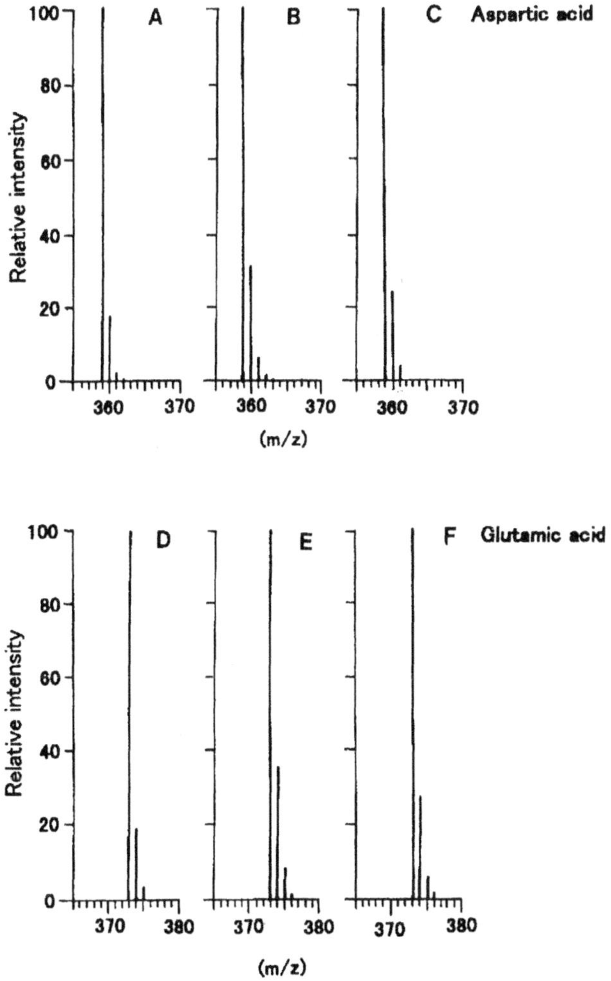

Fig. 3 Chemical ionization mass spectra of *N*–trifluoroacetyl *n*–butyl esters of aspartic acid and glutamic acid of non–incubated and incubated samples for 10 m at Stns. 49 and I–1. A and D :Non–incubated sample from Stn. 49; B and E: Incubated sample from Stn. 49; C and F: incubated sample from Stn. I–1. (modified from Hama (1988).

carbohydrates production.

Based on their functional differences, carbohydrates are usually divided into two groups; structural and reserve. Handa (1969) fractionated the carbohydrates of a marine diatom, Skeletonema costatum, into the two groups by hot-water extraction. He found that more than 90% of hot-water-extractable carbohydrate, which corresponded to the reserve carbohydrates, were constituted by glucose. Hama and Handa (1983) measured the daily variability in the monosaccharide composition of particulate matter in lake water and found that glucose exhibited remarkable changes, increasing in the day and decreasing at night, although other monosaccharides such as xylose, galactose, rhamnose and mannnose showed no significant variations. These observations indicate that glucose in phytoplankton cells mainly functions as a reserve of carbon and energy.

On the other hand, from a determination of the monosaccharide composition of structural carbohydrates of *S. costatum*, which was resistant to hot water extraction, Handa (1969) reported that mannose was the most important species, accounting for 55% of the carbohydrates in the residual fraction followed by xylose, rhamnose and fucose.

Judging from the monosaccharide composition of photosynthetic products at Stn. 49, both reserve and structural carbohydrates were produced in this "maturing" upwelling area. At Stn. I-1, however, most of the carbohydrates produced at the nitrate-depleted "aged" upwelling area were reserve carbohydrates, and only a small amount of the structural carbohydrates were synthesized, indicating

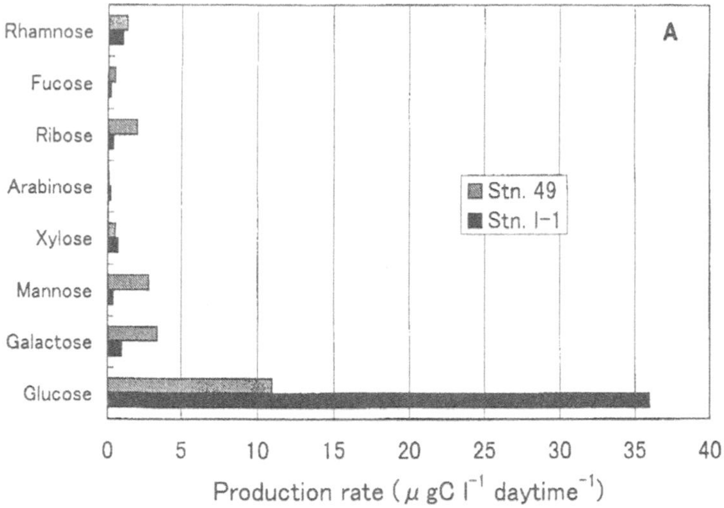

(a)

Fig. 4. Production rates of eight monosaccharides (A), and 12 amino acids (B), constituting carbohydrates and proteins/amino acids, respectively, at 10 m depth of Stns. 49 and I–1.

insignificant production of the cell walls. This finding is consistent with the observation that reserve glucan accumulates under nitrogen-limited conditions in cultured phytoplankton (Myklestad, 1974).

Unlike the monosaccharide composition of carbohydrates, few compositional differences were found for amino acids in the photosynthetic products between the two stations (Fig. 4B). Glutamic acid, aspartic acid, and leucine were found as the main components of the photosynthetic products at both stations. This was comparable to the determination in the amino acid composition of cultured phytoplankton (Chau *et al.*, 1967), marine POM (Siezen and Mague, 1978) and photosynthetic products (Hama, 1992) which showed little variability in amino acid composition among species and different waters. This probably reflects the fact that the amino acid composition does not vary considerably among proteinaceous substances in phytoplankton cells. However, in other studies, variability in the amino acid composition of a free amino acid pool in phytoplankton cells has been observed as a result of varying environmental conditions (Admiraal *et al.*, 1986, Haberstroh and Ahmed, 1985; Goes *et al.*, 1995). Proteins/amino acids were not fractionated into combined and free amino acids in this study. Free amino acid pools account for a minor portion of total proteins/amino acids. Consequently, the determination of the amino acid composition of the total proteins/amino acids applied by Hama (1988) might not detect compositional changes of free amino acid pools, if such a change occurred through the aging of upwelling waters.

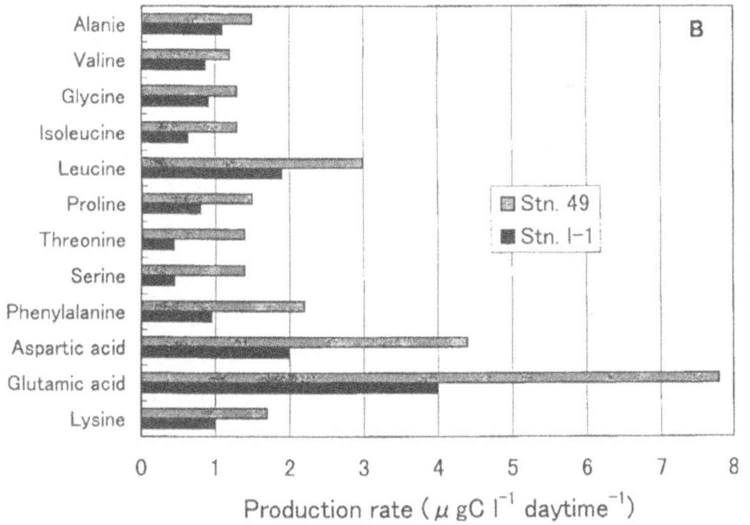

(b)

Fig. 4. (Continued).

The carbohydrate/protein ratios of photosynthetic products were calculated for "maturing" and "aged" upwelling stations and calculated as 0.80 and 2.7, respectively, clearly showing the accumulation of carbohydrates under nitrogen-depleted conditions. These ratios in the regional upwelling area around the Izu Islands, including other stations in addition to Stns. 49 and I-1, were well correlated with glucose/carbohydrate ratios, as shown in Fig. 5 (Hama, 1988). The results plotted in Fig. 5 include the ratio obtained both from daytime and day-night time incubations and glucose/carbohydrate ratio obtained from the day-night time incubation tended to decrease from the ratio with day-time incubation. This day-night change suggests that glucose plays the part of the reserve material and revealed that carbohydrate/protein in the photosynthetic products primarily reflected the degree of production of reserve glucan.

The Chl. *a*-specific productivity of glucose at Stn. I-1 was quite high compare with that at Stn. 49 (5.6 and 48μgCμgChl. a^{-1} daytime^{-1} at Stns. 49 and I-2, respectively), in contrast with the comparable Chl. *a*-specific productivity of seven monosaccharides other than glucose, between two stations (5.6 and 5.2 at Stns. 49 and I-1, respectively). Thus, determination of the photosynthetic products showed that the high Chl *a* specific productivity of organic carbon at

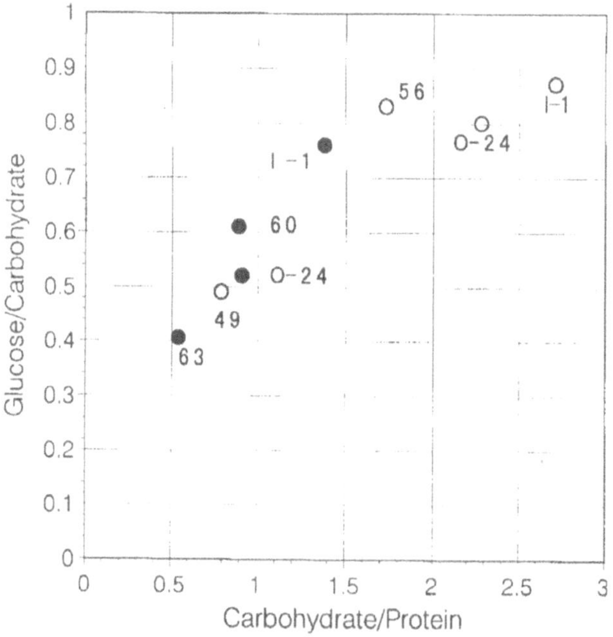

Fig. 5. Relationship between carbohydrate/protein and glucose/carbohydrates in the photosynthetic products of phytoplankton population obtained from a 10 m depth around the Izu Islands, Japan. Stns. 49, 63 and I–1 are inside of upwelling plume and Stns. 56, 60 and O–24 are outside of upwelling plume. ○: daytime incubation; ●: day-night time incubation (after Hama (1988)).

Stn. I-1 was mainly due to the production of reserve material. The high Chl *a* specific productivity at a nitrate-depleted area implied neither a high growth rate nor a healthy physiological state of phytoplankton.

Implications of the composition of photosynthetic products and biogeochemical cycles

The effect of the compositional difference in photosynthetic products is not confined to phytoplankton dynamics. The dynamics of the secondary producers can also be affected because compositional changes in photosynthetic products cause changes in the food sources for these organisms (Jonasdottir *et al.*, 1995). Moreover, considering the variable lability of organic compounds, it is possible that compositional changes in photosynthetic products may result in quantitative and qualitative changes in the biogeochemical cycle.

Variability in the lability of organic constituents has been found in both natural environments and laboratory experiments. Reserve glucan has been reported as one of the most labile constituents of POM. Handa and Yanagi (1969) measured the vertical distribution of hot water extractable carbohydrates, which are mainly composed of reserve glucan, in the northwest Pacific Ocean, and found that their content rapidly decreased from 19–35% of particulate carbohydrate at the surface layer to 6.5–16% at 50 m depth. Further, from the analysis of the monosaccharide composition of particulate carbohydrates in coastal waters, Handa and Tominaga (1969) concluded that glucose and its polymer were preferentially removed from particulate matter. Matsunaga (1981), on the other hand, carried out a decomposition experiment on phytoplankton organic matter under laboratory conditions, and determined the decomposition rate constants for some organic constituents such as total carbohydrates, hot water-extractable carbohydrates and protein/amino acids (Fig. 6). The initial decomposition rate constants obtained for these organic constituents in the first 14 days varied from 0.048 to 0.079 d^{-1}, and were obtained on the assumption that decomposition proceeded as a first-order reaction (Grill and Richards, 1964; Otuki and Hanya, 1972). The highest rate constant was obtained for hot water-extractable carbohydrates, being 1.6 and 1.3 times higher than the total carbohydrates and proteins/amino acids, respectively.

These results on hot water-extractable carbohydrates strongly suggest that the reserve glucan of phytoplankton is one of the most labile organic constituents. The content of reserved glucan in photosynthetic products is highly variable and governs the major composition of photosynthetic products such as the carbohydrate/protein ratio shown in Fig. 5. Thus, it is possible that a varying contribution of reserve glucan to photosynthetic products affects the decomposition rate of POC. POC containing a high level of reserve glucan may have a higher decomposition rate constant than POC poor in reserve material.

TURNOVER OF ORGANIC CONSTITUTENTS

A simple ratio of the standing stock of POC and DOC in oceans to primary productivity, both in particulate and dissolved form, gives the TT of both organic reservoirs, as already discussed (Table 3). The obtained values, however, are only

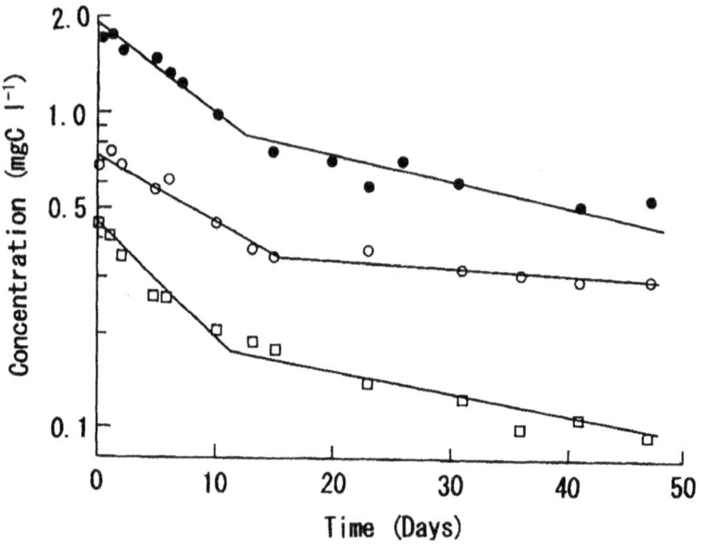

Fig. 6. Changes in the concentration of particulate carbohydrate carbon(○), particulate water extractable carbohydrate carbon(□) and particulate protein carbon(●) during the decomposition process of phytoplankton organic matter (after Matsunaga (1981).

average TT's, regardless of the variability in particulate and dissolved organic constituents. For example, the striking difference between the estimated TT and the "apparent age" of DOC, in the surface oceanic layer implies that there is an exceedingly broad range in biological and chemical labilities of organic compounds in oceanic environments. Organic compounds are likely to have inherent TT's which reflect the biological and chemical functions and lability in aquatic environments. Thus, the estimation of the TT of organic compounds can afford significant information on those functions and on the lability.

When it is possible to superimpose the composition of photosynthetic products onto the composition of ambient organic materials, the spectrum of the TT of compounds can be elucidated. What follows are two summaries of published results. One is an estimation of the TT of whole organic constituents including POM and DOM in lake water (Hama and Handa, 1987a, b). The other is a comparison of TT's of particulate fatty acids constituting marine particulate lipids on the molecular level, by the ^{13}C/GC/MS method (Hama, 1991).

Turnover of particulate and dissolved organic constituents

Production of particulate organic constituents

A ^{14}C tracer experiment in Lake Suwa, which is one of the most eutrophic lakes in Japan, was carried out in summer when the blue-green algae, *Microsystis aeruginosa*, dominated (Hama and Handa, 1987 a, b). Intracellular products were fractionated into biomolecules such as carbohydrates, proteins/amino acids and lipids by extraction with chloroform/methanol and trichloroacetic acid (Morris and Skea, 1978). Carbohydrates were further fractionated into the LMW fraction, the high molecular weight (T) [HMW(T)] fraction (the extractable by TCA), and the high molecular weight (R) [HMW(R)] fraction, which represents the residual carbohydrates against extraction with TCA. Proteins/amino acids were also fractionated into LMW and HMW fractions. Gel-filtration of LMW compounds indicates that the MW of these compounds was centered at 200 Da (Hama and Handa, 1987a).

Carbohydrates were the main products of particulate photosynthesis, accounting for 72% of total organic carbon production during daytime (Fig. 7), but decreasing at night, as has been observed in oceanic phytoplankton (Morris and Skea, 1978). The pattern of production and consumption of total carbohydrates was comparable to HMW(T), indicating that reserve polysaccharides were the main component of HMW(T) carbohydrates (Fig. 8), and that they quantitatively played an important role in the metabolism. Although the LMW carbohydrates showed a rapid increase during the first 6 h, no increase in carbon uptake was found thereafter. No increases in carbon uptake after 6 h from the start of the experiment indicated an equilibrium between the production and consumption of mono- and oligosaccharides rather than an absence of production of these compounds.

Proteins/amino acids and lipids were not important when quantitatively compared with carbohydrates, each accounting for 17 and 8% of the total daytime production, respectively; but they did not decrease at night, and their contribution had increased by the following morning. Nighttime increases in the production rates of proteins indicated synthesis of proteins using carbon which was stored as reserved carbohydrates (Cuhel *et al.*, 1984).

Turnover of particulate organic constituents

Concentrations of biomolecules such as carbohydrates, proteins/amino acids and lipids in each fraction at the start of the Lake Suwa experiments were measured by colorimetric and fluorometric analysis. They corresponded to 75% of the total POC, as measured by an elemental analyzer. A comparison of the production rate and concentration of each organic compound group gave the TT of each biomolecular group (Fig. 9); The result obtained from 3 h incubation was used as the production rate because longer incubations resulted in an underestimation of the production rates of LMW metabolites (Hama and Handa, 1987a).

The TT calculated for bulk POC in Lake Suwa was 38 h, being much shorter than the average TT for POC in the ocean (80 d, Table 3) and roughly comparable with the estimates for phytoplankton carbon in the global ocean (96 h). This was probably because phytoplankton accounted for a major part of the POC in this lake (Hama and Handa, 1982). Apart from the TT of bulk POC, the TT of particulate organic constituents varied from 8 to 58 h, depending on the organic compound group. The shortest TTs were obtained for LMW proteins/amino acids and LMW carbohydrates. The MW of these compounds estimated by gel-filtration (Hama and Handa, 1987a) indicated that mono- and oligosaccharides,

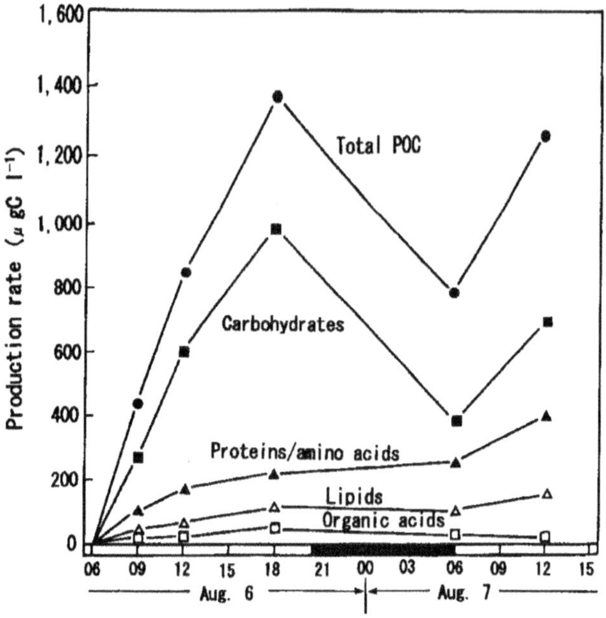

Fig. 7. Production of particulate organic carbon (POC), carbohydrates, proteins/amino acids, lipids and organic acids in the surface layer at Lake Suwa (after Hama and Handa (1987a)).

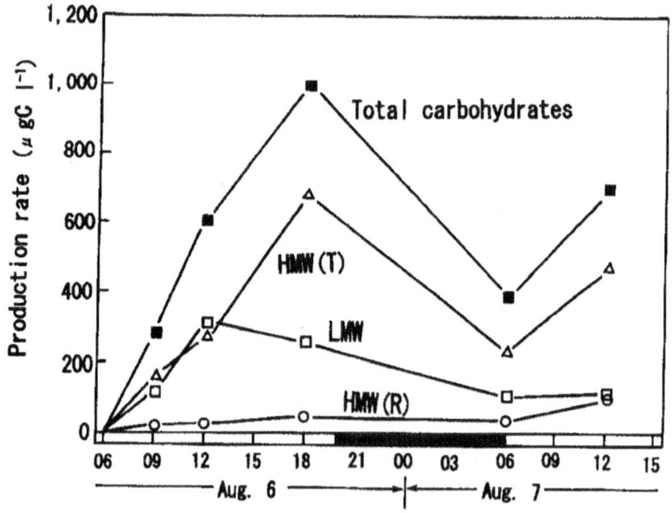

Fig. 8. Production of LMW, HMW(T) and HMW(R) carbohydrates in the surface layer at Lake Suwa (after Hama and Handa (1987a)).

together with free amino acids and oligopeptides, constituted LMW carbohydrates and LMW proteins/amino acids, respectively. These LMW compounds are likely synthesized as the primary compounds of photosynthesis and are then polymerized to polymer. The short TT of LMW compounds probably reflects such an intermediate role for these compounds (Hama and Handa, 1987a; Hama and Handa, 1992).

The TT of HMW(R) carbohydrates and HMW proteins and lipids were roughly comparable, with the range from 51 to 58 h. HMW(R) carbohydrates were considered to correspond to cell-wall materials (Knopka and Shuner, 1980; Hitchcock, 1983), whereas HMW protein and lipid were the main components of phytoplankton cell membranes. The property of these compounds as structural material might account for a longer TT than LMW compounds and HMW(T) carbohydrates.

High MW(T) carbohydrate exhibited intermediate TT in the three carbohydrates fractions, and were nearest to the value of the total carbohydrates. As readily suggested by the time course of carbohydrate production (Fig. 8), polysaccharide in this fraction functioned as the reserve material. The production and consumption of reserve polysaccharides showed a marked diurnal change (Hama, 1983; Hama and Handa, 1992). Thus, the shorter TT for HMW(T) carbohydrates than for HMW(R) carbohydrates is mainly reflected in such a diurnal rhythm.

As discussed above, the TT of POM was mainly caused by differences in the function of organic compounds in the phytoplankton cell. The effect of non-phytoplankton particulate organic material was not obvious in this case, probably

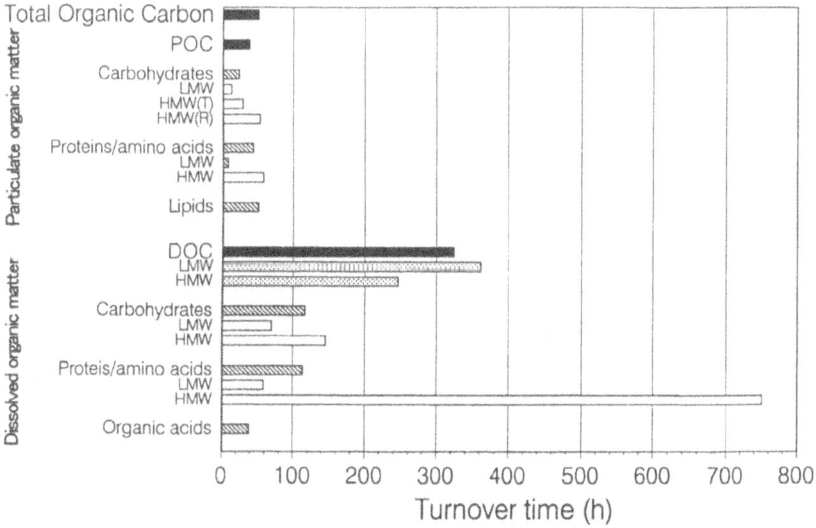

Fig. 9. Turnover times of organic constituents through phytoplankton photosynthesis in Lake Suwa (calculated from the results obtained by Hama and Handa (1987 a, b)).

due to the minor contribution of the non-phytoplankton fraction to POM in a eutrophic lake (Hama and Handa, 1982).

Production of extracellular products

Excreted DOC (EDOC) by phytoplankton photosynthesis also increased with time, accounting for 7.8% of the total (particulate and dissolved) productivity after 12 h (Fig. 10). This value is within the range reported in oceanic environments, as shown in Table 2. The produced dissolved organic compounds were separated by gel-filtration and fractionated into Fractions I (MW > 4,000 Da), II (500 << 4,000 Da), III (200 << 500 Da) and IV(<200 Da) according to the elution volume, as shown in Fig. 11 (the range of MW was estimated by the elution of standard compounds and the absorption of organic compounds to gel material, and might have skewed the estimated MW from the "true" MW (Hama and Handa, 1980)).

Low MW products (<500 Da; fractions III and IV) were the main components of EDOC in every sample, amounting to 69% of total EDOC after a 12 h incubation. High MW products (>500 Da; Fractions I and II) were only minor components in the 3 h incubated sample. However, their contribution increased with incubation length, accounting for 31% after 12 h. The increase with time in the contribution of HMW EDOC was partly due to the increase in its specific activity (Storch and Saunders, 1978; Jensen *et al.*, 1985; Hama and Handa, 1987b), and to the fact that the short term incubation rate did not precisely reflect the production rate of HMW EDOC. Radioactivity in the LMW fraction showed a decrease at night, which indicated the microbial uptake of LMW EDOC.

EDOC was fractionated into organic constituents such as carbohydrates (neutral fraction), proteins/amino acids (basic fraction) and organic acids (acidic fraction) by ion exchange resin. Results obtained from a 12 h incubation are discussed here, since short-term incubation can result in an underestimation of the production of HMW EDOC, as described above. Carbohydrates were uniformly distributed from Fractions I to IV (Fig. 12). Carbohydrates eluted in Fractions I and II were the main components of HMW EDOC, accounting for 84% of the HMW EDOC in those fractions. The contribution of LMW carbohydrates (Fractions III and IV; <500 Da, monosaccharides and oligosaccharides) to the LMW EDOC was much lower than the HMW fraction of 29%.

LMW compounds (free amino acids and oligopeptides) were major products of excreted proteins/amino acids, and 92% of excreted proteins/amino acids were found in the LMW fractions (Fractions III and IV). These free amino acids and oligopeptides showed a higher contribution (37%) to LMW EDOC than carbohydrates and organic acids. Only minor amounts of proteins and polypeptides were found as constituents of HMW EDOC, and the ratio to the total HMW EDOC was quite low. These findings are consistent with observations that free amino acids are the major components of excreted proteins/amino acids (Hellebust, 1974).

Excreted organic acids fractionated into acidic fractions were found in the LMW fraction, and their contribution to the LMW EDOC was slightly lower than that of amino acids and oligopeptides.

Carbohydrates, proteins/amino acids and organic acids accounted for 44, 29 and 27% of EDOC, respectively.

Turnover of dissolved organic constituents

To compare the production rate and ambient concentration of DOM, the rate measured from a 12 h incubation was used (unlike POM production) because short-term incubation resulted in an underestimation of the production rate of HMW compounds as described above. First, the production rate of EDOC was compared with the ambient concentration of DOC in each MW fraction (Fig. 11). The TT calculated for total DOC was 330 h, 8.7 times longer than the TT of POC. The TTs estimated for each MW fraction showed considerable variation. The shortest TT was calculated for Fraction III (150 h) whose MW ranged from 200 to 500 Da, judging from the elution of standard compounds. The TT obtained from the highest MW Fraction I was 2.0 times higher than Fraction III. However, the longest TT was obtained for the lowest MW Fraction IV (510 h). Since the production rate of EDOC in this fraction was highest among the 4 fractions, the longest TT was exclusively due to the high concentration of ambient DOC in this fraction.

When DOC was treated as two fractions; i.e., HMW (combined Fractions I and II, >500 Da) and LMW (combined Fractions III and IV, MW < 500 Da), the calculated TTs were longer in LMW DOC than in HMW DOC (Fig. 9). The comparison of the TTs of both MW fractions indicated the more labile nature of

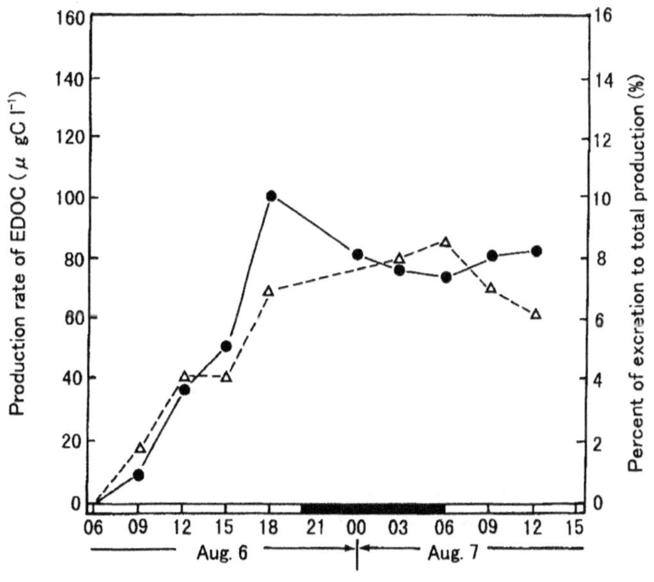

Fig. 10. Production of excreted dissolved organic carbon (●) and the percent of excretion to total (particulate and dissolved) production (△) in the surface layer at Lake Suwa (after Hama and Handa (1987b)).

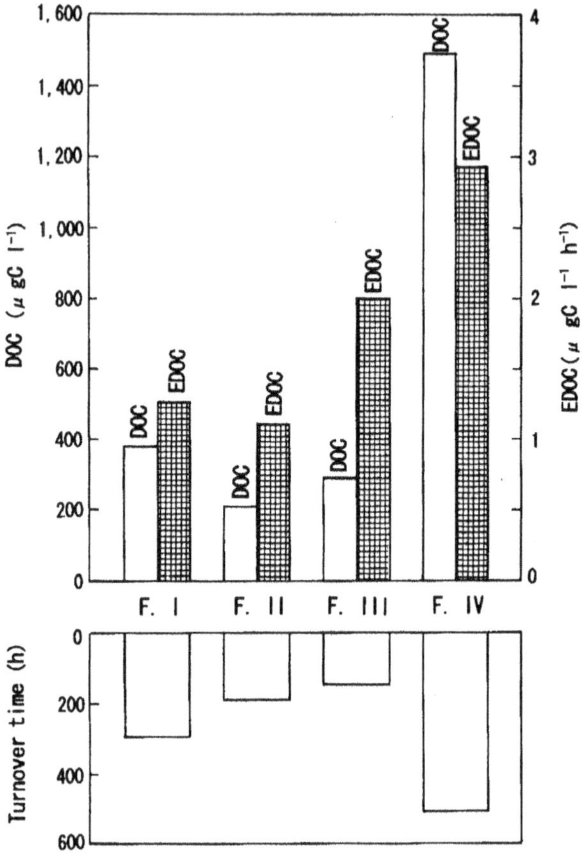

Fig. 11. The ambient concentration of dissolved organic carbon (DOC), the production rate of excreted dissolved organic carbon (EDOC) and the turnover time of DOC through phytoplankton excretion. DOC was fractionated by Sephadex G–25 gel–filtration (modified from Hama and Handa (1987b)).

HMW DOC; a finding inconsistent with the "traditional" concept that LMW compounds were more labile than HMW compounds (Saunders, 1976).

Recently, efforts have been directed to the characterization and estimation of the lability of DOC in seawater (Benner *et al.*, 1992; Guo *et al.*, 1994; Amon and Benner, 1996) as well as to the absolute concentration of DOC (e.g., Sharp *et al.*, 1995). Ammon and Benner (1996) measured the bacterial utilization rate of DOC, which was fractionated into HMW (> 1 kDa) and LMW (< 1 kDa) by tangential flow ultrafiltration. They found that bacterial DOC utilization was higher in the HMW fraction than in the LMW fraction, which is in disagreement with the "traditional" concept. Santschi *et al.* (1995) determined the "apparent" age of DOC fractionated according to their MW and found that HMW DOC (10 > kDa) was contemporary in age, clearly indicating the short TT of HMW DOC. DOC of more than 1kDa in their MW, on the contrary, had 380–4500 y of

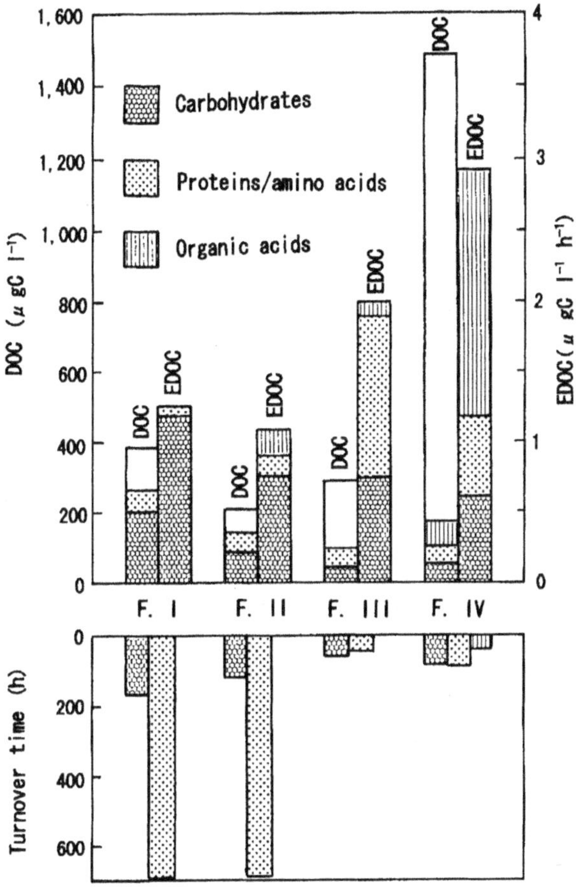

Fig. 12. The ambient concentration of dissolved organic carbon (DOC), the production rate of excreted dissolved organic carbon (EDOC) and the turnover time of dissolved organic matter through excretion. Organic matter was fractionated by Sephadex G–25 gel–filtration and ion exchange resin (modified from Hama and Handa (1987b)).

"apparent" age, indicating that the refractory nature of DOC with 1 kDa << 10 kDa. These recent findings on DOC in seawater are generally consistent with the spectrum of DOC TTs in lacustrine environments reported by Hama and Handa (1987b), consistent with HMW compounds being more labile than those of LMW.

To estimate the TT of DOC on both MW and organic compound levels, the carbohydrates, proteins/amino acids and organic acids in each MW fraction were measured (Fig. 12). The fraction identified by biomolecules was 36% of the total DOC, being much lower than POC. The MW distribution of biomolecules and unidentified DOC was quite different. In HMW Fraction I, carbohydrates were the major components (55% of DOC), and the contribution of uncharacterized DOC was lowest in this HMW fraction. However, in Fraction IV, carbohydrates, proteins/amino acids and organic acids accounted for only 12% of the total DOC, and a major part of the DOC in this fraction was not characterized by these biomolecules. These relationships between molecular weight and biomolecules were comparable with findings in other lake waters (Hama and Handa, 1980).

A comparison of production rate and ambient concentration was made at the organic compound level as well as at the MW level. The spectrum of the TTs of dissolved organic constituents (Fig. 12) was much different from that on the MW level (Fig. 11). The most obvious difference in the two spectra was in the TT of LMW compounds. Biomolecules in the lowest MW Fraction IV such as mono- and oligosaccharides, free amino acids/oligopeptides and organic acids showed short TTs of 86, 89 and 39 h, respectively. These TTs of organic compounds accounted for 0.07–0.18 of the TT of bulk DOC in this fraction. Thus, it appears that the existence of a high concentration of uncharacterized LMW DOC resulted in the long TT of bulk DOC in Fraction IV.

Concerning biomolecules such as carbohydrates and proteins/amino acids, TT decreases with a decrease in MW; LMW carbohydrates (mono- and oligosaccharides) and free amino acids and oligopeptides showed a shorter TT than HMW compounds (Figs. 9 and 12). This relation between the MW of biomolecules and lability is consistent with the "traditional" concept of a relation between lability and MW, but is against the relationship between DOC and MW. The discrepancy in the relation of MW and lability obtained for DOC and biomolecules may reflect the difference in the contribution of biomolecules to DOC in each MW fraction. HMW carbohydrates with a relatively short TT (150 h) were the major constituents of DOC in the HMW fraction (48%); and the shorter TT of DOC in this fraction than in the LMW fraction, reflected the shorter TT of the HMW carbohydrates. Although a very short TT of less than 70 h was estimated for LMW biomolecules, the TT of LMW DOC did not reflect the short TT of LMW biomolecules, because of the low contribution of these identified compounds (15% to LMW DOC).

Turnover of particulate fatty acid

Information such as the isotope ratio or atom %, obtained by the stable isotope tracer method, directly relates to the turnover time of organic molecules (Hama *et al.*, 1987), unlike the radioactive tracer method which needs a separate determination of production rate and concentration to estimate the TT. The ^{13}C/GC/MS method can measure both the concentration and flux of each organic molecule and the TT can be easily estimated at the organic molecule level. However, since the concentration of organic compounds measured by GC includes both phytoplankton and non-phytoplankton organic molecules, the ^{13}C atom % of organic molecules estimated by the ^{13}C/GC/MS method is "diluted" by the presence of non-phytoplankton compounds, which have a natural ^{13}C level of about 1.1%. Thus, TT obtained by this method does not match the actual TT of phytoplankton organic compounds but rather indicates the TT of organic compounds in all particulate matter.

TT determined by the ^{13}C/GC/MS method, in turn, can elucidate the difference in the proportion of the phytoplankton fraction among particulate organic constituents. Further, the chemical composition and lability of the detrital compounds, which are major components of non-phytoplankton POM, can be deduced from the difference in the degree of "dilution" by the non-phytoplankton fraction.

Fatty acid composition varies in different organisms such as phytoplankton, zooplankton, bacteria and higher plants. Additionally, it is generally accepted that the lability of fatty acids differs; unsaturated fatty acid (USFA) is more labile than saturated fatty acid (SAFA). Thus, the spectrum of the TTs of particulate fatty acids can afford information on the differences in the origin and lability of fatty acid species.

The estimation of the production rates and TT of fatty acids in particulate matter was carried out in the regional upwelling area near the Izu Islands, Japan, and was reported in Hama (1991).

Production of particulate fatty acids

The present incubation experiments were carried out in non-upwelling (Stn. I-2) and "maturing" upwelling (Stn. 49) sites. The ^{13}C atom % of fatty acids obtained by the saponification of particulate lipids were estimated for 12 h incubated samples during the daytime.

The concentrations of 15–17 fatty acids in particulate matter from 10 m at both stations are shown in Fig. 13. C16 saturated fatty acid, 16:0, was the most abundant acid and made up 37% of the total fatty acids at Stn. I-2. Other saturated acids, including 14:0 and 18:0, and a 16:1 monounsaturated acid (MUFA) were also abundant, accounting for 12–15% of the total. The contribution of polyunsaturated acids (PUFA) was low in this station and accounted for 11% of the total FA. At upwelling Stn. 49, three fatty acids, 14:0, 16:0 and 16:1 were the main components, as observed at Stn. I-2. However, only 18:0 accounted for 6.6% of the total FA at Stn. 49 (unlike at Stn.I-2). C18 PUFA made higher contributions at Stn. 49, each in the range of 1.4–4.5%. PUFAs are known to be constituents of the membrane lipid of microorganisms (Kayama *et al.*, 1989), and a higher contribution of PUFA in the upwelling Stn. 49 could reflect a greater abundance of microbial biomass than at Stn. I-2.

The composition of photosynthetic products at both stations was estimated and is summarized in Fig. 14. The major constituents of the photosynthetic products were generally comparable with the composition of POM; 16:0, 14:0 and 16:1 acids were found as the main constituents, accounting for 66 and 73% of the total FA production at Stns. I-2 and 49, respectively. This indicates that the FA composition of POM basically reflects that of phytoplankton. However, a distinct difference in the contribution between POM and photosynthetic products was found for some acids. The most striking difference was noticed for 18:0 acid, which accounted for 13% of the total FA in POM at Stn. I-2 but whose contribution to photosynthetic products was only 6.8%. A contribution to photosynthetic products lower than that the POM was also found for other minor acids, including iso 15:0, anteiso 15:0, 15:0, 17:0 and 18:1(n–7).

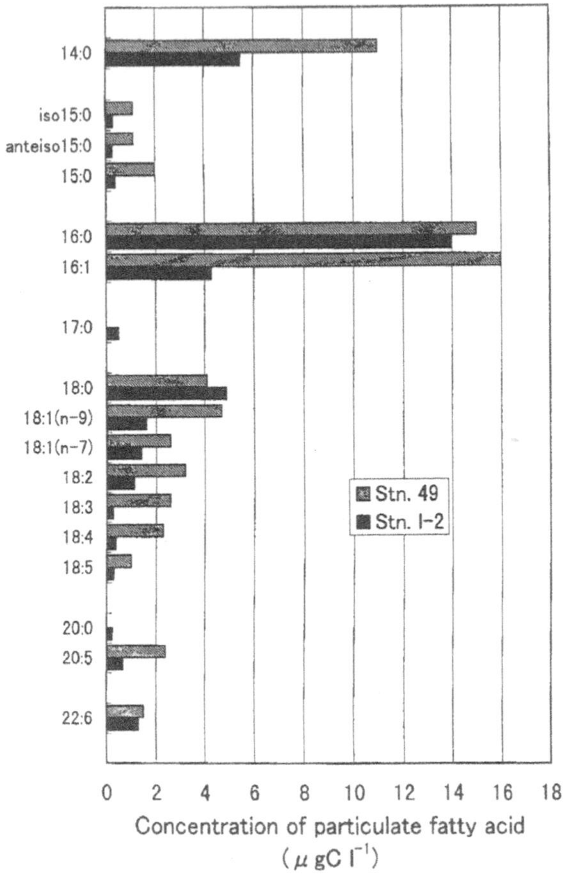

Fig. 13. Concentration of fatty acids in the particulate matter collected from 10 m at Stations 49 and I–2 (modified from Hama, 1991).

One obvious difference in the FA composition of POM and photosynthetic products between the two stations was the contribution of 16:1 acid. This acid is reportedly abundant in diatoms (Wood, 1974; Kayama *et al.*, 1989). This proved to be true of diatoms such as *Nitzschia pungens*, and *Chaetoceros curvisetum* which dominated at upwelling Stn. 49 (Furuya *et al.*, 1986), whereas dinoflagellates were the main species at Stn. I-2, and are known to be poor in 16:1 (Wood, 1974; Kayama *et al.*, 1989). Thus, the difference in the dominant phytoplankton population between the two stations possibly resulted in the difference not only in the FA composition of photosynthetic products but also in that of POM.

Turnover of particulate fatty acids

The TT of particulate fatty acids was calculated from the ^{13}C atom % of each fatty acid of the incubated samples containing ^{13}C at both stations (Fig. 15). The

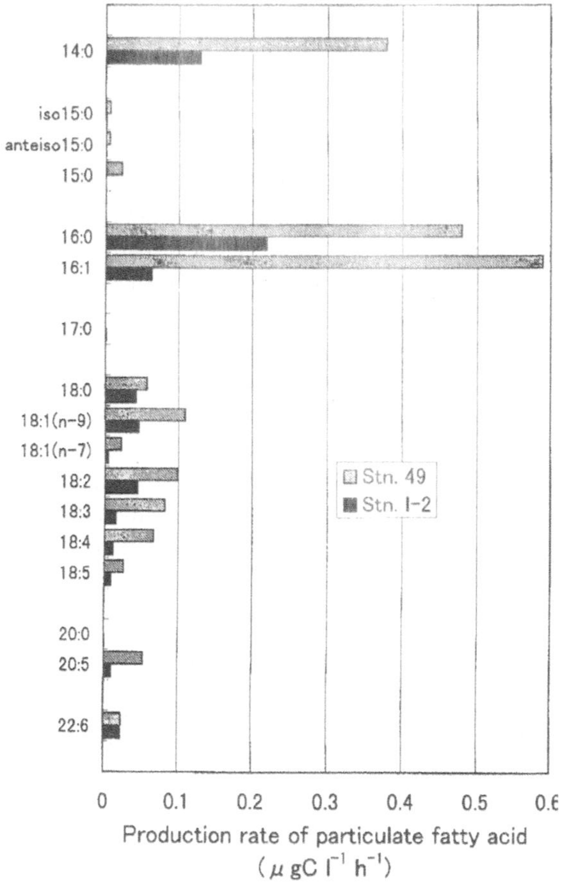

Fig. 14. Production rates of particulate fatty acids at 10 m at Stns. 49 and I–2. Rates for 17:0 and 20:0 at Stn. 49 could not be determined, because of their low concentration (modified from Hama (1991)).

TTs calculated for the total fatty acids were 48 and 17 h at Stns. I-2 and 49, respectively; the TT obtained at upwelling Stn. 49 was about 2.8 times shorter.

The individual TTs of each FA varied greatly at each station, ranging from 6 to 710 h and 6 to 111 h at Stns. I-2 and 49, respectively. PUFA generally exhibited short TTs at both stations. The TT of all C18 PUFA, especially, showed values shorter than 20 h. On the contrary, odd carbon number acids such as 15:0 and 17:0 (only at Stn. I-2), 18:0, 18:1 (n-7) and 20:0 (only at Stn. I-2) showed TTs above 30 h. These characteristics of TT were commonly observed at both stations. However, regional differences in the TT between the two stations were found in compositionally major acids such as 14:0, 16:0 and 16:1; the TT of these acids at Stn. I-2 were 5.3–23 times longer than those at Stn. 49.

The spectrum of the TT of FA obtained here may reflect two factors: the difference in the turnover of FA in phytoplankton cells and the difference in the phytoplankton fraction in particulate FA among FA species (the difference in the degree of "dilution" with the non-phytoplankton fraction among FA). In the case of amino acids and monosaccharides (which are constitutes of proteins/amino acids and carbohydrates, respectively), the first factor may be the main effect on the spectrum of TT because the difference in amino acid and monosaccharide composition among organisms is minor compared with FA (Hama, 1988). On the other hand, FA composition varies with organism groups, and some fatty acids are known to be specific for certain organisms. Thus, the variability of the phytoplankton fraction among fatty acids in POM is likely to govern the variability of TT.

The longer TT of fatty acids, including the branched, 17:0 and 18:1 (n-7) acids, could result from the serious "dilution" effect of a non-phytoplankton fraction, since these acids have been recognized as constituents of bacterial cells (Perry *et al.*, 1979; Joint and Morris, 1982; Wakeham and Canuel, 1988). Other acids, such as 15:0 and 20:0, which likewise showed a longer TT, were also found to be more likely constituents of bacteria than of phytoplankton (Volkman *et al.*, 1980). These facts clearly indicated that the long TT of these FA was mainly due to a minor contribution of phytoplankton.

A distinct difference in TT between the two stations was found in 14:0, 16:0 and 16:1. Although the TTs of these three acids were short at Stn. 49, comparable with the values of PUFA, the values at Stn. I-2 were much longer. As shown in Fig. 14, the production rates of these acids were high and together accounted for 65% of the total FA production even at Stn. I-2, indicating that the long TT of these acids were unlikely to be due to their low production, but were probably caused by the low contribution of phytoplankton.

The proportion of phytoplankton in POM has been discussed repeatedly (Eppley *et al.*, 1977; Furuya *et al.*, 1986; Fuhrman *et al.*, 1987; Kokkinakis and Wheeler, 1987; Dortch and Packard, 1989), and it is commonly accepted that the proportion of phytoplankton is generally higher in eutrophic areas than in oligotrophic areas. A comparison of the concentration of Chl *a* at both stations (1.98 and 0.18μg l^{-1} at Stns. 49 and I-2, respectively) and the Chl. *a*-specific concentration of total fatty acid (36 and 210μgC μg Chl. a^{-1} at Stns. 49 and I-2, respectively) strongly suggests that the proportion of phytoplankton fatty acids to detrital fatty acids was generally higher at upwelling Stn. 49 than at Stn. I-2. The fact that the mean TT for total FA at Stn. 49 was 2.8 times higher than at Stn. I-2 indicates that the degree of "dilution" with detrital FA could be higher at Stn. I-2 than at Stn. 49. Thus, the obviously longer TTs of the three major FA at Stn. I-2, compared with those at Stn. 49, were likely caused by the high proportion of detritus in these FA, indicating that 14:0, 16:0 and 16:1 acids were abundant in detrital lipid. At Stn. 49, on the other hand, little "dilution" was expected because the phytoplankton fraction in particulate lipids should be high due to the increased phytoplankton biomass in the upwelling station. Thus, the TT of major FA at this station did not deviate seriously from the TT in the phytoplankton cell.

In contrast to the marked difference in TT of major FA between the two stations, little difference was found between the TTs of PUFA. The short TTs of

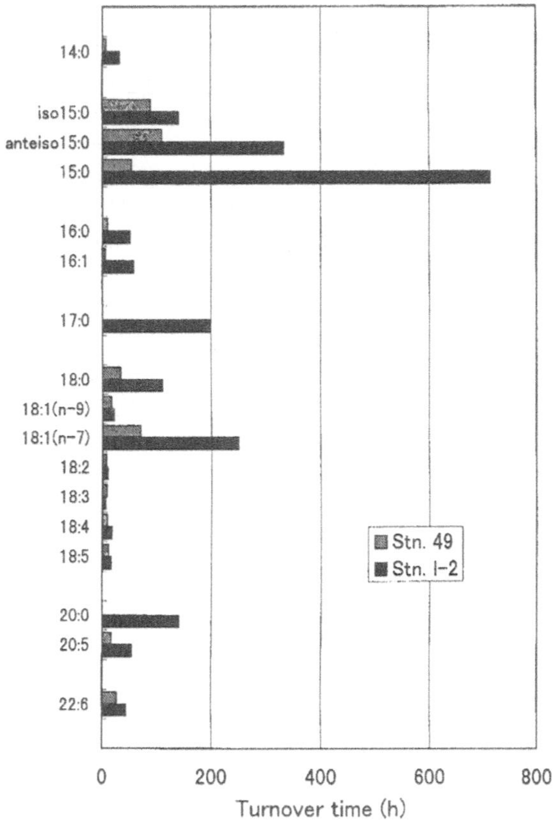

Fig. 15. Turnover times of particulate fatty acids through phytoplankton photosynthesis at 10 m at Stns. 49 and I–2. Rates for 17:0 and 20:0 at Stn. 49 could not be determined, because of their low concentration (modified from Hama (1991)).

PUFA at Stn. 49 were roughly comparable with those of the three major FA, probably reflecting the TT of phytoplankton cellular lipids, because little "dilution" with detrital FA was expected for PUFA as well as for the 14:0, 16:0 and 16:1 acids discussed above. The TTs of PUFA at Stn. I-2, on the other hand, were much shorter than the major three acids. This indicates that the "dilution" effect, which increased the TT of those acids, did not occur in the case of PUFA.

PUFA are more labile than SAFA and MUFA in aquatic environments (de Baar *et al.*, 1983; Prahl *et al.*, 1984; Wakeham and Canuel, 1988), and the PUFA constituting cellular lipids of microorganisms probably degraded rapidly after their death, failing to survive as constituents of detrital organic matter. Tanoue and Handa (1980) compared the fatty acid composition of particulate matter collected from the surface to 2,500 m depth in the northern North Pacific. Although PUFA were found in the particulate matter from the surface layer, they disappeared from particulate matter between the 500 and 2,500 m depths. From the results of vertical flux of fatty acids determined by sediment trap experiments, de Baar *et al.* (1983) estimated the "half depth", which corresponds to "the depth range over which 50% has been lost", of each fatty acid in sinking particles. The "half depths" of 14:0, 16:0 and 16:1 estimated varied from 820 to 960 m, whereas those of PUFA were considerably shallower (440–760 m), representing more the labile nature of PUFA. Thus, the short TT of PUFA, even at oligotrophic Stn. I-2, suggested that the "dilution" effect with detritus cannot be a serious factor for labile organic compounds such as PUFA.

In conclusion, the spectra of TT of fatty acids obtained from the upwelling and non-upwelling stations suggested that the variability of TT mainly reflected the difference in the proportion of the phytoplankton fraction to particulate matter among fatty acids. The TT of compositionally major fatty acids (14:0, 16:0 and 16:1), which were composed of SAFA and MUFA, showed a "dilution" effect of the non-phytoplankton fraction, indicating that these acids were not so labile and existed as components of detrital lipids. On the contrary, a minor "dilution" effect on PUFA by non-phytoplankton was suggested. This was likely due to the fact that small amounts of PUFA exist as constituents of detritus, suggesting the lability of these compounds. Although direct measurements cannot be made of the composition of detrital components, the composition was estimated by the degree of "dilution", and the refractory characteristics appeared to be a result of preferential decomposition.

CONCLUSION

The wide variability in photosynthetically supplied organic matter in the surface layer of aquatic environments, as well as the variability in the TTs of organic compounds discussed in the literature, showed the necessity of a qualitative knowledge of organic constituents in order to understand the relation of organic matter to the biogeochemical cycles in the ocean. The role of oceans in the global carbon cycle is currently of great interest. The transport rate of carbon from the surface layer to the deep layer (Longhurst, 1991; Jahnke, 1996) and from the coastal zone to the open ocean (Wollast, 1991) has been estimated from the viewpoint of its relation to the concentration of atmospheric CO_2. Measurements of the fluxes of bulk carbon are clearly the most basic estimates. However, the role of organic carbon transported as the carbon sink largely depends on lability. When most of the organic constituents transported are labile, they will be decomposed using dissolved oxygen to produce CO_2. Thus, though these compounds are of great importance to the biogeochemical cycles in the ocean, they play no role as a carbon sink. Only when transported organic matter is composed of refractory substance, can they have a role as a carbon sink. Thus, the measurements of bulk carbon flux alone are insufficient to evaluate the role of organic matter transportation to the global carbon budget. The analysis of chemical nature together with temporal parameters is likely necessary to elucidate the role of organic matter in the global carbon budget.

REFERENCES

Admirraal, W., H. Peletier and P. W. P. M. Laane (1986) : Nitrogen metabolism of marine planktonic diatoms; excretion, assimilation and cellular pools of free amino acids in seven species with different cell size. *J. Exp. Mar. Biol. Ecol.*, **98**, 241–263.

Amon, R. M. W. and R. Benner (1996) : Bacterial utilization of different size classes of dissolved organic matter., *Limnol. Oceanogr.*, **41**, 41–51.

Anerson, G. C. and R. P. Zeutschel (1970) : Release of dissolved organic matter by marine phytoplankton in coastal and offshore areas of the Northeast Pacific Ocean., *Limnol. Oceanogr.*, **15**, 402–407.

Barlow, R. G. (1982a) : Phytoplankton ecology in the Southern Benguela Current. II. Carbon assimilation patterns., *J. Exp. Mar. Biol. Ecol.*, **63**, 229–237.

Barlow, R. G. (1982b) : Phytoplankton ecology in the Southern Benguela Current. III. Dynamics of a bloom., *J. Exp. Mar. Biol. Ecol.*, **63**, 239–248.

Benner, R. J. D. Pakulski, M. McCarthy, J. I. Hedges and P. G. Hatcher (1992) : Bulk chemical characteristics of dissolved organic matter in the ocean., *Science*, **255**, 1561–1564.

Berger, W. H., K. Fischer, C. Lai, G. Wu (1987): Ocean productivity and organic carbon flux. Part I. Overview and maps of primary production and export production. p. 67. In *SIO Reference Series, 87–30*, Scripps Institute of Oceanography, University of California, San Diego.

Cauwet, G. (1981) : Non-living particulate matter. pp. 71–89. In *Marine Organic Chemistry*, ed. by E. K. Duursma and R. Dawson, Elsevier Scientific Publishing, Amsterdam.

Chau, Y. K., Chuecas Y. K. L., and J. P. Riley (1967) : The component of combined amino acids of some marine phytoplankton species., *J. Mar. Biol. Ass. U.K.*, **47**, 543–554.

Cuhel, R. L., P. B. Ortner and D. R. S. Lean (1984) : Night synthesis of protein by algae., *Limnol. Oceanogr.*, **29**, 731–744.

Curl, H. and L. F. Small (1965) : Variations in photosynthetic assimilation ratios in natural, marine phytoplankton communities., *Limnol. Oceanogr.*, **10**, suppl., 67–73.

De Baar, H. J. W., J. W. Farrington and S. G. Wakeham (1983) : Vertical flux of fatty acids in the North Atlantic Ocean. *J. Mar. Res.*, **41**, 9–41.

Dortch, Q. and T. T. Packard (1989) : Differences in biomass structure between oligotrophic and eutrophic marine ecosystems., *Deep-Sea Res.*, **6**, 223–240.

Eppley, R. W., W. G. Harrison, S. W. Chisholm and E. Stewart (1977) : Particulate organic matter in surface waters off southern California and its relation to phytoplankton., *J. Mar. Res.*, **35**, 671–696.

Fernandez, E., P. Serret, I. Madariaga, D. S. Harbour and A. G. Davies (1992) : Photosynthetic carbon metabolism and biochemical composition of spring phytoplankton assemblages enclosed in microcosms: the diatom—*Phaeocystis* sp. succession., *Mar. Ecol. Prog. Ser.*, **90**, 89–102.

Fitzwater, S. E., G. A. Knauer and J. H. Martin (1982) : Metal contamination and its effect on primary production measurements., *Limnol. Oceanogr.*, **27**, 544–551.

Fleming, R. H. (1957) : General features of the ocean. p.87–107. In *Treatise on Marine Ecology and Paleoecology*, ed. J. W. Hedgepeth. *Geol. Soc. Amer. Mem.*, **67**.

Fogg, G. E. (1966) : Extracellular products of algae., *Oceanogr. Mar Biol. Ann. Rev.* **4**, 195–212.

Furman, J. A., T. Sleeter and C. Carlson (1987) : Oligotrophic ocean biomass is dominated by non-photosynthetic bacteria, even in the euphotic zone., *EOS*, **68**, 1729.

Furuya, K., M. Takahashi and T. Nemoto (1986) : Summer phytoplankton community structure and growth in a regional upwelling area off Hachijo Island, Japan., *J. Exp. Mar. Biol. Ecol.*, **96**, 43–55.

Gieskes, W. W. C., G. W. Kraay and M. A. Baars (1979): Current ^{14}C methods for measuring primary production: Gross underestimates in oceanic waters., *Neth. J. Sea Res.*, **13**, 58–78.

Goldman, J. C. (1980) : Physiological processes, nutrient availability, and the concept of relative growth rate in marine phytoplankton ecology. pp.179–194. In *Primary Productivity in the Sea*, ed. by P. G. Falkowski, Plenum Press, New York.

Goldman, J. C., J. M. McCarthy and D. G. Peavey (1979) : Growth rate influence on the chemical composition of phytoplankton in oceanic waters., *Nature*, **279**, 210–215.

Gordon, A. E. and A. Frigerio (1972) : Mass fragmentography as an application of gas-liquid chromatography-mass spectrometry in biological research., *J. Chromatogr.*, **73**, 401–417.

Goes, J. I., N. Handa, S. Taguchi, T. Hama and H. Saito (1995) : Impact of UV radiation on the production patterns and composition of dissolved free and combined amino acids in marine phytoplankton., *J. Plankton Res.*, **17**, 1337–1362.

Grill, E. A. and R. A. Richards (1964) : Nutrient regeneration from phytoplankton decomposing in seawater., *J. Mar. Res.*, **22**, 51–69.

Guo, L. C. H. Jr Coleman, P. H. Santschi (1994) : The distribution of colloidal and dissolved organic carbon in the Gulf of Mexico., *Mar. Chem.*, **45**, 105–119.

Haberstroh, P. R. and S. I. Ahmed (1985) : Resolution by high pressure liquid chromatography of intracellular and extracellular free amino acids of a nitrogen deficient marine diatom, Skeletonema costatum (Grev.) Cleve, pulsed with nitrate or ammonium., *J. Exp. Mar. Biol. Ecol.*, **101**, 101–117.

Hama, J. and N. Handa (1986) : Analysis of the production processes of natural phytoplankton population using ^{13}C-gas chromatography-mass spectrometry methods., *Res. Org. Geochem.*, **5**, 41–46. (in Japanese)

Hama, J. and N. Handa (1992) : Diel photosynthetic production of cellular organic matter in natural phytoplankton populations, measured with ^{13}C and gas chromatography/mass spectrometry. I Monosaccharides., *Mar. Biol.*, **112**, 175–181.

Hama, T. (1983) : Carbohydrates in natural waters - especially in Lake Suwa., *La Mer*, **21**, 248–254 (in Japanese).

Hama, T. (1988) ^{13}C-GC-MS analysis of photosynthetic products of the phytoplankton population in the regional upwelling area around the Izu Islands, Japan., *Deep-Sea Res.*, **35**, 91–110.

Hama, T. (1991) Production and turnover rates of fatty acids in marine particulate matter through phytoplankton photosynthesis., *Mar. Chem.*, **33**, 213–227.

Hama, T. (1992) : Primary productivity and photosynthetic products around the Kuroshio warm-core ring., *Deep-Sea Res.*, **39** (Suppl. 1), S279–293.

Hama, T. and N. Handa (1980) : Molecular weight distribution and characterization of dissolved organic matter from lake waters., *Arch. Hydrobiol.*, **90**, 106–120.

Hama, T., and N. Handa (1982) : The seasonal variation of organic constituents in a eutrophic lake, Lake Suwa Japan Part I. Particulate organic matter., *Arch. Hydrobiol.*, **93**, 446–465.

Hama, T., and N. Handa (1983) : The seasonal variation of organic constituents in a eutrophic lake, Lake Suwa Japan Part II. Dissolved organic matter. *Arch. Hydrobiol.*, **98**, 443–462.

Hama, T. and N. Handa (1987a) : Pattern of organic matter production by natural phytoplankton population in a eutrophic lake. 1 Intracellular products., *Arch. Hydrobiol.*, **109**, 107–120.

Hama, T. and N. Handa (1987b) : Pattern of organic matter production by natural phytoplankton population in a eutrophic lake. 2 Extracellular products., *Arch. Hydrobiol.*, **109**, 227–243.

Hama, T., Miyazaki, T., Ogawa, Y., Iwakuma, T., Takahashi, M., Otsuki, A., and Ichimura, S., (1983) : Measurement of photosynthetic production of a marine phytoplankton population using a stable ^{13}C isotope., *Mar. Biol.*, **73**, 31–36.

Hama, T., J. Hama and N. Handa (1987) : Determination of amino acid production rate of a marine phytoplankton population with ^{13}C and gas chromatography-mass spectrometry., *Limnol. Oceanogr.*, **32**, 1144–1153.

Hama, T., J. Hama and N. Handa (1993) : ^{13}C tracer methodology in microbial ecology with special reference to primary production processes in aquatic environments. *Adv. Microbial. Ecol.*, **13**, 39–83.

Handa, N. (1969) : Carbohydrate metabolism in the marine diatom *Skeletonema costatum*., Mar. Biol., 4, 208–214.

Handa, N. and H. Tominaga (1969) : A detailed analysis of carbohydrates in marine particulate matter., *Mar. Biol.*, **2**, 223–235.

Handa, N. and K. Yanagi (1969) : Studies on water-extractable carbohydrates of the particulate matter from the northwest Pacific Ocean., *Mar. Biol.*, **4**, 197–207.

Harris, G. P. (1980) : The measurement of photosynthesis in natural populations of phytoplankton. pp. 129–187. In *The Physiological Ecology of Phytoplankton.* ed. by I. Morris, Blackwell Scientific Publications, Oxford.

Hellebust, J. A. (1965) : Excretion of some organic compounds by marine phytoplankton., *Limnol. Oceanogr.*, **10**, 192–206.

Hellebust, J. A. (1974) : Extracellular products. pp. 838–865. In. *Algal Physiology and Biochemistry*, ed. by W. P. D. Stewart, Blackwell Scientific Publications, Oxford.

Heyward, T. L., E. L. Venrick and J. A. McGowan (1983): Environmental heterogeneity and plankton community structure in the central North Pacific., *J. Mar. Res.*, **41**, 711–729.

Hitchcock, G. L. (1978) : Labeling patterns of carbon-14 in net plankton during a winter-spring bloom. *J. Exp. Mar. Biol. Ecol.*, **31**, 141–153.

Hitchcock, G. L. (1983) : Photosynthate partitioning in cultured marine phytoplankton. I. Dinoflagellates., *J. Exp. Mar, Biol. Ecol.*, **69**, 21–36.

Horne, A. J., G. E. Fogg and D. J. Eagle (1969) : Studies in situ of the primary production of an area of inshore Antarctic Sea. *J. Mar. Biol. Assoc. U.K.*, **49**, 393–405.

Iturriaga, R. and H. -G. Hoppe (1977) : Observations of heterotrophic activity on photoassimilated organic matter., *Mar. Biol.*, **40**, 101–108.

Jahnke. R. A. (1996) : The global ocean flux of particulate organic carbon: A real distribution and magnitude., Global Biogeochem. *Cycles*, **10**, 71–88.

Jenkins (1982) : Oxygen utilization rates in the North Atlantic Subtropical Gyre and primary production in oligotrophic systems., *Nature*, **300**, 246–248.

Jenkins and Goldman (1985) : Seasonal oxygen cycling and primary production in the Sargasso Sea., *J. Mar. Res.*, **43**, 465–491.

Jensen, L. M., N. O. G., Jorgensen and M. Sondergard (1985) : Specific activity: significance in estimating release rates of extracellular dissolved organic carbon (EDOC) by algae., *Verh. Internat. Verein. Limnol.*, **22**, 2893–2897.

Joint, P. A. and R. J. Morris (1982) : The role of bacteria in the turnover of organic matter in the sea., *Oceanogr. Mar Biol. Annu. Rev.*, **20**, 65–118.

Jonasdottir, S. H., D. Fields and S. Pantoja (1995) : Copepod egg production in Long Island Sound, USA, as a function of the chemical composition of seston., *Mar. Ecol. Prog. Ser.*, **119**, 87–98.

Jumars, P. A., D. L. Baross, N. J. Pery and B. W. Ferost (1989) : Closing the microbial loop: dissolved carbon pathway to heterotrophic bacteria from incomplete ingestion, digestion and absorption in animals., *Deep-Sea Res.*, **36**, 483–495.

Kayama, M. S. Araki and S. Sato (1989) : Lipids of marine plants. pp.3–48. In *Marine Biogenic Lipids, Fats, and Oils. Vol. II"*, ed. by R. G. Ackman, CRC Press, Florida

Knauer, G. A. (1993) : Productivity and new production of the ocean system. pp. 221–231 In *Interactions of C, N, P and S Biogeochemical Cycles and Global Change.* Ed. by R. Wollast, Springer-Verlag, Berlin..

Knopka, A. and M. Schnur (1980) : Effect of light intensity on macromolecular synthesis in cyanobacteria., *Microb. Ecol.*, **6**, 291–301.

Koblentz-Mishke, O. I., V. V. Volkovinskiy and Yu. G. Kabanova (1970): Plankton primary production of the World Ocean. pp.183–193 In *Scientific Exploration of the Sough Pacific*, ed. by W. Wooster, National Academy of Sciences., Washington, D. C.

Kokkinakis, S. A. and P. A. Wheeler (1987) : Nitrogen limitation and phytoplankton growth in coastal upwelling regions. *Limnol. Oceanogr.*, **32**, 1112–1123.

Kouchi, H. (1982) : Direct analysis of ^{13}C abundance in plant carbohydrates by gas chromatography-mass spectrometry., *J. Chromatogr.*, **21**, 305–323.

Larsson, U. and A. Hagstrom (1982) : Phytoplankton exudate release as an energy source for the growth of pelagic bacteria., *Mar. Biol.*, **52**, 199–206.

Laws, E. A., G. R. DiTullio and D. G. Redalje (1987) : High phytoplankton growth and production rates in the North Pacific subtropical gyre., *Limnol. Oceanogr.*, **32**, 905–918.

Laws, E. A., G. R. DiTullio, P. R. Betzer, D. M. Karl and K. L. Carder (1989) : Autotrophic production and elemental fluxes at 26°N, 155°W in the North Pacific subtropical gyre., *Deep-Sea Res.*, **36**, 103–120.

Leftley, J. W., D. J. Bonin and S. Y. Maestrini (1983): Problems in estimating marine phytoplankton growth, productivity and metabolic activity in nature: An overview of methodology., *Oceanogr. Mar. Biol. Annu. Rev.*, **21**, 23–66.

Letelier, R. M., J. E. Dore, C. D. Winn, and D. M. Karl (1996) : Seasonal and interannual variations in photosynthetic carbon assimilation at Station ALOHA., *Deep-Sea Res.*, **43**, 467–490.

Li, W. K. W., H. E. Glover and I. Morris (1980) : Physiology of carbon photoassimilation by Oscillatoria thiebautii in the Caribbean Sea., *Limnol. Oceanogr.*, **25**, 447–456.

Lignell, R. (1990) : Excretion of organic carbon by phytoplankton: its relation to algal biomass, primary productivity and bacterial secondary productivity I: the Baltic Sea., *Mar. Ecol. Prog. Ser.*, **68**, 85–99.

Longhurst, A. R. (1991): Role of the marine biosphere in the global carbon cycle., *Limnol. Oceanogr.*, **36**, 1507–1526.

Longhurst, A., S. Sathyendranath, T. Platt and C. Caverhill (1995) : An estimate of global primary production in the ocean from satellite radiometer data., *J. Plankton Res.*, **17**, 1245–1271.

Mague, T. H., E. Friberg, D. J. Hughes and I. Morris (1980) : Extracellular release of carbon by marine phytoplankton; a physiological approach., *Limnol. Oceanogr.*, **25**, 262–279.

Malone, T. C. (1971) : The relative importance of nanoplankton and netplankton as primary producers in tropical oceanic and neritic phytoplankton communities., *Limnol. Oceanogr.*, **16**, 633–639.

Marra, J. and K. R. Heinemann (1987) : Primary production in the North Pacific central gyre: some new measurements based on ^{14}C., *Deep–Sea Res.*, **34**, 1821–1829.

Martin, J. H., G. A. Knauer, D. M. Karl and W. W. Broenkow (1987) VERTEX: carbon cycling in the northeast Pacific., *Deep–Sea Res.*, **34**, 267–285.

Matsunaga, K. (1981) : Studies on the decompositive processes of phytoplanktonic organic matter., *Jap. J. Limnol.*, **42**, 220–229.

Meybeck, M. (1982) : Carbon, nitrogen, and phosphorus transport by world rivers., *Am. J. Sci.*, **282**, 401–450.

Moal, J., V. Martin–Jezequel, R. P. Harris, J–F. Samain, and S. A. Poulet (1987) : Interspecific and intraspecific variability of the chemical composition of marine phytoplankton., *Oceanolo. Acta.*, **10**, 339–346.

Morris, I. And W. Skea (1978) : Products of photosynthesis in natural populations of marine phytoplankton from the Gulf of Maine., *Mar. Biol.*, **47**, 303–312.

Murray, J. W (1992) : The Oceans, pp.175–211. In *Global Biogeochemical Cycles*, ed. by S. S. Butcher, R. J. Charlson, G. H. Orians and G. V. Wolfe, Academic Press Limited, London.

Myklestad, S. (1974) : Production of carbohydrates by marine planktonic diatoms. I. Comparison of nine different species in culture., *J. Exp. Mar. Biol. Ecol.*, **15**, 261–274.

Myklestad, S. (1977) : Production of carbohydrates by marine planktonic diatoms. II. Influence of the N/P ratio in the growth medium on the assimilation ratio, growth rate, and production of extracellular carbohydrate by the diatoms *Chaetoceros affinis var. willei* and *Skeletonema costatum. J. Exp. Mar. Biol. Ecol.*, **29**, 161–179.

Myklestad, S. and Haug, A. (1972) : Production of carbohydrates by the marine diatom *Chaetoceros affinis var. willei* (Gran) Hustedt. I. Effect of the concentration of nutrients in the culture medium., *J. Exp. Mar. Biol. Ecol.*, **9**, 125–136.

Olive, J. H. and J. H. Morrison (1967) : Variations in distribution of ^{14}C in cell extracts of phytoplankton living under natural conditions., *Limnol. Oceanogr.*, **12**, 383–391.

Olive, J. H., D. M. Benton and J. Kishler (1969) : Distribution of C–14 products of photosynthesis and its relationship to phytoplankton composition and rate of photosynthesis., *Ecology*, **50**, 380–386.

Olson, J. S., R. M. Garrels, R. A. Berner, T. V. Armentano, Mi. I. Dyer and D. H. Taalon (1985) : The natural carbon cycle. pp.175–213. In *Atospheric Carbon Dioxide and the Global Carbon Cycle.* ed. by J. R. Trabalka, US Department of Energy, Washington, D.C.

Otsuki, A. and T. Hanya (1972) : Production of dissolved organic matter from dead green algal cells. I. Aerobic microbial decomposition., *Limnol. Oceanogr.*, **17**, 248–257.

Parsons, T. R., K. Stephens and J. D. M. Strickland (1961) : On the chemical composition of eleven species of marine phytoplankton., *J. Fish. Res. Board Can.*, **18**, 1001–1016.

Perry, G. J., J. K. Volkman, R. B. Johns and H. J. Jr Bovor (1979) : Fatty acids of bacterial origin in contemporary marine sediments. *Geochim. Cosmochim. Acta*, **43**, 1715–1725.

Prahl, F. G., G. Eglinton, E. D. S. Corner, S. C. M. OíHara and T. E. V. Forsberg (1984) : Changes in plant lipids during passage through the gut of Calanus., *J. Mar. Biol. Ass. U.K.*, **64**, 317–334.

Priscu, J. C. and L. R. Priscu (1984) : Photosynthate partitioning by phytoplankton in a New Zealand coastal upwelling system., *Mar. Biol.*, **81**, 31–40.

Redfield, A. C., B. H. Ketchum and F. A. Richards (1963) : The influence of organisms on the composition of seawater. pp.26–77. In *The Sea, vol. 2*, ed. by M. N. Hill, Wiley–Interscience, New York.

Rivkin, R. B. (1989) : Influence of irradiance and spectral quality on the carbon metabolism of phytoplankton. I. Photosynthesis, chemical composition and growth., *Mar. Biol. Prog. Ser.*, **55**, 291–304.

Romankovich, E. A. (1984) : Geochemistry of Organic Matter in the Ocean. p.334 Springer–Verlag, New York,.

Sakshaug, E. and O. Holm–Hansen (1977) : Chemical composition of *Skeletonema costatum* and *Pavlova* (*Monochrysis*) *lutheri* as a function of nitrate–, phosphate– and iron–limited growth., *J. Exp. Mar. Biol. Ecol.*, **28**, 109–123.

Samuel, S., N. M. Shah and G. E. Fogg (1971) : Liberation of extracellular products of photosynthesis by tropical phytoplankton., *J. Mar. Biol. Ass. U.K.*, **51**, 793–798.

Santschi, P. H., L. Guo, M. Baskaran, S. Trumbore, J. Southon, T. S. Bianchi, B. Honeyman and L. Cifuentes (1995) : Isotopic evidence for the contemporary origin of high–molecular weight organic matter in oceanic environments., *Geochim. Cosmochim. Acta*, **59**, 625–631.

Saunders, G. W. (1976) : Decomposition in freshwater. pp.341–373. In The role of Terrestrial and Aquatic Organisms in *Decomposition Processes*, ed. by J. M. Anderson and A. Macfadyen, Blackwell Scientific Publications, Oxford.

Sharp, J. H. (1977) : Excretion of organic matter by marine phytoplankton. Do healthy cells do it?, *Limnol. Oceanogr.*, **22**, 381–199.

Sharp, J. H., R. Benner, L. Bennet, C. A. Carlson, S. E. Fizwater, T. T. Peltzer and L. M. Tupas (1995) : Analyses of dissolved organic carbon in seawater: the JGOFS EqPac methods comparison., *Mar. Chem.*, **48**, 91–108.

Siezen, R. J. and T. H. Mague (1978) : Amino acids in suspended particulate matter from oceanic and coastal waters of the Pacific., *Mar. Chem.*, **6**, 215–231.

Slawyk, G., Y. Collos and J. C. Auclair (1977) : The use of the ^{13}C and ^{15}N isotopes for the simultaneous measurement of carbon and nitrogen turnover rates in marine phytoplankton., *Limnol. Oceanogr.*, **22**, 925–932.

Slawyk, G., Y. Collos and J. C. Auclair (1979) : Reply to comment by Fisher *et al.*, *Limnol. Oceanogr.*, **24**, 595–597.

Smith, A. E. and I. Morris (1980) : Pathways of carbon assimilation in phytoplankton from the Antarctic Ocean., *Limnol. Oceanogr.*, **25**, 865–872.

Smith, W. O., R. T. Barber and S. A. Huntsmann (1977) : Primary production off the coast of Northwest Africa: Excretion of dissolved organic matter and its heterotrophic uptake., *Deep–Sea Res.*, **24**, 35–47.

Steemann Nielsen, E. (1952) : The use of radioactive carbon (C 14) for measuring organic production in the sea., *J. Cons. Int. Explor Mer.*, **18**, 117–140.

Storch, T. A. and G. W. Saunders (1978) : Phytoplankton extracellular release and its relation to seasonal cycle of dissolved organic carbon in a eutrophic lake., *Limnol. Oceanogr.*, **23**, 112–119.

Summonds, R. W., W. E. Pereira, W. E. Reynolds, T. C. Rindfleisch and A. M. Duffield (1974) : Analysis of twelve amino acids in biological fluids by mass fragmentography., *Anal. Chem.*, **46**, 582–587.

Takahashi, M., J. Ishizaka, T. Ishimaru, L. P. Atkinson, T. N. Lee, Y. Yamaguchi, Y. Fujita and S. Ichimura (1986) : Temporal change in nutrient concentrations and phytoplankton biomass in short time scale local upwelling around the Izu Peninsula, Japan., *J. Plankton Res.*, **8**, 1039–1049.

Tanoue, E. and N. Handa (1980) : Vertical transport of organic materials in the northern North Pacific as determined by sediment trap experiment. Part 1. Fatty acid composition. *J. Oceanogr. Soc. Japan.*, **36**, 231–245.

Thomas, J. P. (1971) : Release of dissolved organic matter from natural populations of marine phytoplankton., *Mar. Biol.*, **11**, 311–323.

Thomas, W. H. (1970) : On nitrogen deficiency in tropical Pacific oceanic phytoplankton: photosynthetic parameters in poor and rich water., *Limnol. Oceanogr.*, **15**, 380–385.

Van den Heuvel, W. J. A., J. L. Smith and J. S. Cohen (1970) : Gas liquid chromatography and mass spectrometry of carbon–13 enriched and deuterated amino acids as trimethylsilyl derivatives., *J. Chromatogr. Sci.*, **8**, 567–576.

Volkman, K. J., R. B. Johns, F. T. Gillan, G. J. Perry and H. J. Bavor Jr. (1980) : Microbial lipids of an intertidal sediment–1. Fatty acids and hydrocarbons., *Geochim. Cosmochim. Acta.*, **44**, 1133–1143.

Wakeham, S. G. and E. A. Canuel (1988) : Organic geochemistry of particulate matter in the eastern tropical North Pacific Ocean: Implications for particulate dynamics., *J. Mar. Res.*, **46**, 183–213.

Wallen, D. G. and G. H. Geen (1971) : The nature of the photosynthate in natural phytoplankton populations in relation to light quality., *Mar. Biol.*, **10**, 157–168.

Watt, W. D. (1966) : The release of dissolved organic material from the cells of phytoplankton populations., *Proc . Roy. Soc. Lond. B*, **164**, 521–551.

Williams, P. J. LeB. and C. S. Yentsch (1976) : An examination of photosynthetic production, excretion of photosynthetic products, and heterotrophic utilization of dissolved organic compounds with reference to results from a coastal subtropical sea., *Mar. Biol.*, **35**, 31–40.

Williams, P. M. (1971) : The distribution and cycling of organic matter in the ocean. pp.145–163. In *Organic compounds in Aquatic Environments*, ed. by S. J. Faust and J. V. Hunter, Marcel Dekker, New York.

Williams, P. M. and E. R. M. Druffel (1987) : Radiocarbon in dissolved organic matter in the Central North Pacific Ocean., *Nature*, **330**, 246–248.

Wollast, R. (1991) : The coastal organic carbon cycle: fluxes, sources, and sinks. p.365–381. In *Ocean margin Processes in Global Change*, ed. by R. F. C. Mantoura, J. –M. Martin and R. Wollast, John Wiley & Sons, Chichester.

Wood, B. J. B. (1974) : Fatty acids and saponifiable lipids. pp.236–265. In *Algal Physiology and Biochemistry*, ed. by W. D. W. Stewart, Blackwell Scientific Publications, Oxford.

Dynamics and Characterization of Marine Organic Matter, Eds., N. Handa, E. Tanoue and T. Hama, pp. 39–65.

Environmental Control of Nitrate Uptake in Surface Oceanic Waters: an Overview

Jota KANDA

School of Science, Shizuoka University, 836 Ohya, Shizuoka 422, Japan

(Review, Received 3 October 1996; accepted 20 February 1997)

Abstract—Nitrate uptake by natural plankton populations are controlled by external environmental factors. These environmental factors include light intensity, temperature, and concentrations of nitrate, ammonium and other nutrients. Experimental observations on nitrate uptake in oceanic waters are reviewed here. Environmental factors affect nitrate uptake in a complex and indirect way, and the relationship between a certain environmental factor and uptake rates may not be readily identified in field observations. Several characteristic features, however, have been noted. In most cases, larger size species of phytoplankton are dominant in the bloom region where algal biomass is building up, and they contribute to the bulk of nitrate uptake in such waters. Smaller nanoplankton and picoplankton, however, also assimilate nitrate as well as ammonium and urea. In a relatively limited spatial and time scale where other factors are not limiting, a clear temperature dependency of nitrate uptake is demonstrated. Light intensity effectively controls nitrate uptake in most oceanic environments, and results in characteristic diel and vertical variation of the uptake. Low nitrate concentration strongly limits uptake in the surface layer of stratified water columns. The presence of ammonium and urea generally reduce nitrate uptake, although some blooming population shows preference for nitrate over ammonium and urea. The importance of the regenerated forms of nitrogen in the control of nitrate uptake in nitrate-replete waters is suggested, but limited availability of light, iron and silicate is also suggested to suppress nitrate utilization. These observations are yet somewhat qualitative, but will be essential for quantitative provision of nitrate utilization in the world ocean and, hence, new production flux.

INTRODUCTION

Nitrate is one of the major nitrogen sources for natural populations of marine phytoplankton. The supply and utilization of nitrate occurs in a way quite different from those of ammonium and urea, which are being actively regenerated in the euphotic zone. Because of the suppression of nitrification by light, nitrate has been generally assumed not to regenerate in the euphotic zone and, hence, designated as a form of new nitrogen (Dugdale and Goering, 1967). This assumption may not be quantitatively exact as significant nitrification is observed near the bottom of and perhaps even in the upper part of the euphotic zone (Ward *et al.*, 1989). Nevertheless, nitrate dynamics is still unique and will provide insight into the nitrogen cycling in surface ocean. The dominant source of nitrate supplied to the euphotic zone is generally considered to be the deeper layer of the water column (Eppley and Peterson, 1979; Jenkins, 1988), but the supply of nitrate through atmospheric deposition may be significant in some cases (Fanning, 1989; Eppley *et al.*, 1990; DiTullio and Laws, 1991; Owens *et al.*, 1992). Since estimates of new production sometimes largely exceed the postulated nitrate supply (Platt *et al.*, 1989; Laws, 1991; Jenkins and Wallace, 1992; Kanda, 1995), it is frequently claimed that nitrate may not quantitatively dominate new nitrogen supply to the euphotic zone (Hayward, 1987). Given the importance of new production flux in marine biogeochemical cycles (Eppley and Peterson, 1979), quantitative evaluation of supply and utilization of nitrate in surface waters is focused on researches of chemical and biological oceanography (Kanda, 1995).

Environmental factors such as light, temperature, and nutrient concentrations control the rate of nitrate uptake. The environmental control of nitrate uptake is quite complicated. The uptake rate in a natural water is the sum of uptake by each cell of the plankton assemblage. In a relatively shorter time scale, cellular rate of nitrate uptake is physiologically regulated in response to the external environmental signals. In a longer time scale, the growth rate is also under environmental control. Nitrogen demand in a steadily growing population is given as the multiplication of nitrogen cell quota and specific growth rate. The cell quota of nitrogen can be functionally related to the specific growth rate (Droop, 1968), but the relationship dramatically alters depending on which environmental factor limits growth (Droop, 1974; Rhee and Gotham 1981a, b). Furthermore, such functional response of nitrogen uptake to the environmental parameters is likely to be species specific. The uptake rate depends on the biomass of nitrate assimilating plankton as well as species composition of the plankton community. The environmental control over biomass and species composition of plankton are well documented.

Reduction of nitrate to ammonium is essential in nitrate assimilation by planktonic organisms. Because the reduction process must be associated with extra supplies of metabolic energy, it has been postulated that nitrate is the least preferred nitrogenous nutrient in an oceanic environment (Conway, 1977; McCarthy *et al.*, 1977) . This notion has been challenged on several grounds, and the preference and partitioning among various forms of nitrogen are likely to be more complicated than previously assumed (Dortch, 1990). Yet, the availability of other forms of nitrogen certainly affects nitrate uptake.

Field observations of nitrate uptake have been conducted by the use of ^{15}N isotope (Dugdale and Goering, 1967). Recently, highly sensitive analysis of nitrate concentration became available (Garside, 1982; Oudot and Montel, 1988; Raimbault *et al.*, 1990b) and was applied to field observations of nitrate uptake (Eppley *et al.*, 1990; Eppley and Koeve, 1990; Slawyk *et al.*, 1990; Eppley and Renger, 1992). The environmental control of nitrate uptake has been looked for in the relationship between uptake rates and naturally varying or experimentally controlled environmental factors. The influence of environmental factors is more apparent in biomass-specific uptake rates. The uptake rate is commonly normalized to particulate nitrogen (PN) or chlorophyll concentration. Each biomass denominator has its own pros and cons though (Garside, 1991; Dugdale and Wilkerson, 1991, 1992).

We review here the data on correlation between the environmental factors and nitrate uptake rates obtained through field observations at sea. The complicated nature of the environmental control makes it difficult to assess such a relationship in a natural environment. Nevertheless, there are several clear relationships identified in the field and these are certainly worth reviewing. Also, the empirical relationship between variation of an environmental factor and nitrate uptake may produce a quantitative estimate of nitrate uptake from the environmental parameter (Dugdale *et al.*, 1989; Sathyendranath *et al.*, 1991; Probyn *et al.*, 1995). As data on nitrate uptake are usually insufficient due to the labor and cost intensive nature of the field observation, such an estimate could interpolate the lacking observations. Since several environmental factors can be efficiently monitored through remote sensing and various platforms such as mooring buoys, this approach will be more realistic for exploiting a global scheme of nitrate uptake in the sea.

CONTRIBUTION OF SPECIES COMPOSITION

Algae or bacteria?

Phytoplankton, which conventionally includes cyanobacteria and prochlorophytes, has been generally assumed to be responsible for nitrate uptake in marine environment. Marine heterotrophic bacteria, however, may also utilize nitrate under certain conditions (Horrigan *et al.*, 1988). The relative contribution of bacteria to the total rate of nitrate uptake may be assessed by the portion of ^{15}N uptake by a smaller size fraction (typically <1.0 μm). Examining the correlation between the observed nitrate uptake and photosynthetic activity or light intensity will be also useful for this purpose, although the light dependency of nitrate uptake may vary substantially under various environmental conditions (see below). Results of such experiments suggest the bacterial contribution is generally unimportant in various oceanic waters (Harrison *et al.*, 1977; Probyn, 1985; Probyn and Panting, 1985; Le Bouteiller, 1986; Wheeler and Kirchman, 1986; Harrison and Wood, 1988; Probyn *et al.*, 1990; Sambrotto *et al.*, 1993). Lipschultz (1995), working on the ^{15}N tracer experiments combined with the flow cytometry with cell sorting (FCCS), found that nitrate uptake by chlorophyll containing particles (larger than 3 μm) is much larger than the fraction of combined small (<3 μm) particles and non-chlorophyll particles in a coastal water. Nitrate uptake by all of these fractions showed clear light dependency, and he suggested that nitrate uptake by bacteria is fairly small. However, he also noted that a large amount of nitrate taken up was not accounted for in the fractionation process. These results are in sharp contrast to the case of ammonium in which a quite high contribution of heterotrophic bacteria is suggested (Laws *et al.*, 1985; Wheeler and Kirchman, 1986).

Several authors, however, have noted possible uptake of nitrate by heterotrophic bacteria. In the Atlantic, Altabet (1990) found that a significant portion of nitrate was taken up by particles which pass the GF/F glass fiber filters but are retained by aluminum oxide filters (0.2 μm nominal pore size; Anopore). The portion of nitrate uptake by these small particles was higher at the lower portion of the euphotic zone. In the northern North Atlantic, Sambrotto *et al.* (1993) found a significant peak of nitrate uptake below the bottom of the euphotic zone where active photosynthesis was absent. In the subtropical Atlantic, Eppley and Koeve (1990) observed that significant uptake of nitrate (decrease of nitrate concentration) was associated with particles smaller than 0.4 μm. It has been shown that a decrease of nitrate concentration often exceeds the uptake rate determined by ^{15}N in filter-retained particles (Price *et al.*, 1985; Dugdale and Wilkerson, 1986; Ward *et al.*, 1989; Slawyk *et al.*, 1990; Eppley and Koeve, 1990; Eppley and Renger, 1992; Slawyk and Raimbault, 1995; see also Boyd *et al.*, 1995). Transfer of ^{15}N into dissolved organic nitrogen and small particles that pass through the filter was demonstrated (Bronk and Glibert, 1991, 1994; Bronk *et al.*, 1994; Slawyk and Raimbault, 1995). The small particles labeled with ^{15}N may include heterotrophic bacteria. However, small cells of prochlorophyte and other picophytoplankton are also found in this size class (Chisholm *et al.*, 1988). Beside the possible uptake by heterotrophs, the significant fraction of unrecovered ^{15}N requires corrections for the estimation of the uptake rate from the conventional ^{15}N tracer experiments (Kanda *et al.*, 1987, p.89).

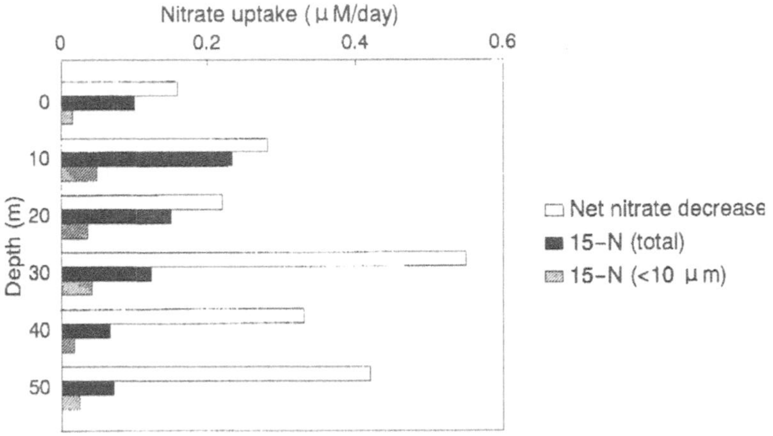

Fig. 1. Vertical profile of nitrate uptake calculated from ^{15}N tracer uptake (total and <10 μm fraction) and of nitrate decrease determined by colorimetric analysis obtained from an *in situ* incubation experiment (unpublished results). The experiment was conducted on 14–15 July 1985 on the KT-85-9 cruise of the R/V Tansei-Maru at 33°15′ N, 139°46′ E near Hachijo Island. Seawater samples were spiked with 0.08 μM of ^{15}N-nitrate at 0830 local time and incubated under *in-situ* condition for 24 hours.

Data from an *in situ* incubation experiment for nitrate uptake are shown in Fig. 1. The experiment was conducted in a regional upwelling area near Izu Islands. The ^{15}N labeled nitrate was added at a concentration of 0.08 μM and incubated for 24 h under *in situ* conditions. For nitrate uptake, both ^{15}N based uptake and decrease of nitrate during the incubation were shown. Both rates showed similar vertical variations in the upper 20 m, although the nitrate decrease was consistently larger. At 30–50 m depth, the nitrate decrease was enigmatically larger than the observed nitrate uptake with ^{15}N. Nitrite concentrations were less than 0.1 μM and did not show significant change over the incubation period. The result may represent a significant transfer of ^{15}N into DON and/or small particulate nitrogen from nitrate-N. Alternatively, isotope dilution effect due to nitrification may have resulted in significant underestimate of nitrate uptake by the ^{15}N method (Ward *et al.*, 1989), or a dissimilatory nitrate reduction may have decreased the nitrate concentration (Omnes *et al.*, 1996). I acknowledge that any conclusive argument will be difficult from the original data set itself. A nitrate decrease much larger than the ^{15}N based nitrate uptake at the lower portion of the euphotic zone was also reported by Boyd *et al.* (1995).

The N partition hypothesis

In oceanic environments, relatively higher algal biomass and primary productivity are often accompanied by enhanced availability of nitrate in the euphotic layer. Such cases are found in upwelling systems, seasonal phytoplankton blooms, and coastal eutrophic waters. It is also suggested that larger size cells of phytoplankton, diatoms in many cases, are dominant in these high biomass conditions. In contrast, the oligotrophic planktonic populations with smaller biomass and productivity are dominated by picoplankton (or ultraplankton) and smaller nanoplankton. In these waters, most of the available nitrogen is in the form of reduced nitrogen (ammonium and urea) which is supplied through *in situ* regeneration. These observations lead to the so-called N partition hypothesis, in which smaller phytoplankton is supported by a regenerated supply of nitrogen while larger phytoplankton depends on nitrate (Malone, 1980; also see Chisholm, 1992).

Numerous studies were conducted for the purpose of verifying this hypothesis. When nitrate uptake is compared to uptake of ammonium and urea, the relative contribution of nitrate is generally higher in the larger size fraction of plankton than in the smaller size fraction (Nalewajko and Garside, 1983; Rönner *et al.*, 1983; Probyn, 1985; Koike *et al.*, 1986; Probyn and Panting, 1985; Le Bouteiller 1986; Wheeler and Kirchman, 1986; Harrison and Wood, 1988; Sahlsten *et al.*, 1988; Owens *et al.*, 1991, 1993; Sambrotto *et al.*, 1993; Probyn *et al.*, 1990; Probyn, 1992; Price *et al.*, 1994; Lipschultz, 1995; Bury *et al.*, 1995), but several exceptions to this general trend have been noted (Furnas, 1983; Harrison and Wood, 1988; Petterson and Sahlsten, 1990). The partitioning of the N source is not exclusive, however. Both nitrate and regenerated forms of nitrogen are assimilated simultaneously in each size class, and the difference in nitrate contribution among different size classes is not as large as generally believed (see Table 1 of Chisholm, 1992). In oligotrophic waters, nitrate is utilized by picoplankton (Harrison and Wood, 1988). In fact, Glover *et al.* (1988) observed that nitrate utilization was associated with small picophytoplankton in the size range of 0.6–1.0 μm. On the other hand, Goldman (1988, 1993) hypothesized that large (>50 μm) diatoms dominate the nitrate uptake near the bottom of the euphotic zone; hence they dominate the new production flux in the oligotrophic open ocean. This hypothesis may be regarded as an extended N partition hypothesis.

In blooming events in mesotrophic and eutrophic waters, however, nitrate is mostly taken up by larger phytoplankton. This could be simply because the larger cells dominate the plankton community. The preference among the available forms of nitrogen may not contribute to the overall trend of the N partition. In Fig. 1, it was shown that most nitrate uptake was accompanied by plankton larger than 10 μm. In the study area, the Kuroshio current and island cause physically perturbed conditions which result in temporal and spatial nitrate entrainment followed by small-scale blooms of phytoplankton. Nitrate uptake was compared between particles retained on a GF/F filter (total) and those passed through a 10 μm pore Nucleopore filter but retained on a GF/F filter (< 10 μm). The ^{15}N results showed most of the nitrate uptake was by larger (> 10 μm) cells, but chlorophyll a and photosynthetic carbon fixation were also high in the larger size fraction (data not shown).

TEMPERATURE

Temperature is one of the prime environmental factors which affect the metabolism of microorganisms. Eppley (1972) summarized growth rate data of microalgae from laboratory and field observations. For each species of planktonic algae, the growth rate generally increases with increasing temperature and shows a maximum at the temperature optima for the species; above the temperature optima the growth rate drops rapidly. Combining data of various species altogether, he showed that the maximum attainable growth rate can be expressed by an exponential function of temperature. The maximum growth rate is presumably attained when nutrient and light do not limit growth. The nitrogen cell quota is known to increase at sub-optimal temperatures for growth (Goldman, 1977; Rhee and Gotham, 1981a), and thus the temperature dependency of nitrate uptake may not be the same as that of the growth rate. Nevertheless, the temperature dependency of nitrate uptake may be observed under experimental conditions where light intensity and nutrient concentration do not limit uptake.

In a relatively limited aerial scale where other factors are not limiting, positive correlation was sometimes found between *in situ* nitrate uptake rate (either chlorophyll or PN specific) and ambient temperature (Slawyk, 1979; Le Bouteiller, 1986; Tobiesen, 1987; Probyn, 1988). A similar correlation with ambient temperature was obtained for the maximum uptake rates of nitrate determined from kinetic experiments under artificially controlled nitrate and light conditions (Glibert *et al.*, 1982b; Kanda *et al.*, 1985a; Harrison *et al.*, 1996). Furthermore, Olson (1980), Mitamura (1986; for a freshwater population), Priscu *et al.* (1989) and Smith and Harrison (1991) demonstrated strong temperature dependency of nitrate uptake in natural planktonic populations at artificially controlled levels of temperature.

The observed correlation of nitrate uptake and temperature is, however, generally less clear than that of ammonium (Kanda *et al.*, 1985a). Light and substrate saturated uptake rates of ammonium (Fisher *et al.*, 1982; Paache and Kristiansen, 1982; Glibert *et al.*, 1982b; Kanda *et al.*, 1985a) and total inorganic nitrogen uptake (Kristiansen, 1983; Carpenter and Dunham, 1985; Pennock, 1987) shows clear temperature dependency when normalized to chlorophyll or PN concentrations. Kanda *et al.* (1985a) observed that maximum rates of nitrate uptake are more clearly covaried with photosynthetic rate than with temperature, and argued that the effect of the prehistory of light conditions was more important. The effect of environmental parameters other than temperature is likely to be more dominant in many cases of field observations.

LIGHT

Nitrate uptake by phytoplankton typically depends on light (MacIsaac and Dugdale, 1972). Light dependency of nitrate uptake is not as straightforward as those of photosynthetic oxygen evolution and carbon dioxide fixation. Assimilatory metabolism of nitrogen in algae is coupled with photosynthesis in some indirect ways (Syrett, 1981). Nitrate uptake rate (ρ) can be expressed as a function of light intensity (I). A typical function used in many studies is in the following form (MacIsaac and Dugdale, 1971), or

$$\rho = \rho_D + \frac{\rho_{\max} I}{K_I + I}.$$

The parameters in the formula represent the rate of uptake in the dark (ρ_D), the maximum rate of the light-dependent portion of uptake ($\rho_{\max}$), and the half-saturating light intensity of the light-dependent portion of uptake (K_I). This approach is similar to the experimental determination of the so-called ρ-I curves in photosynthesis research, and will be referred to as the ρ-I curve approach in this review. The most notable feature of the above function is the inclusion of the term for uptake in the dark. A number of studies reported experimentally determined ρ-I curves for nitrate in the sea (MacIsaac and Dugdale, 1971; MacIsaac *et al.*, 1974, 1985; MacIsaac, 1978; Slawyk, 1979; Nelson and Conway, 1979; Garside, 1981; Nalewajko and Garside, 1983; Kanda *et al.*, 1985a, 1989; Pennock, 1987; Cochlan *et al.*, 1991a; Smith and Harrison, 1991; Bjørnsen *et al.*, 1993; Muggli and Smith, 1993). Cochlan *et al.* (1991a) provides a review for ρ-I curve data in the literature. The light intensity that yields saturation of nitrate uptake is generally lower than those of photosynthesis (e.g. Kanda *et al.*, 1985a).

Light control of nitrate uptake can also be observed in the correlation of the uptake rate with naturally varying light intensity. Nitrate uptake in the sea varies with diel changes of irradiance (Eppley *et al.*, 1971a; McCarthy and Eppley, 1972; Collos and Slawyk, 1976; MacIsaac, 1978; Olson, 1980; Dortch and Maske, 1982; Price *et al.*, 1985; Koike *et al.*, 1986; Tobiesen, 1987; Wheeler *et al.*, 1989; Cochlan *et al.*, 1991b; Glibert and Garside, 1992). Other than the direct effect of light, such diel variation can partly be ascribed to changes in endogenous metabolic activity (e.g. Goering *et al.*, 1964; Eppley *et al.*, 1971b; Miyazaki *et al.*, 1987; Vincent, 1992; Berges *et al.*, 1995). In the oceanic environment, light intensity decreases markedly with depth. In a stratified water column, plankton populations near the bottom of the euphotic zone are chronically light limited. In a vigorously mixed layer, the average light received by the planktonic population would be vertically uniform, but as the mixed layer deepens, the average light intensity received by the planktonic populations decreases concurrently. Vertical profiles of nitrate uptake have been obtained in various regions of the sea (Dugdale and Goering, 1967; Goering *et al.*, 1970; MacIsaac and Dugdale, 1972; McCarthy, 1972; Eppley *et al.*, 1973, 1990; Slawyk *et al.*, 1976; Collos and Slawyk, 1977, 1986; Slawyk, 1979; Conway and Whitledge, 1979; Nelson and Conway, 1979; King and Devol, 1979; Olson, 1980; Glibert *et al.*, 1982a; Harrison *et al.*, 1983, 1985, 1992; Rönner *et al.*, 1983; Blasco *et al.*, 1984; Kanda *et al.*, 1985b; MacIsaac *et al.*, 1985; Lewis *et al.*, 1986; Sambrotto *et al.*, 1986, 1993; Koike *et al.*, 1986; Le Bouteiller, 1986; Nelson and Smith 1986; Owens *et al.*, 1986; Sahlsten, 1987; Tobiesen, 1987; Harrison and Wood, 1988; Probyn, 1988; Sahlsten *et al.*, 1988; Dugdale and Wilkerson, 1989; Murray *et al.*, 1989; Kristiansen and Lund, 1989; Ward *et al.*, 1989; Eppley and Koeve, 1990; Dugdale *et al.*, 1992; Wilkerson and Dugdale, 1992; McCarthy *et al.*, 1992; Peña *et al.*, 1992; Bjørnsen *et al.*, 1993; Boyd *et al.*, 1995; Bury *et al.*, 1995; Probyn *et al.*, 1995; Waldron *et al.*, 1995; Kanda, 1995; Allen *et al.*, 1996). Some of these results show an evident decrease of uptake with depth which likely reflects the effect of light. The decrease of uptake with depth is partly due to the instantaneous effect of light similar to that observed in ρ-I curves (Slawyk *et al.*, 1976), but the effect of light adaptation (see below) and changes in species composition must be accounted for as well. The vertical profiles of nitrate uptake in the oligotrophic open ocean are also affected by nitrate depletion in the upper euphotic zone (e.g. MacIsaac and Dugdale, 1972; Lewis *et al.*, 1986). If nitrate is enriched to a saturation level, the uptake rate (nitrate saturated rates) may also decrease with depth (e.g. Kanda, 1995). Under nitrate replete conditions, however, nitrate uptake is not necessarily high in the upper waters where the light intensity is high, and a lack of light dependency of nitrate uptake is sometimes observed in these waters (e.g. Glibert *et al.*, 1982a; Peña *et al.*, 1990; see below).

Interaction with nitrogen availability

The light dependency is more apparent when the nitrogenous nutrient is sufficient than in the nitrogen-depleted environment (Kanda *et al.*, 1989; Cochlan

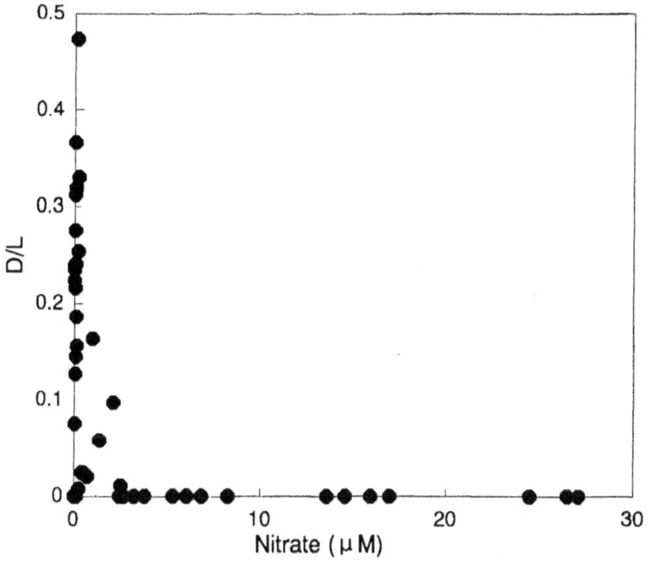

Fig. 2. The D/L ratio (see text) plotted against the ambient nitrate concentration. Data from natural populations in surface seawater in the Pacific (Kanda *et al.*, 1985a) and in Auke Bay (Kanda *et al.*, 1989).

et al., 1991a). The experimentally determined ratio of nitrate uptake in the dark and in the light for a given natural population, hereafter called the D/L ratio, has been used as a conventional index of light dependency (Dugdale and Goering, 1967; Conway and Whitledge, 1979; Nelson and Conway, 1979; Dortch and Maske, 1982; Kanda *et al.*, 1985a; Cochlan *et al.*, 1991a; Smith and Harrison, 1991). The D/L ratio data obtained from surface waters (Kanda *et al.*, 1985a, 1989) are pooled and plotted as a function of the ambient nitrate concentration in Fig. 2. The higher D/L ratio was observed only in nitrate depleted waters. Note, however, these experiments were conducted with surface populations, presumably not limited by light intensity (see below). Kanda *et al.* (1989), working in a small coastal bay in Alaska, found that the shape of the ρ-I curves of nitrate uptake changed dramatically at the point of nitrate exhaustion in a spring bloom. Before and during the development of the bloom where plenty of nitrate was available, nitrate uptake clearly depended on light intensity. Immediately after the nitrate exhaustion, such dependency became less clear and significant uptake was observed in the dark. A similar feature has been found in the laboratory culture of marine planktonic algae, and is commonly postulated as adaptation to low nitrogen availability (Cochlan *et al.*, 1991a). The species-specific response of the light dependency to the variation in nitrogen availability, however, has been well documented (e.g. Eppley *et al.*, 1971b; Harrison, 1976; Laws and Wong, 1978; Paache *et al.*, 1984).

Sahlsten (1987) reported that the diel variation of nitrate uptake was weak and almost insignificant at a depth corresponding to 35% light level in a oligotrophic oceanic water. This observation is consistent with the above-described variation of D/L ratio and ρ-I curve with nitrogen availability. On the other hand, by using high sensitive analysis of nitrate, Eppley and Renger (1988) and Eppley *et al.* (1990) observed that nitrate concentration generally decreased during the day and increased during the night in the oligotrophic open ocean. These data suggest a significant diel cycle of nitrate uptake (and nitrate supply as well). A possible explanation of the significant dark uptake in oligotrophic waters is the contribution of heterotrophic bacteria. As noted in the previous section, the contribution of heterotrophic bacteria in the observed nitrate uptake is still largely unknown. Species-specificity in responding light conditions must be accounted for as well. In nitrate depleted water, Probyn *et al.* (1990) found that the picoplankton population showed insignificant diel variation of nitrate uptake while netplankton population in the same water showed marked diel periodicity

The D/L ratio becomes high in certain cases with nutrient replete waters. MacIsaac (1978) observed a significant diel cycle of nitrate uptake in surface waters off Baja California, but this cycle disappeared or was significantly weakened near the bottom of the euphotic zone (1% light depth) where the nitrate concentration was >5 μM. An essentially similar observation was reported by Cochlan *et al.* (1991b) for nitrate replete waters in the North Pacific. These results suggest that populations conditioned in low light environment exhibit high D/L ratio under nitrogen replete condition. It should be noted here that the data shown in Fig. 2 are obtained exclusively from surface populations and may possibly be adapted to a light-sufficient environment.

In nitrate replete Antarctic waters, Glibert *et al.* (1982a) observed a D/L ratio of about 1.0 as well as several vertical profiles showing larger nitrate uptake in the deeper layer. No significant difference in the nitrate uptake rate during the day and night was reported in nitrate replete waters in the Arctic (Harrison, 1983) and in the North Sea (Petterson and Sahlsten, 1990). For an ice algae population, Harrison *et al.* (1990) demonstrated that nitrate uptake was practically insensitive to light intensity. Smith and Harrison (1991) observed a significant dark nitrate uptake (D/L ratio of up to 0.39) in Arctic waters where nitrate was sufficient. A similar trend was observed in the vertical profile data reported by Smith and Kattner (1989). Smith and Harrison (1991) further reviewed works in high latitude waters where nitrate concentration was high and suggested that dark nitrate uptake was a common phenomenon in these waters. Peña *et al.* (1990) found little light dependency in their vertical profile data sets for nitrate uptake in nitrate-replete waters of the equatorial Pacific. Eppley and Renger (1992) observed that the nighttime decrease of nitrate was even larger than in the daytime in the equatorial Pacific. Some of the high latitude waters belong to regions where intensive vertical mixing occurs, and these results might partially reflect the low light conditioning, as discussed above. In the high latitude and the equatorial waters, the lack of iron and/or the presence of ammonium are suggested as controlling factors of nitrate uptake. The interaction of the light dependency of nitrate uptake with these factors is poorly understood. More recently, Muggli and Smith (1993) observed almost zero nitrate uptake in the dark in the Greenland

Sea. Some other studies in polar regions observed relatively slow nitrate uptakes during the night (Olson, 1980; Koike *et al.*, 1986).

Conditioning by light environment

An important feature of the light dependency of nitrate uptake is its similarity of the light-shade adaptation of the P-I curves (Kanda *et al.*, 1989; Smith and Harrison, 1991; Cochlan *et al.*, 1991a). Populations placed in the deeper part of the euphotic zone or in the vigorously mixing layer are expected to be adapted or conditioned to a low light environment. Such populations generally exhibit a lower rate (especially the chlorophyll-specific rate) at a saturated intensity of light. The half-saturation intensity of light is relatively low and occasionally accompanies a photoinhibition-like pattern of light dependency. These features of the ρ-I curve are consistent with the general pattern of the variation of P-I curves with light-shade adaptations. The uptake in the dark was found to be negligible in the work of Kanda *et al.* (1989) for the postulated shade adapted populations. This lack of dark uptake may be a rather exceptional observation, because populations conditioned in a low light environment show higher D/L ratios, as mentioned above. In either case, the conditioning effect on ρ-I curves, together with a decreasing light field at increasing depth, mainly determines the vertical profile of the nitrate uptake, which decreases with depth.

Dugdale and co-workers (MacIsaac *et al.*, 1985; Wilkerson and Dugdale, 1987; Wilkerson *et al.*, 1987; Dugdale *et al.*, 1989; Dugdale and Wilkerson, 1991, 1992), working with various upwelling regions in the sea, observed a phenomenon called a "shift-up" of nitrate uptake. Nitrate uptake rates denominated by PN (often expressed as V instead of ρ) increase dramatically at a certain time after the upwelling event. The phytoplankton bloom must follow the shift-up of the nitrate uptake. Dugdale and Wilkerson (1991, 1992) postulate that a high nitrate concentration and light availability are necessary for the shift-up, and that the shift-up reflects an induction of metabolic machinery for nitrate assimilation. Garside (1991) criticized the use of PN as the biomass denominator in these observations. Other environmental parameters, such as ammonium, silicate and iron, may enhance or interfere with the development of shift-up conditions (see Section 6. Availability of other nutrients). Kanda *et al.* (1989) observed that the ρ-I curve showed a high maximum uptake rate (chlorophyll denominated rate) and a high half-saturation light intensity during the period of rapid build-up of the phytoplankton biomass. This feature of the ρ-I curve is common with the P-I curves of high-light adapted cells. Kanda *et al.* (1989) postulated that the onset of the spring bloom is accompanied by a high-light adaptation of phytoplankton in the stratified water column.

NITRATE CONCENTRATION

The uptake rate of nitrate follows a saturation type of kinetics and is commonly expressed by the following hyperbolic function which is equivalent to the Michaelis-Menten equation (Dugdale, 1967; Eppley and Coatsworth, 1968;

Goldman and Glibert, 1983), or

$$\rho = \frac{\rho_{max} S}{K_S + S}.$$

The uptake rate (ρ) is a function of the nitrate concentration (S) with two parameters corresponding to the maximum rate of uptake (ρ_{max}) and the half-saturating concentration of nitrate uptake (Ks). The Ks values of nitrate uptake have been experimentally determined in several studies of natural plankton assemblages in the sea (MacIsaac and Dugdale, 1969; Eppley *et al.*, 1977; Paasche and Kristiansen, 1982; Kanda *et al.*, 1985a, 1985b, 1990; Owens *et al.*, 1986; Sahlsten, 1987; Kristiansen and Lund, 1989; Shiomoto and Maita, 1990; Smith and Harrison, 1991; Kristiansen *et al.*, 1994; Shiomoto *et al.*, 1994; Probyn *et al.*, 1995; Harrison *et al.*, 1996). For the natural populations in oceanic waters, Ks is generally determined with an experiment by adding several different levels of ^{15}N labeled nitrate. The interpretation of the Ks value is sometimes difficult; most of the problems arise from the time-scale of the experiment (Goldman and Glibert, 1983; Morel, 1987; Kanda, 1993). If the ambient nitrate concentration is similar to or lower than the Ks value, nitrate uptake may be considered to be limited by the low concentration of nitrate. For oligotrophic waters, however, precise determination of Ks is difficult due to the limitation of the sensitivity of nitrate analysis. For example, Kanda *et al.* (1985a) observed that $Ks+S$ values of nitrate uptake were in the range of 0.00 to 0.12 μM in the oligotrophic waters of the Pacific, but the nitrate concentration was also found in the same range. Given the precision of their nitrate analysis, it was difficult to determined the degree of depression of *in situ* nitrate uptake by the low ambient nitrate concentration.

Recent developments of high precision nitrate analysis allowed this problem to be addressed. Harrison *et al.* (1996) determined the kinetic parameters of nitrate uptake by ^{15}N incubation experiments combined with high-precision analysis of nitrate. They found Ks of nitrate uptake in the oligotrophic Atlantic to be 20–30 nM on average. The Ks values are similar to or larger than the ambient concentrations of nitrate, and thus the uptake is strongly suppressed by the low concentration of nitrate. This does not necessarily mean that the growth rate of plankton is limited by the low nitrate concentration in surface oligotrophic waters. The half-saturation concentration for growth is generally much lower than those for uptake (Morel, 1987; Kanda, 1993). Furthermore, ammonium and urea are generally more important nitrogen sources for plankton populations in these waters, and there is little evidence for severe nitrogen limitation on the growth of phytoplankton in oligotrophic waters (Goldman *et al.*, 1979; Laws *et al.*, 1984, 1987, 1989; Kanda and Hattori, 1988; Kanda *et al.*, 1988). The levels of the ^{15}N nitrate addition in most of the previous researches were generally comparable to or much higher than the Ks values reported by Harrison *et al.* (1996). The elevated rate of uptake generally decreases with depth in the oligotrophic open ocean (Kanda *et al.*, 1985b; Le Bouteiller, 1986; Sahlsten,

1987; Harrison *et al.*, 1992; Kanda, 1995), and the difference between the elevated rate of nitrate uptake and the *in situ* rate of uptake is suggested to be higher in shallower layers (Kanda, 1995). The decrease of the elevated rate of uptake likely reflects the corresponding decrease of light. The high rate of uptake in response to the nitrate addition indicates that the metabolic machinery of nitrate assimilation in plankton cells is readily induced in surface oligotrophic waters. The induction mechanism of nitrate uptake under such conditions is not fully understood, but is apparently species-specific (Glibert and Ray, 1990). Since induction and activation of nitrate assimilation capacity often requires light (Syrett, 1981; Berges *et al.*, 1995), the apparent light dependency in the vertical profile of an elevated rate of nitrate uptake may reflect the light-dependent induction of nitrate assimilation enzymes.

Harrison *et al.* (1996) also observed a positive correlation between *Ks* and the ambient nitrate concentration. This observation is consistent with earlier suggestions that lower *Ks* is an adaptation to an oligotrophic environment (Dugdale, 1967). Although the *Ks* as high as 10 μM is sometimes observed from laboratory cultures of diatoms (Garside and Glover, 1991), the reported range of *Ks* in nitrate replete waters is much smaller than the ambient concentration (Kanda *et al.*, 1985a; Harrison *et al.*, 1996). Thus, the rate of uptake is not generally under suppressions by nitrate concentrations in nitrate replete waters. It should be noted, however, that in such waters, the experimental determination of *Ks* is difficult due to the pre-existing high concentration of nitrate (MacIsaac and Dugdale, 1969).

The high-precision analysis of nitrate showed the ubiquitous distribution of a detectable concentration of nitrate (several to several tens nM) in surface oligotrophic waters (Garside, 1982, 1985; Eppley and Renger, 1988; Eppley *et al.*, 1990; Mantoura *et al.*, 1993; Harrison *et al.*, 1996). This finding led some authors to speculate that there exists a threshold concentration of nitrate uptake in natural populations of such an environment. In the subtropical Atlantic, Eppley *et al.* (1990) found that the observed rates of nitrate decrease in bottle incubation samples are linearly related to a concentration of up to 50 nM of nitrate and that there was a clear threshold of about 4 nM of nitrate for a zero change of nitrate concentration. Based on ^{15}N incubation experiments combined with high-precision nitrate analysis, McCarthy *et al.* (1992) suggested a 16 nM threshold in nitrate uptake, although Harrison *et al.* (1996) found no significant threshold. In laboratory, Raimbault and Gentilhomme (1990) and Raimbault *et al.* (1990a, 1990c) observed that a nitrate-limited culture of the diatom *Paheodactylum tricornutum* can drive down the nitrate concentration to 0 nM in a relatively short time interval in response to a nitrate addition of 15–100 nM. They observed that the uptake rate is a linear function of a nitrate concentration of up to 100 nM without a significant threshold. On the other hand, Garside and Glover (1991) observed that a nitrate limited culture of the diatom *Thalassiosira pseudonana* (clone 3H) was not able to completely use up the added nitrate. Residual nitrate of several tens of nM was observed, and the plot of uptake vs. concentration yielded a linear relationship with a threshold of 50–100 nM. They also conducted similar experiments with a natural plankton assemblage of a neritic origin, and observed a threshold concentration of uptake of about 10 nM. The threshold, however, disappeared if the same population was repeatedly enriched with nitrate.

AVAILABILITY OF OTHER NUTRIENTS

Ammonium and urea

As noted earlier, nitrate is frequently designated as the least preferred nitrogenous nutrient. The experimental addition of ammonium to natural planktonic populations often yields a reduction in the nitrate uptake rate (Conway, 1977; MacIsaac *et al.*, 1979: Blasco and Conway, 1982; Price *et al.*, 1985; Smith, 1991; Muggli and Smith, 1993; Price *et al.*, 1994; Tamminen, 1995; Harrison *et al.*, 1996). These experiments show that the inhibition by ammonium is often incomplete. The uptake rate decreases with an increasing concentration of ammonium, but at a certain concentration, nitrate uptake still proceeds but becomes insensitive to further additions of ammonium. Harrison *et al.* (1996) observed that the addition of ca. 20 nM of ammonium substantially inhibited nitrate uptake. The inhibition observed by Harrison *et al.* (1996) was in a hyperbolic function in that inhibition was reached to the maximum (~50% decrease) at 100 nM. The half-saturation concentration of the inhibition was about 30 nM.

Several indices have been proposed for the quantitative description of the interaction among different nitrogen sources (Dortch, 1990). These include the ratio of nitrate and ammonium uptake (hereafter referred to as the NO_3/NH_4 ratio in this article) , and the ratio of nitrate uptake divided by the total rate of nitrogenous nutrient uptake (f-ratio; Eppley *et al.*, 1979). Also frequently used is the following ratio called the relative preference index (RPI) for nitrate (McCarthy *et al.*, 1977).

$$RPI_{\text{nitrate}} = \frac{(\text{nitrate uptake}) / (\text{total nitrogenous nutrient uptake})}{(\text{nitrate concentration}) / (\text{total nitrogenous nutrient concentration})}$$

The total nitrogenous nutrient includes nitrate, ammonium, urea and nitrite, but the latter two species are sometimes omitted for conventionality. For natural plankton populations in the sea, the nitrate uptake rate, NO_3/NH_4 ratio, f-ratio, or nitrate RPI was shown to be negatively correlated with the ambient concentration of ammonium (MacIsaac *et al.*, 1972; McCarthy *et al.*, 1977; Olson, 1980; Garside, 1981; McCarthy, 1981; Paasche and Kristiansen, 1982; Glibert *et al.*, 1982a; Furnace, 1983; Glibert and McCarthy, 1984; Probyn, 1985, 1988, 1992; Carpenter and Dunham, 1985; Harrison *et al.*, 1987; Pennock, 1987; Tobiesen, 1987; Murray *et al.*, 1989; Kristiansen and Lund, 1989; Wheeler and Kokkinakis, 1990; Smith and Harrison, 1991; Owens *et al.*, 1991; Cota *et al.*, 1992; Muggli and Smith, 1993; Smith, 1993; Probyn *et al.*, 1995; Rees *et al.*, 1995). From these experiments, nitrate uptake is sometimes found to be negligibly small when the ammonium concentration exceeds a certain threshold. McCarthy (1981) found that the contribution of nitrate and nitrite to the total nitrogen uptake became insignificant when the ammonium concentration was 0.5–1 μM or more. Wheeler and Kokkinakis (1990), similarly found that nitrate uptake was negatively correlated with the ambient ammonium concentration and became insignificant at a concentration of 0.1–0.3 μM of ammonium. Such a complete absence of nitrate uptake in the presence of ammonium, however, may not be always observed in the marine environment (see below). The RPI for nitrate is generally lower than 1.0, and thus nitrate is suggested to be not a preferred nitrogen source (McCarthy *et al.*, 1977). Dortch (1990), however, criticizes the use of this index because the effect of both inhibition and preference were combined in this index in field experiments. The term "preference" generally means the disproportionaly high rate of utilization of one nitrogenous nutrient over others through, for example, the high affinity system of uptake for the nutrient (Dortch, 1990).

Urea uptake frequently, but not always, exceeds nitrate uptake in marine environments (McCarthy, 1972; Harvey and Caperon, 1976; McCarthy *et al.*, 1977; Eppley *et al.*, 1977; Furnace, 1983; Kristiansen, 1983; Kaufman *et al.*, 1983; Carpenter and Dunham, 1985; Harrison *et al.*, 1985; Kanda *et al.*, 1985a; Price *et al.*, 1985; Probyn and Painting, 1985; Probyn, 1985, 1988; Kokkinakis and Wheeler, 1987; Sahlsten, 1987; Price and Harrison, 1988; Sahlsten *et al.*, 1988; Kristiansen and Lund, 1989; Harrison *et al.*, 1990; Probyn *et al.*, 1990, 1995; Bury *et al.*, 1995; Waldron *et al.*, 1995). A sequential preference of ammonium > urea > nitrate is suggested (Harvey and Caperon, 1976; McCarthy *et al.*, 1977; Probyn, 1985, 1988), but Kaufman *et al.* (1983) and Price and Harrison (1988) demonstrated cases for preference of urea over ammonium. The direct effect of urea on nitrate uptake is rarely studied. McCarthy and Eppley (1972) found a reduction of nitrate uptake by the addition of urea as well as a reduction of urea uptake by the addition of nitrate. The reduction of nitrate uptake by the addition of urea was smaller than that by the addition of ammonium. Price *et al.* (1985) observed both significant and insignificant reductions of nitrate uptake by the addition of urea. For a freshwater population, Mitamura (1986) observed that nitrate uptake was inhibited by the addition of urea following a hyperbolic function similar to the ammonium inhibition, but the degree of inhibition by urea was much smaller than that by ammonium.

Limitation of nitrate uptake by ammonium and urea has been a focus of intensive research in nitrate-replete waters including polar and high latitude regions, as well as coastal and equatorial upwelling regions. Despite the abundance of nitrate in surface waters, ammonium supplied through the regeneration process limited nitrate uptake significantly in some high latitude waters and thus resulted in a lower f-ratio (Rönner *et al.*, 1983; Koike *et al.*, 1986; Wheeler and Kokkinakis, 1990; Owens *et al.*, 1991). The situation was suggested to be similar in the equatorial Pacific, where the f-ratio is generally low (Murray *et al.*, 1989; Dugdale *et al.*, 1992; Peña *et al.*, 1992). The f-ratio data obtained in polar regions were compiled recently by Smith and Nelson (1990), Kristiansen *et al.* (1994), and Bury *et al.* (1995), but the data suggest a fairly large regional and seasonal

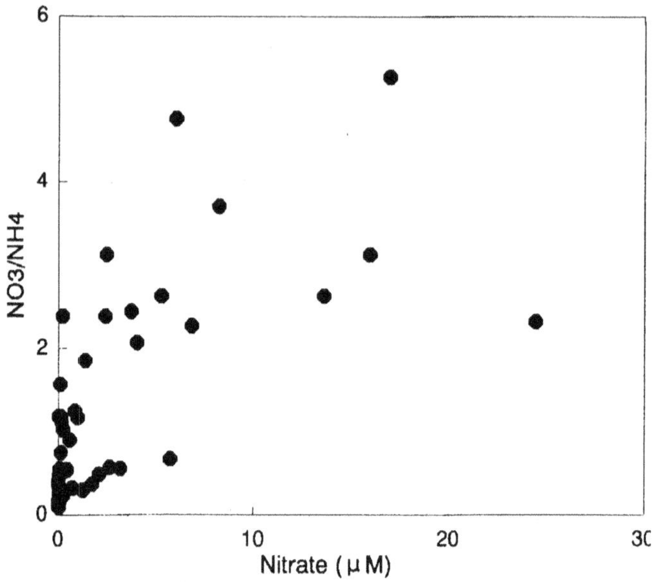

Fig. 3. The NO_3/NH_4 ratio for ρ_{max} (see text) plotted against the ambient nitrate concentration. Data from natural populations in surface seawater in the Pacific (Kanda *et al.*, 1985a), near Hachijo Island (Kanda *et al.*, 1985b), in Auke Bay (Kanda *et al.*, 1990),and in the East China Sea (Kanda *et al.*, in press).

variability. Smith (1991) found a positive correlation of f-ratio and nitrate concentration in Antarctic waters. High f-ratios of around 0.8–0.9 are often reported in polar waters, especially in the blooming period (Nelson and Smith, 1986; Smith and Harrison, 1991; Goeyens *et al.*, 1991; Kristiansen *et al.*, 1994; Bury *et al.*, 1995; Waldron *et al.*, 1995) as well as in several other nitrate-replete waters during bloom (Owens *et al.*, 1986, 1993; Kokkinakis and Wheeler, 1987; Smith and Harrison, 1991; Peña *et al.*, 1992; Rees *et al.*, 1995). Thus, in nitrate replete waters, there will be a clear distinction between the blooming condition (or the "shift-up" condition) where the phytoplankton biomass is rapidly building up and the condition with constantly low biomass (Dugdale and Wilkerson, 1992). The latter is commonly called HNLC (high-nutrient, low chlorophyll) waters. In addition to the light conditions, as discussed above, the presence of ammonium and perhaps urea may likely limit nitrate uptake in HNLC waters. More recently, however, iron was suggested to be a limiting factor (see below).

In an extensive review of the interaction between ammonium and nitrate uptake, Dortch (1990) concluded that the inhibition effect of ammonium on nitrate uptake is sometimes less clear in field observations. For example, natural plankton populations in oyster-ponds utilize nitrate in the presence of >10 μM of ammonium (Collos *et al.*, 1989). Under nitrate replete conditions, a significant nitrate uptake was observed in the presence of a relatively high concentration (several μM) of ammonium (Carpenter and Dunham, 1985). A NO_3/NH_4 ratio calculated with maximum uptake rates (ρ_{max}) for nitrate and ammonium, can be an index of nitrate preference (Dortch, 1990). Since nitrate is first reduced to ammonium during its assimilation process, a NO_3/NH_4 ratio for ρ_{max} higher than 1.0 indicates a strong preference of nitrate over ammonium (Paasche and Kristiansen, 1982; Kanda, 1989). In Fig. 3, the NO_3/NH_4 ratio for ρ_{max} from the data of Kanda *et al.* (1985a, b, 1990, in press) is plotted as a function of the ambient nitrate concentration. A high ratio exceeding 1.0 was observed during the blooming condition in the Auke Bay and the regional upwelling near Hachijo Island. Several other authors observed a high NO_3/NH_4 ratio for ρ_{max} in the bloom condition (MacIsaac *et al.*, 1974; Owens *et al.*, 1986; Dugdale and Wilkerson, 1991; Tamminen, 1995). The RPI values for nitrate exceeding 1.0 is also indicative of nitrate preference. Such cases are rare but have been found on several occasions (Glibert *et al.*, 1982b; Paache and Kristiansen, 1982; Harrison *et al.*, 1982, 1985; Furnace, 1983; Owens *et al.*, 1986; Pennock, 1987; Smith and Kattner, 1989; Smith and Harrison, 1991).

Iron and other nutrients

In HNLC waters, the lack of iron is suggested to depress nitrate uptake. Martin and co-workers observed that the addition of iron substantially stimulated nitrate uptake in such waters (Martin and Fitzwater, 1988; Martin *et al.*, 1990). Banse (1990) and Dugdale and Wilkerson (1990) argued that the addition of iron did not stimulate the average growth rate of the phytoplankton population and the observed increase of biomass and enhancement of nitrate uptake can be attributed to an insufficient loss rate (presumably grazing rate) of phytoplankton. The addition of iron stimulated the growth of a specific group of algae (generally diatoms) in a population which was depressed by the low concentration of iron, but grazing on this group of algae was not able to catch up the biomass increase. The growth rate of the originally dominant group of algae (small size phytoplankton mainly depend on regenerated nitrogen) was also fast but effectively removed by grazing. Such a situation is in favor of the "grazing control" hypothesis for a low biomass (Cullen, 1991). Price *et al.* (1991, 1994) measured nitrate uptake by a ^{15}N tracer method following iron enrichment in the equatorial Pacific. The iron enrichment resulted in a two fold increase of nitrate uptake, but was accompanied by a succession of species of phytoplankton (Price *et al.*, 1991). Price *et al.* (1994) further observed that iron stimulation of nitrate uptake was confined to the larger size fraction of phytoplankton (>3 μm), and that removal of grazing pressure resulted in enhancement of nitrate utilization by small phytoplankton. Iron would be one of several factors that control nitrate uptake in HNLC waters, but apparently not the sole factor.

The effect of iron is thus mostly on species composition. A similar effect was proposed recently for silicate (Dugdale and Wilkerson, 1992; Dugdale *et al.*, 1995). Since diatoms are generally dominant in "shift-up" conditions, silicate deficiency may result in limited nitrate utilization in HNLC waters. Dugdale *et al.* (1995) showed the possibility of enhancing nitrate uptake by an increased supply of silicate. This type of indirect control of nitrate uptake through the changing composition of species may also be brought by other micronutrients. For example, Morel *et al.* (1994) recently proposed that the growth of larger marine phytoplankton cells may be limited by the availability of zinc because a zinc deficiency results in the suppression of bicarbonate utilization. The increase of zinc supply may thus cause a similar enhancement of nitrate uptake.

Perspective for future research

Among the environmental parameters discussed in this article, light intensity, nitrate concentration and ammonium concentration are the three major variables. The correlation of nitrate uptake and these three parameters have been relatively well studied. With recent developments in analytical methods for the nanomolar range of nitrate, efforts in quantitative research on the environmental control of nitrate uptake in oligotrophic waters have been started (e.g. Harrison *et al.*, 1996; Allen *et al.*, 1996). Further accumulation of such data will clarify the role of nitrate nutrition in oligotrophic systems. The combining effect of two or three of these parameters, however, is little understood. Much of the discussion on field observations is focused on this aspect (e.g. the combining effect of light intensity and nitrate availability), but there is still some disagreement among researchers about the interpretation of the data sets.

Some variation of the nitrate uptake rate remains which cannot be accounted for by a variation of the three major parameters. As in HNLC waters, the nitrate uptake rate is sometimes limited to be smaller than that expected from the major environmental parameters. The "shift-up" hypothesis implies that nitrate uptake is under dynamic control and that time is an important variable. The variation of the nitrate uptake rate can be quantitatively described with light intensity, nitrate concentration, and the time elapsed from the onset of the upwelling event in several upwelling systems (Dugdale *et al.*, 1992). Such analysis, however, may not be applicable in other oceanic systems where time cannot be determined or defined as an environmental variable. Moreover, iron, silicate and other micronutrients are suggested to affect the development of the shift-up. Data on the micronutrient effect are rapidly being accumulated, but there is much more to be clarified. The dynamic change in nitrate uptake usually accompanies the succession of species and variation in grazing rates. Thus the composition of species of the plankton community and grazing pressure should be considered in addition to the micronutrient effect. A complete quantitative description of nitrate uptake may require a more general ecosystem dynamic model.

Future research calls for the establishment of a theoretical and physiological framework for dealing with the effect of multiple environmental parameters. The accumulation of field observations and the effort of linking the data with dynamic ecosystem models are also important. Above all, however, it is stressed that there is a need for developing technology in nitrate uptake research in the field. Refinement of the ^{15}N methodology, for example, is necessary for establishing a complete budget for ^{15}N (Bronk *et al.*, 1994) and assigning the contribution by various components in a whole planktonic community.

Acknowledgment—Prof. N. Handa has encouraged me for continuing field work in the oceanic environment. Drs. E. Tanoue and T. Hama provided the opportunity for presenting this review. I thank the Ministry of Education, Japan for financial support (Grants No. 04232206, 04740355, 05216203, 06740420, and 07640655), the MASFLEX project of the Science and Technology Agency, Japan, and the Nissan Science Foundation.

REFERENCES

Allen, C. B., J. Kanda and E. A. Laws (1996): New production and photosynthetic rates within and outside a cyclonic mesoscale eddy in the North Pacific subtropical gyre. *Deep-Sea Res. I*, **43**, 917–936.

Altabet, M. A. (1990): Organic C, N, and stable isotopic composition of particulate matter collected on glass-fiber and aluminum oxide filters. *Limnol. Oceanogr.*, **35**, 902-909.

Banse, K. (1990): Does iron really limit phytoplankton production in the offshore subarctic Pacific? *Limnol. Oceanogr.*, **35**, 772–775.

Berges, J. A., W. P. Cochlan and P. J. Harrison (1995): Laboratory and field responses of algal nitrate reductase to diel periodicity in irradiance, nitrate exhaustion, and the presence of ammonium. *Mar. Ecol. Prog. Ser.*, **124**, 259–269.

Bjørnsen, P. K., H. Kaas, H. Kaas, T. G. Nielsen, M. Olesen and K. Richardson (1993): Dynamics of a subsurface phytoplankton maximum in the Skagerrak. *Mar. Ecol. Prog. Ser.*, **95**, 279–294.

Blasco, D. and H. L. Conway (1977): Effect of ammonium on the regulation of nitrate assimilation in natural phytoplankton populations. *J. Exp. Mar. Biol. Ecol.*, **61**, 157–168.

Blasco, D., J. J. MacIsaac, T. T. Packard and R. C. Dugdale (1984): Relationship between nitrate reductase and nitrate uptake in phytoplankton in the Peru upwelling region. *Limnol. Oceanogr.*, **29**, 275–286.

Boyd, P. W., S. J. Bury, N. J. P. Owens, G. J. Savidge and T. Preston (1995): A comparison of isotopic and chemiluminescent methods of estimating microalgal nitrate uptake in the NE Atlantic. *Mar. Ecol. Prog. Ser.*, **116**, 199–205.

Bronk, D. A. and P. M. Glibert (1991): A ^{15}N tracer method for the measurement of dissolved organic nitrogen release by phytoplankton. *Mar. Ecol. Prog. Ser.*, **77**, 171–182.

Bronk, D. A. and P. M. Glibert (1994): The fate of the missing 15N differs among marine systems. *Limnol. Oceanogr.*, **39**, 189–195.

Bronk, D. A., P. M. Glibert and B. B. Ward (1994): Nitrogen uptake, dissolved organic nitrogen release, and new production. *Science*, **265**, 1843–1846.

Bury, S. J., N. J. P. Owens and T. Preston (1995): ^{13}C and ^{15}N uptake by phytoplankton in the marginal ice zone of the Bellingshausen Sea. *Deep-Sea Res. II*, **42**, 1225–1252.

Carpenter, E. J. and S. Dunham (1985): Nitrogenous nutrient uptake, primary production, and species composition of phytoplankton in the Carmans River estuary, Long Island, New York. *Limnol. Oceanogr.*, **30**, 513–526.

Chisholm, S. W. (1992): Phytoplankton size. p.213–237. In *Primary Productivity and Biogeochemical Cycles in the Sea.* ed. by P. G. Falkowski and A. D. Woodland, Plenum Press, New York.

Chisholm, S. W., R. J. Olson, E. R. Zettler, R. Goericke, J.B. Waterbury and N. A. Welschmeyer (1988): A novel free-living prochlorophyte abundant in the oceanic euphotic zone. *Nature*, **334**, 340–343.

Cochlan, W. P., N. M. Price and P. J. Harrison (1991a): Effects of irradiance on nitrogen uptake by phytoplankton: comparison of frontal and stratified communities. *Mar. Ecol. Prog. Ser.*, **69**, 103–116.

Cochlan, W. P., P. J. Harrison and K. L. Denman (1991b): Diel periodicity of nitrogen uptake by marine phytoplankton in nitrate rich environments. *Limnol. Oceanogr.*, **36**, 1689–1700.

Collos, Y., S. Y. Maestrini and J.-M. Robert (1989): High long-term nitrate uptake by oyster- pond microalgae in the presence of high ammonium concentrations. *Limnol. Oceanogr.*, **34**, 957–964.
Collos, Y. and G. Slawyk (1976): Significance of cellular nitrate content in natural populations of marine phytoplankton growing in shipboard cultures. *Mar. Biol.*, **34**, 27–32.
Collos, Y. and G. Slawyk (1977): Nitrate reductase activity as a function of *in situ* nitrate uptake and environmental factors of euphotic zone profiles. *J. Exp. Mar. Biol. Ecol.*, **29**, 119–130.
Collos, Y. and G. Slawyk (1986): ^{13}C and ^{15}N uptake by marine phytoplankton - IV. Uptake ratios and the contribution of nitrate to the productivity of Antarctic waters (Indian Ocean sector). *Deep-Sea Res.*, **33**, 1039–1051.
Conway, H. L. (1977): Interactions of inorganic nitrogen in the uptake and assimilation by marine phytoplankton. *Mar. Biol.*, **39**, 221–232.
Conway, H. L., and T. E. Whitledge (1979): Distribution, fluxes and biological utilization of inorganic nitrogen during a spring bloom in the New York Bight. *J. Mar. Res.*, **37**, 657–668.
Cota, G. F., W. O. Smith, Jr., D. M. Nelson, R. D. Muench and L. I. Gordon (1992): Nutrient and biogenic particulate distributions, primary productivity and nitrogen uptake in the Weddell-Scotia Sea marginal ice zone during winter. *J. Mar. Res.*, **50**, 155–181.
Cullen, J. J. (1991): Hypothesis to explain high-nutrient conditions in the open sea. *Limnol. Oceanogr.*, **36**, 1578–1599.
DiTullio, G. R. and E. A. Laws (1991): Impact of an atmospheric-oceanic disturbance on phytoplankton community dynamics in the North Pacific Central Gyre. *Deep-Sea Res.*, **38**, 1305–1329.
Dortch, Q. (1990): The interaction between ammonium and nitrate uptake in phytoplankton. *Mar. Ecol. Prog. Ser.*, **61**, 183–201.
Dortch, Q. and H. Maske (1982): Dark uptake of nitrate and nitrate reductase activity of a red-tide population off Peru. *Mar. Ecol. Prog. Ser.*, **9**, 299–303.
Droop, M. R. (1968): Vitamin B_{12} and marine ecology. IV. The kinetics of uptake, growth and inhibition in Monochrysis lutheri. *J. Mar. Biol. Ass. U.K.*, **48**, 689–733.
Droop, M. R. (1974): The nutrient status of algal cells in continuous culture. *J. Mar. Biol. Ass. U.K.*, **54**, 825–855.
Dugdale, R.C. (1967): Nutrient limitation in the sea: Dynamics, identification and significance. *Limnol. Oceanogr.*, **12**, 685–695.
Dugdale, R. C. and J. J. Goering (1967): Uptake of new and regenerated forms of nitrogen in primary productivity. *Limnol. Oceanogr.*, **12**, 196–206.
Dugdale, R. C., A. Morel, A. Bricaud and F.P. Wilkerson (1989): Modeling new production in upwelling centers: A case study of modeling new production from remotely sensed temperature and color. *J. Geophys. Res.*, **94**, 18119–18132.
Dugdale, R. C. and F. P. Wilkerson (1986): The use of ^{15}N to measure nitrogen uptake in eutrophic oceans; experimental considerations. *Limnol. Oceanogr.*, **31**, 673–689.
Dugdale, R. C. and F. P. Wilkerson (1989): New production in the upwelling center at Point Conception, California: temporal and spatial patterns. *Deep-Sea Res.*, **36**, 985–1007.
Dugdale, R. C. and F. P. Wilkerson (1990): Iron addition experiments in the Antarctic: A reanalysis. Global Biogeochem. *Cycles*, **4**, 13–19.
Dugdale, R. C. and F. P. Wilkerson (1991): Low specific nitrate uptake rate: A common feature of high-nutrient, low-chlorophyll marine ecosystems. *Limnol. Oceanogr.*, **36**, 1678–1688.
Dugdale, R. and F. Wilkerson (1992): Nutrient limitation of new production in the sea. p.299–316. In Primary Productivity and Biogeochemical Cycles in the Sea, ed. by P.G. Falkowski and A.D. Woodland, Plenum Press, New York.
Dugdale, R. C., F. P. Wilkerson, R. T. Barber and F. P. Chavez (1992): Estimating new production in the equatorial Pacific Ocean at 150°W. *J. Geophys. Res.*, **97**, 681–686.
Dugdale, R. C., F. P. Wilkerson, and H. J. Minas (1995): The role of silicate pump in driving new production. *Deep-Sea Res.*, **42**, 697–719.
Eppley, R. W. (1972): Temperature and phytoplankton growth in the sea. *Fish. Bull.*, **70**, 1063–1085.
Eppley, R. W., A. F. Carlucci, O. Holm-Hansen, D. Kiefer, J.J. McCarthy, E. Venrick and P. M. Williams (1971a): Phytoplankton growth and composition in shipboard cultures supplied with

nitrate, ammonium, or urea as the nitrogen source. *Limnol. Oceanogr.*, **16**, 741–751.

Eppley, R. W. and J. L. Coatsworth (1968): Uptake of nitrate and nitrite by Ditylum brightwellii - kinetics and mechanisms. *J. Phycol.*, **4**, 151–156.

Eppley, R. W., C. Garside, E. H. Renger and E. Orellana (1990): Variability of nitrate concentration in nitrogen-depleted subtropical surface waters. *Mar. Biol.*, **107**, 53–60.

Eppley, R. W. and W. Koeve (1990): Nitrate use by plankton in the eastern subtropical North Atlantic, March-April 1989. *Limnol. Oceanogr.*, **35**, 1781–1788.

Eppley, R. W. and B. J. Peterson (1979): Particulate organic matter flux and planktonic new production in the deep ocean. *Nature*, **282**, 677–680.

Eppley, R. W. and E. H. Renger (1988): Nanomolar increase in surface layer nitrate concentration following a small wind event. *Deep-Sea Res.*, **35**, 1119–1125.

Eppley, R. W. and E. H. Renger (1992): Nitrate utilization by plankton in the equatorial Pacific March 1988 along 150°W. *J. Geophys. Res.*, **97**, 663–668.

Eppley, R. W., E. H. Renger and W. G. Harrison (1979): Nitrate and phytoplankton production in southern California coastal waters. *Limnol. Oceanogr.*, **24**, 483–494.

Eppley, R. W., E. H. Renger, E. L. Venrick and M. M. Mullin (1973): A study of plankton dynamics and nutrient cycling in the central gyre of the North Pacific Ocean. *Limnol. Oceanogr.*, **18**, 534–551.

Eppley, R. W., J. N. Rogers, J. J. McCarthy and A. Sournia (1971b): Light/dark periodicity in nitrogen assimilation of the marine phytoplankters *Skeletonema costatum and Coccolithus huxleyi* in N-limited chemostat culture. *J. Phycol.*, **7**, 150–154.

Eppley, R. W., J. H. Sharp, E. H. Renger, M.J. Perry and W.G. Harrison (1977): Nitrogen assimilation by phytoplankton and other microorganisms in the surface waters of the central North Pacific Ocean. *Mar. Biol.*, **39**, 111–120.

Fanning, K. A. (1989): Influence of atmospheric pollution on nutrient limitation in the ocean. *Nature*, **339**, 460–463.

Fisher, T. R., P. R. Carlson and R. T. Barber (1982): Carbon and nitrogen primary productivity in three North Carolina estuaries. *Estuar. Coast. Shelf Sci.*, **15**, 621–644.

Furnas, M. J. (1983): Nitrogen uptake dynamics in lower Narragansett Bay, Rhode Island. I. Uptake by size-fractionated phytoplankton populations. *J. Plankton Res.*, **5**, 657–676.

Garside, C. (1981): Nitrate and ammonium uptake in the apex of the New York Bight. *Limnol. Oceanogr.*, **26**, 731–739.

Garside, C. (1982): A chemiluminescent technique for the determination of nanomolar concentrations of nitrate and nitrite in seawater. *Mar. Chem.*, **11**, 159–167.

Garside, C. (1985): The vertical distribution of nitrate in open ocean surface water. *Deep- Sea Res.*, **32**, 723–732.

Garside, C. (1991): Shift-up and the nitrate kinetics of phytoplankton in upwelling systems. *Limnol. Oceanogr.*, **36**, 1239–1244.

Garside, C. and H. E. Glover (1991): Chemiluminescent measurements of nitrate uptake kinetics: I. *Thalassiosira pseudonana* (clone 3H) and neritic assemblages. *J. Plankton Res.*, **13**, s5–s19.

Glibert, P. M., D. C. Biggs and J. J. McCarthy (1982a): Utilization of ammonium and nitrate during austral summer in the Scotia Sea. *Deep-Sea Res.*, **29**, 837–850.

Glibert, P. M. and C. Garside (1992): Diel variability in nitrogenous nutrient uptake by phytoplankton in the Chesapeake Bay plume. *J. Plankton Res.*, **14**, 271–288.

Glibert, P. M., J. C. Goldman and E. J. Carpenter (1982b): Seasonal variations in the utilization of ammonium and nitrate by phytoplankton in Vineyard Sound, Massachusetts, USA. *Mar. Biol.*, **70**, 237–249.

Glibert, P. M. and J. J. McCarthy (1984): Uptake and assimilation of ammonium and nitrate by phytoplankton: indices of nutritional status for natural assemblages. *J. Plankton Res.*, **6**, 677–697.

Glibert, P. M. and R. T. Ray (1990): Different patterns of growth and nitrogen uptake in two clones of marine *Synechococcus spp. Mar. Biol.*, **107**, 273–280.

Glover, H. E., B. B. Prezelin, L. Campbell, M. Wyman and C. Garside (1988): A nitrate-dependent Synechococcus bloom in surface Sargasso Sea water. *Nature*, **333**, 161–163.

Goering, J. J., R. C. Dugdale and D. W. Menzel (1964): Cyclic diurnal variations in the uptake of ammonia and nitrate by photosynthetic organisms in the Sargasso Sea. *Limnol. Oceanogr.*, **9**, 448–451.

Goering, J. J., D. D. Wallen and R. M. Nauman (1970): Nitrogen uptake by phytoplankton in the discontinuity layer of the eastern subtropical Pacific Ocean. *Limnol. Oceanogr.*, **15**, 789–796.

Goeyens, L., F. Sörensson, P. Tréguer, J. Morvan, M. Panouse and F. Dehairs (1991): Spatiotemporal variability of inorganic nitrogen stocks and uptake fluxes in the Scotia- Weddell Confluence area during November and December 1988. *Mar. Ecol. Prog. Ser.*, **77**, 7–19.

Goldman, J. C. (1977): Temperature effects on phytoplankton growth in continuous culture. *Limnol. Oceanogr.*, **22**, 932–936.

Goldman, J. C. (1988): Spatial and temporal discontinuities of biological processes in pelagic surface waters. p.273–296. In *Toward a theory on biological-physical interactions in the world ocean*, ed. by B. J. Rothschild, Kluwer Acad. Publ., Dordrecht.

Goldman, J. C. (1993): Potential role of large oceanic diatoms in new primary production. *Deep-Sea Res.*, **40**, 159–168.

Goldman J. C. and P. M. Glibert (1983): Kinetics of inorganic nitrogen uptake by phytoplankton. p.233–274. In *Nitrogen in the Marine Environment*, ed. by E. J. Carpenter and D. G. Capone, Academic Press, New York.

Goldman, J. C., J. J. McCarthy and D. G. Peavey (1979): Growth rate influence on the chemical composition of phytoplankton in oceanic waters. *Nature*, **279**, 210–215.

Harrison, W. G. (1976): Nitrate metabolism of the red tide dinoflagellate Gonyaulux polyedra Stein. *J. Exp. Mar. Biol. Ecol.*, **21**, 199–209.

Harrison, W. G. (1983): The time-course of uptake of inorganic and organic nitrogen compounds by phytoplankton from the eastern Canadian Arctic: A comparison with temperate and tropical populations. *Limnol. Oceanogr.*, **28**, 1231–1237.

Harrison, W. G., F. Azam, E. H. Renger and R.W. Eppley (1977): Some experiments on phosphate assimilation by coastal marine plankton. *Mar. Biol.*, **40**, 9–18.

Harrison, W. G., G. F. Cota and R. E. H. Smith (1990): Nitrogen utilization in ice algal communities of Barrow Strait, Northwest Territories, Canada. *Mar. Ecol. Prog. Ser.*, **67**, 275–283.

Harrison, W. G., D. Douglas, P. Falkowski, G. Rowe and J. Vidal (1983): Summer nutrient dynamics of the Middle Atlantic Bight: nitrogen uptake and regeneration. *J. Plankton Res.*, **5**, 539–556.

Harrison, W. G., L. R. Harris and B. D. Irwin (1996): The kinetics of nitrogen utilization in the oceanic mixed layer: Nitrate and ammonium interactions at nanomolar concentrations. *Limnol. Oceanogr.*, **41**, 16–32.

Harrison, W. G., L. R. Harris, D. M. Karl, G. A. Knauer and D. G. Redalje (1992): Nitrogen dynamics at the VERTEX time-series site. *Deep-Sea Res.*, **39**, 1535–1552.

Harrison, W. G., E. J. H. Head, R. J. Conover, A. R. Longhurst and D. D. Sameoto (1985): The distribution and metabolism of urea in the eastern Canadian Arctic. *Deep-Sea Res.*, **32**, 23–42.

Harrison, W. G., T. Platt and B. Irwin (1982): Primary production and nutrient assimilation by natural phytoplankton populations of the eastern Canadian Arctic. *Can. J. Fish. Aquat. Sci.*, **39**, 335–345.

Harrison, W. G., T. Platt and M. R. Lewis (1987): f-Ratio and its relationship to ambient nitrate concentration in coastal waters. *J. Plankton Res.*, **9**, 235–248.

Harrison, W. G. and L. J. E. Wood (1988): Inorganic nitrogen uptake by marine picoplankton: Evidence for size partitioning. *Limnol. Oceanogr.*, **33**, 468–475.

Harvey, W. A. and J. Caperon (1976): The rate of utilization of urea, ammonium, and nitrate by natural populations of marine phytoplankton in a eutrophic environment. *Pacific Sci.*, **30**, 329–340.

Hayward, T. L. (1987): The nutrient distribution and primary production in the central North Pacific. *Deep-Sea Res.*, **34**, 1593–1627.

Horrigan, S. G., A. Hagstrom, I. Koike and F. Azam (1988): Inorganic nitrogen utilization by assemblages of marine bacteria in seawater culture. *Mar. Ecol. Prog. Ser.*, **50**, 147–150.

Jenkins, W. J. (1988): Nitrate flux into the euphotic zone near Bermuda. *Nature*, **331**, 521–523.

Jenkins, W. J. and D. W. R. Wallace (1992): Tracer based inferences of new primary production in the sea. p. 299–316. *In Primary Productivity and Biogeochemical Cycles in the Sea*, ed. by P. G. Falkowski and A. D. Woodland, Plenum Press, New York.

Kanda, J. (1989): Control of nitrate uptake during diatom blooms. *Kaiyo Monthly*, **21**, 583–587 [in Japanese].

Kanda, J. (1993): Models of nutrient limitation of uptake and growth of microalgae: variability of half saturating constants. Bull. Jpn. Soc. *Microb. Ecol.*, **8**, 109–123 [in Japanese].

Kanda, J. (1995): Nitrate assimilation and new production in open ocean. p. 227–238. In *Biogeochemical Processes and Ocean Flux in the western Pacific*, ed. by H. Sakai and Y. Nozaki, Terra Scientific Publishing Co., Tokyo.

Kanda, J., and A. Hattori (1988): Ammonium uptake and synthesis of cellular nitrogenous macromolecules in phytoplankton. *Limnol. Oceanogr.*, **33**, 1568–1579.

Kanda, J., E. A. Laws, T. Saino and J. Kanda (1987): An evaluation of isotope dilution effect from conventional data sets of ^{15}N uptake experiments. *J. Plankton Res.*, **9**, 79–90.

Kanda, J., T. Saino and A. Hattori (1985a): Nitrogen uptake by natural populations of phytoplankton and primary production in the Pacific Ocean: regional variability of uptake capacity. *Limnol. and Oceanogr.*, **30**, 987–999.

Kanda, J., T. Saino and A. Hattori (1985b): Variation of carbon and nitrogen uptake capacity in a regional upwelling area around Hachijo Island. *J. Oceanogr. Soc. Japan*, **41**, 373–380.

Kanda, J., T. Saino and A. Hattori (1988): Nitrogen nutrition and physiological state of natural populations of phytoplankton in surface waters of the western North Pacific. *Limnol. Oceanogr.*, **33**, 1580–1585.

Kanda, J., D. A. Ziemann, L. D. Conquest and P.K. Bienfang (1989): Light dependency of nitrate uptake by phytoplankton over the spring bloom in Auke Bay, Alaska. *Mar. Biol.*, **103**, 563–569.

Kanda, J., D. A. Ziemann, L. D. Conquest and P. K. Bienfang (1990): Nitrate and ammonium uptake by phytoplankton populations during the spring bloom period in Auke Bay, Alaska. *Estuar. Coast. Shelf Sci.*, **30**, 509–524.

Kanda, J., T. Itoh, D. Ishikawa and Y. Watanabe (in press): Environmental control of nitrate uptake in the East China Sea. *Deep-Sea Res. II.*

Kaufman, Z. G., J. S. Lively and E. J. Carpenter (1983): Uptake of nitrogenous nutrients by phytoplankton in a barrier island estuary: Great South Bay, New York. *Estuar. Coast. Shelf Sci.*, **17**, 483–493.

King, F. D. and A. H. Devol (1979): Estimates of vertical eddy diffusion through the thermocline from phytoplankton nitrate uptake rates in the mixed layer of the eastern tropical Pacific. *Limnol. Oceanogr.*, **24**, 645–651.

Koike, I., O. Holm-Hansen and D. C. Biggs (1986): Inorganic nitrogen metabolism by Antarctic phytoplankton with special reference to ammonium cycling. *Mar. Ecol. Prog. Ser.*, **30**, 105–116.

Kristiansen, S. (1983): Urea as a nitrogen source for the phytoplankton in the Oslofjord. *Mar. Biol.*, **74**, 17–24.

Kristiansen, S., T. Farbrot and P. A. Wheeler (1994): Nitrogen cycling in the Barents Sea - Seasonal dynamics of new and regenerated production in the marginal ice zone. *Limnol. Oceanogr.*, **39**, 1630–1642.

Kristiansen, S. and B. Aa. Lund (1989): Nitrogen cycling in the Barents Sea -I. Uptake of nitrogen in the water column. *Deep-Sea Res.*, **36**, 255–268.

Laws, E. A. (1991): Photosynthetic quotients, new production and net community production in the open ocean. *Deep-Sea Res.*, **38**, 143–167.

Laws, E. A., G. R. DiTullio, P. B. Betzer, D. M. Karl and K. L. Carder (1989): Autotrophic production and elemental fluxes at 26oN, 155oW in the North Pacific subtropical gyre. *Deep-Sea Res.*, **36**, 103–120.

Laws, E. A., G. R. DiTullio, and D. G. Redalje (1987): High phytoplankton growth and production rates in the North Pacific subtropical gyre. *Limnol. Oceanogr.*, **32**, 905–918.

Laws, E. A., W. G. Harrison and G. R. DiTullio (1985): A comparison of nitrogen assimilation rates based on ^{15}N uptake and autotrophic protein synthesis. *Deep-Sea Res.*, **32**, 85–95.

Laws, E. A., D. G. Redalje, L. W. Haas, P. K. Bienfang, R. W. Eppley, W. G. Harrison, D. M. Karl and J. Marra (1984): High phytoplankton growth and production in oligotrophic Hawaiian coastal waters. *Limnol. Oceanogr.*, **29**, 1161–1169.

Laws, E. A. and D. C. L. Wong (1978): Studies of carbon and nitrogen metabolism by three marine phytoplankton species in nitrate-limited continuous culture. *J. Phycol.*, **14**, 406– 416.

Le Bouteiller, A. (1986): Environmental control of nitrate and ammonium by phytoplankton in the Equatorial Atlantic Ocean. *Mar. Ecol. Prog. Ser.*, **30**, 167–179.

Lewis, M. R., W. G. Harrison, N. S.Oakey, D. Hebert and T. Platt (1986): Vertical nitrate fluxes in the oligotrophic ocean. *Science*, **234**, 870–873.

Lipschultz, F. (1995): Nitrogen-specific uptake rates of marine phytoplankton isolated from natural populations of particles by flow cytometry. *Mar. Ecol. Prog. Ser.*, **123**, 245–258.

MacIsaac, J. J. (1978): Diel cycles of inorganic nitrogen uptake in a natural phytoplankton population dominated by *Gonyaulax poyedra. Limnol. Oceanogr.*, **23**, 1–9.

MacIsaac, J. J. and R. C. Dugdale (1969): The kinetics of nitrate and ammonia uptake by natural populations of marine phytoplankton. *Deep-Sea Res.*, **16**, 45–57.

MacIsaac, J. J. and R. C. Dugdale (1972): Interactions of light and inorganic nitrogen in controlling nitrogen uptake in the sea. *Deep-Sea Res.*, **19**, 209–232.

MacIsaac, J. J., R. C. Dugdale, R. T. Barber, D. Blasco and T. T. Packard (1985): Primary production cycle in an upwelling center. *Deep-Sea Res.*, **32**, 503–529.

MacIsaac, J. J., R. C. Dugdale, S. A. Huntsman and H. L. Conway (1979): The effect of sewage on uptake of inorganic nitrogen and carbon by natural populations of marine phytoplankton. *J. Mar. Res.*, **37**, 51–66.

MacIsaac, J. J., R. C. Dugdale and G. Slawyk (1974): Nitrogen uptake in the Northeast Africa upwelling area: Results from the Cineca-Charcot II cruise. *Tethys*, **6**, 69–76.

Mantoura, R. F. C., C. S. Law, N. J. P. Owens, P. H. Burkill, E. M. S. Woodward, R. J. M. Howland and C. A. Llewellyn (1993): Nitrogen biogeochemical cycling in the northwestern Indian Ocean. *Deep-Sea Res. II*, **40**, 651–671.

Malone, T. C. (1980): Size-fractionated primary production of marine phytoplankton. p.301–319. In *Primary productivity in the sea*, ed. by P. G. Falkowski, Plenum, New York.

Martin, J. H., and S. E. Fitzwater (1988): Iron deficiency limits phytoplankton growth in the north-east Pacific subarctic. *Nature*, **331**, 341–343.

Martin, J. H., S. E. Fitzwater and R. M. Gordon (1990): Iron deficiency limits phytoplankton growth in the Antarctic waters. *Global Biogeochem. Cycles*, **4**, 5–12.

McCarthy, J. J. (1972): The uptake of urea by natural populations of marine phytoplankton. *Limnol. Oceanogr.*, **17**, 738–748.

McCarthy, J. J. (1981): The kinetics of nutrient utilization. p.211–233. In Physiological Bases of *Phytoplankton Ecology*, ed. by T. Platt, *Can. Bull. Fish. Aquat. Sci*, Vol. 210.

McCarthy, J. J. and R. W. Eppley (1972): A comparison of chemical, isotopic, and enzymatic methods for measuring nitrogen assimilation of marine phytoplankton. *Limnol. Oceanogr.*, **17**, 371–382.

McCarthy, J. J., C. Garside and J. L. Nevins (1992): Nitrate supply and phytoplankton uptake kinetics in the euphotic layer of a Gulf Stream warm-core ring. *Deep-Sea Res.*, **39**, S393–S403.

McCarthy, J. J., W. R. Taylor and J. L. Taft (1977): Nitrogenous nutrition of the plankton in the Chesapeake Bay. 1. Nutrient availability and phytoplankton preferences. *Limnol. Oceanogr.*, **22**, 996–1011.

Mitamura, O. (1986): Urea metabolism and its significance in the nitrogen cycle in the euphotic layer of Lake Biwa III. Influence of the environmental parameters on the response of nitrogen assimilation. *Arch. Hydrobiol.*, **107**, 281–299.

Miyazaki, T., H. Suyama and H. Uotani (1987): Diel changes of uptake of inorganic carbon and nitrogen by phytoplankton, and the relationship between inorganic carbon and nitrogen uptake in Lake Nakamura, Japan. *J. Plankton Res.*, **9**, 513–524.

Morel, F. M. M. (1987): Kinetics of nutrient uptake and growth of phytoplankton. *J. Phycol.*, **23**, 137–150.

Morel, F. M. M., J. R. Reinfelder, S. B. Roberts, C. P. Chamberlain, J. G. Lee and D. Yee (1994): Zinc and carbon co-limitation of marine phytoplankton. *Nature*, **369**, 740–742.

Muggli, D. L. and W. O. Smith, Jr. (1993): Regulation of nitrate and ammonium uptake in the

Greenland Sea. *Mar. Biol.*, **115**, 199–208.

Murray, J. W., J. N. Downs, S. Strom, C.-L. Wei and H. W. Jannasch (1989): Nutrient assimilation, export production and ^{234}Th scavenging in the eastern equatorial Pacific. *Deep-Sea Res.*, **36**, 1471–1489.

Nalewajko, C. and C. Garside (1983): Methodological problems in the simultaneous assessment of photosynthesis and nutrient uptake in phytoplankton as functions of light intensity and cell size. *Limnol. Oceanogr.*, **28**, 591–597.

Nelson, D. M. and H. L. Conway (1979): Effects of the light regime on nutrient assimilation by phytoplankton in the Baja California and northwest Africa upwelling systems. *J. Mar. Res.*, **37**, 301–318.

Nelson, D. M. and W. O. Smith, Jr. (1986): Phytoplankton bloom dynamics of the western Ross Sea ice edge -II. Mesoscale cycling of nitrogen and silicon. *Deep-Sea Res.*, **33**, 1389–1412.

Olson, R. J. (1980): Nitrate and ammonium uptake in Antarctic waters. *Limnol. Oceanogr.*, **25**, 1064–1074.

Omnes, P., G. Slawyk, N. Garcia and P. Bonin (1996): Evidence of denitrification and nitrate ammonification in the River Rhone plume (northwestern Mediterranean Sea). *Mar. Ecol. Prog. Ser.*, **141**, 275–281.

Oudot, C. and Y. Montel (1988): A high sensitivity method for the determination of nanomolar concentrations of nitrate and nitrite in seawater with Technicon AutoAnalyzer II. *Mar. Chem.*, **24**, 239–252.

Owens, N. J. P., P. H. Burkill, R. F. C. Mantoura, E.M.S. Woodward, I.E. Bellan, J. Aiken, R. J. M. Howland and C. A. Llewellyn (1993): Size-fractionated primary production and nitrogen assimilation in the northwestern Indian Ocean. *Deep-Sea Res. II*, **40**, 697–709.

Owens, N. J. P., J. N. Galloway and R. A. Duce (1992): Episodic atmospheric nitrogen deposition to oligotrophic oceans. *Nature*, **357**, 397–399.

Owens, N. J. P., R. F. C. Mantoura, P. H. Burkill, R. J. M. Howland, A. J. Pomroy and E. M. S. Woodward (1986): Nutrient cycling studies in Carmarthen Bay: phytoplankton production, nitrogen assimilation and regeneration. *Mar. Biol.*, **93**, 329–342.

Owens, N. J. P., J. Priddle and M. J. Whitehouse (1991): Variations in phytoplankton nitrogen assimilation around South Georgia and in the Bransfield Strait (Southern Ocean). *Mar. Chem.*, **35**, 287–304.

Paasche, E., I. Bryceson and K. Tangen (1984): Interspecific variation in dark nitrogen uptake by dinoflagellates. *J. Phycol.*, **20**, 394–401.

Paasche, E. and S. Kristiansen (1982): Nitrogen nutrition of the phytoplankton in the Oslofjord. *Estuar. Coast. Shelf Sci.*, **14**, 237–249.

Peña, M. A., W. G. Harrison and M. R. Lewis (1992): New production in the central equatorial Pacific. *Mar. Ecol. Prog. Ser.*, **80**, 265–274.

Pennock, J. R. (1987): Temporal and spatial variability in phytoplankton ammonium and nitrate uptake in Delaware Estuary. *Estuar. Coat. Shelf Sci.*, **24**, 841–857.

Pettersson, K. and E. Sahlsten (1990): Diel patterns of combined nitrogen uptake and intracellular storage of nitrate by phytoplankton in the open Skagerrak. *J. Exp. Mar. Biol. Ecol.*, **138**, 167–182.

Platt, T., W. G. Harrison, M. R. Lewis, W. K. W. Li, S. Sathyendranath, R. E. Smith and A. F. Vezina (1989): Biological production of the oceans: the case for a consensus. *Mar. Ecol. Prog. Ser.*, **52**, 77–88.

Price, N. M., B. A. Ahner and F. M. M. Morel (1994): The equatorial Pacific Ocean: Grazer-controlled phytoplankton populations in an iron-limited ecosystem. *Limnol. Oceanogr.*, **39**, 520–534.

Price, N. M., L. F. Andersen and F. M. M. Morel (1991): Iron and nitrogen nutrition of equatorial Pacific plankton. *Deep-Sea Res.*, **38**, 1361–1378.

Price, N. M., W. P. Cochlan and P. J. Harrison (1985): Time course of uptake of inorganic and organic nitrogen by phytoplankton in the Strait of Georgia: comparison of frontal and stratified communities. *Mar. Ecol. Prog. Ser.*, **27**, 39–53.

Price, N. M. and P. J. Harrison (1988): Urea uptake by Sargasso Sea phytoplankton: saturated and *in situ* uptake rate. *Deep-Sea Res.*, **35**, 1579–1593.

Priscu, J. C., A. C. Palmisano, L. R. Priscu and C. W. Sullivan (1989): Temperature dependence of inorganic nitrogen uptake and assimilation in Antarctic sea-ice microalgae. *Polar Biol.*, **9**, 443–446.

Probyn, T. A. (1985): Nitrogen uptake by size-fractionated phytoplankton populations in the southern Benguela upwelling system. *Mar. Ecol. Prog. Ser.*, **22**, 249–258.

Probyn, T. A. (1988): Nitrogen utilization by phytoplankton in the Namibian upwelling region during an austral spring. *Deep-Sea Res.*, **35**, 1387–1404.

Probyn, T. A. (1992): The inorganic nitrogen nutrition of phytoplankton in the southern Benguela: New production, phytoplankton size and implications for pelagic foodwebs. *S. Afr. J. Mar. Sci.*, **12**, 411–420.

Probyn, T. A., B. A. Mitchell-Innes and S. Searson (1995): Primary productivity and nitrogen uptake in the subsurface chlorophyll maximum on the Eastern Agulhas Bank. *Cont. Shelf Res.*, **15**, 1903–1920.

Probyn, T. A. and S. J. Panting (1985): Nitrogen uptake by size-fractionated phytoplankton populations in Antarctic surface waters. *Limnol. Oceanogr.*, **30**, 1327–1332.

Probyn, T. A., H. N. Waldron and A. G. James (1990): Size-fractionated measurements of nitrogen uptake in aged upwelled waters: Implications for pelagic food-webs. *Limnol. Oceanogr.*, **35**, 202–210.

Raimbault, P. and V. Gentilhomme (1990): Short- and long-term responses of the marine diatom Phaeodactylum tricornutum to spike additions of nitrate at nanomolar levels. *J. Exp. Mar. Biol. Ecol.*, **135**, 161–176.

Raimbault, P., V. Gentilhomme and G. Slawyk (1990a): Short-term responses of 24 hour N- starved cultures of Phaeodactylum tricornutum to pulsed additions of nitrate at nanomolar levels. *Mar. Ecol. Prog. Ser.*, **63**, 47–52.

Raimbault, P., G. Slawyk, B. Coste and J. Fry (1990b): Feasibility of using an automated colorimetric procedure for the determination of seawater nitrate in the 0-100 nM range: examples from field and culture work. *Mar. Biol.*, **104**, 347–351.

Raimbault, P., G. Slawyk and V. Gentilhomme (1990c): Direct measurements of nanomolar nitrate uptake by the marine diatom *Phaeodactylum tricornutum* (Bohlin). Implications for studies of oligotrophic ecosystems. *Hydrobiologia*, **207**, 311–318.

Rees, A. P., N. J. P. Owens, M. R. Heath. D. H. Plummer and R. S. Bellerby (1995): Seasonal nitrogen assimilation and carbon fixation in a fijordic sea loch. *J. Plankton Res.*, **17**, 1307– 1324.

Rhee, G.-Y., and I.J. Gotham (1981a): The effect of environmental factors on phytoplankton growth: Temperature and the interaction of temperature with nutrient limitation. *Limnol. Oceanogr.*, **26**, 635–648.

Rhee, G.-Y., and I. J. Gotham (1981b): The effect of environmental factors on phytoplankton growth: Light and the interaction of light with nitrate limitation. *Limnol. Oceanogr.*, 26, 649–659.

Rönner, U., F. Sörensson and O. Holm-Hansen (1983): Nitrogen assimilation by phytoplankton in the Scotia Sea. *Polar Biol.*, **2**, 137–147.

Sahlsten, E. (1987): Nitrogenous nutrition in the euphotic zone of the Central North Pacific Gyre. *Mar. Biol.*, **96**, 433–439.

Sahlsten, E., F. Sorensson and K. Pettersson (1988): Planktonic nitrogen uptake in the south-eastern Kattegat. *J. Exp. Mar. Biol. Ecol.*, **121**, 227–246.

Sambrotto, R. N., J. H. Martin, W. W. Broenkow, C. Carlson and S. E. Fitzwater (1993): Nitrate utilization in surface waters of the Iceland Basin during spring and summer of 1989. *Deep-Sea Res. II.*, **40**, 441–457.

Sambrotto, R. N., H. J. Niebauer, J. J. Goering and R. L. Iverson (1986): Relationships among vertical mixing, nitrate uptake, and phytoplankton growth during the spring bloom in the southeast Bering Sea middle shelf. *Cont. Shelf Res.*, **5**, 161–198.

Sathyendranath, S., T. Platt, E. P. W. Horne, W. G. Harrison, O. Ulloa, R. Outerbridge and N. Hoepffner (1991): Estimation of new production in the ocean by compound remote sensing. *Nature*, **353**, 129–133.

Shiomoto, A. and Y. Maita (1990): Uptake of nitrate and ammonia in the subarctic boundary and adjacent regions of the northwestern Pacific Ocean. *Deep-Sea Res.*, **37**, 1887–1897.

Shiomoto, A., K. Sasaki, T Shimoda and S. Matsumura (1994): Kinetics of nitrate and ammonium uptake by natural populations of marine phytoplankton in the surface water of the Oyashio region during spring and summer. *J. Oceanogr.*, **50**, 515–529.

Slawyk, G. (1979): ^{13}C and ^{15}N uptake by phytoplankton in the Antarctic upwelling area: Results from the Antipod I Cruise in the Indian Ocean Sector. *Aust. J. Mar. Freshwater Res.*, **30**, 431–448.

Slawyk, G., J. J. MacIsaac and R. C. Dugdale (1976): Inorganic nitrogen uptake by marine phytoplankton under *in situ* and simulated *in situ* incubation conditions: Results from the northwest African upwelling region. *Limnol. Oceanogr.*, **21**, 149–152.

Slawyk, G. and P. Raimbault (1995): Simple procedure for simultaneous recovery of dissolved inorganic and organic nitrogen in ^{15}N-tracer experiments and improving the isotopic mass balance. *Mar. Ecol. Prog. Ser.*, **124**, 289–299.

Slawyk, G., P. Raimbault and V. Gentilhomme (1990): On the discrepancies between a colorimetric and isotopic method for measuring nitrate utilization in nutrient depleted waters: implications for the design of experimental protocols in new production studies. *Hydrobiologia*, **207**, 333–339.

Smith, W. O., Jr. (1991): Nutrient distributions and new production in polar regions: parallels and contrasts between the Arctic and Antarctic. *Mar. Chem.*, **35**, 245–257.

Smith, W. O., Jr. (1993): Nitrogen uptake and new production in the Greenland Sea: the spring Phaeocystis bloom. *J. Geophys. Res.*, **98**, 4681–4688.

Smith, W. O., Jr. and W. G. Harrison (1991): New production in polar regions: the role of environmental controls. *Deep-Sea Res.*, **38**, 1463–1479.

Smith, W. O., Jr. and G. Kattner (1989): Inorganic nitrogen uptake by phytoplankton in the marginal ice zone of the Fram Strait. *Rapp. P.-v. Réun. Cons. Int. Explor. Mer.*, **188**, 90– 97.

Smith, W. O., Jr. and D. M. Nelson (1990): Phytoplankton growth and new production in the Weddel Sea marginal ice zone in the austral spring and autumn. *Limnol. Oceanogr.*, **35**, 809–821.

Syrett, P. J. (1981): Nitrogen metabolism of microalgae. p.182–210. In *Physiological Bases of Phytoplankton Ecology*, ed. by T. Platt, *Can. Bull. Fish. Aquat. Sci*, Vol. 210.

Tamminen, T. (1995): Nitrate and ammonium depletion rates and preference during a Baltic spring bloom. *Mar. Ecol. Prog. Ser.*, **120**, 123–133.

Tobiesen, A. (1987): Nitrogen uptake in the phytoplankton of Oslofjorden, Norway: Effects of environmental and biological covariates. *Sarsia*, **72**, 299–311.

Vincent, W. F. (1992): The daily pattern of nitrogen uptake by phytoplankton in dynamic mixed layer environments. *Hydrobiologia*, **238**, 37–52.

Waldron, H. N., C. G. Attwood, T. A. Probyn and M. I. Lucas (1995): Nitrogen dynamics in the Bellingshausen Sea during the austral spring of 1992. *Deep-Sea Res. II*, **42**, 1253–1276.

Ward, B. B., K. A. Kilpatrick, E. H. Renger and R. W. Eppley (1989): Biological nitrogen cycling in the nitracline. *Limnol. Oceanogr.*, **34**, 493–513.

Wheeler, P. A. and D. L. Kirchman (1986): Utilization of inorganic and organic nitrogen by bacteria in marine systems. *Limnol. Oceanogr.*, **31**, 998–1009.

Wheeler, P. A., D. L. Kirchman, M. R. Landry and S. A. Kokkinakis (1989): Diel periodicity of ammonium uptake and regeneration in the oceanic subarctic Pacific: Implications for interactions in microbial food webs. *Limnol. Oceanogr.*, **34**, 1025–1033.

Wheeler, P. A. and S. A. Kokkinakis (1990): Ammonium recycling limits nitrate use in the oceanic subarctic Pacific. *Limnol. Oceanogr.*, **35**, 1267–1278.

Wilkerson, F. P. and R. C. Dugdale (1987): The use of large shipboard barrels and drifters to study the effects of coastal upwelling on phytoplankton dynamics. *Limnol. Oceanogr.*, **32**, 368–382.

Wilkerson, F. P. and R. C. Dugdale (1992): Measurements of nitrogen productivity in the equatorial Pacific. *J. Geophys. Res.*, **97**, 669–679.

Wilkerson, F. P., R. C. Dugdale and R. T. Barber (1987): Effects of El Niño on new, regenerated, and total production in eastern boundary upwelling system. *J. Geophys. Res.*, **92**, 14347–14533.

Dynamics and Characterization of Marine Organic Matter, Eds., N. Handa, E. Tanoue and T. Hama, pp. 67–89.

Distribution and Non-Dominancy of Large Diatoms in Planktonic Algal Communities in the Central Pacific Ocean along 175°E

Masayuki TAKAHASHI[1], Hideyuki KANAZAWA[2],
Hiroshi KIYOSAWA[3,*] and Joji ISHIZAKA[4,**]

[1]*Department of Biology, University of Tokyo,*
3-8-1 Komaba, Meguro-ku, Tokyo 153-8902, Japan
[2]*Department of Comparative Literature and Comparative Culture, University of Tokyo,*
3-8-1 Komaba, Meguro-ku, Tokyo 153-8902, Japan
[3]*Kansai Environmental Engineering Corporation, Chuo, Osaka 541-0052, Japan*
[4]*National Institute for Resources and Environment,*
Onogawa, Tsukuba, Ibaraki 305-8569, Japan

(Original, Received 30 October 1996; accepted 27 December 1996)

Abstract—The abundance and biomass of large diatoms retained in 100 μm mesh nets have been determined from 46°N to 15°S along 175°E in the central Pacific Ocean during August and October, 1991. Twenty-seven diatom species ranging from 13×10^3 to 34×10^6 μm^3 in single cell volume were recognized in the samples obtained by a NORPAC net in the top 200 m of the water column at 13 stations in the area. Numerical abundance of each species within the top 50 m or 100 m which covered most of the euphotic zone, were mostly <10 cells. m^{-3}. Percentage shares of chlorophyll *a* of >100 μm fraction which included large diatoms and dinoflagellates were always less than 5% of the total chlorophyll *a* in the water, and then the fraction of the large diatoms as a group in the total chlorophyll *a* were much less. The distribution of large diatoms in the area was largely classified into three types; distributed only at 44°N or north, at 30°N or north, and at 36°N or south. The second type partly overlapped the distribution of the first type at 44°N or north and the 3rd type at 30–36°N. There was the sub-Arctic front locating between 43°N and 42°N and a subtropical convergence at 28°N, which seems to be important for controlling the large diatom distributions.

INTRODUCTION

The biological community in a given natural environment is generally composed of a variety of species (Whittaker, 1975). Hutchinson (1961) proposed an idea of various micro-environments accommodating each different species in their

---

*Present address: Marine Biological Research Institute of Japan, 4-3-16 Yutaka-cho, Shinagawa-ku, Tokyo 142-0042, Japan.

**Present address: Faculty of Fisheries, Nagasaki University, 1-14 Bunkyo-machi, Nagasaki 852-8521, Japan.

suitable habitats, even in a water which looks homogeneous. However, there is still no complete understanding of such a large biological diversity occurring in the aquatic community (Venrick, 1990).

It is also common knowledge that a given community is generally composed of only a few dominant but many non-dominant species (Odum, 1963). The most probable explanation for the difference in dominancy between species is due to different competitive capabilities of species; species having the superior capability to obtain the necessary resources could dominate less capable species (Tilman, 1977). Consequently, dominant species are justified as well adapted for the environment. Actually, several reports have been produced which demonstrate some supporting evidence of the adaptation of dominant species in the surrounding environment (e.g., Takahashi and Fukazawa, 1982). If this is the only mechanism controlling the abundance of organisms, there will be no suitable environment for non-dominating species unless there occurs a suitable environment for each respective species in some other time or area, otherwise they will be simply classified as defeaters. However, there might be a possibility that non-dominating species have some specific adaptation capabilities in order to maintain the species under natural conditions, even without becoming dominant.

It is, then, worth evaluating the mechanisms not only for the dominancy of abundant species but also for non-dominancy of less abundant species in order to understand the survival of each species under natural conditions. It is also expected that non-dominating species are responsible for maintaining the existing ecosystem, regardless of the differences in the magnitude of dominancy. Therefore, it is necessary to evaluate the life and survival strategies of the non-dominating species.

For the study of non-dominancy, easy extraction of the target species from a natural community will be essential. Since organisms having fewer morphological characteristics require special care, large algae having more morphological characteristics compared to pico- and nano-planktonic algae, will be suitable for the present object. Large diatoms in the open ocean were then focussed on in the study because they were mostly well-described in taxonomy, and were photo-autotrophic and less abundant.

MATERIALS AND METHODS

Samplings and observations in the field

Field observations and samplings were performed from 48°N to 15°S along 175°E in the central Pacific Ocean during the cruise of the R.V. *Hakurei-Maru* between August 9 and October 7 in 1991. Occupied stations are shown in Fig. 1 in which Stns. 1 to 29 were sampled from north to south during August 17 and September 4, and Stns. 30 to 38 were sampled from south to north during September 16 and 21. All field samplings were carried out during daytime. At all the stations, determination of photosynthetically active radiation (PAR), temperature and salinity, and water sampling for nutrients (phosphate, silicic acid and nitrite plus nitrate) was made by a CTD system (Seabird Electronics) having

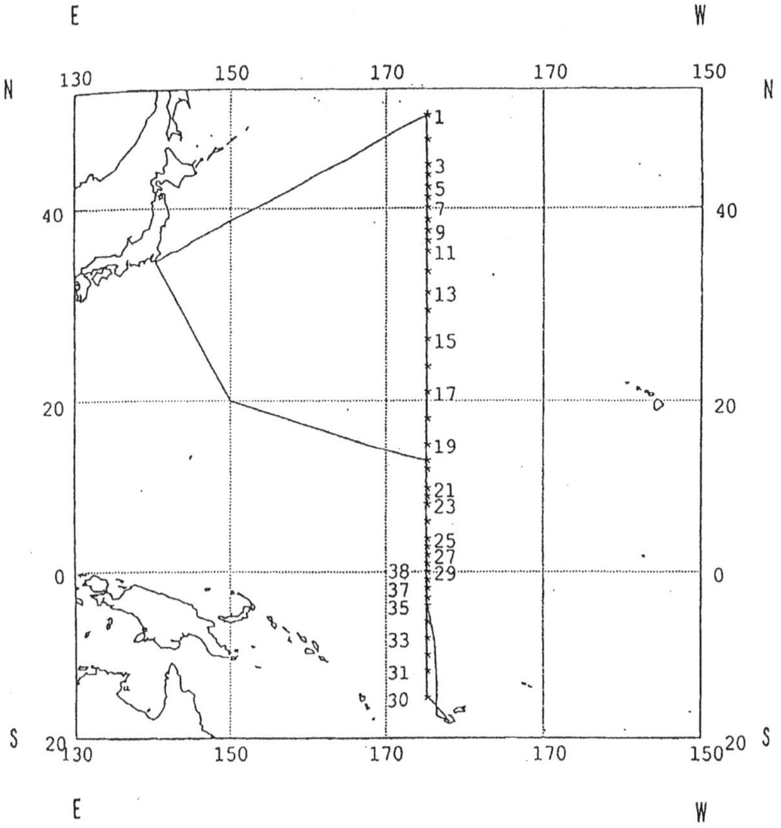

Fig. 1. Cruise track of the study area. Numbers indicate the occupied stations.

12 5 l Niskin bottles and a 4 PAR sensor (Biospherical Instruments). Sampling depths for the water were 10, 20, 30, 40, 50, 75, 100, 125, 150, 200, 250 and 300 m. An additional water sample was collected from 0 m in a plastic bucket. Calibration for salinity measurement by the CTD sensor was made from the water samples collected by the Niskin bottles using an Autosal salinometer (Guideline Instruments, Model 8400A). Nutrients were analyzed on board by an autoanalyzer (Alpkem, Model RFA300).

Plankton samples were collected by vertical hauls of a twin NORPAC type net (mouth diameter, 45 cm) of 100 μm mesh net at 13 stations (Stns. 1, 3, 7, 11, 14, 16, 19, 22, 26, 29, 30, 33 and 35) at a towing speed of 1 m. s^{-1}. The depths of the net tows were 0–100 m and 100–200 m at 36°N or north, and were 0–50 m and 50–200 m at 30°N or south. For the sampling of lower depth intervals, the net was closed by sending a messenger at the depth when the net reached the top depth. The obtained samples were immediately fixed with neutralized formalin (final concentration, 1.2% in v/v). The filtration volume of water during net tow was estimated from the readings of the flow meters which were 0.965 on average with a deviation coefficient of 19.3%.

Estimation of algal biomass

Total algal biomass was estimated by chlorophyll *a*. One liter of each water sample collected by a CTD-rosette sampler at different depths was filtered in series at gravitation vacuum using four Sartorius filtration funnels fitted with filters of 10, 3, and 1 μm of Nuclepore, and of *ca.* 0.5 μm of glass fiber (Whatman GF/F), respectively at all the stations. Another 1 to 2 liters of water samples from different depths were also filtered through 100 and 10 μm of NITEX plankton netting and a glass fiber filter (Whatman GF/F) at 11 stations (Stns. 3, 7, 11, 14, 16, 19, 26, 29, 30, 33 and 35) to estimate biomass of large phytoplankton. Materials retained on the filters were extracted with N,N-dimethylformamide and the chlorophyll *a* was determined by fluoremetry using a fluorometer (Turner Design, model 100-05R) according to Strickland and Parsons (1968).

Variation coefficients for the determination of chlorophyll *a*, checked seven times at two separate stations using surface water samples, were 11.3, 5.8, 8.8 and 10.6% for <1, 1–3, 3–10 and >10 μm fractions, respectively. Those for large fractions were 8.8, 12.7 and 20.7% for >10, 10–100 and >100 μm, respectively.

Identification and biomass estimation of large diatoms

All large diatom cells collected by net tows (100 μm mesh net) were identified and their cell numbers were counted under an inverted microscope (Utermohl, 1958). About 10 to 50 cells for each species were also determined for their cell sizes. The total volume of water applied for microscopic observations for each sample varied approximately between 1500 and 4500 l after filtration through the net tow. A scanning electron microscope was used for detailed observations of some species for identification. Data from both Cupp (1943) and Sundstrom (1986) were used for identification. Each cell was approximated to a suitable geometrical solid and its total volume was determined based upon cell sizes according to Miyai *et al.* (1988). Cell volume was converted into cellular organic carbon using the formula proposed by Strathmann (1967).

RESULTS

Environmental characteristics of the study area

Vertical profiles of temperature, salinity, nutrients, PAR and total chlorophyll *a* in the water column above 300 m along 175°E are graphically shown in Figs. 2, 3 and 4. A marked difference was clearly observed between the north and the south of 42°N particularly in temperature, salinity and nutrients. In the north of 42°N, low salinity water with a moderate temperature of above 10°C and containing moderate concentrations of nutrients was observed above 20 m above the strong seasonal thermocline, and low temperature water below 10°C with high salinity and high concentrations of nutrients occurred below the seasonal thermocline. In the area between 40°N and 30°N, the water temperature in the top 300 m warmed up mostly by more than 10°C and the seasonal thermocline gradually deepened the depths down to 30–40 m southwards. The temperature

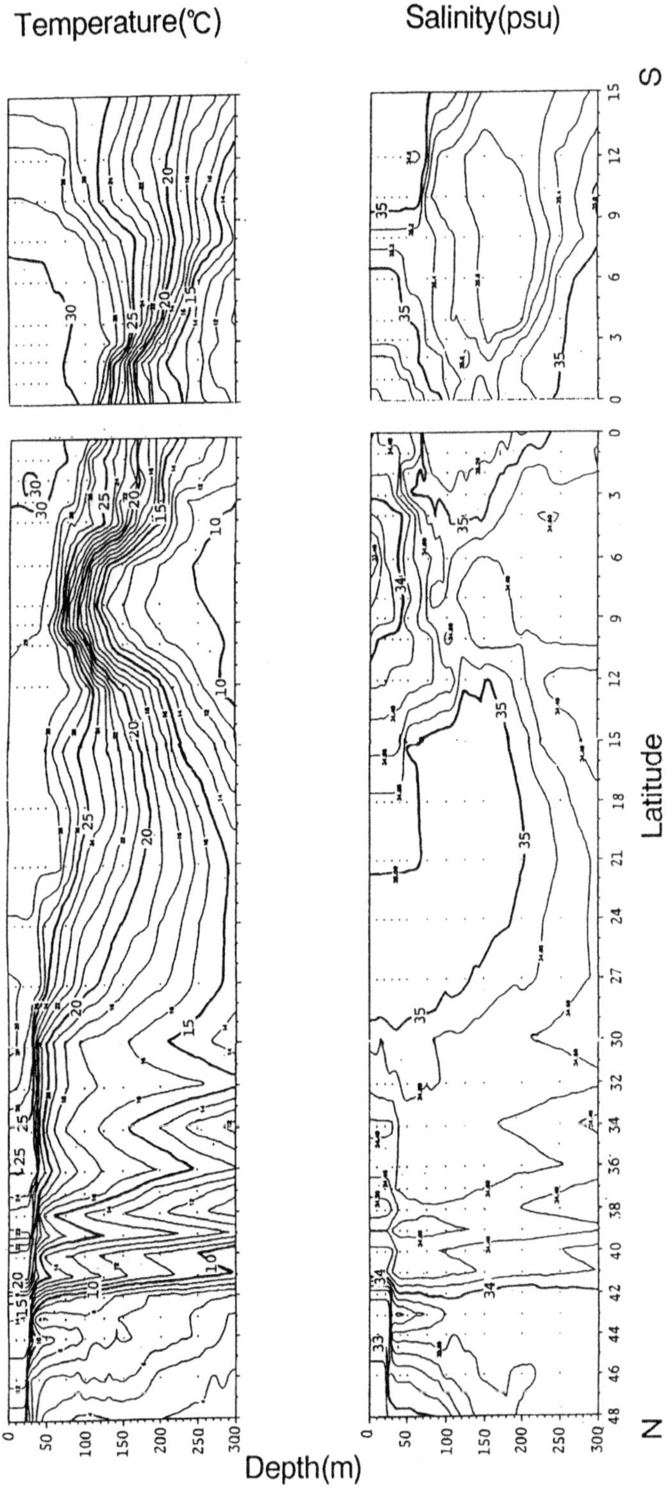

Fig. 2. Vertical profiles of temperature and salinity in the water column above 300 m between 48°N and 15°S along 175°E in the study area.

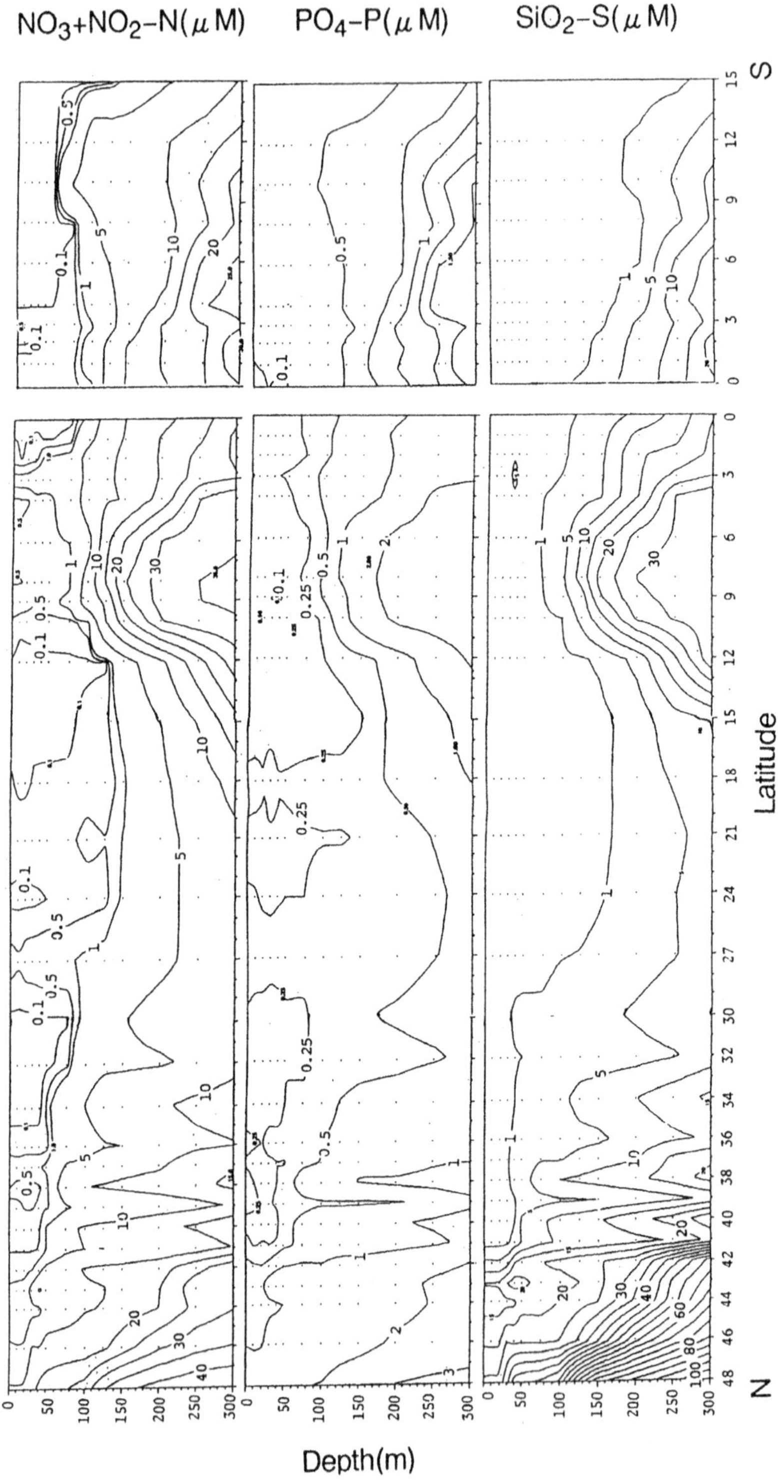

Fig. 3. Vertical profiles of nitrate plus nitrite, phosphate and silicic acid in the water column above 300 m between 48°N and 15°S along 175°E in the study area.

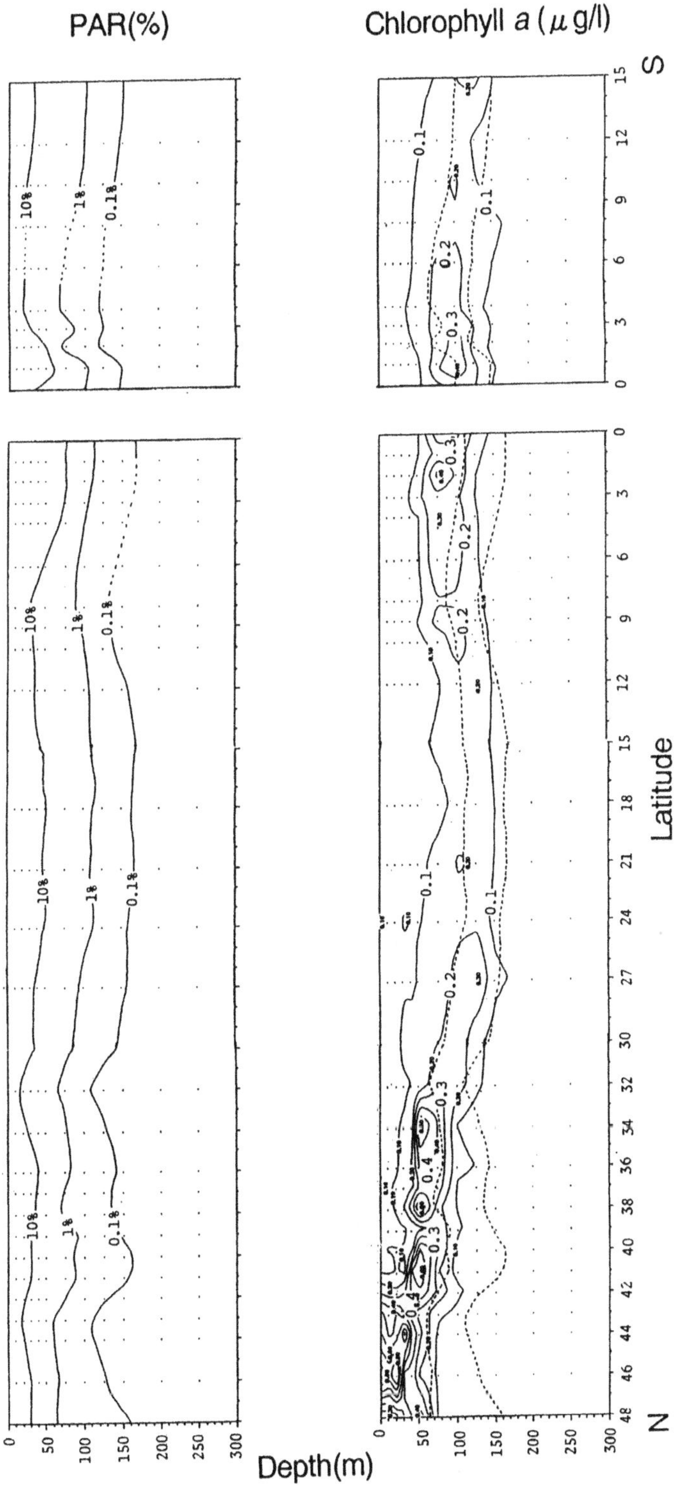

Fig. 4. Vertical profiles of PAR penetration percentages and chlorophyll *a* in the water column above 300 m between 48°N and 15°S along 175°E in the study area.

structure below the seasonal thermocline showed some variations in depth between stations. The temperature in the surface mixing layer varied from 20°C to 26°C. In the area between 30°N and 15°N, the surface water was characterized by high temperatures above 20°C in most of the water column having high salinity, the seasonal thermocline widened greatly, and the nutrient concentrations were low in the entire water column above 300 m. In the area between 12°N and 4°S, the main thermocline became obvious and occurred at 50 to 100 m, which indicated upwelling there. The most strong upwelling occurred between 10°N and 6°N. Along with the upwelling, high concentrations of nutrients also came up to the shallow depths. The surface temperature reached 30°C or over around the equator. In the area between 4°S and 15°S, the vertical gradient of the temperature became less evident, particularly immediately north of the upwelling area. Nutrients again became fairly low in main part of the examined water column.

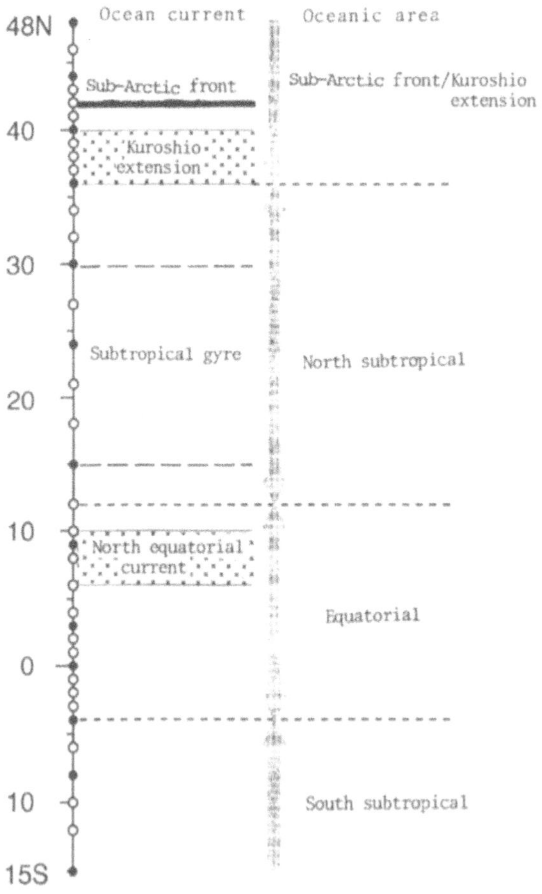

Fig. 5. Ocean currents and oceanic areas recognized along the 175°E study line. Closed circles represent stations where vertical net tows were carried out.

Based upon the physico-chemical characteristics mentioned above and available additional information (Sverdrup *et al.*, 1942), the study area was roughly classified into four different water masses such as the sub-Arctic/Kuroshio extension water north of 36°N, the northern subtropical water occurring between 36°N and 11°N, the equatorial water occurring between 11°N and 4°S, and the southern subtropical water south of 4°S (Fig. 5). Three ocean currents or their effective areas were noticed in the study area, such as the Kuroshio extension between 40°N and 36°N, the subtropical gyre between 30°N and 15°N, and northern equatorial current between 10°N and 6°N. A distinctive discontinuous zone observed at 42°N was justified as the sub-Arctic front (Sverdrup *et al.*, 1942).

Penetration of PAR in the water column became to be 1% of the surface irradiation, mostly above 75 m north of 30°N and 22°S (Fig. 4). The 1% depth mostly occurred between 75 and 100 m in the other study areas. The layer reaching 0.1% of surface PAR, was located at 100–150 m in the former waters and at 150 m or deeper in the latter areas. Considering 0.1% of PAR as the bottom of the euphotic zone, the temperature within the euphotic zone varied from 12°C to 5°C in the sub-Arctic/Kuroshio extension water, 30°C to 20°C in the subtropical water and 30°C to 10°C in the equatorial water (Figs. 2 and 4).

A marked subsurface chlorophyll maximum developed along 175°E where the subsurface maximum appeared near the surface north of 42°N and gradually moved southward into deeper depths, although it appeared in shallower depths in the equatorial water (Fig. 4). Shallow subsurface chlorophyll maxima tended to contain higher chlorophyll concentrations than those at deeper depths. The shallow maxima was located at 1% or higher PAR levels, whereas deep maxima occurred at 1% or lower PAR levels.

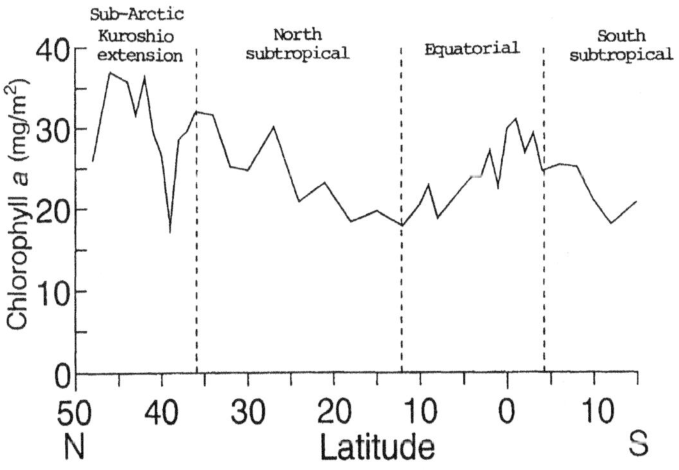

Fig. 6. Total chlorophyll *a* in the top 300 m at each station in the study area.

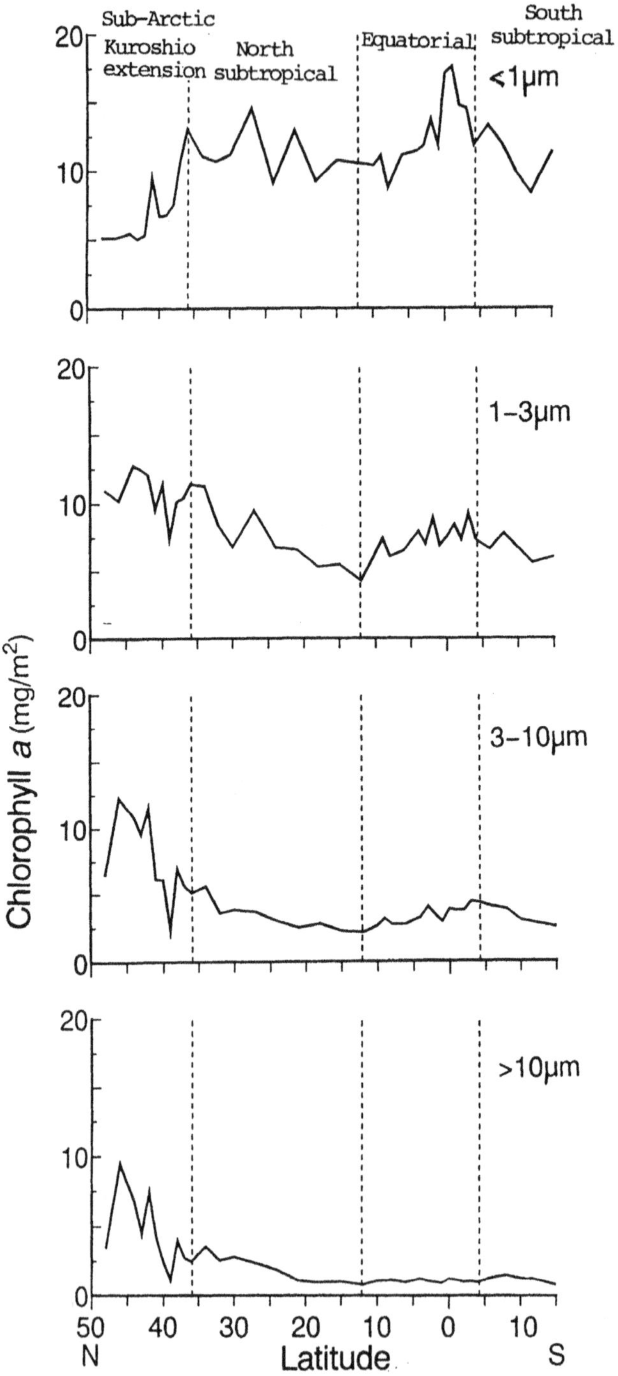

Fig. 7. Total chlorophyll *a* in each size class in the top 300 m at each station in the study area.

Total and size distributions of chlorophyll a in the study area

Total chlorophyll *a* above 300 m in the water column at each station is shown in Fig. 6, and shows a variation of about 2.1 times ranging from a minimum of 17.7 mg chl. *a* m^{-2} at 39°N to a maximum of 37.0 mg chl. *a* m^{-2} at 46°N. The maximum and minimum both occurred in the sub-Arctic/Kuroshio extension water. In the northern subtropical water, total chlorophyll *a* consistently decreased towards the south, and there was a 1.8 times difference in the total chlorophyll *a* changes in the area. Total chlorophyll *a* again increased toward the south in the equatorial water followed by a decrease in the southern subtropical water. Assuming that 0.1% PAR is the bottom of the euphotic zone, 80 to 95% of the total chlorophyll *a* above 300 m was distributed within the euphotic zone.

The total chlorophyll *a* was separated into four different size classes of <1, 1–3, 3–10 and >10 μm by gentle filtration as shown in Fig. 7. The smallest fraction of <1 μm in the sub-Arctic/Kuroshio extension water occupied only a half or less compared to the other areas, whereas the other larger fractions were all highest in the sub-Arctic/Kuroshio extension water compared to the other areas. The trend became pronounced in the larger size fractions; the largest fraction of >10 μm in the sub-Arctic/Kuroshio extension water reached about 5 times compared to the fraction in the other areas. The increase of total chlorophyll *a* in the equatorial water was due mainly to the increases in small size fractions particularly smaller than 3 μm. On the other hand the observed large total chlorophyll *a* in the sub-Arctic/Kuroshio extension water was due to the increase of large size fractions particularly of >3 μm. Fractional share percentages of different size classes at each station showed clear geographical differences (Fig. 8). The fraction of <1 μm became dominant in the south of 36°N, reaching share

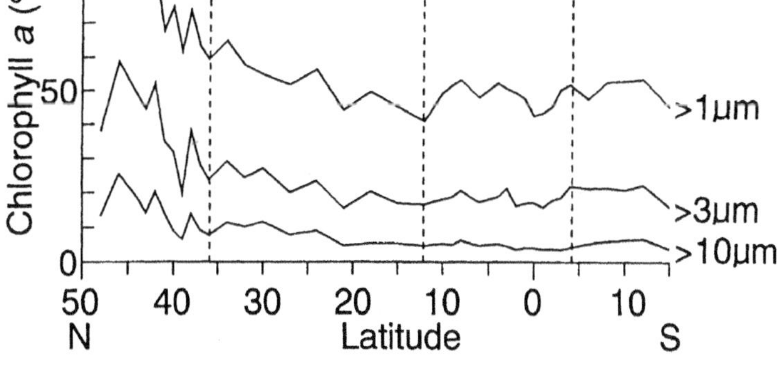

Fig. 8. Fractional components of chlorophyll *a* particularly focussing on the three size categories in the top 300 m in the water column at different stations along 175°E in the study area.

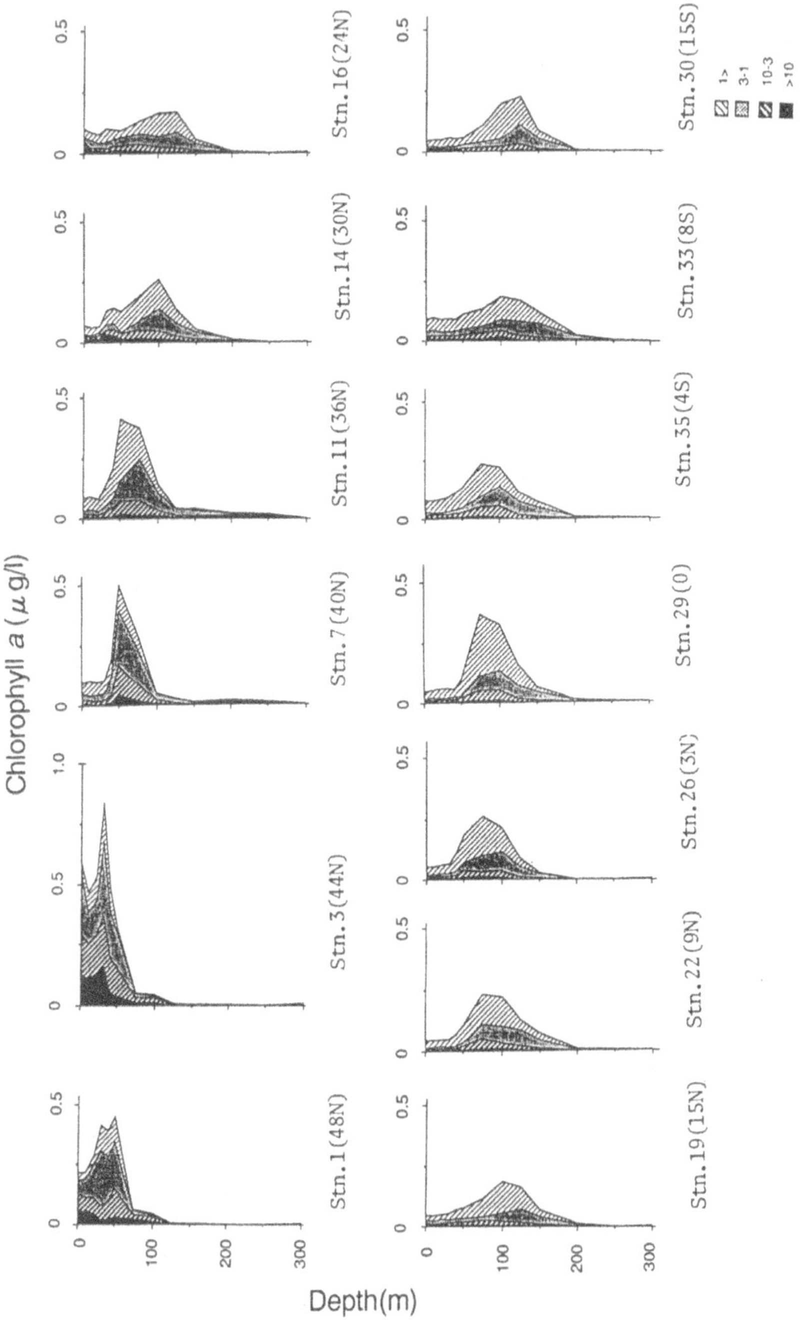

Fig. 9. Vertical distributions of chlorophyll *a* in different particle sizes at various stations where vertical net tows were made along 175°E in the study area.

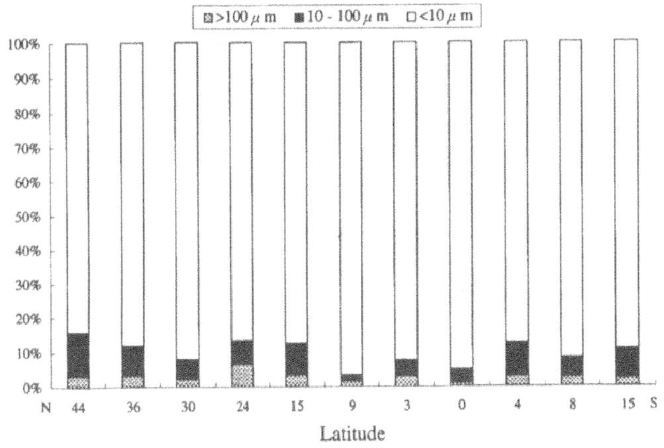

Fig. 10. Fractional components of chlorophyll *a* in the top 300 m of the water column at each station along 175°E in the study area.

percentages of 50% or more in the south of 20°N, and that of 1–3 μm was almost consistent at about 30–45% in the entire area. The fraction of >10 μm was always less predominant, such as 10% or less, except at a few stations in the sub-Arctic/Kuroshio extension water.

Vertical profiles of each size fraction of chlorophyll *a* showed similar depth changes as the depth integrated although the larger fraction tended to show diversified patterns (Fig. 9). For example, the largest fraction of >10 μm showed its maximum at different depths compared with the total chlorophyll *a* at 5 of the 13 stations. Percentage shares of chlorophyll *a* of >10 μm varied from 3.5% at 125 m at 15°S to 25.8% at 30°N and 48°N, which were similar to that of the entire water column mentioned above. The fraction of >10 μm in total chlorophyll *a* at each depth was less predominant at all the stations. The fraction of >10 μm was further separated into two fractions such as 10–100 and >100 μm (Fig. 10). Chlorophyll *a* in the >100 μm fraction was detected at all the stations in which chlorophyll *a* varied from a minimum of 0.3% at 0° to a maximum of 6.3% at 30°N. Then less predominancy became more obvious in the size range of >100 μm.

Large planktonic algae and their sizes in the study area

Planktonic algae collected by 100 μm plankton nets were then identified and quantified for numerical abundance and their biomass in the study area. Because of a large volume of water involved due to a smaller abundance of large phytoplankters, planktonic algae were collected by net tows of 100 μm mesh. The obtained samples were mostly diatoms, dinoflagellates and filamentous blue green algae. All those algae were either in single cells or colonies having more

Table 1. Large diatoms and dinolfagellate species recognized in the study area and their cell volume and approximate cell density throughout the water column in the top 200 m

Species	Average cell volume (μm³)	Approximate cell density (cells/m³)
Diatoms		
Asterolampra marylandica	53,000	<10
Asteromphalus heptactis	60,300	<10
Chaetoceros coarctatus	22,700	<10
Climacodium frauenfeldianum	240,000	<10-100
Corethron hystrix	66,000	<10
Coscinodiscus centralis	1,080,000	<10
C. megalomma	1,020,000	<10
C. noludiefer	77,800	<10
Ethmodiscus rex	19,300,000	<10
Gossleriella tropica	1,230,000	<10
Hemiaulus haukii	14,000	<10
H. membranaceus	104,000	<10-100
Planktoniella sol	26,800	<10-100
Proboscia alata var. curvirostris	311,000	<10
Pr. alata 1	27,900	<10
Pr. alata 2	75,200	<10
Pseudosolenia calcar-avis	112,000	<10
Rhizosolenia acuminata	8,830,000	<10
Rh. bergonii	185,000	<10
Rh. clevei var. clevei	21,200,000	<10
Rh. clevei var. communis	107,000	<10
Rh. cylindrun	34,600	<10
Rh. debyana	34,700,000	<10
Rh. hebetata form hiemalis	61,200	<10
Rh. hebatata form semispina	42,000	<10
Rh. imbricata	65,900	<10
Thalassiothrix longissima	13,300	<10-100
Dinoflagellates		
Ceratium sp. type 1	80,700	<10-100
C. sp. type 2	215,000	<10-100
C. sp. type 3	302,000	<10-100
C. sp. type 4	60,200	<10
C. sp. type 5	352,000	<10
Ceratocorys horrida	64,900	<10
Pyrocystis fusiformis	95,800	<10
P. hamulus var. semicircularis	130,000	<10
P. noctiluca	116,000	<10-100
Pyrophacus horologicum	95,900	<10

than 100 μm in length and/or width. Since some cells or colonies having smaller widths or lengths than 100 μm in a given community might not be collected by the second net in the present study, such small cells and colonies were not collected quantitatively. The large planktonic algae in this study were justified as $>10^4$ μm^3 for their cellular volume. Recognized algal species or groups in the study area are listed in Table 1. The major groups were diatoms and dinoflagellates.

The latter group only included cells which had chlorophyll fluorescence under blue excitation.

Recognized large diatoms were classified into 27 species including 3 varieties and 2 morphologically distinguished cells in 14 genera. All species except for *Th. longissima* of pennate were centric. Two morphological cells of *Pr. alata* examined by SEM observations occurred at different stations. Within 14 genera, 9 species were classified in the genus of *Rhizosolenia*. Since *Proboscia* and *Pseudosolenia* were separated from *Rhizosolenia*, and they had all close morphological shapes, *Rhizosolenia* and its close genera included almost 50% of the total observed large diatom species. Within the large diatom species, *R. debyana* had the largest cell volume of 34.7×10^6 μm^3 followed by *Rh. clevei* var. *clevei* of 21.2×10^6 μm^3 and *Eth. rex* of 19.3×10^6 μm^3. On the other hand, small cells were *Th. longissima* and *H. hauckii* of 0.013×10^6 and 0.014×10^6 μm^3, respectively. Size differences between the largest and the smallest large diatoms were then 2600 times.

Four genera were recognized in the large dinoflagellates. Genus *Ceratium* occupied the most species followed by genus *Pyrocystis*. There were not many differences in size (about 1 order or less) between their largest and smallest cell volumes. They are all in intermediate of cell volumes compared to the large diatoms mentioned above.

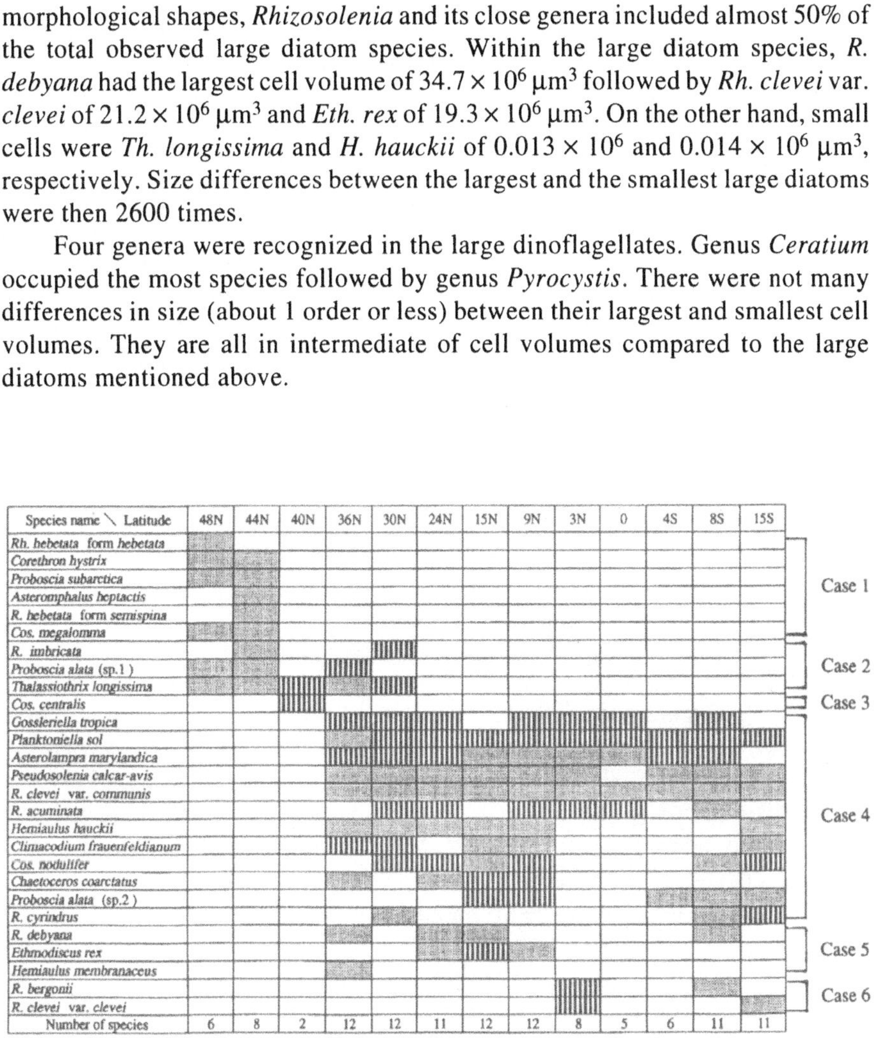

Fig. 11. Occurrences of large diatoms at each station along 175°E in the study area. Dotted area represents predominant occurrence in the upper depths, and vertically lined area represents predominance in the lower depths.

Distributions of the large diatoms in the study area

The large diatoms showed characteristic distributions in the following 6 cases in the study area, as summarized graphically in Fig. 11. Species in case 1 were only distributed at 44°N and 48°N. Six species, such as *As. heptactis, C. hystrix, C. megalomma, Pr. alata* var. *curvirostris, Rh. hebetata* form *hiemalis*, and *Rh. hebetata* form *semispina,* were found to be for case 1. Case 2 species occurred between 48°N and 30°N including 3 species of *Pr. alata* 1, *Rh. imbricata*, and *Th. longissima*. Case 3 species, such as *C. centralis,* only occurred at 40°N. Species in case 4 were distributed widely from 36°N to 15°S and included *As. marylandica, Ch. coarctatus, Cl. frauenferdianum, C. noludifer, C. tropica, H. haukii, Pl. sol, Pr. alata* 2, *Ps. calcar-avis, Rh. acuminata, Rh. cylindrus*, and *Rh. clevei* var. *communis*. Case 5 species occurred from 36°N to 9°N and included *Eth. rex. H. membranaceus* and *Rh. debyaba*. Case 6 species occurred from 3°N to 15°S, and included *Rh. bergonii* and *Rh. clevei* var. *clevei*.

Six species belonging to case 1 and 17 species to cases 4, 5 and 6 did not have common habitats where both species coexisted. These two large groups had then segregated distributions in the north of 40°N for case 1 and in the south for cases 4, 5 and 6. Species belonging to case 2 were only distributed in both habitats but in the north of 30°N. Therefore, there were no overlappings in the species distributions in the north of 44°N and in the south of 24°N. Case 4 covered both cases 5 and 6, although cases 5 and 6 did not overlap each other's distributions.

It is highly possible that the species in cases 1 and 2 could extend the distribution range in the north of 48°N although there was no data available in the present study. On the other hand, the northern limit of distribution for cases 4, 5 and 6 could be in the area of 40°N and 36°N, at least during the study period in the study area. Within the 17 species of cases 4, 5 and 6 occurring in the south of 36°N, 5 species in case 4 were observed at 9 to 10 stations of the 11 total stations, and were expected to have wide distributions in the subtropical and tropical waters. Seven other species in case 4 occurred only at 3 to 7 stations. Three species of case 5 were recognized at 1 to 3 stations, and 2 species of case 6 occurred at 2 stations. Except for *Rh. acuminata,* of the 12 species of the less widely distributed 7 species and of 5 species in cases 5 and 6, none was distributed at the equator and fewer numbers of species were distributed at stations around the equator.

Comparing the shallow depth samples collected from 0–50 m or 0–100 m with the deep samples from 50–200 m or 100–200 m at each station, significant differences of cell density per unit volume of water were noticed statistically by t-test evaluation in some species. *Ps. calcar-avis* and *Rh. clevei* var. *communis* were both significantly abundant in the shallow depth ($P < 0.025$), and *Pl. sol* was abundant in the deep depth ($P < 0.025$). Considering the wide distributions of these 3 species occurring at 9 stations, the observed differences in their vertical distributions were considered to be generous.

There were many species of large diatoms at 36°N–9°N and 8°S–16°S, i. e., 11 and 12 species, respectively. Those were 5 species at 0°, 2 species at 40°N, 8 species at 3°N, and 6 species at 48°N.

Estimated biomass of the large diatoms in the study area

Average cell numbers of the large diatoms and dinoflagellates in the water column above 300 m were mostly <10 cells. m^{-3}. Four diatom species of *Cl. frauenfeldianum*, *H. membranaceus*, *Pl. sol*, and *Th. longissima* showed large variations such as 10 to 100 cells m^{-3}, and 5 dinoflagellate species of *C.* spp. types 1, 2, 3 and 5, and *P. noctiluca* showed similar large variations (Table 1).

Total biomass of each large diatom and dinoflagellate collected by net tows were estimated from the cellular volume based upon cell numbers multiplied by the cell volume determined by microscopic observations. Chlorophyll *a* in the

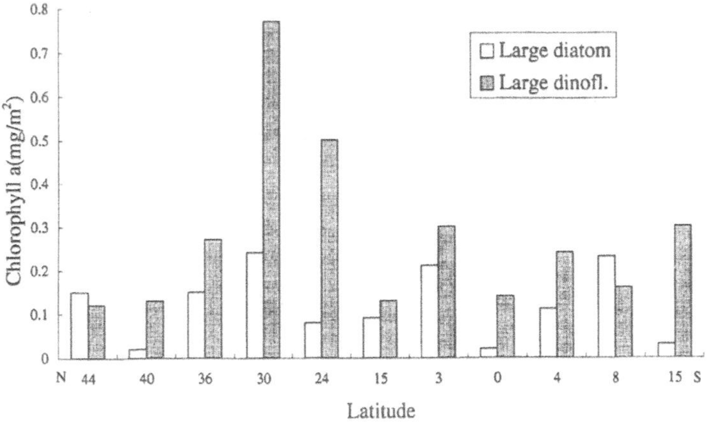

Fig. 12. Estimated chlorophyll *a* of large diatoms and dinoflagellates in the total chlorophyll *a* above 300 m at each station along 175°E in the study area.

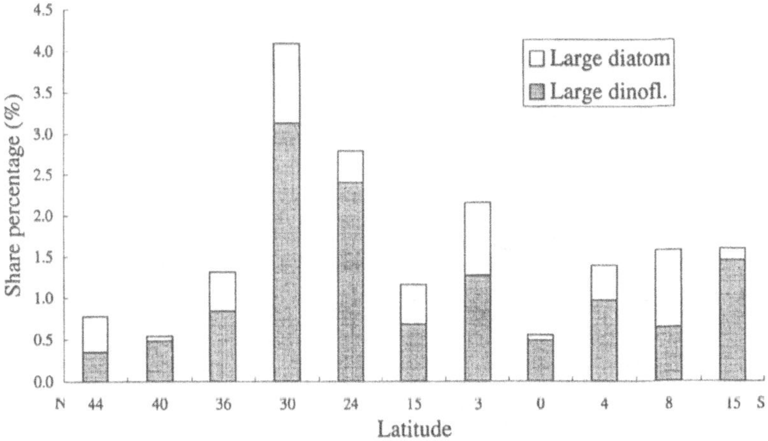

Fig. 13. Estimated percentage shares of large diatoms and dinoflagellates in the total algal population above 300 m at each station along 175°E in the study area.

size fraction of >100 μm was then fractionated for those two algal groups in proportion to the cell volume under the assumption that the chlorophyll *a* was composed mostly of those two observed algal groups in the fraction (Fig. 12). Chlorophyll *a* of the large diatoms in the water column above 300 m varied from a low of 0.16 mg chl. *a* m^{-2} at 40°N to a maximum of 0.24 mg chl. *a* m^{-2} at 30°N, and those of large dinoflagellates were from about 0.12 mg chl.*a* m^{-2} at 15°N, 40°N and 44°N to 0.77 mg chl. *a* m^{-2} at 30°N. Even though the total chlorophyll *a* in the fraction of >10 μm increased in the sub-Arctic water (Fig. 7), neither large diatoms nor large dinoflagellates showed any such corresponding increase there. Consequently, the increase of chlorophyll *a* of >10 μm was due to a possible increase of intermediate size algae in 10–100 μm rather than the large algae concerned in the present study. This is also understandable from the actual large increase of chlorophyll *a* of >10 μm, e. g. about 5 mg chl. *a* m^{-2}, compared to low levels of such as less than 1 mg chl. *a* m^{-2} (Fig. 7).

Table 2. biomass percentages of large diatom species occupying more than 10% in the large diatom community

Station	Species	Biomass percentage
Stn. 1 (48°N)	C. megalomma	74
	Pr. alata var. curvirostris	19
Stn. 3 (44°N)	Th. longissima	32
	C. megalomma	28
	Pr. alata var. curvirostris	20
Stn. 7 (40°N)	C. centralis	99
Stn.11 (36°N)	Rh. debyana	83
	G. tropica	10
Stn.14 (30°N)	Rh. acuminata	54
	G. tropica	15
Stn.16 (24°N)	Rh. debyana	73
	Eth. rex	16
Stn.19 (15°N)	Rh. debyana	57
	H. haukii	21
	Eth. rex	15
Stn.22 (9°N)	Eth. rex	62
	Rh. acuminata	28
Stn.26 (3°N)	G. tropica	37
	Pl. sol	21
	Rh. clevei var. clevei	21
	Rh. acuminata	18
Stn.29 (0°)	Pl. sol	56
	Rh. acuminata	30
Stn.35 (4°S)	Pl. sol	87
Stn.33 (8°S)	Rh. acuminata	69
	Pl. sol	23
Stn.30 (15°S)	Cl. frauenferdianum	89

Share percentages of the large diatoms and dinoflagellates in the water column above 300 m were then only 4.1% of the total algal biomass at maximum at 30°N and 0.5% at minimum at 40°N (Fig. 13), and they were then obviously fairly less predominant as a group in the algal community in the study area. Both large diatoms and dinoflagellates showed similar variations in the share percentages between stations. The large diatoms almost always showed a lower biomass, reaching only 0.1–1% of the total algal biomass, than the dinoflagellates except at 8°S.

Table 2 represents species which occupied more than 10% of the biomass in the total large diatoms. It is clear that different species became to be dominant at each respective station. Three species of *Th. longissima*, *H. haukii* and *Pl. sol* were the dominant components at 44°N, 15°N and 3°N-4°S, respectively, although their cell size was fairly small within the large diatoms. Actual biomass of these large diatoms could be more because of possible loss during net tows and filtration procedures for chlorophyll measurements. Huge diatom species, such as *Rh. debyana*, *Rh. clevei* var. *clevei*, *Rh. acuminata* and *Eth. rex*, often became the major biomass component at certain stations, but their occurrences were fairly random in the present net tow samplings. On the other hand, *Ceratium* was always occupying more than 50% of the total biomass of large dinoflagellates.

DISCUSSION

Rabinowitz (1981) proposed two aspects for defining the rarity of species occurring in natural environments; one is a limited size of habitat for a given species in a given environment, and the other is low individual density in a given habitat of a given species for some other reason. The former was duc to a possible limited size of habitat and/or possible narrow environmental requirements of the species. These aspects of rarity were applied for further consideration in the present study as follows.

The large diatoms collected vertically by 100 μm mesh net tows in the water column above 300 m existed at all the stations during the present study. No species showed wide distribution throughout stations, and there were some characteristic distribution patterns in several groups of species. Six species were only distributed in the sub-Arctic water in the north of 44°N, and 17 other species occurred only in the subtropical and tropical waters in the south of 36°N. There was a marked discontinuity zone justified as the sub-Arctic front around 42°N where there were distinctive differences at least in the profiles of temperature, salinity and nutrients throughout the water column. The above-mentioned two groups of large diatom species then inhabit either the sub-Arctic waters or in the subtropical and tropical waters. This suggested that each group of large diatoms could have different requirements for environmental conditions, such as temperature and salinity. Then two waters offering different environmental conditions could separate the large diatoms into two groups.

Even within the group observed in the south of 36°N, actual occurrences varied depending upon species from only occurring at 1 station to many stations. Five species of *G. tropica*, *Pl. sol*, *As. marylandica*, *Ps. calcar-avis*, and *Rh.*

clevei var. *communis* were commonly recognized at almost all stations, i. e. 9 to 10 stations ranging from 36°N to 15°S. Considering the wide range of more than 5000 km in the direct north-south distance for the area, they were not attributed to a limited size of habitat for their rare occurrences. The other 12 species in the same area occurred only at a limited number of stations, i. e. 1 to 7 stations out of 10. These species equally showed their occurrences in the subtropical waters both in the north and the south of the equatorial water, but not many in the equatorial water (3°N, 0° and 4°S). The latter area was characterized by high temperatures and low salinities compared with high salinities and high temperatures in the surrounding waters because of possible influences by upwelling and ocean currents. It is therefore concluded that the 5 species mentioned above extended their distributions even in equatorial waters with high environmental variabilities but the other 12 species were not always favored for such environments within the 17 species in the subtropical and tropical waters. Furthermore, *Eth. rex* and *Rh. debyana* centered their distributions only in the north of 9°N, and *Rh. bergonii*, and *Rh. clevei* var. *clevei* never appeared north of 9°N. These results suggested that the different distributions between groups, as mentioned, were probably due to differences in the environmental requirements of each respective group. Three species of *Rh. imbricata*, *Pr. alata* and *Th. longissima* occurred in both the sub-Arctic and subtropical waters between 48°N and 30° but not in the south of 30°N. These three species could have wide ranges for salinity and temperature compared to the other species mentioned above, but they were unable to extend their distribution to the south of 24°N because of a possible failure of the temperature and salinity environments there. In addition to the temperature and salinity environments, winter mixing of the surface zone could have possible effects on the distributions of these species.

Considering the size of the habitat range, the distribution of large diatoms varied from a wide range of several 1000 km in the north-south direction to a small range of a few 100 km depending upon the differences in species. It was expected that the observed differences in the areal distributions of large diatoms were due to possible differences in habitat requirements between each species, not only for single but also for multiple environmental parameters.

In terms of the rarity of the biomass of the large diatoms, it was common that all species were extremely less pronounced, i. e. 1% of the total algal community, as a group. Share percentages of single species for the large diatoms then became less pronounced, even though *Rh. debyana* and *Pl. sol* occupied more than half of the large diatom biomass at some stations. Since the cell sizes of large diatoms were all extremely large compared with the other algae, biomass density in terms of cell numbers became fairly small. Huge species such as *Rh. debyana* and *Eth. rex* showed a little variation in occurrence with very small cell densities. On the other hand, small size species in large diatoms such as *H. haukii* and *Pl. sol,* had greatly varying cell densities at different stations. These results showed that there were some differences in occurrence even within the large diatom species group.

The followings are further interpretations for evaluating the rarity of the large diatoms in the study area. The first possibility was defeaters of the large diatoms compared to other small algae. The rarity of the large diatoms was then due to possible depression by other algae. Under these circumstances, it is necessary for the large diatoms to have suitable environments with less depressive pressure from other algae (spatial resource isolation) or at some other time (temporal resource isolation). Otherwise, those large diatoms will be distinct because of the competitive segregation (Gause, 1935). Even though there were some heterogeneity in the environment in the study area, it is impossible to explain the rarity of large diatoms only through the spatial and temporal heterogeneity of the environments (cf. Venrick, 1990). Rarity could reduce grazing pressure on the large diatoms due to increased energy requirements of grazers for searching. Share percentages of the large diatoms were always less than 1%, and there was no place where the large diatoms were abundant in the study area. Differences in the environment can give some explanations which support differences in the species distributions, but are not capable of explaining the observed wide distributions of large diatoms with less abundant biomass.

There were two other possibilities to explain the existing rarity of the large diatoms. One was that there were suitable habitats for the large diatoms in areas which possibly supplied the large diatoms to the surrounding areas. The other was that a suitable environment for some of the large diatoms was created at a different time in the study area (Iriarte and Fryxell, 1995). Recently, some large diatom species such as *Rhyzosolenia* sp. (tentatively identified as *R. castracanei*) were found to formulate a bloom coloring the blue water into yellowish brown and having chlorophyll *a* concentrations ranging from 5 to 29 mg m^{-3} in the equatorial Pacific Ocean (Yoder *et al.*, 1994). The peculiar characteristics of this *Rhyzosolenia* sp. could contribute to making a bloom in the oligotrophic tropical water (Hayward, 1993; Villareal *et al.*, 1993) as well as the possible hydrographic features of the area (Yoder *et al.*, 1994). This is evidence for the latter case.

Since the large diatoms observed in the subtropical and tropical waters never occurred in the sub-Arctic waters and those observed in the latter water were not in the former waters, it was hard to believe that the large diatoms were transported in such large areas. This interpretation was also supported by past observations (Smayda, 1958; Guillard and Kilham, 1977; Sundstrom, 1986) in which those species observed in the subtropical and tropical waters were only observed in the same type of waters. Therefore, it was expected that there was no active transport of species exchanging between the sub-Arctic and subtropical and tropical waters. The possible supply of cells from coastal areas was also unlikely because most species observed in the present study did not occur or no species was found with large biomass in coastal waters. Past observations reported the cell density of *Ps. calcar-avis* and *Eth. rex* decreased toward the coastal waters (McHugh, 1954; Sundstrom, 1986).

In the subtropical and tropical waters in the north Pacific Ocean, it was found that environmental conditions did not vary greatly from field observations carried out in summer and in winter (McGowan and Walker, 1985; Rodriguez and Mullin, 1986). However, it was possible that vertical mixing occurring seasonally created suitable environments for the large diatoms in the sub-Arctic waters and waters where seasonal thermocline was created. It was possible that species occurring at 48°N and 30°N had temporal variations in their biomass. However, it was unlikely that the biomass sizes of species occurring in the south of 24°N were affected by seasonal thermocline.

The above-mentioned considerations suggested that it was hard to explain the rarity of large diatoms only by the less competitive capability of each species.

Total numbers of large diatom species were less in waters having active water mixing such as in the Kuroshio extension, equatorial current and equatorial upwelling, compared to the waters under less perturbation. This suggested that the occurrences of large diatoms were not due to a possible inbalanced condition created by perturbation.

The present study confirmed a wide distribution of large diatoms in the central Pacific Ocean as well as their extremely low abundance and biomass.

Acknowledgments—This study was carried out as a part of NOPACCS (Northwest Carbon Cycle Study) which was supported by the New Energy Industry Development Organization. Support was also given by Grant-in-Aid for Scientific Research (No. 06454009) from the Ministry of Education, Science and Culture, Japan. We thank Dr. K. Ishikawa (chief scientist for the cruise), the captain and crew members of the R.V. *Hakurei-Maru* for their help during the cruises. Special thanks are given to Drs. Y. Watanabe and H. Ishida for their great help for this study.

REFERENCES

Cupp, E. E. (1943): Marine plankton diatoms of the west coast of North America. *Bull. Scripps Inst. Oceanogr. Tech. Ser.*, **5**, 1–238.

Gause, G. F. (1935): *The Struggle for Existence.* Dover Publications, London, England.

Guillard, R. R. and P. Kilham (1977): The ecology of marine planktonic diatoms. In: *The Biology of Diatoms*, ed. by D. Werner, Univ. Calif. Press, Berkeley, pp. 498.

Hayward, T. L. (1993): The rise and fall of *Rhizosolenia*. *Nature*, **363**, 675–676.

Hutchinson, G. E. (1961): The paradox of the plankton. *Amer. Nat.*, **95**, 137–145.

Iriarte, J. L. and G. A. Fryxell (1995): Micro-phytoplankton at the equatorial Pacific (140°W) during the JGOFS EqPac time series studies: March to April and October 1992. *Deep-Sea Res.*, **42**, 559–583.

McGowan, J. A. and P. A. Walker (1985): Dominance and diversity maintenance in an oceanic ecosystem. *Ecol. Monogr.*, **55**, 103–118.

McHugh, J. L. (1954): Distribution and abundance of the diatom *Ethimodiscus rex* off the west coast of North America. *Deep-Sea Res.*, **1**, 216–222.

Miyai, H., K. Matsuzaki, K. Ogawa and T. Sugihara (1988): A simple method for the estimation of phytoplankton biomass based on cell morphology. *Bull. Plankton Soc. Japan*, **35**, 121–126.

Odum, E. P. (1963): *Ecology*. Holt, Reinhart and Winston, New York, U.S.A., 152 pp.

Rabinowitz, D. (1981): Seven forms of rarity. pp. 205–217. In: *The Biological Aspects of Rare Plant Conservation*, ed. by H. Synge, John Wiley, Chichester, England.

Rodriguez, J. and M. M. Mullin (1986): Diel and interannual variation of size distribution of oceanic zooplankton biomass. *Ecology*, **67**, 215–222.

Smayda, T. (1958): Biogeochemical studies of marine phytoplankton. *OIKOS*, **9**, 160–189.

Strathmann, R. R. (1967): Estimating the organic carbon content of phytoplankton from cell volume or plasma volume. *Limnol. Oceanogr.*, **12**, 411–418.

Strickland, J. D. H. and T. R. Parsons (1968): A practical handbook of seawater analysis. *Fish. Res. Bd. Can. Bull.*, **167**, 311, Ottawa, Canada.

Sundstrom, B. G. (1986): The marine diatom *Rhyzosolenia*. A new approach to the taxonomy. Ph.D. Thesis, Lund University, 196 pp.

Sverdrup, H. U., M. W. Johnson and R. H. Fleming (1942): *The Oceans, Their Physics, Chemistry and General Biology*. Prentice-Hall, New York, 1087 pp.

Takahashi, M. and N. Fukazawa (1982): A mechanism of "red tide" formation. II. Effect of selective nutrient stimulation on the growth of different phytoplankton species in natural water. *Mar. Biol.*, **70**, 267–273.

Tilman, D. (1977): Resource competition between planktonic algae: an experimental and theoretical approach. *Ecology*, **58**, 338–348.

Utermohl, H. (1958): Zur Verkommung der quantitativen Phytoplanktonmethodik. *Mitt. int. Ver. Limnol.*, **9**, 1–38.

Venrick, E. L. (1990): Phytoplankton in an oligotrophic ocean: species structure and interannual variability. *Ecology*, **71**, 1547–1563.

Villareal, T. A., M. A. Altabet and K. Culver-Rymsza (1993): Nitrogen transport by vertically migrating diatom mats in the North Pacific Ocean. *Nature*, **363**, 709–712.

Whittaker, R. H. (1975): *Communities and Ecosystem*. 2nd ed., The Macmillan Company, New York, U.S.A.

Yoder, J. A., S. G. Ackleson, R. T. Barber, S. Flament and W. M. Balch (1994): A line in the sea. *Nature*, **371**, 689–692.

Dynamics and Characterization of Marine Organic Matter, Eds., N. Handa, E. Tanoue and T. Hama, pp. 91–104.

The Relation between Primary Production and the Flux of Particulate Organic Matter in the Sub-Arctic Coastal Sea

Mitsuru YANADA and Yoshiaki MAITA

Department of Fishery and Marine Science, Faculty of Fisheries, Hokkaido University, Hakodate 041-8611, Japan

(Original, Received 3 June 1996; accepted 1 November 1996)

Abstract—The relation between primary production and the precipitating processes of particulate organic materials was manifested by measuring *in situ* some size-fractionated (No Screen, <690-μm and <350-μm) organic material fluxes and amino acids composition in the precipitating particles during a spring bloom and autumn in Funka Bay, the sub-Arctic coastal sea. The total flux (No Screen) of particulate organic material, in term of carbon, at a depth of 74 m during the spring bloom and the autumn accounted for, on average, 51% and 57% of primary production, respectively. However, the fluxes of particulate organic carbon in the <690-μm and <350-μm fractions were only, on average, 19% and 9% of primary production during the spring bloom, and 30% and 12% during the autumn, respectively. The composition of protein amino acids, the existence of glucosamine and detectable degrees of non-protein amino acids (γ-aminobutyric acid, β-alanine and ornithine) in the precipitating particles obtained from our sediment trap experiment demonstrated that the precipitating organic particles were dominated by some microzooplankton source materials even in the<350-μm fraction. At the coastal sea, the precipitating organic particles are as few as the phytoplankton itself produced in the euphotic zone, the bulk of the phytoplankton might be removed from a water column by the predation of zooplankton, and a part of the removed phytoplankton might change in larger-sized precipitating zooplankton source materials.

INTRODUCTION

In recent years, some sediment trap experiments have been performed in various coastal and open seas by some investigators. The results on the vertical flux and the composition of organic particles obtained from the experiments have offered important information to explain global organic material cycling processes (production, distribution, transportation and decomposition) of the ocean. In open oceans, it has been thought that the vertical flux of organic particles increases with increasing primary production *in situ* (Eppley and Peterson, 1979; Deuser and Ross, 1980, Deuser *et al.*, 1981, 1983; Pace *et al.*, 1987), and also decreases with increasing depth as a consequence of predation by zooplankton or decomposition depending on bacteria (Knauer *et al.*, 1979; Lorenzen *et al.*, 1983; Karl *et al.*, 1988). Suess (1980) and Betzer *et al.* (1984) proposed that the vertical organic carbon fluxes below a depth of 50 m in open oceans could be expressed as a function with primary production and depth *in situ*. However, it is still not evident whether the function can be adopted in coastal seas including complicated processes for biological activity.

Protein amino acids, a major component of organic nitrogen compounds in particulate organic matter, have been used as an important indicator in order to predict biochemical decomposition and/or sources of precipitating organic matter (Lee *et al.*, 1983; Wakeham *et al.*, 1984). Non-protein amino acids (e.g., glucosamine, β-alanine, γ-aminobutyric acid, ornithine) also have been used to predict biochemical decomposition of organic matter depending on bacteria (Ittekkot *et al.*, 1984a, 1984b; Müller *et al.*, 1986). The many investigations on protein or non-protein amino acids in precipitating particles, however, have been performed at below the productive layer of the open ocean, but a few in coastal seas (e.g., Lee and Cronin, 1982).

Then, the objective of this study is to speculate the relation between primary production and the flux of particulate organic matter in the sub-Arctic coastal sea by measuring *in situ* primary production, the size-fractionated fluxes of particulate organic matter, and the composition of protein and non-protein amino acids in precipitating particles during both a spring bloom and autumn in Funka Bay.

MATERIALS AND METHODS

Sediment trap experiments

A moored sediment trap array was deployed at a depth of 74 m in Station 30 (42°16.2′ N, 140°6.0′ E; water depth 92 m) of central Funka Bay. The sediment trap designed by Maita *et al.* (1986) was composed of eight cylinders (the length: 45 cm, the inside diameter: 12 cm). One part of the cylinders was covered *in situ* with a 690-μm mesh screen of polyethylene (<690-μm) to remove some large zooplankton (e.g., swimmer), the other part was covered with a 350-μm mesh screen (<350-μm) to catch smaller particles (for example, phytoplankton). Another cylinder was not covered with any screens to collect all settling particles (No Screen). The each cylinder was added a 1.5 M sodium chloride solution containing formalin (final concentration: 1%) as a preservative. The sediment trap was developed during spring 1985 and 1986, and during autumn 1984. The one-time development periods ranged from 10 days to 27 days (Table 1).

After recovery of the trap, the samples were immediately collected on a Millipore filter (HA type, 47 mmø) by filtration. The remains on the filter were dried in a desiccator under vacuum, then were measured for total dry weight. Aliquots from the trap samples were analyzed for determination of the contents of carbon and nitrogen using a CHN analyzer. The other parts of the trap samples were analyzed for determination of the contents of protein amino acids (17 types) and non-protein amino acids (glucosamine, β-alanine, γ-aminobutyric acid, ornithine) by a ninhydrin reaction method using a high performance liquid chromatograph. The homogenized dry sample (about 20 mg) was accurately weighted into a glass ampoule, and hydrolyzed with 6N HCl under nitrogen gas

Table 1. The deploying periods of the sediment trap during spring and autumn in Funka Bay (Station 30)

Period No.	Initial Date	Final Date	During (days)	Depth	Remark†
*A (b)	Aug. 21 -	Sep. 3, 1984	13	74m	Autumn
S (a)	Sep. 3 -	Sep. 21, 1984	17	74m	Autumn
S (b)	Sep. 21 -	Oct. 1, 1984	10	74m	Autumn
F (a)	Feb. 1 -	Feb. 18, 1985	17	74m	Pre-bloom
F (b)	Feb. 18 -	Feb. 28, 1985	10	74m	Pre-bloom
M (a)	Feb. 27 -	Mar.14, 1986	15	74m	Bloom
M (b)	Mar. 14 -	Mar.26, 1986	12	74m	Bloom
A (a)	Mar. 26 -	Apr. 17, 1986	22	74m	Bloom
A (b)	Apr. 17 -	Apr. 30, 1986	13	74m	Post-bloom
M (c)	May, 10 -	May,23, 1985	13	74m	Post-bloom

†See in the text.

at 110°C for 22 hr. The hydorolysate was evaporated to dryness at a temperature below 50°C using a rotary evaporator. The residue was redissolved in 0.02N NaOH to neutralize the sample and to remove ammonia in the hydrolysate and evaporated to dryness again at a temperature below 50°C. The residue was taken up in sodium citrate buffer solution of pH 2.2. An aliquot (100 μl) of the solution was injected into an Amino Acid Analyzer (Hitachi 835).

Measurements of primary production and chlorophyll a content in seawater

The primary production and chlorophyll *a* content in seawater were determined at the times of deployment and recovery of the sediment trap. Seawater samples for measurement of primary production were collected at depths accounting for respective light intensities of 100%, 30%, 15% and 2%. Analysis of primary production was performed with suspected ^{14}C-method under surface temperature and natural light according to Parsons *et al.* (1984).

Seawater samples for the measurement of chlorophyll *a* were collected at the same depths for measurement of primary production and at depths of 10 m interval until 70 m. The content of chlorophyll *a* was determined by fluorometric analysis with 90% acetone extraction according to Parsons *et al.* (1984).

RESULTS

Temporal variations in primary production and chlorophyll a content

During deployment of the sediment trap in spring, primary production estimated from ^{14}C-method started increasing in late February, reaching the largest value of about 2000 mgC $m^{-2}d^{-1}$ in late March, and afterward decreasing

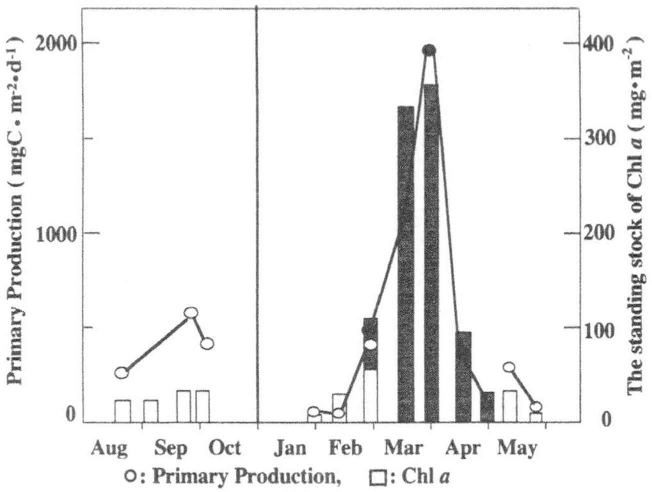

Fig. 1. Temporal variations in primary production using the ^{14}C-method and the standing stock(0–70 m) of chlorophyll *a* in Funka Bay (Sta. 30).

dramatically (Fig. 1). The average primary production was 950 mgC $m^{-2}d^{-1}$ during the spring in 1986, continuously deploying the sediment trap (Feb. 27–Apr. 30, 1986; 62 days). On the other hand, the average primary production during the autumn (Aug. 21–Oct. 1; 41 days) was 420 mgC $m^{-2}d^{-1}$, accounting for about one-half of that during the spring.

The temporal variation in the standing stock of chlorophyll *a* (Chl *a*) in the water column (0–74 m) integrated from the vertical distribution of Chl *a* concentration was almost consistent with that in the primary production ($r = 0.92$, $n = 14$). The average standing stock of Chl *a* was approximately 10-fold higher during the spring in 1986 than during the autumn.

Temporal variations in the flux of organic materials

The mass fluxes including both organic and inorganic particles in the respective size-fractions (No Screen, <690-μm and <350-μm), showed maximum value in February (Fig. 2). After that, the mass fluxes in the <690-μm and <350-μm fractions decreased and increased again in the middle of May. In contrast, the flux in the "No Screen" fraction tended to increase from March to April. The mass fluxes in the respective size-fractions during the autumn were lower than those in the middle of May.

The temporal variations in the fluxes of particulate organic carbon and nitrogen in <350-μm fraction were approximately consistent with those in the mass flux in the same fractions (carbon: $r = 0.96$, nitrogen: $r = 0.89$). That is, the fluxes of particulate organic carbon and nitrogen in the <350-μm fraction also revealed a maximum value in February, a secondary peak in the later half of March, a decrease after that, and an increase in the middle of May. The temporal

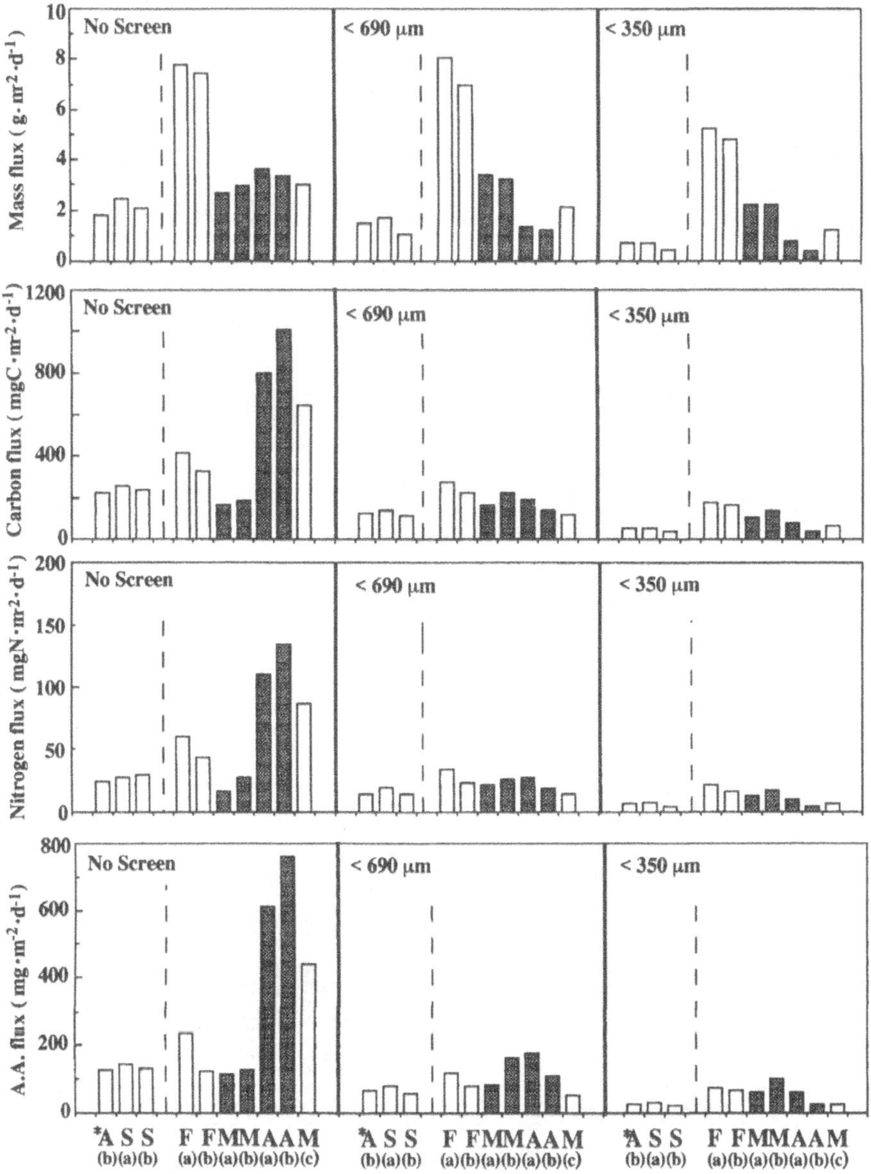

Fig. 2. Temporal variations in mass flux, organic carbon flux, organic nitrogen flux and total protein amino acid flux size-fractionated *in situ* by the screens at a depth of 74 m in Funka Bay (Sta. 30).

Table 2. The relative proportions of individual protein amino acids to total protein amino acids and the contents of non-protein amino acids in the three size-fractions of sediment trap samples at a depth of 74 m in Funka Bay (Station 30)

Period	Asp	Thr	Ser	Glu	Gly	Ala	Cys	Val	Met	Ile	Leu	Tyr	Phe	Lys	His	Arg	Pro	Total AA	GlcN†	γ-ABA	Orn	β-Ala
	(weight %)																	($mg\ g^{-1}$)				
No Screen																						
*A(b)	11.9	6.0	6.0	13.0	7.0	7.1	1.7	5.4	1.3	4.1	7.5	4.2	4.9	6.3	2.2	6.0	5.4	70	6.1	0.48	-	-
S(a)	11.8	6.0	6.0	14.9	6.8	7.1	1.7	5.3	1.3	4.1	7.9	2.9	4.6	5.5	2.3	6.4	5.5	58	5.7	0.21	-	-
S(b)	11.9	5.9	5.9	14.5	6.9	7.2	1.5	6.0	1.2	4.7	8.3	0.9	4.9	5.7	2.3	6.8	5.4	63	4.8	0.16	-	-
F(a)	13.7	6.3	7.3	14.2	8.7	7.7	0.3	5.7	<0.1	4.4	7.6	<0.1	4.9	5.5	2.3	6.8	4.4	30	2.1	0.05	0.23	<0.01
F(b)	13.7	6.5	7.1	13.9	9.3	7.3	0.6	5.8	<0.1	4.4	7.4	0.6	4.9	5.3	2.0	6.7	4.4	17	1.6	0.05	0.04	0.10
M(a)	12.1	5.7	5.9	13.4	7.0	7.0	1.4	5.6	1.8	4.9	8.5	2.8	5.1	6.1	2.0	6.3	4.4	43	5.1	0.06	0.03	0.04
M(b)	11.8	5.8	5.9	13.0	6.7	7.3	1.2	5.7	1.9	5.2	8.9	4.0	5.6	5.4	1.8	5.4	4.2	43	5.0	0.08	0.05	0.04
A(a)	11.7	5.5	5.3	14.0	6.4	6.9	1.1	5.6	2.8	5.0	9.2	2.4	5.2	5.3	2.2	6.8	4.5	168	13.4	0.05	<0.01	<0.01
A(b)	11.2	5.4	5.0	14.1	6.2	6.7	1.3	5.8	2.8	5.0	9.0	3.1	4.9	5.8	2.2	6.9	4.5	225	17.7	0.05	<0.01	<0.01
M(c)	11.4	5.4	5.3	14.8	6.2	7.2	1.1	5.9	2.5	4.7	8.4	2.2	4.8	7.0	2.5	6.2	4.4	145	16.5	0.18	0.06	0.12
<690 μm																						
*A(b)	12.8	6.4	6.3	12.6	7.2	7.4	0.8	5.6	0.9	4.2	7.7	3.3	4.8	5.8	2.1	6.0	5.3	43	3.3	0.31	-	-
S(a)	13.0	6.4	6.3	13.9	7.1	7.5	0.6	5.6	1.0	4.2	8.3	1.1	4.7	5.7	2.4	6.7	5.5	47	3.4	0.11	-	-
S(b)	13.0	6.2	6.2	14.3	7.1	7.6	0.7	6.0	0.5	4.5	8.3	0.4	5.0	6.2	2.4	6.2	5.4	52	3.9	0.11	-	-
F(a)	14.2	6.9	7.7	13.3	9.3	7.9	0.5	5.8	<0.1	4.3	7.1	0.8	4.9	5.0	1.8	6.1	4.1	15	1.1	0.06	0.04	<0.01
F(b)	14.0	6.6	7.5	13.2	9.2	7.5	1.0	5.7	0.8	4.2	6.9	0.9	4.8	4.9	1.9	6.1	4.9	12	0.8	0.04	0.04	0.08
M(a)	13.1	5.9	6.5	12.7	8.0	7.3	1.4	5.4	1.4	4.7	8.3	2.9	5.2	4.8	1.7	5.9	4.8	24	3.4	0.04	0.03	0.03
M(b)	11.6	5.7	5.9	12.5	6.8	7.0	1.4	5.6	1.6	5.2	8.9	3.8	5.6	6.1	1.6	6.4	4.3	50	6.0	0.08	0.04	0.04
A(a)	11.7	5.6	5.6	13.2	6.4	7.1	1.3	5.6	2.4	5.2	9.2	2.7	5.4	5.6	1.8	6.9	4.3	127	10.3	0.06	0.05	<0.01
A(b)	11.6	5.8	5.7	13.7	6.6	6.9	1.2	5.5	1.7	4.9	8.7	3.4	5.0	6.0	1.8	6.7	4.6	89	5.2	0.08	0.05	<0.01
M(c)	13.3	6.7	7.4	13.9	8.5	8.3	0.3	5.9	<0.1	4.4	7.3	0.4	4.8	5.5	2.0	5.9	5.4	25	2.4	0.15	0.14	<0.01
<350 μm																						
*A(b)	12.9	6.4	6.4	12.8	7.4	7.3	1.0	5.5	1.2	4.3	7.8	2.9	4.9	5.8	2.2	6.1	5.2	37	2.3	0.18	-	-
S(a)	12.6	6.2	6.3	13.5	7.2	7.2	1.4	5.9	1.4	4.5	8.1	1.1	4.8	5.7	2.3	6.6	5.2	41	2.5	0.13	-	-
S(b)	12.6	6.2	6.4	13.9	7.2	7.0	1.2	6.0	1.0	4.8	8.3	0.3	5.0	6.2	2.3	6.7	4.9	51	2.6	0.10	-	-
F(a)	14.5	6.9	7.8	13.5	9.5	8.1	0.5	5.9	<0.1	4.4	7.2	<0.1	4.9	5.0	1.9	6.0	4.0	14	1.1	0.05	0.07	<0.01
F(b)	14.8	6.9	8.0	13.2	9.6	7.6	0.3	5.8	<0.1	4.4	7.8	0.5	4.1	4.9	1.9	6.2	4.2	14	0.9	0.06	0.05	0.01
M(a)	12.9	5.9	6.3	12.3	7.9	7.2	1.5	5.5	1.3	4.8	8.2	3.7	5.2	5.2	1.8	6.0	4.5	27	3.4	0.06	0.03	0.04
M(b)	12.5	5.9	6.0	13.3	6.9	7.6	1.0	5.6	1.6	5.0	9.3	2.5	5.6	5.0	1.5	6.3	4.4	44	3.0	0.04	0.03	0.04
A(a)	12.4	5.6	5.8	13.6	6.6	7.6	1.0	5.1	2.3	4.6	9.1	3.1	5.5	5.4	1.4	6.5	4.5	83	4.3	0.06	0.04	<0.01
A(b)	11.8	5.8	5.9	13.4	6.6	6.9	1.4	5.6	1.5	5.1	8.8	3.6	5.1	6.1	1.6	6.5	4.4	70	4.7	0.09	0.05	<0.01
M(c)	13.3	6.7	7.4	13.7	8.6	8.2	0.4	5.7	<0.1	4.3	7.3	1.0	4.9	5.6	1.8	5.7	5.4	23	1.9	0.14	0.05	<0.01

†The count of GlcN was estimated by multiplying the measured value by the correction factor of 1.4 (Müller *et al.*, 1986).

variations in the fluxes of particulate organic carbon and nitrogen in the <690-μm fraction agreed roughly with those in the <350-μm fraction (carbon: $r = 0.85$, nitrogen: $r = 0.67$), although the flux of particulate organic nitrogen showed the maximum value in the first half of April. In "No Screen", the temporal variations in the fluxes of particulate organic carbon and nitrogen, however, remarkably differed from that in the mass flux. The fluxes of particulate organic carbon and nitrogen revealed the maximum value in April rather than in February when the mass flux revealed its maximum value. In autumn, the fluxes of particulate organic carbon and nitrogen in the respective size-fractions were lower than those in the middle of May. On the other hand, the temporal variations in the flux of total protein amino acids nearly agreed with those of particulate organic carbon or nitrogen in the respective size-fractions.

The composition of protein amino acids

In the 17 protein amino acids identified, the relative proportions of Asp and Glu to the total protein amino acids were higher with more than 10% in the respective size-fractions, whereas the proportions of Cys, Met, Tyr and His were lower with less than 5% (Table 2). The compositions of amino acids in both February and the middle of May were slightly different from those during March–April and autumn in the respective size-fractions. In particular, the quantitative relation between Asp and Glu showing relatively high values represented the relation of Asp ≥ Glu in both February and the middle of May, and the relation of Glu ≥ Asp during the other periods conversely. The proportions of Cys, Met and Tyr showing relatively low values, were lower in both February and the middle of May, and revealed an increasing tendency from March to April. Further, the proportion of Gly represented relatively higher values in February rather than in the other periods.

The contents of non-protein amino acids

The content of glucosamine (GlcN) was low in February, then increased, and reached the maximum value in April in the respective size-fractions (Table 2). The content of GlcN in "No Screen" fraction showed a higher tendency in comparison with those in <650-μm and <350-μm fractions during all the periods. The content of γ-ABA was low from February to April (below 0.1 mg g^{-1}), but high during autumn. On the other hand, the contents of β-Ala and Orn were as low as one detectable degree (below 0.1 mg g^{-1}).

DISCUSSION

The relation between the primary production and the size-fractionated organic matter fluxes

The spring bloom in Funka Bay, judging from primary production measurements by the ^{14}C-method and the standing stock of Chl *a* in the water column, started from early March, reached maximum in late March, and finished in the middle of April. In the pre-bloom (February) when primary production is extremely low, the fluxes of particulate organic materials (e.g., carbon, nitrogen and amino acids) in the respective size-fractions are high in comparison with those in the beginning of spring bloom (earlier March). The fluxes of particulate organic carbon in the pre-bloom exceed primary production at the same period, even in <350-μm fraction. The mass flux in the pre-bloom was higher by far than that in the other periods. Moreover, the mass flux in <350-μm fraction accounted for about 70% of the flux in <690-μm fraction in the pre-bloom. These suggest that precipitating particles collected in the sediment trap during the pre-bloom comprise smaller particles with a low organic material content. During this period the active vertical mixing of water column in Funka Bay happened. Therefore, the maximum mass flux and relatively high organic matter flux found in the pre-bloom may be overestimated by vertically blowing up some particles existing in the water column or sediment. Namely, the fluxes in the pre-bloom cannot necessarily express the sinking flux of particles from the surface.

The temporal variations of particulate organic carbon flux in the <690-μm and <350-μm fractions are significantly positively correlated to that of primary production ($r = 0.93$ and $r = 0.85$, $n = 8$, respectively), except February when the vertical water mixing is active. This means that the temporal variations of organic matter fluxes in both the <690-μm and <350-μm fractions coincide with that of primary production during the spring bloom. In contrast, there is no correlation between the temporal variation of particulate organic carbon flux in the "No Screen" fraction and the temporal variation of primary production. Especially, the flux of particulate organic carbon in the "No Screen" fraction at the post-bloom (later April and the middle of May) was higher than that obtained at the peak of the spring bloom, and far exceeded the primary production *in situ*. This implies that the precipitating large-sized particles with higher organic contents are caught in the sediment trap at the post-bloom and the particulate organic particles produced during the spring bloom precipitate with some time lags at the post-bloom.

Here, we calculated average fluxes of organic materials during autumn in 1994 and the spring bloom in 1996 when the trap was deployed successively (Table 3). The average fluxes of organic materials in <690-μm and <350-μm fractions during the spring bloom were approximately 30% and 15% of that in the "No Screen" fraction, respectively. This suggests that precipitating organic particles during the spring bloom do not contribute to smaller particles, but contribute to large-sized organic particles of more than 690-μm size (approximately 70%). In contrast, the fluxes of the organic materials in <690-μm and <350-μm fractions during the autumn account for approximately 50–60% and 20–25% of that in the "No Screen" fraction, respectively. This suggests that the contribution of the precipitating smaller organic particles is slightly higher during the autumn than during the spring bloom, although the fluxes of organic materials during the autumn reveal low values in the respective size-fractions in comparison to the spring bloom.

Table 3. The average fluxes of organic carbon, organic nitrogen and protein amino acids for the three size-fractions during the spring bloom and autumn

Period	During (days)	Depth (m)	Screen	Carbon Flux ($mgC \cdot m^{-2} \cdot d^{-1}$)	Nitrogen Flux ($mgN \cdot m^{-2} \cdot d^{-1}$)	A.A. Flux ($mg \cdot m^{-2} \cdot d^{-1}$)
Spring	62	74	No Screen	572	76.7	428
860227			<690 μm	180	24.2	136
- 860430			<350 μm	87	10.9	62
Autumn	41	74	No Screen	242	27.1	135
840821			<690 μm	127	16.5	69
- 841001			<350 μm	50	6.5	27

Table 4. The average proportions of organic carbon flux and organic nitrogen flux to primary production for the three size-fractions during the spring bloom and autumn

Period	Deployed Time (days)	Depth (m)	Screen condition	Primary Production ($mgC \cdot m^{-2} \cdot d^{-1}$)	C-Flux / P.P. (%)	N-Flux / P.P.* (%)
Spring	62	74	No Screen	948	60	46
860227			<690 μm		19	14
- 860430			<350 μm		9	7
Autumn	41	74	No Screen	421	57	37
840821			<690 μm		30	22
- 841001			<350 μm		12	9

*The primary production in terms of nitrogen was calculated from primary production using an atomic ratio of C/N = 6.6.

At a depth of 74 m in Funka Bay, the proportion of organic carbon flux in the "No Screen" fraction for primary production is equivalent to about 60% during both the spring bloom and the autumn (Table 4), despite the obvious difference of primary production which is recognized between the spring bloom and the autumn. Knauer *et al.* (1979) reported that the proportion of carbon flux for primary production was 53% at a depth of 50 m in coastal high productivity and 50% at a depth of 75 m in the open sea. Bishop and Marra (1984) also reported that mean euphotic zone production and particulate carbon flux estimated for the Panama Basin were 261 and 122 mgC $m^{-2}d^{-1}$, respectively. This particulate carbon flux is estimated to be 47% of primary production. Further, using the equation suggested by Suess (1980), the particulate carbon flux at a depth of 74 m in the open ocean was calculated to be 51% of primary production. That is, the proportions of particulate carbon flux to primary production in the "No Screen" fraction in Funka Bay during both the spring bloom and autumn appear to be nearly consistent with the proportion calculated from Suess's equation representing the relation of the flux, primary production and depth. However, the fluxes of particulate organic carbon in the <690-μm and <350-μm fractions at a depth of 74 m during the spring bloom accounted for only about 20% and 10% of primary production, respectively, and at a depth of 74 m during the autumn, only about 30% and 12% of primary production, respectively. This means that precipitating organic particles are as few as the phytoplankton itself (probably <350-μm fraction) produced in the euphotic zone, the most part of the produced phytoplankton changed in the larger organic particles of more than 350-μm size and/or is removed from the water column.

The source of precipitating organic particles

Among the 17 kinds of protein amino acids identified in the precipitating particles in Funka Bay, the relative contents of Asp, Glu, Cys, Met, Tyr and Gly represent the temporal or size-fractionation characteristic variations. It is possible that a part of their amino acids can be used to distinguish either phytoplankton origin materials or zooplankton origin materials for the sources of precipitating organic particles. Therefore, the amino acids of Asp, Ala, Gly and Ser are contained with relatively high values in the phytoplankton as compared with zooplankton, whereas the amino acids of Glu, Lys and Tyr are contained with relatively high values in the zooplankton, conversely (Maita, 1973).

During the spring bloom, Glu typical of zooplankton is only higher in the "No Screen" fraction than in <690-μm and <350-μm fractions, whereas Asp, Ser and Gly typical of phytoplankton are lower. The relative contents of individual amino acids in the <690-μm fraction are similar to those in the <350-μm fraction. On the other hand, relative contents of individual amino acids during the autumn coincide with that during the spring bloom in respective size-fractions. This correspondence between the autumn and the spring bloom for the amino acids composition demonstrate the same supply source for precipitating organic particles during both periods. In comparison with the amino acids composition of phytoplankton or zooplankton reported by Maita (1973), the relative content of individual amino acids in the precipitating organic particles was consistent with the contents in the zooplankton rather than in the phytoplankton.

It is also thought that glucosamine (GlcN) is an amino sugar included in the kitchenaceous matter of a crust zooplankton as an indicator of zooplankton source materials (Ittekkot *et al.*, 1984a, 1984b; Müller *et al.*, 1986; Haake *et al.*, 1993). Therefore, the GlcN content and the relative ratio of GlcN to amino acids (GlcN/AA) can be also used as a good indicator for estimating the relative contribution between phytoplankton source material and zooplankton source material. The GlcN content and the ratio of GlcN/AA in the "No Screen" fraction

Table 5. The average value of AA content, GlcN content and GlcN/Total AA for the three size-fractions during spring bloom and autumn

Period	Deployed Time (days)	Depth (m)	Screen condition	Total A.A. content ($mg \cdot g^{-1}$)	GlcN content ($mg \cdot g^{-1}$)	GlcN / Total AA (%)
Spring	62	74	No Screen	125	7.4	5.9
860227			<690 µm	79	4.4	5.7
- 860430			<350 µm	59	2.8	4.7
Autumn	41	74	No Screen	63	4.0	6.3
840821			<690 µm	47	2.5	5.3
- 841001			<350 µm	42	1.8	4.3

reveals higher value in comparison with <690-μm and <350-μm fractions in both periods (Table 5). This means the intrusion of some crust zooplankton materials into the "No Screen" fraction. The high contribution of the large-sized organic particle and the time lag of the organic matter flux in the "No Screen" fraction recognized during the spring bloom may depend on some crust zooplankton materials. On the other hand, the ratios of GlcN/AA in the <690-μm and <350-μm fractions in both periods are close to that of the "No Screen" fraction and reveal a higher value in comparison with the ratio for phytoplankton (hexosamine/amino acid: <1.25%; Haake *et al.*, 1993). This result suggests that the precipitating organic particles even in the <350-μm fraction during the spring bloom are directly influenced by some crust zooplankton materials. The majority of precipitating organic particles obtained from our sediment trap experiment may comprise some microzooplankton source materials including crust zooplankton materials and detritus or fecal materials passing once in the zooplankton body (Bishop *et al.*, 1977; Dilling and Alldredge, 1993; Marty *et al.*, 1994; Buck and Newton, 1995). In other words, the precipitating organic particles are as few as phytoplankton itself produced in the euphotic zone, even during the spring bloom.

The decomposition of precipitating organic particles

The proportions of protein amino acid carbon to total organic carbon (AA-C/TOC) and protein amino acid nitrogen to total organic nitrogen (AA-N/TON) can be used as an indicator of biochemical decomposition in particulate organic matter. Their proportions are almost equivalent to all size-fractions in Funka Bay, but slightly higher during the spring bloom than during autumn (Table 6). Their values in Funka Bay appear to be higher than those in the open sea (Lee and Cronin, 1982; Lee *et al.*, 1983; Ittekkot *et al.*, 1984a, 1984b).

Table 6. The average values of AA content, AA-C/TOC and AA-N/TON for the three size-fractions during the spring bloom and autumn

Period	Deployed Time (days)	Depth (m)	Screen condition	AA-Con. ($mg \cdot g^{-1}$)	AA-C/TOC (%)	AA-N/TON (%)
Spring	62	74	No Screen	125	33	78
860227			<690 μm	79	34	76
- 860430			<350 μm	59	32	81
Autumn	41	74	No Screen	63	25	64
840821			<690 μm	47	24	53
- 841001			<350 μm	42	24	59

Table 7. The average values of the ratios of γ-ABA/Glu, β-Ala/Asp and Orn/Arg for the three size-fractions during the spring bloom and autumn

Period Time	Deployed (days)	Depth (m)	Screen condition	γ-ABA/Glu (%)	β-Ala/Asp (%)	Orn/Arg (%)
Spring	62	74	No Screen	0.7	0.4	0.8
860227			<690 μm	0.9	0.4	1.2
- 860430			<350 μm	1.0	0.5	1.2
Autumn	41	74	No Screen	3.2	-	-
840821			<690 μm	3.0	-	-
- 841001			<350 μm	2.5	-	-

It has been suggested that the non-protein amino acids of γ-ABA, β-Ala and Orn are decompositional products of Glu, Asp and Arg, respectively. Therefore, their non-protein amino acids contents and the ratios of γ-ABA/Glu, β-Ala/Asp and Orn/Arg are also effective indices for predicting the process of biochemical decomposition in precipitating organic particles (Lee and Cronin, 1982, 1984; Ittekkot *et al.*, 1984a, 1984b). The content of γ-ABA and the ratio of γ-ABA/Glu were low during the spring bloom, although their values were high during the pre-bloom and autumn (Table 7). On the other hand, the contents of β-Ala and Orn are detectable in precipitating organic particles with <0.1% of total protein amino acids in respective size-fractions during the spring bloom. Their values are lower than those reported by Ittekkot *et al.* (1984a, 1984b) and Lee and Cronin (1982), suggesting no evidence of remarkable biochemical decomposition. Therefore, the precipitating organic particles in Funka Bay may not receive so much biochemical decomposition, although the precipitating organic particles during the autumn give an impression of degradation products in comparison with the spring bloom.

CONCLUSION

The data on protein and non-protein amino acids in precipitating particles showed that the precipitating organic particles were chiefly composed of zooplankton source materials (crust zooplankton materials or fecal materials) that did not receive very much the biochemical decomposition. In other words, at coastal sea regions like Funka Bay, the precipitating organic particles are as few as the phytoplankton itself produced in the euphotic zone, the bulk of the phytoplankton was removed from the water column by the predation of zooplankton, and a part changed in larger-sized precipitating organic particles. It was also concluded that the fluxes of particulate organic materials below 50 m depth can be at least expressed as a function of primary production and the *in situ* depth and adopted the equation proposed by Suess (1980) even sub-Arctic coastal sea.

Acknowledgements—We would like to thank Drs. A. Handa and A. Tanoue for their encouragement and useful discussions.

REFERENCES

Betzer, P. R., W. J. Showers, E. A. Laws, C. D. Winn, G. R. DiTullio and P. M. Kroopnick (1984): Primary productivity and particle fluxes on a transect of the equator at 153°W in the Pacific Ocean. *Deep-Sea Res.*, **31**, 1–11.

Bishop, J. K. B. and J. Marra (1984): Variations in primary production and particulate carbon flux through the base of the euphotic zone at the site of the Sediment Trap Intercomparison Experiment (Panama Basin). *J. Mar. Res.*, **42**, 189–206.

Bishop, J. K. B., J. M. Edmond, D. R. Ketten, M. P. Bacon and W. B. Silker (1977): The chemistry, biology, and vertical flux of particulate matter from the upper 400 m of the equatorial Atlantic Ocean. *Deep-Sea Res.*, **24**, 511–548.

Buck, K. R. and J. Newton (1995): Fecal pellet flux in Dabob Bay during a diatom bloom: Contribution of microzooplankton. *Limnol. Oceanogr.*, **40**, 306–315.

Deuser, W. G. and E. H. Ross (1980): Seasonal change in the flux of organic carbon to the deep Sargasso Sea. *Nature*, **283**, 364–365.

Deuser, W. G., E. H. Ross and R. F. Anderson (1981): Seasonality in the supply of sediment to the deep Sargasso Sea and implications for the rapid transfer of matter to the deep ocean. *Deep-Sea Res.*, **28**, 495–505.

Deuser, W. G., P. G. Brewer, T. D. Jickells and R. D. Commeau (1983): Biological control of the removal of abiogenic particles from the surface ocean. *Science*, **219**, 388–391.

Dilling, L. and A. L. Alldredge (1993): Can chaetognath fecal pellets contribute significantly to carbon flux?. *Mar. Ecol. Prog. Ser.*, **92**, 51–58.

Eppley, R. W. and B. J. Peterson (1979): Particulate organic matter flux and planktonic new production in the deep ocean. *Nature*, **282**, 677–680.

Haake, B., V. Ittekkot, S. Honjo and S. Manganini (1993): Amino acid, hexosamine and carbohydrate fluxes to the deep Subarctic Pacific (Station P). *Deep-Sea Res.*, **40**, 547–560.

Ittekkot, V., W. G. Deuser and E. T. Degens (1984a): Seasonality in the fluxes of sugars, amino acids, and amino sugars to the deep ocean: Sargasso Sea. *Deep-Sea Res.*, **31**, 1057–1069.

Ittekkot, V., E. T. Degens and S. Honjo (1984b): Seasonality in the fluxes of sugars, amino acids, and amino sugars to the deep ocean: Panama Basin. *Deep-Sea Res.*, **31**, 1071–1083.

Karl, D. M., G. A. Knauer and J. H. Martin (1988): Downward flux of particulate organic matter in the ocean: a particle decomposition paradox. *Nature*, **332**, 438–441.

Knauer, G. A., J. H. Martin and K. W. Bruland (1979): Fluxes of particulate carbon, nitrogen, and phosphorus in the upper water column of the northeast Pacific. *Deep-Sea Res.*, **26**, 97–108.

Lee, C. and C. Cronin (1982): The vertical flux of particulate organic nitrogen in the sea: decomposition of amino acids in the Peru upwelling area and the equatorial Atlantic. *J. Mar. Res.*, **40**, 227–248.

Lee, C. and C. Cronin (1984): Particulate amino acids in the sea: effects of primary productivity and biological decomposition. *J. Mar. Res.*, **42**, 1075–1097.

Lee, C., S. G. Wakeham and J. W. Farrington (1983): Variations in the composition of particulate organic matter in a time-series sediment trap. *Mar. Chem.*, **13**, 181–194.

Lorenzen, C. J., N. A. Welschmeyer and A. E. Copping (1983): Particulate organic carbon flux in the subarctic Pacific. *Deep-Sea Res.*, **30**, 639–643.

Maita, Y. (1973): Proteinous substances and amino acids in sea water. *Kaiyo Monthly*, **5**, 55–61.

Maita, Y., M. Yanada and A. Shiomoto (1986): Fluxes of particulate organic carbon and nitrogen in Funka Bay in autumn. *Bull. Fac. Fish. Hokkaido Univ.*, **37**, 124–133.

Marty, J. C., E. Nicolas, J. C. Miquel and S. W. Fowler (1994): Particulate fluxes of organic compounds and their relationship to zooplankton fecal pellets in the northwestern Mediterranean Sea. *Mar. Chem.*, **46**, 387–405.

Müller, P. J., E. Suess and C. A. Ungerer (1986): Amino acids and amino sugars of surface particulate and sediment trap material from waters of the Scotia Sea. *Deep-Sea Res.*, **33**, 819–838.

Pace, M. L., G. A. Knauer, D. M. Karl and J. H. Martin (1987): Primary production, new production and vertical flux in the eastern Pacific Ocean. *Nature*, **325**, 803–804.

Parsons, T. R., Y. Maita and C. M. Lalli (1984): *A Manual of Chemical and Biological Methods for Seawater Analysis*. Pergamon Press, New York, 173 pp.

Suess, E. (1980): Particulate organic carbon flux in the oceans—surface productivity and oxygen utilization. *Nature*, **288**, 260–263.

Wakeham, S. G., C. Lee, J. W. Farrington and R. B. Gagosian (1984): Biogeochemistry of particulate organic matter in the oceans: results from sediment trap experiments. *Deep-Sea Res.*, **31**, 509–528.

Dynamics and Characterization of Marine Organic Matter, Eds., N. Handa, E. Tanoue and T. Hama, pp. 105–115.

Fatty Acid Composition of Sinking Particles in the Western North Pacific

Kazuhide HAYAKAWA[1] and Nobuhiko HANDA[2]

[1]*Lake Biwa Research Institute, 1-10 Uchidehama, Otsu, Shiga 520-0806, Japan*
[2]*Aichi Prefectural University, Nagakute-cho, Aichi 480-1198, Japan*

(Original, Received 12 December 1996; accepted 23 January 1997)

Abstract—Fatty acids were measured in sinking particles collected using a time-series sediment trap from the continental slope off Hokkaido, Japan from September 1988 to May 1989. The fatty acids were mainly composed of saturated ones. However, remarkable components of fatty acids were obtained in the sinking particles in September 1988 and April-May 1989. The sinking particles in September 1988 showed a high abundance of 9-octadecenoic acid (18:1 *n*-9), suggesting a contribution of zooplankton derived materials. On the other hand, the sinking particles in April-May 1989 had a large content of 7-hexadecenoic acid (16:1) and eicosapentaenoic acid (20:5), suggesting a sedimentation of diatom cells during a spring bloom. The low contents of long-chain fatty acids at the same periods indicated a relatively small contribution of terrestrial inputs to the mass fluxes, in comparison to other periods. High values of unsaturated fatty acids to saturated fatty acids ratios (UFA/SAFA ratios) were found in the sinking particles in deep water. During a spring bloom, the sinking particles of large mass flux can transport unsaturated fatty acids, which are one of the most labile compound classes, from surface to deep water. It suggested that sedimentation of intact cells of phytoplankton occurred in the coastal ocean during a spring bloom.

INTRODUCTION

Vertical fluxes of sinking particles to the deep ocean reflected the primary productivity in the surface ocean (Suess, 1980; Betzer *et al.*, 1984; Pace *et al.*, 1987). In a high latitude of the world ocean, where sunlight and stable stratification govern the primary productivity in the euphotic zone, the vertical mass flux of sinking particles have the seasonal fluctuation. Early studies on sediment trap showed episodic fluxes of sinking particles during spring or summer in annual observation (Deuser and Ross, 1980; Honjo, 1982; Honjo *et al.*, 1988; Wefer *et al.*, 1988). These studies reported that the transport form (e.g. fecal pellets, aggregates) of sinking particles to the deep ocean was variable in each seasons. Vertical transport of particulate organic matter to deep water also varies with the flux and chemical composition (Lee *et al.*, 1983; Ittekkot *et al.*, 1984; Handa and Nakatsuka, 1993). Large seasonal variability in sinking carbon flux could be related to biological processes in the surface waters (Wefer, 1989).

Coastal and continental margin areas are of importance since high nutrient supply and large export production comparing open ocean occur (Wassmann,

1990; Bienfang and Ziemann, 1992). However, it is difficult to estimate the sources of sinking organic matter in the continental margin areas since the sinking organic matter in this area originated from various sources such as terrestrial inputs, export production from euphotic zone and resuspension from sediment.

Fatty acids are useful biomarker, providing source information of organic materials (e.g. phytoplankton, zooplankton, bacteria, terrestrial inputs) as well as transformations of particulate matter (Tanoue and Handa, 1980; Wakeham *et al.*, 1984). In the present study, vertical fluxes of fatty acids were measured with time-series sediment trap samples collected from the continental slope of the western North Pacific for one year. Here, we deduced the sources of sinking organic matters in this oceanic area based on the fatty acid composition in the sinking particles.

MATERIALS AND METHODS

A time-series sediment trap was moored from September 1988 to May 1989 at a depth of 5200 m at station NNWP-01 in the continental slope off Hokkaido, Japan (144°13′ E and 40°56′ N, water depth of 5369 m), as shown in Fig. 1. The sediment trap was cone-shaped with 13 cup-collectors and a cone-mouth of 0.5 m^2 (PARFLUX Mark-VI, McLane Res. Lab. Inc.). The samples were collected at a time interval of 20 days. Formaldehyde was added to the cup water as a preservative. After the recovery of the samples, swimmers in the sampling cups were removed by forceps, and the cup water removed by centrifugation. All the samples were kept frozen at –20°C until analysis.

Organic carbon and total nitrogen of the sinking particles were analyzed using a CHN analyzer (Handa and Nakatsuka, 1993; Nakatsuka *et al.*, 1995).

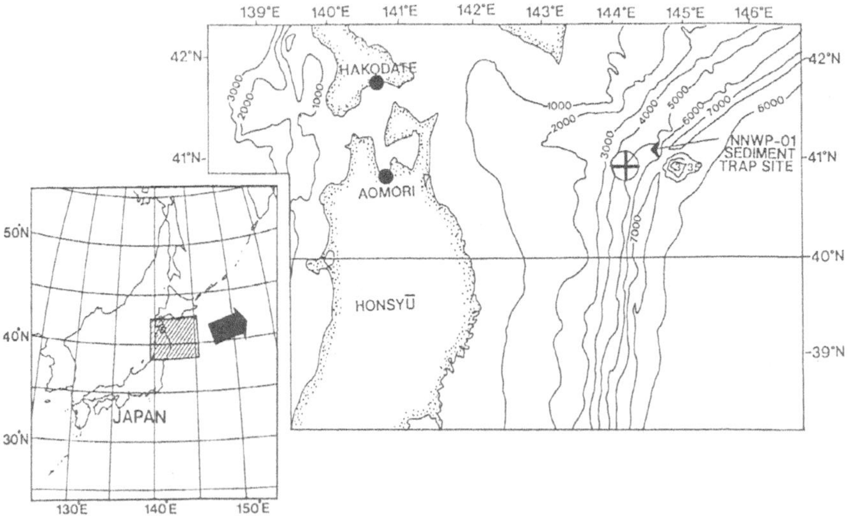

Fig. 1. Location of the time-series sediment trap in the western North Pacific.

Fatty acids in the sinking particles were analyzed by gas chromatography, as described by Hayakawa *et al.* (1996). Lipid materials were extracted from the sinking particles with chloroform/methanol (2/1, v/v) (Folch *et al.*, 1957). Heneicosanoic acid (21:0) as an internal standard was added to the extracts. The extracts were saponified with 0.5 N KOH/methanol. Subsequently, the fatty acids were esterified to their methyl esters using 14 % BF_3/methanol. Fatty acid methyl esters were cleaned up by thin-layer chromatography (TLC) using TLC plates coated with Silica gel G (Analtech Inc.) and mixture solvent of benzene/hexane (1/1, v/v). The fatty acid methyl esters were quantified by a Shimadzu GC-9A gas chromatograph equipped with an FID and a fused silica capillary column (FFS ULBON HR-Thermo-3000B, 25 m length, 0.25 mm i.d., Shinwa Industry Co. Ltd.). N_2 was used as a carrier gas at 0.9 ml/min. The column temperature was programmed from 130°C at 2°C/min, held at 180°C for 15 min, raised to 210°C at 2°C/min and held isothermally at 210°C for 15 min.

Each of the GC peaks was identified by comparing their retention times with those of authentic standards as well as mass spectra obtained with a gas chromatograph/mass spectrometer (GC/MS: JEOL JMS-DX302). A fused silica capillary column (FFS ULBON HR-SS-10, 25 m length, 0.25 mm i.d., Shinwa Industry Co. Ltd.) was equipped with the GC/MS instrument.

RESULTS AND DISCUSSION

Source strength for sinking organic matter based on fatty acid composition

Vertical fluxes of organic carbon in the continental slope off Hokkaido were in the range of 4.44 to 24.4 mgC $m^{-2}day^{-1}$ with two maxima in September (sample #1) and April-May (sample #12, #13), as shown in Fig. 2 (after Handa and Nakatsuka (1993)). These high fluxes of organic carbon can be caused by a high productivity in the surface water during autumn and spring. In the north Pacific Ocean, there are two maxima of standing stock of plankton communities on

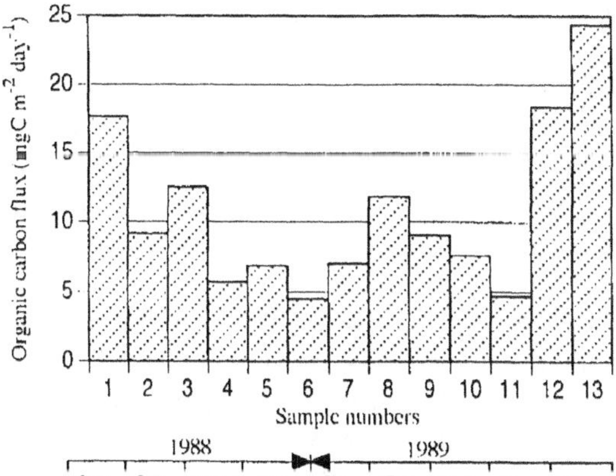

Fig. 2. Vertical fluxes of organic carbon in the western North Pacific

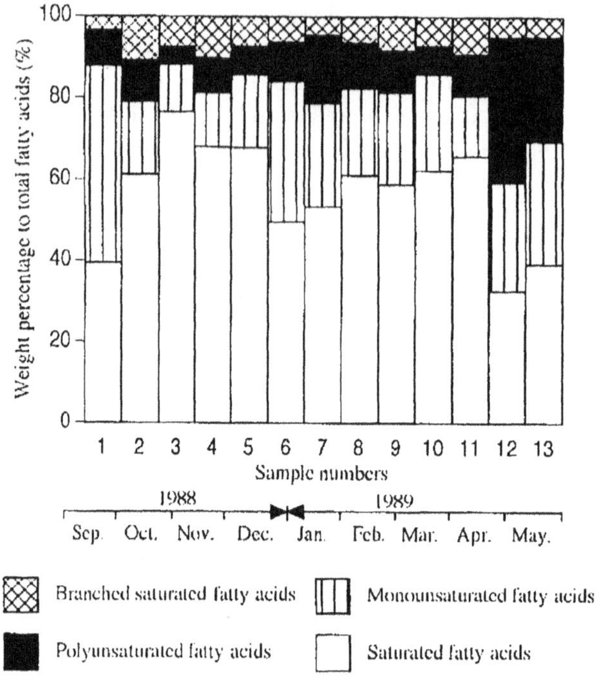

Fig. 3. Temporal compositional changes of each fatty acid classes (saturated, monounsaturated, polyunsaturated and branched fatty acids) in the sinking particles in the western North Pacific.

spring and autumn as a typical seasonal cycle (Parsons *et al.*, 1984).

Fatty acids in the sinking particles were detected in a range of carbon numbers from C_{13} to C_{28}. Carbon contents of fatty acids to organic carbon in the sinking particles were in the range of 2.9–15.4 %. The data of fatty acid fluxes will be available elsewhere. Saturated fatty acids as tetradecanoic acid (14:0), hexadecanoic acid (16:0) and octadecanoic acid (18:0) were the major components throughout the mooring periods (Fig. 3). From October 1988 to March 1989, saturated fatty acids accounted for over 50 % of the total fatty acids in the sinking particles. The fatty acid composition in the sinking particles is similar to those reported in previous studies of sinking particles (Tanoue and Handa, 1980; de Baar *et al.*, 1983; Matsueda *et al.*, 1986; Wakeham and Lee, 1989; Reemtsma *et al.*, 1990)

On the other hand, 9-octadecenoic acid (18:1 *n*-9) was remarkably dominant in the September sample, #1, and 7-hexadecenoic acid (16:1) and eicosapentaenoic acid (20:5) were dominant in the April-May samples, #12, #13 (Fig. 4). Wakeham *et al.* (1984) suggested that a dominance of 18:1 *n*-9 in marine particulate matter is an indicator of zooplankton materials, because crustacean zooplankton have 18:1 *n*-9 which exceeds 50 % of total fatty acids (Morris, 1971). The relative high content of 18:1 *n*-9 of the fatty acid composition suggested that the sinking

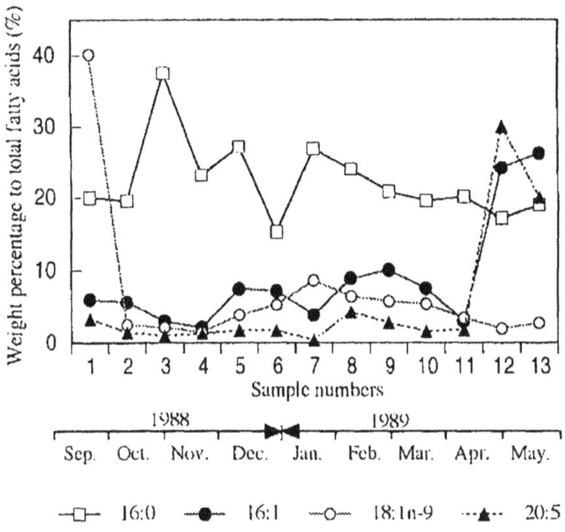

Fig. 4 . Temporal compositional changes of 16:0, 16:1, 18:1n-9 and 20:5 asids in the sinking particles in the western North Pacific.

particles in the September sample, #1 were largely affected by zooplankton. The 20:5 is a typical marine fatty acid of zooplankton and fish origin (Ackman, 1989). However, some studies of the fatty acid composition of diatom showed that 14:0, 16:0, 16:1 and 20:5 were the dominant fatty acids (Ackman *et al.*, 1968; Orcutt and Patterson, 1975; Volkman *et al.*, 1989; Dunstan *et al.*, 1994). A microscopic observation of the samples in April-May, #12, #13 indicated the large contribution of diatoms. This result and the high contents of 16:1 and 20:5 in the sinking particles suggested that most of the sinking organic particles in the April-May samples, #12, #13 were mainly produced by diatoms.

Weight percentages of the branched fatty acids (mainly br-15:0 and br-17:0) to the total fatty acids were shown in Fig. 5. They were relatively low in September 1988 (sample #1), January 1989 (sample #7), and April-May 1989 (sample #12, #13). Br-15:0, br-17:0 and 18:1 *n*-7 indicated bacterial influence on the sinking particles, because these acids are generally considered to be bacterial origin (Perry *et al.*, 1979; Volkman *et al.*, 1980) and normally represent less than 10 % of total fatty acids in sinking particles (de Baar *et al.*, 1983; Reemtsma *et al.*, 1990; Najedek, 1993). The relative abundance of branched fatty acids in the present study were similar to those reported in the previous studies.

The contents of long-chain saturated fatty acids over C_{24} were relatively low in sample #1, #7, #12, and #13 (Fig. 5). Long-chain ($\geq C_{24}$) saturated fatty acids with a predominance of even over odd carbon number are derived from terrestrial higher plants (Saliot *et al.*, 1991). Marine diatoms have been found long-chain fatty acids (Volkman *et al.*, 1989). However, in the experimental area, large amounts of terrestrial materials would input into sediment trap (Saito *et al.*, 1992). Thus, the presence of long-chain fatty acids in the sinking particles

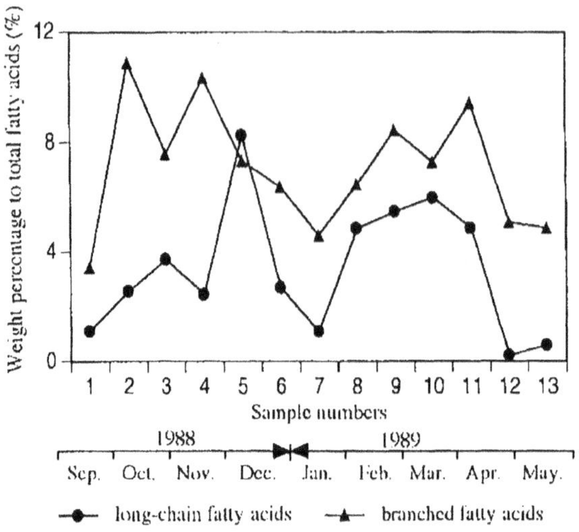

Fig. 5. Temporal variation of the composition of branched fatty acids and long-chain fatty acids in the sinking particles in the western North Pacific.

probably indicate terrestrial inputs. The low contents of long-chain saturated fatty acids over C_{24} in sample #1, #7, #11, and #12 suggested that terrestrial inputs were relatively small during these periods, comparing to the inputs of marine organisms.

UFA/SAFA ratio of sinking organic matter

Ratios of unsaturated fatty acids (UFA) to saturated fatty acids (SAFA) are shown in Fig. 6. These ratios were high values in the September sample, #1 and the April-May samples, #12, #13. Living marine organisms (e.g. phytoplankton, zooplankton) have a large content of unsaturated fatty acids (more than 60 % of the total fatty acids) (Ackman, 1989). This fact means that the UFA/SAFA ratio of living organisms is over 1. On the other hand, in detrital materials, the content of unsaturated fatty acids was low (Shultz and Quinn, 1973). Fecal pellets have large amounts of saturated fatty acids (Tanoue *et al.*, 1982; Harvey *et al.*, 1987). The high UFA/SAFA ratios in the September sample, #1 and the April-May samples, #12, #13 suggested a rapid transport of fresh organic matter to deep water. In the September sample, #1, the sinking particles contain zooplankton-derived materials, which were indicated by a large amount of 18:1 *n*-9. In the April-May samples, #12, #13, fresh diatom cells containing 16:1 and 20:5 was supposed to contribute to the high values of the UFA/SAFA ratios of the sinking particle.

Relationship between UFA/SAFA ratio and vertical transport of organic matter

Several studies on fatty acid composition of sinking particles have reported that the major fatty acids of sinking particles were saturated fatty acids (Table 1).

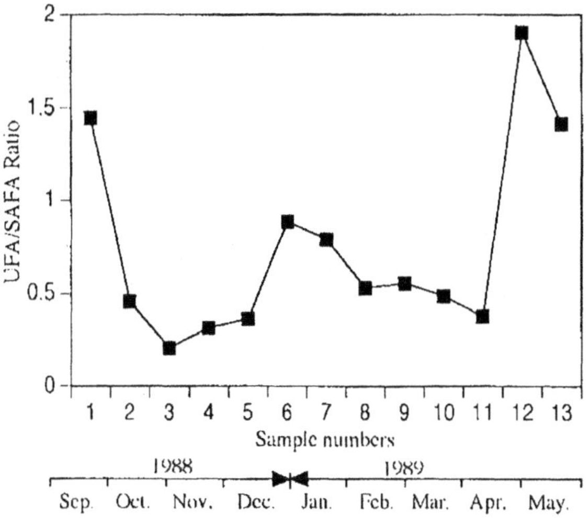

Fig. 6. Temporal variation of the ratios of unsaturated fatty acids (UFA) to saturated fatty acids (SAFA) in the sinking particles in the western North Pacific.

Table 1. The ratios of unsaturated fatty acids to saturated fatty acids of sinking particles in the reports of sediment trap studies.

UFA/SAFA* ratio	Depth (m) of setting sediment trap	Area	Reference
0.33	1100	Northern North Pacific Ocean	Tanoue and Handa (1980)
0.26	2200		
0.29	4400		
0.41	5250		
1.39	389	North Atlantic Ocean	de Baar *et al.* (1983)
0.76	988		
0.78	3755		
0.94	5068		
0.92	360	Sagami Bay, Japan	Matsueda (1986)
1.71	560		
1.28	1060		
0.74	740	Antarctic Ocean	Matsueda (1986)
0.66	940		
0.31	3440		
0.46	4240		
1.64	630	Antarctic Ocean	Yagi (1986)
0.55	1430		
0.41	3230		
0.34 – 0.47	732	Central Arabian Sea	Reemtsma *et al.* (1990)
0.24 – 0.34	2914		
0.44 – 0.72	35	Northern Adriatic Sea	Najdek *et al.* (1994)
0.34 – 1.30	110	Breid Bay, Antarctica	Hayakawa *et al.* (1996)
0.21 – 1.90	5200	This study	

*: Unsaturated fatty acids / Saturated fatty acids

Three-dimensional plots of sampling depths, organic carbon fluxes and UFA/SAFA ratios of sinking particles are shown in Fig. 7, which are based on the present study and early reports of the fatty acid composition of sinking particles collected by the sediment trap. Vertical distribution of organic carbon flux indicated a decrease with increasing sampling depth. Most of the UFA/SAFA ratio in the sinking particles were lower than a unity, because unsaturated fatty acids are more degradable than saturated fatty acids in the water column during vertical transport to deep water (Tanoue and Handa, 1980; de Baar *et al.*, 1983). The high UFA/SAFA ratios, being greater than a unity, were found at a shallow depth in the open ocean (e.g. North Atlantic Ocean, Antarctic Ocean) and at deep depths in the coastal ocean (e.g. Sagami Bay, present study). The sinking particles of the high UFA/SAFA ratio contain labile compounds of organic matter exported from euphotic layer. Some reports documented that sedimentation of labile organic materials occurred during bloom periods (Davies and Payne, 1984; Ittekkot *et al.*, 1984; Skjoldal and Wassmann, 1986; Honjo and Manganini, 1993). During a spring bloom, large and intact cells of phytoplankton settle down from the euphotic layer and can transport the labile organic compounds to deep water. Additionally, in the coastal ocean, large export production occur from the high nutrient supply, and rapid sinking of large and intact cells play an important role in the large export production (Smetacek *et al.*, 1985; Berger *et al.*, 1989; Bienfang and Ziemann, 1992). In the coastal ocean, transportation of labile organic compounds to deep water is more effective than that in the open ocean, as indicated by the high UFA/SAFA ratios at deep water in the present study.

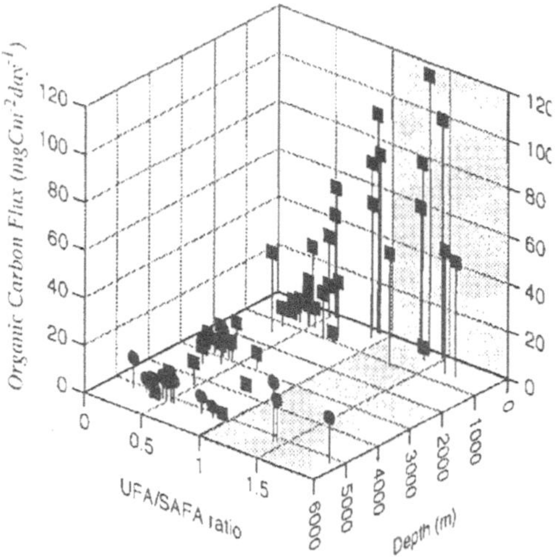

Fig. 7. Relationship between sampling depth, organic carbon flux and the UFA/AFA ratio of sinking particles in this study and early studies (Table 1) of fatty acid composition of sinking particles. The solid square means the data of the early studies. The solid circle means the data of this study.

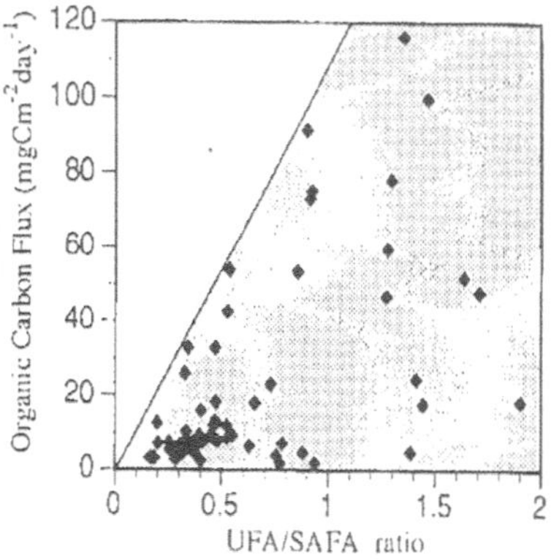

Fig. 8. Relationship between organic carbon flux and the UFA/SAFA ratio of sinking particles in this study and early studies of fatty acid composition of sinking particles.

The relationship between the UFA/SAFA ratios and organic carbon fluxes is shown in Fig. 8. The ratios are scattered against organic carbon fluxes (shadow zone in Fig. 8). However, sinking particles with large organic carbon flux do not show a low ratio of the UFA/SAFA (clear zone in Fig. 8). Such a large flux of sinking particles is closer to living organisms than detritus and fecal pellets which are low UFA/SAFA ratio. This result supports the fact that the rapid sinking of large and intact cells, which is often observed in high latitudinal or coastal ocean, plays an important role in the large export production.

CONCLUSIONS

The result that the fatty acids in the sinking particles mainly consist of saturated acids is similar to those reported in the early studies of fatty acid. On the other hand, the remarkabley high component of 18:1 *n*-9 in autumn indicated that the sources of sinking particles were mainly derived from zooplankton, and high components of 16:1 and 20:5 in spring also indicated that the sources of the sinking particles were mainly diatoms. The presence of long-chain fatty acids also indicated the contribution of terrestrial inputs to the sinking particles. UFA/SAFA ratios are useful for inferring the freshness of sinking organic matters. The high UFA/SAFA ratios in deep water suggested that the sinking particles of large mass flux during a spring bloom can transport fresh organic matter to deep water.

REFERENCES

Ackman, R. G. (1989): Fatty acids. p103–138. In: *Marine Biogenic Lipids, Fats, and Oils*, Vol. I, ed. by R. G. Ackman, CRC Press, Orlando, FL.

Ackman, R. G., C. S. Tocher and J. McLachlan (1968): Marine phytoplankter fatty acids. *J. Fish. Res. Bd. Canada*, **25**, 1603–1620.

Berger, W. H., V. S. Smetacek and G. Wefer (1989): Ocean Productivity and Paleoproductivity-An Overview. In: *Productivity of the Ocean: Present and Past*, ed. by W. H. Berger, V. S. Smetacek and G. Wefer, John Wiley and Sons, Chichester.

Betzer, P. R., W. J. Showers, E. A. Laws, C. D. Winn, G. R. DiTullio and P. M. Kroopnick (1984): Primary productivity and particle fluxes on a transect of the equator at 153°W in the Pacific Ocean. *Deep-Sea Res.*, **31**, 1–11.

Bienfang, P. K. and D. A. Ziemann (1992): The role of coastal high latitude ecosystems in global export production. p. 285–297. In: *Primary Productivity and Biogeochemical Cycles in the Sea*, ed. by P. G. Falkowski and A. D. Woodhead, Plenum Press, New York.

Davies, J. M. and R. Payne (1984): Supply of organic matter to the sediment in the northern North Sea during a spring phytoplankton bloom. *Mar. Biol.*, **78**, 315–324.

de Baar, H. J. W., J. W. Farrington and S. G. Wakeham (1983): Vertical flux of fatty acids in the North Atlantic Ocean. *J. Mar. Res.* **41**, 19–41.

Deuser, W. G. and E. H. Ross (1980): Seasonal change in the flux of organic carbon to the deep Sargasso Sea. *Nature*, **283**, 364–365.

Dunstan, G. A., J. K. Volkman, S. M. Barrett, J. M. Leroi and S. W. Jeffrey (1994): Essential polyunsaturated fatty acids from 14 species of diatom (Bacillariophyceae). *Phytochem.*, **35**, 155–161.

Folch, J., M. Lees and G. H. S. Stanley (1957): A simple method for the isolation and purification of total lipides from animal tissues. *J. Biol. Chem.*, **226**, 497–509.

Handa, N. and T. Nakatsuka (1993): Vertical fluxes of organic materials in the northern North West Pacific and Breid Bay, Antarctica, with special reference in the effect of phytoplankton bloom. p. 221–233. In: *Deep Ocean Circulation, Physical and Chemical Aspects*, ed. by T. Teramoto, Elsevier, Amsterdam.

Harvey, H. R., G. Eglington, S. C. M. O'Hara and E. D. S. Corner (1987):Biotransformation and assimilation of dietary lipids by Calanus feeding on a dinoflagellates. *Geochim. Cosmochim. Acta*, **51**, 3031–3040.

Hayakawa, K., N. Handa, N. Ikuta and M. Fukuchi (1996): Downward fluxes of fatty acids and hydrocarbons during a phytoplankton bloom in the austral summer in Breid Bay, Antarctica. *Org. Geochem.*, **24**, 511–523.

Honjo, S. (1982): Seasonality and interaction of biogenic and lithogenic particulate flux at the Panama Basin. *Science*, **218**, 883–884.

Honjo, S. and S. J. Manganini (1993): Annual biogenic particle fluxes to the interior of the North Atlantic Ocean; studied at 34°N 21°W and 48°N 21°W. *Deep-sea Res*, **40**, 587–607.

Honjo, S., S. J. Manganini and G. Wefer (1988) Annual particle flux and a winter outburst of sedimentation in the northern Norwegian Sea. *Deep-Sea Res.*, **35**, 1223–1234.

Ittekkot, V., W. G. Deuser and E. T. Degens (1984): Seasonality in the fluxes of sugars, amino acids, and amino sugars to the deep ocean: Sargasso Sea. *Deep-sea Res.*, **31**, 1057–1069.

Lee, C., S. G. Wakeham and J. W. Farrington (1983): Variations in the composition of particulate organic matter in a time-series sediment trap. *Mar. Chem.*, **13**, 181–194.

Matsueda, H. (1986): Doctor Dissertation, Nagoya Univ. (in Japanese).

Matsueda, H., N. Handa, I. Inoue and H. Takano (1986): Ecological significance of salp fecal pellets collected by sediment traps in the eastern North Pacific. *Mar. Biol.*, **91**. 421–431.

Morris, R. J. (1971): Variations in the fatty acid composition of oceanic euphausiids. *Deep-Sea Res.*, **18**, 525–529

Najdek, M. (1993): Factors influencing fatty acid and hydrocarbon composition of sedimenting particles in the northeastern Adriatic Sea. *Mar. Chem.*, **41**, 299–310.

Najdek, M., S. Puskaric and A. B. Bochdansky (1994): Contribution of zooplankton lipids to the flux of organic matter in the northern Adriatic Sea. *Mar. Ecol. Prog. Ser.*, **111**, 241–249.

Nakatsuka, T., N. Handa and S. Imaizumi (1995): Spatial and temporal variation of $\delta^{15}N$ in sinking particles in deep waters: Its implication for the origin and transport of particulate organic matter. p355–374. In: *Biogeochemical Processes and Ocean Flux in the Western Pacific*. ed. by H. Sakai and Y. Nozaki, Terra Sci. Pub. Co., Tokyo.

Orcutt, D. M. and G. W. Patterson (1975): Sterols, fatty acid and elemental composition of diatoms growth in chemically defined media. *Comp. Biochem. Physiol.*, **50B**, 579–583.

Pace, M. L., G. A. Knauer, D. M. Karl and J. H. Martin (1987): Primary production, new production and vertical flux in the eastern Pacific Ocean. *Nature*, **325**, 803–804.

Parsons, T. R., M. Takahashi and B. Hargrave (1984): *Biological Oceanographic Processes. 3rd edn.*, Pergamon Press, Oxford, 330pp.

Perry, G. J., J. K. Volkman, R. B. Johns and H. J. Bavor Jr (1979): Fatty acids of bacterial origin in contemporary marine sediments. *Geochim. Cosmochim. Acta*, **43**, 1715–1725.

Reemtsma, T., B. Haake, V. Ittekkot, R. R. Nair and U. H. Brockmann (1990): Downward flux of particulate fatty acids in the central Arabian Sea. *Mar. Chem.* , **29**, 183–202.

Saito, C., S. Noriki and S. Tsunogai (1992): Particulate flux of Al, a component of land origin, in the western North Pacific. *Deep-Sea Res.*, **39**, 1315–1327.

Saliot, A., J. Laureillard, P. Scribe and M. A. Sicre (1991): Evolutionary trends in the lipid biomarker approach for investigating the biogeochemistry of organic matter in the marine environment. *Mar. Chem.*, **36**, 233–248.

Schultz, D. M. and J. G. Quinn (1973): Fatty acid composition of organic detritus from *Spartina alterniflora. Estuarine Coastal Mar. Sci.*, **1**, 177–190.

Skjoldal, H. R. and P. Wassmann (1986): Sedimentation of particulate organic matter and silicium during spring and summer in Lindåspollene, Western Norway. *Mar. Ecol. Prog. Ser.*, **30**, 49–63.

Smetacek, V. S. (1985): Role of sinking diatom life-history cycles: ecological, evolutionary and geological significance. *Mar. Biol.* , **84**, 239–251

Suess, E. (1980): Particulate organic carbon flux in the oceans - surface productivity and oxygen utilization. *Nature*, **288**, 260–263.

Tanoue, E. and N. Handa (1980): Vertical transport of organic materials in the northern North Pacific as determined by sediment trap experiment. Part 1. Fatty acid composition. *J. Oceanogr. Soc. Japan*, **36**, 231–245.

Tanoue, E., N. Handa and H. Sakugawa (1982): Difference of the chemical composition of organic matter between fecal pellet of *Euphausia superba* and its feed, *Dunaliella tertiolecta. Trans. Tokyo Univ. Fish.*, **5**, 189–196.

Volkman, J. K., S. W. Jeffrey, P. D. Nichols, G. I. Rogers and C. D. Garland (1989): Fatty acid and lipid composition of 10 species of microalgae used in mariculture. *J. Exp. Mar. Biol. Ecol.* , **128**, 219–240.

Volkman, J. K., R. B. Johns, F. T. Gillan, G. J. Perry and H. J. Bavor Jr (1980):Microbial lipids of an intertidal sediment—I. Fatty acids and hydrocarbons. G*eochim. Cosmochim. Acta* , **44**, 1133–1143.

Wakeham, S. G. and C. Lee (1993): Organic geochemistry of particulate matter in the ocean: The role of particles in oceanic sedimentary cycles. *Org. Geochem.*, **14**, 83–96.

Wakeham, S. G., C. Lee, J. W. Farrington and R. Gagosian (1984): Biogeochemistry of particulate organic matter in the oceans: results from sediment trap experiments. *Deep-Sea Res.* , **31**, 509–528.

Wassmann, P. (1990): Relationship between primary and export production in the boreal coastal zone of the North Atlantic. *Limnol. Oceanogr.*, **35**, 464–471.

Wefer, G. (1989): Particle Flux in the Ocean: Effects of Episodic Production. p. 139–154. In: *Productivity of the Ocean: Present and Past*, ed. by W. H. Berger, V. S. Smetacek and G. Wefer, John Wiley and Sons, Chichester.

Wefer, G., G. Fischer, D. Füetterer and R. Gersonde (1988): Seasonal particle flux in the Bransfield Strait, Antarctica. *Deep-Sea Res.* , **35**, 891–898.

Yagi, K. (1986): Master Dissertation, Nagoya Univ. (in Japanese).

Dynamics and Characterization of Marine Organic Matter, Eds., N. Handa, E. Tanoue and T. Hama, pp.117–140.

Fluxes of Major Biochemicals in the Equatorial Pacific Ocean

Stuart G. WAKEHAM[1], Cindy LEE[2] and John I. HEDGES[3]

[1]*Skidaway Institute of Oceanography, 10 Ocean Science Circle, Savannah, GA 31411, U.S.A.*
[2]*Marine Sciences Research Center, State University of New York, Stony Brook, NY 11794, U.S.A.*
[3]*School of Oceanography, Box 357940, University of Washington, Seattle, WA 98195-7940, U.S.A.*

(Synthesis, Received 21 October 1996; accepted 20 January 1997)

Abstract—A coordinated study of the geochemistry of major biochemical classes has been carried out in the equatorial Pacific Ocean (EqPac) as part of the U.S. Global Ocean Flux Study (JGOFS). Amino acids, carbohydrates, pigments, and lipids were measured at the molecular level in plankton, sediment trap material, and sediments along a transect from 12°N to 12°S along 140°W to address questions related to the efficiency and dynamics of the carbon pump in the equatorial Pacific and how processes in the water column affect the burial of biogenic components. Here we present an overview of our findings, discussing fluxes and relative reactivities of the biochemical classes. We conclude with an inventory of the major particulate biochemicals in the equatorial Pacific, in which the proportion of organic carbon amenable to conventional molecular-level analyses decreases from ~82% in plankton to ~20% in sediments, demonstrating the need for new analytical approaches to characterize residual macromolecular material.

INTRODUCTION

The equatorial zone from New Guinea to South America in the Pacific Ocean is characterized by elevated levels of new production, perhaps representing about 50% of global new production (Eppley and Peterson, 1979; Chavez and Barber, 1987). The transfer of biogenic carbon-containing particles from surface waters to the ocean's interior, the so-called "biological pump" (Longhurst and Harrison, 1989), is an important component of the global mechanism for regulating the atmospheric and oceanic carbon reservoirs. Remineralization of particulate organic matter in the ocean's interior provides the nutrients which are upwelled through physical processes near the equator and support new production. The eventual sequestration of organic matter in sediments provides a long-term sink for elements as well as a record of global climate change on geologic time scales. A major goal of the Global Ocean Flux Study (JGOFS) Equatorial Pacific

(EqPac) program was therefore to quantify the role of this region in the global biogeochemical cycles.

Although the ocean stores and removes carbon largely in the inorganic form (i.e. bicarbonate and calcium carbonate), particulate organic material is the primary vehicle for the vertical redistribution of bioactive elements within the water column. Thus it is appropriate that any study of carbon fluxes and biogeochemical processing include direct, detailed investigations of the organic component. Organic compounds of different origins and molecular structural types are degraded with high selectivity throughout the ocean. Measurements of fluxes of particulate organic carbon alone reveal none of the complex compositional trends that accompany this degradation and are diagnostic for the processes at work. Without molecular-level tags which trace organic materials of different origins and reactivity, it is difficult to relate bulk fluxes of organic carbon to the processes controlling organic matter cycling in the water column and sediments.

Our organic geochemical study focusses on the compositions and corresponding fluxes of major biochemical classes in order to address the two major goals of the EqPac study (Murray *et al.*, 1992), specifically 1) to determine the efficiency and dynamics of the carbon pump in the equatorial Pacific and 2) to establish how mid-water and deep-water processes affect the burial of biogenic components. To define the sources, quality and flux of particulate organic compounds, we intercepted organic material at sequential stages of transport and diagenesis from the surface ocean to the underlying sediments (Fig. 1). These stages include 1) the standing crop of particles in the euphotic zone (net-plankton); 2) particles exiting the euphotic zone (floating sediment traps at 105 m); 3) particles transiting the ocean's interior (moored traps at 1000 m below the sea surface and 1000 m above the sea floor); 4) material raining down onto the seafloor (sediment floc); and 5) material being buried in the sediment column (sediment cores). Samples were subjected to bulk elemental analyses and molecular-level quantification of individual carbohydrates, amino acids, lipids and pigments. Our choice of these biochemicals is based on their importance in living organisms, their utility as tracers of source and alteration processes which affect the composition of particulate matter, and the fact that they are readily amenable to analysis at the molecular level. As we will show, a large fraction of particulate and sedimentary organic carbon cannot be accounted for by these biochemical classes, and much of this material remains uncharacterized with respect to molecular composition.

METHODS

Samples

Samples were collected on four cruises of the EqPac program along the transect from 12°N to 5°S along 140°W and extending to 12.5°S, 135°W. Net-plankton was collected on two survey cruises (Survey 1: February–March, 1992; Survey 2: August–September, 1992) using oblique tows of a 26 μm mesh net. Samples were passed through a 850 μm stainless steel sieve, and split for multiple

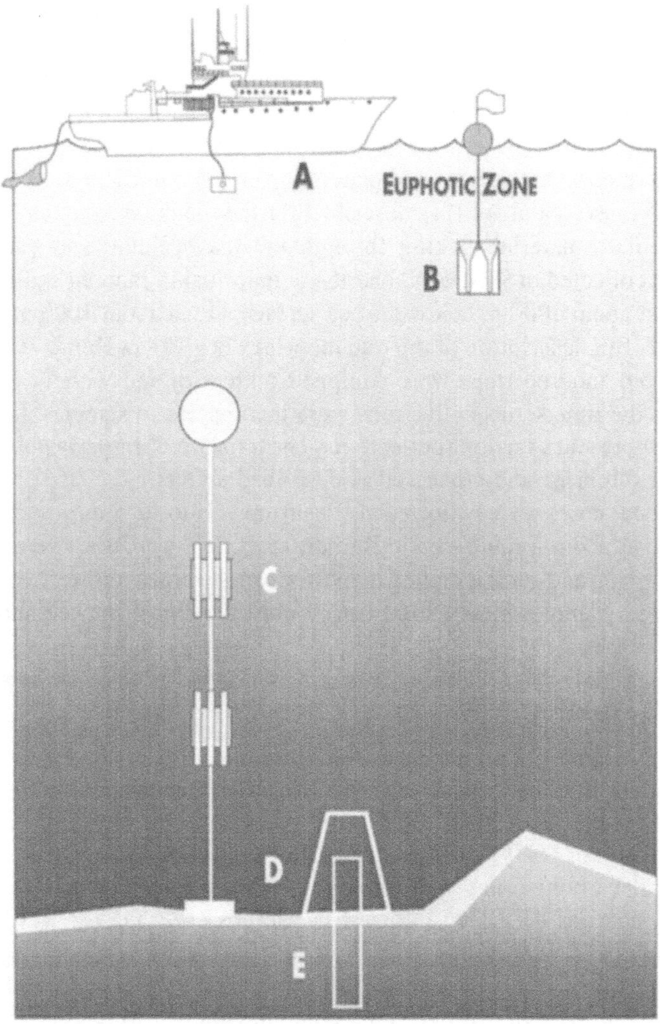

Fig. 1. Schematic diagram of the sample collection strategy used in this study. A: net-plankton; B: 105-m floating sediment traps; C: moored traps at 1000 m below the sea surface and 1000 m above the sea floor; D: surficial sediment; E: buried sediment.

analysis, where 1/2 was used for elemental and carbohydrate analyses, 1/4 for lipids, and 1/16 each for amino acids, pigments, archive, and other miscellaneous analyses. The elemental/carbohydrate split was centrifuged and the pellet frozen. The other splits were filtered onto glass fiber filters and frozen. Both phytoplankton and zooplankton were included.

Material exiting the photic zone (the export flux) was collected in floating sediment traps deployed for 1.5–3 days at 105 m along the survey transects. The floating sediment traps were equipped with Indented Rotating Sphere (IRS;

Peterson *et al.*, 1993) valves to minimize collection of zooplankton swimmers and to prevent washout of poison and preservative treatments. The mooring consisted of a single array containing four sediment traps with a total collecting surface area of ~1.5 m^2. Prior to deployment, 0.5 l of a solution of 50 mg $HgCl_2$ and 50 g NaCl in filtered seawater was introduced into the collection cup of each trap. Upon retrival, the sedimented material from the four traps was composited, and treated (sieved and split) as described for the plankton samples.

Particulate material sinking through the mesopelagic and bathypelagic zones was collected at 9°N, 5°N, and the equator using moored sediment traps deployed at about 1000 m below the sea surface and at about 1000 m above the sea floor. A full description of the trap moorings is given in Honjo *et al.* (1995). Although our moored traps were equipped both with and without time-series collectors, the time series collections were incomplete so samples discussed in this paper represent year-long collections. The traps were similarly poisoned with $HgCl_2$ in a salt brine and processed as described above.

Sediment cores were collected at sites from 9°N to 12°S during the benthic/ coring cruise using a multi-corer (Barnett *et al.*, 1984). Cores were sectioned onboard the ship and corresponding intervals from several cores were composited. The sediment samples were stored frozen until they were thawed and split for organic analyses.

Analytical methods

Organic carbon, total nitrogen, and inorganic carbon (by difference) were determined in duplicate using a Carlo Erba CHN analyser according to the method of Hedges and Stern (1984). The procedure was modified (Cowie and Hedges, 1991) for analysis of the carbonate-rich sediments.

Neutral carbohydrates were analysed using the technique of Cowie and Hedges (1984a) as described in detail by Hernes *et al.* (1996). Samples were pretreated with 72 wt% H_2SO_4 for 2 hr, diluted to 1.2 M H_2SO_4 and hydrolysed for 3 hr at 100°C. Following neutralization with $Ba(OH)_2$ and passage of the supernatant through an ion exchange column, samples were equilibrated in a 0.2% $LiClO_4$ in pyridine solution, derivatized with BSTFA + 1% TMCS, and analysed by gas chromatography. Adonitol was used as an internal standard.

Amino acids were measured by fluorescence-high pressure liquid chromatography after hydrolysis according to Lee and Cronin (1982) and Lee *et al.* (submitted). Samples were hydrolysed with 6N HCl under N_2 at 110°C for 19 hr. Hydrolysates were dried *in vacuo*, taken up in water, and amino acids were analysed by HPLC using a modification of the Lindroth and Mopper (1979) *o*-phthaldialdehyde derivative HPLC technique.

Lipid analyses were described in Wakeham *et al.* (1997). Lipids were Soxhlet extracted with CH_2Cl_2:MeOH (2:1) and partitioned into the organic phase with a 5% NaCl solution. An aliquot of the extract was saponified with 0.5N KOH in methanol and the neutral and acidic fractions extracted from basic (pH > 13) and acidic (pH < 2) solution, respectively. The neutral lipids, (sterols, fatty alcohols, hydrocarbons, and long-chain alkenones) were treated with BSTFA to derivatize hydroxyl functions to trimethylsilyl ethers. Acids were methylated with diazomethane and each fraction was analysed by gas chromatography and gas chromatography-mass spectrometry. Cholestane and *n*-methylnonadecanoate were used as internal standards for the neutral and acid fractions, respectively.

The major chlorophyl pigments and their degradation products were analysed by HPLC (Lee *et al.*, submitted) following extraction with 100% acetone. Pigments were separated by gradient reverse-phase chromatography using a methanol/acetone/ion pairing solution (tetrabutylammonium acetate-ammonium acetate; Mantoura and Llewellyn, 1983), followed by fluorescence detection.

RESULTS AND DISCUSSION

Export fluxes of organic materials out of the euphotic zone

The fluxes of organic carbon (C_{org} (units of mgC_{org}/m^{2}d)), amino acids, carbohydrates, lipids, and pigments (all biochemical classes expressed as mg-class/m^{2}d) exported out of the euphotic zone to sediment traps deployed at 105 m varied along the EqPac transect from 12°N to 12°S (Fig. 2). Fluxes near the equatorial region (2°N to 2°S) were 3–5-fold higher than at the ends of the transects (12°N and 12°S). This latitudinal trend is consistent with synoptic primary production data (Barber *et al.*, 1996) showing production to be higher near the equator (±2°) than at poleward locations (Fig. 2). C_{org} fluxes generally are comparable to fluxes determined independently by a concurrent combined thorium isotope-sediment trap experiment (Murray *et al.*, 1996) and by thorium tracer techniques (Buesseler *et al.*, 1995; Luo *et al.*, 1995; Bacon *et al.*, 1996). There are no comparable biochemical flux data from the EqPac study. A comparison of the relative magnitudes of biochemical export fluxes gives the order amino acids > carbohydrates ≈ lipids > pigments, as would be expected based on abundances of these compound classes in plankton (our analyses, see below).

The trends in export flux for C_{org} and the biochemical classes are, in part, compatible with the physical and biogeochemical characterization of this region during the EqPac study. Equatorial upwelling, driven by local trade winds, results in enhanced primary productivity in the zone from about 5°N to 5°S, hence the locally elevated organic matter fluxes. On the other hand, weakening of the trade winds and subsequent reduction in upwelling during El Niño conditions of Survey 1 (February–March) resulted in lower primary production rates than during non-El Niño conditions of Survey 2 (August–September; Barber *et al.*, 1996). Equatorial productivity was ~50% higher during Survey 2 (up to 1800 mgC/m^{2}d; Fig. 2) than during Survey 1 (~1000 mgC/m^{2}d). However, our 105-m sediment traps collected lower fluxes in the equatorial region during non-El Niño Survey 2 than during El Niño Survey 1. This result contrasts with C_{org} export flux estimates by other JGOFS studies which were based on thorium tracer measurements rather than sediment traps (Buesseler *et al.*, 1995; Luo *et al.*, 1995; Bacon *et al.*, 1996; Murray *et al.*, 1996) in which C_{org} fluxes during the non-El Niño period were up to 2-fold were higher than during the El Niño period. We are

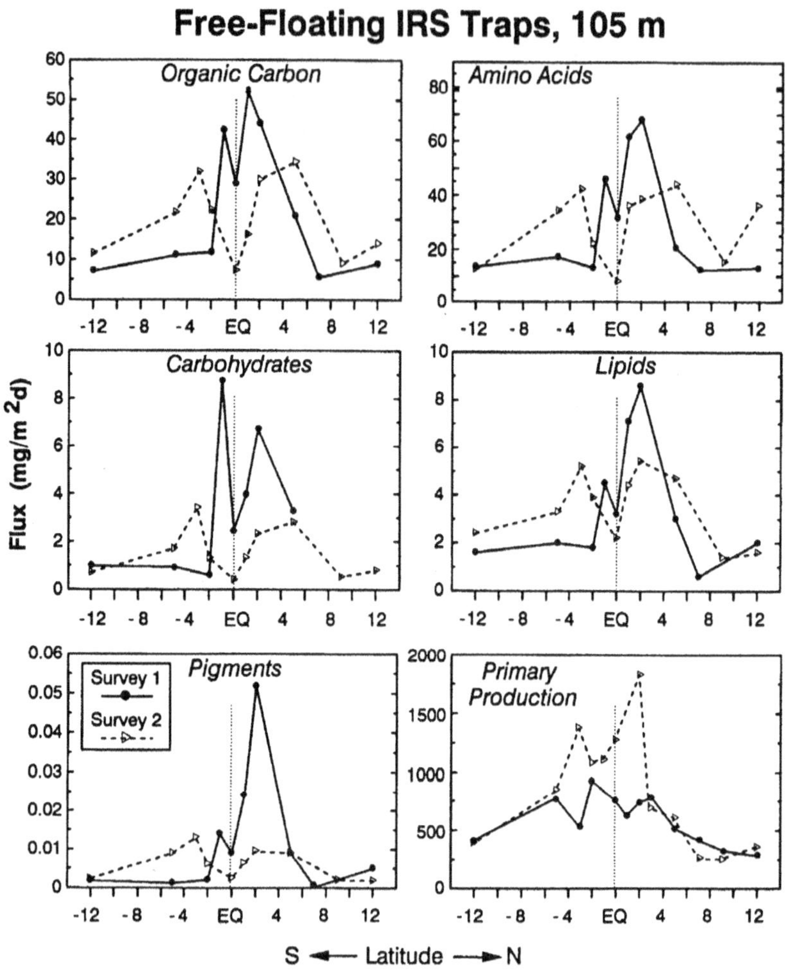

Fig. 2. Fluxes for organic carbon (mgC_{org}/m^2d) and major biochemical classes (mg class/m^2d) measured in the floating sediment traps along the N-S transects of EqPac Survey 1 and Survey 2. An estimate of the carbon flux represented by each class may be obtained by assuming that amino acids are 45% carbon, carbohydrates 40% C_{org}, lipids 85% C_{org}, and pigments 90% C_{org}. Primary production data from Barber *et al.* (1996) are shown for comparison.

uncertain as to why our 105-m traps gave lower fluxes but under trapping of sinking particles in the high energy regime of the equatorial region may be involved. During the non-El-Niño Survey 2 period, lateral shear of the Equatorial Countercurrent was stronger (>100 cm/sec) than during El-Niño conditions of Survey 1 (~20 cm/sec) (Kessler and McPhaden, 1995), and this condition could enhance trapping biases or advect particulate material laterally away from the equatorial region. Spatial and temporal variations in zooplankton grazing might also be involved. In general mesozooplankton biomass and grazing rates were highest near the equator than at higher latitudes and were lower during Survey 1 than Survey 2 (Zhang *et al.*, 1996; White *et al.*, 1996). Synoptic studies of microzooplankton grazing (Landry *et al.*, 1995; Verity *et al.*, 1996) showed that microzooplankton are the major grazers of the small (<5 μm) phytoplankton that dominate in the equatorial Pacific (Chavez *et al.*, 1996; Bidigare and Ondrusek, 1996). It is, however, the larger zooplankton which produce fecal pellets of sufficient size to be collected in sediment traps, and indeed our trap samples contained abundant zooplankton fecal pellets.

Fluxes in the interior of the equatorial Pacific

Sediment traps moored at 9°N, 5°N, and the equator collected year-long samples of material passing 1000 m below the sea surface and 1000 m above the sediment. Fluxes of C_{org} and biochemicals to 1000 m depth showed the same latitudinal trend as was observed at 105 m depth, being considerably higher at the equator than at 9°N (Fig. 3). While primary production rates varied by about 50% along the transect from 9°N to the equator, fluxes to the 1000 m traps varied to a much greater extent along the same transect, with the greatest range seen for lipids (~20-fold) and the least for carbohydrates (~5-fold). In general, fluxes to 1000 m were about 10% of the export flux measured at 105 m. Organic matter fluxes were further attenuated between the two moored traps at each site, but the magnitude of this attenuation varied depending on trap location. Fluxes at 9°N were significantly less attenuated than at the equator, and in fact lipid fluxes at 4200 m depth at 9°N were slightly—but perhaps not significantly—higher than at 1000 m depth. Overall, fluxes of C_{org} to our moored traps compare favorably both in magnitude and spatial and temporal trends with other mid-water EqPac trap (Honjo *et al.*, 1995) and thorium tracer-based (Luo *et al.*, 1995) flux data.

Fluxes to the sediment

Latitudinal variability in primary production in surface waters and export flux of organic carbon, and biochemicals out of surface waters in the equatorial Pacific is, in general, faithfully recorded in the flux of organic materials accumulating in the surface sediments along the transect. Accumulation rates of organic materials in surface (0–0.5) cm sediments were estimated using sediment accumulation rates of DeMaster *et al.* (in prep.) and sedimentary concentrations of C_{org} and biochemicals. Sediment (mass) accumulation rates range from 0.004 g/m^2d at 9°N and 0.013 g/m^2d at 12°S to 0.05 g/m^2d at 2°S (DeMaster *et al.*, in prep.). Accumulation rates for C_{org} and biochemicals varied accordingly with latitude (Fig. 4), and relative accumulation rates of the major biochemical classes were consistent with relative abundances of these compound classes in plankton and sediment trap materials. Pigments were detected only in surface sediments from 9°N, 5°N, and the equator. As was observed for the floating and moored traps, the range in sediment accumulation rates for C_{org} and biochemicals (6-fold for C_{org}; up to 20-fold for lipids) was significantly greater than was the range in primary production. This further strengthens the case for selective degradation

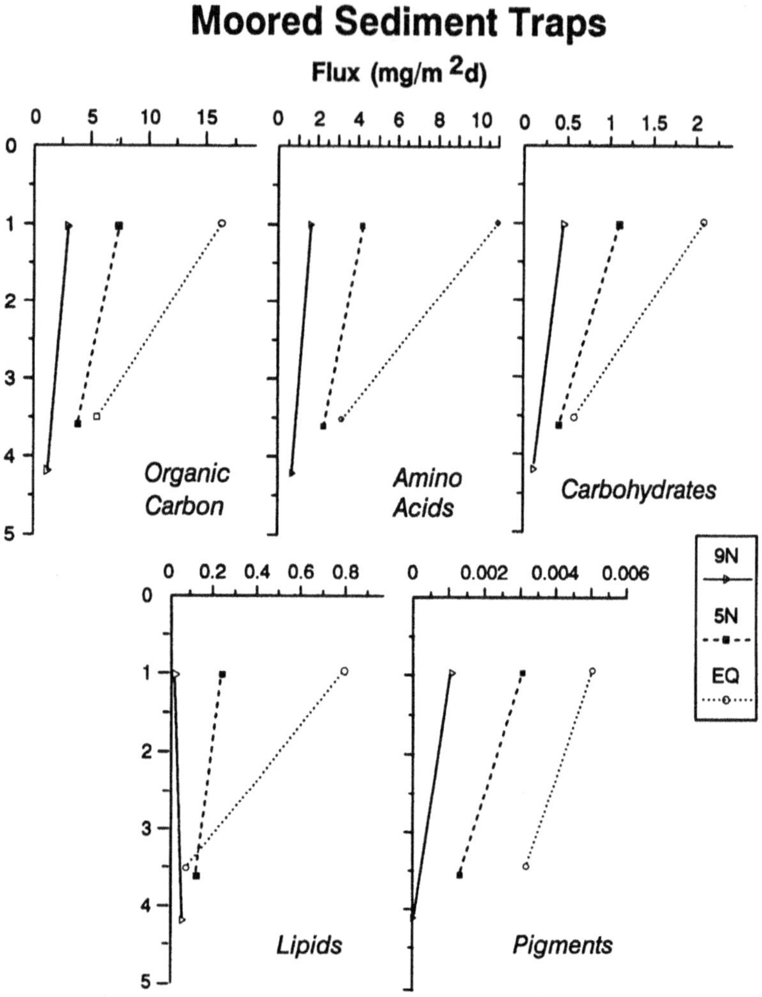

Fig. 3. Fluxes for organic carbon (mgC_{org}/m^2d) and major biochemical classes (mg class/m^2d) to moored sediment traps at 9°N, 5°N and the equator.

and alteration of organic matter during transport between surface waters and its burial in underlying sediments. Accumulation rates ranged from 0.1% to 0.001% of export flux for C_{org} and pigments, respectively.

Comparison of fluxes from plankton source to burial in the sediment

Systematic analyses of particulate organic matter from its site of production in surface waters to the site of its burial in sediments allows us to make a comprehensive comparison of fluxes at three sites (9°N, 5°N, and the equator) in

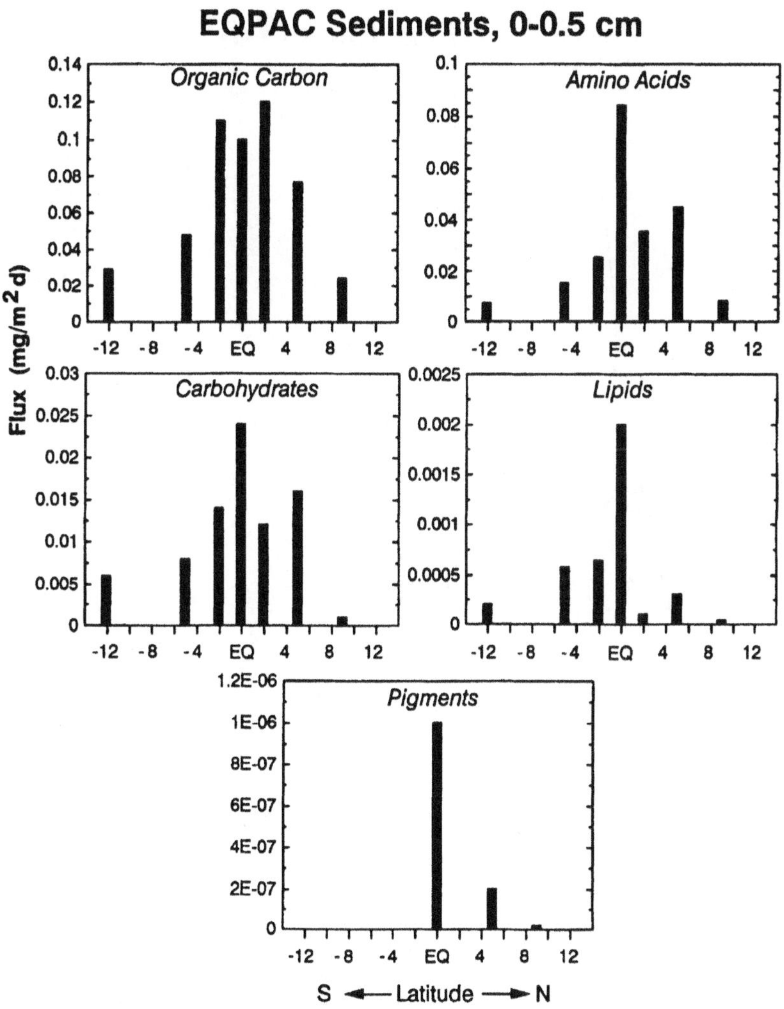

Fig. 4. Fluxes for organic carbon (mgC_{org}/m^2d) and major biochemical classes ($mg class/m^2d$) measured in surface (0–0.5 cm) sediments along the N-S transect.

the equatorial Pacific (Fig. 5). The data in Fig. 5 must be viewed with care because the sample types represent a very wide range time scales, from a day for primary productivity, to days, a year, and centuries for floating traps, moored traps and surface sediments, respectively. This range of time scales notwithstanding, fluxes of organic constituents were greatly attenuated during transit of particles down the water column and into surface sediments. Two regions of maximum flux attenuation are identified: the epipelagic zone and the benthic boundary layer. Export fluxes estimated from 105-m sediment traps were 1–3 orders of magnitude lower, depending on compound class, than "production fluxes" in

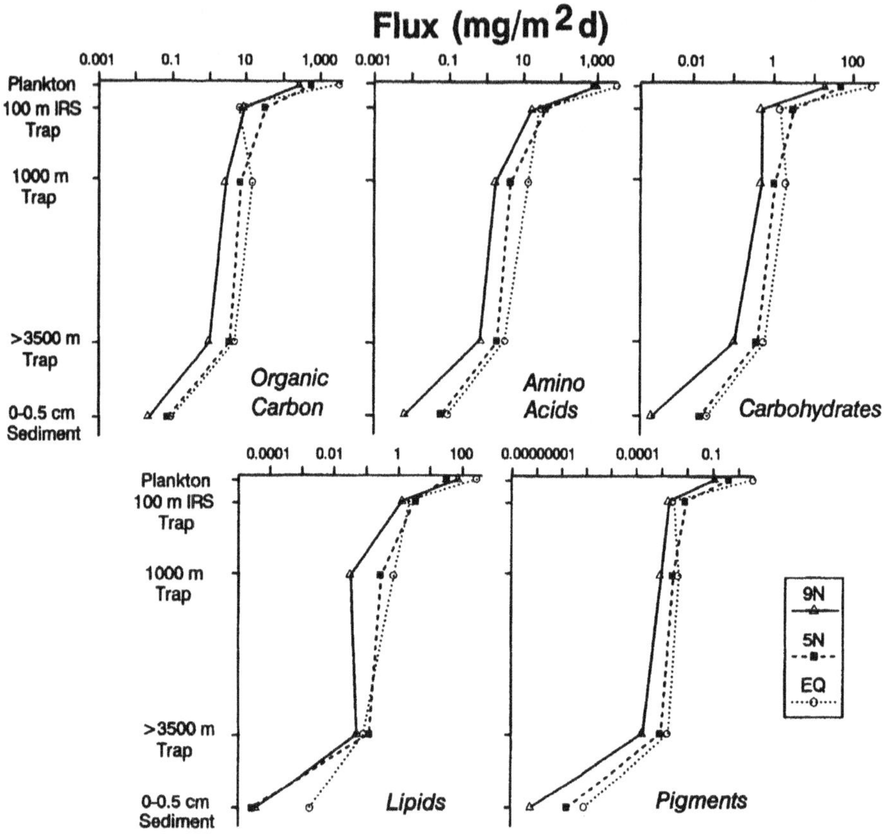

Fig. 5. Fluxes for organic carbon (mgC_{org}/m^{2}d) and major biochemical classes (mg class/m^{2}d) estimated for primary production and measured in the floating sediment traps, moored traps, and the surface sediment (0–0.5 cm) at 9°N, 5°N, and the equator. The floating sediment trap (export) flux at 5°N and the equator are means of the two surveys; for 9°N, only a single measurement is available. Fluxes of biochemicals into the surface sediment (0–0.5 cm) were calculated using the biochemical concentrations in the sediment and sediment accumulation rates from DeMaster *et al.* (in prep.). Carbon equivalent fluxes may be calculated as described in the caption for Fig. 2.

surface waters. "Production fluxes" for C_{org} are simply primary production rates at the corresponding latitudes (Barber *et al.*, 1996), while "production fluxes" for the biochemical classes are estimated from rates of primary production and the fraction of C_{org} represented by each biochemical class as determined by our analyses of net-plankton. Comparisons of "production fluxes" and trap-measured export fluxes show, for example, that at the equator during Survey 2, fluxes of C_{org} and amino acids into the 105 m traps were ~3% of their respective "production flux", vs. 1–2% for carbohydrates, ~1% for lipids and 0.3% for pigments. Little of the particulate organic matter produced in surface waters was collected in the traps at 105-m. Pigments were the most labile fraction and C_{org} and amino acids the least labile, at least on a net basis. Fluxes between the 105-m floating traps and the moored traps at 1000 m depth and 1000 m above the sea floor decreased by another 2–3 fold, signifying continued degradation in the oceans interior but at a lower intensity than in the epipelagic zone. About 1% of C_{org} from primary production reached our moored traps, and again considerably less of the individual biochemical classes survive transit in the ocean's interior. The second major step-decrease in flux of organic materials occurred between the deepest traps, deployed at about 1000 m above the sea floor, and the 0–0.5 cm interval of sediment, a decrease in flux of an additional 1–3 orders of magnitude. Organic matter accumulation in surface sediments therefore represents a small fraction of the organic material collected in traps located at 1000 m above the bottom and a tiny fraction of material originating in surface waters (at the equator, for instance: 0.01% for C_{org}, amino acids and carbohydrates; 0.0007% for lipids; and 0.00004% for pigments). It is probable that this decrease in flux occurs primarily at the water-sediment interface. In general, the flux of organic materials at the equatorial station was the highest down the entire water column and into the sediment for the sites for which we have comparable data, in line with the relative primary productivities at the three sites.

Attenuation in flux of organic matter in the ocean may result from degradation or alteration as particles sink vertically through the interior of the ocean and across the water-sediment interface, or from particles being advected laterally away from their site of production and not being collected in traps. We cannot quantitatively evaluate the relative importance of degradation/alteration vs lateral advection in the equatorial Pacific. Many studies (reviewed by Wakeham and Lee, 1993) have concluded that organic matter degradation is intense in the epipelagic ocean, continues at a lower intensity in the deep ocean, and is again intense at the sediment surface. At the same time, lateral advection of organic matter in the equatorial Pacific has been invoked as a mechanism responsible for low export flux during EqPac, as discussed above in light of the strong, but temporally and spatially variable lateral sheer of the Equatorial Countercurrent.

In the case of export flux in the epipelagic zone, we measured export ratios (ratios of particulate flux to 105-m traps to primary production) for C_{org} of ~3%. Complementary thorium tracer-based measurements during EqPac by Buesseler *et al.* (1995), Luo *et al.* (1995), Bacon *et al.* (1996), and Murray *et al.* (1996) yielded export ratios of ~2–20%. There are no comparable data for export ratios of individual biochemical classes. McCarthy *et al.* (1996) report new production rates based on ^{15}N techniques (f-ratios) for the equatorial Pacific of ~12%. Using McCarthy *et al.*'s data, Buesseler *et al.* (1995), Luo *et al.* (1996) Bacon *et al.* (1996) argue that their low export ratios (2–5%) show an imbalance between new production and export flux and that advection of organic matter, largely DOC, out of the region is responsible for the discrepancy. Murray *et al.* (1996), on the other hand, obtained higher export ratios (3–28%) and thus suggest that there is in fact no imbalance and DOC advection is not needed.

Compositions of biochemicals

Molecular analyses of the major biochemical classes also reveal important information about the behavior of organic matter in the equatorial Pacific. Detailed discussion of the molecular compositions of the major biochemical classes is beyond the scope of this paper, but is available elsewhere (Hernes *et al.*, 1996; Wakeham *et al.*, 1997; Lee *et al.*, submitted). Selected sets of samples from the equator are used to illustrate the salient trends. Amino acid, carbohydrate, and lipid compositions of net-plankton and 105-m trap material were generally quite similar and for simplicity only the trap material is shown in the respective figures.

Amino acids

Amino acid composition in floating traps, moored traps and surface sediment from the equatorial station are shown in Fig. 6. Compared to the large decreases in flux with increasing depth, only subtle changes in composition are noticeable. The most abundant amino acids in all samples were aspartic and glutamic acids, glycine and alanine, as is commonly found in all marine plankton (Degens, 1970; Siezen and Mague, 1978). With increasing depth, aspartic acid, glycine and alanine became even more predominant. A relative enrichment with depth of serine and threonine did not occur. This was somewhat surprising since preferential preservation of these two compounds with increasing depth has been observed previously in suspended and sinking particles (Lee and Cronin, 1984) and sediments (Cowie and Hedges, 1992). Preferential preservation is most likely due

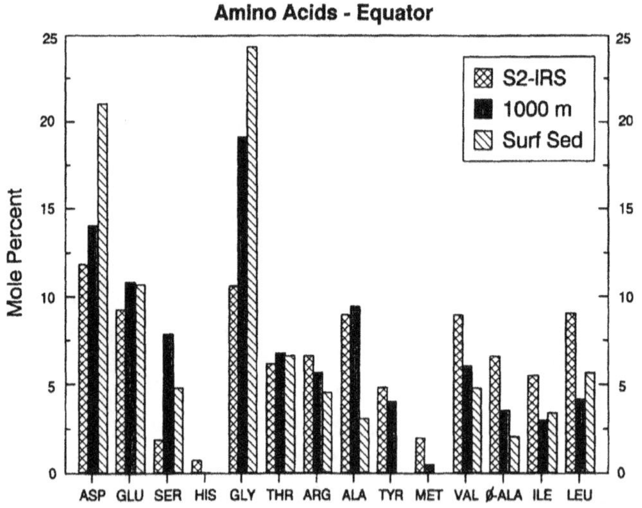

Fig. 6. Amino acid distributions in selected samples from the equator station. S2-IRS: Survey 2 floating sediment trap; 1000 m: moored sediment trap; surf sed: 0–0.5 cm sediment. ASP: aspartic acid; GLU: glutamic acid; SER: serine; HIS: histidine; GLY: glycine; THR: threonine; ARG: arginine; ALA: alanine; TYR: tyrosine; MET: methionine; VAL: valine; ϕ-ALA: phenylalanine; ILE: isoleucine; LEU: leucine.

to the enrichment of these compounds in a resistant protein-silica complex found in diatom cell walls (Hecky *et al.*, 1973). Diatoms were relatively abundant during the EqPac experiment during Survey 2 (12% of total phytoplankton chlorophyll), although less so (3%) during Survey 1 (Bidigare and Ondrusek, 1996). The relative proportion of glutamic acid decreased markedly with depth as did proportions of histidine and the aromatic compounds tyrosine and phenylalanine. Most of these losses occur at the sediment surface rather than in the water column. Although loss of both acidic amino acids is usually reported, only glutamic acid decreased in the EqPac stations. However, Burdige and Martens (1988) reported that glutamic acid decreased relative to aspartic acid in organic-rich sediments of Cape Lookout Bight. They also found that serine did not increase with depth, while glycine did, as we observe here. Both tyrosine and histidine were totally absent from EqPac surface sediments. These compositional changes are consistent with those observed in previous studies of particle and sediment diagenesis (Lee and Cronin, 1984; Burdige and Martens, 1988; Cowie and Hedges, 1992).

Carbohydrates

Typical carbohydrate compositions (as aldoses) are shown in Fig. 7. The common features of aldose distributions were similar throughout the sample set, from plankton to the sediment, with glucose as the major monomer, followed by mannose and galactose. At the three locations where we obtained samples thoughout the entire water column (9°N, 5°N, and the equator), the most striking feature of the neutral carbohydrate composition was the decrease in wt% glucose,

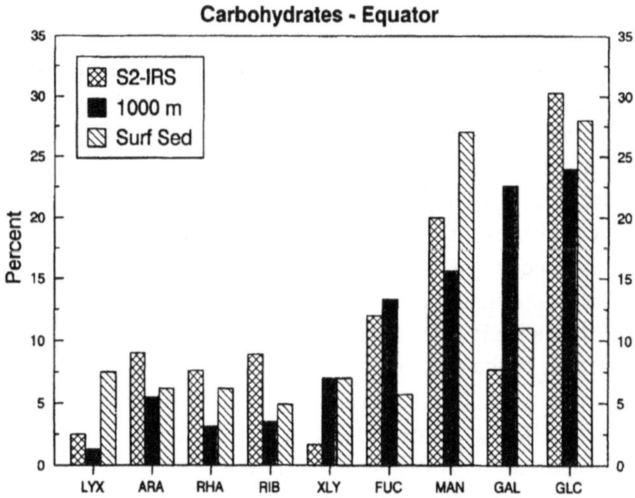

Fig. 7. Carbohydrate distributions (wt percent of total carbohydrates) in selected samples from the equator station. LYX: lyxose; ARA: arabinose; RHA: rhamnose; RIB: ribose; XLY: xlyose; FUC: fucose; MAN: mannose; GAL: galactose; GLC: glucose.

from 35–45% in the plankton tows to ~30 wt% in the floating traps, ~25 wt% in the moored traps and 20 wt% in the surface sediments (Hernes *et al.*, 1996). This trend likely reflects preferential loss of water-soluble storage carbohydrates, for instance β-1,3 glucan (Handa and Tominaga, 1969). A similar, but less strong, trend was observed for ribose. On the other hand, relative abundances of carbohydrates involved in structural polymers, including galactose, mannose, xylose, and rhamnose (Tanoue and Handa, 1987), tended to increase from plankton to the sediment, consistent with their being more resistant to degradation in the water column. In the case of galactose, however, degradation does appear to become important in the sediment. The compositional data indicate that carbohydrates in the equatorial Pacific exist in three components: 1) a relatively labile storage component consisting of glucose; 2) a group of structural components of intermediate lability enriched in galactose and perhaps arabinose; and 3) a relatively refractory structural component comprised of rhamnose, fucose, xylose, and mannose. Caution is required in interpreting trends for persistent deoxysugars, in particular fucose and rhamnose, as these may also result from *in situ* production from a secondary source, such as bacteria (Hedges *et al.*, 1994). Thus, compared to plankton, the sediment is enriched in rhamnose and xylose and depleted in glucose and ribose; distributions of these carbohydrates may be useful indicators of organic matter "freshness" (Cowie and Hedges, 1992).

Lipids

Lipids were operationally separated into neutral and acidic compounds during analysis. In all samples regardless of location or depth, acidic lipids (defined here as fatty acids released by alkaline hydrolysis of the lipid extract) were approximately 3–4-fold more abundant than neutral lipids (including sterols, fatty alcohols, long-chain alkanediols, long-chain alkenones, and hydrocarbons). This is consistent with a predominance of fatty acid-containing acyl esters (triacylglycerols, wax esters and polar lipids) in marine organisms (Sargent, 1976), and hence in particulate matter derived from them. The fatty acids (Fig. 8(a)) in plankton and 105-m traps were generally dominated by 16:0 (where 16 = number of carbon atoms and 0 = number of double bonds) and polyunsaturated fatty acids (PUFAs; e.g. 20:5 and 22:6). As particulate material sank to moored traps, polyunsaturated constituents were selectively degraded, leading to a distribution dominated by 16:0 and $18{:}1\Delta^9$ but with <2% PUFAs. The increased abundance of $18{:}1\Delta^9$ in the moored traps is probably indicative of inputs from deep-water zooplankton (Wakeham *et al.*, 1984b). Polyunsaturated acids were all but absent from the 0–0.5 cm sediment, typically leaving 16:0, $18{:}1\Delta^9$ and $18{:}1\Delta^{11}$ as the dominant constituents. The ratio of $18{:}1\Delta^{11}/18{:}1\Delta^9$ is used as an indicator of bacterial alteration of organic matter (Sargent *et al.*, 1987) and increased from 0.1–0.3 in plankton and 105-m trap material to 0.5–2.5 in the sediment. Likewise, the sediments were relatively enriched in *iso*- and *anteiso*-branched C_{15} and C_{17} fatty acids, further indicating bacterial alteration (Kaneda, 1991).

The neutral lipids were dominated by sterols, in particular cholest-5-en-3β-ol [27(5) in Fig. 8(b)]. Sterol distributions are typical for mixed plankton

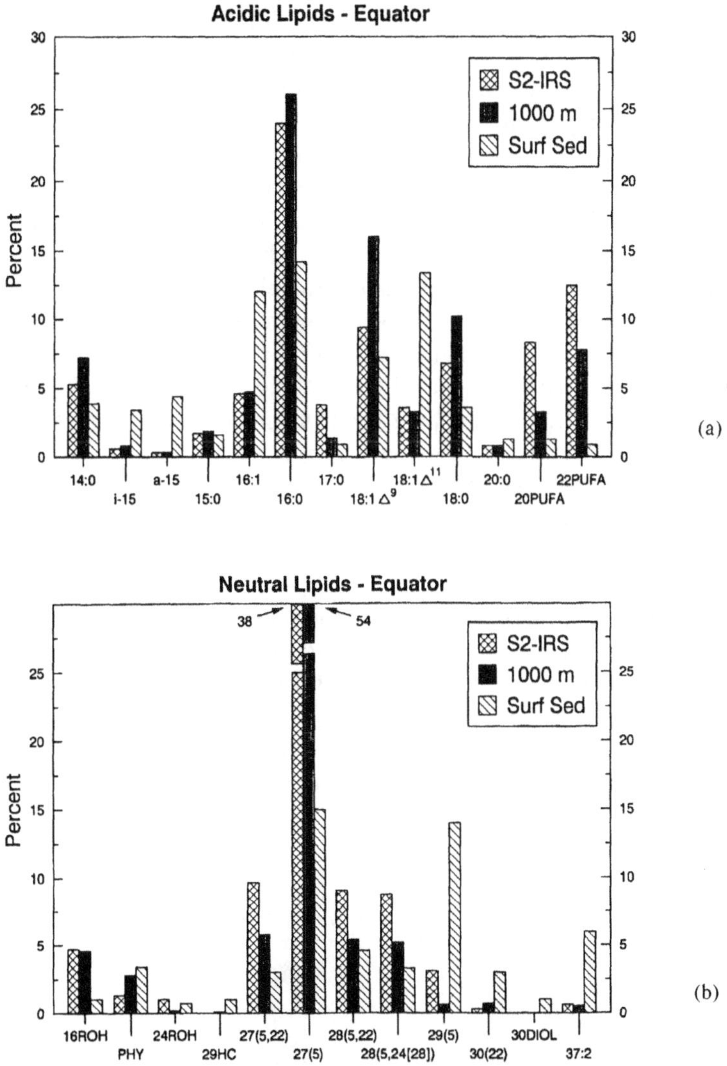

Fig. 8. Major fatty acid (a) and neutral lipid (b) distributions (wt percent of total acids or neutral lipids) in selected samples from the equator station. Fatty acids — X:Y = carbon number: double bond number; *i*- and *a*-: *iso* and *anteiso*; PUFA : polyunsaturated. Neutral lipids — XROH: x-carbon number fatty alcohol; PHY: phytol; 29HC: *n*-nonacosane; 27(5,22):cholesta-5,22E-dien-3β-ol; 27(5): cholest-5-en-3β-ol; 28(5,22): 24-methylcholesta-5,22E-dien-3β-ol; 28(5,24[28]): 24-methylcholesta-5,24(28)-dien-3β-ol; 29(5): 24-ethylcholest-5-en-3β-ol; 30(22): 4α,23,24-trimethyl-5α(H)-cholest-5-en-3β-ol; 30diol: C_{30}-alkan-1,15-diol; 37:2: heptatriacontadien-2-one.

(Volkman, 1986), with the dominance of cholest-5-en-3β-ol most likely derived from both phytoplankton and zooplankton. The fatty alcohol, hexadecanol (16ROH in Fig. 8(b)), and its less abundant homolog, tetracontanol (24ROH), were probably derived from wax esters contributed by zooplankton (Sargent,

1976), while the isoprenoid alcohol, phytol (PHY), originates from chlorophyll-*a* of phytoplankton. Long-chain C_{37} and C_{38} alkenones (e.g. 37:2) that are biosynthesized by haptophytes such as *Emiliania huxleyi* (Volkman *et al.*, 1980) were minor components in samples from the upper water column. Neutral lipid distributions in the moored traps were similar to the distributions in the floating traps, except for slightly increased abundances of cholest-5-en-3β-ol in the moored traps. The pattern of neutral lipids in surface sediments was markedly different than in water column particles. Although sterols as a class remained dominant, distributions of individual sterols changed, with 24-ethylcholest-5-en-3β-ol [29(5)] considerably more abundant in the 0–0.5 cm sediment than in the water column. Likewise, abundances of long-chain alkanes (e.g. 29HC in Fig. 8(b)), 4α,23,24-trimethyl-5α(H)-cholest-22E-en-3β-ol [30(22)], a C_{30}-alkan-1,15-diol (30DIOL), and the long-chain alkenones (e.g. 37:2) were significantly higher in the sediment. Thus, while *n*-alkanes, alkanediols, alkenones and 4α,23,24-trimethyl-5α(H)-cholest-22E-en-3β-ol were only of minor importance in the water column materials, they became selectively enriched in the sediments as more labile components were preferentially degraded.

Pigments

The pigment distribution in plankton samples was, as would be expected for fresh phytoplankton, dominated by chlorophyll-*a* (95% of total pigments) with only minor amounts of the chlorophyll degradation products, phaeophorbide, pyrophaeophorbide, and phaeophytin (Fig. 9). Chlorophyll-*a*/phaeophorbide ratios were in the 20–400 range. Further details on both phytoplankton chlorin and carotenoid pigment distributions during the EqPac program are given by Bidigare *et al.* (1995). Sinking of particles was accompanied by preferential loss

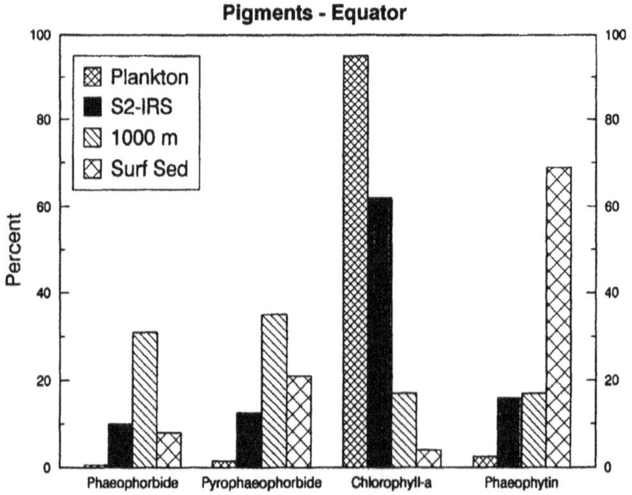

Fig. 9. Pigment distributions (wt percent of total pigments) in selected samples from the equator station.

of Chl-*a*, so that in the floating traps it accounted for ~60% of the pigments while phaeopigments increased in relative abundance. Chl-*a*/phaeophorbide ratios in the floating traps were in the range of 2–20. Degradation of Chl-*a* continued down the water column, yielding a predominance of phaeophorbide and pyrophaeophorbide and Chl-*a*/phaeophorbide ratios of about 0.5 in the moored traps at both depths. Chlorophyll-*a* was the only pigment detected in the sediments, and there only at the equator. Presumably, only at the equator was the flux of pigments through the water column high enough to allow a small fraction of the Chl-*a* to survive degradation and mixing and become incorporated into the surface sediment. In the surface sediment at the equator, the chlorophyll-*a*/phaeophorbide ratio was 0.5. Phaeophytin was usually the dominant pigment in the surface sediments.

Inventory of major biochemicals

Our data clearly indicate that in the equatorial Pacific the fluxes and compositions of biochemical classes, and hence the behavior of organic matter in general, are markedly more dynamic than would be perceived from measurements of bulk organic carbon alone. A major goal of our multi-class, molecular analysis scheme was to obtain a mass balance between that fraction of organic carbon which could be characterized at the molecular level and the remaining organic material which we did not characterize. Results of this inventory are illustrated in Fig. 10, summarizing mean percentages of C_{org} for the entire data set.

Amino acids, carbohydrates and lipids cumulatively account for ~84% of the organic carbon in the plankton samples, with amino acids clearly dominating at ~67% of C_{org}, carbohydrates at ~4% and lipids at ~15%. Pigments, while important constituents of phytoplankton in terms of their biochemical function and certainly useful for taxonomic studies, account for only 0.2% or less of the planktonic organic carbon. Material uncharacterized at the molecular level comprised an average of 16% of the planktonic C_{org} and likely includes other major biochemical classes such as nucleic acids, uronic acids, and amino sugars. In the 105-m sediment traps, the major biochemical classes account for ~80% of C_{org} and the proportions of biochemical classes comprising C_{org} are similar to those in the plankton. About 18% of C_{org} was uncharacterized in the floating traps.

It is between the floating traps and the moored traps that the amount of characterizable material decreases dramatically versus uncharacterizable C_{org}. Amino acids and lipids account for most of the loss, with amino acids decreasing to about 24% of C_{org} and lipids decreasing to about 3% of C_{org} in both shallow and deep moored traps. Carbohydrates were generally ~5% of the C_{org}, not too dissimilar from their percentage in both plankton and floating traps. The result of this significant but preferential loss of amino acids and lipids is that only ~30% of the C_{org} in the moored traps, and hence the sinking particulate matter in the interior of the equatorial Pacific, can be accounted for by our molecular level biochemical analyses. Even less of the organic carbon in the sediments, about

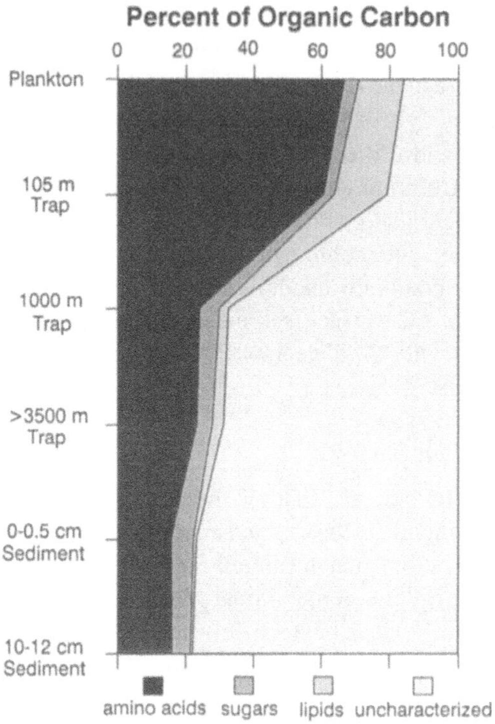

Fig. 10. Cumulative biochemical class distributions (compound class-C_{org} as percent of total C_{org}) in the equatorial Pacific. Compound class-carbon contents were calculated directly for amino acids and carbohydrates, while for lipids they were estimated assuming lipids to be 85% C_{org}. Pigments are not included as their contribution is negligible. The residual, uncharacterized material was obtained as the difference between total C_{org} and the sum of (amino acid-C + carbohydrate-C + lipid-C).

20%, can be identified, with the contributions from amino acids and lipids decreasing yet again relative to the deep water column trap material, whereas the carbohydrate-C actually increasing slightly.

A comparison of vertical trends shown in Fig. 5 with trends in Fig. 10 suggests an interesting paradox that may provide key new information on the fate of particulate organic matter in the equatorial Pacific. Figure 5 indicates a large drop in flux occurring between net-plankton production in surface waters and the sediment traps deployed at 105 m, while fluxes between 105 m and 1000 m were less strongly attenuated. This decrease in flux to 105-m traps could be interpreted either in terms of degradation of organic matter between its production by plankton and its collection at 105-m or as loss due to lateral advection away from the 105-m traps. Yet Fig. 10 shows the relative proportions of the organic matter identified as the major biochemical classes in plankton and 105-m traps to be remarkably similar. It is between the 105-m traps and the 1000-m traps that significant compositional changes occur, but the corresponding drop in flux is less severe than in the epipelagic region. These results would suggest that if significant degradation does occur in the upper 105 m of the equatorial Pacific, then that degradation is nonselective with respect to biochemical classes. Perhaps a more plausible scenario is that advective processes remove most the plankton-derived particulate organic matter before it can sink into the 105-m traps, reducing the flux, but that material which is collected at 105 m has yet to be greatly degraded and thus is compositionally similar to its planktonic source. Assuming that the material in the 1000-m traps is derived from overlying surface waters, and we cannot rule out that advection brought it from elsewhere, then it is between 105 m and 1000 m that degradation alters the relative proportions of identifiable biochemicals comprising C_{org} (Fig. 10). Midwater degradation thus leads to the observed decrease in flux. Clearly these issues raise questions that can be answered only through further research.

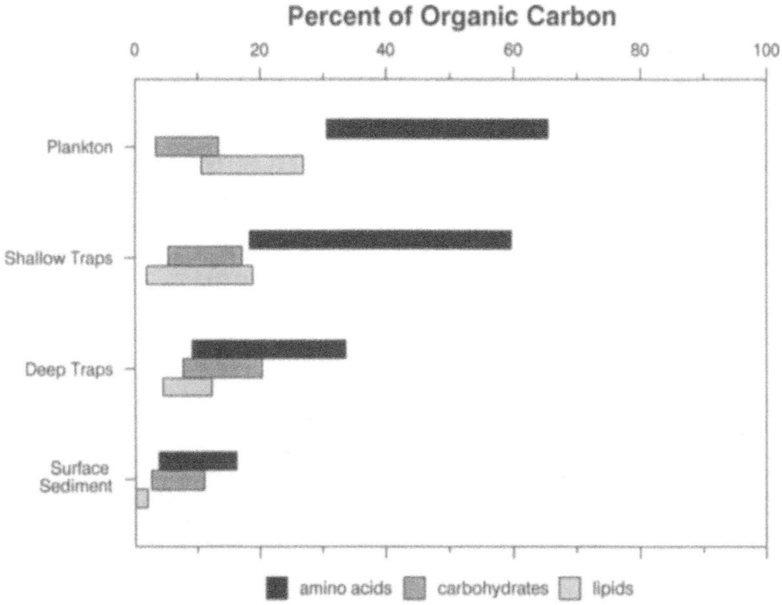

Fig. 11. Ranges of amino acid, carbohydrate and lipid contributions to C_{org} in plankton, shallow sediment traps deployed near the base of the euphotic zone, deep traps in the mesopelagic and bathypelagic zones, and deep-sea surface sediments. References — plankton: Degens and Mopper, 1976; Smetacek and Hendrikson, 1979; Tanoue *et al.*, 1982; Wefer *et al.*, 1982; Cowie and Hedges, 1984b, 1996; Parsons *et al.*, 1984; Montani and Okaichi, 1985; Sargent and Henderson, 1986; Tanoue and Handa, 1987; Cowie and Hedges, 1992; shallow traps: Wakeham *et al.*, 1993, 1984a, b; Lee and Cronin, 1982, 1984; Hedges *et al.*, 1988; Cowie and Hedges, 1992; deep traps: Lee and Cronin, 1984; Wefer *et al.*, 1982; Ittekkot *et al.*, 1984a, b; Wakeham *et al.*, 1984b; Müller *et al.*, 1986; Wakeham and Lee, 1989; Haake *et al.*, 1992; sediments: Whelan, 1977; Farrington and Tripp, 1977; Tanoue and Handa, 1987; Venkatesan *et al.*, 1987; Steinberg *et al.*, 1987.

This study is, to the best of our knowledge, the first to systematically determine the fraction of organic carbon characterized at the molecular level as major biochemical classes in particulate matter from the sea surface to the sediment. There are, however, data available from different stages through the water column and sediment which may be compared with our EqPac results, although these literature data come from a wide range of oceanic regimes and were obtained by a variety of analytical techniques. Figure 11 summarizes typical ranges of amino acid, carbohydrate, and lipid contributions to total C_{org} that have been reported in plankton, sediment traps near the base of the euphotic zone, traps in the meso- and bathy-pelagic zones, and abyssal surface sediments. The generalized organic carbon balance we derived for the equatorial Pacific is consistent with these literature ranges. It is possible to assign an order of relative reactivity for the biochemical classes during early diagenesis in the water column and surface sediment: pigments >> lipids > amino acids > carbohydrates. In part, the relative reactivity of carbohydrates may be due to their function as structural components, for while the storage carbohydrates (e.g. glucose and ribose) are preferentially lost as particles sink, other components (rhamnose, fucose, xylose, and mannose) tend to survive either because of their structural function or because they are resynthesized by bacteria. Both our EqPac data and the literature compilation clearly show the increasing abundance of the fraction of C_{org} which is uncharacterized at the molecular level, rising from ~10–20% in plankton and shallow traps to ~80% in surface sediments. Much of this uncharacterized material is undoubtedly macromolecular material that is intractable with respect to our molecular analyses and may include material that is crosslinked during early diagenesis in the water column or that represents resistant biopolymers that survive diagenesis (Tegelaar *et al.*, 1989). However, this does not imply that this material cannot be utilized by mesopelagic, bathypelagic or benthic organisms, although clearly the most "nutritious" material has been consumed long before the resistant material is buried in the sediments.

CONCLUSIONS

A systematic survey of organic carbon and major biochemicals in the equatorial Pacific Ocean reveals a wide range of export fluxes, fluxes through the oceans interior, and burial fluxes in the sediments. Overall, the magnitude of fluxes through the water column are closely coupled with spatial variations in primary production in surface waters. Two sites of enhanced organic matter remineralization are indentified: the epipelagic zone and the benthic boundary layer. Molecular level analyses show that as organic material sinks through the ocean and is buried in the sediments, the proportion of C_{org} which can be characterized decreases dramatically. Only about 20% of C_{org} in sediments can be identified, demonstrating clearly that there is a great need for improving analytical capablities for the study of this uncharacterized, macromolecular material that plays such a key role in the marine carbon cycle.

Acknowledgements—We thank our many EqPac colleagues for their assistance at many stages of this work. M. L. Peterson, P. J. Hernes, S. Pantoja and B. Bergamaschi helped with sampling and analysis. This research was supported by National Science Foundation collaborative research grants OCE90-22238, OCE-9022319 and OCE-9024372.

REFERENCES

Bacon, M. P., J. K. Cochran, D. Hirschberg, T. R. Hamar and A. P. Fleer (1996): Export flux of carbon at the equator during the EqPac time-series cruises estimated from ^{234}Th measurements. *Deep-Sea Res. II*, **43**, 1133–1153.

Barber, R. T., M. P. Sanderson, S. T. Lindley, F. Chai, J. Newton, C. C. Trees, D. G. Foley and F. P. Chavez (1996): Primary productivity and its regulation in the equatorial Pacific during and following the 1991–92 El Niño. *Deep-Sea Res. II*, **43**, 933–969.

Barnett, R. P. O., J. Watson and D. Connelly (1984): A multiple corer for taking virtually undisturbed samples from shelf, bathyal, and abyssal sediments. *Oceanol. Acta*, **7**, 399–408.

Bidigare, R. R. and M. E. Ondrusek (1996): Influence of the 1992 El Niño on phytoplankton pigment distributions in the equatorial Pacific Ocean. *Deep-Sea Res. II*, **43**, 809–833.

Buesseler, K. D., J. A. Andrews, M. C. Hartman, R. Belastock and F. Chai (1995): Regional estimates of the export flux of particulate organic carbon derived from thorium-234 during the JGOFS EqPac program. *Deep-Sea Res. II*, **42**, 777–804.

Burdige, D. J. and C. S. Martens (1988): Biogeochemical cycling in an organic-rich coastal marine basin: 10. The role of amino acids in sedimentary carbon and nitrogen cycling. *Geochim. Cosmochim. Acta*, **52**, 1571–1584.

Chavez, F. P. and R. T. Barber (1987): An estimate of new production in the equatorial Pacific. *Deep-Sea Res.*, **34**, 1229–1243.

Chavez, F. P., K. R. Buck, S. K. Service, J. Newton and R. T. Barber (1996): Phytoplankton variability in the central and eastern tropical Pacific. *Deep-Sea Res. II*, **43**, 835–870.

Cowie, G. L. and J. I. Hedges (1984a): Determination of neutral sugars by capillary gas chromatography of equilibrated isomeric mixtures. *Anal. Chem.*, **56**, 497–504.

Cowie, G. L. and J. I. Hedges (1984b): Carbohydrate sources in a coastal marine environment. *Geochim. Cosmochim. Acta*, **48**, 2075–2087.

Cowie, G. L. and J. I. Hedges (1991): Organic carbon and nitrogen geochemistry of Black Sea surface sediments from stations spanning the oxic:anoxic boundary. pp. 343–349. In: *Black Sea Oceanography*, ed. by E. Izdar and J. W. Murray, Kluwer Academic Publishers.

Cowie, G. L. and J. I. Hedges (1992a): Sources and reactivities of amino acids in a coastal marine environment. *Limnol. Oceanogr.*, **37**, 703–724.

Cowie, G. L. and J. I. Hedges (1992b): Biochemical indicators of diagenetic alteration in natural organic mixtures. *Nature*, **369**, 304–307.

Cowie, G. L. and J. I. Hedges (1996): Degestion and alteration of the biochemical constituents of a diatom (*Thalassiosira weissflogii*) ingested by a herbivorous zooplankton (*Calanus pacificus*). *Limnol. Oceanogr.*, **41**, 581–594.

Cowie, G. L., J. I. Hedges, F. G. Prahl and G. J. De Lange (1995): Elemental and major biochemical changes across an oxidation front in a relict turbidite: An oxygen effect. *Geochim. Cosmochim. Acta*, **59**, 33–46.

Degens, E. T. (1970): Molecular nature of nitrogenous compounds in sea water and recent marine sediments. pp. 77–106. In: *Organic Matter in Natural Waters*, ed. by D. W. Hood, Univ. Alaska Pub.

Degens, E. T. and K. Mopper (1976): Factors controlling the distribution and diagenesis of organic materials in marine sediments. pp. 59–113. In: *Chemical Oceanography*, ed. by J. P. Riley and R. Chester, Academic Press.

DeMaster, D. J., R. H. Pope, O. Rageuneau and C. R. Smith (1997): Burial rates of biogenic material along the EqPac transect: Holocene variability and paleoflux indicators. *Deep-Sea Res. II* (in prep.).

Eppley, R. W. and B. J. Peterson (1979): Particulate organic matter flux and planktonic new production in the deep ocean. *Nature*, **282**, 677–680.

Farrington, J. W. and B. W. Tripp (1977): Hydrocarbons in western North Atlantic sediments. *Geochim. Cosmochim. Acta*, **41**, 1627–1641.

Haake, B., V. Ittekkot, V. Ramaswamy, R. R. Nair and S. Honjo (1992): Fluxes of amino acids and hexosamines to the deep Arabian Sea. *Mar. Chem.*, **40**, 291–314.

Handa, N. and H. Tominaga (1969): A detailed analysis of carbohydrates in marine particulate matter. *Mar. Biol.*, **2**, 228–235.

Hecky, R. E., K. Mopper, P. Kilham and E. T. Degens (1973): The amino acid and sugar composition of diatom cell walls. *Mar. Biol.*, **19**, 323–331.

Hedges, J. I. and J. H. Stern (1984): Carbon and nitrogen determinations of carbonate containing solids. *Limnol. Oceanogr.*, **29**, 657–663.

Hedges, J. I., W. A. Clark and G. L. Cowie (1988): Fluxes and reactivities of organic matter in a coastal marine bay. *Limnol. Oceanogr.*, **33**, 1137–1152.

Hedges, J. I., G. L. Cowie, J. E. Richey, P. D. Quay, R. Benner, M. Strom and B. R. Forsberg (1994): Origins and processing of organic matter in the Amazon River as indicated by carbohydrates and amino acids. *Limnol. Oceanogr.*, **39**, 743–761.

Hernes, P. J., J. I. Hedges, M. L. Peterson, S. G. Wakeham and C. Lee. (1996): Neutral carbohydrate geochemistry of particulate matter in the central Equatorial Pacific. *Deep-Sea Res. II*, **43**, 1181–1204.

Honjo, S., J. Dymond, R. Collier and S. J. Manganini (1995): Export production of particles to the interior of the equatorial Pacific Ocean during the 1992 EqPac experiment. *Deep-Sea Res. II*, **42**, 831–870.

Ittekkot, V., W. G. Deuser and E. T. Degens (1984a): Seasonality in the fluxes of sugars, amino acids, and amino sugars to the deep ocean: Sargasso Sea. *Deep-Sea Res.*, **31**, 1057–1069.

Ittekkot, V., W. G. Deuser and E. T. Degens (1984b): Seasonality in the fluxes of sugars, amino acids, and amino sugars to the deep ocean: Panama Basin. *Deep-Sea Res.*, **31**, 1071–1083.

Kaneda, T. (1991): Iso- and anteiso-fatty acids in bacteria: biosynthesis, function, and taxonomic significance. *Microbiol. Rev.*, **55**, 288–302.

Kessler, W. S. and M. J. McPhaden (1995): The 1991–1993 El Niño in the central Pacific. *Deep-Sea Res. I*, **42**, 295–333.

Lee, C. and C. Cronin (1982): The vertical flux of particulate organic nitrogen in the sea: decomposition of amino acids in the Peru upwelling area and the equatorial Atlantic. *J. Mar. Res.*, **40**, 227–251.

Lee, C. and C. Cronin (1984): Particulate amino acids in the sea: effects of primary productivity and biological decomposition. *J. Mar. Res.*, **42**, 1075–1097.

Lee, C., J. I. Hedges, S. G. Wakeham, M. L. Peterson and P. J. Hernes (submitted): Amino acids and pigments in the equatorial Pacific Ocean.

Lindroth, P. and K. Mopper (1979): High performance liquid chromatographic determination of subpicomole amounts of amino acids by precolumn fluorescence derivatization with *o*-phthaldialdehyde. *Anal. Chem.*, **51**, 1667–1674.

Longhurst, A. R. and W. G. Harrison (1989): The biological pump: Profiles of plankton production and consumption in the upper ocean. *Prog. Oceanogr.*, **22**, 47–123.

Luo, S., T.-L. Ku, M. Kusakabe, J. K. B. Bishop and Y.-L. Yang (1995): Tracing particle cycling in the upper ocean with ^{230}Th and ^{228}Th: An investigation in the equatorial Pacific along 140°W. *Deep-Sea Res. II*, **42**, 805–829.

Mantoura, R. F. C. and C. A. Llewellyn (1983): The rapid determination of algal chlorophyll and carotenoid pigments and their breakdown products in natural waters by reverse-phase high-performance liquid chromatography. *Analyt. Chim. Acta*, **151**, 297–314.

McCarthy, J. J., C. Garside, J. L. Nevins and R. T. Barber (1996): New production along 140°W in the equatorial Pacific during and following the 1992 El Niño event. *Deep-Sea Res. II*, **43**, 1065–1093.

Montani, S. and T. Okaichi (1985): Amino acid variations in marine particles during sinking and sedimentation in Harima-Nada, the Seto Inland Sea. pp. 15–27. In: *Marine and Estuarine*

Geochemistry, ed. by A. C. Sigleo and A. Hattori, Lewis Publishers.

Müller, P. J., E. Suess and C. A. Ungerer (1986): Amino acids and amino sugars of surface particulate and sediment trap material from waters of the Scotia Sea. *Deep-Sea Res.*, **33**, 819–838.

Murray, J. W., M. W. Leinen, R. A. Feely, J. R. Toggweiler and R. Wanninkof (1992): EqPac: a process study in the central equatorial Pacific. *Oceanography*, **5**, 134–142.

Murray, J. W., K/ Young, J. Newton, J. Dunne, T. Chapin and B. Paul (1996): Export flux of particulate organic carbon from the central equatorial Pacific Ocean using a combined drifting trap ^{234}Th approach. *Deep-Sea Res. II*, **43**, 1095–1132.

Parsons, T. R., M. Takahashi and B. Hargrave (1984): *Biological Oceanographic Processes*. Pergamon Press, 330 pp.

Peterson, M. L., P. J. Hernes, D. S. Thoreson, J. I. Hedges, C. Lee and S. G. Wakeham (1993): Field evaluation of a valved sediment trap. *Limnol. Oceanogr.*, **38**, 1741–1761.

Sargent, J. R. (1976): The structure, function, and metabolism of lipids in marine organisms. pp. 149–212. In: *Biochemical and Biophysical Perspectives in Marine Biology, Vol. 3*, ed. by D. C. Malins and J. R. Sargent, Academic Press, New York.

Sargent, J. R. and R. J. Henderson (1986): Lipids. pp. 59–108. In: *The Biological Chemistry of Marine Copepods*, ed. by E. D. S. Corner and S. C. M. O'hara, Clarendon Press, Oxford.

Sargent, J. R., R. J. Parkes, I. Mueller-Harvey and R. J. Henderson (1987): Lipids in marine ecology. pp. 119–138. In: *Microbes in the Sea*, ed. by M. A. Sleigh, John Wiley and Sons, New York.

Siezen, R. J. and T. H. Mague (1978): Amino acids in suspended particulate matter from oceanic and coastal waters of the Pacific. *Mar. Chem.*, **6**, 215–231.

Smetacek, V. and P. Hendrikson (1979): Composition of particulate organic matter in Kiel Bight in relation to phytoplankton succession. *Oceanol. Acta.*, **2**, 287–298.

Steinberg, S. M., M. I. Venkatesan and I. R. Kaplan (1987): Organic geochemistry of sediments from the continental margin off southern New England, U.S.A.—Part I. Amino acids, carbohydrates and lignin. *Mar. Chem.*, **21**, 249–265.

Tanoue, E. and N. Handa (1987): Monosaccharide composition of marine particles and sediments from the Bering Sea and northern North Pacific. *Oceanol. Acta*, **10**, 91–99.

Tanoue, E., N. Handa and H. Sakugawa (1982): Difference in chemical composition of organic matter between fecal pellet of *Euphausia superba* and its feed, *Duniella tertiolecta. Trans. Tokyo Univ. Fish.*, **5**, 189–196.

Tegelaar, E. W., J. W. de Leeuw, S. Derenne and C. Largeau (1989): A reappraisal of kerogen formation. *Geochim. Cosmochim. Acta*, **53**, 3103–3106.

Venkatesan, M. I., E. Ruth, S. Steinberg and I. R. Kaplan (1987): Organic geochemistry of sediments from the continental margin off southern New England, U.S.A.—Part II. Lipids. *Mar. Chem.*, **21**, 267–299.

Verity, P. G., D. K. Stoecker, M. E. Sieracki and J. R. Nelson (1996): Microzooplankton grazing of primary production at 140°W in the equatorial Pacific. *Deep-Sea Res. II*, **43**, 1227–1255.

Volkman, J. K. (1986): A review of sterol markers for marine and terrigenous organic matter. *Org. Geochem.*, **9**, 83–99.

Volkman, J. K., G. Eglinton, E. D. S. Corner and T. E. V. Forsberg (1980a): Long-chain alkenes and alkenones in the marine coccolithophoprid *Emiliania huxleyi. Phytochem.*, **19**, 2619–2622.

Wakeham, S. G. and C. Lee (1989): Organic geochemistry of particulate organic matter in the ocean: The role of particles in oceanic sedimentary cycles. *Org. Geochem.*, **14**, 83–96.

Wakeham, S. G., J. W. Farrington and R. B. Gagosian (1984a): Variability in lipid flux and composition of particulate matter in the Peru upwelling region. *Org. Geochem.*, **6**, 203–215.

Wakeham, S. G., C. Lee, J. W. Farrington and R. B. Gagosian (1984b): Biogeochemistry of particulate organic matter in the oceans: results from sediment trap experiments. *Deep-Sea Res.*, **31**, 509–528.

Wakeham, S. G., R. B. Gagosian, J. W. Farrington and C. Lee (1984c): Biogeochemistry of particulate organic matter in the oceans—results from sediment trap experiments. *Deep-Sea Res.*, **31**, 509–528.

Wakeham, S. G., J. I. Hedges, C. Lee and T. K. Pease (1993): Effects of poisons and preservatives on the composition of organic matter in a sediment trap experiment. *J. Mar. Res.*, **51**, 669–696.

Wakeham, S. G., J. I. Hedges, C. Lee, M. L. Peterson and P. J. Hernes (1997): Compositions and fluxes of lipid biomarkers through the water column and surficial sediments of the equatorial Pacific Ocean. *Deep-Sea Res. II*, **44**, 2131–2162.

Wefer, G., E. Suess, W. Balzer, G. Leibezeit, P. J. Müller, C. A. Ungerer and W. Zenk (1982): Fluxes of biogenic components from sediment trap deployments in circumpolar waters of the Drake Passage. *Nature*, **299**, 145–147.

Whelan, J. K. (1977): Amino acids in a surface sediment core of the Atlantic abyssal plain. *Geochim. Cosmochim. Acta*, **41**, 803–810.

White, J. R., X. Zhang, L. A. Welling, M. R. Roman and H. G. Dam (1995): Latitudinal gradients in zooplankton biomass in the tropical Pacific at 140°W during the JGOFS EqPac study: Effects of El Niño. *Deep-Sea Res. II*, **43**, 715–733.

Zhang, X., H. G. Dam, J. R. White and M. R. Roman (1995): Latitudinal gradients in mesozooplankton grazing and metabolism in the central equatorial Pacific during the U.S. JGOFS EqPac study. *Deep-Sea Res. II*, **42**, 695–714.

Dynamics and Characterization of Marine Organic Matter, Eds., N. Handa, E. Tanoue and T. Hama, pp. 141–168.

Source of Organic Matter in Sinking Particles in the Japan Trench: Molecular Composition and Carbon Isotopic Analyses

R. ISHIWATARI[1], K. YAMADA[1], K. MATSUMOTO[1], H. NARAOKA[1], S. YAMAMOTO[2] and N. HANDA[3]

[1]*Department of Chemistry, Faculty of Science, Tokyo Metropolitan University, Minami Ohsawa, Hachioji, Tokyo 192-03, Japan*
[2]*Faculty of Education, Soka University, Tangicho, Hachioji, Tokyo 192, Japan*
[3]*Research Institute for Atmospheric and Hydrospheric Science, Nagoya University, Chikusa, Nagoya 464, Japan*

(Synthesis, Received 12 August 1996; accepted 14 November 1996)

Abstract—Molecular and stable carbon isotopic compositions were obtained for the organic matter (OM) in sediment trap samples from the Japan Trench at 4500 and 8500–8800 m in depth. The analytical results show that OM in sinking particles are composed of both freshly-deposited and resuspended OM. The relative abundance of resuspended OM in the total OM in shallower sediment trap samples is generally lower than that in deeper trap samples. However, no marked differences both in py-GCMS characteristics and stable carbon isotopic compositions of TOC and biomarkers (n-alkanes, fatty acids) are recognized. All sinking particles examined indicate that the major source of OM is plankton-derived. Terrestrial OM contribution is minor.

Molecular and carbon isotopic analyses of individual lipid compounds in sinking particles revealed that they are a mixture of compounds derived from algae, bacteria and terrestrial higher plants. The analytical data show the usefulness of the biomarker and carbon isotopic analytical approach to draw a picture of biogeochemical processes of the carbon cycle in the ocean.

INTRODUCTION

Sinking of large particles in the ocean is an important process for the transportation of organic carbon from the surface waters to the sea-floor (e.g. Lee and Wakeham, 1992; Wakeham and Lee, 1993). Therefore, the assessment of sources of organic matter (OM) in sinking particles is necessary for a better understanding of the carbon cycle in the ocean. In this respect, sediment trap experiments have been conducted extensively to elucidate organic molecular composition of sinking particles in the ocean (e.g. Wakeham *et al.*, 1980; Wakeham, 1982, 1985; Ittekkot *et al.*, 1984a, b). These studies have revealed that lipid compounds in sinking particles in the ocean are predominantly of planktonic origin, with a lesser contribution of those from microbial and terrestrial sources. These studies have also indicated that the relative lipid composition changes dramatically during deposition through the water column (Wakeham *et al.*, 1980). Moreover, lipid fluxes fluctuate strongly from season to season depending on the sources (Najdek, 1993).

Mid- and long-chain normal hydrocarbons have been detected in sinking particles in the ocean (Wakeham *et al.*, 1980). Wakeham *et al.* (1980) believe that n-C_{17} and n-C_{19} alkanes are of planktonic origin, whereas C_{25}–C_{31} alkanes with odd carbon number predominance are of land-derived. This idea is commonly accepted in the scientific community. A similar idea was stated by Simoneit *et al.* (1986) in the study of upward and downward flux of particulate OM in the Pacific Ocean, where they found n-C_{16}–C_{35} alkanes in sediment trap samples. Matsueda and Handa (1986) observed n-C_{17} alkane as a major component with lesser amounts of long-chain n-alkanes (C_{21}–C_{32}) in sinking particles collected in the eastern North Pacific Ocean. These long-chain n-alkanes, however, did not show an odd-even carbon number predominance, unlike to those reported by Simoneit *et al.* (1986). Matsueda and Handa (1986) claimed that the likely source of the long-chain n-alkanes (C_{21}–C_{32}) is bacteria grown on the sinking particles. They also detected unsaturated hydrocarbons (n-$C_{21:6}$ and branched $C_{25:2}$, $C_{25:3}$, $C_{25:4}$), which were thought to be derived from phytoplankton. These hydrocarbons may be highly branched isoprenoid alkenes and they are believed to be mainly derived from diatoms (Volkman *et al.*, 1994). The works cited above indicate the importance of molecular approach in assessment of the source of organic compounds in sinking particles.

The compound specific isotope analysis (CSIA) using GC-IRMS (Gas chromatography-isotope ratio monitoring mass spectrometry) is another important approach to assessing the source of organic compounds in sinking particles in the ocean. This technique has been recently extensively developed and successfully applied to various environmental and geochemical samples (e.g. Hayes *et al.*, 1990; Ishiwatari *et al.*, 1994; Schoell and Hayes, 1994). The principle of this method is completely different from that of the molecular approach. It provides an important criteria for deciphering the source of individual organic compounds in sinking particles on the basis of stable carbon isotopes.

This paper is not intended to review the whole aspect of organic geochemical studies of sinking particles in the ocean, but intended to describe molecular characteristics and stable carbon isotopic composition of organic compounds in sinking particles obtained from the Japan Trench (site JT), where sediment-trap experiments have been conducted by a group of Japanese scientists since 1986. This paper includes published work (Ishiwatari *et al.*, 1995) and carbon isotopic data which will be published elsewhere as an original work.

STUDY AREA AND MATERIALS

The first sinking particle samples used for our study (JT-01) were collected at site JT (maximum water depth 9,200 m; 34°10.4′ N, 141°58.9′ E) in the Japan Trench (Fig. 1). In this case, the sampling device was deployed at the depth of 8,798 m from Aug. 30, 1986 to May 4, 1987, and totally 13 sinking particle samples (JT-06) were taken in this period. Another set of sinking particle samples

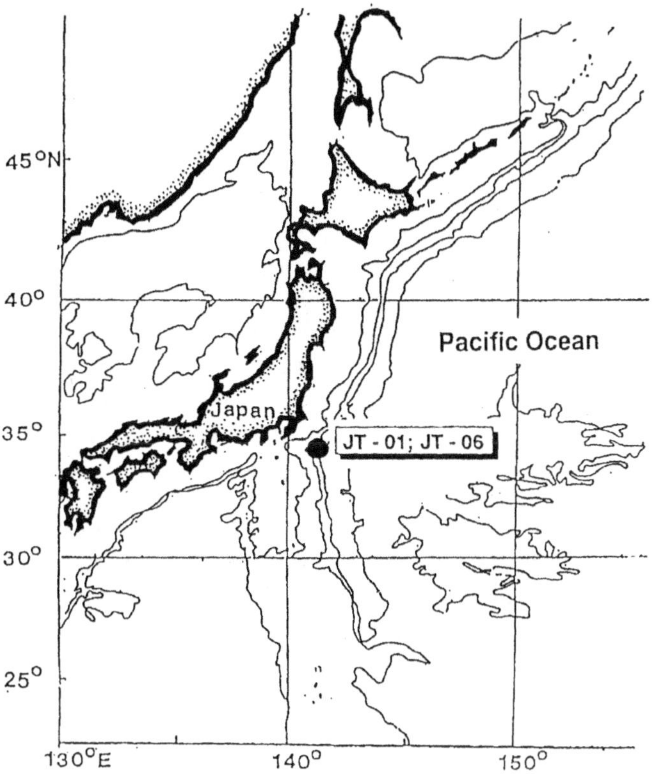

Fig. 1. Location of the sediment trap-experiments.

was collected at site JT (maximum water depth 8,930 m; 34°10.4′ N, 142°00.4′ E) in the Japan Trench. The sampling device was deployed at water depths of 4,519 m and 8,519 m from in March 1990 to Feb. 1991, and in total, 13 sinking particle samples were collected in this period.

Portions of these samples were subjected to pyrolysis-GCMS analysis. In the case of molecular and isotopic analyses, several selected samples were combined because of sample limitation, and subjected to analyses. The following two sets of sinking particle sample were prepared: #4001, #9001, #4002 and #9002. Sample #4001 is the sinking particle sample collected from 4,519 m in water depth during the period of March 25 to June 9, 1991 (Cup Nos. 1–3), while #9001 is that collected from 8,519 m in water depth during the same period. #4002 is the sample collected from 4,519 m in water depth during the period of December 31, 1991 to January 25, 1992 (Cup Nos. 12 and 13) and #9002 is that from 8,519 m in water depth in the same period. The sinking particle samples were pulverized after freeze-drying and subjected to molecular composition and carbon isotopic analyses. At the procedure of freeze-drying, a trap containing liquid N_2 was used to prevent any possible contamination of traces of pump oil.

PROCEDURES FOR MOLECULAR COMPOSITION AND CARBON ISOTOPIC ANALYSES

Pyrolysis-gas chromatography-mass spectrometry (py-GCMS)

The py-GCMS was conducted by using a CDS Pyroprobe Model 100 connected to a Varian Model 3400 GC-Finnigan Model INCOS 50 quadrupole mass spectrometer. The gas chromatograph was equipped with a fused silica capillary column whose wall was coated with DB-5 30 m × 0.25 or 0.32 mm i.d. In a typical operation, 2–3 mg of powdered sample was loaded in a small quartz tube and pyrolyzed at 750°C (set temperature) for 20 sec. The actual pyrolysis temperature may be lower by *ca.* 200°C than the set temperature. The interface was heated at 250°C or 300°C. The GC column oven temperature was first maintained at 40°C for 5 min. and programmed from 40°C to 300°C at 5°C min^{-1}. Mass spectrometric data were acquired in the electron impact mode (70 eV), with a range of scanning from 50–650 mass units at 1.5 sec. per decade.

Identification of the pyrolysates was based on a comparison of their mass spectra with those of standard compounds or literature data, or on mass spectral interpretations. The pyrolysis products were quantified from a calibration curve established between the area of a characteristic mass fragment of a compound and the area of m/z 71 of normal octadecane used as an external standard.

Biomarker analyses

Hydrocarbons

A freeze-dried sample was extracted with n-hexane by ultrasonication. The neutrals were fractionated into aliphatic hydrocarbons and other fractions by silica gel column chromatography (silica gel: Mallinckrodt Inc. 100 mesh; column size: 50 mm × 5 mm i.d.). The aliphatic hydrocarbons including normal alkanes and hopanoid hydrocarbons were obtained as the first n-hexane eluate (2 ml) and analyzed by gas chromatography (GC) or gas chromatography-mass spectrometry (GCMS).

Normal alkanes were isolated from the total alkane fraction by the procedure reported by Yamada *et al.* (1994). Briefly, the hexane fraction was contacted with molecular sieves 5A (Wako Chemicals Co., Ltd.) in isooctane solvent at room temperatures for 3 hr. Normal alkanes were recovered from the molecular sieves by dissolution with HF solution followed by extraction with n-hexane of n-alkanes from the resultant HF solution.

GC analysis of n-alkanes was conducted with a HP 5890 instrument equipped with a flame ionization detector (FID) and a splitless/split injector at a splitless mode. A DB-5 fused silica capillary column (30 m × 0.32 mm i.d.) was used. Helium was used as the carrier gas. The column oven temperature was programmed from 60 to 120°C at 30°C min^{-1} and then to 310°C at 5°C min^{-1}. Injector temperature was maintained at 300°C. Quantitation was achieved by integration of FID peak areas of n-alkanes. 1-Docosene was used as a reference (injection) standard for quantitation and the response factor of individual n-alkanes relative to the standard was assumed to be 1.0. Analytical error for n-alkanes is within

10%.

GCMS analysis was carried out using a Finnigan INCOS 50 mass spectrometer/Varian 3400 GC equipped with a 30 m × 0.25 mm i.d. DB-5 fused silica capillary column. Helium was used as the carrier gas. The other GC conditions were almost the same as those for GC analysis. Hopanoid hydrocarbons were identified by comparison of mass spectra and relative GC retention times with those of standards and those reported in the literature. Quantitation of hopanoid hydrocarbons was done by comparing the peak area of 17β(H), 21β(H) hopane (C_{30}) which had been quantified, with peak areas of each hopanoid on a mass chromatogram scanned at m/z 191, which is a fragment ion characteristic to hopanoid structures.

Sterols

An air-dried sample remaining after hydrocarbon extraction was extracted with benzene/methanol (6:4) mixture by ultrasonication. The extracts were saponified with 0.5N NaOH/methanol at 95°C for 2 hr. After saponification, the neutrals were extracted with hexane/ethyl ether (9:1) mixture. The sterol fraction was obtained as a benzene/methanol (4:1) fraction by silica gel column chromatography (silica gel: Mallinckrodt, Inc. 100 mesh; column: 50 mm × 5 mm i.d.). The sterol fraction was analyzed as TMS-esters by GCMS and as acetates by GC-IRMS.

Stable carbon isotopic analysis

Stable carbon isotopes of individual organic compounds in sinking particle samples were determined using a Finnigan MAT GC-IRMS system. The system consists of a HP Model 5890 GC connected to a Finnigan MAT delta-S mass spectrometer via a combustion interface. In a typical analysis, 40–80 ng of individual organic compounds in 1 μl hexane was injected into a DB-5 chemically bonded fused silica capillary column (60 m × 0.32 mm i.d.) with on-column mode. Helium was used as the carrier gas. The GC oven temperature was programmed from 50 to 150°C at 30°C min^{-1}, from 120 to 310°C at 5°C min^{-1}, and then held at 310°C for 25 min. The combustion furnace was maintained at 940°C. Deuterated n-alkane ($C_{24}D_{50}$), whose $\delta^{13}C$ value had been independently determined by a conventional method, was employed as the isotopic standard. $\delta^{13}C$ values are expressed in units of $\delta^{13}C$ relative to the PDB standard. The precision for the GC-IRMS measurements is within 0.4‰.

GENERAL FEATURES OF ORGANIC MATTER IN SINKING PARTICLES IN THE JAPAN TRENCH

C/N ratio and carbon isotopic characteristics

Table 1 gives total organic carbon, total nitrogen and C/N weight ratios in the JT-01 sinking particles taken at 8,798 m of water depth in 1986–1987. Table 2 gives TOC, total nitrogen, stable carbon isotopic composition and $\Delta^{14}C$ values of the JT-06 sinking particle samples taken at both 4,519 m and 8519 m in 1990–1991. The latter data are quoted from Handa and Sugimoto (unpublished results,

Table 1. Total organic carbon (TOC), total nitrogen and the weight ratios of TOC to total nitrogen (C/N ratios) in the JT-01 sinking particles (water depth: 8798 m; Ishiwatari *et al.*, 1995).

Sample No.	Date of sampling	TOC (mg/g)	Total N (mg/g)	C/N ratio (weight)
1	Aug.30 - Sep.18,1986	20.9	2.6	8.1
2	Sep.18 - Oct. 07,	19.2	2.3	8.2
3	Oct.07 - Oct.26,	19.6	2.5	7.9
4	Oct.26 - Nov.14,	17.1	2.2	7.8
5	Nov.14 - Dec.03,	16.9	2.0	8.5
6	Dec.03 - Dec.22,	16.6	2.0	8.3
7	Dec.22 - Jan.10, 1987	13.3	1.8	7.5
8	Jan.10 - Jan.29,	20.4	2.6	7.9
9	Jan.29 - Feb.17,	20.3	2.7	7.6
10	Feb.17 - Mar.08,	18.5	2.0	9.4
11	Mar.08 - Mar.27,	19.7	1.8	10.7
12	Mar.27 - Apr.15,	20.0	2.2	9.0
13	Apr.15 - May.04.	27.1	2.7	10.1
	average	19.2	2.3	8.54
	st. dev.	3.2	0.3	0.99

1993).

In the case of JT-06 samples, $\delta^{13}C$ values of TOC ranging from –22.9 to –23.3‰ (average: –23.2 ± 0.1‰) in the shallow sinking particle samples are similar to those in the deep sinking particle samples having –22.6 to –24.2‰ (average: –23.4 ± 0.4‰). $\Delta^{14}C$ values of TOC in the sinking particle samples are considerably low as compared with a post-nuclear bomb $\Delta^{14}C$ value ($\Delta^{14}C$ = +120‰: Bauer *et al.*, 1992). $\Delta^{14}C$ value of TOC photosynthesized before nuclear-bomb experiments (1956) should be –50‰. Therefore, TOC in these sinking particle samples is considerably affected by the resuspension of sedimentary OC. $\Delta^{14}C$ values indicate that TOC in the sinking particle samples at the shallow trap depth (4519 m) is, on an average, younger than that at the deep sediment trap depth (8,519 m) (–115‰ vs. –244‰). Nakatsuka *et al.* (1996) estimated that resuspended sedimentary OC accounts for 47% of TOC in the shallow sinking particle samples and for 84% of TOC in the deep samples, respectively. If we assume that OC in sinking particles is a mixture of freshly photosynthesized organic carbon ($\Delta^{14}C$ = +120‰) with resuspended OC with $\Delta^{14}C$ = –1,000‰, resuspended OC accounts for 21% of TOC for the shallow sinking particles and for 32% of TOC for the deep ones, respectively. Temporal fluctuation in $\Delta^{14}C$ values of OM suggests that resuspended OM is supplied to the sediment trap cups

Table 2. Chemical and carbon isotopic characteristics of the sediment-trap samples from the Japan Trench (JT-06) (modified from Handa and Sugimoto, unpublished data, 1993). *4001;9001:samples collected in 25 Mar. 1990-9 June; 4002; 9002; samples collected in 31 Dec. 1990-25 Feb. 1991.

		Water Depth of 4519 m					Water Depth of 8519 m				
Sample No.	Date of Sampling	TOC mg/g	N mg/g	C/N weight	$\delta^{13}C$ of TOC(‰)	$\Delta^{14}C$ of TOC(‰)	TOC mg/g	N mg/g	C/N weight	$\delta^{13}C$ of TOC(‰)	$\Delta^{14}C$ of TOC(‰)
1	25Mar. 90 - 20Apr.90	40.39	4.51	8.96	-22.93	-92.76	30.08	3.78	7.96	-22.94	-239.68
2	20Apr90 - 14May90	36.75	4.07	9.03	-23.39	-88.65	32.22	4.03	8.00	-22.80	-139.78
3	15May 90 - 9June90	37.70	3.61	10.44	-23.19	-62.16	31.27	4.01	7.80	-22.73	-118.67
4	10June 90 - 4July90	35.48	3.96	8.96	-23.24	-152.71	32.75	4.36	7.51	-23.25	-201.45
5	5July90 - 30July90	34.39	3.49	9.85	-23.19	-91.19	29.11	3.79	7.68	-23.44	-196.38
6	31July 90 - 24Aug.90	36.20	3.67	9.86	-23.15	-74.99	29.58	3.86	7.66	-23.85	-391.51
7	25Aug. 90 - 19Sep..90	37.69	4.42	8.53	-23.26	-66.53	27.99	3.52	7.95	-23.58	-308.86
8	20Sep. 90 - 14Oct.90	36.42	4.24	8.59	-23.26	-160.86	29.36	3.70	7.94	-23.62	-357.30
9	15Oct. 90 - 9Nov.90	38.19	4.09	9.34	-23.27	-155.90	31.06	4.31	7.21	-23.71	-330.53
10	10Nov. 90 - 4Dec.90	41.01	4.20	9.76	-23.07	-97.13	31.36	4.12	7.61	-24.16	-262.54
11	5Dec. 90 - 30Dec.90	34.85	3.77	9.24	-23.20	-66.53	28.94	3.80	7.62	-23.65	-216.50
12	31Dec. 90 - 24Jan.91	30.85	3.82	8.08	-23.01	-93.67	30.46	3.96	7.69	-23.16	-166.44
13	25Jan.91 - 25Feb.91	30.66	3.88	7.90	-23.15	-295.69	30.87	3.94	7.84	-23.19	-246.17
Whole	average	36.20	3.98	9.12	-23.18	-115.29	30.39	3.94	7.73	-23.39	-244.29
samples	st. dev.	3.09	0.31	0.74	0.12	64.21	1.37	0.24	0.22	0.42	83.70
4001;9001*	average	38.28	4.06	9.48	-23.17	-81.19	31.19	3.94	7.92	-22.82	-166.04
	st. dev.	1.89	0.45	0.84	0.23	16.61	1.07	0.14	0.10	0.11	64.64
4002:9002*	average	30.76	3.85	7.99	-23.08	-194.68	30.67	3.95	7.76	-23.18	-206.31
	st. dev.	0.13	0.04	0.12	0.10	142.85	0.29	0.01	0.10	0.02	56.38

intermittently, as far as JT-06 samples are concerned.

C/N ratios (weight ratio: 9.1 ± 0.7) of OM in the JT-06 shallow sinking particles are higher than those (7.7 ± 0.2) in the JT-06 deep ones. Although a clear-cut explanation for this unusual result cannot be given at present because chemical forms of nitrogen are not known in those samples, this result suggests that microbially labile organic fractions such as carbohydrates are preferentially lost from sinking particles during sedimentation or N-enriched OM such as microbial biomass contributes to the deeper samples.

The downward trend in C/N ratios for JT-06 sinking particles is different from that reported for sinking particles from the Ligurian Sea, north-western Mediterranean Sea (Peulve *et al.*, 1996). According to Peulve *et al.* (1996), C/N ratio of sinking particles increases with water depth from 7 at 100 m to 11 at 2,000 m.

Molecular characteristics from pyrolysis-GCMS

Py-GCMS results

Figure 2 shows the representative total ion chromatograms of pyrolysates of JT-01 sinking particle samples. The pyrolysates are composed of a complicated mixture of organic compounds. A series of alkylbenzenes, and phenols and alkylphenols are present as major components. These products may be mainly derived from proteins and related materials. Since methoxy phenols were not detected in the pyrolysates, lignin is practically absent in the sinking particle samples.

As shown in mass chromatograms (Fig. 3(a)), pyrroles and, pyridines are present in the pyrolysates. These compounds are typical pyrolysates of proteins, polypeptides and amino acids (Bracewell and Robertson, 1984; Munson and Vick, 1985; Tsuge and Matsubara, 1985). However, since the yield of pyridine for sinking particle samples is relatively high as compared with that observed for common proteins (Munson and Vick, 1985), proteins in the sinking particles might be partially modified. Indoles, phenylalkanenitriles and furan-derivatives are detected in the pyrolysates (Fig. 3(b)). Indole and phenylalkanenitriles are also typical pyrolysates of polypeptides. Furan-derivatives may be derived from polysaccharides (Helleur *et al.*, 1985).

A series of n-alkanes/n-alkenes is observed in the mass chromatograms (Fig. 3(c)). These hydrocarbons are derived from lipids materials with long methylene chains. Isoprenoid hydrocarbons (pristenes) may be derived from the phytol side chain of chlorophyll pigments.

The similar molecular characteristics are obtained for the JT-06 sinking particle samples (#4001: shallow trap sample) collected at site JT in 1990–1991. Pyrroles, pyridines, indoles, and phenylalkanenitriles are clearly seen in the pyrolysates. Furfural, which is derived from carbohydrates, is also present in the pyrolysates, as shown in Fig. 4(a). In samples similar to the JT-01 as samples, major pyrolytic products appeared on the gas chromatogram alkylbenzenes and alkylphenols. A series of n-alkane and n-alkene doublets is also present in the

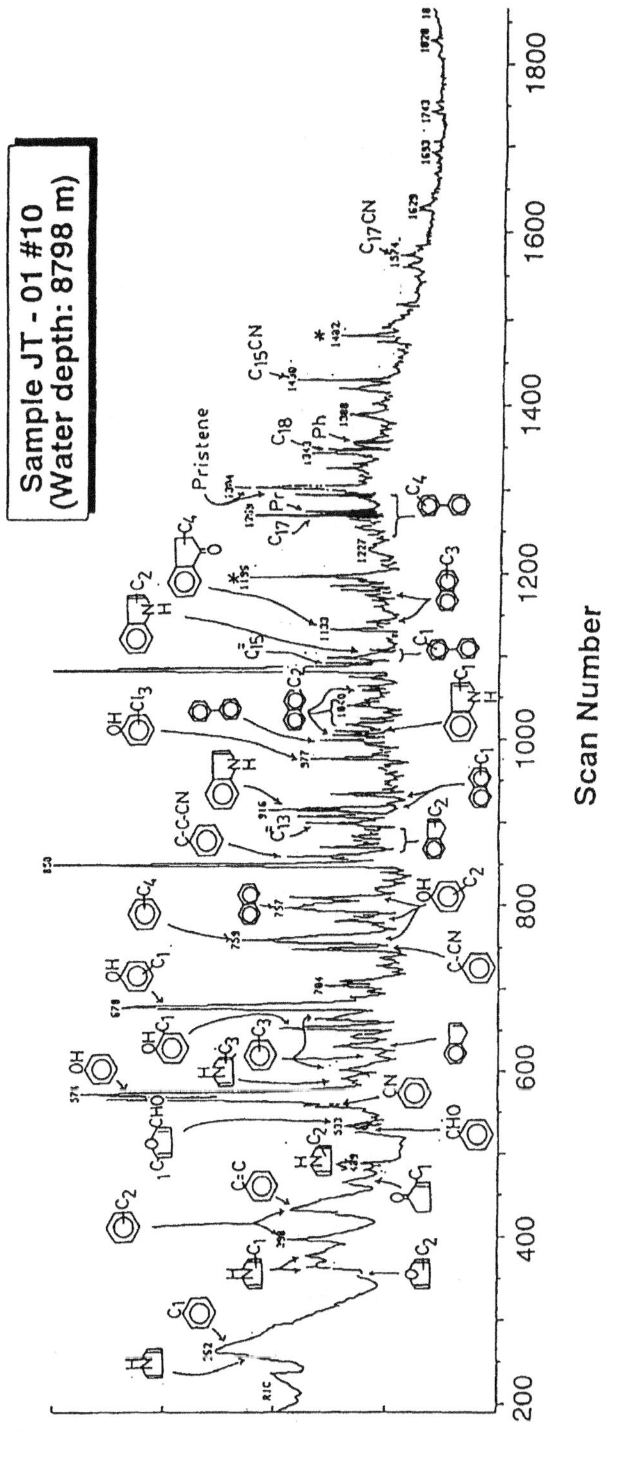

Fig. 2. Total ion chromatogram of pyrolysates of sinking particles from the Japan Trench (Sample: JT-01 #10).

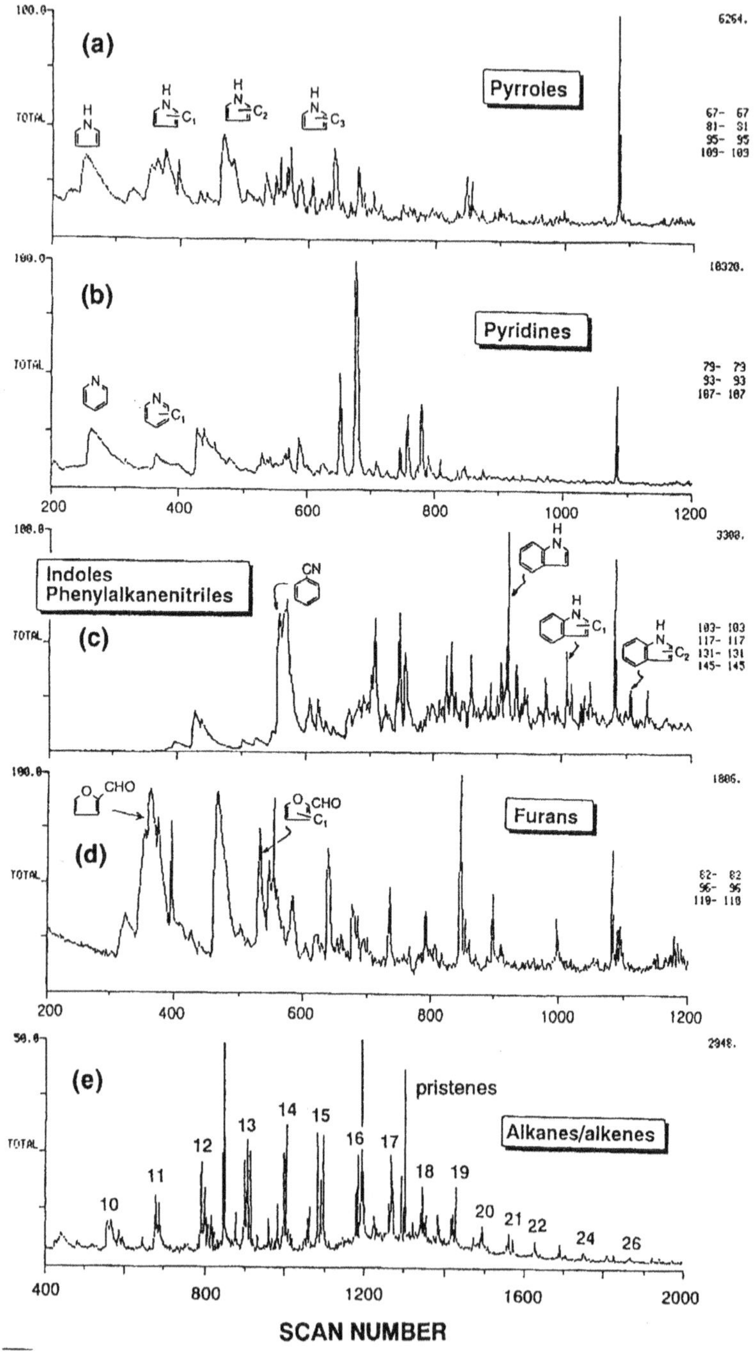

Fig. 3. Mass chromatograms of pyrolysates of sinking particles from the Japan Trench (Sample: JT-01 #11): (a) pyrroles (m/z 67 + 81 + 95 + 109), (b) pyridines (m/z 79 + 93 + 107), (c) indoles and phenylalkanenitriles (m/z 103 + 117 + 131 + 145), (d) furans (m/z 82 + 96 + 110), and (e) alkanes/alkenes (m/z 69 + 71).

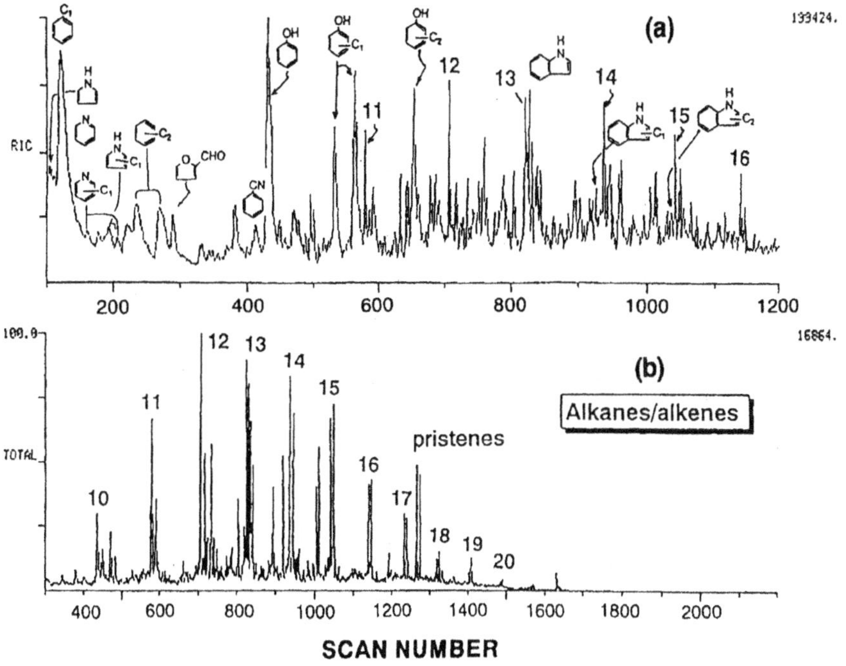

Fig. 4. (a) Total ion and (b) Mass chromatograms at m/z 69 + 71 of pyrolysates of sinking particles from the Japan Trench (Sample: JT-06 #4001).

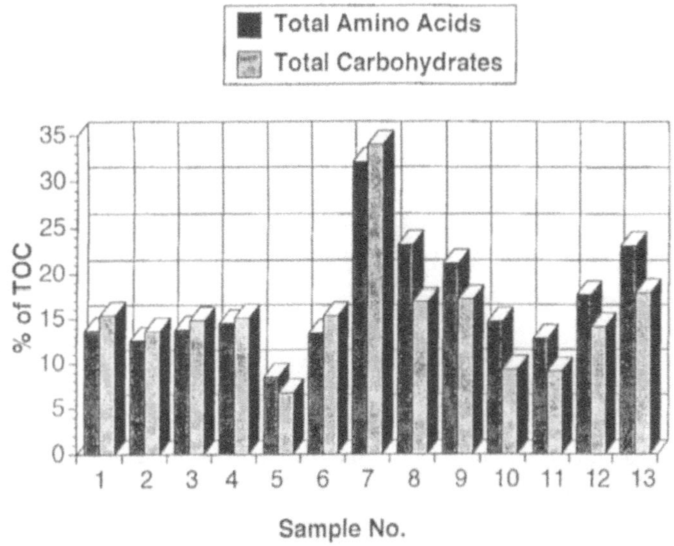

Fig. 5. Temporal variations in the content of total amino acids and carbohydrates determined by py-GCMS for JT-01 sediment trap samples. Sample numbers correspond to those given in Table 1.

Table 3. Assignment of the organic compounds in the pyrolysates of sinking particles from the Japan Trench (JT-01 and JT-06).

Abbreviation	Pyrolysis product	Carbon range	Mass fragment for analysis	Possible precursor
Aliphatic hydrocarbons				
n	n-alkanes	n = 9 ~ 26	71	lipid
n=	n-alkenes	n = 9 ~ 26	69	lipid
i-n	isoprenoid	n = 19, 20	71	phytol
i-n=	isoprenoid	n = 19	69	phytol
Aliphatic compounds				
ACn	fatty acids	n = 12 ~ 18	73	fatty acid
CNn	alkanenitrile	n = 12 ~ 18	110	fatty acid
Aromatic hydrocarbons				
BZn	C_n-benzenes	n = 1 ~ 4	92 + 14n	protein (n=1~2)
INn	C_n-indenes	n = 0 ~ 3	116 + 14n	?
NAn	C_n-naphthalenes	n = 0 ~ 4	128 + 14n	?
BIn	C_n-biphenyls	n = 0 ~ 4	154 + 14n	?
APn	C_n-anthracenes	n = 0 ~ 3	178 + 14n	?
APn	C_n-phenanthrenes	n = 0 ~ 3	178 + 14n	?
Phenols				
PHn	C_n-phenols	n = 0 ~ 3	94 + 14n	protein (n=0~2)
Furans				
FRn	C_n-furans	n = 1 ~ 3	68 + 14n	carbohydrate
EAn	C_n-furylaldehydes	n = 0 ~ 1	96 + 14n	carbohydrate
BFn	C_n-benzofurans	n = 0 ~ 3	118 + 14n	?
Nitrogenous compounds				
PRn	C_n-pyrroles	n = 0 ~ 4	67 + 14n	protein (n=0~2)
PDn	C_n-pyridines	n = 0 ~ 2	79 + 14n	protein
BCn	phynyl-C_n-cyanides	n = 0 ~ 4	103 + 14n	protein
IDn	C_n-indoles	n = 0 ~ 3	117 + 14n	protein (n=0~1)

pyrolysates (Fig. 4(b)).

Table 3 lists the compounds identified in the pyrolysates of sinking particles and their possible precursors. Consequently, the py-GCMS study revealed that OM in sinking particles from the Japan Trench is categorized largely into three classes of precursor: proteins (nitrogenous compounds such as pyrroles, pyridines, phenylalkanenitriles and indoles as pyrolysates), carbohydrates (furan-derivatives as pyrolysates) and lipids (alkenes/alkanes, alkanenitriles/fatty acids as pyrolysates).

Estimation of the amounts of carbohydrates and amino compounds in sinking particles

The amounts of polypetides and carbohydrates in sinking particle samples can be estimated by using the quantitative relationship between the results from py-GCMS and those obtained by conventional methods. In the case of amino compounds, a good relationship was obtained between total amino acids determined by a wet chemical method and pyrrole as a pyrolysate for sinking particles and sediment samples from coastal areas and lakes (Ishiwatari *et al.*, 1995). A good correlation is also present between total carbohydrates (determined by GC as monosaccharides) and furfural as a pyrolysate (Ishiwatari *et al.*, 1995). Thus, temporal variations in the content of total amino acids and carbohydrates estimated from py-GCMS data for JT-01 sinking particle samples, as shown in Fig. 5. It shows that (1) both amino acids and carbohydrates in the sinking particles account for approximately 16% (7–30%) of the total OM.

Comparison between shallow and deep sediment trap samples in py-GCMS characteristics

Figure 6 gives pyrolysi gas chromatograms (mass chromatograms at m/z 50–72 plus m/z 74–146) for both shallow and deep sinking particle samples (#4001 vs. #9001). The py-GCMS was obtained for the samples prepared after 6M HCl-hydrolysis. Therefore, carbohydrates and protenaceous components must be absent in the samples. Ion at m/z 73 was excluded from the mass chromatograms to remove the effect of contamination by Si-derivatives (m/z = 73). As shown in Fig. 6, the py-gas chromatograms for both #4001 and #9001 are roughly the same, suggesting that molecular composition of the hydrolysis-resistant part (parts other than carbohydrates and protenaceous materials) of TOM in these samples is similar.

It is interesting to note that peaks of alkane/alkene doublets, pristene and phenols are marked in the py-GC chromatograms. Fatty acids must have been removed by the procedure of HCl-hydrolysis. A similar py-GC chromatogram is obtained for the #4001 sample, even after KOH-hydrolysis following HCl-hydrolysis. This result indicates that aliphatic structures are present in the hydrolysis-resistant portion of TOM in the sinking particles, as suggested by Peulve *et al.* (1996).

Freshness Index

It is known that amino acids and carbohydrates tend to degrade faster than lipid materials during the early stage of diagenesis of OM in sediments (Tissot and Welte, 1984). A similar selective degradation of amino acids and carbohydrates is thought to take place in the water column. If this process takes places, the ratios of the amounts of amino acids and carbohydrates with that of lipids in sinking particles may decrease with increasing degradation of OM. However, this is not necessarily the case for the degradation of OM in the water column (Wakeham and Lee, 1993; Wakeham, personal communication). For example, in a sediment trapexperiment conducted in equatorial North Atlantic Ocean, hydrocarbon flux decreases by a factor of 14, amino acids by 25, sterols by 45 and fatty

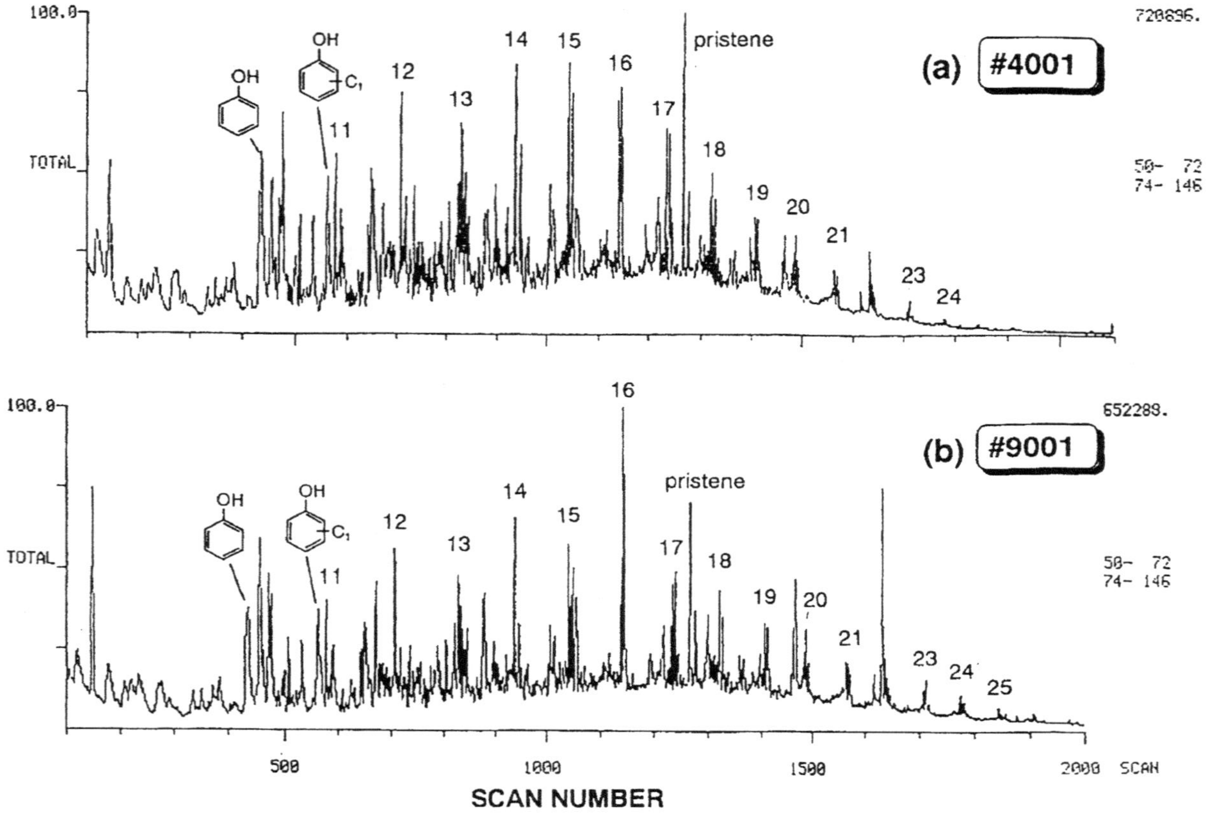

Fig. 6. Mass chromatograms (m/z 50-72 + m/z 74-146) of pyrolysates of sinking particles from the Japan Trench. Sample: (a) JT-06 #4001, (b) JT-06 #9001. This py-GCMS was conducted for the samples obtained after HCl-hydrolysis.

Table 4. Freshness Indexes proposed from py-GCMS

No.	Ratio	Fresh ↔ Old	
1.	$\frac{\text{Indoles}}{\text{Alkanes/Alkenes}}$	High	Low
2.	$\frac{\text{Pyrroles}}{\text{Alkanes/Alkenes}}$	High	Low
3.	$\frac{\text{Furans}}{\text{Alkanes/Alkenes}}$	High	Low
4.	$\frac{\text{Alkenes}}{\text{Alkanes}}$	High	Low
5.	$\frac{\text{Pyridines}}{\text{Alkanes/Alkenes}}$	Low	High

acids by 60 between 389 m and 5068 m (Wakeham and Lee, 1993).

In a previous study (Ishiwatari *et al.*, 1995), we proposed a "Freshness Degree (FD)" or "Freshness Index" which is defined as a ratio of pyrolysis product x (normalized by the sum of n-alkanes and n-alkenes with a carbon number ranging from 10 to 26 in a sample compared to that in a reference material; where x = pyrroles, indoles (as a proxy of polypeptides), pyridines and furans (as a proxy of polysaccharides) (Table 4). FD is calculated by the following equation.

$$\text{FD}\left(\text{Product } x\right) = \frac{\left(\dfrac{\text{Product } x}{\text{Alkanes} + \text{Alkenes}}\right)_{\text{sample}}}{\left(\dfrac{\text{Product } x}{\text{Alkanes} + \text{Alkenes}}\right)_{\text{reference}}} \times 100.$$

However, it should be noted that both n-alkanes and n-alkenes are not only produced from lipids such as fatty acids but also from hydrolysis-resistant organic fractions on pyrolysis, as already shown. Therefore, the "Freshness Index" proposed above should mean the ratios of labile organic fractions with lipids plus hydrolysis-resistant organic fraction. These ratios would decrease with increasing the extent of the degradation of OM.

Another Freshness Index, FD (pyridine), is proposed to characterize the diagenetic changes of nitrogenous compound composition, because a percentage of pyridine in the total nitrogenous pyrolysates appears to increase with diagen-

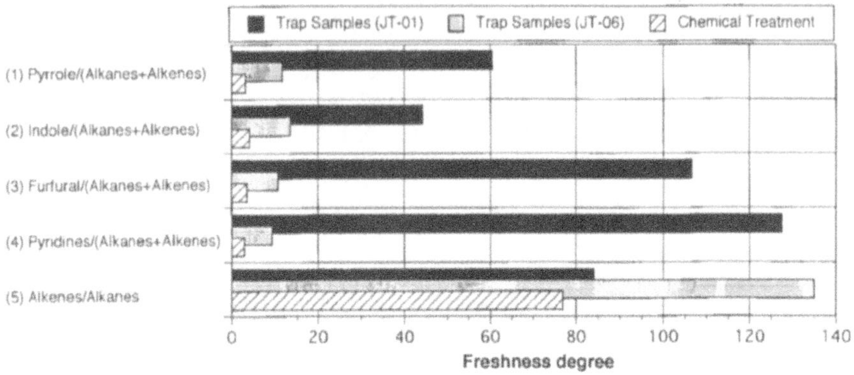

Fig. 7. "Freshness Degree" values calculated for the JT-01 and JT-06 sinking particle samples and artificially-hydrolyzed JT-06 samples. Chemical treatment consists of hydrolysis with 6 M HCl for 20 hours followed by 0.5 M KOH in methanol for 3 hours.

esis:

$$\mathrm{FD}\,(\text{pyrrole / pyridine}) = \frac{\left(\dfrac{\text{pyrrole}}{\text{pyridine}}\right)_{\text{sample}}}{\left(\dfrac{\text{pyrrole}}{\text{pyridine}}\right)_{\text{reference}}} \times 100,$$

$$\mathrm{FD}\,(\text{alkene / alkane}) = \frac{\left(\dfrac{\text{n - alkene}}{\text{n - alkane}}\right)_{\text{sample}}}{\left(\dfrac{\text{n - alkene}}{\text{n - alkane}}\right)_{\text{reference}}} \times 100$$

where n-alkanes/n-alkenes = C_{10}–C_{26}. FD (alkene/alkane) is also proposed because the ratio of n-alkenes/n-alkanes in matured OM (e.g. OM in sedimentary rocks) appears to be low as compared with those for freshly deposited OM. Giant sinking particles (>1 mm in diameter) collected at site JT are used as reference materials, because the giant particles are believed to be composed of fecal pellets and rather freshly formed organic debris.

Figure 7 gives FD values calculated for the JT-01 and JT-06 sinking particle samples. An artificially-hydrolyzed JT-06 (#4001) sample is shown for comparison in Fig. 7. The artificially-hydrolyzed JT-06 (#4001) sample was prepared as follows: After the removal of organic solvent-soluble materials, the sediment trap sample was hydrolyzed first by 6 M HCl at 100°C for 20 hours and then by 0.5

M KOH-methanol under reflux for 3 hours.

The following points can be raised as interesting from the biogeochemical point of view: (1) FD values for the hydrolyzed-#4001 sample are the lowest among the three samples given in Fig. 7. This result is reasonable, because acid and alkaline hydrolysis would remove most of the proteins and carbohydrates from the sinking particle samples, resulting in lower FD values; (2) The FD (pyrrole) and FD (indole) for the three samples are below 60, while FD (furfural), and FD (pyridine) are around 100 or higher for the JT-01 samples; (3) FD values, except for FD (alkene/alkane) for the JT-06 (#4001) sample, are small as compared with those for the JT-01 samples. This indicates that the JT-06 OM (#4001) is older than JT-01 OM; and (4) FD (alkene/alkane) are similar among three samples.

As described above, pyrolytic composition clearly indicates that OM in sinking particles in the Japan Trench is essentially autochthonous. Marked evidence showing the contribution of terrestrial OM (e.g. lignin) to the sinking particles is not detected by py-GCMS.

MOLECULAR AND CARBON ISOTOPIC COMPOSITIONS OF SELECTED ORGANIC COMPOUNDS IN SINKING PARTICLES AND BOTTOM SEDIMENT IN THE JAPAN TRENCH

OM in sinking particle samples from the Japan Trench were characterized by lipid biomarker and GC-IRMS analyses (Yamada and Ishiwatari, unpublished results, 1996; Matsumoto and Ishiwatari, unpublished results, 1996), as described below.

Normal alkanes

As already described, n-alkanes can provide information on the sources of OM in sinking particles. Figure 8 shows gas chromatograms (RIC) of n-hexane eluate of JT-06 sinking particle samples. Normal alkanes range from C_{14} to C_{35} in all the sinking particle samples. The distribution in the region of mid-molecular weight hydrocarbons centers at n-C_{17} with a significant hump. This may indicate planktonic and bacterial inputs. n-C_{27}–C_{33} alkanes show an odd carbon number predominance, indicating terrestrial plant wax input. Carbon preference index (CPI) values of n-C_{27}–C_{33} alkanes range from 2.4 to 2.7. These values are slightly smaller than those in common coastal sediments with considerable terrestrial hydrocarbon input. But, the molecular range of n-alkanes is similar to those reported for sinking particles in the north-west Pacific Ocean (Simoneit *et al.*, 1986) and in the equatorial Atlantic Ocean (Wakeham *et al.*, 1980). n-C_{35} alkane peak is superimposed by lycopane, as described later.

One might think here that the result described above is inconsistent with the results of py-GCMS because n-C_{27}–C_{33} alkanes clearly indicate the contribution of terrestrial OM to the sinking particles. The discrepancy is not strange because py-GCMS is insensitive to very small amounts of components such as n-alkanes.

Figure 9 gives the stable carbon isotopic composition of individual n-C_{15}–

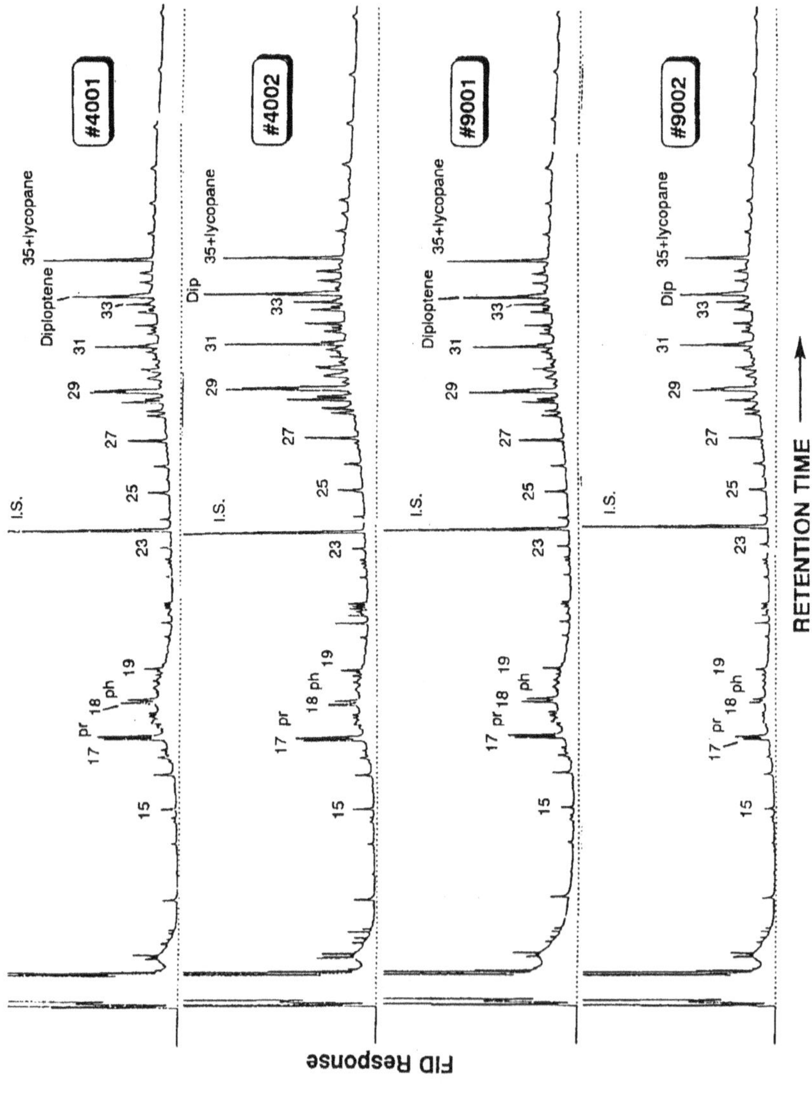

Fig. 8. Gas chromatograms of n-hexane eluate of sinking particle samples from the Japan Trench (JT-06).

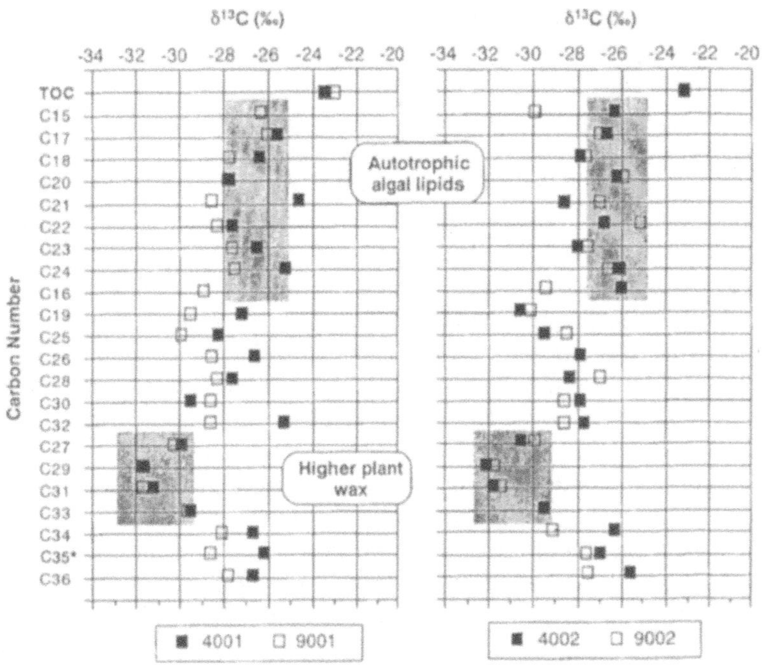

Fig. 9. Stable carbon isotopic composition of individual n-C_{15}–C_{36} alkanes from the JT-06 sinking particles. *: n-C_{35} + lycopane.

C_{36} alkanes from the sinking particles (JT-06). Carbon isotopic values of n-alkanes lie in the range from –25.6 to –32.1‰. $\delta^{13}C$ values of n-C_{14}–C_{20} alkanes except for n-C_{16} and n-C_{19} alkanes fall in the range of –25 to –27‰, which is probably indicative of marine planktonic input. $\delta^{13}C$ values of n-C_{16} and n-C_{19} alkanes are variable (-26.0 ~ –30.6‰: –28.1 on average for n-C_{16} and –29.4 for n-C_{19}, respectively). This suggests a complicated origin of these n-alkanes. Odd numbered n-C_{27}–C_{33} alkanes give the lowest $\delta^{13}C$ values (–29.5 ~ –32.1‰) lying in the range of those observed for n-C_{27}–C_{33} alkanes of higher (C3) plants (Rieley *et al.*, 1991; Collister *et al.*, 1994; Yamada and Ishiwatari, unpublished results). Therefore, the odd numbered n-C_{27}–C_{33} alkanes in the sinking particle samples may be of terrestrial higher (mainly C3) plant origin, mixed with minor amounts of n-alkanes of other sources (e.g. CAM plants, plankton, bacteria, etc.).

Normal C_{34}–C_{36} alkanes give relatively heavy $\delta^{13}C$ values ranging from –26.3 to –28.6‰. These alkanes might originate from terrestrial higher plants, because their $\delta^{13}C$ values resemble those of cedar n-C_{33}–C_{35} alkanes with $\delta^{13}C$ of –28 ~ -29‰ (Yamada, unpublished data). In particular, the dominant alkane is n-C_{35} for both sinking particle and cedar samples, and its $\delta^{13}C$ value in the sinking particles (–27.4 ± 1.0‰) resembles that in the cedar samples (–28‰). However, by a detailed examination of GC peaks, the n-C_{35} alkane peak in sediment trap samples was found to be superimposed by lycopane which was obtained from an aliphatic hydrocarbon fraction after molecular sieve separation (using molecular sieve 5A) and identified by GCMS. Since the $\delta^{13}C$ value of lycopane was determined to be –24.5‰ (Yamada, unpublished results), the true $\delta^{13}C$ value of n-C_{35} alkane should be much lower than the apparent value (–27.4‰).

$\delta^{13}C$ values of n-alkanes show a considerably wide range of distribution. This result gives a marked contrast to that of TOC, which shows fairly constant values (–23.2 ± 0.1‰ for shallow sinking particle samples: n = 13 and –23.4 ± 0.4‰ for deep sinking particle samples: n = 13). According to the literature (e.g. Degens *et al.*, 1968; Hayes, 1993; Lichtfouse *et al.*, 1994), lipid compounds in living organisms are usually ^{13}C-depleted by 3–5‰ (or more in some cases) relative to biomass (TOC). The present result that n-C_{14}–C_{20} alkanes excluding n-C_{16} and n-C_{19} alkanes are ^{13}C-depleted by 3.7‰ relative to TOC for the sinking particles indicates both n-alkanes and TOC are largely of phytoplankton origin. The odd numbered n-C_{27}–C_{33} alkanes are ^{13}C-depleted by 7.5‰ relative to TOC, indicating them to be of terrestrial higher (mainly C3) plant origin. n-C_{34}–C_{36} alkanes seem to be ^{13}C-enriched by around 3‰ relative to n-C_{27}–C_{33} alkanes, suggesting the former being phytoplankton origin. $\delta^{13}C$ values of the other n-alkanes fall between –27 and –30‰, being ^{13}C-depleted by 4–7‰ relative to TOC. These results suggest these n-alkanes are of mixed (terrestrial plant-wax and phytoplankton) origin.

Hopanoid hydrocarbons

Figure 10 gives GCMS selected ion chromatograms at m/z 191 of aliphatic hydrocarbons in the sinking particles from the Japan Trench (JT-06). 17β(H), 21β(H)-hop-22(29)-ene (Diploptene: see Appendix) is the most abundant in the hopanoids in the all sinking particle samples. Hop-22(29)-ene is derived from living prokaryotic organisms, such as bacteria and cyanobacteria (e.g. Rohmer *et al.*, 1984; Venkatesan, 1988). The other $\beta\beta$-type [17β(H), 21β(H)-configurations] hopanoid hydrocarbons (hop-17(21)-ene, $C_{30}\beta\beta$ and $C_{31}\beta\beta$R: Numbers indicate the carbon number in a molecule, and R indicates R-configuration at C-22 chiral center) are also prokaryotic organism-derived. $C_{30}\beta\beta$ and $C_{31}\beta\beta$R hopanes are believed to be produced by degradation of bacterial hopanoids with an extended side chain (Rohmer *et al.*, 1980). This result indicates significant contribution of bacterial (both heterotrophic and autotrophic) lipids in the sinking particles. Moreover, a series of hopanoid hydrocarbons which are typical for fossil hopanoid hydrocarbons ($\alpha\beta$-type [17α(H), 21β(H)-configurations] ($C_{29}\alpha\beta$, $C_{30}\alpha\beta$, $C_{31}\alpha\beta$R, $C_{31}\alpha\beta$S, $C_{32}\alpha\beta$R, $C_{32}\alpha\beta$S, $C_{33}\alpha\beta$R and $C_{33}\alpha\beta$S) are present in all sinking particle samples, although their concentration level is very low. This indicates significant petroleum oil contamination in the samples.

Diploptene shows almost the same $\delta^{13}C$ values (–23.2 ± 0.3‰) for all the sinking particles examined, suggesting that it is from the similar source.

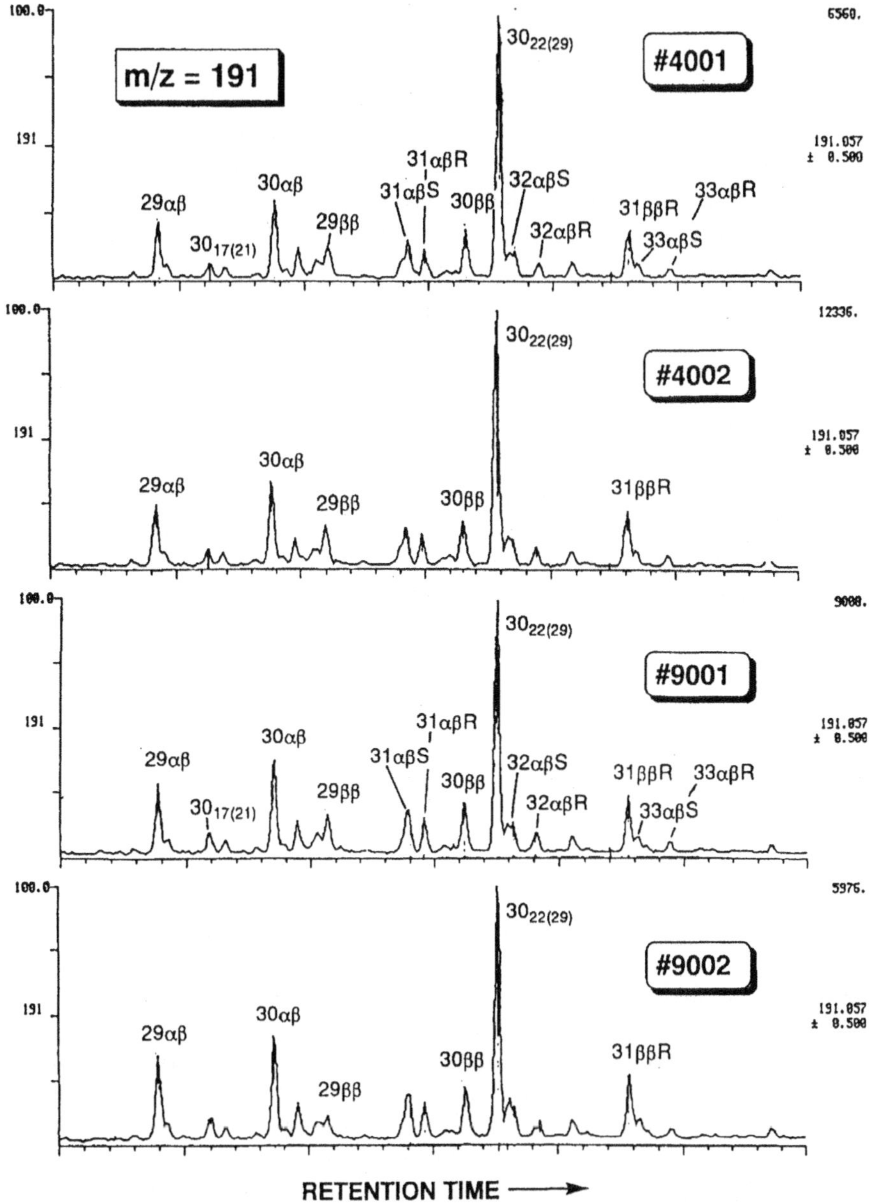

Fig. 10. GCMS mass chromatograms at m/z 191 of aliphatic hydrocarbons in the sinking particles from the Japan Trench (JT-06). Peak assignments: 29$\alpha\beta$, 17α(H),21β(H)-30-norhopane; $30_{17(21)}$,hop-17(21)-ene; 30$\alpha\beta$, 17α(H),21β(H)-hopane; 29$\beta\beta$, 17β(H),21β(H)-30-norhopane; 31$\alpha\beta$S, 17β(H),21β(H)-30-homohopane (22S); 31$\alpha\beta$R, 17α(H),21β(H)-30-homohopane(22R); 30$\beta\beta$, 17β(H),21β(H)-hopane; $30_{22(29)}$, hop-22(29)-ene (Diploptene); 32$\alpha\beta$S, 17α(H),21β(H)-30,31-bishomohopane (22S); 32$\alpha\beta$R, 17α(H),21β(H)-30,31-bishomohopane (22R); 31$\beta\beta$R, 17β(H),2β(H)-30-homohopane (22R); 33$\alpha\beta$S, 17α(H),21β(H)-30,31,32-trishomohopane (22S); 33$\alpha\beta$R, 17α(H),21β(H)-30,31,32-trishomohopane (22R).

Lycopane

Lycopane (a C_{40}-isoprenoid alkane: 2,6,10,14,19,23,27,31-octamethyldotriacontane: see Appendix) in the sinking particles (JT-06) was identified by comparison with the literature (Wakeham, 1990) with respect to mass spectrum and GC-retention time datum. A likely source of lycopane is thought to be marine phytoplankton from the circumstantial evidence that this compound has been found in particulate matter in the Cariaco Trench (Wakeham, 1990) and sediments in the Peru upwelling area (Farrington *et al.*, 1988).

$\delta^{13}C$ value of lycopane from the sinking particles in the Japan Trench gives -24.5 ± 0.3‰ on average (Yamada, unpublished result). This $\delta^{13}C$ value is close to that for photosynthetic C_{28}-sterol (24-methylcholesta-5,22-dien-3β-ol: –24.1 ± 1.1‰: Matsumoto and Ishiwatari, unpublished result). The present result indicates lycopane to be of marine phytoplankton origin. The remarkably similar $\delta^{13}C$ values for lycopane to that in the Japan Trench were reported by Wakeham *et al.* (1993) for water columns and sediment samples from the central Pacific gyre and the Cariaco Trench. They concluded that lycopane is of photoautotrophic origin, and postulated an additional route of formation of lycopane by anaerobic microbial action on an algal precursor in anoxic zones.

Sterols

Cholesterol (Cholest-5-en-3β-ol: Δ^5C_{27}-sterol: see Appendix), 5α-cholestan-3β-ol (5αC_{27}-sterol) and 24-methylcholesta-5,22-dien-3β-ol ($\Delta^{5,22}C_{28}$-sterol) are found to be present in sinking particles from the Japan Trench (JT-06) as major sterols. According to Volkman (1986), cholest-5-en-3β-ol (Δ^5C_{27}-sterol) are generally distributed in phytoplankton, zooplankton and other marine fauna, while 24-methylcholesta-5,22-dien-3β-ol ($\Delta^{5,22}C_{28}$-sterol) is present in most of diatom species as a major sterol. Therefore, the former has often been used as a biomarker for phytoplankton plus zooplankton, and the latter as a biomarker for diatoms. 24-Methylcholesta-5,22-dien-3β-ol must come from Prymnesiophytes in oligotrophic waters more likely than from diatoms.

Sterols in the sinking particles in the Japan Trench (JT-06) give $\delta^{13}C$ values of –22.5 ± 0.3‰ on average for cholesterol (Δ^5C_{27}-sterol) and –24.1 ± 1.1‰ for 24-methylcholesta-5,22-dien-3β-ol ($\Delta^{5,22}C_{28}$-sterol) (Matsumoto and Ishiwatari unpublished results). Relative enrichment in ^{13}C of cholesterol to algal sterol (24-methylcholesta-5,22-dien-3β-ol) by 1.6‰ indicates that zooplankton is a more likely source of cholesterol than phytoplankton.

Factors controlling $\delta^{13}C$ values of individual compounds in sinking particles

Table 5 summarizes the $\delta^{13}C$ values of various types of organic compounds in sinking particles at site JT. It is clear from Table 5 that the relative enrichment in ^{13}C increases in the following order: (1) n-alkanes (higher plant-derived: –31‰) < (2) n-alkanes (algae-derived: –27‰) < lycopane, $\Delta^{5,22}C_{28}$-sterol (phytoplankton-derived: -24 ~ -25‰) < diploptene (heterotrophic bacteria-derived: -

Table 5. Analytical results of sinking particles and bottom sediments from the Japan Trench (sinking particles: JT-06; sediments: core LM6 34°11.89′ N, 141°55.81′ E; water depth: 9200 m)

	Sinking Particles				Sediment	
	WD=4519 m 4001; 4002	WD=8519 m 9001; 9002	Total Samples	Deviation from TOC		Deviation from TOC
TOC ($\delta^{13}C$ ‰)	-23.2 ± 0.1	-23.4 ± 0.4	-23.3 ± 0.3	0.0	-21.0	0.0
TOC (mg/g-dry sediment)	36.2 ± 3.1	30.4 ± 1.4	33.3 ± 0.3	-	7.7	-
C/N (weight ratio)	9.1 ± 0.7	7.7 ± 0.2	8.4 ± 0.9	-	8.0	-
n-Alkanes						
CPI (C27-C33)	-	-	2.4 - 2.7	-		-
Algae-derived* ($\delta^{13}C$ ‰)	-26.7 ± 1.1	-27.3 ± 1.1	-27.0 ± 0.5	-3.7	-28.3 ± 0.9	-7.3
Higher plant-derived** ($\delta^{13}C$‰)	-30.8 ± 1.1	-30.7 ± 1.0	-30.8 ± 1.1	-7.5	-31.3 ± 0.1	-10.3
Diploptene ($\delta^{13}C$ ‰)	-23.0 ± 0.6	-23.4 ± 0.3	-23.2 ± 0.3	+0.1	-	-
Lycopane ($\delta^{13}C$ ‰)	-24.5 ± 0.4	-24.5 ± 0.5	-24.5 ± 0.3	-1.2	-	-
C_{27}-sterol*** ($\delta^{13}C$ ‰)	-22.7 ± 1.0	-22.2 ± 1.0	-22.5 ± 0.3	+0.8	-	-
C_{28}-sterol**** ($\delta^{13}C$ ‰)	-24.9 ± 0.9	-23.3 ± 0.2	-24.1 ± 1.1	-0.8	-	-
Fatty acids						
n-C14 - C20 ($\delta^{13}C$ ‰)	-	-	-		-23.4 ± 1.2	-2.4
n-C21 - C22 ($\delta^{13}C$ ‰)	-	-	-		-26.0 ± 0.0	-5.0
n-C23 - C26 ($\delta^{13}C$ ‰)	-	-	-		-27.2 ± 0.8	-6.2
n-C27 - C32 ($\delta^{13}C$ ‰)	-	-	-		-30.4 ± 1.1	-9.4

* Alkanes of C15, C17, C18, C20, C21, C22, C23, C24

** Alkanes of C27, C29, C31, C33

*** Cholest-5-en-3β-ol

**** 24-Methylcholesta-5,22-dien-3β-ol

23‰), TOC (-23‰) < $\Delta^5 C_{27}$-sterol (zooplankton-derived: –23‰).

Detailed discussion on the relationship in $\delta^{13}C$ values of different organic compounds in sinking particles is impossible at present, because little information exists regarding the quantitative relationship in carbon isotopic composition among lipid biomarkers in the same organisms and effect of grazing on their $\delta^{13}C$ values. However, the following relationships in ^{13}C composition among organic compounds in the sinking particles can be pointed out:

(1) $\delta^{13}C$ values of lycopane and $\Delta^{5,22}C_{28}$-sterol (organic compound biosynthesized by primary producer organisms) are similar (–24.5‰ vs. –24.1‰), although these two compounds may not be synthesized by the same organisms.

(2) n-Alkanes (a mixture of n-C_{15}, C_{17}, C_{18}, C_{20}, C_{21}, C_{22}, C_{23}, C_{24}-alkanes) of probable algal origin are depleted in ^{13}C by 2.5–2.9‰ relative to polyisoprenoids (lycopane and $\Delta^{5,22}C_{28}$-sterol). This depletion of n-alkanes seems to be reasonable in the light of previous studies (Hayes, 1993), since n-alkyl species are depleted in ^{13}C relative to isoprenoid lipids in anorganism. In fact, fatty acids are depleted in ^{13}C by 1–2‰ relative to polyisoprenoids (Galimov and Shirinskiy,

1975).

(3) The enrichment of Δ^5C_{27}-sterol in ^{13}C by 1.6‰ relative to $\Delta^{5,22}C_{28}$-sterol may be due to heterotrophy, because it is known that biomass is enriched in ^{13}C relative to food source by *ca.* 1‰ (Hayes, 1993). However, a study should be done to test that the heterotrophic shift in ^{13}C is valid at molecular level.

(4) Diploptene is enriched in ^{13}C by *ca.* 1‰ relative to algae-originated compounds (lycopane and $\Delta^{5,22}C_{28}$-sterol). This result suggests that diploptene is synthesized by heterotrophic bacteria.

(5) n-C_{27}, C_{29}, C_{31}, C_{33}-alkanes are depleted in ^{13}C relative to any other organic compounds. The conclusion of these alkanes to be largely of terrestrial higher-plant origin may be reasonable, because n-alkanes (n-C_{27}, C_{29}, C_{31}, C_{33}) extracted from soils (A_o layer) show $\delta^{13}C$ values of -31.4 ± 1.3 (Yamada, unpublished results), which is close to n-alkanes in both sinking particles and bottom sediments. However, a possibility of algal (micro algae) lipid contribution rather the higher plant waxes cannot be ruled out (Lichtfouse *et al.*, 1994).

As already described, the relative abundance of resuspended OM in the total OM in the deeper (8519 m) trap samples is generally higher than that in the shallower (4519 m) ones. Nevertheless, no practical difference is observed in $\delta^{13}C$ of both TOC and lipid biomarkers (n-alkanes). This fact suggests that the nature of OM in resuspended particles is nearly the same as that in freshly sinking particles.

Normal Alkanes and fatty acids in bottom sediment

TOC in the underlying bottom sediments is enriched in ^{13}C by *ca.* 2‰ relative to TOC in the sinking particles, but $\delta^{13}C$ values of terrestrial n-alkanes show no change among these materials. This fact indicates that the terrestrial n-alkanes are more stable than the other organic fractions and are a sensitive indicator of a terrestrial OM input.

Planktonic n-alkanes (a mixture of n-C_{15}, C_{17}, C_{18}, C_{20}, C_{21}, C_{22}, C_{23}, C_{24}-alkanes) in the bottom sediments appear depleted in ^{13}C by *ca.* 1‰ relative to those in the sinking particles. This might be a result of reworking of n-alkanes in the course of sinking particles to bottom sediments. $\delta^{13}C$ of higher plant-derived n-alkanes (n-C_{27}, C_{29}, C_{31}, and C_{33}) are practically the same as those from sinking particles. This indicates that n-alkanes of higher plant wax origin indeed survives during sedimentation.

Fatty acids (FAs) in JT bottom sediments show a wide range of $\delta^{13}C$ values from −23‰ to −30‰, indicating complicated sources. n-C_{14}–C_{20} FAs are enriched in ^{13}C by 3.6‰ relative to algae-derived n-alkanes, suggesting that this enrichment might be due to loss of ^{13}C-depleted CH_4 from zooplanktonic gut communities (Hayes, 1993). Since the $\delta^{13}C$ of n-C_{21}–C_{26} FAs resemble those of algal n-alkanes, these FAs are possibly derived from phytoplankton, although this idea is at a stage of speculation at present (Naraoka *et al.*, 1995). In the study of Naraoka *et al.* (1995), similar $\delta^{13}C$ values (−26 ~ −27‰) were obtained for n-C_{20}–C_{26} (even numbered) FAs from pelagic sediments in north western Pacific

Ocean.

n-C_{27}–C_{32} FAs in JT bottom sediments are slightly enriched in ^{13}C by 1‰ relative to odd-carbon numbered n-C_{27}–C_{33} alkanes which are thought to be of higher plant origin. Therefore, n-C_{27}–C_{32} FAs are probably derived mainly from terrestrial higher plants with minor contribution of planktonic organisms. Elucidating factors controlling these $\delta^{13}C$ values is a challenging problem for further investigation.

CONCLUSION

The present examination of sinking particles from the Japan Trench indicates that the major source of OM is marine plankton-derived, and terrestrial OM contribution is minor. OM in sinking particles are composed of both freshly-deposited and resuspended OM. The relative abundance of resuspended OM in the total OM in the shallower sediment trap samples is generally lower than that in the deeper trap samples. However, no marked differences both in py-GCMS characteristics and stable carbon isotopic compositions of TOC and biomarkers are recognized between the shallower and the deeper samples. Therefore, one of the features of the sedimentation system of OM in the Japan Trench might be that considerable amounts of resuspended OM contribute to sinking particles throughout a year.

The results described above also confirm that the first drastic change in organic matter composition in sinking particles is finished in the water column shallower than 4,500 m in depth. This is in accord with previous findings from sediment-trap experiments that the most drastic diagenetic change in organic matter composition takes place in the water column shallower than 1,000 m (e.g. Wakeham and Lee, 1993).

Biomarker and carbon isotopic analysis of individual organic compounds revealed that lipid compounds in sinking particles are a mixture of compounds derived from algae, bacteria and terrestrial higher plants. These analyses also suggest that long-chain n-alkanes are surviving during sedimentation and in bottom sediments. However, the present analytical data are too limited to make further discussion on the transformation of OM in the marine water column, and further works are clearly needed. Nevertheless, this study demonstrates that the biomarker and CSIA approach play an important role in drawing a picture of biogeochemical processes of the carbon cycle in the ocean.

Acknowledgements—The authors thank Stuart Wakeham for his constructive review of the manuscript. The authors also thank the crews and scientists on the cruises of R/V *Tansei Maru* and *Hakuho Maru* for the collection of sediment trap samples. This work was partly funded by Grant-in-Aid for Scientific Research of the Japanese Ministry of Education, Science, Sports and Culture.

Appendix

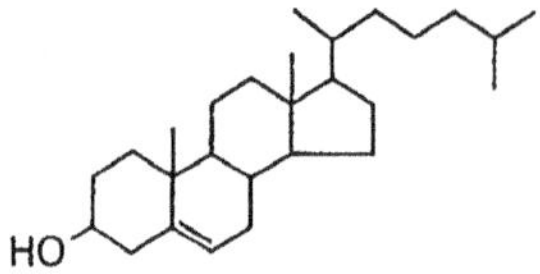

Cholest-5-en-3β-ol

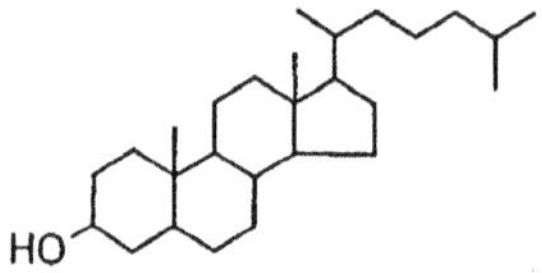

5α-Cholestan-3β-ol

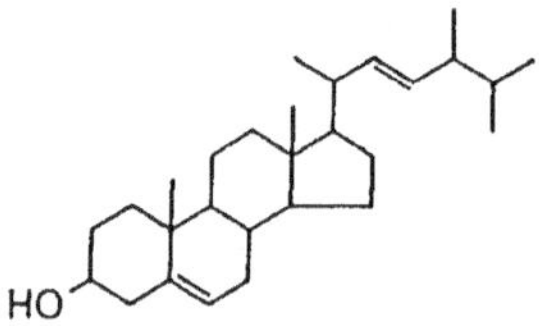

24-Methylcholesta-5,22E-dien-3β-ol

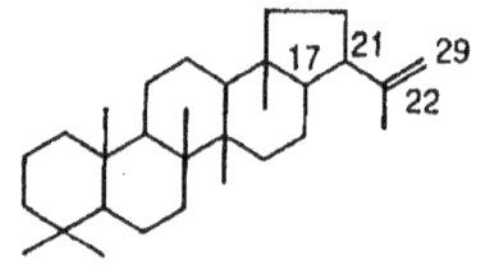

17β(H),21β(H)-hop-22(29)-ene
(Diploptene)

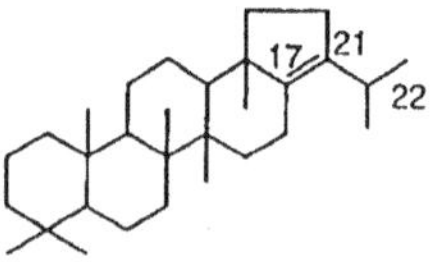

Hop-17(21)-ene

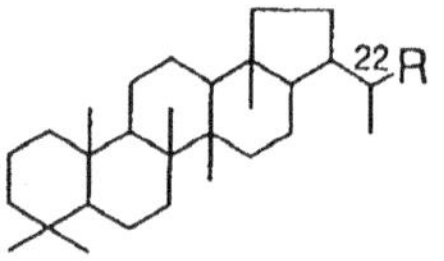

C_{29}: R = H
C_{30}: R = CH_3
C_{31}: R = C_2H_5
C_{32}: R = C_3H_7

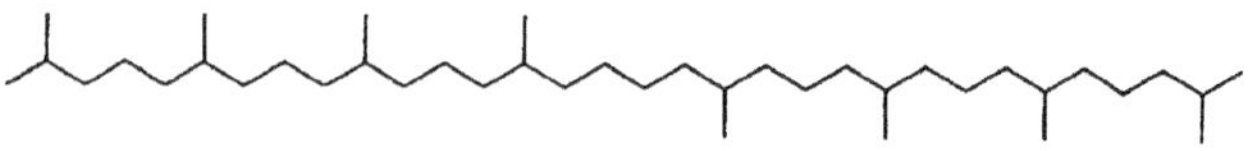

2,6,10,14,19,23,27,31-octamethyldotriacontane
(Lycopane)

REFERENCES

Bauer, J. E., Williams, P. M. and Druffel, E. R. M. (1992): ^{14}C activity of dissolved organic carbon fractions in the north-central Pacific and Sargasso Sea. *Nature*, **357**, 667–670.

Bracewell, J. M. and Robertson, G. W. (1984): Quantitative comparison of the nitrogen-containing pyrolysis products and amino acid composition of soil humic acids. *J. Anal. Appl. Pyrol.*, **6**, 19–29.

Collister, J. W., Rieley, G., Stern, B., Eglinton, G. and Fry, B. (1994): Compound-specific $\delta^{13}C$ analyses of leaf lipids from plants with differing carbon dioxide metabolisms. *Org. Geochem.*,

21, 619–627.

Degens, E. T., Behrendt, M., Gotthardt, B. and Reppmann, E. (1968): Metabolic fractionation of carbon isotopes in marine plankton. II. Data on sample collected off the coasts of Peru and Ecuador. *Deep-Sea Res.*, **15**, 1–9.

Farrington, J. W., Davis, A. C., Sulanowski, J., McCaffrey, M. A., McCarthy, M., Clifford, C. H., Dickinson, P. and Volkman, J. K. (1988): Biogeochemistry of lipids in surface sediments from the Peru Upwelling Area at 15°S. In *Advances in Organic Geochemistry 1987*, ed. by L. Mattavelli and L. Novelli, *Org. Geochem.*, **13**, 607–617.

Galimov, E. M. and Shirinskiy, V. G. (1975): Ordered distributions of carbon isotopes in compounds and components of the lipid fraction in organisms. *Geochem. Int.*, **12**, 157–180.

Hayes, J. M. (1993): Factors controlling ^{13}C contents of sedimentary organic compounds: principles and evidence. *Mar. Geol.*, **113**, 111–125.

Hayes, J. M., Freeman, K. H., Popp, B. N. and Hoham, C. H. (1990): Compound-specific isotopic analyses: A novel tool for reconstruction of ancient biogeochemical processes. *Org. Geochem.*, **16**, 1115–1128.

Helleur, R. J., Hayes, E. R., Jamieson, W. D. and Craigie, J. S. (1985): Analysis of polysaccharide pyrolysate of red algae by capillary gas chromatography-mass spectrometry. *J. Anal. Appl. Pyrol.*, **8**, 333–347.

Ishiwatari, R., Uzaki, M. and Yamada, K. (1994): Carbon isotope composition of individual n-alkanes in recent sediments. *Org. Geochem.*, **21**, 801–808.

Ishiwatari, R., Yamamoto, S. and Handa, N. (1995): Characterization of sinking particles in the ocean by pyrolysis-gas chromatography/mass spectrometry. *J. Anal. Appl. Pyrol.* **32**, 75–89.

Ittekkot. V., Deuser, W. C. and Degens, E. T. (1994a): Seasonality in the fluxes of sugars, amino acids, and amino sugars to the deep ocean: Sargasso Sea. *Deep-Sea Res.*, **31**, 1057–1069.

Ittekkot. V., Degens, E. T. and Honjo, S. (1994b): Seasonality in the fluxes of sugars, amino acids, and amino sugars to the deep ocean: Panama Basin. *Deep-Sea Res.*, **31**, 1071–1083.

Lee, C. and Wakeham, S. G. (1992): Organic matter in the water column: future research challenges. *Mar. Chem.*, **39**, 95–118.

Lichtfouse, E., Derenne, S., Mariotti, A. and Largeau, C. (1994): Possible algal origin of long chain odd n-alkanes in immature sediments as revealed by distribution and carbon isotope ratios. *Org. Geochem.*, **22**, 1023–1027.

Matsueda, H. and Handa, N. (1986): Vertical flux of hydrocarbons as measured in sediment traps in the eastern North Pacific Ocean. *Marine Chem.*, **20**, 179–195.

Munson, T. O. and Vick, J. (1985): Comparison of human hair by pyrolysis-capillary column gas chromatography and gas chromatography-mass spectrometry. *J. Anal. Appl. Pyrol.*, **8**, 493–501.

Najdek, M. (1993): Factors influencing fatty acid and hydrocarbon composition of sedimenting particles in the northeastern Adriatic Sea. *Mar. Chem.*, **41**, 299–310.

Nakatsuka, T., Handa, N., Harada, N., Sugimoto, T. and Imaizumi, S. (1998): Origin and decomposition of sinking particulate organic matter in the deep water column inferred from the vertical distributions of its $\delta^{15}N$, $\delta^{13}C$ and $\Delta^{14}C$. *Deep-Sea Res.*(in press).

Peulve, S., Sicre, M.-A., De Leeuw, J. W., Saliot, A. and Baas, M. (1996): Molecular characterization of macromolecular organic matter in sediment traps from the Northwestern Mediterranean Sea. *Geochim. Cosmochim. Acta.*

Rieley, G., Collier, R. J., Jones, D. M., Eglinton, G., Eakin, P. A. and Fallick, A. E. (1991): Sources of sedimentary lipids deduced from stable carbon-isotope analyses of individual compounds. *Nature*, **352**, 425–427.

Rohmer, M., Dastillung, M. and Ourisson, G. (1980): Hopanoids from C_{30} to C_{35} in Recent muds. *Naturwissenschaften*, **67**, 456–458.

Rohmer, M., Bouvier-Nave, P. and Ourisson, G. (1984): Distribution of hopanoid triterpenes in prokaryotes. *J. General Microbiol.*, **130**, 1137–1150.

Schoell, M. and Hayes, J. M. (1994): *Compound-Specific Isotope Analysis in Biogeochemistry and Petroleum Research* (*Org. Geochem.*, **21**, Nos. 6/7), Pergamon.

Simoneit, B. R. T., Grimalt, J. O., Fischer, K. and Dymond, J. (1986): Upward and downward flux of particulate organic material in abyssal waters of the Pacific Ocean. *Naturwissenschaften*, **73**,

S322–S325.

Tissot, B. T. and Welte, D. H. (1984): *Petroleum Formation and Occurrence*, 2nd edn., Springer-Verlag, Berlin, Heidelberg, New York, Tokyo.

Tsuge, S. and Matsubara, H. (1984): High-resolution pyrolysis-gas chromatography of proteins and related materials. *J. Anal. Appl. Pyrol.* **8**, 49–64 (1985).

Venkatesan, M. I. (1988): Diploptene in Antarctic sediments. *Geochim. Cosmochim. Acta*, **52**, 217–222.

Volkman, J. K. (1986): A review of sterol markers for marine and terrigenous organic matter. *Org. Geochem.*, **9**, 83–99.

Volkman, J. K., Barret, S. M. and Dunstan, G. A. (1994): C_{25} and C_{30} highly branched isoprenoid alkenes in laboratory culture of two marine diatoms. *Org. Geochem.*, **21**, 407–413.

Wakeham, S. G. (1982): Organic matter from a sediment trap experiment in the equatorial North Atlantic: wax esters, steryl esters, triacylglycerols, and alkyldiacylglycerols. *Geochim. Cosmochim. Acta*, **46**, 2239–2257.

Wakeham, S. G. (1985): Wax esters and triacylglycerols in sinking particulate matter in the Peru upwelling area (15°S, 75°W). *Marine Chem.*, **17**, 213–235.

Wakeham, S. G. (1990): Algal and bacterial hydrocarbons in particulate matter and interfacial sediment of the Cariaco Trench. *Geochim. Cosmochim. Acta*, **54**, 1325–1336.

Wakeham, S. G. (1993): A photoautotrophic source for lycopane in marine water columns. *Geochim. Cosmochim. Acta*, **57**, 159–165.

Wakeham, S. G. and Lee, S. (1993): Production, transport, and alteration of particulate organic matter in the marine water column.pp. 145–169. In *Organic Geochemistry*, ed. by M. H. Engel and S. A. Macko, Plenum Press, New York.

Wakeham, S. G., Farrington, J. W., Gagosian, R. B., Lee, C., DeBaar, H., Nigrelli, G. E., Tripp, B. W., Smith, S. O. and Frew, N. M. (1980): Organic matter fluxes from sediment traps in the equatorial Atlantic Ocean. *Nature*, **286**, 798–800.

Yamada, K., Uzaki, M., Kon, M., Naraoka, H. and Ishiwatari, R. (1994): Molecular sieve isolation technique for use in stable carbon isotope analysis of individual long-chain n-alkanes in crude oil. *J. Mass Spectrom. Soc. Jpn.*, **42**, 237–246.

Dynamics and Characterization of Marine Organic Matter, Eds., N. Handa, E. Tanoue and T. Hama, pp. 169–186.

^{14}C Budget of Sinking Particulate Organic Matter in the Japan Trench: A New Approach to Estimate the Contribution from Resuspended Particles in Deep Water Column

Takeshi NAKATSUKA*, Ayako HOSOKAWA, Nobuhiko HANDA**, Eiji MATSUMOTO and Toshiyuki MASUZAWA

Institute for Hydrospheric-Atmospheric Sciences, Nagoya University, Nagoya 464-01, Japan

(Original, Received 27 September 1996; accepted 27 December 1996)

Abstract—The ^{14}C contents of organic carbon ($\delta^{14}C$) were measured in sinking particles collected by year-long time-series sediment trap systems moored at four different depths on the deepest axis of the Japan Trench. The annual mean $\delta^{14}C$ values in sinking particles decreased with depth consistently, from +39‰ at 1171 m to −39‰ at 8681 m, while there were also some seasonal changes in $\delta^{14}C$ of sinking particles at each depth. Because the $\delta^{14}C$ value of sinking particulate organic carbon (POC) originated from surface water cannot change significantly during its settling process in deep water column due to its relatively long half life (5720 yrs), this nearly 80‰ downward decrease in $\delta^{14}C$ value inevitably indicated that some "old" carbon is entrained into sinking particles in the deep water column. Considering that the $\delta^{14}C$ values of sinking POC in the central regions of oceans do not show such large vertical changes, this "old" carbon must come from surface sediments on the continental slope area, adjacent of the trap site, through resuspension of sedimentary particles. The $\delta^{14}C$ value of sinking POC at each trap depth was compared with the $\delta^{14}C$ values of surface sediments and the hypothesized $\delta^{14}C$ values of sinking POC exported from surface water, in order to calculate the proportion of the resuspended sediment fraction in each sinking POC sample. While the results showed that 24–56% of sinking organic carbon at 8681 m depth may come from the surface sediment beneath the trench, this flux of resuspended "old" surface sediment could not solely explain the increase in sinking POC flux near bottom of water. Consequently, it is also suggested that there is not only the resuspension of "old" sedimentary particles but also the "rebound" of fresh sinking particles on the bottom which induces the anomalous increase in sinking particle flux near sediment surface.

INTRODUCTION

In the ocean, sinking particulate organic matter from surface water acts as a main route of food supply to the deep water and benthic ecosystems. On the other hand,

*Present address: Institute of Low Temperature Science, Hokkaido University, Sapporo 060, Japan.
**Present address: Aichi Prefectural University, Nagoya 467, Japan.

it also makes the steep vertical gradients of nutrients and carbonate materials in the oceanic water columns through the active uptake of those dissolved materials by phytoplankton in surface water to form sinking particles, and the successive regeneration of those dissolved materials in deep water columns during its settling process. This mechanism of sinking POC to remove the carbon dioxide from surface to deep water consequently keeps the partial pressure of CO_2 at a low level in the atmosphere too. The sinking POC is thus a key pathway in the material cycling on Earth which controls our living environments, and has been called the "biological pump" (Berger *et al.*, 1989) in the ocean.

Exact measurements of the spatio-temporal variations in the flux of sinking POC are, therefore, very important in understanding the Earth's surface environment, and many experiments to collect the sinking particles have been carried out all over the world during the past two decades using mooring systems equipped with sediment traps. However, especially in the regions near continental margins, the anomalous vertical profiles in the flux of sinking particulate organic matter have been sometimes reported in deep water columns, which do not decrease but increase downward (Honjo, 1982; Honjo *et al.*, 1982). Because the usual degradation processes of sinking particles cannot explain this downward increasing trend in their flux, the anomalous profiles near the bottom of the water have been attributed to the resuspended particles from sediment surfaces, especially from the sediment on the continental slopes. The input of resuspended sedimentary particles largely affect the apparent rates of downward decreases in the sinking POC flux, and it makes the calculation of vertical transport of organic carbon difficult there. On the other hand, the resuspension and diffusive outflow of sedimentary particles from continental slope areas are also evaluated as one of the major pathways of horizontal transport of organic matter from coastal regions, where very high primary production usually occurs due to nutrient input from land and coastal upwelling, to the open ocean (Biscaye *et al.*, 1988; Walsh *et al.*, 1991). Therefore, it is very important to quantitatively estimate the contribution from resuspended sedimentary particles in sinking POC in continental margins, and make clear the mechanism of resuspension of sedimentary particles from continental slope to open ocean.

In this study, the $\delta^{14}C$ values in sinking POC are applied to estimate the contribution from the resuspended sedimentary particles to sinking POC. The $\delta^{14}C$ measurement for such a small amount of carbon as the sinking POC sample has become possible through the recent development of tandem-accelerator mass spectrometry (AMS) (Nakamura *et al.*, 1987). Because the ^{14}C has a very long half life (5720 yrs), the $\delta^{14}C$ of sinking POC from surface water cannot change significantly during its settling periods, which is in the order of 30–50 days (Deuser *et al.*, 1991), and must keep a high value as well as that in the surface water (about +100‰). On the other hand, the organic carbon in the surface sediment on the bottom of deep water has very low $\delta^{14}C$ values (Druffel *et al.*, 1992) due to the bioturbation effect there. A simple mass balance calculation of ^{14}C in sinking POC may, therefore, provide a clear identification of the contribution of resuspended sedimentary particles from the sea bottom.

HOW TO ESTIMATE THE FRACTION OF RESUSPENDED SEDIMENT IN SINKING PARTICLES?

So far, the addition of resuspended sedimentary particles into sediment traps has been suggested as follows (Fig. 1; Walsh *et al.*, 1988b etc.). Firstly, the vertical profiles in annual mean fluxes of sinking particulate organic matter are measured using year-long time-series sediment traps moored at several different depths. Secondly, the downward decreasing rates in their fluxes are calculated, using the data from "upper traps" where the resuspended sedimentary particles may not affect the apparent fluxes, on the assumption of the first-order kinetics for the decomposition process of organic matter during their settling in the deep water column. Thirdly, the fluxes of sinking particulate organic matter are extrapolated to "lower trap depths", using the estimated first-order decomposition rate and the assumption of the constant sinking speed of particles in the deep water column. Finally, the differences between the measured and extrapolated fluxes of sinking particulate organic matter are explained as a resuspended fraction from the bottom of the water in the sinking particles.

However, this traditional procedure to estimate the resuspended effect has several disadvantages, as follows. 1) This procedure requires two or more sediment traps at upper water depths which are not affected by resuspended sedimentary particles. In the area of continental margins, however, even the upper traps deployed at shallow depths may also be affected by resuspended sedimentary particles, because the horizontal transport of resuspended sedimentary particles from the continental slope may occur at various water depths. 2) The extrapolation of flux of sinking particulate organic matter from upper traps to a lower trap is based on the assumption that the decomposition rates of organic matter in sinking particles are constant during their settling process in the deep water column. However, the rates of decomposition usually become smaller after the removal of labile components in the upper water column (Walsh *et al.*, 1988a; Banse, 1990) and, therefore, the calculation of flux for the lower trap depth by the extrapolation procedure may result in an under-estimation of the actual flux of sinking particulate organic matter which originated from surface water. 3) In this procedure, the flux at each trap depth must be measured very precisely in order to obtain a comparison between different trap depths. However, there are also several other factors besides the resuspended sedimentary particles which may affect the estimation of the sinking particle flux in sediment trap experiments; i.e. water currents, especially in a shallow trap depth (Gardner, 1980). Nevertheless, if there is no input of additional particles from outside the sinking particles, this under- or over-estimation of the fluxes of sinking particulate organic matter may be calibrated using some refractory components in sinking particles such as aluminum (Walsh *et al.*, 1988a) and/or clays (Noriki and Tsunogai, 1986), on the assumption that Al or clay fluxes must be constant vertically and the apparent vertical changes in the Al or clay fluxes can be attributed to some artificial problems in sediment trap experiments. However, if the resuspended sedimentary particles also affect the measured flux of sinking particulate organic matter, this calibration cannot be carried out.

$\delta^{14}C$ MASS BALANCE MODEL FOR SINKING POC

In contrast to the traditional estimation of resuspended sedimentary particles in the deep water column, the ^{14}C mass balance method presented in this paper consists of very simple procedures. The $\delta^{14}C$ values of sinking POC originating from surface water does not change significantly during its settling process, except for the small isotope fractionation effect during the decomposition of sinking particulate organic matter (Nakatsuka *et al.*, 1997). This is the great advantage of the $\delta^{14}C$ values, compared to the sinking POC flux or its concentration which vary greatly during the settling process. If the dissolved organic carbon (DOC) is largely transferred into particulate form in the deep water column, the apparent $\delta^{14}C$ values of sinking POC may become lower with increasing depth, because the DOC in the deep water column has a very "old" $\delta^{14}C$ age such as 6000 BP (Williams and Druffel, 1987). However, a recent report on $\Delta^{14}C$ of sinking POC in the central gyre of the North Pacific Ocean does not show significant vertical changes in their $\Delta^{14}C$ values at all (Druffel *et al.*, 1992). Therefore, we can conclude that only the resuspension of sedimentary particles in the continental margin areas can probably make the $\delta^{14}C$ values of sinking POC lower in the deep water column, as already reported from the Atlantic Bright of the North American Continent (Anderson *et al.*, 1994).

On the basis of a simple two end-member model, the measured $\delta^{14}C$ values of sinking POC ($\delta^{14}Cm$) can be represented as follows:

$$\delta^{14}Cm = \delta^{14}Cp\cdot(1 - r) + \delta^{14}Cs\cdot r, \tag{1}$$

where $\delta^{14}Cp$ and $\delta^{14}Cs$ are the $\delta^{14}C$ values of the sinking POC exported from the surface water and the resuspended sedimentary POC which is incorporated into sinking particles in the deep water column, respectively. And, r is the proportion of the resuspended sediment fraction in sinking POC. After the arrangement of Eq. (1), the proportion of resuspended sediment fraction (r) in each sinking POC sample can be obtained as follows:

$$r = (\delta^{14}Cp - \delta^{14}Cm)/(\delta^{14}Cp - \delta^{14}Cs). \tag{2}$$

Compared to the traditional method described in Fig. 1, the estimation of the resuspended sedimentary fraction (r) according to Eq. (2), has several great advantages. Because this procedure need not extrapolate the vertical trend in sinking POC fluxes observed at shallower traps to deeper depths, it is not necessary to assume the first-order kinetics for the decompositional process of sinking organic matter during their settling process. And, moreover, the estimation of the fraction of resuspended sedimentary particles in each sinking POC sample can be carried out using only one sediment trap, moored in one layer of water. And if the efficiency of the traps in collection of sinking particles is affected by some artificial problems, it does not influence the $\delta^{14}C$ values of sinking POC itself or the estimation by Eq. (2) at all.

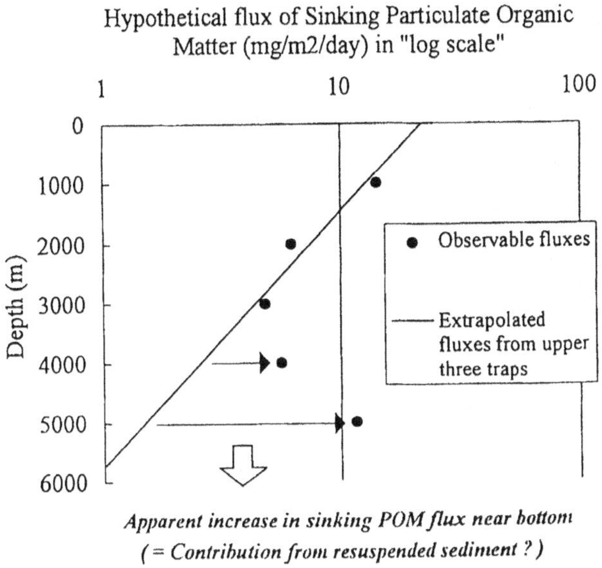

Fig. 1. Schematic illustration of the traditional concept of the contribution from resuspended sedimentary particles into sinking particle flux in the near-bottom sediment trap.

On the other hand, however, the most critical subject in order to apply Eq. (2) is the determination of the end-member values ($\delta^{14}Cp$ and $\delta^{14}Cs$). Because we do not have exact knowledge about where the resuspension of sedimentary particles actually occurs on the bottom of the sea, it is not easy to determine the $\delta^{14}Cs$ values now. In the case of $\delta^{14}Cp$, the situation is a little easier because the $\delta^{14}Cp$ can be estimated from the $\Delta^{14}C$ values of total carbonate materials of surface water in the region where a sediment trap is deployed (Oslund *et al.*, 1985; Toggweiler, 1989). However, the $\Delta^{14}C$ values vary about 30‰ seasonally in temperate surface water (Broecker and Peng, 1980), and the vertical changes in $\Delta^{14}C$ of ΣCO_2 in surface and subsurface waters may also affect the $\delta^{14}Cp$ if the primary production not only occurs in surface mixed layer but also in thermocline waters. Moreover, eolian dusts deposited onto the sea surface may also influence the $\delta^{14}C$ values of sinking POC from surface water. In this paper, therefore, we discuss those problems in details, based on the $\delta^{14}C$ measurements of sinking POC and sediment samples.

SEDIMENT TRAP AND SEDIMENT SAMPLES

Sediment trap experiment was carried out from March 20, 1992 to March 8, 1993 on the deepest axis of the Japan Trench (9181 m; 34°10.27′ N; 141°58.36′ E) off the Boso Peninsula, Japan (Fig. 2) using the research vessels *Tansei-Maru* (Ocean Research Institute, University of Tokyo, Japan) and *Shinyo-Maru* (Tokyo University of Fishery, Japan). Time-series sediment traps (McLean Mark VI and

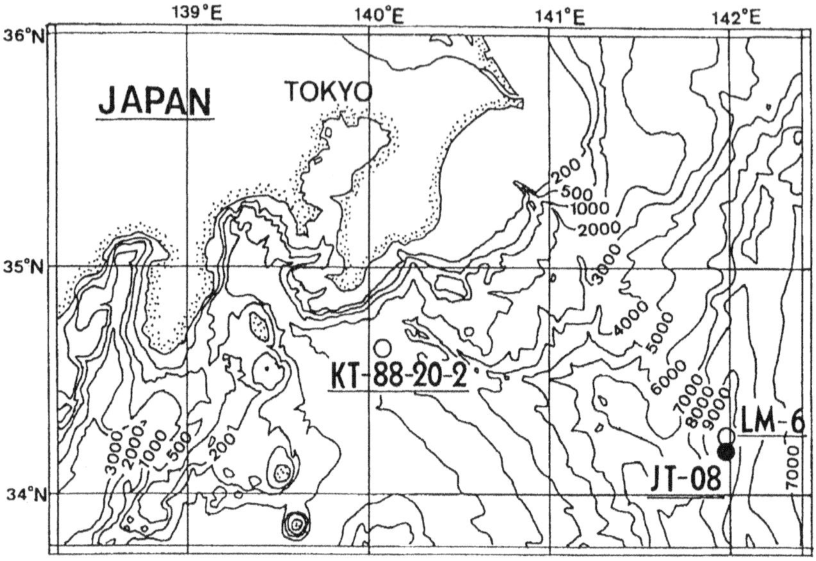

Fig. 2. Site locations of sediment trap experiment and deep sea sediment samplings.

VII) were moored at four different depths (1171, 3673 , 5679 and 8681 m), and 13 cups were attached to each sediment trap to collect sinking particles. Each sampling interval was set at 27 days. Before deployment of traps, each sample collecting cup was filled with a buffered 5% formalin solution of filtrated sea surface water to prevent in-situ degradation of samples until the recovery of the sediment traps. Unfortunately, some samples from 1171 and 8681 m could not be obtained due to mechanical troubles.

The sieving of the recovered samples by 1 mm or 0.5 mm meshes to remove swimmer were not carried out in this study, but it does not seem to significantly affect the results of chemical and isotopic analyses in this study, because the largest particles were no more than 0.5 mm in the samples, at least from the deeper three traps. Each sample was concentrated by a centrifuge to separate sea water from the sinking particles, and the particle samples were stored at –20°C until chemical and isotopic analyses.

The surface sediment samples were collected in the Japan Trench below the trap site (LM-6; 9200 m; 34°13.09′ N; 141°59.10′ E) and on the continental slope area adjacent to trap site (KT-88-20-2; 2574 m; 34°39.30′ N; 140°05.20′ E) (Fig. 2) using a multiple corer and a box corer, respectively. The recovered core samples were immediately subdivided into 1 cm sections, and stored at –20°C until chemical and isotopic analyses.

In order to check effect of the preservative (formalin) on the $\delta^{14}C$ values of the sinking POC samples, about 10 g of fresh sediments collected from 0–1 and 2–3 cm depths below trap site (LM-6) were fixed by buffered 5% formalin solutions of filtrated sea water in 20 ml glass vials, and stored in a refrigerator

(4°C) for two weeks. The size-fractionation effect on the $\delta^{14}C$ values of resuspended sedimentary particles was also estimated using a sieved sample (< 12.5 mm mesh) for the 0–1 cm sediment collected at the continental slope area (KH-88-20-2). These fixed and sieved samples were also analyzed for their ^{14}C (and ^{13}C) contents by the following methods as well as all other samples of sinking particles and surface sediments.

ANALYTICAL METHODS

Samples of sinking particles and surface sediments (0–1 and 2–3 cm at LM-6, and 0–1 cm at KT-88-20-2) were first homogenized in glass vials, and parts of them were acidified in 1N HCl solution to remove carbonate materials for about 12 hours. (In the case of the LM-6 sediment samples, the acidification was not carried out, because carbonate material was not contained there at all due to its very deep location, 9200 m.) Parts of the carbonate-free samples were dried up at 200°C and analyzed for their total carbon and nitrogen contents using an elemental analyzer (Yanaco CHN-corder MT-1S). The remaining the carbonate-free samples were washed with enough distilled water to remove the salts, dried at 110°C for about 8 hours and then ground. (In the formalin contamination test to LM-6 sediments, half of the test samples were not washed but only dried, in order to check whether the formaldehyde can be removed by evaporation or not.)

The carbonate-free sediment powders were sealed in quartz tubes in a vacuum with 1 g Cu-oxide, 1 g-Cu and a fragment of Ag-foil, and combusted at 850°C for two hours. Those combusted tubes were cooled down gradually for about 16 hours in order to convert all the carbon and nitrogen in the samples to CO_2 and N_2 gases (Minagawa *et al.*, 1984), and produced CO_2 and N_2 gases were separated and purified using the cryogenic method. Aliquots of the purified CO_2 gases were directly analyzed for their $^{13}C/^{12}C$ ratios using an isotope ratio mass spectrometer (MAT 251). The remaining CO_2 gases were sealed again in quartz tubes with H_2 gas and 1 mg Fe-powder, and heated at 650°C to reduce all the CO_2 to graphite carbon (Kitagawa *et al.*, 1993). The produced graphite carbons were packed into aluminum holders, and used as targets for measurements of the $^{14}C/^{13}C$ ratios by tandem accelerator mass spectrometer (AMS) in Nagoya University (Nakamura *et al.*, 1987). Due to a limitation of machine time, the $^{14}C/^{13}C$ measurements were carried out only for about one sample in two at each depth. After measuring the $^{14}C/^{13}C$ ratios, the graphite targets were combusted again using almost the same method as used by Minagawa *et al.* (1984), and the CO_2 produced gases were analyzed again for their $^{13}C/^{12}C$ ratios by a MAT-251 mass spectrometer in order to calculate the $^{14}C/^{12}C$ ratios. These $^{13}C/^{12}C$ ratios before and after ^{14}C measurements were also used to check the isotope fractionation effect during the production of graphite targets and to calibrate it.

The measured $^{14}C/^{12}C$ and $^{13}C/^{12}C$ ratios are represented using following notations:

$$\delta^{1X}C = [(^{1X}C/^{12}C)_{sam}/(^{1X}C/^{12}C)_{std} - 1]\cdot 1000,$$

where X is 4 or 3, and the standard materials for $\delta^{14}C$ and $\delta^{13}C$ values are the ideal tree carbon (whose $\delta^{13}C$ values were –25‰) grown at A.D.1950 (Stuiver and Polach, 1977) and Pee Dee Belemnite (PDB), respectively. Usually, the $^{14}C/^{12}C$ ratios are shown by the notation of $\Delta^{14}C$ for geochemical purposes, because $\Delta^{14}C$ has been normalized to $\delta^{13}C$ of –25‰, and we can cancel out the isotope fractionation effect and consider only the "age" of the carbon materials using $\Delta^{14}C$. In this study, however, we do not estimate the "age" of sinking POC but discuss the "mixing ratio" of resuspended sediment in sinking POC using Eqs. (1) and (2). Therefore, the notation of $\delta^{14}C$ is used in this paper in order to exclude the effect from variations of ^{13}C contents in samples. However, in practice, the $\Delta^{14}C$ and $\delta^{14}C$ values are not so different in this study, because all of the samples analyzed are organic matter which have $\delta^{13}C$ values near –25‰, and the isotope fractionation effects during sinking processes of POC are very small (Wada *et al.*, 1987; Nakatsuka *et al.*, 1992; Nakatsuka *et al.*, 1997), far less than the measurement error of $\delta^{14}C$ by AMS (about +/–10‰ for 1σ).

RESULTS AND DISCUSSIONS

Temporal changes in $\delta^{14}C$ values of sinking POC and its flux

The $\delta^{14}C$ values of sinking POC consistently decrease with depth, suggesting a gradual increase in the contribution from "old" resuspended sediment in the traps with depth, but they do not show large seasonal variations except for 8681 m (Fig. 3(a)). Because the ranges of seasonal variations in $\delta^{14}C$ values at 3673, 5679 and 1171 m are all within about 30‰, we cannot exclude the possibility that these temporal changes are not caused by sediment resuspension, but derived from the seasonal variations in the $\Delta^{14}C$ values of ΣCO_2 in surface water (Broecker and Peng, 1980). However, the range of temporal change in $\delta^{14}C$ values at 8681 m is far beyond 50‰, and this inevitably suggests that the seasonal variation in $\delta^{14}C$ of sinking POC at 8681 m is largely affected by changes in the mixing ratios of resuspended "old" carbon into sinking POC from surface sediment. The sinking POC flux (Fig. 3(b)) at 8681 m also varies greatly although at three shallower traps, they do not change very much. This fact also suggests the possibility that the near-bottom trap suffers from very large variations in the sedimentary environment.

However, the relationships between the $\delta^{14}C$ values of sinking POC and their fluxes are not simple. Because the sample of the highest sinking POC flux at 8681 m, in October 1992 (Fig. 3(b)), corresponds to the highest $\delta^{14}C$ value, this highest sinking POC flux cannot be explained by an episodic input of resuspended "old" sedimentary particles into the trap, rather it must be derived from the large amount of "new" sinking POC produced in sea surface water. But, this high sinking POC flux in October 1992 cannot be supported either by the increase in export production from sea surface, because the sinking POC fluxes in October and/or September did not show such episodic increases at the shallower trap depths, as observed at 8681 m. This fact suggests the existence of another mechanism for the transport of sinking particles which is neither the direct

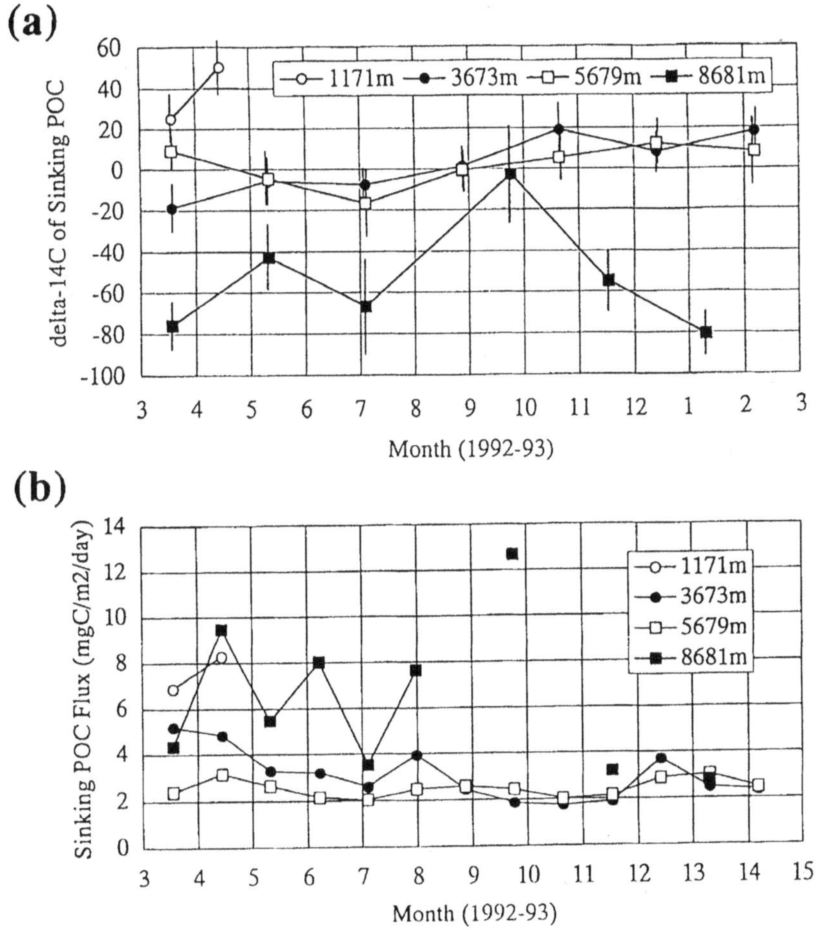

Fig. 3. Seasonal variations in the $\delta^{14}C$ value of sinking POC (a) and its flux (b) at the JT-08 site.

sinking of "new" POC from surface water nor the resuspension of "old" sediment from the sea floor.

The temporal patterns of $\delta^{14}C$ values in sinking POC, however, partly contain the effect from the seasonal variation in the $\Delta^{14}C$ values of ΣCO_2 in surface water (Broecker and Peng, 1980), and we cannot proceed to a more detailed and quantitative discussion about the source of the sinking POC using these temporal patterns. Therefore, we would like to discuss firstly the annual weighted mean values of $\delta^{14}C$ in sinking POC at the four trap depths in order to cancel the effect of the seasonal variation in the $\delta^{14}C$ of sinking POC from surface water, and then come back again to these temporal characteristics on the sinking POC flux and its $\delta^{14}C$ value at the deepest trap depth.

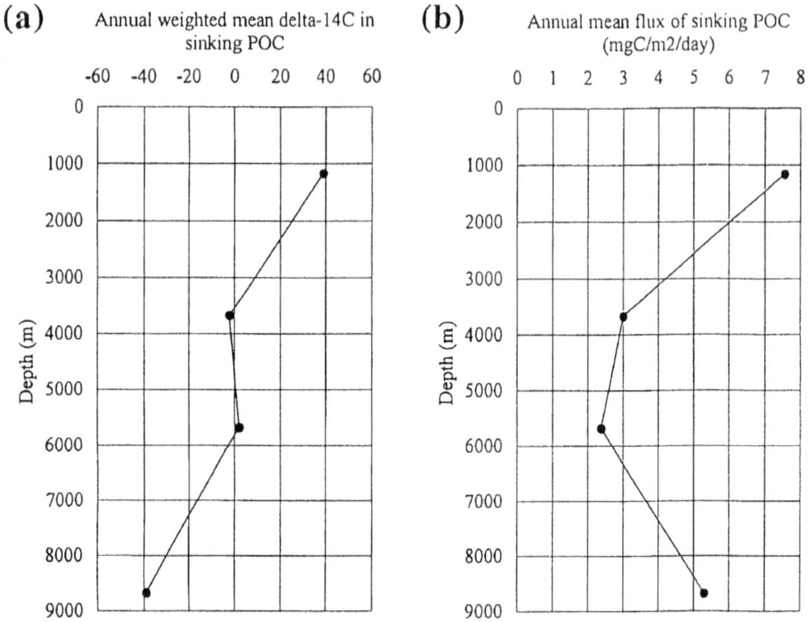

Fig. 4. Vertical profiles of the annual weighted mean $\delta^{14}C$ value of sinking POC (a) and its flux (b) at the JT-08 site. For calculation of the mean sinking POC flux at each depth, only the samples, where the $\delta^{14}C$ values were actually measured, were used.

Vertical change in $\delta^{14}C$ of sinking POC and its flux

The profiles of annual weighted mean values of $\delta^{14}C$ in sinking POC and its flux are shown in Fig. 4. The calculations for average fluxes of sinking POC were carried out only for the samples where the $\delta^{14}C$ values were measured. The weighted mean $\delta^{14}C$ values decreased downward from +39‰ at 1171 m to −39‰ at 8681 m. Because the calculation procedure for annual averaged $\delta^{14}C$ values cancelled the effect of the seasonal changes in surface water $\Delta^{14}C$, this downward decreasing trend in $\delta^{14}C$ of sinking POC inevitably indicates the fact that the mixing ratio of resuspended "old" sedimentary POC in the sediment traps increases together with depth. This seems consistent with the trend that the sinking POC flux increases downward very much at the trap near the sea bottom (Fig. 4(b)). In contrast to the shallowest (1171 m) and deepest (8681 m) traps, the two traps from the intermediate layers (3673 m and 5679 m) showed almost identical $\delta^{14}C$ values (about 0‰), suggesting that the entrainment of the resuspended "old" sedimentary POC had not occurred much between the two trap depths. Although, at present, it is not clear why the resuspension did not occur only at the intermediate depth interval, it may be controlled by the geographical situation of the trap site and/or the depth profile of the intensity of the Kuroshio current above the trap.

Table 1. Results of formalin contamination tests for $\delta^{14}C$ values using sediments in surface bioturbation zone (LM-6)

	(a)	(b)	(c)
LM-6 (0–1 cm)	−291 +/− 10‰	−268 +/− 9‰	−264 +/− 9‰
LM-6 (2–3 cm)	−283 +/− 11‰	−277 +/− 12‰	−296 +/− 12‰

*Numbers in this table are $\delta^{14}C$ values determined by the same method used for sinking particle samples.

*(a), (b) and (c) mean "samples of non-formalin treatment", "samples washed with distilled water after formalin treatment" and "samples not-washed with distilled water after formalin treatment", respectively.

For this downward decreasing trend in $\delta^{14}C$, however, the adsorptive contamination of formalin (preservative for sediment trap samples) may partly affect the deeper trap samples. Because these samples contain higher amount of clay minerals (Hosokawa, 1996) and the formalin must have a very low $\delta^{14}C$ value (probably equal to −1000‰ = dead carbon), the adsorption of formalin to clay mineral, if possible, may have the $\delta^{14}C$ values lower at the deepest trap. We, therefore, considered the effect of formalin contamination on sinking POC by some tests using surface sediment samples (Table 1). The treatment of surface sediment by formalin did not show any distinct decreases in the $\delta^{14}C$ values of POC at all. Moreover, the samples, which was not washed but just dried after formalin treatment, also showed almost the same $\delta^{14}C$ values as the non-formalin treatment of the original sample, suggesting that most of formaldehyde molecules can be removed only by evaporation procedures.

(Strictly speaking, the expression "annual mean" is not correct in this study, because the $\delta^{14}C$ values were not measured for about one in every two samples at the deeper three traps, and at the shallowest trap no sample could be collected after No. 3 sample. However, the estimated $\delta^{14}C$ values seem to be almost equal to the "real" annual mean values at the middle two trap depths because there are only small seasonal variations in both the sinking POC flux and its $\delta^{14}C$ value there. For the trap from 8681 m, the calculated average of sinking POC flux is slightly lower than the total average and closer to the values at the middle trap depths, because some samples of a higher POC flux at 8681 m (No. 2, 4 and 6 in Fig. 3(b)) were not included for average calculation. The source identification of sinking POC at 8681 m using $\delta^{14}C$ values in this paper, therefore, may be biased closer to those at the middle trap depths.)

Endmember values of $\delta^{14}C$ for the source identification of sinking POC

In order to calculate the mixing ratio of resuspended sedimentary particles (r) into each trap depth using Eq. (2), it is necessary to predetermine the endmember values of $\delta^{14}C$ for the resuspended sedimentary POC ($\delta^{14}Cs$) and the sinking POC exported from the surface water ($\delta^{14}Cp$), besides the annual mean

Table 2. Four extreme cases on the $\delta^{14}C$ endmember values for the exported particles from surface water ($\delta^{14}Cp$) and the resuspended sedimentary particles ($\delta^{14}Cs$)

	$\delta^{14}Cp$	$\delta^{14}Cs$
Case 1	+39‰	−291‰
Case 2	+39‰	−149‰
Case 3	+100‰	−291‰
Case 4	+100‰	−149‰

*The hypothesized values for $\delta^{14}Cp$ and $\delta^{14}Cs$ are estimated as the upper and/or lower limit values (see text for the details).

$\delta^{14}C$ value for each trap depth ($\delta^{14}Cm$). As discussed above, it is not easy to determine the endmember values. We considered the following four extreme cases for those values (Table 2).

For the $\delta^{14}Cp$ value, two extreme cases can be considered. One is that the $\delta^{14}Cp$ is equal to $\delta^{14}C$ at 1171 m, +39‰. If there is no input from resuspended "old" sediment to sinking particles throughout the water column above 1171 m depth, the $\delta^{14}C$ values at 1171 m must be equal to $\delta^{14}Cp$. But if there are any "old" POC inputs in the upper water column, the real $\delta^{14}Cp$ value must be higher than $\delta^{14}C$ at 1171 m. Therefore, the $\delta^{14}C$ values at 1171 m can be set as the lower limit for $\delta^{14}Cp$. The other extreme value for $\delta^{14}Cp$ is $\Delta^{14}C$ of surface water ΣCO_2, about +100‰ in this area (Toggweiler, 1989). Because all the surface water planktons are produced from ΣCO_2 in surface water via primary production, $\delta^{14}Cp$ must be almost equal to $\Delta^{14}C$ of surface water ΣCO_2 there. However, if the primary production occurs below the surface mixed layer, and/or if there is a significant eolian flux of "old" terrestrial POC to the sea surface, the endmember value of $\delta^{14}C$ in sinking POC from the surface and subsurface water, which is by definition equal to $\delta^{14}Cp$, must become lower than $\delta^{14}C$ of the surface water ΣCO_2. Therefore, the $\delta^{14}C$ of ΣCO_2 in surface water, +100‰, can be set as the upper limit for $\delta^{14}Cp$.

For the $\delta^{14}Cs$ value, we present the following two extreme values as the hypothetical upper and lower limits. Because the resuspension of sedimentary particles must occur at some place in the region from the upper continental slope to the bottom of the trench, we use the $\delta^{14}C$ values of POC in 0–1 cm layers of the surface sediments obtained at the upper slope of the Sagami Trough (KT-88-20-2), −149‰, and the bottom of the Japan Trench (LM-6), −291‰, as the upper and lower limits for $\delta^{14}Cs$ value, respectively (Fig. 2). The $\delta^{14}C$ value of surface sediment decreases monotonously with depth from Sagami Bay to the Japan Trench in this region (T. Nakatsuka, unpublished data), probably reflecting a decrease in sedimentation rate and an increase in the resuspended sediment fraction on the sediment surface with depth. Therefore, if the resuspension in fact occurs above the contour line of 2500 m depth (Fig. 2), the real $\delta^{14}Cs$ value must be higher than those assumed in Table2, and the calculated results of the resuspended fraction in sinking POC will be under-estimated in this study.

The $\delta^{14}C$ value of real resuspended sedimentary POC may be shifted from the hypothesized $\delta^{14}Cs$ values, determined from the $\delta^{14}C$ values of the organic matter in surface sediment, if the size fractionation of particle occurs at the time of resuspension. In order to check this effect, we measured the $\delta^{14}C$ value of a size-fractionated sample of 0–1 cm sediment at KT-88-20-2. The $\delta^{14}C$ value of POC in the fine fraction (<12.5 mm) of that sediment was –153 +/– 14‰, which is almost as same as that of the total sediment (–149 +/– 15‰). This suggests that the size fractionation of particle at the time of resuspension does not significantly affect the $\delta^{14}C$ value of resuspended sediment, while the TOC concentration itself must largely change during the size-fractionation process (Tanoue and Handa, 1979).

Source of sinking POC in deep water inside the Japan Trench

According to Eq. (2), the mixing ratio of resuspended sedimentary POC (r) in each trap depth can be calculated on the basis of the hypothesized endmember values for $\delta^{14}Cp$ and $\delta^{14}Cs$ (Table 2). The calculated percentage of the resuspended fraction (Fig. 5) increases with depth and it ranges from 0 to 24% at 1171 m, and from 24 to 56% at 8681 m. Because the $\delta^{14}C$ values for the endmembers of mixing were estimated as the upper and lower limit values for $\delta^{14}Cp$ and $\delta^{14}Cs$, respectively, the real percentage of the resuspended fraction in each sediment trap depth must be also covered by the estimated ranges of the resuspended fraction at each trap depth (Fig. 5).

Using the estimated percentages of the resuspended POC fraction and the profile of total sinking POC flux (Fig. 4(b)), we further calculated the profile of the vertical flux of the "non-sediment-origin" fraction in sinking POC (Fig. 6), which must be equal to POC from surface water, in order to check whether or not

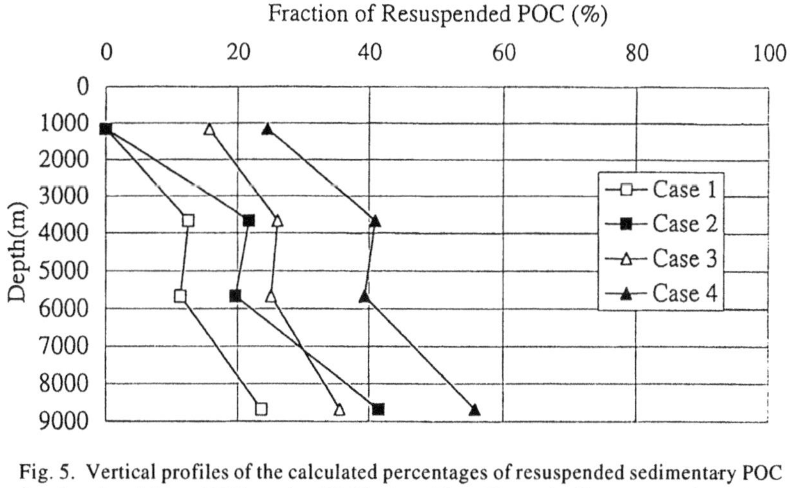

Fig. 5. Vertical profiles of the calculated percentages of resuspended sedimentary POC fraction in sinking POC at the JT-08 site. The "Cases 1–4" correspond to those determined in Table 2.

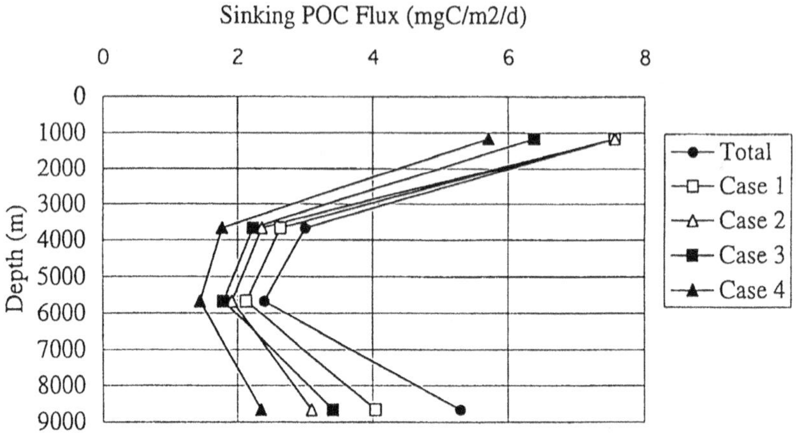

Fig. 6. Vertical profile of the calculated fluxes of the non-sedimentary sinking POC at the JT-08 site. The "Cases 1–4" correspond to those determined in Table 2, and "total" fluxes are equal to that in Fig. 4(b).

it decreases steadily with depth. In all of the four cases of hypothesized endmember values, however, the "non-sediment-origin" sinking POC flux still increased significantly at 8681 m, which cannot be explained by the resuspension of surface sediment. From the viewpoint of $\delta^{14}C$, we must conclude that the increase in sinking POC flux near the sea bottom (Fig. 4(b)) must be attributed not only to the resuspension of surface sediment but also to the lateral input of fresh particles which have just come from the surface water.

Where do those fresh particles come from? One possible mechanism for this near-bottom increase of the fresh particles in sinking POC is the "rebound" of newly deposited fresh particles on the sediment surface (Walsh *et al.*, 1988b). Billet *et al.* (1983) and Lampitt *et al.* (1985) observed, using a deep-sea camera, that large amounts of phytoplankton aggregates deposit on the sea floor during a short time interval just after a phytoplankton bloom, probably a diatom bloom (Alldredge *et al.*, 1995), at surface water, and that newly deposited materials quickly disappear again by resuspension due to the bottom-water current. Because the newly deposited fresh materials can "rebound" (=resuspended again quickly) on the sediment-water interface without being ingested into the bioturbation zone, whose "old" $\delta^{14}C$ values we analyzed for $\delta^{14}C_s$ endmembers, the $\delta^{14}C$ values of the "rebound" fraction must be almost equal to those of sinking POC which directly come from surface water ($\delta^{14}C_p$).

Under the assumptions that, in the deep water column below 3500 m depth, first-order kinetics governs the downward degradation process of the refractory fraction in the exported sinking POC from surface water, and the "rebound" particles are not included in the middle two sediment traps, we can calculate the mixing ratios of the three sources in sinking POC (A: resuspended sedimentary particles, B: exported particles from surface water and C: "rebound" fresh

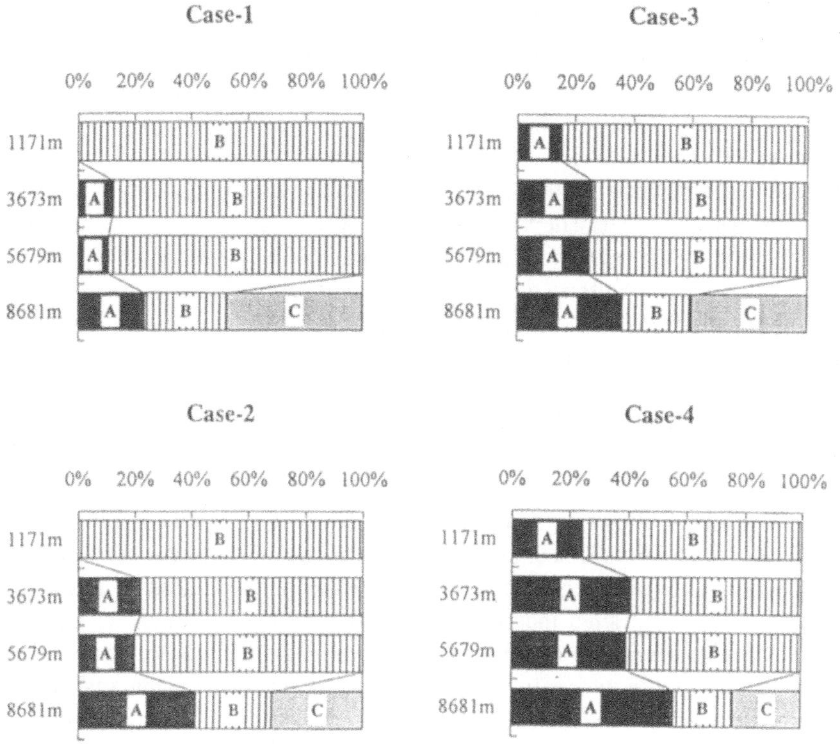

Fig. 7. Vertical changes in the proportions of the source of sinking POC at the JT-08 site. "A", "B" and "C" correspond to the POC sources; "resuspended sedimentary particles", "exported particles from surface water" and "rebound fresh particles from water bottom", respectively. The "Cases 1–4" correspond to those determined in Table 2.

particles from water bottom) quantitatively at each trap depth (Fig. 7). The "A" fraction at each trap depth has been already determined by Eq. (2) as in Fig. 5 and, the "B" fraction in the sinking POC at 8681 m can be calculated, using a log-plot extrapolation of the estimated profiles of "non-sediment-origin" flux of sinking POC at the middle two trap depths (Fig. 6). According to Fig. 7, we can infer that the sinking POC at 8681 m comprises about 24–47% of the fresh "rebound" particles which have just been deposited on the sea bottom and immediately resuspended again there. (In this case, the procedure to determine the "B" fraction at 8681 m is almost the same as that described in Fig. 1, for which we previously pointed out several disadvantages. While the removal of resuspended sediment fractions from 3673 m and 5679 m before this procedure must make the calculated results of "B" fraction more reliable ones, we should also notice the possibility that this result may be distorted partly by some of the problems discussed for Fig. 1.)

Mechanisms of particle transport in continental slope area

Now we discuss again the temporal variations in $\delta^{14}C$ values of sinking POC and its flux at each trap depth (Fig. 3). At the middle trap depths (3673 and 5679 m), the $\delta^{14}C$ values do not change temporally very much although the $\delta^{14}C$ values themselves are always less than the estimated $\delta^{14}Cp$ values in Table 2. This suggests that the contributions from resuspended sedimentary particles to sinking POC at these middle trap depths are almost constant, and the significant amounts of sedimentary particles are always resuspended from the continental slope area to the open ocean. Contrary to these middle depths, the $\delta^{14}C$ values at the deepest trap (8681 m) near the water bottom temporally change greatly. And, moreover, the highest flux of sinking POC in October was not accompanied by the lowest $\delta^{14}C$ value, which indicates the episodic resuspension of "old" sedimentary particles, but by the highest $\delta^{14}C$ value, which apparently suggests an increase in "new" POC flux due to the enhanced export production from surface water. As discussed above, however, there were no signs of the corresponding increases in the sinking POC fluxes at the upper trap depths (Fig. 3(b)). Therefore, the high $\delta^{14}C$ value with a high flux of sinking POC at the deepest trap means the episodic resuspension of newly deposited sinking particles on the bottom (="rebound" as mentioned above), rather than the resuspension of aged sedimentary particles.

In consequence, we can image two kinds of characteristic mechanisms of particle transport in the continental slope area. One is the constant resuspension of fine sedimentary particles on the continental slope. The resuspended fine sedimentary particles can be horizontally transported into the open ocean area on the isopycnal surfaces. Then, they are steadily scavenged by the exported large particles from the surface water, and entrained into sinking flux throughout the deep water column. The other is the episodic resuspension of newly deposited fresh particles (="rebound" particles) from the sea bottom. This "rebound" particles may not go up to higher levels in the deep water column and settle down again immediately. Consequently, it may result in an increase in sinking POC flux only at the near-bottom sediment trap as observed at 8681 m. The reason why the "rebound" occurs episodically at some particular periods is not clear at present, and it is a future subject for study.

CONCLUSION

The annual mean $\delta^{14}C$ values in sinking POC consistently decreased with depth, from +39‰ at 1171 m to −39‰ at 8681 m at the deepset site of the Japan Trench. Using the two source $\delta^{14}C$ model of sinking POC, consisting of resuspended sedimentary particles (low $\delta^{14}C$) and the exported particles from surface water (high $\delta^{14}C$), we could calculate the proportion of the resuspended sedimentary fraction in sinking POC at each trap depth. Although the uncertainties about the hypothesized endmember values for $\delta^{14}C$ of the resuspended sediment ($\delta^{14}Cp$) and the exported particle from surface water ($\delta^{14}Cs$) make the ranges of calculated results quite large, we could at least find the following facts from the $\delta^{14}C$ mass balances. The percentages of resuspended sedimentary POC increased with depth, indicating that large amounts of resuspended "old" sedimentary POC are entrained into sinking particles, even above the middle trap depths (~3500, 5500 m). The estimated vertical flux of non-sediment-origin POC, however, still increased largely at the near-bottom sediment trap. This inevitably suggested the existence of another source of sinking POC at the near-bottom water layer besides the previous two sources. It could be inferred from the temporal pattern of $\delta^{14}C$ variation in the sinking POC at 8681 m that the episodic resuspension of newly deposited fresh particles from the sediment-water interface causes an anomalous increase in the flux of "non-sediment-origin POC" near bottom.

Acknowledgements—We thank all the scientists, especially Y. Nozaki, captains and crews in the research cruises of *Tansei-Maru*, *Hakuho-Maru* and *Shinyo-Maru* for their help in collecting sediment trap samples and deep sea sediments. Thanks are also due to T. Nakamura and A. Ikeda for the facility to use the tandem accelerator mass spectrometer at Nagoya University. We are also grateful to T.Yoshioka and K.Oguri for the facilities to prepare the CO_2 gases and graphite targets from POC samples. This study was partly supported by a grant-in-aid No. 07740438 of the Japanese Ministry of Education, Science, Sports and Culture.

REFERENCES

Alldredge, A. L., C. Gotschalk, U. Passow and U. Riebesell (1995): Mass aggregation of diatom blooms: Insights from a mesocosm study. *Deep-Sea Res. II*, **42**, 9–27.

Anderson, R. F., G. T. Rowe, P. F. Kemp, S. Trumbore and P. E. Biscaye (1994): Carbon budget for the mid-slope depocenter of the Middle Atlantic Bight. *Deep-Sea Res. II*, **41**, 669–703.

Banse, K. (1990): New views on the degradation and disposition of organic particles as collected by sediment traps in the open sea. *Deep-Sea Res.*, **37**, 1177–1195.

Berger, W. H., V. S. Smetacek and G. Wefer (1989): Ocean productivity and paleoproductivity—an overview. pp. 1–34. In *Productivity of the Ocean: Present and Past*, ed. by W. H. Berger, V. S. Smetacek and G. Wefer, John Wiley & Sons, Chichester.

Billet, D. S. M., R. S. Lampitt, A. L. Rice and R. F. C. Mantoura (1983): Seasonal sedimentation of phytoplankton to the deep-sea benthos. *Nature*, **302**, 520–522.

Biscaye, P. E., R. F. Anderson and R. L. Deck (1988): Fluxes of particles and constituents to the eastern United States continental slope and rise: SEEP-I. *Cont. Shelf Res.*, **8**, 885–904.

Broecker, W. S. and T.-H. Peng (1980): Seasonal variability in the $^{14}C/^{12}C$ ratio for surface ocean water. *Geophys. Res. Lett.*, **7**, 1020–1022.

Druffel, E. R. M., P. M. Williams, J. E. Bauer and J. R. Ertel (1992): Cycling of dissolved and particulate organic matter in the open ocean. *J. Geophys. Res.*, **97**, 15,639–15,659.

Gardner, W. D. (1980): Sediment trap dynamics and calibration: a laboratory evaluation. *J. Mar. Res.*, **38**, 17–39.

Honjo, S. (1982): Seasonality and interaction of biogenic and lithogenic particulate flux at the Panama Basin. *Science*, **218**, 883–884.

Honjo, S., D. W. Spencer and J. W. Farrington (1982): Deep advective transport of lithogenic particles in the Panama Basin. *Science*, **216**, 516–518.

Hosokawa, A. (1996): Characteristics of material transport system in coastal regions. Master Thesis, Nagoya University, Japan (in Japanese).

Kitagawa, H., T. Masuzawa, T. Nakamura and E. Matsumoto (1993): A batch preparation method of graphite targets with low background for AMS ^{14}C measurements. *Radiocarbon*, **35**, 295–300.

Lampitt, R. S. (1985): Evidence for the seasonal deposition of detritus to deep-sea floor and its subsequent resuspension. *Deep-Sea Res.*, **22**, 885–897.

Minagawa, M., D. A. Winter and I. R. Kaplan (1984): Comparison of Kjeldahl and combustion methods for measurement of nitrogen isotope ratios in organic matter. *Anal. Chem.*, **56**, 1859–1861.

Nakamura, T., N. Nakai and S. Ohishi (1987): Techniques of tandem accelerator mass spectrometry and their applications to ^{14}C measurement. *Nuc. Inst. Meth. Phys. Res.*, **B29**, 335–360.

Nakatsuka, T., N. Handa, E. Wada and C.-S. Wong (1992): The dynamic changes of stable isotopic ratios of carbon and nitrogen in suspended and sedimented particulate organic matter during a phytoplankton bloom. *J. Mar. Res.*, **50**, 267–296.

Nakatsuka, T., N. Handa, N. Harada, T. Sugimoto and S. Imaizumi (1997): Origin and decomposition of sinking particulate organic matter in the deep water column inferred from the vertical distributions of its $\delta^{15}N$, $\delta^{13}C$ and $\delta^{14}C$. *Deep-Sea Res. I*, **44**, 1957–1979.

Noriki, S. and S. Tsunogai (1986): Particulate fluxes and major components of settling particles from sediment trap experiments in the Pacific Ocean. *Deep-Sea Res.*, **33**, 903–912.

Ostlund, H. G., H. Craig, W. S. Broecker and D. Spencer (1987): *GEOSECS Atlantic, Pacific and Indian Ocean Expeditions Vol. 7, Shorebased Data and Graphics*. National Science Foundation, Washington, D.C., 200 pp.

Stuiver, M. and H. A. Polach (1977): Discussion reporting of ^{14}C data. *Radiocarbon*, **19**, 355–363.

Tanoue, E. and N. Handa (1979): Differential sorption of organic matter by various sized sediment particles in recent sediment from the Bering Sea. *J. Oceanogr. Soc. Japan*, **35**, 199–208.

Toggweiler, J. R., K. Dixon and K. Bryan (1989): Simulations of radiocarbon in a coarse-resolution world ocean model 2. Distributions of bomb-produced carbon 14. *J. Geophys. Res.*, **94**, 8243–8264.

Wada, E., M. Terasaki, Y. Kabaya and T. Nemoto (1987): ^{15}N and ^{13}C abundances in the Antarctic Ocean with emphasis on the biogeochemical structure of the food web. *Deep-Sea Res.*, **34**, 829–841.

Walsh, I., J. Dymond and R. Collier (1988a): Rates of recycling of biogenic components of settling particles in the ocean derived from sediment trap experiments. *Deep-Sea Res.*, **35**, 43–58.

Walsh, I., K. Fischer, D. Murray and J. Dymond (1988b): Evidence for resuspension of rebound particles from near-bottom sediment traps. *Deep-Sea Res.*, **35**, 59–70.

Walsh, J. J., D. A. Dieterle and J. R. Pribble (1991): Organic debris on the continental margins: a simulation analysis of source and fate. *Deep-Sea Res.*, **38**, 805–828.

Williams, P. M. and E. R. M. Druffel (1987): Radiocarbon in dissolved organic matter in the Central North Pacific Ocean. *Nature*, **330**, 246–248.

Dynamics and Characterization of Marine Organic Matter, Eds., N. Handa, E. Tanoue and T. Hama, pp. 187–211.

Productivity of Gephyrocapsacean Algae Revealed from Long-Chain Alkenones and Alkyl Alkenoates in the Northwestern Pacific off Japan

Ken Sawada[1]*, Hisatake Okada[2], Yoshihiro Shiraiwa[3] and Nobuhiko Handa[1]**

[1]*Institute for Hydrospheric-Atmospheric Sciences, Nagoya University, Chikusa-ku, Nagoya 464-01, Japan*
phone: +81-52-789-3475, FAX: +81-52-789-3436
[2]*Department of Earth and Planetary Sciences, Graduate School of Science, Hokkaido University, N10 W8, Sapporo 060, Japan*
[3]*Department of Biology, University of Tsukuba, Tsukuba, Ibaraki 305, Japan*

(Original, Received 31 January 1997; accepted 17 April 1997)

Abstract—Long-chain (n C_{37}–C_{39}) alkenones and (n C_{37}–C_{38}) alkyl alkenoates (A&A), which are derived mainly from coccolithophorids specifically, the family Gephyrocapsaceae such as *Emiliania huxleyi* and *Gephyrocapsa oceanica*, were analyzed in, 1) sinking particles collected by year-long time-series sediment traps at 1674, 4180, 5687 and 8688 m depths, 2) in the underlying bottom sediment at 9200 m depth and 3) in a sediment core of KT92-17 St. 14 (3252 m water depth) from in the northwestern Pacific off Japan. Pronounced maxima of sinking A&A fluxes in sediment trap samples at 1674 m were observed in late spring to summer. Seasonal patterns of alkenone temperature records in sediment trap samples from 1674 to 8688 m were similar to sea surface temperature (SST) signals with a time delay of one half to two months. Thus, A&A in the particles sinking in deep sea water column could be mainly derived from the primary products of surface layer and reflected strongly the state of productivity of their source organisms. Furthermore, significant amounts of A&A were contained in the underlying bottom sediment, in which no coccoliths were observed. However, A&A fluxes tended to decrease with water depths due to decomposition of these compounds in the water column and sediment-water interface. A & A and coccolith records at 1674 m suggested that A&A fluxes were not always correlated with the coccolith fluxes of *E. huxleyi* and *G. oceanica*. The relative abundances of *E. huxleyi* and *G. oceanica* estimated by A&A fingerprints were not necessarily comparable to the estimated values from coccolith observations.

Paleo-productivity estimations of the Gephyrocapsaceae based on A&A mass accumulation rates for a core St. 14 were generally higher during the last glacial period than the Holocene. This result disagrees with previous studies on coccolith carbonates in the middle to high latitudinal North Atlantic sediments, in which coccolithophorid productions increased from the last glacial period to

---

*Present address: Department of Chemistry, University of Tsukuba, Tsukuba, Ibaraki 305-8571, Japan. e-mail: sawadak@staff.chem.tsukuba.ac.jp(Ken Sawada)

**Present address: Aichi Prefectural University, Ibagakaima 1522-3, Nagakute, Aichi 480-1198, Japan.

the Holocene. The millenarian-scale fluctuation in productivity of coccolithophorids including the Gephyrocapsaceae in the northwestern Pacific off central Japan was presumably different from that in the middle to high-latitudinal areas of the North Atlantic.

INTRODUCTION

A knowledge of long-term changes in productivity and spatial distributions of individual types of phytoplanktons, such as calcareous and siliceous phytoplanktons, in the sea surface layer is important for understanding the interaction between carbon cycling and marine biogeochemical processes (Dymond and Lyle, 1984; Taylor *et al.*, 1991; Holligan, 1992). Coccolithophorid is the most dominant calcareous phytoplankton, and especially *Emiliania huxleyi* (Lohman) Hay and Mohler of the family Gephyrocapsaceae often occur dominantly in the calcareous nannoplankton community and sometimes form large blooms (Balch *et al.*, 1991; Holligan *et al.*, 1993). *E. huxleyi* is a ubiquitous species and is distributed from low to high latitudinal areas of the world's oceans (McIntyre and Bé, 1967; Okada and Honjo, 1973), and production and export of this species might affect the air-sea exchange of CO_2 due to the production of a large amount of coccolith carbonates and organic matter (Holligan *et al.*, 1993). In addition, the genus *Gephyrocapsa* of the Gephyrocapsaceae, especially *Gephyrocapsa oceanica* Kampther, predominates exclusively in coastal, inland and marginal sea areas of the western tropical to subtropical Pacific and the Indian Ocean (Houghton and Guptha, 1991; Blackburn and Cresswell, 1993). Coccolithophorid productions (or productivity) have been generally estimated from the abundance of coccoliths in marine sinking particles (e. g. Honjo, 1976; Samtleben and Bickert, 1990; Knappertsbusch and Brummer, 1995) and the concentrations of fine carbonates (e. g. $<32\mu m$) in sediments (e. g. McIntyre *et al.*, 1972; van Kreveld *et al.*, 1995). However, the sedimentary coccolith carbonate record is not a good proxy of production from marine particles and sediments in the deep sea below the lysocline depth, because of coccolith carbonate dissolution.

C_{37}–C_{39} alkenones and C_{37}–C_{38} alkyl alkenoates (A&A) appear to be derived from the families Gephyrocapsaceae and Isochrysidales (Marlowe *et al.*, 1990), but the Isochrysidales is coastal species and probably minor components in the marine nannoplankton community. In recent years, A&A in marine sediments have been used for reconstructing past changes in the productivity of the Gephyrocapsaceae (Prahl *et al*, 1989a; McCaffrey *et al.*, 1990; Kennedy and Brassell, 1992; Martinez *et al.*, 1996).

It is known that lipid compounds are generally labile in early diagenesis within the sea water column and benthic boundary layer, and only a small fraction in these compounds survives and its composition is altered in transit from the euphotic layer to the bottom sediments (Crip *et al.*, 1979; Wakeham *et al.*, 1980; Gagosian *et al.*, 1982; Matsueda and Handa, 1986). In addition, the fluxes of specific organic compounds (e.g. fatty acids) are not necessarily correlated with primary productivity of total phytoplanktons in the sea surface layer (Wakeham and Canuel., 1988), despite a good correlation between bulk particulate organic carbon (POC) fluxes and primary productivity (Suess, 1980; Pace *et al.*, 1987). The relationships between productivity and organic compound fluxes could vary due to environmental and/or physiological variables, and preferential decomposition within the water column and sediment (reviewed in Wakeham and Lee, 1993). Thus, it should be confirmed that the fluxes and accumulation rates of A&A reflect the productivity of their source organisms in the euphotic layer.

A&A, in particular C_{37} alkenones, also provide a record of changes in water temperature. The number of double bonds of these alkenones tends to vary depending on the growth temperatures of their source organisms in cultures (Marlowe *et al.*, 1984; Prahl and Wakeham, 1987). This relationship was expressed by the alkenone unsaturation index as follows; $U^{k'}_{37}$ = ([37:2] – [37:4])/ ([37:2] + [37:3] + [37:4]) (Brassell *et al.*, 1986) or $U^{k'}_{37}$ = [37:2]/([37:2] + [37:3]) (Prahl and Wakeham, 1987), where [37:x] denotes the concentration of alkenones consisting of 37 of carbon with x numbers of double bonds. It has been reported that $U^{k'}_{37}$ or $U^{k'}_{37}$ in field samples were related to the water temperature of the surface euphotic layer where A&A were produced (Conte *et al.*, 1992), and were not altered during the diagenetic degradation within the water column (Conte *et al.*, 1992) and after deposition (Prahl *et al.*, 1989b; Madureira *et al.*, 1995).

The main aims of this work are to develop the use of A&A as a practical method for estimating substantial productivity of Gephyrocapsacean algae, and to reconstruct the fluctuations of the productivity of these microalgae at centennial and millenarian-scales from these compounds in the northwestern Pacific off central Japan. We report here changes in downward fluxes of A&A within the water column and compare these biomarker fluxes with coccolith fluxes by a time-series sediment trap investigation at the Japan Trench (JT-07). Furthermore, we report the variation of A&A accumulation rates in a single core of KT92-17 St. 14 collected from the Nishishichitou Ridge off central Japan over the last 30,000 calendar years (30 kyr).

MATERIALS AND METHODS

Samples

Sinking particles were collected using a year-long time-series sediment trap system (Maclane, Mark VI and VII) at a station JT-07 (34°10′ N, 142°E; Fig. 1), located at the triple junction site of the Pacific, North American and Philippine Sea Plates in the northwestern Pacific off central Japan. Four sediment traps were moored at 1674, 4180, 5687 and 8688 m from March 1991 to March 1992. Each of the collecting cups for particles fixed in the sediment traps were automatically changed every 28 days, and then 13 samples were obtained for each of the depths, while only seven samples were obtained at 8688 m due to the mechanical trouble of the sediment trap system. Bottom sediment samples were collected from 9200 m at the sediment trap site at JT-07 using a multiple corer (LM-6). Chemical analyses in this study were restricted to the surface (from 0 to 1.0 cm thick).

Sediments were collected by means of a piston core at KT92-17 St. 14 (32°40.1′ N, 138°27.3′ E, water depth 3252m; Fig. 1) in the Nishishichitou Ridge

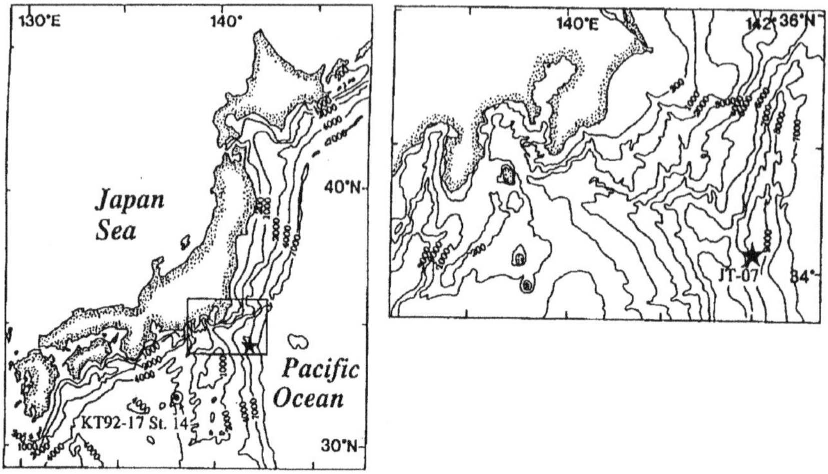

Fig. 1. Maps of the northwestern Pacific off Japan showing the sampling lacations for time-series sediment traps at JT-07 and the underlying bottom sediment of LM-6 (star), and a sediment core, KT92-17 St. 14 (double circle).

off central Japan. The sediments were carefully sectioned into 2.4 cm thick slices for chemical analyses.

Chemical analyses

Portions of the sediment trap samples or sediment samples were acidified with 1M HCl and left for half a day to remove carbonates. Carbonate-free samples were vacuum dried and analyzed for particulate organic carbon (POC) by a Yanaco MS-1 CHN analyzer.

Detailed procedures for long-chain alkenones and alkyl alkenoates (A&A) analyses have been described elsewhere (Sawada *et al.*, 1996). Briefly, A&A were extracted from sediment trap and sediment samples by high energy ultrasonication and by successive treatment with CH_2Cl_2: MeOH (1:3 v/v), CH_2Cl_2: MeOH (1:1 v/v) and CH_2Cl_2: MeOH (3:1 v/v). Distilled water was added to the combined extracts, shaken and centrifuged to separate lipid fraction. The resulting organic solvent phase with lipids was carefully siphoned out and passed through a column packed with anhydrous Na_2SO_4. The lipid extract was passed through a column packed with 95% activated silica gel and the ketone and ester fractions were eluted with a hexane-ethyl acetate mixtures of 90:10 (v/v) and 95:5 (v/v), respectively. These fractions were analyzed by a Shimadzu GC-9A capillary gas chromatograph (50 m × 0.32 mm i.d., 0.12 μm film thickness, CPSil 5CB fused silica column, Chrompack) equipped with a flame-ionization detector. Identification of A&A was confirmed on a Shimadzu GC-QP2000A quadrupole mass spectrometer, using identical GC conditions.

Graphite targets for radiocarbon (^{14}C) datings of planktonic foraminifera, *Neogloboquadorina dutertrei* samples, from the sediment core were prepared by the hydrogen reduction method (Kitagawa *et al.*, 1991). ^{14}C ages using the graphite targets were determined by an accelerator mass spectrometer (AMS; GENERAL IONIC, 1.8 MeV, 25 nA) of the Dating and Materials Research Center, Nagoya University (Nakamura and Nakai, 1988). A 380 year reservoir correction at 34°N in the Pacific (Bard, 1988) was subtracted from the ^{14}C ages. The corrected ages were converted to calendar ages using the equation given by Bard *et al.* (1993).

Counting of coccolith

Detailed procedures for counting coccoliths have been described elsewhere (Okada, 1992; Sawada *et al.*, 1996). The coccoliths on the slide were counted under a transmitted light microscope using cross-polarized light at 1000 to 1500 X magnification with immersion oil. Identifications of the coccoliths of *E. huxleyi*, *G. oceanica* and other coccoliths were carried out as per Okada (1992). Around 300 to 400 coccolith specimens were usually counted in each sample.

QUANTITATIVE CHANGES IN ALKENONES AND ALKENOATES DURING TRANSPORT AND DEPOSITION

Downward fluxes of alkenones and alkenoates in deep sea off central Japan

The temporal variations of the fluxes of sinking POC and A&A in the sediment trap samples of JT-07 are shown in Figs. 2b–e. Pronounced maxima of A&A fluxes were observed in sediment trap samples from 1674 m in late spring to summer (from June to August of 1991). The fluxes of POC and particulate organic nitrogen (PON) at 1674 m were also highest in spring to summer (Nakatsuka *et al.*, 1995), indicating that the seasonal trend of A&A flux was concordant with those of POC and PON.

Alkenone unsaturation indices ($U^{k'}_{37}$) in sediment trap samples at 1674 m ranged from 0.59 to 0.89, and were converted to the temperature using a equation as follows,

$$U^{k'}_{37} = 0.033\ T + 0.001\ (°C)\ (r^2 = 0.946,\ n = 20) \qquad (1)$$

which was obtained from the relationships between $U^{k'}_{37}$ of the suspended particle samples collected from the sea surface in several areas off Japan and *in situ* measured sea surface temperatures (SST) (Sawada *et al.*, 1998; Fig. 3). $U^{k'}_{37}$-based temperatures in sediment trap samples at 1674 m almost coincided with the SST delivered by the Japan Maritime Safety Agency, with a time delay of a half to one month (Fig. 2a). However, $U^{k'}_{37}$-based temperatures were about 2 to 3°C lower than the observed SSTs during June to September of 1991. Such lower $U^{k'}_{37}$-based temperatures might be attributed to the formation of seasonal thermocline in the upper euphotic layer during these months. The time delay between $U^{k'}_{37}$-based temperatures of trap samples at 1674 m and the observed SSTs was

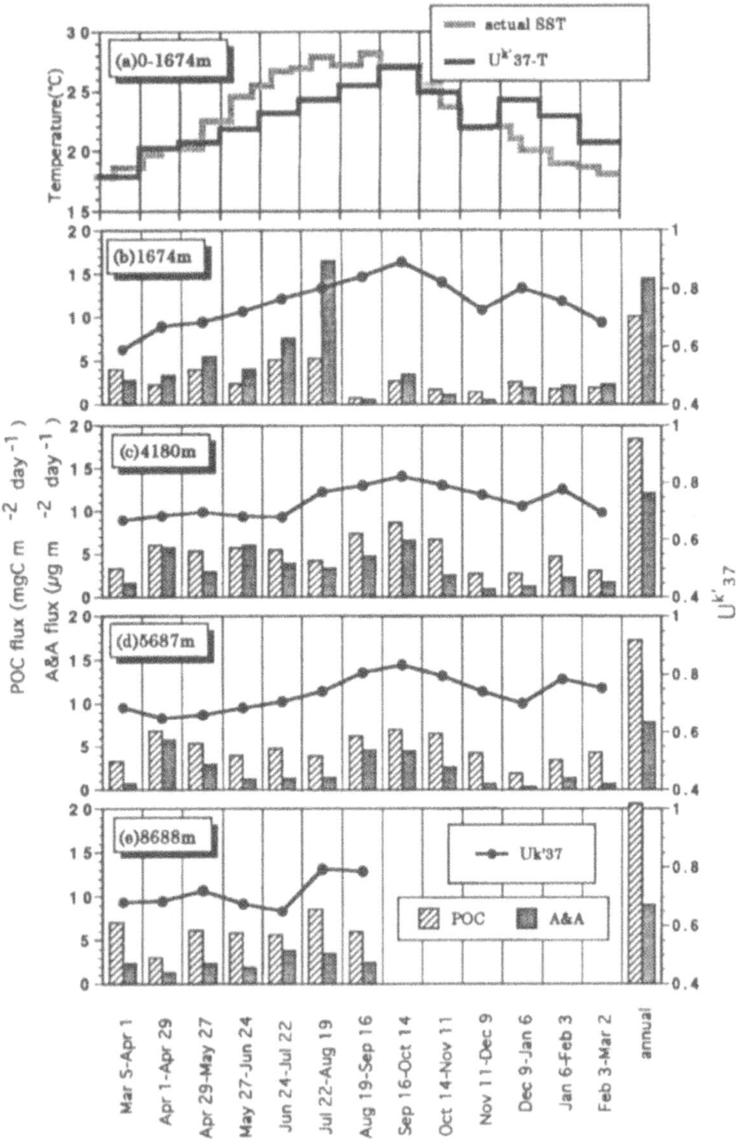

Fig. 2. (a) Seasonal variations of water temperature estimated from $U^{k'}_{37}$ in sinking particles collected from the time-series sediment traps JT-07 at 1674 m (solid line) and observed sea surface temperatures in the mooring trap site of JT-07 (dashed line) reported by Japan Maritime Safety Agency (Fast report on the ocean) (Sawada *et al.*, 1998). The temporal variations of the fluxes of particulate organic carbon (POC; mgC m^{-2} day^{-1}; hatched bar) and A&A fluxes(μg m^{-2} day^{-1}; dotted bar), and alkenone unsaturation indices ($U^{k'}_{37}$) in sinking particles collected from time-series sediment traps of station JT-07 at 1674 (b), 4180 (c), 5687 (d), and 8688 m (e) from March 1991 to March 1992 (Sawada *et al.*, 1998). annual: annual fluxes of POC (gC m^{-2} $year^{-1}$) and A&A (mg m^{-2} $year^{-1}$)

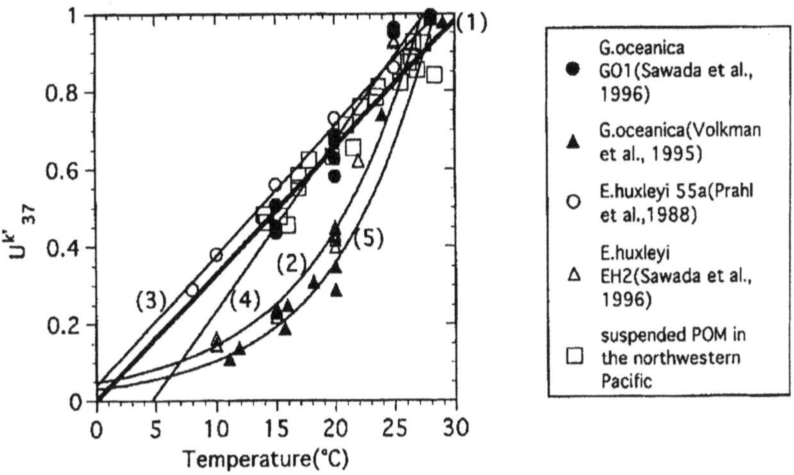

Fig. 3. Plot of $U^{k'}_{37}$ against water temperature for suspended particulate organic matter (POM) collected from the sea surface in several areas of the northwestern Pacific off Japan (open square; Sawada *et al.*, 1998). Values of $U^{k'}_{37}$ for laboratory cultured *E. huxleyi* strain 55a (open circle; Prahl *et al.*, 1988), *E. huxleyi* strain EH2 (open triangle; Sawada *et al.*, 1996), *G. oceanica* strain GO1 (solid circle; Sawada *et al.*, 1996) and *G. oceanica* strain reported by Volkman *et al.* (1995) (solid triangle) are also plotted. Line (1), (3) and (4) are linear fits for the data of suspended POM ($U^{k'}_{37} = 0.033T$), cultures of *E. huxleyi* 55a ($U^{k'}_{37} = 0.034T + 0.039$) and *G. oceanica* GO1 ($U^{k'}_{37} = 0.044T - 0.204$). Curve (2) and (5) are nonlinear fits for the cultured *E. huxleyi* EH2 ($U^{k'}_{37} = 0.047 \exp(0.112T)$) and *G. oceanica* reported by Volkman *et al.* (1995) ($U^{k'}_{37} = 0.03\exp(0.123T)$).

probably the time taken for the downward transport of the sinking particles containing A&A from the sea surface layer to the depth of the mooring traps. Thus, the maximum A&A fluxes during June to August of 1991 might result from the maximum A&A productions by their source organisms in the surface euphotic layer during May to July of 1991, including the period of the spring bloom.

The fluxes of POC tended to increase from 1674 to 8688 m (Figs. 2b–e), suggesting that secondary products such as resuspended materials were possibly supplied to the deep sea. Thus, the source of these sinking particles cannot be explained as only directly sinking particles derived from their source organisms in the sea surface layer. Nevertheless, seasonal variations of $U^{k'}_{37}$ in sediment trap samples from 4180 to 5687 m were also concordant with that of SSTs with a time delay of one to two months (Fig. 2b–e). From this fact, the secondary input of A&A from secondary resuspended particles could have hardly contributed to the A&A fluxes in deep water. Annual fluxes of A&A tended to decrease with depth from 1674 to 8688 m (Fig. 2b–e). The relationships between annual A&A fluxes ($F_{A\&A}$; mg m^{-2} $year^{-1}$) and the water depth (z; km) from 1684 to 5687 m was

$$\ln F_{A\&A} = -0.144\,z + 0.663 \; (r^2 = 0.86,\ n = 3) \qquad (2)$$

(Sawada *et al.*, 1998; Fig. 2b–e).

This decrease of A&A fluxes is probably due to the decomposition and/or attenuation of these compounds during downward transport to the deep sea. The decreasing constant of annual A&A fluxes was –0.14 km^{-1} from the slope of equation (2) in these oceanic regions. It has been believed that A&A were more resistant compounds against biological degradation as compared with the other solvent-extractable lipids (Volkman *et al.*, 1980; Conte *et al.*, 1992), but our results in sediment trap samples indicated that A&A were not refractory compounds within the sea water column.

We, however, confirmed that A&A significantly contained (1.50 $\mu g\ g^{-1}$ dry weight) in the bottom sediment at 9200 m, despite there being no preservation of coccolith carbonates. In addition, $U^{k'}_{37}$ obtained from the underlying bottom sediment was similar to the flux-weighted annual mean values of $U^{k'}_{37}$ in the sediment trap samples, which are similar to samples from 1674 to 8688 m and estimated to be 0.72 to 0.76 (Sawada *et al.*, 1998). These results provided further evidence that A&A in particles within deep sea waters and bottom sediments could be mainly derived from primary products in the sea surface layer, and, therefore, A&A must be a reliable indicator of the productivity of their source organisms.

Diagenetic loss at sediment-water interface

At the sediment-water interface, a few centimeters of the surface bottom sediment, most organic materials arriving at the seafloor are rapidly remineralized by benthic communities (Smith and Baldwin, 1984). Such a diagenetic loss of organic compounds in the sediment-water interface can be evaluated by a comparison of vertical fluxes in the sea water column with the mass accumulation rates (MAR) in the top of bottom sediment. However, determination of MAR of primary products was perturbed by reworking of the resuspended materials from continental slope. The reworking resuspended lithogenic materials and POC might occur frequently near the seafloor in the area mooring the sediment traps of JT-07, as mentioned previously. Even if the resuspended A&A in this area might be minor components, the sedimentary A&A were attenuated by the other resuspended materials and the accumulation rates of A&A could be underestimated. Thus, annual fluxes of A&A in sediment trap samples above 5687 m were compared with the MAR of A&A obtained from surface bottom sediment samples of St. 14 in the Nishishichitou Ridge, in which secondary supplies of resuspended materials might scarcely occur (Table 1). MARs (TOC: mg $cm^{-2}\ yr^{-1}$; A&A: ng $cm^{-2}\ yr^{-1}$) were calculated using the following equation:

$$\mathrm{MAR} = \mathrm{a} \cdot \mathrm{X} \cdot \mathrm{SR} \cdot \mathrm{DBD} \tag{3}$$

where X is organic carbon content (%) or concentration of A&A (ng g^{-1}), SR is the sedimentation rate (cm yr^{-1}) based on calendar ages calibrated from AMS ^{14}C datings (Table 2) and DBD is the dry bulk density (g cm^{-3}). α in TOC and A&A are 10.0 and 1.0, respectively.

Table 1. Annual fluxes in long-chain alkenones and alkyl alkenoates (A&A) in sinking particles collected with time-series sediment traps at station JT-07, and mass accumulation rates (MARs) of A&A in a sediment core, KT92-17 St. 14. Fluxes and MARs of long-chain alkenones in the Washington Coast of the northeastern Pacific are also shown (Prahl *et al.*, 1993).

Location	Water depth (m)	Flux ($ng\ cm^{-2}yr^{-1}$)	MAR ($ng\ cm^{-2}yr^{-1}$)	Preservation efficiency(%)[a]
Off central Japan				
JT-07	1674	145		
	3200	122[b]		
St. 14	3252		26.1	21.39
Washington coast				
Nearshore	1000	150		
	2712		33	22
Midway	1000	110		
	3111		11	10
Gyre	1000	140		
	3680		0.35	0.25

a: the ratios of MAR to flux within the water column of A&A.

b: a flux value calculated using equation (2).

Table 2. Radiocarbon ages of a sediment core, KT92-17 St. 14 (after Sawada and Handa, 1998).

^{14}C datings/Aira-Tanzawa volcanic ash			
Depth from the top(cm)	Conventional age (yr BP)	Corrected age[b] (yr BP)	Calendar age[c] (yr BP)
24.8-27.3	3810±116	3430	3413
58.4-60.9	6745±209	6365	7053
150.6-153.1	14055±143	13675	16117
203.4-205.9	16509±264	16129	19160
285-299.5[a]	24330±225[a]	23950	28858

a: Aira-Tanzawa (AT) ash layers: the age has been previously determined (Murayama *et al* ., 1993)

b: Correction of reservoir effect assumes 380 years in the area of 33°N of the North Pacific (Bard, 1988).

c: Corrected ages were converted to calendar ages using the equation given by Bard *et al* . (1993).

The MAR of A&A was 26.1 ng cm^{-2} yr^{-1} in the bottom sediment of St. 14 at 3252 m, whereas, A&A flux at 3200 m is calculated to be 122 ng cm^{-2} yr^{-1} by using equation (2). Preservational efficiency, which was defined by the ratio of MAR to trap flux (Prahl *et al.*, 1993), was calculated to be 21.4% (Table 1). The MAR of A&A in the bottom sediment was one-fifth lower than the flux of these compounds within the deep water column at the same depth. It was presumed that such a decrease corresponded to the diagenetic loss in the sediment-water interface off central Japan.

Prahl *et al.* (1993) compared the annual fluxes of long-chain alkenones from sediment trap samples at 1000 m, with the MAR of these compounds in bottom sediments at three locations (Nearshore, Midway and Gyre) of the northeastern North Pacific (Table 1). They demonstrated that the magnitude of diagenetic loss (or preservational efficiency) of long-chain alkenones from 1000 m to the bottom sediment was different between these three locations. These differences were caused by the spatial variations of redox conditions at the seafloor. Hence, it needs to take into account the variations in the redox condition in seafloor for substantial estimate of productivity and export production of A&A from MAR in bottom sediments.

Comparison between the fluxes of alkenones and alkenoates, and coccoliths

Fluxes and species compositions of coccoliths in the sediment trap samples of JT-07 were determined by counting the coccoliths under a transmitted light microscope (Figs. 4a and b). Coccolith of *E. huxleyi* accounted for 30 to 65% of the total coccoliths, and was the most abundant component of coccolith assemblage in the sediment trap samples of JT-07. *Florisphaera profunda* Okada and Honjo and *Gladiolithus flabellatus* Halldal and Markali were also quite abundant in these samples, but do not presumably synthesize A&A (Marlowe *et al.*, 1990). Coccolith of *G. oceanica* accounted for 5 to 15% of the total coccoliths. Coccoliths for other potential A&A sources, such as the placoliths of *Gephyrocapsa ericsonii* McIntyre and Bé, were present but were only minor components (1 to 5% in total coccoliths). The coccolith fluxes of total coccolithophorids and A&A-producing algae ranged from 0.1 to 1.8×10^9 m^{-2} day^{-1} and from 0.05 to 1.0×10^9 m^{-2} day^{-1}, respectively. Maximal peaks of these coccolith fluxes were observed in June to July 1992, and the timings of these maxima were a month earlier than those of the A&A fluxes.

The A&A fluxes were plotted against the coccolith fluxes in the sediment trap samples of JT-07 at 1674 m (Fig. 5). It was found that a large scatter of a dataset during the period of the highest A&A fluxes (from July 22 to August 19 of 1992), so that the A&A fluxes were not simply correlated with the A&A-producing coccolith fluxes. However, the coccolith fluxes of the A&A-producing algae gave a better linear regression fit when plotted against the A&A fluxes which preclude the dataset during the period of the highest A&A fluxes. The flux ratios of A&A to coccoliths of their producing algae was the highest in July 22 to August 19 of 1992 and was 24 μg /(number $\times 10^9$), whereas, those in the other months were 4 to 16 μg /(number $\times 10^9$). The A&A in sediment trap samples at 1674 m during July to August might be produced in the euphotic layer during the period of spring bloom as mentioned previously, suggesting that the A&A fluxes increased remarkably as compared with the coccolith fluxes during the period of spring bloom. This fact suggested that A&A fluxes cannot be directly converted to coccolith carbonate fluxes.

Such variations of the flux ratios of A&A to coccoliths could be attributed to the variations of intercellular production ratios of A&A to coccoliths and/or the

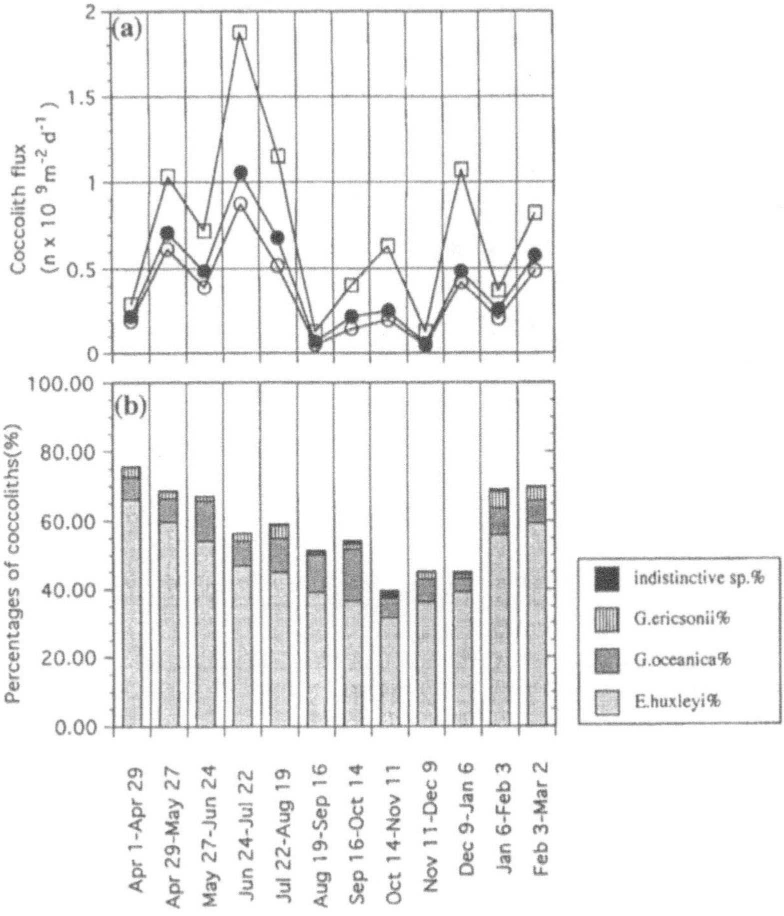

Fig. 4. The temporal variations of coccolith fluxes of total coccolithophorids (open square), A&A-producing species (*E. huxleyi, G. oceanica, G. ericsonii* and indistinctive species; (solid circle) and *E. huxleyi* (open circle) (a), and the species compositions of A&A-producing algae (b). in the sediment-trap samples of JT-07 at 1674 m from March 1991 to March 1992. The indistinctive species are Gephyrocapsacean algae which cannot be distinguished *G. oceanica* from *G. ericsonii* and other species.

differences of transport and decomposition mechanisms to the deep ocean. A&A are presumed to be components of membrane lipids (Marlowe *et al.*, 1984) and their productions are independent of coccolith productions (Volkman *et al.*, 1980). A&A / coccolith production ratios may be variable in response to changing environmental and physiological states. In general, intercellular concentrations of membrane lipids do not vary depending on changing physiological state, but it was reported that those of A&A significantly increased from the exponential

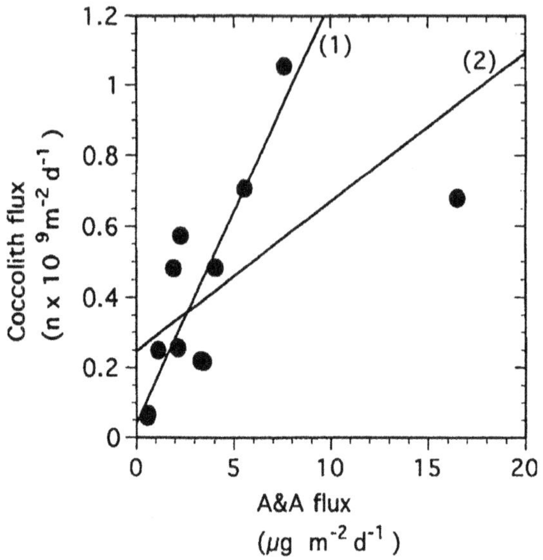

Fig. 5. Plot of A&A fluxes ($F_{A\&A}$; μg m^{-2} day^{-1}) against coccolith fluxes of A&A-producing species (F_{lith}; n × 10^{9} m^{-2} day^{-1}) in sinking particles collected with the time-series sediment traps of JT-07 at 1674 depth.
Lines (1) and (2) are linear fits for the datasets precluding a dataset during the period of the highest A&A fluxes (from July 22 to August 19 of 1992) ($F_{A\&A} = 0.120*F_{lith} + 0.042$, $r^2 = 0.74$) and all datasets ($F_{A\&A} = 0.042*F_{lith} + 0.247$, $r^2 = 0.39$), respectively.

phase to the stationary phase (Bell and Pond, 1996). Also, A&A / coccolith production ratios can be variable in genetic and taxonomical differences of A&A producing algae. Four morphotypes of *E. huxleyi* (A, B, C and var. corona) were recognized and the numbers of coccolith per cell for type C cell were remarkably higher than that for type A cells (Young and Westbroek, 1991). It is also known that *E. huxleyi* has a completely naked (non-coccolith bearing) cell type in its life cycle (Klaveness, 1972). Dominance of the naked cell type in the community of A&A-producing algae in the euphotic layer must result in an increase of A&A / coccolith production ratios. Balch *et al.* (1991) found that *E. huxleyi* detaches a large amount of coccoliths into the water column under late bloom conditions. Field observations of an extensive bloom of *E. huxleyi* in the northeastern North Atlantic (Holligan *et al.*, 1993) also reported that concentrations of detached coccoliths and non-coccolith-bearing cells of *E. huxleyi* increased in the central area of the bloom, where the growth phase of the phytoplankton communities corresponds to a late stationary phase under low nutrient conditions. If A&A-containing algal cells are grazed by zooplanktons and are successively transported to the deep ocean by their fecal pellets more efficiently than free coccoliths, the export flux ratios of A&A to coccolith can increase, as seen in our observations of sediment traps.

EVALUATION OF SPECIES' COMPOSITIONS OF GEPHYROCAPSACEAN ALGAE FROM ALKENONE AND ALKENOATE FINGERPRINTS

Taxonomic dependence on the composition of alkenones and alkenoates

The productivity of the Gephyrocapsaceae could potentially vary through the taxonomical difference of the A&A source organisms, possibly attributed to the variations of intercellular concentrations of these lipids and the differences of transport and decomposition mechanisms within the water column due to the differences of cell sizes and other factors. It is important for the estimation of Gephyrocapsaceae productivity to evaluate the contribution of individual species within A&A fluxes.

It was found that the ratios of alkenones to alkyl alkenoates or C_{37} alkenones to C_{38} alkenones were significantly different between the laboratory cultured algal cells of *E. huxleyi* and *G. oceanica* (Conte *et al.*, 1995; Volkman *et al.*, 1995). Systematical investigations of A&A using laboratory cultures and sediment samples (Sawada *et al.*, 1995, 1996) suggested that the difference in the composition of A&A could be applied to determine relative abundances of *E. huxleyi* and *G. oceanica* in natural samples. Nonlinear relationships between the ratios of C_{38} di-unsaturated ethyl alkenoates to C_{37} alkenones (EE/K37) and $U^{k'}_{37}$ were found for both algal cultures of the *E. huxleyi* strain EH2 collected from the Great Barrier Reef and the *G. oceanica* strain GO1 collected from Mutsu Bay (41°N, 141°E) for growth temperatures, and the correlation curves were clearly distinct for these two species (Fig. 6a). The relationships between EE/K37 and $U^{k'}_{37}$ (EE/K37–$U^{k'}_{37}$) for the *E. huxleyi* strain EH2 and the *G. oceanica* strain GO1 were

$$EE/K37 = 0.307 \exp(-3.315\ U^{k'}_{37}) \quad (r^2 = 0.959, n = 18) \qquad (4)$$

and

$$EE/K37 = 1.162 \exp(-3.068\ U^{k'}_{37}) \quad (r^2 = 0.910, n = 11) \qquad (5)$$

respectively. For a simple mathematical model using EE/K37–$U^{k'}_{37}$, the relative abundance of *E. huxleyi* and *G. oceanica* in a certain sample was expressed by A_{EH} and A_{GO} respectively, which is shown as follows:

$$GO/EH = A_{GO} / (A_{GO} + A_{EH})\cdot 100 = (E_{SED} - E_{EH}) / (E_{GO} - E_{EH})\cdot 100 \quad (\%) \qquad (6)$$

where E_{SED} is the value of EE/K37 measured for a natural sediment and particulate organic matter (POM) sample, and E_{EH} and E_{GO} are the endmembers of EE/K37 which are given by equations of the curves EH*f* and GO*f*, as shown in Fig. 6a for a given $U^{k'}_{37}$ in *E. huxleyi* and *G. oceanica*, respectively (Model 1 in Sawada *et al.*, 1996; Fig. 7a). For the calculation of the ratios of these algal species using this model, it is necessary that the $U^{k'}_{37}$-temperature (T) relationships for these two algal species mixed in natural samples are similar. However,

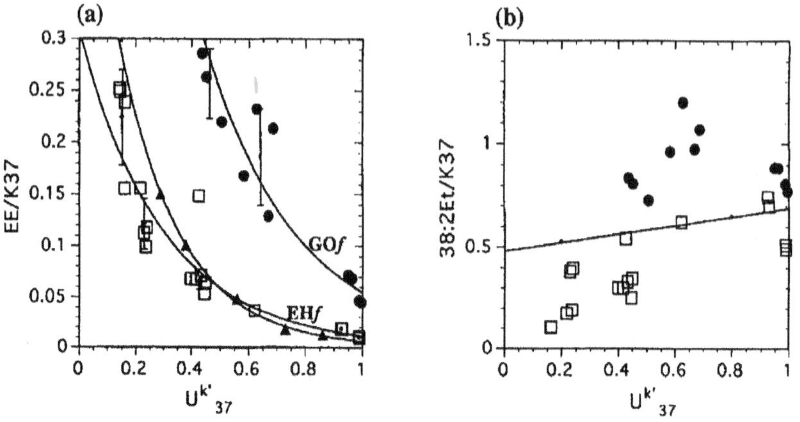

Fig. 6. (a) Nonlinear exponential relationships between EE/K37 and $U^{k'}_{37}$ in laboratory strains of *E. huxleyi* cultured at Nagoya University (open square), *E. huxleyi* cultured at Niigata University (□) and *G. oceanica* (solid circle) (Sawada *et al.*, 1996). EE/K37-$U^{k'}_{37}$, relationship reported by Prahl *et al.* (1988) are also shown. Error bars indicate standard deviation (±2σ) from the mean. (b) Plots of 38:2Et/K37 against $U^{k'}_{37}$. Line represents the boundary between datasets of *E. huxleyi* and *G. oceanica* based on the linear discriminant function (after Sawada *et al.*, 1995).

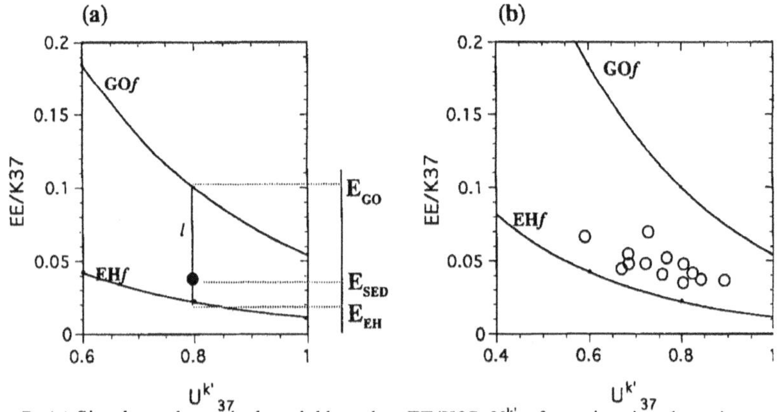

Fig. 7. (a) Simple mathematical model based on EE/K37–$U^{k'}_{37}$ for estimating the ratios of *G. oceanica* to *E. huxleyi* (Sawada *et al.*, 1996; See text). (b) Plots of EE/K37 against $U^{k'}_{37}$ for the sediment-trap samples of JT-07 at 1674 m on the diagrams produced from laboratory cultured *E. huxleyi* EH2 (curve EHf) and *G. oceanica* GO1 (curve GOf).

it was found that there were large deviations between the $U^{k'}_{37}$-temperature (T) calibration lines or curves based on cultures of *E. huxleyi* (Prahl *et al.*, 1988; Conte *et al.*, 1995) and *G. oceanica* (Volkman *et al.*, 1995; Sawada *et al.*, 1996) as shown in Fig. 3. For example, if a mixture of *E. huxleyi* EH2 and *G. oceanica* GO1 grown at 20°C were considered, and if the values of $U^{k'}_{37}$ for *E. huxleyi* EH2 and *G. oceanica* GO1 were 0.40 and 0.65 respectively, the $U^{k'}_{37}$ for the mixture of these species would be the 'objective' mean value. In such a situation, a mathematical model considering the difference in the $U^{k'}_{37}$-T relationships for these two algal species must be applied (Model 2 in Sawada *et al.*, 1996). Unfortunately, there are still no reports for the $U^{k'}_{37}$-T calibration of the *E. huxleyi* strain collected from the northwestern Pacific. $U^{k'}_{37}$-T relationships, however, are probably similar between *E. huxleyi* and *G. oceanica* for the sediment samples in the sea off Japan, because the linear regression of $U^{k'}_{37}$ on temperature in suspended POM collected from sea surface water off Japan, in which *E. huxleyi* is the major constituent, was not significantly different from that of *G. oceanica* strain GO1 (Fig. 3). Therefore, it seems likely that Model 1 can be applied to the sediment samples of the northwestern Pacific off Japan.

The ratios of C_{38} di-unsaturated ethyl alkenones to C_{37} alkenones (38:2Et/K37) against $U^{k'}_{37}$ for cultured strains of *E. huxleyi* and *G. oceanica* were also separable and these two species could be distinguished (Sawada *et al.*, 1995; Fig. 6b). Applying a discriminatory analysis to these 38:2Et/K37 data, a linear discriminant function, which indicates the boundary of each domain, can be obtained as follows:

$$38{:}2\ \mathrm{Et/K37} = 0.209\ U^{k'}_{37} + 0.479 \qquad (7)$$

and the correct discrimination in this classification is high (96.3%). The 38:2 Et/K37-$U^{k'}_{37}$, however, hardly gave a regression curve and, therefore, is not applicable to quantitative separation of *E. huxleyi* from *G. oceanica*, although it can be used for the qualitative separation of these two species.

Applications to sinking particles

EE/K37 *vs.* $U^{k'}_{37}$ were plotted for sediment trap samples of JT-07 at 1674 m (Fig. 7b). All data points fell in between the two curves of EE/K37-$U^{k'}_{37}$ obtained for *E. huxleyi* EH2 and *G. oceanica* GO1 in the culture experiments reported by Sawada *et al.* (1996), suggesting that the A&A were derived from a mixture of these two species. Counting coccoliths under a microscope indicated that the coccoliths of *E. huxleyi* and *G. oceanica* accounted for 80 to 95% in the total coccoliths of A&A-producing species (Fig. 4b), which was in agreement with the results of the A&A fingerprints. The ratios of *G. oceanica* to *E. huxleyi* (GO/EH) were estimated using Model 1, based on equation (6), from these data, ranging from 9 to 43% (Fig. 8). The general seasonal trends of GO/EH based on EE/K37-$U^{k'}_{37}$ indicated that the percentages of *G. oceanica* were lower in winter to spring and increased during summer to fall. This trend was basically similar to that of GO/EH estimated by the microscopic counting of coccoliths. However, GO/EH estimated by microscopic counting were 2 to 3 times higher than those estimated by EE/K37-$U^{k'}_{37}$ during June to August 1991 and during October to December 1991. The months of June to August 1991 corresponded to the period of the highest A&A fluxes (Fig. 2b), which was a response to the spring bloom. Furthermore, a pronounced higher value of the flux ratio of A&A to coccolith was

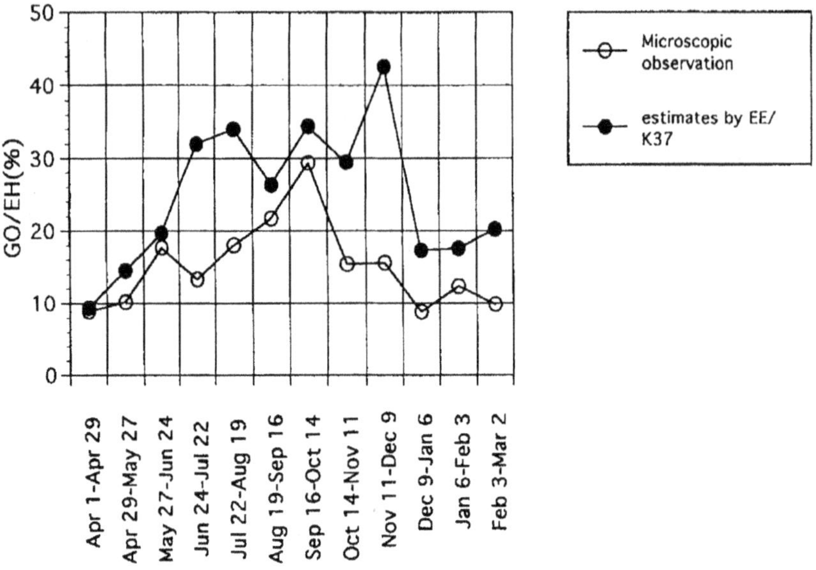

Fig. 8. The temporal variations of the relative abundances of *G. oceanica* to *E. huxleyi* estimated by EE/K37–$U^{k'}_{37}$ (solid circle) and microscopic coccolith counting (open circle) in sinking particles collected the sediment traps of JT-07 at 1674 m.

observed during these months, as mentioned above (Fig. 5). It was presumed that an 'anomalously' high A&A/coccolith flux ratio was connected with the discrepancy between GO/EH-estimated A&A fingerprints and microscopic counting. These implied that the cells of *G. oceanica* were possibly produced in the euphotic layer and/or transported to the deep sea more efficiently than coccolith of its species under a bloom condition. An anomalous high value of EE/K37 for *E. huxleyi* could result in an 'objectively' high value of GO/EH under a bloom condition. However, EE/K37 for *E. huxleyi* appeared to be insignificantly affected by changes of physiological states such as the growth phase (Prahl *et al.*, 1988; K. Sawada and Y. Shiraiwa, unpublished data). From these results, it is possible to conclude that the contribution of *E. huxleyi* and *G. oceanica* within A&A fluxes was not necessarily comparable to that within coccolith fluxes.

RECONSTRUCTION OF THE PRODUCTIVITY OF GEPHYROCAPSACEAN ALGAE IN A SEDIMENT CORE

Sedimentary records in a core of St. 14

Vertical profiles of concentrations of A&A and TOC, and the ratios of organic carbon to nitrogen (C/N) for a sediment core KT92-17 St. 14 are shown in Figs. 9b and c. The concentrations of A&A and TOC ranged from 0.7 to 4.5 μg/g dry weight and from 3.5 to 8.5 mg/g dry weight (TOC content: 0.35 to 0.85%), respectively. There were several negative spikes of TOC concentration corresponding to the coarse turbidite and volcanic ash layers described by Murayama *et al.* (1994). The C/N ratios in a core St. 14 range from 6 to 9, indicating that organic matter in the sediment samples of St. 14 is almost of marine origin (Fig. 9b).

Calendar ages based on AMS ^{14}C ages (Table 2) were plotted against core depth (cm) in homogeneous silty clay layers (Fig. 9a). Sedimentation rates based on the calendar ages were estimated to be 5.3 to 13.4 cm kyr^{-1}.

The MARs of A&A and TOC for the homogeneous silty clay layers of a core St. 14 were calculated by equation (3) and ranged from 20 to 150 $\mu g\ cm^{-2}\ kyr^{-1}$ and from 60 to 290 mg $\mu g\ cm^{-2}\ kyr^{-1}$ over the last 30,000 calendar years (30 kyrs), respectively (Fig. 10a). The smooth trend observed showed that the MARs of A&A exhibited high values from 30 kyr BP to the middle of the last glacial period (about 15 kyr BP) and a gradual decrease from the middle of the last glacial period to the Holocene (after 12.5 kyr BP). The MARs of TOC were slightly higher than those of the present-day from the last glacial to early Holocene (7 kyr BP), and

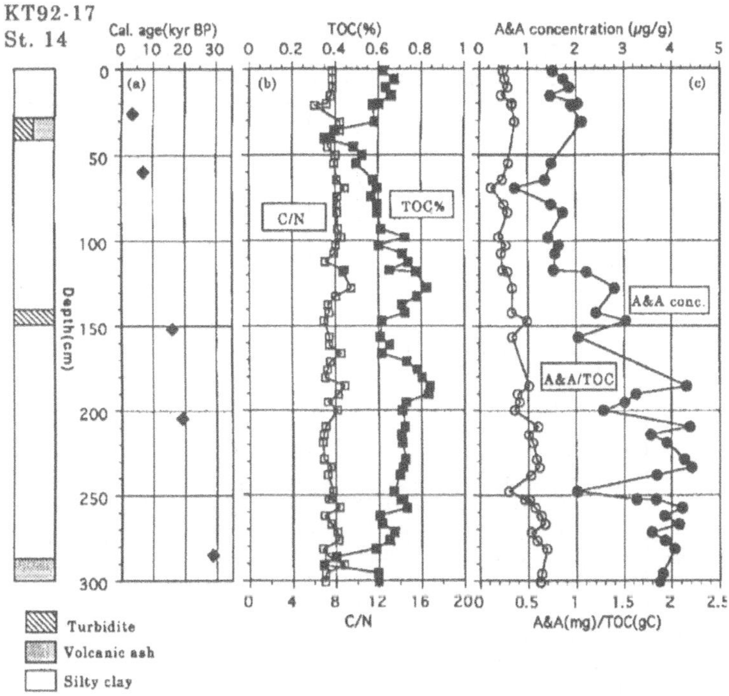

Fig. 9. (a) Profiles of calendar ages based on AMS^{14}C ages of planktonic foraminifera in a sediment core KT92–17 St. 14. Sediments of this core consisted of homogeneous olive greenish silty clay with several volcanic ash (dotted bars) and turbidite layers (hatched bars) (Murayama *et al.*, 1994). Aira-Tanzawa (AT) ash layers were identified and its age has been determined previously as 24330 ± 140 yr BP by AMS^{14}C datings (Murayama *et al.*, 1993). Profiles of TOC content (solid square; %) and organic C/N ratios (open square) (b), and concentrations of A&A (solid circle; μg/g dry weight) and A&A/TOC ratios (open circle) (c) in a sediment core KT92–17 St. 14.

then were the lowest values during the middle Holocene (7 to 4 kyr BP). High peaks in both A&A and TOC MARs were distinctively discernible from 16 to 19 kyrs BP.

Time series records of GO/EH were reconstructed using a equation (6) based on EE/K37-$U^{k'}_{37}$ and microscopic coccolith counting in St. 14 over the last 30 kyr (Fig. 10b). It was found that the absolute values of GO/EH estimated by EE/K37–$U^{k'}_{37}$ were about 50% lower than those estimated by microscopic coccolith counting. This discrepancy was contrasted with that observed in sediment trap samples of JT-07, in which GO/EH estimated by EE/K37–$U^{k'}_{37}$ were about 2 to 3 times higher than by microscopic counting (Fig. 8). From these facts, such a difference of GO/EH estimations observed in a core St. 14 could be attributed to diagenetic alterations after deposition, rather than to differences of A&A/coccolith production ratios and the mechanisms of transport and decomposition (or dissolution) within the water column. It is possible that the sedimentary diagenesis alters the primary GO/EH of coccoliths in abyssal sediments such as

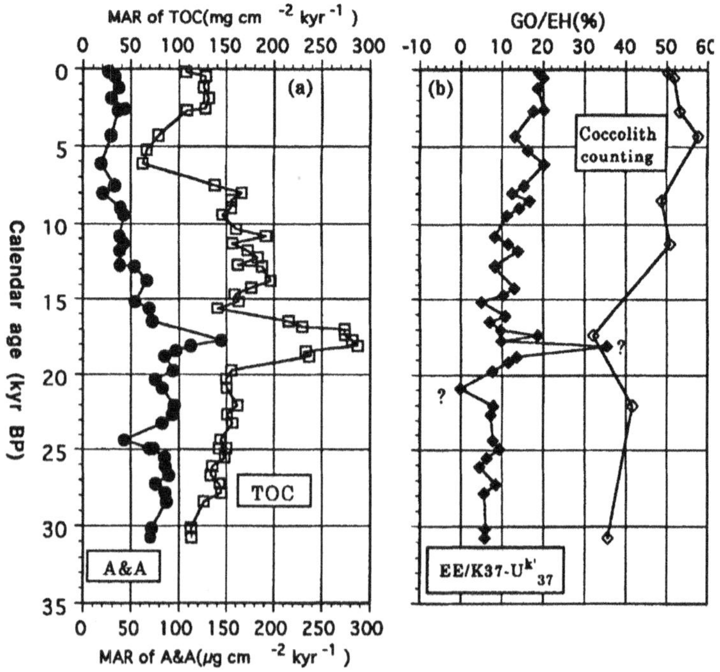

Fig. 10. (a) Variations in mass accumulation rates (MARs) of TOC (mg µg cm^{-2} kyr^{-1}; open square) and A&A (µg cm^{-2} kyr^{-1}; solid circle) in a core KT92-17 St. 14 over 30 kyr BP. The ages of all sediment samples analyzed were calculated by linear interpolations on the basis of sediments of which the calendar ages were known. (b) Variations in the relative abundances of *G. oceanica* to *E. huxleyi* (GO/EH) estimated by Model 1 based on EE/K37–$U^{k'}_{37}$ (solid diamond) and microscopic coccolith counting (open diamond) in a core KT92-17 St. 14 over the last 30 kyr.

St. 14 is due to the preferential dissolution of the coccoliths of *E. huxleyi* because its coccolith is more fragile than that of *G. oceanica* (McIntyre and McIntyre, 1970). However, the trends of variations in GO/EH estimations were similar between these chemical and microscopic methods, which indicated that the percentage abundances of *E. huxleyi* from 30 to 15 kyr BP were higher than at present and then decreased after 15 kyr BP. Thus, it was presumed that *E. huxleyi* was more dominant in the community of the Gephyrocapsaceae off central Japan for the last glacial than the Holocene.

Implication of variations in paleoproductivity of Gephyrocapsacean algae

The variations in the MARs of A&A (Fig. 10a) showed that the productivity of the Gephyrocapsaceae off central Japan during the glacial period was higher than that in the present, and then gradually decreased from the middle of the last glacial period to the Holocene. The productivity of these algae arrived at the present level at about 13 kyr BP. This trend for productivity of the Gephyrocapsaceae in St. 14 agreed with that reconstructed from the MARs of long-chain alkenones in a core SO50-31KL (18°45.3′ N; 115°52.39′ E) collected from the South China Sea, a mid-latitudinal area of the North Pacific (Huang *et al.*, 1997). However, the variations of coccolithophorid productivity deduced from sedimentary coccolith records in mid- to high-latitudinal area of North Atlantic Ocean indicated that coccolithophorid productivity was lower than that of the present and then increased in the Holocene (McIntyre *et al.*, 1972; van Kreveld *et al.*, 1995), which clearly disagreed with the results obtained from mid-latitudinal North Pacific sediments. It was found that *E. huxleyi* accounted for 50 to 80% of the coccolithophorid community in the mid- to high latitudinal area of the North Atlantic Ocean in spring to early fall (Okada and McIntyre, 1979), so that the productivity of the total coccolithophorids possibly to a large extent reflects the productivity of *E. huxleyi*.

It was reported that the decrease of coccolithophorid productivity in the North Atlantic for the last glacial period was possibly caused by, the lower water temperature of the surface layer due to the southward movement of the polar front (McIntyre *et al.*, 1972). The difference of SST between the last glacial maximum and the present was estimated to be –10.0 to –4.0°C in the oceanic areas of 35 to 45°N of North Atlantic Ocean (CLIMAP project members, 1976). The cooling of the sea surface layer during the last glacial period was also observed in $U^{k'}_{37}$-based temperature records of the St. 14 site, but the difference of water temperatures between the last glacial and the present by $U^{k'}_{37}$ in St. 14 was small (–2.5 to –1.5°C: averaging 20.5°C for the last glacial period and 22.5°C for the Holocene) (Sawada and Handa, 1998). Analytical data for marine microfossil assemblages and foraminiferal d18O obtained from a sediment core of C-4 off central Japan (33°9.0′ N, 137°41.9′ E) suggested that the cold water mass associated with the Oyashio probably intruded into the Kuroshio from the north and a mixed water region developed off central Japan during the last glacial period (Chinzei *et al.* 1987; Oba and Yasuda, 1992). The mixing region of the Kuroshio and Oyashio currently exists at 35 to 37°N and 143 to 145°E, which corresponds to the polar front in the northwestern Pacific. However, the cold water of the Oyashio generally penetrates below the euphotic layer when it extends south of 35°N (Yang *et al.*, 1993). The microfossil assemblage analyzed in the C-4 core included planktonic foraminifera inhabiting the intermediate layer (~400 m depth), whereas, Gephyrocapsaceae such as *E. huxleyi* grows in the upper euphotic layer (Balch *et al.*, 1991). Thus, the cold water bearing the Oyashio might have less influence on the surface layer off central Japan, even during the last glacial age.

It is noticeable that the MAR of A&A increased with the dominance of *E. huxleyi* for the last glacial period (Figs. 10a and b), implying that these things were possibly linked. Slight lower SST than the present (averaged annual SST was estimated to be 20.5°C; Sawada and Handa, 1998) rendered the environment favorable for *E. huxleyi*, although this possibility is entirely speculative. We presumed that the high values of the MAR of A&A for the last glacial period resulted from the high productivity of *E. huxleyi* off central Japan.

Organic compound contents possibly depend not only on their export production from the sea surface layer but also on the preservation efficiency controlled by the benthic redox condition as mentioned above. The oxygen concentrations in deep sea waters would vary off central Japan from the last glacial period to the Holocene, although there are no systematic reports for the variations of oxygen concentration in this region. The glacial cooling increased the ventilation of the North Pacific Intermediate Water, so that young and well-oxygenated waters could be supplied to the deep sea of the North Pacific during the glacial period (Ohkouchi *et al.*, 1994; Kennett and Ingram, 1995). This inference implies that the oxygen concentrations of the deep and bottom waters in the North Pacific for the last glacial period were possibly higher than those in the Holocene, and biodegradation under oxic conditions could be activated on the seafloor of the North Pacific off Japan during the last glacial period. Hence, it is plausible that the millenarian scale variations of the sedimentary organic preservation were superimposed on those of export production of biogenic organic materials off Japan.

It is widely accepted that atmospheric CO_2 concentrations decreased and the greenhouse effect declined during the glacial period (e. g. Broecker and Denton, 1987). Several hypotheses for the mechanisms of a decrease of atmospheric CO_2 concentrations have been proposed. Recently, several studies demonstrated that the formations of carbonate in the surface ocean reduce alkalinity and result in a decline of capacity in the uptake of CO_2 in surface ocean (e. g. Tsunogai and Noriki, 1991; Taylor *et al.*, 1991). Holligan *et al.*(1993) reported that a decline of pCO_2 was observed in the sea surface of an extensive bloom area of *E. huxleyi*. It was suggested that the decline of atmospheric pCO_2 for the glacial period can be strongly linked to reducing the formation of biogenic carbonate in the surface ocean (Dymond and Lyle, 1985). Indeed, sedimentary coccolith carbonate records obtained from the middle to high latitudinal area of the North Atlantic indicated that productivity of coccolithophorids during the last glacial was lower than that in the Holocene (McIntyre, 1972; van Kreveld *et al.*, 1995). However, our data indicated that productivity of the Gephyrocapsaceae decreased in the northwestern Pacific off central Japan from the last glacial period to the present. If the hypothesis proposed by Dymond and Lyle (1985) and the other investigators is accepted, it is possible that the capacity of uptake of CO_2 increased in the surface ocean off central Japan during the last glacial period. However, for clarifying the linkage between variations of pCO_2 in the atmosphere and the formation of biogenic carbonates from sedimentary A&A investigations, the percent abundances of coccolith carbonates produced by the Gephyrocapsaceae in total biogenic carbonates containing foraminiferal and other carbonates and the production ratios of carbonate carbon to organic carbon should be examined in future studies.

CONCLUSIONS

Sinking particles collected from year-long time-series sediment traps at 1674, 4180, 5687 and 8688 m, the underlying bottom sediment at 9200 m and a sediment core of KT92-17 St. 14 at 3252 m, in the northwestern Pacific off Japan were analyzed for long-chain alkenones and alkyl alkenoates (A&A).

A&A fluxes tended to decrease with water depth in sediment trap samples from 1674 to 8867 m, attributed to the decomposition of these compounds within the water column. Nevertheless, A&A were significantly contained in the underlying bottom sediment at 9200 m, despite no preservation of coccolith carbonates. Alkenone unsaturation indices ($U^{k'}_{37}$) in sediment-trap and bottom-sediment samples were well-preserved water temperature signals of the surface euphotic layer. From these results, it could be confirmed that sedimentary A&A can be an applicable and reliable proxy for reconstructing the productivity of the Gephyrocapsaceae, even in sediments below the lysocline depth. However, the A&A fluxes were not simply correlated with the coccolith fluxes of A&A producing algae (*E. huxleyi* and *G. oceanica*) determined by microscopic counting of the coccolith at 1674 m. Also, assemblages of A&A-producing algae estimated using the relationship between the ratios of C_{38} di-unsaturated ethyl alkenoates to C_{37} alkenones (EE/K37) and $U^{k'}_{37}$ (EE/K37–$U^{k'}_{37}$) were not necessarily compared to those determined by microscopic counting of the coccolith at 1674 m and a core KT92-17 St. 14. These results suggested that the production, transport and decomposition (or dissolution) mechanisms within the water column and sediment-water interface were different between the A&A and coccolith of their source organisms. Hence, the limitations were recognized and taken into account when the MAR and species composition of coccolith carbonate are estimated by using those of A&A.

The MARs of A&A in a core KT92-17 St. 14 for the last glacial period were higher than those in the Holocene, implying that the productivity of the Gephyrocapsaceae during the glacial period was higher than that in the Holocene off central Japan. This result disagreed with previous studies on coccolith carbonates in North Atlantic sediments, in which coccolithophorid productions increased from the last glacial period to Holocene. Hence, it was presumed that the millenarian-scale fluctuation in coccolithophorid productivity in the northwestern Pacific off central Japan was possibly different from that in the middle to high-latitudinal areas of the North Atlantic from the last glacial period to the Holocene.

It is expected that systematic investigations of A&A and coccolith carbonates are conducted for a wider range of marine particulate organic matter and sediments. We believe that it will be possible to evaluate the long-term fluctuations in the productivity of the Gephyrocapsaceae species, and biogenic carbonates in the world oceans can be estimated from A&A analyses. These informations are essential for understanding the roles of the marine biogeochemical process for material cycling and climatic conditions.

Acknowledgements—We thank the all cruise members of the R/V Tansei-maru (KT92-3 and KT92-17) and the R/V Hakuho-maru (KH94-3) for collecting sediment traps and sediment cores. Helpful comments on the work were provided by Drs. T. Nakatsuka, T. Saino, T. Hama, K. Ohta, N. Harada, T. Oba, N. Ohkouchi, T. Masuzawa, T. Nakamura, M. Murayama and N. Ahagon.

REFERENCES

Balch, W. M., Holligan, P. M., Ackleson, S. G. and Voss, K. J. (1991): Biological and optical properties of mesoscale coccolithophore blooms in the Gulf of Maine. *Limnol. Oceanogr.*, **36**, 629–643.

Bard, E., Arnold, M., Fairbanks, R. G. and Hamelin, B. (1993): 230Th–234U and ^{14}C ages obtained by mass spectrometry on corals. *Radiocarbon*, **35**, 191–199.

Bard, E. (1988): Correction of accelerator mass spectrometry ^{14}C ages measured in planktonic foraminifera: paleoceanographic implications. *Paleoceanogr.*, **3**, 635–645.

Bell, M. V. and Pond, D. (1996): Lipid composition during growth of motile and coccolith forms of Emiliania huxleyi. *Phytochemistry*, **41**, 465–471.

Blackburn S. I. and Cresswell G. (1993): A coccolithophorid Bloom in Jervis Bay, Australia. Aust. *J. Mar. Freshwater Res.*, **44**, 253–260.

Brassell S. C., Eglinton G., Marlowe I. T., Pflaumann U., and Sarnthein M. (1986): Molecular stratigraphy: a new tool for climatic assessment. *Nature*, **320**, 129–133.

Broecker, W. S. and Denton, G. H. (1980): The role of ocean-atmosphere reorganizations in glacial cycles. *Geochim. Cosmochim. Acta.*, **53**, 2465–2501.

Chinzei, K., Fujioka, K., Kitazato, H., Koizumi, I., Oba, T., Oda, M., Okada, H., Sakai, T. and Tanimura, Y. (1987): Postglacial environmental change of the Pacific Ocean off the coasts of central Japan, *Mar. Micropaleontol.*, **11**, 273–291.

CLIMAP project members (1976): The surface of the ice-age earth. *Science*, **191**, 1131–1137.

Conte M. H., Thompson A. and Eglinton G. (1995): Lipid biomarker diversity in the coccolithophorid Emiliania huxleyi (Prymnesiophyceae) and related species *Gephyrocapsa oceanica*. *J. Phycol.*, **31**, 272–282.

Conte M. H., Eglinton G., and Madureira L. A. (1992): Long-chain alkenones and alkyl alkenoates as paleotemperature indicators: Their production, flux and early diagenesis in the eastern North Atlantic. *Org. Geochem.*, **19**, 287–298.

Crip, T. P., Brenner, S., Venkatesan, M. I., Ruth, E. and Kaplan, I. R. (1979): Organic chemical characterization of sediment trap particles from San Nicholas, Santa Barbara, Santa Monica, and San Pedro Basins, California. Geochim. *Cosmochim. Acta.*, **43**, 1791–1801.

Dymond, J. and Lyle, M. (1984): Flux comparisons between sediments and sediment traps in the eastern tropical Pacific: Implications for atmospheric CO_2 variations during the Pleistocene.

Limnol. Oceanogr., **30**, 699–712.

Fairbanks, R. G., and Wiebe, P. H. (1980): Foraminifera and chlorophyll maximum: vertical distribution, seasonal succession and paleoceanographic significance. *Science*, **209**, 1524–1526.

Gagosian, R. B., Smith, S. O. and Nigrelli, G. E. (1982) Vertical transport of steroid alcohols and ketones measured in a sediment trap experiment in the equatorial Atlantic Ocean. *Geochim. Cosmochim. Acta.*, **46**, 1163–1172.

Holligan, P. M., Fernandez, E., Aiken, J. A., Balch, W. M., Boyd, P., Burkill, P. H., Finch, M., Groom, S. B., Malin, G., Muller, K., Purdie, D. A., Robinson, C., Trees, C., Turner, S. M. and van der Wal, P. (1993): A biogeochemical study of the coccolithophore *Emiliania huxleyi, in the North Atlantic. Global Biogeochem. Cycle.*, **7**, 879–900.

Holligan, P. M. (1992): Do marine phytoplankton influence global climate? In: Falkowski, P. G. and Woodhead, A. D. (Editors), Primary productivity and biogeochemical cycle in the sea, Plenum Press, New York, pp. 487–501.

Honjo, S (1976): Coccoliths: Production, transportation and sedimentation. *Mar. Micropaleontol.*, **1**, 65–79.

Houghton S. D. and Guptha M. V. S. (1991): Monsoonal and fertility controls on recent marginal sea and continental shelf coccolith assemblages from the western Pacific and northern Indian oceans. *Mar. Geol.*, **97**, 251–259.

Huang, C.-Y., Liew, P.-M., Zhao, M., Chang, T.-C., Kuo, C.-M., Chen, M.-T., Wang, C.-H. and Zheng, L.-F. (1997): Deep sea and lake records of the Southeast Asian paleomonsoons for the last 25 thousand years. *Earth Planet. Sci. Lett.*, **146**, 59–72.

Kennedy, J. A. and Brassell, S. C. (1992): Molecular stratigraphy of the Santa Barbara basin: comparison with historical records of annual climate change. *Org. Geochem.*, **19**, 235–244.

Kennett, J. P. and Ingram, B. L., (1995): A 20,000-year record of ocean circulation and climate change from the Santa Barbara basin. *Nature*, **377**, 510–514.

Kitagawa, H., Masuzawa, T., Matsumoto, E., Yamaguchi, K. and Nakamura, T. (1991) A preparation method of graphite target by reduction of CO_2 with H2 for AMS^{14}C measurement. In *Summaries of Research Usinf AMS ^{14}C Nagoya University (II).* (Edited by Dating and Materials Research Center Nagoya University). pp. 113–121 (in Japanese).

Klaveness, D. (1972): Coccolithus huxleyi (Lohm.) Kamptn. II. The flagellate cell, aberrant cell types, vegetative propagation and life cycles. *Br. Phycol. J.*, **7**, 309–318.

Knappertsbusch, M. and Brummer, G. J. A. (1995): A sediment trap investigation of sinking coccolithophorids in the North Atlantic. *Deep-Sea Res.*, 42, 1083–1109.

Madureira, L. A. S., Conte, M. H. and Eglinton, G. (1995): Early diagenesis of lipid biomarker compounds in North Atlantic sediments. *Paleoceanogr.*, **10**, 627–642.

Marlowe I. T., Brassel S. C., Eglinton G. and Green J. C. (1990): long-chain alkenones and alkyl alkenoates and fossil coccolith record of marine sediments. *Chem. Geol.*, **88**, 349–375.

Marlowe I. T., Green J. C., Neal A. C., Brassell S. C., Eglinton G. and Course P. A (1984): long chain (n–C_{37}–C39) alkenones in the Prymnesiophyceae. Distribution of alkenones and other lipids and taxonomic significance. *Br. Phycol. J.*, **19**, 203–216.

Martinez, PH., Bertrand, PH., Bouloubassi, I., Bareille, G., Shimmield, D., Vautravers, B., Grousset, F., Guichard, S., Ternois, Y. and Sicre, M.-A. (1996): An integrated view of inorganic biogeochemical indicators of palaeoproductivity changes in a coastal upwelling area. *Org. Geochem.*, **24**, 411–420.

Matsueda, H. and Handa, N. (1986): Vertical flux of hydrocarbons as measured in sediment traps in the eastern North Pacific Ocean. *Mar. Chem.*, **20**, 179–195.

McCafferey, M. A., Farrington, J. W. and Repeta, D. J. (1990): The organic geochemistry of Peru margin surface sediments: I. A comparison of the C_{37} alkenone and historical El Niño records. *Geochim. Cosmochim. Acta.*, **54**, 1671–1682.

McIntyre, A., Ruddiman, W. F. and Jantzen, R. (1972): Southward penetrations of the North Atlantic polar front: Faunal and floral evidence of large-scale surface water mass movements over the last 225,000 years, *Deep-Sea Res.*, **19**, 61–77.

McIntyre A. and McIntyre R. (1970): Coccolith concentrations and differential solution in oceanic sediments. In *Micropaleontology of the Oceans* (Edited by Funnell B. H. and Riedel W. R.). pp. 253–261, Cambridge University Press, Cambridge, England.

McIntyre A. and Be A. W. H. (1967): Modern coccolithpphores of the Atlantic Oceans-1. Placolith and cyrtoliths. *Deep-Sea Res.*, **14**, 561–597.

Murayama, M., Ahagon, N., Hyong, S., Kanamatsu, T. and Taira, A. (1994): Lithology in sediment cores collected during the cruises of KT92–17 and KT93-7 (IGBP). *Kaiyou Monthly*, **26**, 434–439 (in Japanese).

Murayama, M., Matsumoto, E., Nakamura, T., Okamura, M., Yasuda, H. and Taira, A. (1993): Re-examination of the eruption age of Aira-Tn Ash (AT) obtained from a piston core off Shikoku-determined by AMS ^{14}C dating of planktonic foraminifera. *J. Geol. Soc. Japan (Chishitsugaku Zasshi)*, **99**, 787–798 (in Japanese with English abstract).

Nakamura, T. and Nakai, N. (1988): Fundamentals of radiocarbon datings with accelarator mass spectrometry. *Mem. Geol. Soc. Japan*, **29**, 83–106.

Nakatsuka, T., Handa, N. and Imaizumi, S. (1995): Spatial and temporal variation of d15N in sinking particles in deep waters: Its implication for the origin and transport of particulate organic matter: In: Sakai, H. and Nozaki, Y. (Editors), *Biogeochemical processes and ocean flux in the western Pacific*, Terra Scientific Publishing Company, Tokyo, pp. 355–374.

Oba, T. and Yasuda, H. (1992): Paleoenvironmental change of the Kuroshio region since the last glacial age (in Japanese with English abstract). *The Quarternary Research (Daiyonki Kenkyu)*, **31**, 329–339 (in Japanese with English abstract).

Ohkouchi, N., Kawahata, H., Murayama, M., Okada, M., Nakamura, T. and Taira, A. (1994): Was deep water formed in the North Pacific during the late Quaternary? Cadmium evidence from the northwest Pacific. *Earth Planet. Sci. Lett.*, **124**, 185–194.

Okada H. (1992): Biogeographic control of modern nannofossil assemblages in surface sediments of Ise Bay, Mikawa Bay and Kumano-nada, off the coast of central Japan. *Memorie Di Scienze Geologiche*, **13**, 431–449.

Okada and McIntyre (1979): Seasonal distribution of modern coccolithophores in the western North Atlantic Ocean. *Mar. Biol.*, **54**, 319–328.

Okada H. and Honjo S. (1973): The distribution of oceanic coccolithophorids in the Pacific. *Deep-Sea Res.*, **20**, 355–374.

Pace, M. L., Knauer, G. A., Karl, D. M. and Martine, J. H. (1987): Primary production, new production and vertical flux in the eastern Pacific Ocean. *Nature*, **325**, 803–805.

Prahl F. G., Collier R. B., Dymond J., Lyle M. and Sparrow M. A. (1993): A biomarker perspective on prymnesiophyte productivity in the northeast Pacific Ocean. *Deep-Sea Res.*, **40**, 2061–2076.

Prahl, F. G., Muehlhausen, L. A. and Lyle, M. (1989a): An organic geochemical assessment of oceanographic conditions at MANOP Site C over the past 26,000 years. *Paleoceanogr.*, **4**, 495–510.

Prahl, F. G., de Lange, G. J., Lyle, M. and Sparrow, M. A. (1989b): Post-depositional stability of long-chain alkenones under contrasting redox conditions. *Nature*, **341**, 434–437.

Prahl F. G., Muehlhausen L. A. and Zahnle D. L. (1988): Further evaluation of long-chain alkenones as indicators of paleoceanographic conditions. *Geochim. Cosmochim. Acta.*, **52**, 2303–2310.

Prahl F. G. and Wakeham S. G. (1987): Calibration unsaturation patterns in long-chain ketone composition for paleotemperature assessment. *Nature*, **330**,367–369.

Samtleben, C. and Bickert, T. (1990): Coccoliths in sediment traps from the Norwegian Sea. *Mar. Micropaleontol.* **16**, 39–64.

Smith, K. L., Jr. and Baldwin, R. J. (1984): Seasonal fluctuations in deep-sea sediment community oxygen consumption: Central and eastern North Pacific. *Nature*, **307**, 624–626.

Sawada, K. and Handa, N. (1998): Variability of the path of the Kuroshio oceanic current over the past 25,000 years. *Nature*, **392**, 592–595.

Sawada, K., Handa, N., Shiraiwa, Y., Danbara, A. and Montani, S. (1996): Long-chain alkenones and alkyl alkenoates in the coastal and pelagic sediments of the northwest North Pacific with special reference to the reconstruction *of Emiliania huxleyi and Gephyrocapsa oceanica* ratios. *Org. Geochem.*, **24**, 751–764.

Sawada, K. and Handa, N., Nakatsuka, T. (1998): Production and transport of long-chain alkenones and alkyl alkenoates in sea water column in the northwestern North Pacific off central Japan. *Mar. Chem.*, **59**, 219–234.

Sawada, K., Handa, N., Shiraiwa, Y. and Danbara, A. (1995): An attempt at reconstructing prymnesiophyte assemblages revealed from sedimentary long-chain alkenones and alkyl alkenoates: An approach from laboratory cultures. *Res. Org.Geochem.*, **10**, 21–26 (in Japanese).

Suess, E. (1980): Particulate organic carbon flux in the oceans- surface productivity and oxygen utilization. *Nature*, **288**, 260–263.

Taylor, A. H., Watson, A. J., Ainsworth, M., Robertson, J. E. and Turner, D. R. (1990): A modelling investigation of the role of phytoplankton in the balance of carbon at the surface of the North Atlantic. *Global Biogeochem. Cycles*, **5**, 151–171.

Tsunogai, S. and Noriki, S. (1991): Particulate fluxes of carbonate and organic carbon in the ocean, Is the marine biological activity working as a sinking of the atmospheric carbon? *Tellus*, **43B**, 256–.

van Kreveld, S. A., Knappertsbusch, M., Ottens, J., Ganssen, G. M. and van Hinte, J. E. (1996): Biogenic carbonate and ice-rafted debris (Heinrich layer) accumulation in deep-sea sediments from a Northeast Atlantic piston core. *Mar. Geol.*, **131**, 21–46.

Volkman J. K., Barrett S. M., Blackburn S. I. and Sikes E. L. (1995): Alkenones in *Gephyrocapsa oceanica*: implications for studies of paleoclimate. *Geochim. Cosmochim. Acta.*, **59**, 513–520.

Volkman J. K., Eglinton G., Corner E. D. S. and Forsberg T. E. V. (1980): Long-chain alkenes and alkenones in the marine coccolithophorid *Emiliania huxleyi*. *Phytochemistry*, **19**, 2619–2622.

Wakeham, S. G. and Lee, C. (1993): Production, transport, and alteration of particulate organic matter in the marine water column. In *Organic Geochemistry: Principles and Applications* (Edited by Engel, M. H. and Macko, S. A.). pp. 145–169, Plenum Press, New York.

Wakeham, S. G. and Canuel, E. A. (1988): Organic geochemistry of particulate matter in the eastern tropical North Pacific Ocean: Implications for particle dynamics. *J. Mar. Res.*, **46**, 183–213.

Wakeham, S. G., Farrington, J. W., Gagosian, R. B., Lee, C., De Baar, H., Nigrelli, G. E., Tripp, B. W., Smith, S. O. and Frew, N. M. (1980): Fluxes of organic matter from a sediment trap experiment in the equatorial Atlantic Ocean. *Nature*, **286**, 798–800.

Yang, S.-K., Nagata, Y., Taira, K. and Kawabe, M. (1993): Southward intrusion of the Intermediate Oyashio Water along the east coast of the Boso Peninsula, Japan II. Intrusion events into Sagami Bay. *J. Oceanogr.*, **49**, 173–191.

Young, J. R. and Westbroek, P. (1991): Genotypic variation in the coccolithophorid species Emiliania huxleyi. *Mar. Micropaleontol.*, **18**, 5–23.

Dynamics and Characterization of Marine Organic Matter, Eds., N. Handa, E. Tanoue and T. Hama, pp. 213–229.

Photochemical Degradation of Dissolved Organic Carbon to Carbon Monoxide in Coastal Seawater

Keiichi OHTA[1], Yayoi INOMATA[2], Atsushi SANO[1] and Koichi SUGIMURA[1]

[1]*Institute for Hydrospheric-Atmospheric Sciences, Nagoya University, Nagoya 464-01, Japan.*
[2]*Solar-Terrestrial Environment Laboratory, Nagoya University, Nagoya 464-01, Japan*

(Original, Received 27 November 1996; accepted 14 April 1997)

Abstract—Carbon Monoxide concentrations in the Seto Inland Sea (Stns. 3 and 6) and Ise Bay (Sugashima) were measured by the headspace method during the 95-1 cruise of the R/V Toyoshio Maru (Hiroshima University) and at the Sugashima Marine Laboratory (Nagoya University), respectively.

The vertical distribution of CO concentrations displayed a marked diurnal variation in sunlit water layers (photochemical depths), i.e., 0–10 m at Stn. 3, 0–20 m at Stn. 6, and 0–3 m at Sugashima. The sea-to-air CO fluxes were estimated to be 2–11 μmole/m^2/d at these observation sites. The CO concentrations in the photochemical depth were integrated to give a CO column burden. Diurnal variations in the CO column burden were analyzed for daily oxidation and production rates of CO in water. Carbon monoxide in water was suggested to be rapidly oxidized by high microbial activities, showing the turnover times of 2–3 h in the Seto Inland Sea and 2 h at Sugashima. The CO production rates estimated were 0.16 (Stn. 3), 0.63 (Stn. 6) and 0.28 (Sugashima) mmole C/m^2/d.

On the basis of daily CO production rates and DOC concentrations in water, the lifetimes of DOC were estimated to be 19, 10 and 5 yr at Stns. 3 and 6 and Sugashima, respectively. These values which were obtained for the first time for coastal oceanic DOC are about two order of magnitude as small as those reported for DOC in the Sargasso Sea (500–2100 yr).

INTRODUCTION

The oceanic surface water is generally supersaturated with respect to atmospheric carbon monoxide (CO) mixing ratios, indicating the ocean is one of natural sources for atmospheric CO (Swinnerton *et al.*, 1969, 1970; Seiler and Junge, 1970; Lamontagne *et al.*, 1971; Linnenbom *et al.*, 1973; Seiler and Schmidt, 1974; Wilkniss *et al.*, 1979).The global sea-to-air flux of CO was estimated by Conrad and Seiler (1986), Erickson (1989) and Bates *et al.* (1995) to be 100 ± 90, 165 ± 80 and 13 Tg/yr, respectively, although a great difference exists among these estimates.

In the water column, CO concentrations decrease with depth in the euphotic zone to approach relatively constant, low levels occurring in deeper waters

(Swinnerton *et al.*, 1968; Seiler and Schmidt, 1974; Swinnerton and Lamontagne, 1974). Carbon monoxide in the euphotic zone shows a marked diurnal variation on sunny days with amaximum concentration in the early morning and a minimum level just before sunrise (Conrad *et al.*, 1982; Johnson and Bates, 1996; Ohta, 1996).The vertical CO profiles and their diurnal variations resemble in pattern those of in situ light intensities.

Photochemical degradation of dissolved organic matter (DOM) is known to be most responsible for the production of CO in the euphotic zone (Wilson *et al.*, 1970; Bauer *et al.*, 1980; Conrad and Seiler, 1980; Conrad *et al.*, 1982; Valentine and Zepp, 1993). As removal processes for CO, in situ oxidations, mainly by bacteria, and a sea-to-air transfer are important (Linnenbom *et al.*, 1973; Conrad and Seiler, 1980, 1982; Conrad *et al.*, 1982; Jones, 1991; Jones and Amador, 1993). Among bacteria, ammonium oxidizing bacteria are identified as a CO oxidizer (Jones and Morita, 1983; Jones *et al.*, 1984).

Mopper *et al.* (1991) and Miller and Zepp (1995) suggested that the photochemical degradation would largely contribute to the loss of dissolved organic carbon (DOC) and thereby to the oceanic carbon cycle. Their suggestion was based on the data obtained from experiments of irradiation of sunlight or artificial light on water samples. Field studies on the flux of photochemical degradation of DOC have therefore been required. Recently, Johnson and Bates (1996) reported the rates of photochemical production and consumption of CO from the field observation in the Tropical Pacific Ocean. However, the magnitude of the DOC loss due to CO production was not described.

We made field measurements of CO concentrations in the Seto Inland Sea and Ise Bay, Japan, in May and July, respectively. Based on the data on diurnal changes in CO concentration, we estimated the production and removal rates of CO as well as the magnitude of DOC degradation to CO.

MATERIAL AND METHODS

Field studies in the Seto Inland Sea were carried out during the TY-95-01 cruise (May 15–19, 1995) of the research vessel *Toyoshio Maru* of Hiroshima University at two stations (Stn. 3: 34°05′ N, 13°221′ E, 22 m deep, and Stn. 6: 33°40′, 13°153′ E, 50 m deep) (Fig. 1). Depth profiles of temperature, salinity and irradiance in water, wind velocity and atmospheric solar radiation were measured on board ship (Sakugawa *et al.*, 1996). The DOC concentrations were determined by S. Hirata of the Chugoku National Industrial Research Institute by use of Shimadzu TOC-5000 (Sakugawa *et al.*, 1996). Water samples from different depths were collected in a 12-l GO-FLO sampler. The surface water was sampled in a polyethylene bucket.

A field study in Ise bay was conducted during July 13–16, 1994, at the Sugashima Marine Biological Laboratory (34°29.5′ N, 136°50.5′ E), Nagoya University, on Sugashima Island 3.3 km offshore (Fig. 1). Water samples were collected from a floating pier near the Laboratory using a 5-l Van Dorn sampler. Water temperature, salinity, and irradiance in water were measured at the sampling site. Wind velocities monitored at a meteorological station locating

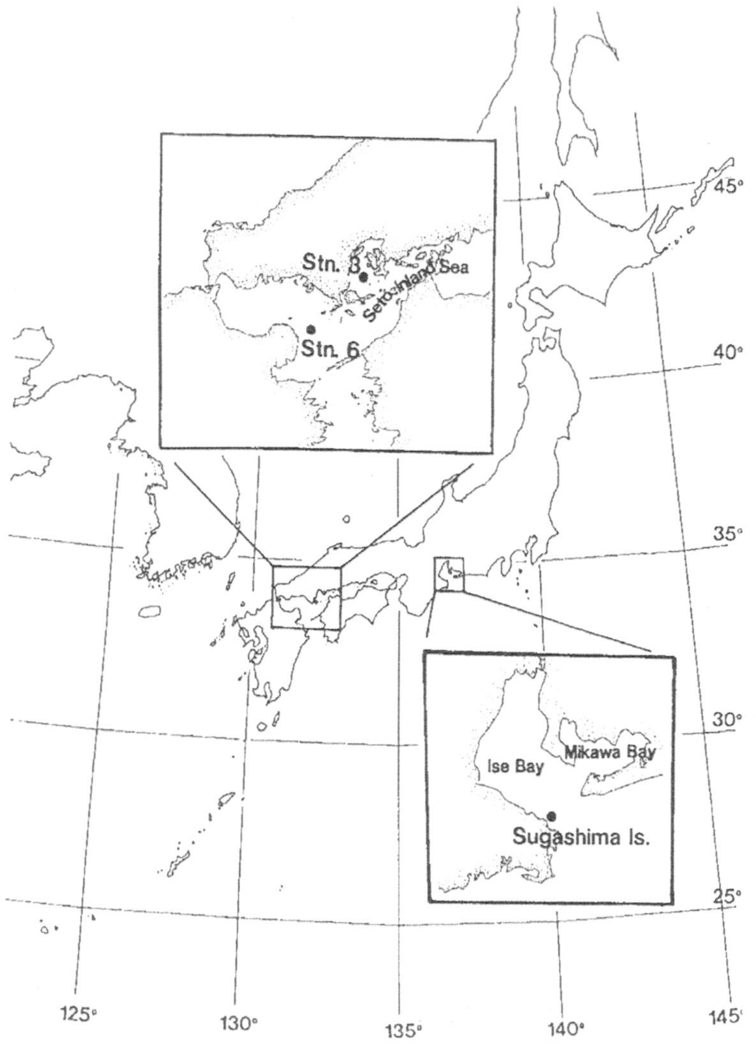

Fig. 1. Locations of observation sites.

about 5 km west of Sugashima Island were used for estimating sea-to-air fluxes of CO.

Measurements of CO dissolved in water were made according to the method reported by Ohta (1996).The procedure is described briefly below. Each water sample was siphoned into a 100 ml glass flask with a greaseless glass stopcock. A 10 ml portion of the water sample in the flask was replaced by CO-free synthetic air to make a headspace. The flask was then shaken vigorously by hand for 1 min. and kept standing for 5 min. in the dark in order to complete the equilibration between aqueous and gaseous phases. A 1 or 2 ml gas sample was taken from the headspace and subjected to a gas chromatographic analysis.

As a gas chromatograph, a GS-12 model (Sensor Tech Co., Japan) fitted with a semiconductor detector (Ohta *et al.*, in preparation) was used for on board analyses in the Seto Inland Sea. A model of GC-9A (Shimadzu, Japan) fitted with a methanizer (Shimadzu MTN-1) and a flame ionization detector was used at Sugashima Island in Ise Bay. A chromatographic column (2.6 mm in ID x 1 m) packed with molecular sieve 5A was used at 80°C to separate CO from hydrogen and methane. The detection limit was 0.05 nMCO dissolved in the water sample, and the precision of the analysis was ±2.5%.

The standard CO gases (1.02 and 4.94 ppm) used in the present study were purchased from Taiyo-Toyo Sanso Corporation, Japan. A preparation method for them was described elsewhere (Ohta, 1996).

The short-term incubation experiments for examining bacterial CO oxidations were conducted as follows. The 100 ml glass flask for headspace preparation was filled with a water sample taken from the sea surface. As a reference, a water sample without microbial activities was prepared by adding a $HgCl_2$ solution to the surface seawater sample in the flask (0.5 mM, final concentration). The flasks were kept standing at 26–27C in the dark. The CO concentrations were determined before and after the incubation period (1 to 3 h) by the same procedure described above.

The CO concentrations thus measured were corrected at first based on the equilibrium constants (Henry's constants) of CO between the aqueous and gaseous phases in the flask for headspace preparation. The concentrations were further corrected for water vapor contained in the headspace gas (Smithsonian Meteorological Table, 1958).

RESULTS AND DISCUSSION

Diurnal variation in CO concentration

Gentle winds (2–6 m/s) prevailed during the field observation periods in both the Seto Inland Sea (Stns. 3 and 6) and Ise Bay. Solar radiation was intense at Stn. 6 and Ise Bay. At Stn. 3, it was fine till 10:00 and cloudy thereafter (Fig. 2). There were two cycles of the ebb and flow of the tide around Sugashima Island on July 14. The high tide appeared at 0:00 and 14:26, and low tide at 7:57 and 19:00, resulting in a change in water depth by about 1 m at the observation site.

The CO concentrations in the surface water displayed a marked diurnal variation (Fig. 2). At Stn. 3, the highest CO concentration appeared at 13:30, while the solar radiation reached its maximum level at 10:00 (Fig. 2). A time lag of about 3.5 hours was also found at Stn. 6 between the maxima of CO concentration and solar radiation. Similar time lags have been observed in the open ocean (Conrad and Seiler, 1980; Conrad *et al.*, 1982; Ohta, 1996).At Sugashima, the highest level of CO appeared at 13:00; only 1 hour after the solar radiation showed a peak (Fig. 2). A turn of the tide, which occurred around 14:30, might contribute to this shorter time lag and also to a sharp CO decrease in the early afternoon as physical processes (mixing, transport or replace of water),

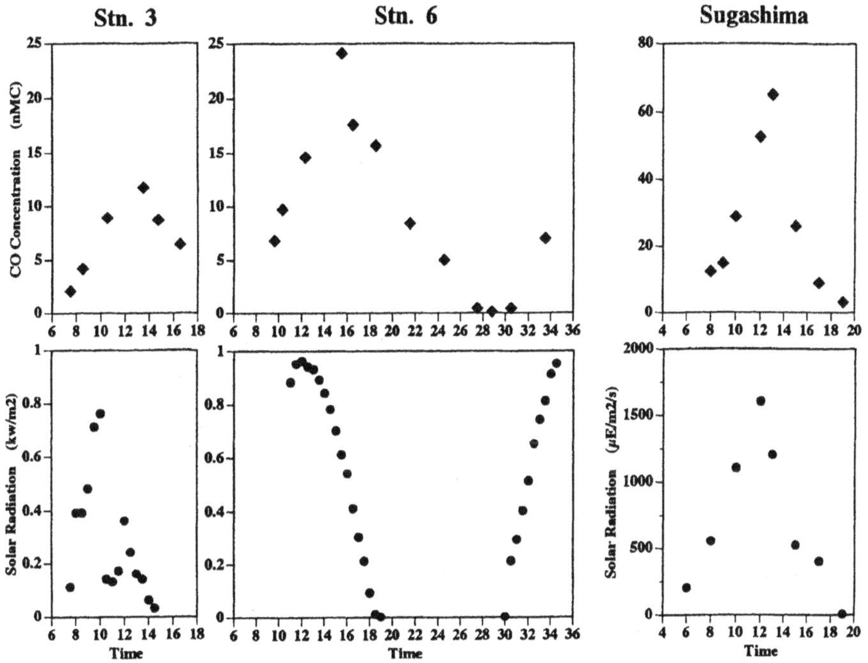

Fig. 2. Diurnal variations in surface CO concentrations (upper) and solar radiation (bottom) at Stns. 3 and 6 and Sugashima.

which can modify the CO concentration principally controlled by the photochemical and biological processes.

Carbon monoxide concentrations were highest at the sea surface and gradually decreased with depth in the euphotic zone throughout all the three observation sites. Figure 3 shows a time series of vertical CO distribution at Stn. 6 where the depth of the euphotic zone was about 35 m. Although the diurnal variation in CO concentration was pronounced in the upper water layers, it was still discernible at a depth of 20 m where the light intensity reached a range of 2–7% of the surface levels. A value of 2.5%, which was reported as a minimum light level causing a discernible diurnal CO variation in the equatorial Pacific Ocean (Ohta, 1996), falls within this range. As shown in Fig. 4, marked diurnal variations were observed at Stn. 3 down to depths of 10 m (the light intensity: 8–10%) in water and at Sugashima in the entire water column (the light intensity at the depth of 3 m: 20%).These suggest that photochemical production of CO could occur within the water layers where light intensities higher than 2% are available.

It is worthy to note that CO was removed so extensively in the afternoon that only a trace amount of it could be found in water late at night (St. 3 and Sugashima, Fig. 4) or early in the morning (Stn. 6, Fig. 3).In the open ocean, a considerable concentration of CO is maintained in water throughout the nighttime (Conrad *et al.*, 1982; Jones, 1991, Johnson and Bates, 1996; Ohta, 1996).

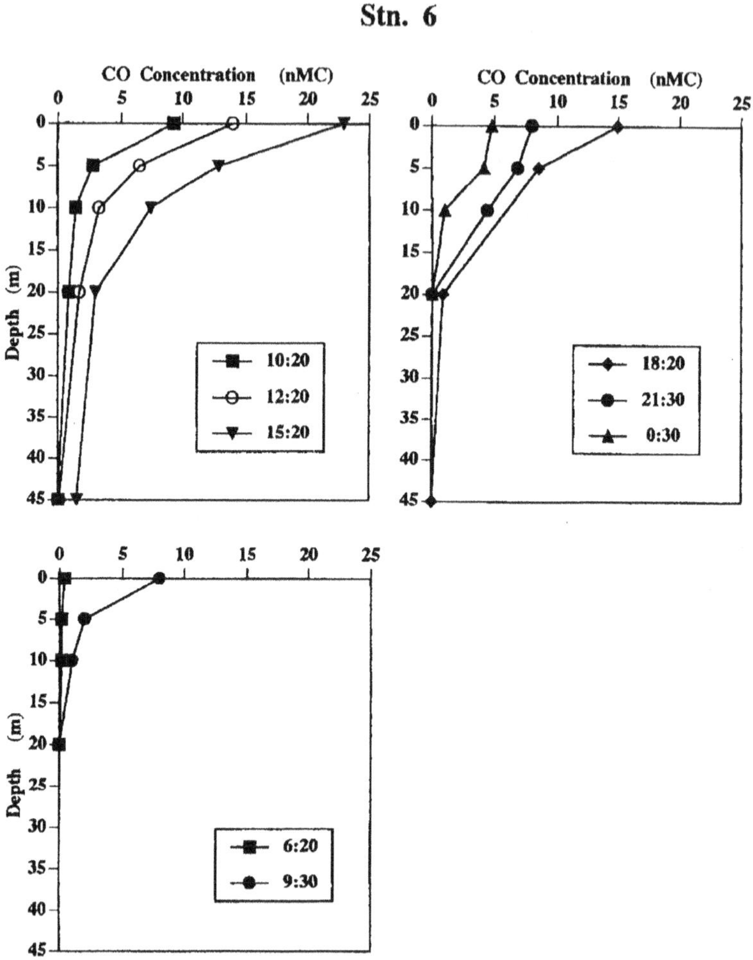

Fig. 3. Vertical CO distribution at Stn. 6 at different daytimes (local time).

This difference would largely result from higher microbial activities oxidizing CO in the coastal water than open ocean, although some contribution of complicated water movements to disperse CO in the coastal areas cannot be precluded.

Using the above-described data on diurnal variations in CO concentration, we estimated the production and removal rates of CO, as well as degradation rates of DOC into CO, according to the methods which will be discussed hereafter. A similar estimation was given by Johnson and Bates (1996), who made CO measurements at 4–h intervals in the Tropical Pacific Ocean and discussed the dynamics of CO based on the data thus obtained.

The same authors integrated CO concentrations in the water column to obtain a column burden. Then they analyzed diurnal variations in the column burden for the production and oxidation rates of CO in order to avoid an influence

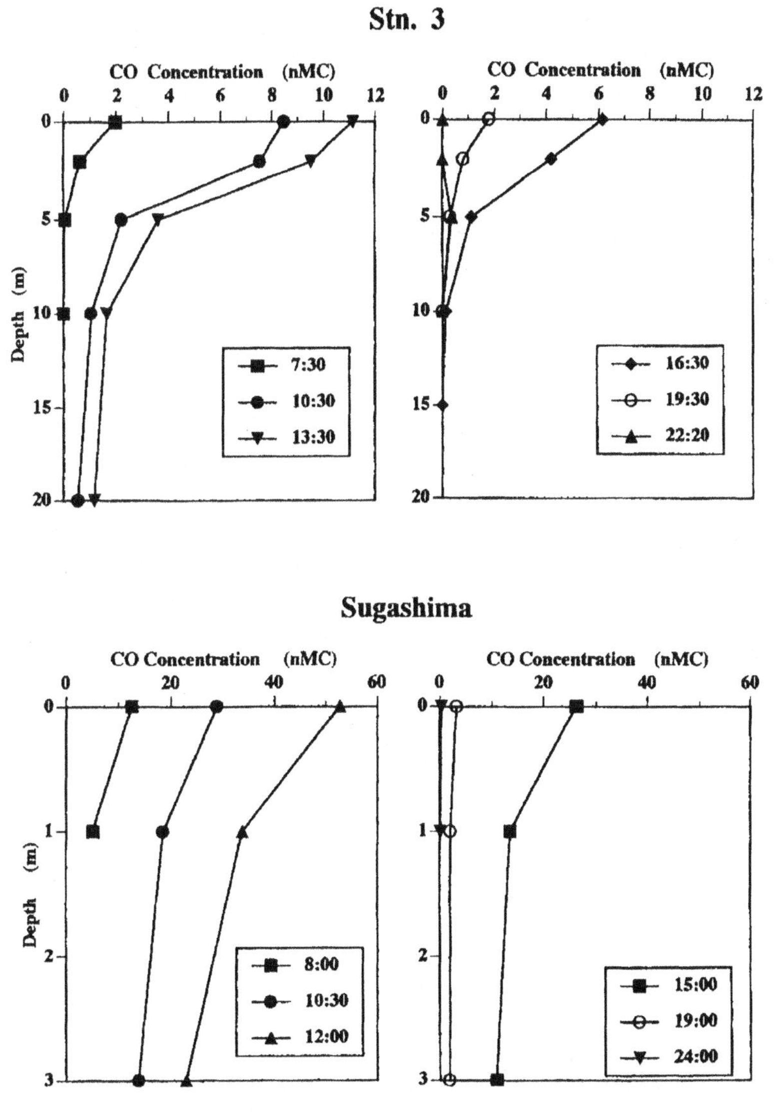

Fig. 4. Vertical CO distribution at Stn. 3 and Sugashima at different daytimes (local time).

of vertical water mixing on the CO distribution, using the following equation.

$$dB / dt = P - (F + \lambda) \qquad (1)$$

where B is the column burden of CO, t is time, P is the integrated production rate in the water column (column production rate), F is the sea-to-air flux, and λ is the integrated oxidation rate in the water column (column oxidation rate).

In the present study, we calculated dB/dt and λ in a similar manner as Johnson and Bates (1996), and F was estimated from the wind velocity and CO surface concentrations. The calculation procedure and result of each item in this equation will be described in the following sections.

Sea-to-air flux of CO

The model of Liss and Slater (1974), described below, was used for estimating the sea-to-air flux (F) of CO.

$$F = k_w (C_l - C_g / H) \tag{2}$$

where k_w is the gas transfer velocity, C is the concentration in the liquid or gas phase, and H is the Henry's constant. In the present study, the CO concentrations measured in the fields (Fig. 2) were interpolated onto 30-min. intervals, assuming a linear change in CO concentration with time during each time interval of measurement, and used as C_l. The atmospheric CO concentration was measured on board ship in the Seto Inland Sea by introducing 5 ml of the air sample into the GC. The concentrations fluctuated within the range of 85–161 ppb. The mean value of 117 ppb was used as C_g at Stns. 3 and 6. We used the same 117 ppb for Sugashima where the atmospheric CO measurement was not made, assuming that this concentration is common to the atmospheric CO over coastal marine areas under the influence of the urban air.

The transfer velocity kw was calculated from the equations given by Liss and Merlivat (1986) and modified by Gabric *et al.* (1995).

$$kw = 0.17\alpha w \text{ for } w \leq 3.6 \text{ (Sugashima)}, \tag{3}$$

$$kw = \beta(2.85w - 10.26) + 0.61\alpha \text{ for } 3.6 < w \leq 13 \text{ (Stns. 3 and 6)} \tag{4}$$

$$\alpha = (600 / Sc)^{2/3} \text{ and } \beta = (600 / Sc)^{1/2},$$

where w is a wind velocity at 10 m above sea level and Sc is a Schmidt number of CO. Wind velocities were measured every 2 hours on board at 10 m in the Seto Inland Sea (Stns. 3 and 6).For Sugashima, the wind velocities monitored every hour at the meteorological station (about 20 m above sea level) were used after being reduced by 6% for conversion to the values at 10 m (Ohta, 1996). The wind velocity changed in the range of 2–6 m/s at Stn. 3, 1–8 m/s at Stn. 6, and 0–3 m/s at Sugashima. In the present study, a daily average of the velocity at each site was used as w, because the response of the CO flux to the wind would not occur so quickly as to follow hourly changes in velocity (Harn, 1981).

The Schmidt number Sc can be calculated by the following equation,

$$Sc = \eta / \rho / D, \tag{5}$$

where η and ρ were the viscosity and density of seawater, respectively, and D is

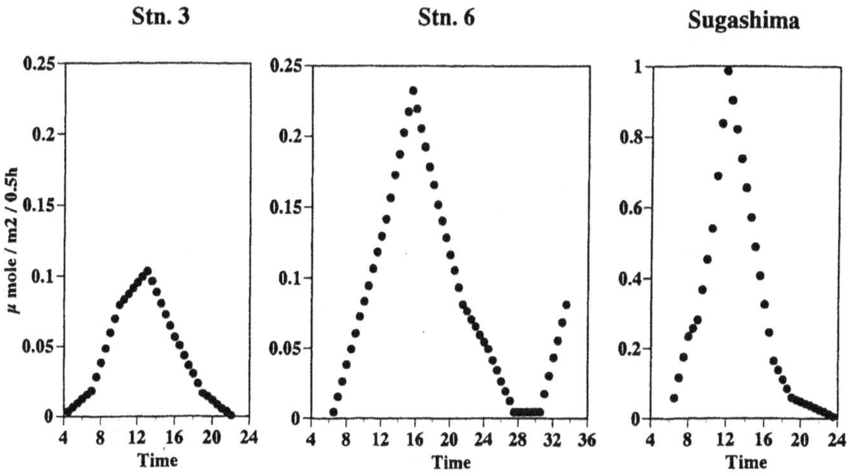

Fig. 5. Sea-to-air fluxes of CO at 0.5-hours intervals at Stns. 3 and 6 and Sugashima.

Table 1. Sea-to-air CO fluxes at Stns. 3 and 6 (the Seto Inland Sea) and Sugashima (Ise Bay). Average CO wind velocities at the height of 10 m (w), Schmidt numbers (Sc), and gas transfer velocities (k_w) are also listed.

Station	w (m/s)	Sc	k_w (cm/h)	F (μ mole/m^2/d)
Stn. 3	4.1	770	1.83	1.57
Stn. 6	4.2	770	2.03	4.51
Sugashima	2.0	402	3.74	11.0

the gas diffusivity. The values of η, ρ, and D were calculated according to Jähne *et al.* (1987), Millero (1974), and Millero and Poisson (1981), respectively. The Schmidt number obtained from daily averages of these three parameters was used as Sc in the present study.

The CO fluxes thus calculated are shown in Fig. 5, and daily fluxes are in Table 1 together with parameters used.

CO column burden and column oxidation

The column burden of CO was obtained by integrating CO concentrations from the surface down to the depth (photochemical depth) where a diurnal CO variation was still observed, that is, 0–10 m at Stn. 3, 0–20 m at Stn. 6, and 0–3 m at Sugashima. The results are shown in Fig. 6(a).

The CO oxidation rate was estimated based on the assumption that there is no CO production at night and, therefore, the nocturnal loss of the CO column burden depends only on microbial oxidations and sea-to-air fluxes. For Stn. 6 and Sugashima, the CO column burdens after 19:30 were used as nocturnal data. For Stn. 3, we used the column burdens after 16:30, since it was cloudy in the

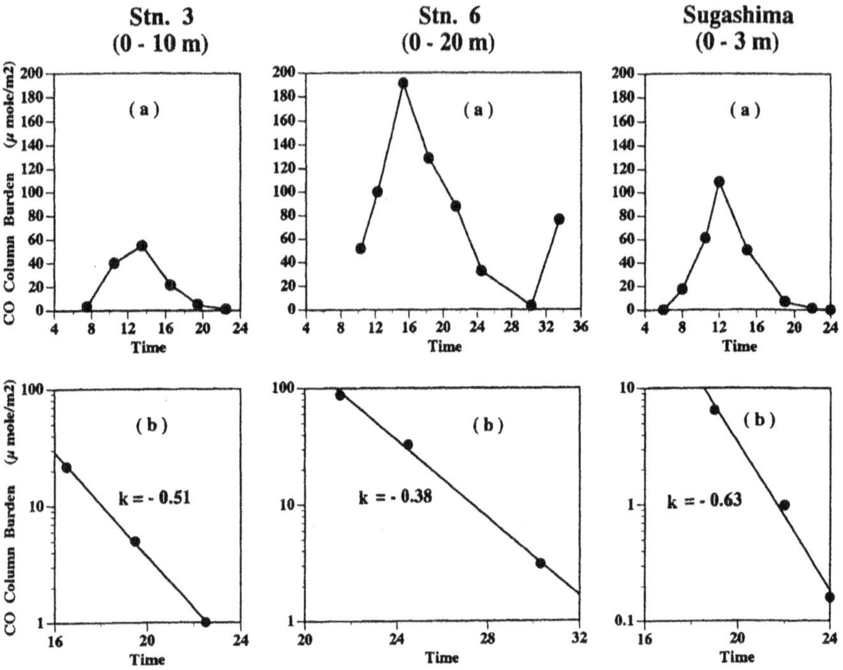

Fig. 6. The CO column burdens (a) and the first-order kinetics plot of column burdens versus time (b) for Stns. 3 and 6 and Sugashima.

afternoon and dark after 14:00 enough to expect no photochemical CO production (Fig. 2).

The column burdens were corrected at first for the loss due to sea-to-air fluxes, using the flux F at 30-min. intervals, then the logarithm of the column burdens thus obtained were plotted against the time (Fig. 6b). A straight line could be fitted to the plots, indicating the oxidation of CO followed first-order kinetics expressed by the following equation, and also the slope gave the value of the rate constant k.

$$-\ln(B_t / B_0) = kt \tag{6}$$

or

$$B_t = B_0 e^{-kt}, \tag{7}$$

where t is time, and B_0 and B_t are column burdens at time zero and after t, respectively. The values of k for Stns. 3 and 6, and Sugashima are –0.51, –0.38, and –0.63 (1/h), respectively (Fig. 6b).

Another estimate of oxidation rate constant was given by the incubation experiment. Several times of incubation were performed under the dark at each

Table 2. Turnover times of CO in water at different ocean sites.

Ocean site	Season	Turnover time (h)	Method	Reference
Coastal region				
Gulf of Paria, Venezuela (surface water)				
	spring	16-19	^{14}CO incubation	Jones and Amador (1993)
	fall	3-7	^{14}CO incubation	Jones and Amador (1993)
Olinoco delta region, Venezuela (surface water)				
	spring	2-15	^{14}CO incubation	Jones and Amador (1993)
	fall	3-7	^{14}CO incubation	Jones and Amador (1993)
Seto Inland Sea (Stns. 3 and 6), Japan (euphotic zone)				
	May	2-3	field observation	present study
Ise Bay (Sugashima), Japan (euphotic zone)				
	July	2	field observation	present study
Open ocean				
Sargasso Sea (euphotic zone)				
	June	27-93	^{14}CO incubation	Jones (1991)
	September	89-392	^{14}CO incubation	Jones (1991)
South Pacific (mixed layer)				
19°S, 149.5°W	April	53	field observation	Johnson and Bates (1996)
4°S, 140°W	December	259	field observation	Johnson and Bates (1996)
Eastern Caribbean (surface water)				
	spring	34-88	^{14}CO incubation	Jones and Amador (1993)
	fall	8-98	^{14}CO incubation	Jones and Amador (1993)

of Stns. 3 and 6. The CO concentration in the water sample contained in a flask decreased rapidly during the short term incubation (1–3 h), while the water sample spiked with $HgCl_2$ maintained a original CO concentration, indicating a principal contribution of biological oxidation to the CO decrease. If the reaction is first-order, the rate constants of -0.48 ± 0.06 and -0.35 ± 0.05 (1/h) can be given to Stns. 3 and 6, respectively. Each value is close to the above-described constant k of each site. This is indicative of the validity of analyzing nocturnal variation in CO column burden for the oxidation rate constant.

The turnover time of CO, which is defined as the inverse of the first-order rate constant of CO oxidation, was reported to be short (3–19 h) in the coastal region and much longer (several days) in the open ocean (Table 2). The turnover times estimated from the above-described values of k for CO at our observation sites were shortest (2–3 h) among those reported so far (Table 2). This indicates that the microbial activities oxidizing CO are at very high levels in the Seto Inland Sea and Ise Bay in summer.

The column burden oxidized (B_{oxi}) during time t can be calculated using following equations,

$$B_{oxi} = B_0 - B_t \tag{8}$$

$$= B_0 - B_0\, e^{-kt} \tag{9}$$

$$= B_0\,(1 - e^{-kt}). \tag{10}$$

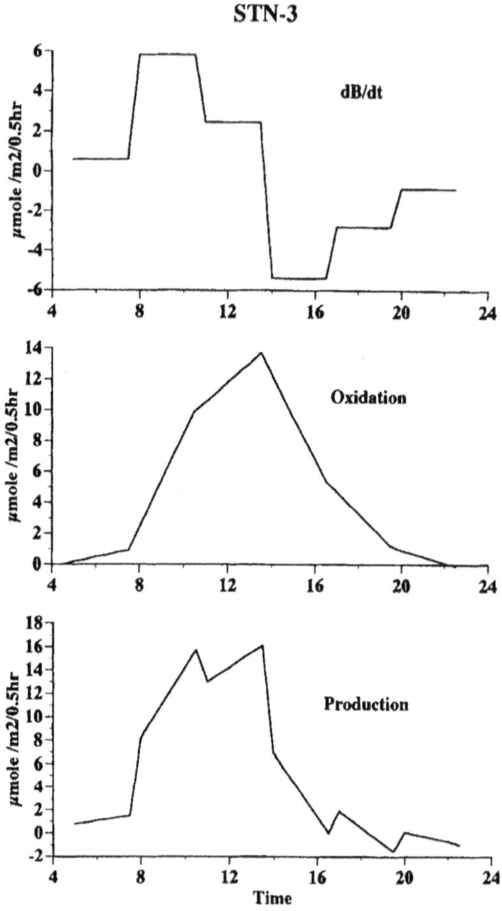

Fig. 7. Diurnal variations of *dB/dt*, CO column oxidation rates and CO column production rates at Stn. 3.

In the present study, the value of B_{oxi} was calculated at 30-min. intervals to yield a half-hourly column oxidation rate (λ).The diurnal pattern of λ (Figs. 7–9) well resembles that of the CO column burden (Fig. 6a) at each observation site, because B_{ox} depends on the column burden.

CO column production

The CO column production rate (P) was calculated from

$$P = dB / dt + (F + \lambda). \tag{11}$$

The column burdens obtained at about 3-h intervals (Fig. 6a) at the three observation sites were interpolated onto 30-min. intervals, assuming a linear change in the column burden with time, and then the derivatives *dB/dt* were

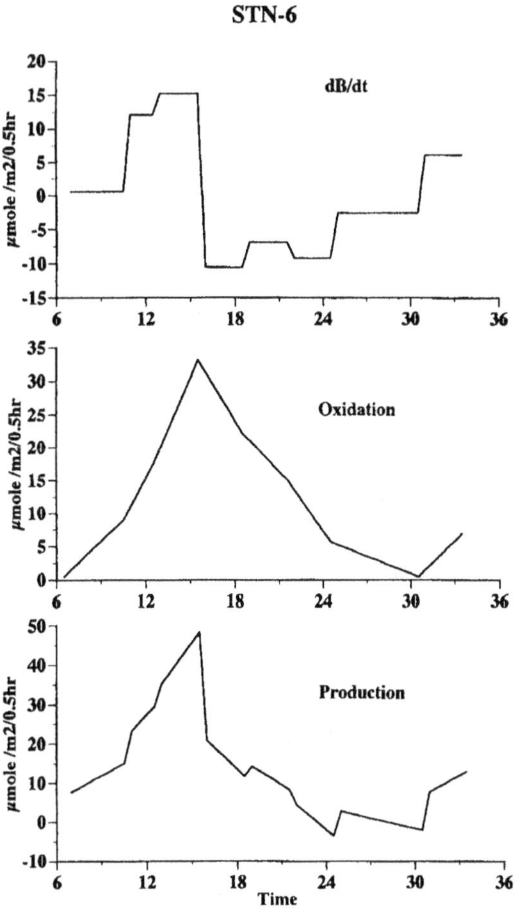

Fig. 8. Diurnal variations of *dB/dt*, CO column oxidation rates and CO column production rates at Stn. 6.

calculated for every interval. The column production rates *P* at every 30-min. intervals were estimated based on the assumption that the rate constant *k* of the CO column oxidation is constant night and day. The results of calculation for *dB/dt* and *P* of each observation site are illustrated in Figs. 7–9.

The diurnal pattern of *P* at Stn. 3 displayed two peaks during 10:30–13:30 (Fig. 7), presumably reflecting the fluctuation of solar radiation (Fig. 2). At Stn. 6, extremely high CO concentrations were observed at 15:20 in the entire euphotic zone (Figs. 2 and 3). They made up a steep increase in the CO column burden between 12:20 and 15:20 (Fig. 6), resulting in the highest *dB/dt* during the same period and further in an appearance of a sharp peak of *P* at 15:20 (Fig. 8). We cannot explain at present why the extremely high CO concentrations appeared at 15:20 and whether any other high CO concentrations occurred between 12:20 and 15:20.A detailed figure of a CO concentration change during this period

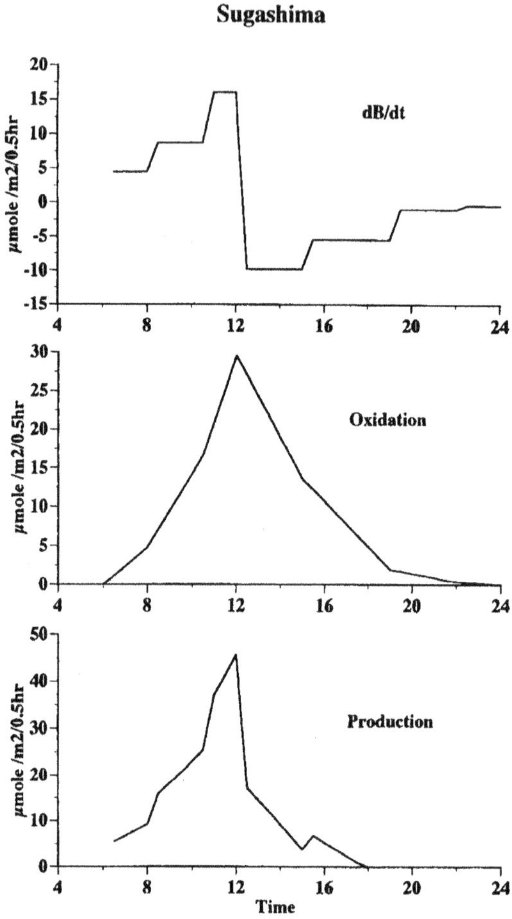

Fig. 9. Diurnal variations of *dB/dt*, CO column oxidation rates and CO column production rates at Sugashima.

might be followed if we had carried out the observation at shorter time intervals in the early afternoon.

The daily CO column productions obtained were 0.16, 0.63 and 0.28 mmoleC/m^2 at Stn. 3 (0–10 m), Stn. 6 (0–20 m) and Sugashima (0–3 m), respectively (Table 3). It is obvious that the sea-to-air fluxes (1.6–11 µmole/m2/d, Table 1) contribute as only a minor process to the removal of CO produced photochemically in the water column. Thus more than 96% of CO is oxidized in water due to high microbial activities suggested by the short turnover times of CO (Table 2). This is in a contrast to the Sargasso Sea where the microbial CO oxidation proceeds very slowly and therefore the air-sea exchange acts as almost the sole process removing CO from the water (Jones, 1991).

The values of daily CO column production (0.16–0.63 mmole/m^2/d) at our three sites are more than five times as large as those in the Tropical Pacific Ocean

Table 3. Photochemical CO production rates, DOC column burdens and DOC lifetimes at Stns. 3 and 6 (the Seto Inland Sea) and Sugashima (Ise Bay).

Site		Stn. 3	Stn. 6	Sugashima
Photochemical depth	(m)	10	20	3
CO column production	(mmoleC/m^2/d)	0.16	0.63	0.28
DOC column burden	(moleC/m^2)	1.13	2.31	0.51
DOC lifetime	(y)	19.3	10.0	5.0

(Johnson and Bates, 1996). The coastal marine may be of much importance as a photochemical field, although the field observation for CO production has not been performed at other ocean sites.

Photochemical lifetime of DOC

The lifetime of DOC photochemically degraded into CO was estimated by dividing the integrated DOC concentration within the photochemical depths (DOC column burden) by the CO column production rate *P*. The DOC column burdens at Stn. 3 (1.13 moleC/m^2, 0–10 m) and Stn. 6 (2.31 moleC/m^2, 0–20 m) were obtained from the data on vertical distribution of DOC concentrations measured at these sites (Sakugawa *et al.*, 1996). The DOC measurement was not made at Sugashima. If we can use an average (171 μMC) of DOC concentrations (141–207 μMC) measured near the mouth of Mikawa Bay being 25 km far from Sugashima (Inomata, unpublished data), the DOC column burden of 0.51 moleC/m^2 (0–3 m) is assumed.

The photochemical lifetimes thus obtained for DOC at Stns. 3 and 6 and Sugashima were 19.3, 10 and 5 y, respectively (Table 3). Mopper *et al.* (1991) estimated the lifetime of photochemically reactive DOC in the Sargasso Sea to be 500–2100 y, which are two orders of magnitude as long as ours. Although their estimate was based on data obtained not from the field observation but via sunlight irradiation experiments using seawater samples, it is obvious that DOC in the coastal region can be degraded photochemically far faster than that in the open ocean.

Several carbonyl compounds with low molecular weights are known to be produced photochemically from DOC, although their production rates are about 3/7 of that of CO (Mopper *et al.*, 1991). If we take the production rates of carbonyl compounds into the account, the lifetime of DOC in our observation sites may be reduced to 4–14 y.Miller and Zepp (1995) reported that an irradiation of artificial light on river waters resulted in a production of dissolved inorganic carbon 17–24 times as much as CO. Such data are not available at present on coastal waters. If photochemical decarboxylation takes place easily on DOC in coastal waters, the above-described lifetimes will be much reduced.

Acknowledgements—The authors appreciate the support of Drs. K. Fujiwara, H. Sakugawa and K. Takeda of Faculty of Integrated Arts and Sciences, Hiroshima University, and officers and crew of R/V Toyoshio Maru of Hiroshima University. The authors express their thanks to Professor H. Hayashi and stuffs of the Sugashima Marine Biological Laboratory, Nagoya University, for their kind assistance in this study.

REFERENCES

Bauer, K., R. Conrad and W. Seiler (1980): Photooxidative production of carbon monoxide by phototrophic microorganisms. *Biochim. Biophys. Acta*, **589**, 46–55.

Bates, T. S., L. C. Kelly, J. E. Johnson and R. H. Gammon (1995): Regional and seasonal variation in the flux of oceanic carbon monoxide to the atmosphere. *J. Geophys. Res.*, **100**, 23093–23101.

Conrad, R. and W. Seiler. (1980): Photooxidative production and microbial consumption of carbon monoxide in water. *FEMS Microbial. Lett.*, **9**, 61–64.

Conrad, R. and W. Seiler (1982): Utilization of traces of carbon monoxide by aerobic oligotrophic microorganisms in ocean, lake and soil. *Arch. Microbial.*, **132**, 41–46.

Conrad, R., W. Seiler, G. Bunse and H. Giehl (1982): Carbon monoxide in seawater (Atlantic Ocean). *J. Geophys. Res.*, **87**, 8839–8852.

Conrad, R. and W. Seiler (1986): Exchange of CO and H2 between ocean and atmosphere. pp. 269–282. In *The Role of Air-Sea Exchange in Geochemical Cycling*, ed. by P. Buat-Menard, D. Reidel Publishing Company, Dordrecht.

Gabric, A. J., G. P. Ayers and G. C. Sander (1995): Independent marine and atmospheric model estimates of the sea-air flux of dimethylsulfide in the Southern Ocean. *Geophys. Res. Lett.*, **22**, 3521–3524.

Erickson, D. J. III (1989): Ocean to atmosphere carbon monoxide flux: global inventory and climate implications. *Global Biogeochem. Cycles*, **3**, 305–314.

Harn, J. (1981): Nitrous oxide in the ocean. pp. 191–227. I*n Denitrification, Nitrification, and Atmospheric Nitrous Oxide*, ed. by C. C. Delwiche, John Wiley and Sons, New York.

Jhne, B., G. Heinz and W. Dietrich (1987): Measurement of the diffusion coefficients of sparingly soluble gases in water. *J. Geophys. Res.*, **92**, 10767–10776.

Johnson, J. E. and T. S. Bates (1996): Sources and sinks of carbon monoxide in the mixed layer of the tropical South Pacific Ocean. *Global Biogeochem. Cycles*, **10**, 347–359.

Jones, R. D. and Y. Morita (1983): Carbon monoxide oxidation by chemolithotrophic ammonium oxidizers. *Can. J. Microbiol.*, **29**, 1545–1551.

Jones, R. D., R. Y. Morita, and R. P. Griffiths (1984): Methods for estimating *in situ* chemolithotrophic ammonium oxidation using carbon monoxide oxidation. *Mar. Ecol. Prog. Ser.* **17**, 259–269.

Jones, R. D. (1991): Carbon monoxide and methane distribution and consumption in the photic zone of the Sargasso Sea. *Deep-Sea Res.*, **38**, 625–635.

Jones, R. D. and J. A. Amador (1993): Methane and carbon monoxide production, oxidation, and turnover times in the Caribbean Sea as influenced by the Orinoco River. *J. Geophys. Res.*, **98**, 2353–2359.

Lamontagne, R. A., J. W. Swinnerton and V. J. Linnenbom (1971): Non equilibrium of carbon monoxide and methane at the air-sea interface. *J. Geophys. Res.*, **76**, 5117–5121.

Linnenbom, V. J., J. W. Swinnerton and R. A. Lamontagne (1973): The ocean as a source for atmospheric carbon monoxide. *J. Geophys. Res.*, **78**, 5333–5340.

Liss, P. S. and P. G. Slater (1974): Flux of gases across the air-sea interface. *Nature*, **247**, 181–184.

Liss, P.S. and L. Merliavat (1986): Air-sea gas exchange rates: introduction and synthesis. pp. 113–127. In *The Role of Air-Sea Exchange in Geochemical Cycling*, ed. by P. Buat-Menard, D. Reidel Publishing Company, Dordrecht.

Miller, W. L. and R.G. Zepp (1995): Photochemical production of dissolved inorganic carbon from terrestrial organic matter: significance to the oceanic organic carbon cycle. *Geophys. Res. Lett.*, **22**. 417–420.

Millero, F. J. (1974): Seawater as a multicomponent electrolyte solution. pp.3–80. In *The Sea*, Vol. 5, ed. by E. D. Goldberg, John Wiley & Sons Ltd., New York.

Millero, F. J. and A. Poissoon (1981): International one-atmosphere equation of state of seawater. *Deep-Sea Res.*, **28A**, 625–629.

Mopper, K., X. Zhou., R. J. Kieber, D.J. Kieber, R.J. Sikorski and R.D. Jones (1991): Photochemical degradation of dissolved organic carbon and its impact on the oceanic carbon cycle. *Nature*, **353**, 60–62.

Ohta, K. (1997): Diurnal variations of carbon monoxide concentration in the equatorial Pacific upwelling region. *J. Oceanogr.*, **53**, 173–178.

Sakugawa, H., K. Takeda, K. Ohta, S. Hirata, A. Takami, H. Kawai and K. Fujiwara (1997): Characterization of marine photochemical processes in the Seto Inland Sea, Japan. *Umi no Kenkyu* (to be submitted).

Seiler, W. and C. Junge (1970): Carbon monoxide in the atmosphere. *J. Geophys. Res.*, **75**, 2217–2226.

Seiler, W. and U. Schmidt (1974): Dissolved nonconservative gases in seawater. pp. 219–243. In *The Sea*, Vol. 5, ed. by E. D. Goldberg, John Wiley and Sons Ltd., New York.

Smithsonian Meteorological Table (1958): Density of pure water vapor at saturation. p. 381–385. Smithsonian Institution, Washington, D.C.

Swinnerton, J. W., V. J. Linnenbom and C. H. Cheek (1968): A sensitive gas chromatographic method for determining carbon monoxide in seawater. *Limnol. Oceanogr.*, **13**, 193–195.

Swinnerton, J. W., V. J. Linnenbom and C. H. Cheek (1969): Distribution of methane and carbon monoxide between the atmosphere and natural waters. *Environ. Sci. Technol.*, **3**, 836–838.

Swinnerton, J. W., V. J. Linnenbom and R. A. Lamontagne (1970): The ocean: a natural source of carbon monoxide. *Science*, **167**, 984–986.

Swinnerton, J. W. and R. A. Lamontagne (1974): Carbon monoxide in the South Pacific Ocean. *Tellus*, **26**, 136–142.

Valentine, R. L. and R. G. Zepp (1993): Formation of carbon monoxide from the photodegradation of terrestrial dissolved organic carbon in natural waters. *Environ. Sci. Technol.*, **27**, 409–412.

Wilkniss, P. E., E. B. Rodgers, J. W. Swinnerton, R. E. Larson and R. A. Lamontagne (1979): Trace gas concentrations, intertropical convergence, atmospheric fronts, and ocean currents in the tropical Pacific. *J. Geophys. Res.*, **84**, 7023–7033.

Wilson, D. F., J. W. Swinnerton and R. A. Lamontagne (1970): Production of carbon monoxide and gaseous hydrocarbons in seawater: relation to dissolved organic carbon. *Science*, **168**, 1577–1579.

Dynamics and Characterization of Marine Organic Matter, Eds., N. Handa, E. Tanoue and T. Hama, pp. 231–240.

The Occurrence of Organic Peroxides in Seawater

Hiroshi SAKUGAWA[1], Akira TAKAMI[1], Hiroshi KAWAI[1], Kazuhiko TAKEDA[1], Kitao FUJIWARA[1] and Shizuko HIRATA[2]

[1]*Faculty of Integrated Arts and Sciences and School of Biosphere Sciences, Hiroshima University, 1-7-1 Kagamiyama, Higashi-Hiroshima 739, Japan*
[2]*Chugoku National Industrial Research Institute, 2-2-2, Hirosuehiro, Kure 737-01, Japan*

(Original, Received 5 November 1996; accepted 27 December 1996)

Abstract—Organic peroxides (ROOH), as well as hydrogen peroxide (H_2O_2), dissolved in seawater were determined by fluorometry. Seawater samples were collected at Hiroshima Bay and the Iyo-Nada in the Seto Inland Sea, western Japan in May 1995. Using an enzymatic destruction method, ROOH and H_2O_2 were descriminated from each other. The concentration of ROOH measured was in the range of 38–389 nM in seawater samples, whereas the H_2O_2 concentration was in the range of 0–195 nM. The vertical distribution (0–45 m depth) of ROOH in seawater was found to be almost uniform or a little higher at the bottom layer waters, whereas the H_2O_2 concentration was highest at surface waters and decreased with depth. Linear regression analysis was conducted to examine the correlations between peroxide concentrations and other analytical items such as CTD data, nutrients, fluorescent organic matter and dissolved organic carbon. The results showed that the ROOH concentration has almost no correlation with other parameters, whereas the H_2O_2 concentration highly correlates with the salinity in the Hiroshima Bay samples but mostly correlates with water temperature in the Iyo-Nada samples. These results clearly indicate that the generation and decomposition mechanism of ROOH is different from that of H_2O_2.

INTRODUCTION

During the last decade, the concentration of hydrogen peroxide (H_2O_2) in seawater was measured by several of investigators to understand the marine photochemistry of chemical species such as redox reactions involving transition metals and organic compounds (Zika *et al.*, 1985; Moore *et al.*, 1993). Its concentration is reportedly tens to hundreds of nanomolar levels in the surface waters of coastal and open ocean. Fujiwara *et al.* (1993) reported H_2O_2 concentrations ranging from 0 to 400 nM in the Seto Inland Sea, western Japan. The vertical distribution of H_2O_2 in seawater was found to be highest in the surface and decreasing with depth, suggesting the importance of photochemical reactions.

Organic peroxides (ROOH) such as hydroperoxides and peroxyacids occur in atmospheric samples (gas or precipitation) and they have similar behaviors to H_2O_2 (Hellpointner and Gab, 1989, Hewitt and Kok, 1991). The concentrations of ROOH (as total values) are in the range of 0 to hundreds of nanomolar levels in atmospheric wet samples. As far as we know, no study has been carried out about the distribution of ROOH in seawater, although might play a similar role as H_2O_2 such as redox control of chemical species.

The aim of this study is to determine organic peroxide concentrations as well as H_2O_2 in seawater and to clarify the distribution patterns in coastal and open ocean waters. We measured the concentration of ROOH as well as H_2O_2 in seawaters of a coastal region, of Japan by a technique, using enzymes, that separates ROOH from H_2O_2 (Sakugawa *et al.*, 1995). Here we report the results of the separation of ROOH from H_2O_2 by this method and discuss on distribution patterns of ROOH in coastal seawaters.

MATERIALS AND METHODS

Collection of seawater samples

Seawater samples were collected by Go-Flo samplers (General Oceanic, USA) and surface waters (0 m) were collected in polyethylene containers at Hiroshima Bay (Station 3, water depth 22 m, 34°05′ N, 132°21′ E) and the Iyo-Nada (Station 6, water depth 50 m, 33°40′ N, 131°53′ E), in the Seto Inland Sea during the cruise of the R.V. *Toyoshio Maru*, Faculty of Applied Biological Science, Hiroshima University, May, 1995 (Fig. 1). Seawater sample were filtered with GF/F glass fiber filters before analysis. H_2O_2 and ROOH concentrations were immediately (within 15 min) measured by fluorometry, which is described later.

Analytical procedure of H_2O_2 and ROOH

ROOH and H_2O_2 concentrations were measured by the method used by Sakugawa *et al.* (1995). Briefly, ROOH and H_2O_2 react with p-hydroxyphenylacetic acid under the existence of peroxidase to produce 6,6′-dihydroxy-3,3′-biphenyldiacetic acid which was detected by a fluorometer (320 nm excitation and 400 nm emission: JASCO FP920). Catalase was used for selective decomposition of ROOH and H_2O_2 before the analysis of the total peroxides. The detection limit of ROOH and H_2O_2 in seawater is 17 and 25 nM, respectively, and the analytical reproducibility is ±0.76 and ±4.1%, respectively, at 500 nM. Commercial H_2O_2 and peracetic acid solution were used as the standard solution of H_2O_2 and ROOH, respectively. We assume that the sensitivities of other organic hydorperoxides to the fluorescent reagents used are same as that of peracetic acid. H_2O_2, peroxyacetic acid and p-hydroxyphenylacetic acid used were purchased from Aldrich. Catalase and peroxidase were purchased from Sigma.

Interference of nitrite and aldehydes occurring in seawater with the measurement of ROOH and H_2O_2 was examined by adding known amounts of $NaNO_2$, formaldehyde and acetaldehyde standard solutions to the seawater samples. The result showed no positive or negative interferences of these reductants in the

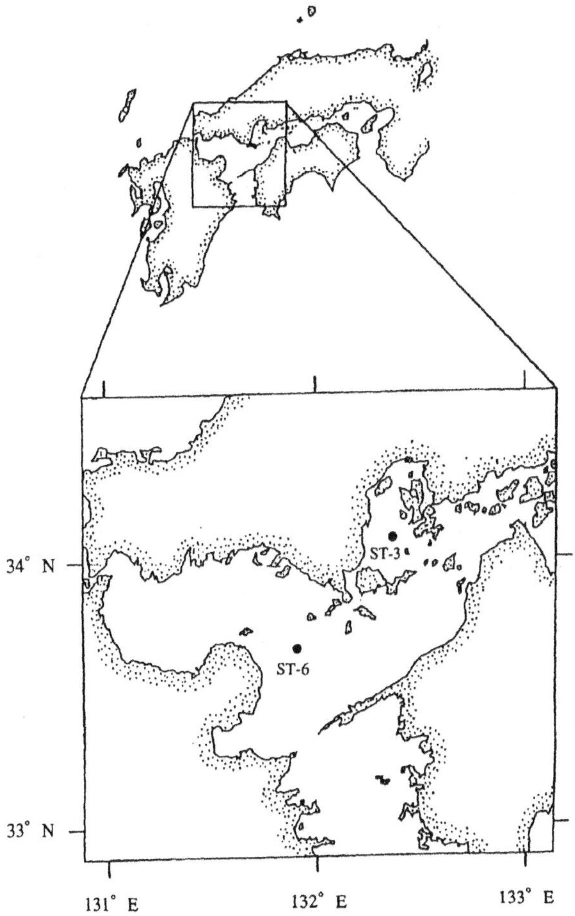

Fig. 1. Map of the locations where seawaters were collected during the cruise of the R. V. *Toyoshio-Maru* in May, 1995.

analysis of peroxides. The interference of fluorescent organic matter occurring in seawater, such as the humic substance in the measurement of peroxides, should be negligible because the background fluorescence value of each seawater sample was subtracted from the fluorescence values of ROOH and H_2O_2. Although our method measures only the total amount of organic peroxides in seawater (not specific to individual organic peroxides), this method would be useful for rough by estimating the concentrations of organic peroxides occurring in seawater.

Measurements of environmental parameters

Light intensity was measured by a quantum/radiometer/photometer (LI-185A, LAMBDA Instruments) using an underwater quantum sensor (LI-192S). Dissolved oxygen was measured by a DO sensor (model 8500, Nester) which was calibrated by the Winkler method. Water temperature, density, salinity and fluorescence were measured by a CTD sensor (model SBE-9/11, Sea-Bird Electronics). Fluorescent organic matter dissolved in seawater was measured by a fluorometer (excitation: 320 nm, emission 420 nm). Quinine sulfate solution (pH 2), purchased from Nacalai Tesque, Inc., was used for the standard solution of the fluorescence intensity. The concentration of dissolved organic carbon (DOC) in the seawater was measured by a Shimadzu TOC 5000 analyzer.

RESULTS AND DISCUSSION

Measurement of ROOH and H_2O_2 concentrations in the Seto Inland Sea

Table 1 summarizes the observed concentrations of ROOH and H_2O_2 at Hiroshima Bay (Station 3) and the Iyo-Nada (Station 6), the Seto Inland Sea in May, 1995. ROOH and H_2O_2 concentrations measured were in the range of 32–389 and 0–195 nM, respectively, at both stations. Figure 2 shows a temporal variation of the ROOH and H_2O_2 depth profiles between 0400 and 2200, on May 16, 1995 at Hiroshima Bay. The weather was cloudy on May 16. Average solar radiation between 1000 and 1600 was about 0.3 Kw/m^2. The ROOH and H_2O_2 concentrations measured were in the ranges of 144–389 and 0–195 nM, respectively. In general, the ROOH concentrations were almost uniform from the surface to the bottom waters, whereas the H_2O_2 concentration was highest at the surface or in near-surface water (0 or 2 m), decreasing with depth.

Figure 3 also shows a temporal variation of ROOH and H_2O_2 depth profiles between 0920 and 0600, on May 17–18 at the Iyo-Nada. The weather was fine during May 17–18. Average solar radiation during 1000–1600 in both days was about 0.8 Kw/m^2. ROOH and H_2O_2 concentrations measured were in the ranges of 38–296 and 7–146 nM, respectively. In general, ROOH concentrations were relatively lower at the surface or almost uniform from the surface to the bottom waters, whereas the H_2O_2 concentration was highest at the surface or in near-surface water (0 or 2 m) decreasing with depth.

Correlation of ROOH and H_2O_2 with environmental parameters measured

The correlation between peroxide concentrations and other analytical items such as CTD data, nutrients, fluorescent organic matter and DOC was examined (Table 2). The correlation between ROOH and these environmental parameters was very weak at Station 3, suggesting that the generation and decomposition mechanism of ROOH is independent of these parameters or may be too complex to elucidate. The H_2O_2 concentration strongly correlated with salinity (negatively), suggesting that the H_2O_2 levels in Hiroshima Bay are controlled by the influx of river water. In other words, river waters supply large amounts of H_2O_2 (or its precursor) into the Bay and the mixing of river waters with seawaters dilutes H_2O_2 concentrations. Similar results were obtained by Fujiwara *et al.* (1993), by measuring the H_2O_2 concentrations in Hiroshima Bay indicating that H_2O_2 is highest at the estuary of the Ohta River and decreases with distance from the

Table 1. H_2O_2 and ROOH concentrations in the water samples collected at Stations 3 and 6 in the Seto Inland Sea

Sampling Site	Date & Time	Water Depth	H_2O_2	ROOH	Sampling Site	Date & Time	Water Depth	H_2O_2	ROOH
		m	nM	nM			m	nM	nM
ST-3	5/16/95	0	195	162	ST-6	5/17/95	0	99	166
	4:00	2	105	312		9:20	2	118	161
		5	89	261			5	115	147
		10	37	244			10	66	151
		15	9	222			20	43	161
		20	9	278			45	23	170
	7:00	0	117	231		12:00	0	138	38
		2	96	239			2	76	127
		5	69	231			5	62	203
		10	16	218			10	53	208
		15	23	205			20	56	142
		20	7	201			45	13	184
	10:00	0	153	254		15:00	0	146	72
		2	134	254			2	97	281
		5	82	226			5	111	220
		10	22	268			10	73	281
		15	41	259			20	35	286
		20	38	245			45	69	253
	13:00	0	88	236		18:00	0	122	154
		2	109	286			2	108	176
		5	107	240			5	97	182
		10	38	231			10	104	187
		15	19	254			20	69	259
		20	8	268			45	45	198
	16:00	0	113	302		21:00	0	112	172
		2	84	319			2	90	226
		5	15	325			5	112	190
		10	6	325			10	99	243
		15	0	360			20	48	220
		20	0	389			45	45	226
	19:00	0	96	269		5/18/95	0	106	125
		2	91	263		0:00	2	83	202
		5	58	332			5	77	178
		10	47	276			10	58	154
		15	41	282			20	22	214
		20	41	276			45	26	214
	22:00	0	82	144		3:00	0	22	103
		2	58	251			2	92	116
		5	91	169			5	84	122
		10	32	257			10	99	109
		15	6	282			20	95	193
		20	32	238			45	11	212
						6:00	0	95	32
							2	77	212
							5	59	199
							10	37	270
							20	44	296
							45	7	277

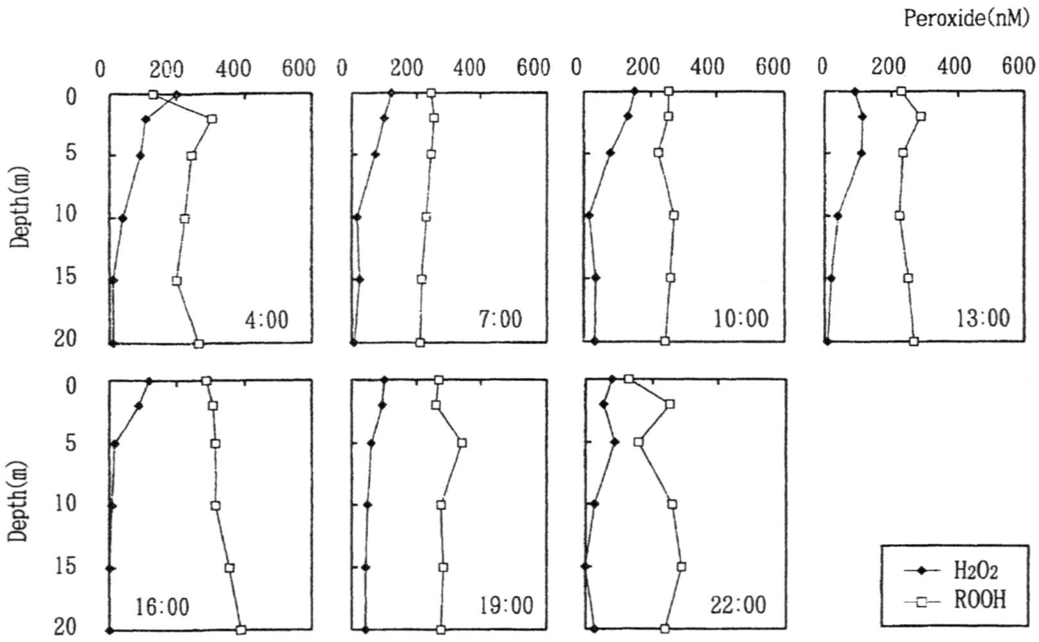

Fig. 2. Temporal variations of the ROOH and H_2O_2 depth profiles observed at Station 3 in the Seto Inland Sea.

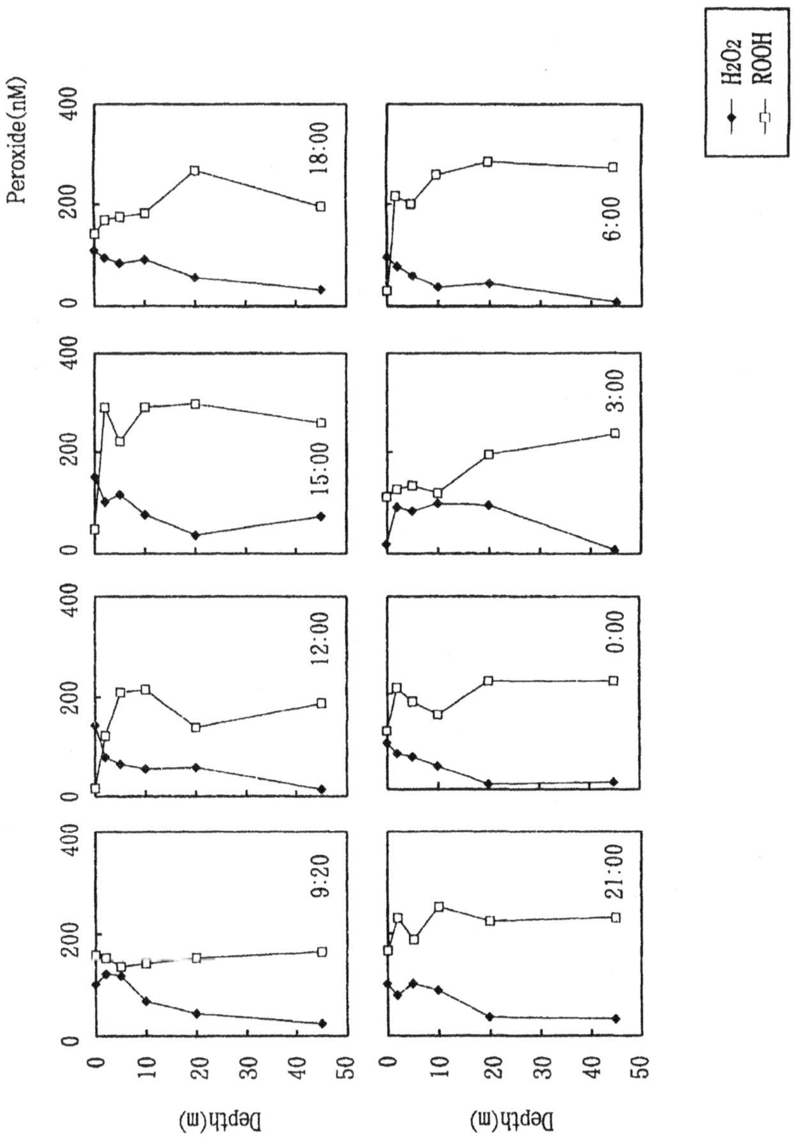

Fig. 3. Temporal variations of the ROOH and H_2O_2 depth profiles observed at Station 6 in the Seto Inland Sea.

Table 2. Correlations between peroxide concentrations and other analytical items

ST-3 (n=42)

	H_2O_2	ROOH	FOM	DOC	NO_2+NO_3	PO_4	temperature	light intensity	fluorescence	DO	Salinity
H_2O_2	1.000										
ROOH	-0.315	1.000									
FOM*	-0.045	-0.184	1.000								
DOC**	0.376	0.067	-0.284	1.000							
NO_2+NO_3	-0.342	-0.060	0.292	-0.403	1.000						
PO_4	-0.362	-0.030	0.214	-0.341	0.437	1.000					
temperature	0.799	-0.077	-0.212	0.442	-0.421	-0.317	1.000				
light intensity	0.488	-0.010	-0.209	0.327	-0.118	-0.043	0.758	1.000			
fluorescence	-0.031	0.131	-0.011	0.095	-0.345	-0.438	-0.101	-0.597	1.000		
DO***	-0.328	0.144	-0.129	-0.045	0.117	-0.529	-0.326	-0.405	0.415	1.000	
Salinity	-0.923	0.149	-0.017	-0.370	0.376	0.368	-0.858	-0.535	0.052	0.381	1.000

ST-6 (n=48)

	H_2O_2	ROOH	FOM	DOC	NO_2+NO_3	PO_4	temperature	light intensity	fluorescence	DO	Salinity
H_2O_2	1.000										
ROOH	-0.432	1.000									
FOM*	-0.217	-0.175	1.000								
DOC**	0.243	-0.105	-0.327	1.000							
NO_2+NO_3	-0.619	0.156	0.291	-0.359	1.000						
PO_4	-0.543	0.234	0.151	0.001	0.538	1.000					
temperature	0.744	-0.407	-0.158	0.379	-0.709	-0.565	1.000				
light intensity	0.442	-0.467	-0.063	-0.100	-0.230	-0.453	0.348	1.000			
fluorescence	-0.379	0.249	0.106	0.247	0.064	0.377	-0.171	-0.572	1.000		
DO***	0.418	0.114	-0.126	0.094	-0.583	-0.356	0.427	-0.186	0.204	1.000	
Salinity	-0.468	0.508	-0.066	-0.300	0.606	0.420	-0.779	-0.102	-0.123	-0.416	1.000

*Fluorescent organic matter. **Dissolved organic carbon. ***Dissolved oxygen.

estuary. Moore *et al.* (1993) also reported that the distribution pattern of H_2O_2 concentrations in the eastern Caribbean is largely affected by the Orinoco River input into the sea.

The correlation between ROOH and the environmental parameters measured was generally very weak at Station 6, but there were some correlation with salinity (positively) and light intensity (negatively) at a statistically significant level. This result, combined with the observed vertical profiles of ROOH as described above, suggests that photochemical decomposition of ROOH occurs in surface layer waters and thus ROOH concentration tends to be lower at surface layer waters. Our results show that there was a diurnal variation of ROOH concentrations in the surface waters (0 and 2 m depths), which indicates that the ROOH concentration is lowest during the afternoon and highest at nighttime. This supports the occurrence of photochemical decomposition of ROOH in surface waters. Our laboratory experiments indicate that organic peroxides are produced by the irradiation of artificial light (using a xenon lamp) to seawater containing low molecular weight organic compounds such as acetaldehyde and acetone (unpublished data). However, a high light intensity may reduce the production of organic peroxides or may decompose organic peroxides.

The H_2O_2 concentration mostly correlated with water temperature (positively), $NO_2 + NO_3$ (negatively), and PO_4 (negatively) at Station 6. There was a diurnal variation of the H_2O_2 concentration at surface waters; highest during the afternoon and lowest at nighttime. The thermocline was present around 10–15 m depth at Station 6 during our sampling period. These results, combined with the observed vertical profiles of H_2O_2 as described above, suggest that photochemical production of H_2O_2 occurs only in the mixing layer (nutrient-poor, warm water mass), whereas, in deep layer waters (nutrient-rich, cool water mass), no photochemical production of H_2O_2 occurs. Photochemically produced H_2O_2 may be quickly decomposed by chemical reactions or microorganisms in mixing layer waters and thus its influx (supply) to deep layer waters may be negligible. Similar H_2O_2 vertical profiles were obtained by Miller and Kester (1994) by measuring H_2O_2 concentrations in the Sargasso Sea. Their data show that the H_2O_2 concentrations dropped rapidly to almost zero below the thermocline.

SUMMARY

As far as we know, we were the first to analyze organic peroxides dissolved in seawater in this study. The concentrations of organic peroxides were in the range of 32–389 nM in the Seto Inland Sea, western Japan during May, 1995. The concentration levels were comparable with those of H_2O_2. The vertical profile of ROOH was, however, quite different from that of H_2O_2, e.g. the ROOH concentration was almost uniform from surface through bottom waters, or was higher at deep waters, whereas the H_2O_2 concentration was highest at surface waters and decreased with depth. Although H_2O_2 appears to be photochemically produced and quickly decomposed by chemical reactions or microorganisms in the mixing layer waters, the generation and decomposition mechanism of ROOH is still unknown (while the photochemical decomposition of ROOH was suggested to occur in the surface waters of the Iyo-Nada). It is likely that dissolved organic matter such as aldehydes, ketons and double bond-containing compounds is involved in the formation process of ROOH, as well as H_2O_2. More study is clearly required to elucidate its distribution and behavior in the marine environment.

Acknowledgements—We are grateful to the captain Akio Gou and crew of the R.V. *Toyoshio-Maru*, Faculty of Applied Biological Science, Hiroshima University for helping with the seawater sampling.

REFERENCES

Fujiwara, K., T. Ushiroda, K. Takeda, Y. Kumamoto and H. Tsubota (1993): Diurnal and seasonal distribution of hydrogen peroxide in seawater of Seto Inland Sea. *Geochem. J.*, **27**, 103–115.

Hellpointner, E. and S. Gab (1989): Detection of methyl, hydroxymethyl and hydroxyethyl hydroperoxides in air and precipitation. *Nature*, **337**, 631–634.

Hewitt, C. N. and G. L. Kok (1991): Formation and occurrence of organic hydroperoxides in the troposphere: laboratory and field observations. *J. Atmos. Chem.*, **12**, 181–194.

Miller, W. L. and D. R. Kester (1994): Peroxide variations in the Sargasso Sea. *Mar. Chem.*, **48**, 17–29.

Moore, C. A., C. T. Farmer and R. G. Zika (1993): Influence of the Orinoco River on hydrogen peroxide distribution and production in the Eastern Caribbean. *J. Geophys. Res.*, **98**, 2289–2298.

Sakugawa, H., T. Yamashita and K. Fujiwara (1995): Determination of hydrogen peroxide and organic peroxides in seawater. pp. 452–457. In *Global Fluxes of Carbon and Its Related Substances in the Coastal Sea—Ocean-Atmosphere System*, ed. by S. Tsunogai, K. Iseki, I. Koike and T. Oba, M&J International, Yokohama.

Zika, R. G., J. W. Moffett, R. G. Petasne, W. J. Cooper, and E. S. Saltzman (1985): Spatial and temporal variations of hydrogen peroxide in Gulf of Mexico waters. *Geochim. Cosmochim. Acta*, **49**, 1173–1184.

Dynamics and Characterization of Marine Organic Matter, Eds., N. Handa, E. Tanoue and T. Hama, pp. 241–256.

"Picopellets" Produced by Phagotrophic Nanoflagellates: Role in the Material Cycling within Marine Environments

Toshi NAGATA

Ocean Research Institute, The University of Tokyo,
1-15-1 Minami-dai, Nakano, Tokyo 164-8639, Japan

(Review, Received 5 August 1996; accepted 23 October 1996)

Abstract—Recent studies have suggested that marine phagotrophic protists can release large numbers of small fecal pellets (<1 μm in linear size dimension). These "picopellets" may contribute substantially to the turnover of submicron and colloidal particles in the sea. This paper reviews recent studies that examined release of colloidal and dissolved organic matter by phatotrophic nanoflagellates, dominant grazers of picophytoplankton and bacteria in diverse marine environments. Difficulties in quantitative recovery of flagellate egesta have seriously restricted the analysis of the mass balance and stoichiometry of ingestion/egestion processes in flagellates, but available evidence suggests that a significant fraction of ingested prey materials is egested as colloidal and dissolved organic matter particularly when prey abundance is high. Biochemical characterization of picopellets has revealed that some of these particles have liposome-like structures, which may partly explain the formation of the semi-labile and refractory pool of dissolved organic matter in seawater. Future challenges include recovery, detection and characterization of picopellets from natural environments and evaluation of their roles in major biogeochemical fluxes including particle sinking, organic matter decomposition and nutrient regeneration.

INTRODUCTION

Recent evidence indicates that excretion by protists of inorganic nutrients including ammonium and phosphate is a dominant mechanism of N and P regeneration in diverse aquatic environments (Sherr *et al.*, 1983; Goldman *et al.*, 1985; Caron and Goldman, 1990; Harrison, 1992; Ferrier-Pagès and Rassoulzadegan 1994; Haga *et al.*, 1995). Models have been proposed to predict general relationships between nutrient excretion and carbon consumption by protists (Caron and Goldman, 1990; Caron, 1991), providing a guide to better understanding the cycling of C, N and P in the upper oceans. Previous models, however, have generally neglected egestion of unassimilated organic materials by protists, in part because of the lack of reliable data on assimilation efficiencies and egestion rates of protists (Landry, 1993; Caron and Goldman, 1993).

Recent studies have suggested that protists can release a large amount of submicron-sized or colloidal organic materials (in short "picopellets" as coined by Fereidoun Rassoulzadegan and his colleagues) and dissolved organic matter

(Koike *et al.*, 1990; Nagata and Kirchman, 1992a, b). The need to formulate models that include the egestion parameters has become increasingly apparent (Jumars *et al.*, 1989; Landry, 1993; Caron and Goldman, 1993). Novel hypotheses have been proposed to explain roles of protist egesta and submicron particles in the storage and transport of dissolved organic matter in the oceans (Nagata and Kirchman, 1992b, 1997; Nagata and Koike, 1995). Given the significance of dissolved and colloidal organic carbon in seawater as one of the largest active reservoirs of organic carbon on Earth (Williams and Druffel, 1988; Ogawa and Ogura, 1992; Benner *et al.*, 1992; Kepkay, 1994), it is fundamentally important to identify parameters of protist egestion and to understand the contribution of protists to the turnover of organic matter pools in the oceans.

Here we review recent data concerning the egestion process of protist grazers in marine environments. we focus on phagotrophic (heterotrophic and mixotrophic) nanoflagellates, which are dominant consumers of picophytoplankton and bacteria in diverse marine environments (Fenchel, 1987; Sanders *et al.*, 1992). The egestion processes of ciliates and larger phagotrophic flagellates (e.g., dinoflagellates) have been discussed by Stoecker (1984), Gowing and Silver (1985), Nothig and von Dodungen (1989) and Buck and Newton (1995).

GENERAL FEATURES OF INGESTION, DIGESTION AND EGESTION IN NANOFLAGELLATES

Electron microscopic observations have revealed the sequence of food consumption by phagotrophic nanoflagellates (Aaronson, 1973; Cole and Wynne, 1974; Dubowsky, 1974; Fenchel, 1982a). Although digestive strategies of flagellates are diverse (Gaines and Taylor, 1984; Fenchel, 1987), the dominant mode of feeding in planktonic nanoflagellates involves phagocytotic engulfment, followed by enzymatic digestion of prey in acidic food vacuoles, assimilation of digested molecules, and finally egestion of incompletely digested or unassimilated materials. Some flagellate digestive enzymes (e.g., acid phosphatase) are released along with egestion (Dubowsky, 1974; Nagata and Kirchman, 1992b).

Several investigators have noted that "fragments of membranes" (Cole and Wynne, 1974) or "cell wall material" (Fenchel, 1986) accumulate as prey residues in food vacuoles at the later stage of digestion, and that "membranous vesicles" (Aaronson, 1971) or "amorphous material" (Twiss and Campbell, 1995), which are apparently released by flagellates, are observed abundantly in the culture of flagellates. These observations indicate that flagellates may release fragile, colloidal particles as egesta, a process suggested by more recent data obtained by particle counting and biochemical analyses (Koike *et al.*, 1990; Nagata and Kirchman 1992b; see below).

MASS BALANCE CONSIDERATIONS

The carbon budget of food consumption by flagellates can be described by the following mass balance equation:

$$I = G + R + E, \tag{1}$$

where I = ingested organic carbon; G = organic carbon used for growth; R = loss of organic carbon to respiration; E = egested organic carbon. Nitrogen and phosphorus budgets can be described by similar formulas, where R represents the release of ammonium and phosphate, respectively.

Two ecologically important efficiencies, growth yield (Y; or gross growth efficiency) and assimilation efficiency (A), are defined as follows:

$$Y = G/I, \tag{2}$$

$$A = (G + R)/I. \tag{3}$$

Batch culture experiments have been conducted to examine the mass balance of flagellates in terms of carbon (Fenchel, 1982b; Caron *et al.*, 1985; Geider and Leadbeater, 1988; Caron *et al.*, 1990a, b), nitrogen (Goldman *et al.*, 1985; Caron *et al.*, 1990b; Nagata and Kirchman, 1991; Hoch *et al.*, 1996), and phosphorus (Andersen *et al.*, 1986; Caron *et al.*, 1990b) and energy (Kopylov *et al.*, 1980). Most previous studies have generally focused on growth yield and nutrient regeneration efficiencies (R/I in terms of N and P), which are important for discussing roles of flagellates in fluxes of matter and energy within aquatic ecosystems (Ducklow *et al.*, 1986; Sherr *et al.*, 1987; Caron, 1991). Much less attention has been paid to the parameter of egestion (reviewed by Nagata and Kirchman 1992a). To our knowledge, there is no previous study in which the rate of egestion (E) was directly determined. Instead, E has been estimated as a "residual" in mass balance calculation. Here we summarize some estimates of parameter E along with other variables (Table 1), and discuss some problems inherent to previous approaches.

Fenchel (1982b) determined Y from the relation between initial prey bacterial carbon and the total flagellate carbon produced. He also determined respired carbon from oxygen consumption assuming a respiration quotient of 1. Egested carbon was calculated as the missing part of the budget (E = I – G – R). According to his calculation of total carbon ingested, 30–40% is egested either as dissolved (DOC) or particulate organic carbon (POC) (Table 1). Similarly, Nagata and Kirchman (1991) observed that the missing part of the nitrogen budget is high (40–50%), suggesting that a significant portion of ingested nitrogen was egested (Table 1).

Caron *et al.* (1985, 1990a) found that the values of Y estimated from POC budget [Y_{POC} = (POC in the culture after all the prey carbon is consumed by flagellates)/(POC initially present in the culture as prey biomass)] was systematically higher than those estimated from biovolume [Y_{vol} = (flagellate biovolume produced)/(prey biovolume consumed)]. The difference between Y_{POC} (53–71%) and Y_{vol} (31–62%) was about 10%, which Caron *et al.* (1985, 1990a) ascribed to egested POC (Table 1). A similar argument was presented by Geider and

Table 1. Growth yield (Y), assimilation effeciency (A) and egestion as a percentage of ingestion (E) in phagotrophic nanoflagellates.

Flagellates	Prey	C, N or P	Y (%)	A (%)	E (% of ingestion) POM/DOM*	POM**	DOM***	References
Ochromonas sp.	Bacteria (*Pseudomonas*)	C	34	57	43			Fenchel (1982b)
Pleuromonas jaculans	Bacteria (*Pseudomonas*)	C	43	72	28			Fenchel (1982b)
Paraphysomonas imperforata	Bacteria (mixed)+ diatom (*Dactylum tricornutum*)	C	44			10	10 (DOC)	Caron et al. (1985)
Stephanoeca diplocostata	Bacteria (pseudomonad)	C	27			13		Geider and Leadbeater (1988)
Paraphysomonas imperforata	Bacteria (Mixed)	N	25	61	39		22 (DFAA)	Nagata and Kirchman (1991)
Paraphysomonas imperforata	Bacteria (Mixed)	N	30	50	50		5 (DFAA+ DCAA)	Nagata and Kirchman (1991)
Ochromonas sp.	Bacteria (*Eschericia coli*)	N					7 (DFAA)	Anderssen et al. (1985)
Poterioochromonas malhamensis	Bacteria (*Pasteurella* sp.)	P	40-49				10 - 20 (DOP)	Caron et al. (1990b)

*Egestion is estimated assuming that total ingestion (I) that was not explained by growth (G) and respiration (R, or regeneration in the case of nitrogen) is egestion (E); i.e., $E = I - (G + R)$. Note that E is not a direct measurement, but a residual of the mass balance equation (see text).
**Egestion as POM is estimated from the difference between biovolume-based growth yield and POC-based growth yield (see text).
***Egestion as DOM is estimated as an accumulation of DOM, except that the release of DFAA is corrected for uptake by bacteria (see text).

Leadbeater (1988) who estimated the egestion of POC to be 13% of ingestion in choanoflagellates (Table 1).

An obvious limitation of the "residual" method is that these estimates intrinsically bear large uncertainties due to propagated errors associated with multiple measurements and assumptions. Also, the validity of estimating the difference between Y_{POC} and Y_{vol} (Caron *et al.*, 1985; Geider and Leadbeater, 1988) hinges on conversion factors (e.g., C to volume ratio for prey and flagellates), which may vary depending on the method of fixation and sizing (see Choi and Stoecker, 1989; Nagata and Watanabe, 1990). Direct measurements of egested materials are required for discussing the digestion-egestion mass balance of flagellates in a more coherent fashion.

Quantitative recovery of egested materials is a difficult, if not impossible, task (Landry, 1993; Caron and Goldman, 1993). As a first approximation, the release rate of dissolved organic matter (DOM) is probably a good, albeit conservative, estimate of E because picopellets produced by nanoflagellates are likely to be colloidal particles that can pass through glass fiber filters (nominal retention 0.7 μm) (see Koike *et al.*, 1990). Furthermore, picopellets may release low molecular weight DOM [e.g., dissolved free amino acids (DFAA)] by diffusion (Jumars *et al.*, 1989) or after enzymatic hydrolysis of macromolecules (Nagata and Kirchman, 1991). In fact, previous studies have found that the accumulation of DOM represents a significant fraction (5–20%) of the total ingestion in terms of carbon (Caron *et al.*, 1985), nitrogen (Hoch *et al.*, 1996) and phosphorus (Andersen *et al.*, 1986; Caron *et al.*, 1990b) (Table 1). However, we should interpret the values of "DOM release" very carefully, because DOM may be released artificially from flagellates due to breakage of cells by filtration as discussed by Nagata and Kirchman (1990, 1992a). Also, as will be discussed below, the "accumulation" rate of DOM does not necessarily represent the "release" rate of DOM.

Rapid bacterial degradation of labile DOM may result in serious underestimation of the release rate of DOM. This possibility was examined by Nagata and Kirchman (1991) who measured turnover rates of ^{3}H-DFAA in the culture of flagellates (*Paraphysomonas imperforata*) grazing on bacteria. Their data clearly indicated that a large fraction of DFAA released by flagellates was rapidly taken up by prey bacteria. After correcting this bacterial uptake of DFAA, Nagata and Kirchman (1991) estimated that the release of DFAA accounted for 4–22% of total nitrogen ingestion by flagellates. In an earlier study with *Ochromonas* grazing on bacteria, Anderssen *et al.* (1985) reported that DFAA release, after correcting for bacterial uptake, was 7% of total ingestion. It should be noted that the correction by the above method can be applied to only a limited class of organic components (such as DFAA) of which chemical form is identified and a tracer available. The use of metabolically inactive prey (e.g., heat-killed bacteria) is potentially useful to minimize degradation of egested organic matter and underestimation of E (cf. Caron *et al.*, 1990b; Ferrier-Pagès and Rassoulzadegan, 1994).

One possible, but so far not demonstrated, process that may influence the estimation of E is coprophagy, i.e., reingestion of picopellets. Recent studies have demonstrated that flagellates can ingest a wide variety of colloidal particles (Sherr, 1988; Tranvik *et al.*, 1993). If flagellates in fact ingest their own picopellets, it is obvious that E would be underestimated.

As a summary of mass balance considerations, note that most previous estimates of E (Table 1) are byproducts of studies in which the parameters I, G, R, rather than E, were the prime concern. As a result, authors have tended to discount the significance of the egestion term, even though, as summarized in Table 1, some studies have indicated that E can be a significant fraction (10–50%) of the total mass balance. We should also notice that the mass budget is far less complete because of practical problems in direct determinations of E (see above). In future, it is particularly important to estimate C:N:P ratio of egested organic matter in order to evaluate stoichiometric implications of egestion (Landry 1993).

RELATIONSHIP BETWEEN INGESTION RATE AND ASSIMILATION EFFICIENCY

As emphasized by Fenchel (1982b), balanced growth is never attained in batch cultures, which seriously restricts the analysis of relationship among feeding parameters of flagellates (see Eqs. (1), (2), and (3)). Practically, it is difficult to measure ingestion rates precisely from changes of prey-predator numbers alone (Fenchel, 1982b; Goldman *et al.*, 1985). Rapid recycling of egested organic matter is another complication associated with batch culture approaches (see above, Hoch *et al.*, 1996). To solve some of these problems, Nagata and Kirchman (1991) determined grazing rates by using fluorescent labeled-bacteria (Sherr *et al.*, 1987) and turnover rates of DFAA and ammonium by using isotope tracers (^{3}H-DFAA and ^{15}N-ammonium) over the course of flagellates growth in batch cultures. They found that both ammonium and DFAA release rates increased with the increase of ingestion rates, but the ratio of DFAA release relative to ammonium release was much higher when the ingestion rate was higher (Fig. 1). At the maximum level of the ingestion rate, DFAA release exceeded ammonium release. Although DFAA probably comprises only a part of egested materials by flagellates, the results of Nagata and Kirchman (1991) strongly suggest that egestion relative to ammonium regeneration (E:R in Eq. (1)), and therefore assimilation efficiency (A), substantially changes depending on the ingestion rate. We may predict that A is lower when prey is more abundant and I is higher.

Physiological basis of the possible relationship between A and I in nanoflagellates is unknown. A decrease of assimilation efficiency with an increase of ingestion rate has been observed in some crustacean zooplankton species (Gaudy, 1974 and references therein), suggesting that the inverse relation between A and I is a general phenomenon among planktonic grazers. Penry and Jumars (1986) and Jumars *et al.* (1989) have developed a model to describe digestion process of grazers. According to their "optimum digestion model", the

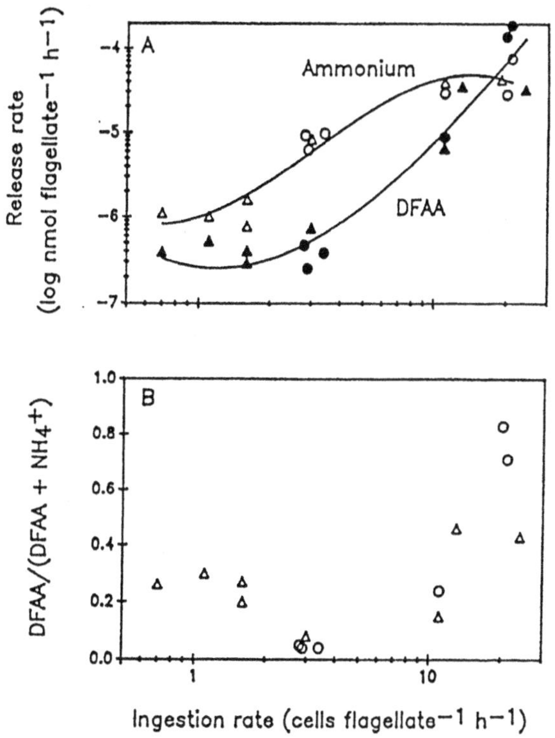

Fig. 1. (A) The relationship between release rates of ammonium and DFAA by phagotrophic nanoflagellates grazing on mixed bacteria (circles) or *Vibrio splendidus* (triangles). Release rates are plotted against ingestion rate determined by the fluorescently-labeled bacteria method. (B) The relationship between ratios of DFAA release rate to total (DFAA + ammonium) release and ingestion. From Nagata and Kirchman (1991).

greater the rate of ingestion, the greater will be the release of food remains relative to ingestion. This model prediction is apparently consistent with the experimental results reported by Nagata and Kirchman (1991, Fig. 1).

Although we know very little about the release of DOM by flagellate assemblages in natural environments and its relationship with nutritional conditions of flagellates, the available data from experiments and models just discussed above suggest that the release of DOM by flagellates could be important in envrionments where prey abundance is high; e.g., during picophytoplankton blooms or in microenvironments of organic aggregates where bacteria densely colonize.

SUBMICRON PARTICLES IN SEAWATER AND PICOPELLETS PRODUCED BY FLAGELLATES

Koike *et al.* (1990) used a resistive pulse counter (Elzone 80XY) to discover many small (size 0.4–1 μm) particles in seawater. These particles are highly

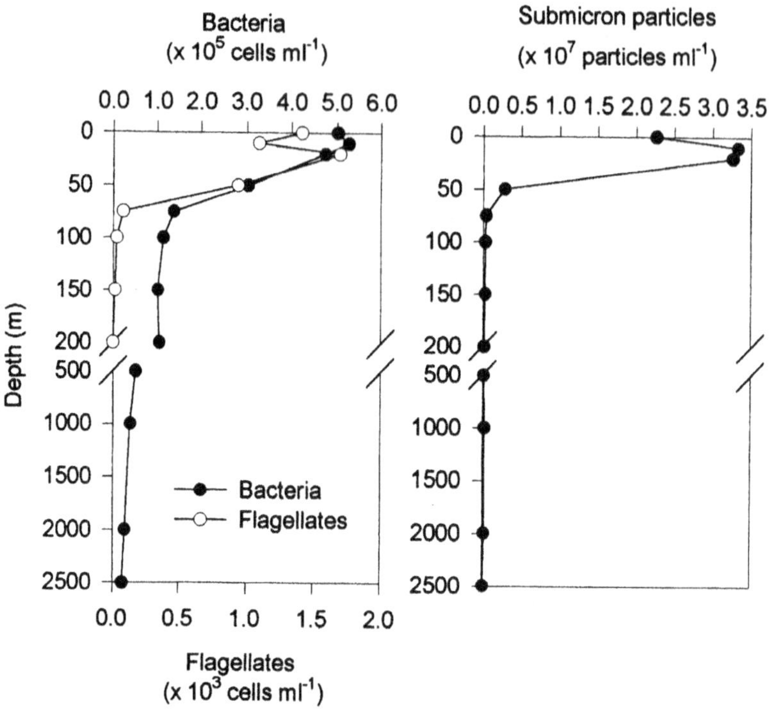

Fig. 2. Vertical distribution of bacteria, nanoflagellates (both pigmented and nonpigmented forms are included), and submicron particles in the Western Pacific (off Sagami Bay, 34°29.0′ N, 138°33.4′ E). Bacteria and nanoflagellates were counted by the DAPI method (Porter and Feig, 1980). Submicron particles were counted with an Elzone particle counter according to Kogure and Koike (1987). Samples were collected from CTD-Niskin bottles on July 10, 1992 during the R/V *Tansei-maru* cruise (KT-92-10). From T. Nagata (unpublished).

abundant (typical abundance, 10^6–10^7 particles ml^{-1}) in the upper ocean and may play substantial roles in oceanic material cycling (Koike *et al.*, 1990; Longhurst *et al.*, 1992; Nagata and Koike, 1995; Nagata and Kirchman, 1997). Oceanographers are keen to characterize submicron particles and other colloids, and to identify sources and sinks of small particles.

Flagellate picopellets are suggested to be one of the major sources of submicron particles in seawater (Koike *et al.*, 1990). Koike *et al.* (1990) observed that the number of submicron particles increased substantially with the increase of flagellates grazing on bacteria. To confirm and emphasize possible relations between flagellate egestion and submicron particles, some data from my own observations in the western Pacific off Sagami Bay and those from a seawater culture experiment are presented in Figs. 2 and 3. As depicted in Fig. 2, the abundance of submicron particles was high in the euphotic zone and low in the deeper layer, covarying well with biological parameters such as numbers of bacteria and nanoflagellates. This result is consistent with previous observations

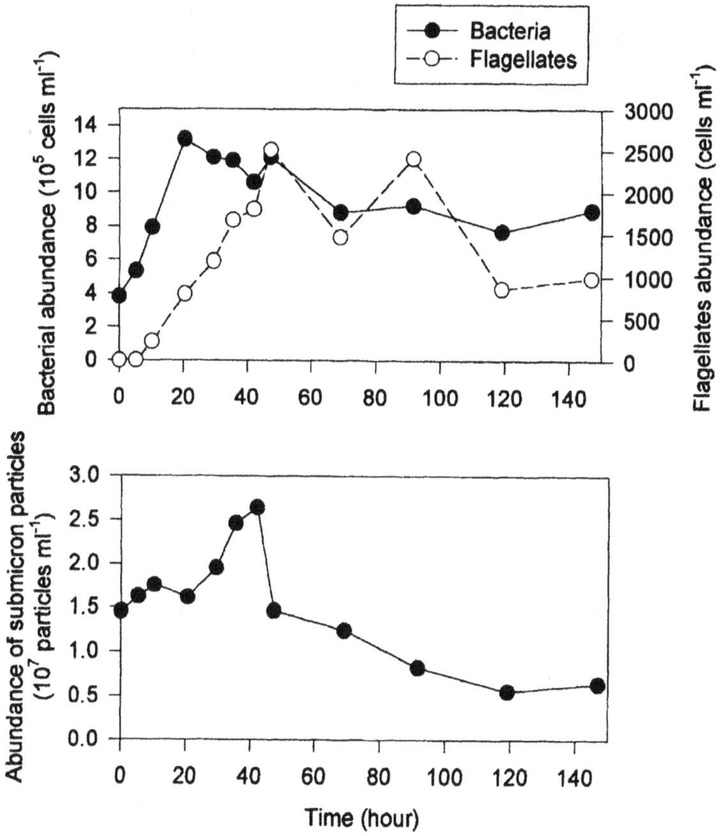

Fig. 3. Changes of the abundances of bacteria, nanoflagellates (A) and submicron particles (B) in a seawater culture. The water sample collected at a 20 m-depth (35°1.1′ N 139°23.3′ E) was filtered either through a 5 μm-mesh-size plankton net or 0.2 μm-pore-size Nuclepore filters. The filtrates were mixed (<5 μm:<0.2 μm = 4:5) and contained in a 1 liter capacity polycarbonate bottle. To stimulate bacterial growth, glucose was added at a final concentration of 1 μM. The sample was incubated in the dark at 20°C, and subsamples were withdrawn over the time for counting bacteria, nanoflagellates and submicron particles, as described above (see legend of Fig. 2). From T. Nagata (unpublished).

in the subarctic Pacific (Koike *et al.*, 1990), suggesting that the dynamics of submicron particles are closely coupled with biological processes.

Results from a seawater culture experiment also indicated that large numbers (in the order of 10^7 particles ml^{-1}) of submicron particles (mostly 0.4–0.6 μm in size) are released along with an increase of nanoflagellates grazing on bacteria (Fig. 3). Again, this result is consistent with that of Koike *et al.* (1990), suggesting that some flagellate pellets are bacteria-sized particles (0.4–0.6 μm) and can contribute substantially to the dynamics of bulk submicron particles in seawater. However, we should also notice that the number of submicron particles decreased rapidly during the stationary growth phase of flagellates, a phenomenon which was not observed by Koike *et al.* (1990). The mechanism of this reduction in the number of submicron particles is unclear. One possibility is biological degradation due to coprophagy (see above) or hydrolysis by bacteria. Alternatively, and more likely, aggregation to larger forms or adsorption to the wall of the incubation bottle could explain the disappearance of submicron particles. Much remains to be examined of the chemical and physical characteristics of submicron particles released by flagellates.

BIOCHEMICAL CHARACTERIZATION OF FLAGELLATE EGESTA

Nagata and Kirchman (1992b) suggested that some of the picopellets produced by flagellates (*P. imperforata*) are liposome-like particles. Liposomes are phospholipid vesicles containing a water phase inside, and are formed spontaneously when the concentration of phospholipids exceed a critical micelle concentration. Nagata and Kirchman (1992b) suggested that partially digested prey bacterial membranes in food vacuoles should spontaneously form liposomes, entrapping other components of prey cells and flagellate digestive enzymes, and that these liposome-like organic complexes are released to seawater along with egestion. The data in support of this hypothesis include: (1) lipid content of colloidal DOM released by flagellates was much higher than that of prey bacterial cells, and (2) a flagellate digestive enzyme (acid phosphatase) associated with picopellets was not degraded by proteases, but it was susceptible to proteases after treatment with polymyxin B, which destroys bacterial membranes. In addition, electron-microscope observations showed that liposome-like vesicles accumulate abundantly in food vacuoles of flagellates at the later stage of digestion (e.g., Cole and Wynne, 1974), suggesting that bacterial wall materials are more slowly digested by flagellates than other prey components are. Based on these experimental data, Nagata and Kirchman (1992b) suggested that liposome-like picopellets may protect labile DOM from enzymatic degradation and these particles play a role in the formation of semi-labile and refractory DOM pools in seawater (see below). Obviously, future studies should examine in greater detail structures, compositions and physico-chemical stability of micellar vesicles in natural seawaters.

Tranvik (1994) examined the release of DOM by mixotrophic flagellates, *Poterioochromonas malhamensis*. In his experiment, flagellates were incubated either with ^{14}C-bicarbonate or with ^{14}C-labeled bacteria, in order to compare release of $DO^{14}C$ from photosynthetic production and that from ingestion of bacteria (i.e., egested DOC). Size-fractionation experiments revealed that about half of the egested $DO^{14}C$ was in the fraction of large colloidal particles (0.02–0.45 μm), whereas a major fraction of photosynthetic $DO^{14}C$ consisted of smaller compounds in the <0.02 μm fraction. It was also observed that a natural bacterial assemblage degraded egested $DO^{14}C$ more slowly than photosynthetically produced $DO^{14}C$. Based on these results, the author suggested that flagellate bacterivory and egestion may contribute to a transformation of labile organic matter into more refractory forms, which is consistent with the liposome hypothesis of Nagata and Kirchman (1992b).

Turk *et al.* (1992) found that flagellates, *Ochromonas* sp., release a large amount of prey bacterial DNA. In a continuous culture experiment, substantial accumulation of dissolved DNA was observed in the culture with nanoflagellates and bacteria, whereas a high amount of dissolved DNA was not observed in the bacteria-alone control. From mass balance calculations, they estimated that the majority of bacterial DNA ingested by flagellates was egested as dissolved DNA. The authors suggested that the release of unassimilated DNA by flagellates may contribute substantially to phosphorus cycling in marine systems. It is somewhat paradoxical that flagellates do not digest DNA, even though DNA is a potentially important source of C, N, and P. Nonetheless, the significant release of bacterial DNA by flagellates, and the possible association of egested DNA with liposome-like complexes (see above), may explain high concentrations of "coated" DNA that have been found in the dissolved fraction ($<0.2\ \mu m$) of seawater (Maruyama *et al.*, 1993).

ROLE OF PICOPELLETS IN BIOGEOCHEMICAL CYCLING IN THE SEA

There is a rapidly growing list of studies suggesting important roles of organic colloids and flagellate picopellets in marine systems (reviewed by Lee and Wakeham, 1992; Kepkay, 1994; Nagata and Koike 1995; Nagata and Kirchman, 1997). Briefly, organic colloids may be important (1) in the formation of macroaggregates (marine snow) (Alldredge *et al.*, 1993; Wells and Goldberg, 1993, 1994; Kepkay, 1994; Passow *et al.*, 1994), (2) as nutritional sources of protists and zooplankton (Sherr, 1988; Flood *et al.*, 1992; Tranvik *et al.*, 1993), (3) in the transport of radionuclides and contaminants (Rees, 1991; Baskaran *et al.*, 1992; Moran and Buesseler, 1992) and (4) in the formation of semi-labile and refractory organic matter (Nagata and Kirchman 1992b, 1997; Tranvik, 1994). In addition, recent studies have suggested a possible involvement of flagellate ingestion and egestion in the transformation of iron (Barbeau *et al.*, 1996), and trace metals (Twiss and Campbell, 1995) in aquatic systems.

Here we discuss one potential role of flagellate picopellets in the formation of semi-labile and refractory organic matter in seawater (Nagata and Kirchman, 1992b, 1997). Based on its turnover time, oceanic DOM can be classified into three categories, i.e., refractory, semi-labile and labile (Carlson and Ducklow, 1995). The refractory pool dominates deep water DOM which has an average age in the order of 1,000–6,000 years (estimated from ^{14}C dating; Williams and Druffel, 1987; Bauer *et al.*, 1992). The surface DOM pool consists of roughly 50% refractory and 50% semi-labile components; this semi-labile pool turns over in time scales of months (Carlson *et al.*, 1994; Carlson and Ducklow, 1995; Williams, 1995). Highly labile DOM, which turns over as fast as minutes to hours (e.g., DFAA and monomeric sugars), also exists in the upper water, but the concentration of labile DOM is very low (nanomolar), representing a negligible (<1%) portion of the total DOM pool.

Pathways producing semi-labile and refractory DOM are controversial (Hedges, 1988; Francois, 1990). Reactions involving condensation of organic compounds including amino acids and sugars have been proposed to transform labile compounds into less labile compounds (Yamamoto and Ishiwatari, 1989; Keil and Kirchman, 1993). However, source materials (amino acids and sugars) are generally thought to be quickly degraded by bacteria, before the concentration of these compounds becomes high enough so that geochemical reactions can proceed (Hedges, 1988). Nagata and Kirchman (1992b) hypothesized that flagellate picopellets, by virtue of their liposome-like structures (see above), protect labile organic matter from rapid bacterial attack. Moreover, concentrations of organic matter may be high on the surface and inside of picopellets, providing "micro-reaction sites" where geochemical reactions which cannot occur in diluted conditions may proceed rapidly. In support of this hypothesis, Nagata *et al.* (1997) found that proteins associated with bacterial membranes and walls are much less easily degraded than the proteins in soluble fractions (free proteins). Moreover, results from model experiments have suggested that adsorption of free proteins to submicron particles and other colloids may substantially reduce the degradation rate of proteins (Keil and Kirchman, 1994; Nagata and Kirchman, 1996). One can speculate that temporal storage of organic matter owing to protection by picopellets against bacterial degradation may result in a large-scale transport of organic matter, which is driven by convective mixing and horizontal advection (Carlson *et al.*, 1994; Pelzer and Hayward, 1996).

CONCLUDING REMARKS

Transformations of materials along the trophic transfer through complex food web pathways including the "classical food chain" (phytoplankton-zooplankton-fish pathway) and the "microbial loop" (DOM-bacteria-protists pathway), to a large extent, determine rates of nutrient recycling and particle sinking in the oceans (Longhurst, 1991). Although the major role of the microbial loop in these processes has been well recognized during the past two decades (Pomeroy, 1974; Williams, 1981; Azam *et al.*, 1983; Cole *et al.*, 1988; Ducklow and Carlson, 1992; Sanders *et al.*, 1992; Azam *et al.*, 1994; Kirchman, 1994), the potential significance of egestion by small protist grazers in material cycling had been largely neglected until very recently (Nagata and Kirchman 1992a, b, 1996b). This situation is contrasted with general interests among oceanographers in the formation of large pellets by mesozooplankton which may affect vertical fluxes of particulate organic matter in the sea (Shrader, 1971; Honjo and Roman, 1977; Matsueda *et al.*, 1986; Handa, 1991). In this review, we have summarized recent experimental results showing production of picopellets by flagellates, and emphasized that these small particles can play a significant role in biogeochemical cycles in the oceans. Future studies should be aimed at better describing the mass and energy balance of flagellates; we need studies that explicitly examine egestion. More detailed biochemical and physical characterization of picopellets is required for understanding the behavior of these particles in seawater. Finally, one of the major future challenges is to clarify the role of picopellets in the formation of semi-labile and refractory DOM, which has substantial implications for storage and transport of organic matter in the oceans.

Acknowledgements—I thank Nobuhiko Handa for his encouragement and stimulating my oceanographic work. Thanks are also due to Eiichiro Tanoue and Takeo Hama for inviting me to contribute to this synthesis. David Kirchman improved the manuscript. David Kirchman, Isao Koike, Kazuhiro Kogure, and Farooq Azam provided valuable discussions on "picopellets" in seawater. This review was written and some of the unpublished data were collected with the support by grants from the MESC, Japan (Nos. 04232105, 06680489, and 08458142).

REFERENCES

Aaronson, S. (1971): The synthesis of extracellular macromolecules and membranes by a population of the phytoflagellate *Ochromonas danica*. *Limnol. Oceanogr.*, **16**, 1–9.

Aaronson, S. (1973): Digestion in phytoflagellates. pp. 18–37. In *Lysosomes in Biology and Pathology*, ed. by J. T. Dingle, North-Holland Publishing Co., Amsterdam.

Alldredge, A. L., U. Passow and B. E. Logan (1993): The abundance and significance of a class of large, transparent organic particles in the ocean. *Deep-Sea Res.*, **6**, 1131–1140.

Andersen, O. K., J. C. Goldman, D. A. Caron and M. R. Dennett (1986): Nutrient cycling in a microflagellate food chain: III. Phosphorus dynamics. *Mar. Ecol. Prog. Ser.*, **31**, 47–55.

Anderssen, A., C. Lee, F. Azam and A. Hagstrom (1985): Release of amino acids and inorganic nutrients by heterotrophic marine microflagellates. *Mar. Ecol. Prog. Ser.*, **23**, 99–106.

Azam, F., T. Fenchel, J. G. Field, J. S. Gray, L. A. Meyer-Reil and F. Thingstad (1983): The ecological role of water column microbes in the sea. *Mar. Ecol. Prog. Ser.*, **10**, 257–263.

Azam, F., D. C. Smith, G. F. Steward and A. Hagstrom (1994): Bacteria-organic matter coupling and its significance for oceanic carbon cycling. *Microb. Ecol.*, **28**, 167–179.

Barbeau, K., J. W. Moffett, D. A. Caron, P. L. Croot and D. L. Erdner (1996): Role of protozoan grazing in relieving iron limitation of phytoplankton. *Nature*, **380**, 61–64.

Bauer, J. E., P. M. Williams and E. R. M. Druffel (1992): ^{14}C activity of dissolved organic carbon fractions in the north-central Pacific and Sargasso Sea. *Nature*, **357**, 667–670.

Baskaran, M., P H. Santschi, G. Benoir and B. D. Honeyman (1992): Scavenging of thorium isotopes by colloids in seawater of the Gulf of Mexico. *Geochim. Cosmochim. Acta.*, **56**, 3375–3388.

Benner, R., J. D. Pakulski, M. Macarthy, J. T. Hedges and P. G. Hatcher (1992): Bulk chemical characteristics of dissolved organic matter in the ocean. *Science*, **255**, 1561–1564.

Buck, K. R. and J. Newton (1995): Fecal pellet flux in Dabob Bay during a diatom bloom: Contribution of microzooplankton. *Limnol. Oceanogr.*, **40**, 306–315.

Carlson, C. A. and H. W. Ducklow (1995): Dissolved organic carbon in the upper ocean of the central equatorial Pacific Ocean, 1992: Daily and finescale vertical variations. *Deep-Sea Res.*, **42**, 639–656.

Carlson, C. A., H. W. Ducklow and A. F. Michaels (1994): Annual flux of dissolved organic carbon from the euphotic zone in the northwestern Sargasso Sea. *Nature*, **371**, 405–408.

Caron, D. A. (1991): Evolving role of protozoa in aquatic nutrient cycles. pp. 387–415. In *Protozoa and Their Role in Marine Processes*, NATO ASI series, Vol. G25, ed. by P. C. Reid, Springer-Verlag, Berlin, Heidelberg.

Caron, D. A. and J. C. Goldman (1990): Protozoan nutrient regeneration. pp. 283–306. In *Ecology of Marine Protozoa*, ed. by G. M. Capriulo, Oxford University Press, NewYork.

Caron, D. A. and J. C. Goldman (1993): Predicting excretion rates of protozoa: Reply to the comment by Landry. *Limnol. Oceanogr.*, **38**, 472–474.

Caron, D. A., J. C. Goldman, O. K. Andersen and M. R. Dennett (1985): Nutrient cycling in a microflagellate food chain: II. Population dynamics and carbon cycling. *Mar. Ecol. Prog. Ser.*, **24**, 243–254.

Caron, D. A., J. C. Goldman and M. R. Dennett (1990a): Carbon utilization by the omnivorous flagellate *Paraphysomonas imperforata*. *Limnol. Oceanogr.*, **35**, 192–201.

Caron, D. A., K. G. Porter and R. W. Sanders (1990b): Carbon, nitrogen and phosphorus budgets for the mixotrophic phytoflagellate *Poterioochromonas malhamensis* (Chrysophyceae) during bacterial ingestion. *Limnol. Oceanogr.*, **35**, 433–443.

Cole, G. T. and M. J. Wynne (1974): Endocytosis of Microcyctis aeruginosa by *Ochromonas danica*. *J. Phycol.*, **10**, 397–410.

Cole, J. J., S. Findlay and M. L. Pace (1988): Bacterial production in fresh and saltwater ecosystems: a cross-system overview. *Mar. Ecol. Prog. Ser.*, **43**, 1–10.

Choi, J. W. and D. K. Stoecker (1989): Effects of fixation on cell volume of marine planktonic protozoa. *Appl. Environ. Microbiol.*, **55**, 1761–1765.

Dubowsky, N. (1974): Selectivity of ingestion and digestion in the chrysomonad flagellate *Ochromonas malhamensis*. *J. Protozool.*, **21**, 295–298.

Ducklow, H. W. and C. A. Carlson (1992): Oceanic bacterial production. *Adv. Microb. Ecol.*, **12**, 113–181.

Ducklow, H. W., D. A. Purdie, P. J. le B. Williams and J. M. Davies (1986): Bacterioplankton: a sink for carbon in a coastal marine plankton community. *Science*, **232**, 865–867.

Fenchel, T. (1982a): Ecology of heterotrophic microflagellates. I. Some important forms and their functional morphology. *Mar. Ecol. Prog. Ser.*, **8**, 211–223.

Fenchel, T. (1982b): Ecology of heterotrophic microflagellates. II. Bioenergetics and growth. *Mar. Ecol. Prog. Ser.*, **8**, 225–231.

Fenchel, T. (1986): The ecology of heterotrophic microflagellates. *Adv. Microb. Ecol.*, **9**, 57–97.

Fenchel, T. (1987): *Ecology of Protozoa: The Biology of Free-Living Phagotrophic Protists*. Springer-Verlag, Berlin, Heidelberg

Ferrier-Pagès, C. and F. Rassoulzadegan (1994): N remineralization in planktonic protozoa. *Limnol. Oceanogr.*, **39**, 411–419.

Flood, P. R., D. Deibel and C. C. Morris (1992): Filtration of colloidal melanin from sea water by planktonic tunicates. *Nature*, **355**, 630–632.

Francois, R. (1990): Marine sedimentary humic substances: Structure, genesis and properties. *Rev. Aquat. Sci.*, **3**, 41–80.

Gaines, G. and F. J. R. Taylor (1984): Extracellular digestion in marine dinoflagellates. *J. Plankton Res.*, **6**, 1057–1061.

Gaudy, R. (1974): Feeding four species of pelagic copepods under experimental conditions. *Mar. Biol.*, **25**, 125–141.

Geider, R. J. and B. S. C. Leadbeater (1988): Kinetics and energetics of growth of the marine choanoflagellate Stephanoeca diplocostata. *Mar. Ecol. Prog. Ser.*, **47**, 169–177.

Goldman, J. C., D. A. Caron, O. K. Andersen and M. R. Dennett (1985): Nutrient cycling in a microflagellate food chain: I. Nitrogen dynamics. *Mar. Ecol. Prog. Ser.*, **24**, 231–242.

Gowing, M. M. and M. W. Silver (1985): Minipellets: A new and abundant size class of marine fecal pellets. *J. Mar. Res.*, **43**, 395–418.

Haga, H., T. Nagata and M. Sakamoto (1995): Size-fractionated NH_4^+ regeneration in the pelagic environments of two mesotrophic lakes. *Limnol. Oceanogr.*, **40**, 1091–1099.

Handa, N. (1991): Chemical studies on organic matter and carbon cycle in the ocean—Lecture by the member awarded the Oceanographic Society of Japan Prize for 1990. *J. Oceanogr. Soc. Japan*, **47**, 49–61 (in Japanese).

Harrison, W. G. (1992): Regeneration of nutrients. pp. 385–407. In *Primary Productivity and Biogeochemical Cycles in the Sea*, ed. by P. G. Falkowski and A. D. Woodhead, Plenum, New York.

Hedges, J. T. (1988): Polymerization of humic substances in natural environments. pp. 45–58. In *Humic Substances and Their Role in the Environment*, ed. by F. H. Frimmel and R. F. Christman, John Wiley & Sons Limited, NewYork.

Hoch, M. P., R. A. Snyder, L. A. Cifuentes and R. B. Coffin (1996): Stable isotope dynamics of nitrogen recycled during interactions among marine bacteria and protists. *Mar. Ecol. Prog. Ser.*, **132**, 229–239.

Honjo, S. and M. R. Roman (1977): Marine copepod fecal pellets: production, preservation and sedimentation. *J. Mar. Res.*, **36**, 45–57.

Jumars, P. A., D. L. Penry, J. A. Baross, M. J. Perry and B. W. Frost (1989): Closing the microbial loop: dissolved carbon pathway to heterotrophic bacteria from incomplete ingestion, digestion and absorption in animals. *Deep-Sea Res.*, **36**, 483–495.

Keil, R. G. and D. L. Kirchman (1993): Dissolved combined amino acids: chemical form and utilization by marine bacteria. *Limnol. Oceanogr.*, **38**, 1256–1270.

Keil, R. G. and D. L. Kirchman (1994): Abiotic transformation of labile protein to refractory protein in seawater. *Mar. Chem.*, **45**, 187–196.

Kepkay, P. E. (1994): Particle aggregation and the biological reactivity of colloids. *Mar. Ecol. Prog. Ser.*, **109**, 293–304.

Kirchman, D. L. (1994): The uptake of inorganic nutrients by heterotrophic bacteria. *Microb. Ecol.*, **28**, 255–271.

Koike, I., S. Hara, K. Terauchi and K. Kogure (1990): The role of submicrometer particles in the ocean. *Nature*, **345**, 242–244.

Kogure, K. and I. Koike (1987): Particle counter determination of bacterial biomass in seawater. *Appl. Environ. Microbiol.*, **53**, 274–277.

Kopylov, A. I., T. I. Mamayeva and S. F. Batsanin (1980): Energy balance of the colorless flagellate *Parabodo attenuatus* (Zoomastigophora, Protozoa). *Oceanology*, **20**, 705–708.

Landry, M. R. (1993): Predicting excretion rates of microzooplankton from carbon metabolism and elemental ratios. *Limnol. Oceanogr.*, **38**, 468–472.

Lee, C. and S. G. Wakeham (1992): Organic matter in the water column: future research challenges. *Mar. Chem.*, **39**, 95–118.

Longhurst, A. R. (1991): Role of the marine biosphere in the global carbon cycle. *Limnol. Oceanogr.*, **36**, 1507–1526.

Longhurst, A. R., I. Koike, W. K. W. Li, J. Rodriguez, P. Dickie, P. Kepkay, F. Partensky, B. Bautista, J. Ruiz, M. Wells and D. Bird (1992): Sub-micron particles in northernwest Atlantic shelf water. *Deep-Sea Res.*, **39**, 1–7.

Maruyama, A., M. Oda and T. Higashihara (1993): Abundance of virus-sized non-DNase-digestible DNA (coated DNA) in eutrophic seawater. *Appl. Environ. Microbiol.*, **59**, 712–717.

Matsueda, N., N. Handa, I. Inoue and H. Takano (1986): Ecological significance of salp fecal pellets collected by sediment traps in the eastern north Pacific. *Mar. Biol.*, **91**, 421–431.

Moran, S. B. and K. O. Buesseler (1992): Short residence time of colloids in the upper ocean estimated from ^{238}U-^{234}Th disequilibria. *Nature*, **359**, 221–223.

Nagata,T.,R. Fukuda, I. Koike, K.Kogure and D.L. Kirchman(1998) Degradation by bacteria of membrane and soluble protein in seawater. *Aquat. Microb. Ecol.*, **14**, 29–37.

Nagata, T. and D. L. Kirchman (1990): Filtration-induced release of dissolved free amino acids: Application to cultures of marine protozoa. *Mar. Ecol. Prog. Ser.*, **68**, 1–5.

Nagata, T. and D. L. Kirchman (1991): Release of dissolved free and combined amino acids by bacterivorous marine flagellates. *Limnol. Oceanogr.*, **36**, 433–443.

Nagata, T. and D. L. Kirchman (1992a): Release of dissolved organic matter by heterotrophic protozoa: implications for microbial food webs. *Arch. Hydrobiol. Beih. Ergebn. Limnol.*, **35**, 99–109.

Nagata, T. and D. L. Kirchman (1992b): Release of macromolecular organic complexes by heterotrophic marine flagellates. *Mar. Ecol. Prog. Ser.*, **83**, 233–240.

Nagata, T. and D. L. Kirchman (1996): Bacterial degradation of protein adsorbed to model submicron particles in seawater. *Mar. Ecol. Prog. Ser.*, **132**, 241–248.

Nagata, T. and D. L. Kirchman (1997): Roles of submicron particles and colloids in microbial food webs and biogeochemical cycles within marine environments. *Adv. Microb. Ecol.* **15**, 81-103.

Nagata, T. and I. Koike (1995): Marine colloids: Their roles in food webs and biogeochemical fluxes. pp. 275–292. In *Biogeochemical Processes and Ocean Flux in the Western Pacific*, ed. by H. Sakai and Y. Nozaki, Terra Scientific Pub. Co., Tokyo.

Nagata, T. and Y. Watanabe (1990): Carbon- and nitrogen-to-volume ratios of bacterioplankton grown under different nutritional conditions. *Appl. Environ. Microbiol.*, **56**, 1303–1309.

Nothig, E.-M. and B. von Bodunge (1989): Occurrence and vertical flux of faecal pellets of probably protozoan origin in the southeastern Weddell Sea (Antarctica). *Mar. Ecol. Prog. Ser.*, **56**, 281–289.

Ogawa, H. and N. Ogura (1992): Comparison of two methods for measuring dissolved organic carbon in the sea water. *Nature*, **356**, 696–698.

Passow, U., A. L. Alldredge and B. E. Logan (1994): The role of particulate carbohydrate exudates in flocculation of diatom blooms. *Deep-Sea Res.*, **41**, 335–357.

Peltzer, E. T. and N. A. Hayward (1996): Spatial and temporal variability of total organic carbon along 140°W in the equatorial Pacific Ocean in 1992. *Deep-Sea Res.* **43**: 1155-1180.

Pomeroy, L. R. (1974): The ocean's food web, a changing paradigm. *BioScience*, **24**, 499–504.

Penry, D. L. and P. A. Jumars (1986): Chemical reactor analysis and optimal digestion. *BioScience*, **36**, 310–315.

Porter, K. and Y. S. Feig (1980): The use of DAPI for identifying and counting aquatic microflora. *Limnol. Oceanogr.*, **25**, 943–948.

Rees, T. F. (1991): Transport of contaminants by colloid-mediated processes. pp. 25–44. In *The Handbook of Environmental Chemistry, Vol. 2F*, ed. by F. O. Hutzinger, Springer-Verlag, Heidelberg.

Sanders, R.W., D. A. Caron and U.-G. Berninger (1992): Relationships between bacteria and heterotrophic nanoplankton in marine and fresh waters: an inter-ecosystem comparison. *Mar. Ecol. Prog. Ser.*, **86**, 1–14.

Sherr, B. F., E. B. Sherr and T. Berman (1983): Grazing, growth, and ammonium excretion by a heterotrophic microflagellate fed with four species of bacteria. *Appl. Environ. Microbiol.*, **45**, 1196–1201.

Sherr, E. B. (1988): Direct use of high molecular weight polysaccharide by heterotrophic flagellates. *Nature*, **335**, 348–351.

Sherr, E. B., B. F. Sherr and L. J. Albright (1987a): Bacteria: Link or sink? *Science*, **235**, 88.

Sherr, E. G., B. F. Sherr and R. D. Fallon (1987b): Use of monodispersed, fluorescently labeled bacteria to estimate in situ protozoan bacterivory. *Appl. Environ. Microbiol.*, **53**, 958–965.

Shrader, H. J. (1971): Fecal pellets: Role in sedimentation of pelagic diatoms. *Science*, **174**, 55–57.

Stoecker, D. K. (1984): Particle production by planktonic ciliates. *Limnol. Oceanogr.*, **29**, 930–940.

Tranvik, L. J. (1994): Colloidal and dissolved organic matter excreted by a mixotrophic flagellate during bacterivory and autotrophy. *Appl. Environ. Microbiol.*, **60**, 1884–1888.

Tranvik, L. J., E. B. Sherr and B. F. Sherr (1993): Uptake and utilization of "colloidal" DOM by heterotrophic flagellates in seawater. *Mar. Ecol. Prog. Ser.*, **92**, 301–309.

Turk, V., A.-S. Rehnstam, E. Lundberg and A. Hagstrom (1992): Release of bacterial DNA by marine nanoflagellates, an intermediate step in phosphorus regeneration. *Appl. Environ. Microbiol.*, **58**, 3744–3750.

Twiss, M. R. and P. G. C. Campbell (1995): Regeneration of trace metals from picoplankton by nanoflagellate grazing. *Limnol. Oceanogr.*, **40**, 1418–1429.

Wells, M. L. and E. D. Goldberg (1993): Colloid aggregation in seawater. *Mar. Chem.*, **41**, 353–358.

Wells, M. L. and E. D. Goldberg (1994): The distribution of colloids in the North Atlantic and Southern Ocean. *Limnol. Oceanogr.*, **39**, 286–302.

Willams, P. J. leB (1981): Incorporation of microheterotrophic processes in the classical paradigm of the planktonic food web. *Kieler Meeresforsh.* **5**, 1–28.

Williams, P. J. leB (1995): Evidence for the seasonal accumulation of carbon-rich dissolved organic material, its scale in comparison with changes in particulate material and consequential effect on net C/N assimilation ratios. *Mar. Chem.*, **51**, 17–29.

Williams, P. M. and E. R. M. Druffel (1987): Radiocarbon in dissolved organic matter in the central North Pacific Ocean. *Nature*, **330**, 246–248.

Williams, P. M. and E. R. M. Druffel (1988): Dissolved organic matter in the ocean: comments on a controversy. *Oceanography*, **1**, 14–17.

Yamamoto, S. and R. Ishiwatari (1989): A study of the formation mechanisms of sedimentary humic substances II. Protein-based melanoidin model. *Org. Geochem.*, **14**, 479–489.

Dynamics and Characterization of Marine Organic Matter, Eds., N. Handa, E. Tanoue and T. Hama, pp.257–276.

Feeding of Carnivorous Zooplankton, Chaetognaths in the Pacific

Makoto TERAZAKI

Ocean Research Institute, University of Tokyo,
1-15-1 Minamidai, Nakano, Tokyo 164-8639, Japan

(Review, Received 2 August 1996; accepted 27 December 1996)

INTRODUCTION

The phylum chaetognatha consists of some 100 species arranged in 22 genera (Bieri, 1991). Chaetognaths are found in every marine habitat, from the benthos to all zones of coastal waters and the open oceans. Although small (2–120 mm), they are often abundant and play an important role in the marine food web as the primary predators of copepods. The biomass of chaetognaths has been estimated as 10–30 per cent of that of copepods in the world oceans, so that they are of great significance and Grey (1930) also described a similar event. Parry (1944) provided the first report of feeding in the laboratory other than by chance observation. Though he found *Sagitta setosa* difficult to keep in the laboratory, he was able to report on the difference between day and night feeding.

The first report of controlled laboratory feeding was by Reeve (1964) working with *S. hispida* in Miami, Florida. Reeve fed *Artemia* nauplii to *S. hispida* in small plastic compartment and presented estimates of the daily ration and weight-specific daily ration for the species. Later, Reeve (1970) reported the complete cycle of development of *S. hispida* in the laboratory for a planktonic species for the first time. Additional information on feeding of this hardy species was provided in these and subsequent papers by Reeve and his co-workers. About the same time, Murakami (1966) successfully kept *S. crassa* alive for 34 days indoors in glass jars and for 106 days outdoors in large concrete tanks. Protozoans were provided as food for young chaetognaths, and small copepods for the adults. One of the earliest reports of the chaetognath diet was by Busk (1856). He found fragments of minute fish, crustaceans and other chaetognaths in their guts. In another early report, Scott (1892) noted that *Sagitta* lives on larval and postlarval fishes, copepods and small amphipods. Scott (1893) also observed that *Sagitta* is "not fastidious, providing the object is of convenient size to suit the capacity of its jaws". Although we agree with this comment today, this view was not always held in the intervening years.

Other early reports of chaetognath feeding come from Shipley (1910), who included diatoms and infusorians in the diet, Steur (1910) who presented a sketch of *S. lyra* eating an amphipod, and Lebour (1922, 1923), who made extensive observations on the gut contents of *S. bipunctata* (probably *S. setosa*). Lebour recognized the problems caused by feeding in the net or cod-end jar. Her report that chaetognaths were probably responsible for the many herring larvae found cut in two or decapitated is, however, probably in error. Her reproduced figures of *Sagitta* grabbing prey may represent artifacts of net feeding, because the chaetognaths did not live in the laboratory. Writing on the plankton of the Gulf of Maine, Bigelow (1924) said "*Sagitta* are strictly carnivorous and so active, fierce, and well armed that it is no wonder they are recorded as feeding on things as far apart as tintinnids, crustaceans, other *Sagitta*, and young fish".

Wimpenny (1936) made the first attempt to systematically analyze the gut contents of chaetognaths. He looked at the percentage of *S. setosa* and *S. elegans* containing food by location and time of year in the south-west North Sea. He was, however, misled by "green matter" in the guts of young chaetognaths and believed diatoms were a direct source of food. He concluded that diatom patches were nursery areas, and centres of reproduction for *S. setosa* and tried to correlate his feeding data with algal concentrations.

In recent studies, the gut contents of chaetognaths have been quantitatively analyzed to allow for estimation of the daily feeding and energetic considerations (Pearre, 1981; Terazaki, 1995).

THE FEEDING PROCESS

Chaetognaths detect prey by sensing their movement. In the laboratory, attacks have been experimentally induced against low frequency vibrating probes placed within millimeters of the animal's body (Horridge and Boulton, 1967; Feigenbaum and Reeve, 1977). Small organisms such as copepods, barnacle nauplii and appendicularians, which make a distinct beat while moving, appear to be the most susceptible to attack. Fish larvae and other chaetognaths may be detected by their tail beat, although lateral motion alone may be sufficient (Feigenbaum and Reeve, 1977).

Planktonic chaetognaths initiate an attack with a sudden flexing and flick of their tail. Benthic species, which are generally adherent to the substrate, make a rapid upward jerk of the forward part of their body when a prey swims overhead. The hold on the substrate is maintained. In both chaetognath groups, attacks cover short distances; the prey is not pursued if it is missed or struggles free (Parry, 1944).

The prey is grabbed by the chaetognath's hooks in a quick action during which the hood is folded backward and the mouth projected forward. The chitinous hooks act like rigid figures, moving individually to manipulate and stuff the prey down the chaetognath's "throat" (Parry, 1944). In *Spadella schizoptera*, the lateral plates and musculature associated with the hooks take on the appearance of a forearm when the chaetognath is feeding. Prey are rarely pierced by the hooks, yet seem to be immobilized before they are ingested (Nagasawa, 1985).

The rigid prey is manipulated by the hooks and eaten endwise, sticking out of the mouth for long periods of time in some cases. Soft-bodied prey such as fish larvae are sometimes folded over (Kuhlmann, 1977). Thus, the orientation of prey in the gut cannot be used to indicate whether the prey was moving towards or away from the chaetognath at the time of capture.

Prey size is limited to organisms which are large enough to be handled by the hooks and small enough, in at least one dimension, to pass through the mouth. On rare occasions, large prey become lodged in the forward gut and the chaetognath dies (Nagasawa, 1985).

The role of the one or two rows of small teeth is still conjectural. It is unlikely that they are used to grasp or hold prey, as some have suggested (John, 1933; Parry, 1944; Hyman, 1959; Furnestin, 1977), because prey are easily handled by the hooks. Furthermore, some genera lack teeth. The teeth are not hollow (Cosper and Reeve, 1970; Bieri *et al.*, 1983), but several investigators have nevertheless suggested that they are used to inject a toxin into the prey, to paralyze them and prevent damage to the chaetognaths gut lining (Burfield, 1927; Bieri *et al.*, 1983). However, Nagasawa (1985) could not confirm this hypothesis in her scanning electron microscope study of captured copepod exoskeletons.

Once ingested, the prey are wrapped in a peritrophic membrane (Reeve *et al.*, 1975; Sullivan, 1977) and passed to the posterior gut by pelistaltic movements of the gut wall (Parry, 1944). This process may take from a few seconds to more than 30 minutes, depending on the species and temperature (Reeve *et al.*, 1975; Feigenbaum, 1982). Once in the posterior gut, the prey may be moved back and forth throughout the gut or may remain near the anus until defecation. Observations of both have been made, sometimes for the same species (Grey, 1930; Reeve *et al.*, 1975; Canino and Grant, 1985; Feigenbaum, 1991). The peritrophic membrane may be thick (*Sagitta hispida*) or thin (*S. enflata*). Sheader and Evans (1975) believe that its function is to protect the gut lining from the sharp spines of crustaceans.

After *Spadella cephaloptera* has swallowed a copepod, it gulps water, which may serve to carry enzymes liberated by the oesophagus to the ingested prey (Parry. 1944).

Many prey can be recognized, and often identified to species, during the digestive process because some parts of their anatomy resist digestion. These parts include the hooks and lateral plates of prey chaetognaths, the mouth parts of crustaceans and the fecal pellets of appendicularians (Sullivan, 1977; Nagasawa and Marumo, 1979; Feigenbaum, 1982).

After digestion, the prey is defecated as a soft pellet. If several prey are in the gut they are usually defecated together, even if some were ingested considerably later than the other (Reeve *et al.*, 1975; Nagasawa and Marumo, 1984; Nagasawa, 1985; Feigenbaum, 1991). Digestive efficiency for *Sagitta hispida*, feeding on at least three copepods at a time, averaged 80%(Reeve and Cosper, 1975). This is similar to efficiencies reported for *S. crassa* (Nagasawa 1985) and typically of canivores (Conover, 1978). Usually, however, chaetognaths contain one prey at a time (Mironov, 1960; Newbury, 1978; Feigenbaum, 1979). For example, a total number of 163 out of 1149 *S. zetesios* were found to have one or more food organisms in their gut. Of the 163 animals, 24 had consumed two items, 3, 1 and 1 had consumed 3, 4 and 5 items, respectively (Terazaki and Marumo,

1982). In some instances, the frequency of multiple prey can be predicted by a random distribution. In others it cannot, indicating that either the prey distributions are patchy or chaetognath movement patterns are not random (Sullivan, 1977). Given the opportunity, hungry chaetognaths will eat several prey items in rapid succession (Cosper and Reeve, 1975).

FOOD ITEMS

Chaetognaths begin feeding a few days after hatching; they feed on proportionately small prey. Copepod nauplii are presumably the mainstay of the young chaetognath's diet, but tintinnids, and sometimes rotifers and small meroplanktons, are also eaten (Mironov, 1960; Reeve, 1966, 1970, 1980; McLaren, 1969; Pearre, 1981).

Chaetognatha feed on Copepoda, other Chaetognatha, Ostracoda, Mesdusae, Salps and the larvae of other marine animals. The food of the Chaetognaths is considered heterogeneous, including cannibalism, and representantives of every group of plankton; diatoms, ciliates, Tintinnids, Medusae, Crustacea, Copepoda, Amphipoda, Cladocera, larval stages of Crustacea, Mollusca, Pteropoda, Heteropoda, Chaetognatha, Tunicata, fish eggs and larvae (Alvariño, 1985). A listing of the food of chaetognaths collected from the Pacific is given in Table 1.

Studies of the Chaetognatha population of Southeastern Australia (Thomson, 1947) show that they feed mainly on Copepoda, fish larvae and other chaetognaths, and are considered to play an important part in the food cycle of the pelagic communities because of their role as competitors and prey. Chaetognatha eat larvae of fish and of small crustacea (Johnson and Snook, 1927), young fish (Graham *et al.*, 1954). Fish larvae, copepods ciliates and cannibalism was more outstanding in *Sagitta enflata* and *S. hexaptera* (Suarez-Caabro, 1955; Alvariño, 1985); preying also on fish larvae, copepods, euphausiids, cladocerans, amphipods, other crustacea and the larvae of other animals of commercial importance (Alvariño, 1962, 1965, 1974).

The distributional characteristics of plankton in the oceanic fishing grounds of salmon in the North Pacific (Tsuruta, 1963) showed in 80% of cases that high concentrations of *S. elegans* or medusae concurred with the absence of Pleuronectidae and Gadidae fish larvae. In the tuna fishing grounds of the Marshall, Gilbert and Phoenix Islands, the abundance of *S. bipunctata* coincided with the scarcity or absence of fish; and in waters around Bikini Atoll, high concentrations of any of the chaetognaths (mainly *S. pacifica*, *S. hexaptera*, *S. enflata*, *P. draco*) coincided with the lowest occurrence of fish larvae.

David (1955) has pointed out that although in bulk plankton collections, *S. gazellae* is frequently observed to have seized numbers of large animals (medusae, fishes, or their own tails). This is not necessarily indicative of their feeding habits, but may rather be the result of the convulsive movements when a large number of individuals are concentrated in the bucket of the net. The food organisms found in the gut of *S. gazellae* were *Calanus propinquus*, *Rhincalanus gigas*, *Pleuromamma robusta*, *Metridiagerlachii*, other copepods, *Thysanoessa macrura*, *Thysanoessa vicina*, other euphausiids and their larvae, *Eukrohnia hamata* and other chaetognaths, and crustaceans such as *Primmo macropa*. No mature *S. gazellae* was found, however, with food in the gut.

The food organisms found in the digestive tract of chaetognaths from the Hilutangan Channel, Philippines, in January 1972 (Jumao-as and Westernhagen, 1975) included 401 Copepoda (74.8%), 105 Chaetognatha (19.6%), 7 Foraminifera (1.3%), 6 Echinoidlarvae (1.1%), 6 Decapod larvae (1.1%), 4 fish larvae (0.8%), 3 Tunicate (0.6%), 1 Polychaeta larvae (0.2%), 1 Isopoda (0.2%) and 1 veliger larvae (0.2%). The species observed were *S. enflata*, *S. neglecta*, *S. robusta*, *S. regularis*, *Krohnittapacifica*, *S. bedoti*, *S. decipienens*, *S. pulchra*, *S. ferox*, *K. subtilis*, *S. pacifica*, *P. draco*, *S. hexaptera*. Jumao-as and Westernhagen (1975) did not find considerable variations in dietary composition with season and standing crop of the plankton communities.

At Ocean Station "P" in the Gulf of Alaska, *S. elegans* lives and feeds primarily in the upper water column (0–25 m) and the vertical distribution of this species may be explained primarily by its requirement of high prey densities. *Eukrohnia hamata* apparently needs less food and is found mainly below the mixed layer with a vertical distribution restricted by temperature. The same prey species, those with the highest densities at Station "P", make up the diet of both chaetognaths, although *E. hamata* eats significantly small prey. The major prey of juvenile chaetognaths are small cyclopoid copepods, a group frequently overlooked in plankton surveys. These species also comprise 41% of the diet of older chaetognaths, which, in addition, on prey stage V copepodites of the largest copepods. The development of larval chaetognaths depends on abundant small prey but the older chaetognaths can handle prey of a very wide range of sizes.

The largest component of the gut contents of *S. elegans* (9.0–23.0 mm: 17.0 mm in mean body length) collected from the sub-arctic Pacific (45°N, 160°10′E), was copepods such as *Pseudocalanus*, *Metridiapacifica* and *Neocalanus flemingeri*. The size of food organisms ranged from 1.0 mm to 3.5 mm (mean = 2.0 mm). In coastal waters off Otsuchi Bay, *S. elegans* chiefly consumed small-sized copepods such as *Acartia*, *Oithona* and *Pseudocalanus*. Therefore, *S. elegans* inhabiting the open sea, consumes large copepods compared to those in coastal water (Terazaki, 1995).

In the Japan Sea, a total of 1510 out of 6053 individual *S. elegans* were found to have one or more prey organisms in their gut. The predominant prey were copepods, and the proportions of Copepoda, Chaetognatha, Ostracoda and unidentified items in the gut of *S. elegans* were 74.6, 1.7, 3.4 and 20.3%, respectively. Most frequent were calanoid copepods such as species of *Neocalanus* and *Paracalanus*, *Scaphocalanus* sp. and *Aetideus* sp. which were also found in *S. elegans* gut contents, are meso- and bathypelagic calanoids. There was no diet variation in the proportions of the various prey ingested (Terazaki, 1993).

In the Central Equatorial Pacific, a total of 200 out of 1854 *S. enflata* were found to have one or more food organisms in their gut. Of these 200 animals, 24 had consumed two items and two had consumed three items. The percentage of Copepoda, Foraminifera, crustacean larvae, Chaetognatha, Pteropoda, Ostracopda

Table 1. Food organisms of pelagic chaetognaths

Species	Location	Food	Reference
Eukrohnia fowleri	N. W. Pacific	Copepoda; Radiolaria	1
	North Pacific	*Calanus pacificus, Lucicutia bicornuta* (Copepoda)	2
E. hamata	Sagami Bay	*Euchaeta* sp. (Copepoda)	3
	N.W. Pacific	Copepoda; Ostracoda; Chaetognatha; Radiolaria	1
	British Columbia	Herring larvae	4
	Gulf of Alaska	*Oithona similis, Metridia pacifica, Oncaea borealis Calanus plumchrus* (Copepoda)	5
Sagitta crassa	Seto Inland Sea	*Acartia clausi, A. erythraea, Calanus, Tortanus Centropages, Paracalanus* (Copepoda)	6
	Seto Inland Sea	Larval red sea bream	7
	Tokyo Bay	*Oithona aruensis, Pseudodiaptomus marinus* (Copepoda), *Sagitta crassa* (Chaetognatha)	8
S. elegans	off British Columbia	Fish larvae; *Calanus plumchrus* (Copepoda)	4
	Dabob Bay	Copepoda	9
	Gulf of Alaska	*Calanus cristatus, Oithona similis, Metridia pacifica Calanus plumchrus* (Copepoda); Chaetogmatha	5
	N. E. Pacific	Fish larvae; Copepoda; Euphausiacea; Medusae; Crustacea larvae; Chaetognatha	10
	Japan Sea	*Neocalanus spp., Paraclanus parvus, Acrocalanus grtacilis, Scaphocalanus* sp., *Oithona* sp., *Aetideus* sp., *Undinopsis* sp.(Copepoda); *Sagitta elegans* (Chaetognatha); *Conchoecia pseudodiscophora* (Ostracoda)	11
	N.W. Pacific	*Pseudocalanus* sp., *Metridia pacifica, Neocalanus flemingeri* (Copepoda)	12
	Otsuchi Bay	*Parathemisto japonica* (Amphipoda); *Acartia clausi, Oithona* spp., *Pseudocalanus minutus* (Copepoda); Euphausiacea	12
S. enflata	Kaneohe Bay, Hawaii	Copepoda; Chaetognatha; Appendicularia; Ciliata	13
	N.E. Pacific	Fish larvae; Copepoda; Euphausiscea; Chaetognatha; Hydromedusae; Siphonophorae	10
	Central Equatorial Pacific	*Microsetella* sp. *Calanoida* (Copepoda); Ostracoda; Foraminifera; Chaetognatha; Crustacean larvae; Pteropoda	14
S. ferox	N.E. Pacific	Fish larvae; Copepoda; Euphausiacea; Crustacean larvae; Medusae; Siphonophorae	10
S. gazellae	Antarctic	*Calanus propinguos, Rincalanus gigas, Calanus* sp. *Pleuromamma robusta, Metridia gerlachi* (Copepoda); *Tysanoessa sp., T. macrura, T. vicina* (Euphausiacea); Amphipoda; Chaetognatha; Primo macropa and other Crustacea	15

1. Tchindonova (1959); 2. Terazaki *et al.* (1977); 3. Tokioka (1939); 4. Lee (1966); 5. Sullivan (1980); 6. Murakami (1959); 7. Fukuhara and Fukunaga (1984); 8. Nagasawa and Marumo (1984); 9. King (1979); 10. Alvarinõ (1985); 11. Terazaki (1993); 12. Terazaki (1995); 13. Szyper (1978); 14. Terazaki (1996); 15. David (1955); 16. Nagasawa and Marumo (1972); 17. Nagasawa and Marumo (1976); 18. Terazaki (unpublished data); 19. Terazaki and Marumo (1982).

Table 1. (continued).

S. hexaptera	N.E. Pacific	*Sardina pilchardus*, *Engraulis encrasicholus*, *Sardinops caerulea*, *Engraulis mordax* and other fish larvae; Copepoda; Euphausiacea; Amphipoda; other Crustacea; Hydromedusae; Siphonophorae; Chaetognatha	10
S. lyra	N.W. Pacific	Copepoda, Ostracoda	1
	N.E. Pacific	Fishlarvae; Copepoda; Euphausiacea; Amphipoda; Crustacea larvae; Medusae; Siphonophorae; Chaetognatha	10
S. macrocephala	N.W. Pacific	Chaetognatha	1
	North Pacific	*Calanus pacificus*, *Lucicutia bicornuta* (Copepoda)	2
S. maxima	off British Columbia	Fish larvae	4
	N.E. Pacific	Fish larvae; Copepoda; Euphausiacea; Crustacea larvae	10
S. minima	N.E. Pacific	Small Copepoda; Invertebrate larvae	10
S. nagae	Suruga Bay	*Candacia pinnata*, *Calanus pacificus*, *Pareuchaeta russelli* (Copepoda)	16
	Suruga Bay	*Calanus pacificus*, *Corycaeus affinis*, *C. dahli*, *C. dubius*, *Clausocalanus arcuicorris*, *Paracalanus parvus*, *Ctenocalanus vanus* (Copepoda)	17
	Sagami Bay	Copepoda; Anchovy larvae	18
	N.E. Pacific	Fish larvae; Copepoda; Crustacea larvae; Mollusca larvae; Hydromedusae	10
S. pacifica	N.E. Pacific	Copepoda, Hydromedusae, Chaetognatha	10
S. pulchra	N.E. Pacific	Fish larvae; Copepoda; Crustacea larvae; Polychaeta larvae	10
S. robusta	N.E. Pacific	Fish larvae; Copepoda; Crustacea larvae; Euphausiacea	10
S. scrippsae	N.E. Pacific	Fish larvae; Copepoda; Euphausiacea; Medusae; Chaetognatha	10
S. serratodentata	N.E. Pacific	Copepoda; Medusae; Chaetognatha	10
S. zetesios	Sagami Bay	*Calanus cristatus*, *C. plumchrus*, *C. pacificus*, *Undinula vulgaris*, *Eucalanus bungii*, *E. crassus*, *E. subtenuis*, *Rhincalanus nasutus*, *R. cornutus*, *Clausocalanus arcuicornis*, *Aetideus armatus*, *Gaetanus armiger*, *Euchaeta concinna*, *Oncaea conifera*, *Pareuchaeta russelli*, *P. scaphula*, *P. rubra*, *Scaphocalanus echinatus*, *Scolecithrix danae*, *S. nicobarica*, *Scolecithricella valida*, *Pleuromamma abdominalis*; *Heterorhabdus pacificus*, *Candacia bipinnata* (Copepoda); *Sagitta nagae*, *S. layra*, *S. neodecipiens*, *S. macrocephala*, *S. zetesios* (Chaetognatha); *Conchoecia elegans* (Ostracoda)	19

and unidentified materials in the gut were 51.9, 6.7, 3.8, 2.9, 1.9, 1.9 and 30.9, respectively. The predominant food organisms were small-sized copepods such as *Microsetella* sp. (Terazaki, 1996). *S. enflata* mainly consume food organisms ranging from 0.4 to 1.8 mm (mean = 0.8 mm) in body length; 63.4% of all prey items were in this range (Fig. 1). The regression equation of body length (mm, X) versus prey length (mm, Y) was

$$Y = 0.1X - 0.003 \ (r^2 = 0.127).$$

Therefore, there was no strong correlation between body length and prey length.

In Suruga Bay, *Calanus pacificus* was the most common species and its copepodite V stage was frequently found in the gut of *S. nagae* collected from the near-bottom habitat. Second to *C. pacificus*, the predominant species in the gut were *Corycaeus affinis*, *C. dahli*, *C. dubius*, *Clausocalanus arcuicornis*, *Paraclanus parvus* and *Ctenocalanus vanus*. In the free water, *S. nagae* also fed on the dominant species, *Calanus pacificus*, *Candacia bipinnata* and *Pareuchaeta russelli*. *S. nagae*, both in the free water and the near-bottom water, takes copepods which are abundantly distributed in the two areas. *S. nagae* migrates vertically and are found in copepod-rich waters, although the dominant species are different depending on the area (Nagasawa and Marumo, 1972, 1976: Fig. 2).

In Tokyo Bay, food organisms of *Sagitta crassa* are *Oithona aruensis* (Copepoda; 95.4%), *Pseudodiaptomus marinus* (Copepoda; 3.9%) and *S. crassa*

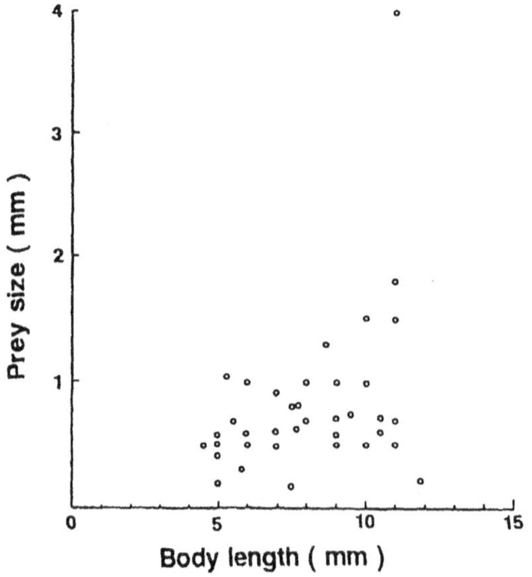

Fig. 1. Relationship between body length of *Sagitta enflata* and food size (after Terazaki, 1995).

(0.7%). Some specimens contained both *O. aruensis* and *P. marinus* (Nagasawa and Marumo, 1984).

In comparison with epipelagic chaetognaths, not much is known about the feeding habits of meso- and bathypelagic chaetognaths (Table 1), because their sampling takes much time and they are often damaged during net sampling from the deep sea. The predominant food organisms of meso- and bathypelagic species, *Sagitta zetesios* collected from Sagami Bay, were copepods and the percentages of Copepoda, Chaetognatha, Euphusiacea, Ostracoda and unidentified organisms in the gut, was 71.6%, 16.2%, 1.4%, 1.4% and 9.4%, respectively. The 24 species of copepods, 5 of chaetognaths and 1 of ostracods were identified and 20 species were meso- and bathypelagic and another 10 species were epipelagic. The food organisms frequently found in the gut were *Calanus pacificus*, *Pareuchaeta russelli* and *S. nagae* and the major prey of adults were large-sized meso- and bathypelagic copepods. There was also cannibalism in stage II and III individuals (Terazaki and Marumo, 1982). *S. zetesios* mainly consumed food organisms ranging from 2 to 3 mm in body length and this size accounted for 43.8% of all their prey. Generally, the large-sized *S. zetesios* have a tendency to consume large food organisms.

Most pelagic chaetognaths are colourless and transparent milky white, although Bieri (1966, 1977) reported pale blue and deep blue epipelagic chaetognaths. However, some deep sea species have yellow, red or orange pigments, for example, *Sagitta macrocephala* (Tokioka, 1939), *Eukrohnia fowleri* (Owre, 1960), *Heterokrohnia bathybia* (Marumo and Kitou, 1966), H. Iongicaudata

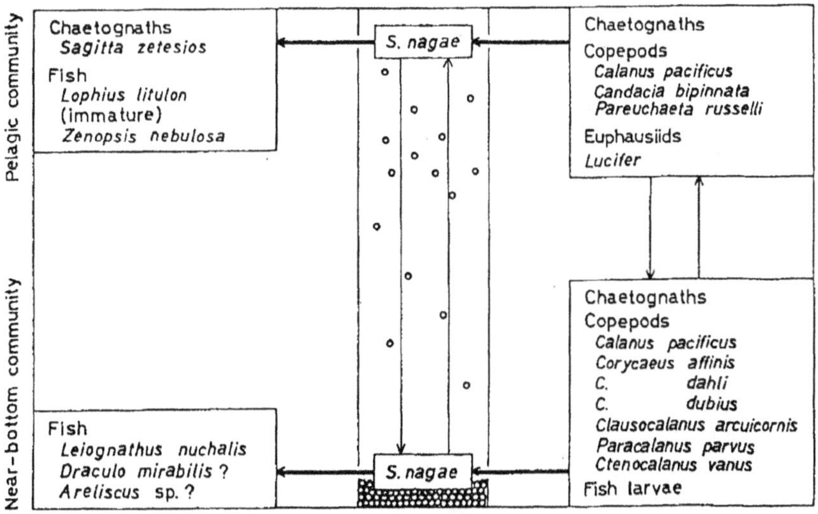

Fig. 2. A scheme of food chains and vertical distribution of *Sagitta nagae* in Suruga Bay. Thick and thin arrows indicate food transfer and vertical migration, respectively (after Nagasawa and Marumo, 1976).

Table 2. Comparison of major carotenoids in deep sea chaetognaths and copepods

Species	Absorption maxima (nm)				R_f value
	Petroleum ether	Hexane	Benzine	Chloroform	(1% acetone in petroleum ether)
Chaetognaths					
Sagitta macrocephala	457–460	460	475–477	478–480	1.0
Eukrohnia fowleri	457–460	459–460	473	478–480	1.0
Copepods					
Calanus pacificus	465	465	ND	484	0.4
Lucicutia bicornuta	468	465	485	483	0.3

ND, Not determined. Reproduced, with permission of the publishers, from Terazaki *et al.* (1977)

(Hagen and Kapp, 1986), and *H. rubra* (Casanova, 1986). The pigment generally occurred in the gut, but occasionally extended to the head and fins in *E. fowleri*. The gut of *S. macrocephala* and *E. fowleri* was found to contain a carotenoid pigment with many small oil droplets. Tokioka (1939) supposed that the carotenoids of the prey were transferred to the predator body without any chemical change. However, Terazaki *et al.* (1977) studied the pigments chromatographically and found that the characteristics of the carotenoids in chaetognaths were different from those of dietary pigments (Table 2). There was no difference between the pigmentation of the intestine of mature and immature specimens; all pigments were carotenoids. Like all carotenes, the major carotenoid was very soluble in polar solvents, although its absorption spectrum formed a single broad peak around 460 nm. Thus, it is inferred that carotenoid in chaetognaths are synthesized by chaetognaths themselves, rather than being dietary pigments deposited in the intestinal tissue (Terazaki, 1991).

Chaetognaths prey on chaetognaths of their own, and other species. By number, chaetognath prey usually forms a small part of the diet. However, by weight the contribution can be significant, particularly for larger-size classes, because they feed on larger prey when available (Mironov, 1960; Feigenbaum, 1979).

Some species are evidently more cannibalistic than others. Reeve and Walter (1972) suggested that cannibalism is related to copulation among mature *S. hispida*. Cannibalism occurs in immature chaetognaths as well. Pearre (1982) pointed out that cannibalism had potential energetic benefits and might be necessary for the existence of large species. If so, the benefits are complex because cannibalism can account for a sizeable part of population mortality in some ecosystems (Szypel; 1978; Øresland, 1987).

Pearre (1982) analyzed quantitative reports of chaetognath cannibalism in the literature. He found chaetognath head width to be the best predictor of the degree of cannibalism in a given species and arrived at the following best-fit multiple regression for cannibalism as a function of chaetognath size and abundance:

$$C_S = 1.531e^{2.431}P \cdot U^{0.455},$$

where, C = arcsine transformation of proportion of chaetognaths among the diet items, P = chaetognath head width, and U = number of chaetognaths/m^3.

FOOD SELECTION

Chaetognath show selectivity in their feeding behaviour and, as a result, influence the distribution of prey populations. Fulton (1984) illustrated this potential with a series of experimental manipulations. In the experiments, predation by *S. hispida* resulted in a significant decline in the abundance of the larger, active copepod *Acartia tonsa*, and an increase in the abundance of the small, less active copepod *Oithona colcarva*.

Although chaetognaths depend on prey movement to detect food, they respond to a broad range of low frequency vibrations (Bone and Goto, 1991) and are likely to select specific prey species on this basis (Feigenbaum, 1977; Feigenbaum and Reeve, 1977). Instead, selection may be based on size, or may result from differential movement patterns or the escability of prey. Large chaetognaths eat larger prey than small individuals of the same species (Reeve, 1966; Rakusa-Suszczewski, 1969; Reeve and Walter, 1972; Szyper, 1976; Feigenbaum, 1982; Terazaki and Marumo, 1982; Nagasawa and Marumo, 1984). This will often result in a shift in prey species as the chaetognath grows. However, there is a difference in size preference among chaetognath species (Pearre, 1980).

Electivity indices obtained from natural gut contents must be interpreted cautiously because many chaetognaths undergo vertical migrations and may not have fed in the layer in which they were caught. Pearre (1973) used the electivities of *S. elegans* to show that the chaetognath fed on the surface and then transported the food downwards. Even where they have been applied to non-migrating stages or species, electivity indices are variable, reflecting the difficulty of estimating the actual availability of prey with a tow net on a scale important to the predator (Sullivan, 1980).

DIGESTION TIME

Digestion time, defined here as the time from ingestion of prey until its defecation, is important in chaetognath biology. It allows feeding rates to be estimated from static values obtained from the gut content analysis of preserved samples. Digestion time can be estimated in several ways: (1) direct observation - getting the chaetognath to feed in the laboratory and timing the process; (2) partial observation - collecting chaetognaths with food in their guts and using the longest time to defecation or an average of the five longest times as an estimate; (3) the same as (2), but using twice the mean of all times so obtained; and (4) Szyper's (1978) methods. In this last method, groups of chaetognaths caught in good condition are preserved at different intervals after capture. The proportion of those with prey, p, is then regressed on the "time to preservation". The $p = 0$ intercept of the regression line provides the estimate. Either the chaetognaths have to be isolated or all potential prey removed while waiting for preservation.

Of the above, method (1) is the most straightforward and has the advantage of only using animals which are healthy enough to feed in the laboratory. The difficulties faced with (1) lie in getting the chaetognaths to feed and in providing a prey-species mix similar to a natural diet. Chaetognaths that feed in the laboratory must be isolated to prevent them from further feeding. As a practical matter, the chaetognaths are generally starved for a day to induce them to feed on demand. Methods (2), (3) and (4) avoid the necessity of achieving laboratory feeding, but require the chaetognaths to be healthy enough for digestion to proceed at its normal pace.

Table 3 lists published digestion times. They range from 40 to 614 min. Being a metabolic process, digestion time varies with temperature (Mironov, 1960; Pearre, 1981) and there is some evidence that it also varies with prey size (Nagasawa and Marumo, 1984; Øresland, 1987). Multiple prey tend to increase digestion time and make it more variable (Reeve, 1980; Canino and Grant, 1985). Szyper (1978) and Sullivan (1980) found digestion time to be unrelated to size. Szyper hypothesized that this was because large chaetognaths digest prey faster than small ones, but eat larger prey that take longer to digest. However, Reeve (1980) found that digestion time increased with chaetognath size and Feigenbaum (1982) was unable to find a significant relationship between digestion time and the ratio of chaetognath weight to prey weight. Individual digestion time tends to be quite variable, which may be responsible for the lack of a clear relationship with other variables. Digestion time may also vary with different populations of

Table 3. Chaetognath digestion times

Species	Temp. (°C)	Food organisms	DT (min)	Reference
Sagitta crassa	-	*Tigriopsis japonicus*	360	Takano (1971)
	-	*Oithona aruensis*	60	Nagasawa and Marumo (1984)
	-	*Acartia clausi*	120-180	Nagasawa and Marumo (1984)
	22 - 24	*A. clausi*	102	Nagasawa (1985)
	22 - 24	*T. japonicus*	168	Nagasawa (1985)
S. elegans	15	Copepods	147	Kuhlmann (1976)
	-	-	210	Reeve (1980)
	0	Copepods	614	Feigenbaum (1982)
	6	*Calanus*	560	Øresland (1987)
	6	Small copepods	294	Øresland (1987)
S. enflata	40	Chaetognath	40	Grey (1931)
	24 - 26	Natural zooplankton	60	Szyper (1978)
	23	Copepods	190	Feigenbaum (1979)
	-	Copepods	100	Terazaki (1996)
S. nagae	-	Copepods	120-300	Terazaki (unpublished data)
Pterosagitta draco	-	Copepods	165	Terazaki (unpublished data)

the same species. For example, Szyper (1978) estimated digestion time for the *S. enfata* population of Kaneohe Bay, Hawaii to be 60 min, while Feigenbaum (1979) arrived at 190 min for the Gulf Stream population of this species. Temperatures were similar in both studies.

Mironov (1960) estimated the digestion time of 2 hr at 11.5°C and 1.5 hr at 15.5°C for *S. setosa*. Through extrapolation, he arrived at an estimate of 1 hr for 20°C. Data are also available for *S. elegans* at different temperatures. Kuhlmann (1977) estimated a time of 147 min at 15°C, while Feigenbaum (1982) found it to be 614 min at 0°C. Based on these, Pearre (1981) developed the following equation for the relationship of digestion time and temperature for this species:

$$\mathrm{DT} = 10.24\ \mathrm{e}^{-0.095T},$$

where DT = digestion time in hours, and T = temperature in °C.

FEEDING ACTIVITY

Natural feeding rates can be estimated from an analysis of gut contents if an estimate of the digestion time is available. The appropriate equations are:

$$\mathrm{FR}_n = \mathrm{NPC}(24)/\mathrm{DT}, \quad \mathrm{FR}_w = (\mathrm{FR}_n)(\mathrm{MPW}),$$

where DT = digestion time in hours; $\mathrm{FR_n}$ = daily feeding rate in number of prey per day, $\mathrm{FR_w}$ = daily feeding rate by weight; MPW = mean prey weight; NPC = number of prey per chaetognath, from gut content analysis.

It is easy to count and measure the prey in a chaetognath's gut because the animal swallows its prey whole and its body is fairly transparent. Quantitative analysis of gut contents makes the following assumptions:

1) Prey species are equally identifiable in the gut.
2) All prey are digested at the same rate.
3) Cod-end feeding is either negligible or can be accounted for in subsequent analysis.
4) The chaetognath neither regurgitates nor defecates prey during the handling and preservation processes.

Assumption 1) is probably false (Pearre, 1974); however, this will have little effect on the FR_n equation as long as the prey can be identified as such, and is not confused with parasites or organisms that may have been inadvertently ingested (e.g. diatoms). Data is also accumulating to prove assumption 2) false (Pearre, 1974; Øresland, 1987), although the assumption can be accounted for in the same way that "mean prey weight" accounts for differences in prey sizes. Cod-end feeding (assumption 3)) can be handled in a number of ways - principally by eliminating from the analysis of food that is forward in the gut or by making adjustments in the collection methods; a detailed discussion is provided in Feigenbaum and Maris (1984). Assumption 4) appears to be valid under limited conditions. Sullivan (1980) found no loss during preservation in her study.

Feigenbaum (1977) used cold formalin to minimize preservation losses in his study.

The food containing ratio (FCR) is the percentage of chaetognaths with food in their gut. When multiple prey are rare, FCR is sometimes used to estimate NPC. Some studies have used NPC or FCR to compare feeding rates.

Chaetognaths display a diurnal feeding rhythm, with a sizeable increase during night. This has been demonstrated in the laboratory (Parry, 1944; Reeve, 1964; Nagasawa, 1985) and supported in field studies by high FCR and NPC (Rakusa-Suszczewski, 1969; Nagasawa and Marumo, 1972; Pearre, 1973; Newbury, 1978; Szyper, 1978; Feigenbaum, 1979, 1982; Bushing and Feigenbaum, 1984; Kimmerer, 1984). However, due to variability in the data, the differences between day and night levels are often statistically significant.

The feeding of *Sagitta nagae*, the most predominant chaetognath species in Suruga Bay, was studied by Nagasawa and Marumo (1972). The FCR of *S. nagae* was higher at night and lower in the daytime, and almost constant vertically in the upper 100 m layer. The FCR was generally neither influenced by food density nor by the copepod numbers per *Sagitta*. *S. nagae* took food throughout life and food ingested by *S. nagae* per day was calculated as 37.6% of *S. nagae* in dry weight.

Sagitta crassa occurs abundantly, particularly in summer, in the inner part of Tokyo Bay, and the FCR of this chaetognath reached a peak of over 60% at 2100 and 2200 hr. FR_n and FR_w are 7.1 and 5.2, respectively. *S. crassa* are calculated 14.9 prey/l/day (Nagasawa and Marumo, 1984).

The FCR of juvenile, stages 1, 2, 3 and 4 individuals of meso- and bathypelagic species, *Sagitta zetesios* collected from Sagami Bay, were 15.5%, 20.7%, 13.2%, 1.5% and 3.2%, respectively. It is clear that the young *S. zetesios* had a high FCR compared with maturing and adult animals. *S. zetesios* contained food organisms in the gut more frequently during daytime than during the night, and 16.6–22.3% of *S. zetesios* collected from the upper 500 m layer had the prey in their gut, but the FCR was under 8.7% in the deep water below 500 m (Terazaki and Marumo, 1982).

The most abundant species at 5°S, 160°W, was *Sagitta enflata* which comprised 32.4–61.1% of the individuals collected from the 0–500 m layer (Terazaki, 1996). The FCR of *S. enflata* collected in the 0–50 m layer ranged from 4.8 to 12.5% (mean = 10.8%). The variation in FCR in the 50–150 m layer ranged from 6.2 to 23.1%, according to the following degrees of digestion: I) those containing food organisms in the anterior part of the gut near the mouth; II) food organisms in the mid-gut, not completely digested and shapes still clearly recognized; III) food organisms in the posterior part of the gut or near the anus. Activity of *S. enflata* was high after sunrise and throughout the day in the 0–50 m layer, but was almost constant after sunset in the 50–150 m layer because prey items were found in the digestion conditions of I and II in >50% of the animal (Fig. 3). Copepods were abundant in the upper 50 m depth both during the day and at night, and their abundance decreased with depth. Feeding of *S. enflata* was very active when a high standing stock of copepods (400 ind.m^{-3}) was observed in the 0–150 m layer in the morning. The NPC, MPW (Copepoda), FR_n and FR_w were

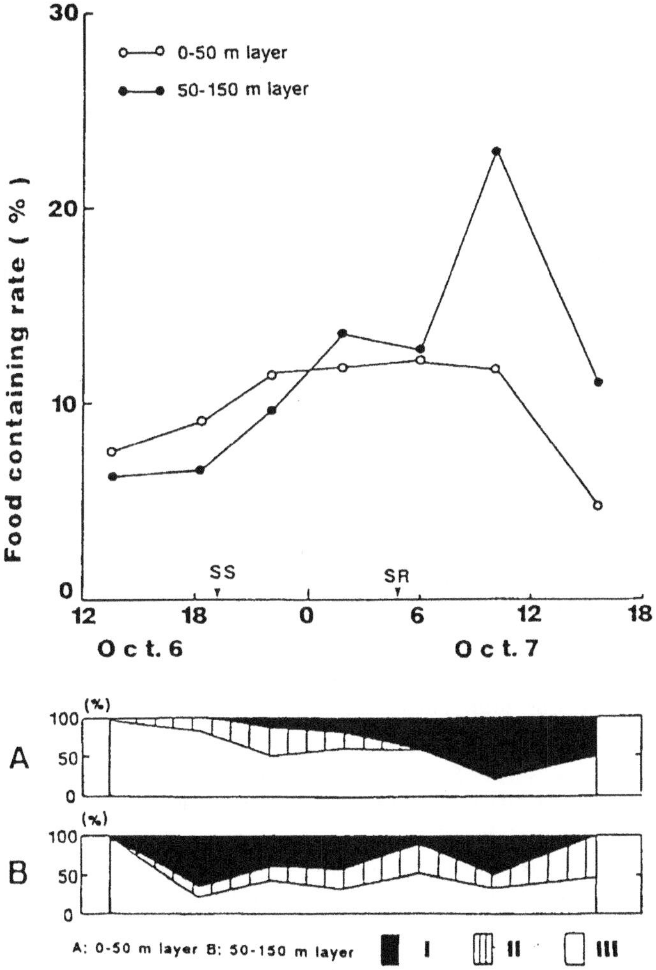

Fig. 3. Daily variation in the FCR (%) of *Sagitta enflata* in the 0–50 m and 50–150 m layers (top). Percentage of individuals in each digestion condition (I, II and III) in both layers (bottom). SS: sunset, SR: sunrise (after Terazaki, 1995).

0.126 prey ind.$^{-1}$, 10 μg, 1.81 prey ind.$^{-1}$day^{-1} and 18.1 μg ind.$^{-1}$day^{-1}, respectively.

The numbers of prey in the gut of *S. enfata* in the Western Mediterranean (Pearre, 1976), Florida Current (Feigenbaum, 1979), Kaneohe Bay (Szyper, 1978) and Virginia continental shelf (Bushing and Feigenbaum, 1984) were 0.77, 0.295, 0.288 and 0.162, respectively. Thus, the NPC (0.126) reported from the Central Equatorial Pacific, was very low compared with results from elsewhere. The mean body length of *S. enflata* collected from 5°S, 160°W was 8.2 mm and this animal consumes 18.1% of its weight per day on a dry weight basis. Beers (1966) reported that the carbon content (% dry body weight) of copepods collected from the Sargasso Sea off Bermuda ranged from 35.2 to 47.6% (mean = 41.6%). Similar values (38.2–48.0%: mean = 42.8%) were reported by Ikeda (1974), who studied the carbon content of copepods from tropical waters. Using the mean value (42%), the carbon equivalent feeding rate of *S. enflata* was ~7.6 μg C $ind.^{-1}day^{-1}$ This translates to a rate of 8.06 mg C $m^{-2}day^{-1}$ in the 0–300 m water column at a mean density of 1060 individuals m^{-2}. The mean primary productivity level in the Central Equatorial Pacific has been reported as 510 mg C $m^{-2}day^{-1}$ (Betzer *et al.*, 1984). According to Ryther (1969), the ecological efficiency of food conversion from phytoplankton (primary production) to herbivores (secondary production) in the upwelling areas is ~20%. Therefore, carnivorous plankton, *S. enflata* at 5°S, 160°W, consumed ~7.9% of the daily secondary production in the Central Equatorial Pacific.

The FCR of *Sagitta elegans* from the surface layer in the Japan Sea, was very high (44.7 and 73.6%) compared with that in other strata. FCR decreased with depth and was 0% in the layer below 1000 m. There were no marked differences in the FCR between day and night and the nighttime FCR in the layer between 200 and 500 m at 42°N, 136°E, was 14.0%. Bodies of *S. elegans* collected from the meso- and bathypelagic layer below 200 m contained large numbers of yellow oil droplets (Terazaki, 1993).

The FCR of *S. elegans* collected from the 0–300 m layer in the sub-Arctic Pacific (45°N, 160°10′ E) ranged from 0% to 13.5% (mean = 9.6%) and the feeding activity was high during the nighttime hours. The NPC was 0.344 (Terazaki, 1995). High feeding activity during night was recognized at Otsuchi Bay (Terazaki *et al.*, 1985) and the NPC was very high (0.94). The NPC of *S. elegans* in the Northeast Pacific (Sullivan, 1977) and Vineyard Sound, Massachusetts (Feigenbaum, 1982) was 0.458–0.550 and 0.400–0.564, respectively. The DT of *S. elegans* collected from 45°N, 165°10′ E, was 6.7 hr as the mean water temperature in the 0–300 m layer was 4.5°C. The FR_n was 0.344 and the FR_w was 0.069 mg $ind.^{-1}day^{-1}$ The population density of *S. elegans* in the 0–300 m layer was *ca.* 150 ind. m^{-2}. They consumed 6.3 mg C m^{-2} day^{-1} from the 0–300 m layer of the water column (Terazaki, 1995). The mean primary productivity level of this area during spring was 1700 mg C m^{-2} day^{-1} (Taniguchi, 1972).

Assuming that the conversion efficiency from primary production to secondary production at 45°N, 165°10′ E is 10% (Ryther, 1969), *S. elegans* would consume 4% of the total secondary production daily. The percentage of secondary production consumed daily by *S. elegans* was more than 100% in St. Margaret's Bay, Nova Scotia during the winter and early spring (Sameoto, 1972), 36% in Bedford Basin, Nova Scotia (Sameoto, 1973), and 10% in the Bering Sea during the summer (Kotori, 1976).

In virtually every case, these estimates rely on significant assumptions or broad generalizations that probably have caused them to be low in some instances and certainly too high in others. It is interesting that hardly any really good quantitative estimates of the impact of chaetognath feeding exist. Thus it is certainly an important area for further research.

REFERENCES

Alvariño, A. (1962): Two new Pacific chaetognaths, their distribution and relationship to allied species. *Bull. Scripps Inst. Oceanogr.*, **8**, 1–50.

Alvariño, A. (1965): Chaetognaths. *Oceanogr. Mar. Biol. Ann. Rev.*, **3**, 115–194.

Alvariño, A. (1974): Chaetognatha. p. 18–19. In *Encyclopaedia Britannica*, 15th edn., Encyclopaedia Britannica Inc.

Alvariño, A. (1985): Predation in the plankton realm; mainly with reference to fish larvae. *Inv. Mar. CICIMAR*, **2**, 1–122.

Beers, J. R. (1966): Studies on the chemical composition of the major zooplankton groups in the Sargasso Sea off Bermuda. *Limnol. Oceanogr.*, **11**, 520–528.

Betzer, P. R., W. J. Showers, E. A. Laws, C. D. Winn, G. R. Ditullio and P. M. Kroopnick (1984): Primary productivity and particle fluxes on a transect of the equator at 153°W in the Pacific Ocean. *Deep-Sea Res.*, **31**, 1–11.

Bieri, R. (1966): A pale blue chaetognath from Tanabe Bay. *Publ. Seto Mar. Biol. Lab.*, **14**, 21–22.

Bieri, R. (1977): A third blue chaetognath and notes on the distribution of hyponeuston observed *in situ*. *Publ. Seto Mar. Biol. Lab.*, **24**, 27–28.

Bieri, R. (1991): 11. Systematics of the Chaetognatha. pp. 122–136. In *The Biology of Chaetognaths*, ed. by Q. Bone, H. Kapp and A. C. Pierrot-Bults, Oxford Univ. Press, Oxford.

Bieri, R., D. Bonilla and F. Arcos (1983): Function of the teeth and vestibular organ in the Chaetognatha as indicated by scanning electron microscope and other observations. Proc. *Biol. Soc. Wash.*, **96**, 110–114.

Bigelow, H. B. (1926): Plankton of the offshore waters of the Gulf of Maine. *Bull. U.S. Bur. Fish.*, **40**, 1–509.

Bone, Q. and T. Goto (1991): 3. The nervous system. pp. 18–31. In *The Biology of Chaetognaths*, ed. by Q. Bone, H. Kapp and A C. Pierrot-Bults, Oxford Univ. Press, Oxford.

Burfield, S. T. (1927): *Sagitta. Liverpool Mar. Biol. Comm. Mem.*, **28**, 1–104.

Bushing, M. and D. Feigenbaum (1984): Feeding by an expatriate population of *Sagitta enflata*. *Bull. Mar. Sci.*, **34**, 240–243.

Busk G. (1856): An account of the structure and relations of *Sagitta bipunctata*. *Quart. J. Microsc. Soc.*, **4**, 14–27.

Canino, M. F. and G. C. Grant (1985): The feeding and diet of *Sagitta tenuis* (Chaetognatha) in the lower Chesapeake Bay. *J. Plankton Res.*, **7**, 175–188.

Casanova, J. P. (1986): *Archeterokrohnia rubra* n. gen. n. sp. nouveau Chaetognathe abyssai de l'Atlantique nord-africain: description et position systematique, hypothese phylogenetique. *Bull. Mus. Natn. Hist. Nat., Paris 4th Ser.*, **8**, 185–194.

Conover, R. J. (1978): Transformation of organic matter. p. 221–499. In *Marine Ecology 4*, ed. by O. Kinne, Wiley, New York.

Cosper, T. C. and M. R. Reeve (1970): Structural details of the mouthparts of a chaetognath, as revealed by scanning electron microscopy. *Bull. Mar. Sci.*, **20**, 441–445.

Cosper, T. C. and M. R. Reeve (1975): Digestive efficiency of the chaetognath *Sagitta hispida* Conant. *J. Exp. Mar. Biol. Ecol.*, **17**, 33–38.

David, P. M. (1955): The distribution of *Sagitta gazellae* Ritter-Zahony. *Discovery Repts.*, **27**, 235–278.

Feigenbaum, D. L. (1979): Daily ration and specific daily ration of the chaetognath *Sagitta enflata*. *Mar. Biol.*, **54**, 75–82.

Feigenbaum, D. L. (1982): Feeding by the chaetognath, *Sagitta elegans*, at low temperatures in Vineyard Sound, Massachusetts. *Limnol. Oceanogr.*, **27**, 699–706.

Feigenbaum, D. L. (1991): 5. Food and feeding behavior. p. 45–54. In *The Biology of Chaetognaths*, ed. by Q. Bone, H. Kapp and A. C. Pierrot-Bults, Oxford Univ. Press, Oxford.

Feigenbaum, D. L. and R. C. Maris (1984): Feeding in the Chaetognatha. *Oceanogr. Mar. Biol. Ann. Rev.*, **22**, 343–392.

Feigenbaum, D. L. and M. R. Reeve (1977): Prey detection in the Chaetognatha: response to a vibrating probe and experimental determination of attack distance in large aquaria. *Limnol.*

Oceanogr., **22**, 1052–1058.

Fukuhara, O. and T. Fukunaga (1984): Predation by *Sagitta* on larval fish in earthen pond. *Bull. Nansei Reg. Fish. Lab.*, **17**, 151–153.

Fulton, R. S. (1984): Effects of chaetognath estuarine copepod communities. *Oecologia*, **62**, 97–101.

Furnestin, M. L. (1977): Les dents de Chaetognathes au microscope electronique à balayage. Relations avec la nutrition. *Rapp. Comm. Int. Mer. Medit.*, **24**, 141–142.

Graham, M., G. C. Trout, R. J. H. Beverton, J. Corlett, A. J. Lee and R. W. Blacker (1954): Report on research from Ernest Holt into the fishing near Bear Island, 1949 and 1950. *Fish. Invest. London, Ser.*, **2**, 18, 1–87.

Grey, B. B. (1930): Chaetognatha from the Society Islands. *Proc. R. Soc. Queensl.*, **42**, 62–67.

Hagen, W. and H. Kapp (1986): Heterokrohnia longicaudata, a new species of chaetognatha from Antarctic waters. *Polar Biol.*, **5**, 181–183.

Horridge, G. A. and P. S. Boulton (1967): Prey detection by Chaetognatha via vibration sense. *Proc. R. Soc. B*, **168**, 413–419.

Hyman, L. H. (1959): The enterocoelous coelomates-phylum Chaetognatha. p. 1–71. In *The Invertebrates. Vol. 5*, Smaller Coelomate Groups, McGraw-Hill Book Co., Inc., New York.

Ikeda, T. (1974): Nutritional ecology of marine zooplankton. *Mem. Fac. Fish. Hokkaido Univ.*, **22**, 1–97.

John, C. C. (1933): Habits, structure, and development of *Spadella cephaloptera*. *Quart. J. Microsc. Sci.*, **75**, 625–696.

Johnson, M. E. and H. J. Snook (1927): *Seashore Animals of the Pacific Coast*. McMillan and Co., New York, 659 pp.

Jumao-as, D. and H. von Westernhagen (1975): Vertical distribution of epiplanktonic chaetognaths in the upper 100 m of the Hilutangan Channel, Cebu, the Philippines. *Mar. Biol.*, **29**, 201–210.

Kimmerer, W. J. (1984): Selective predation and its impact on prey of *Sagitta enflata* (Chaetognatha). *Mar. Ecol. Progr. Ser.*, **15**, 55–62.

King, K. R. (1979): The life history and vertical distribution of the chaetognath, *Sagitta elegans*, in Dabob Bay, Washington. *J. Plankton Res.*, **1**, 153–167.

Kotori, M. (1976): The biology of chaetognatha in the Bering Sea and the northern North Pacific Ocean, with emphasis on *Sagitta elegans*. *Mem. Fac. Fish. Hokkaido Univ.*, **23**, 95–183.

Kuhlmann, D. (1977): Laboratory studies on the feeding behavior of the chaetognaths *Sagitta setosa* Müller and *S. elegans* Venill, with special reference to fish eggs and larvae as food organisms. *Meeresforsch.*, **25**, 163–171.

Lebour, M. V. (1922): The food of plankton organisms. *J. Mar. Biol. Ass. U.K.*, **12**, 644–677.

Lebour, M. V. (1923): The food of plankton organisms, II. *J. Mar. Biol. Ass. U.K.*, **13**, 70–92.

Lee, J. Y. (1966): Oeufs et larves planctoniques de poisons. *Inst. Sci. Peches Maritimes Spec. Publ.*, **3**, 59–96.

Marumo, R. and M. Kitou (1966): A new species of *Heterokrohnia* (Chaetognatha) from the western North Pacific. *La Mer*, **4**, 178–183.

McLaren, I. A. (1969): Population and production ecology of zooplankton in Ogac Lake, a landlocked fjord on Baffin Island. *J. Fish. Res. Bd. Canada*, **26**, 1485–1559.

Mironov, G. N. (1960): The food of plankton predators. 2. Food of *Sagitta*. *Trudy Sevastopol'skoi Biol. Stantsii*, **13**, 78–88.

Murakami, A. (1959): Marine biological study on the planktonic chaetognaths in the Seto Inland Sea. *Bull. Naikai Reg. Fish. Res. Lab.*, **12**, 1–186.

Murakami, A. (1966): Rearing experiments of a Chaetognatha *Sagitta crassa*. *Inf. Bull. Planktol. Jap.*, **13**, 62–65.

Nagasawa, S. (1985): The digestive efficiency of the chaetognath *Sagitta crassa* Tokioka, with observations on the feeding process. *J. Exp. Mar. Biol. Ecol.*, **77**, 271–281.

Nagasawa, S. and R. Marumo (1972): Feeding of a pelagic chaetognath, *Sagitta nagae* Alvariño in Suruga Bay, Central Japan. *J. Oceanogr. Soc. Japan*, **28**, 181–186.

Nagasawa, S. and R. Marumo (1976): Further studies on the feeding habits of *Sagitta nagae* Alvariño in Suruga Bay, Central Japan. *J. Oceanogr. Soc. Japan*, **32**, 209–218.

Nagasawa, S. and R. Marumo (1979): Identification of Chaetognatha based on the morphological

characteristics of hook. *La Mer*, **17**, 178–188.
Nagasawa, S. and R. Marumo (1984): Feeding habits and copulation of the chaetognath *Sagitta crassa*. *La Mer*, **22**, 8–14.
Newbury, T. K. (1978): Consumption and growth rates of chaetognaths and copepods in subtropical oceanic waters. *Pacific Sci.*, **32**, 61–78.
Øresland, V. (1987): Feeding of chaetognaths *Sagitta elegans* and *S. setosa* at different seasons in Gullmarsfjorden, Sweden. *Mar. Ecol. Prog. Ser.*, **39**, 69–79.
Owre, H. B. (1960): Plankton of the Florida Current. Part VI. The Chaetognatha. *Bull. Mar. Sci. Gulf Carib.*, **10**, 255–322.
Parry, D. A. (1944): Structure and function of the gut in *Spadella cephaloptera* and *Sagitta setosa*. *J. Mar. Biol. Ass. U.K.*, 26, 16–36.
Pearre, S., Jr. (1973): Vertical migration and feeding in *Sagitta elegans* Verrill. *Ecology*, **54**, 300–314.
Pearre, S., Jr. (1974): Ecological studies of three West-Mediterranean chaetognaths. *Invest. Pesq.*, **38**, 325–369.
Pearre, S., Jr. (1976): A seasonal study of the diet of three sympatric chaetognaths. *Investigacion Preguera, Barcelone*, **40**, 1–16.
Pearre, S., Jr. (1980): Feeding by Chaetognatha: The relation of prey size to predator size in several species. *Mar. Ecol. Prog. Ser.*, **3**, 125–134.
Pearre, S., Jr. (1981): Feeding by Chaetognatha: Energy balance and importance of various components of the diet of *Sagitta elegans*. *Mar. Ecol. Prog. Ser.*, **5**, 45–54.
Pearre, S., Jr. (1982): Feeding by Chaetognatha: Aspects of inter- and intraspecific production. *Mar. Ecol. Prog. Ser.*, **7**, 33–45.
Rakusa-Suszczewski, S. (1969): The food and feeding habits of Chaetognatha in the seas around the British Isls. *Pol. Arch. Hydrobiol.*, **16**, 213–232.
Reeve, M. R. (1964): Feeding of zooplankton, with special reference to some experiments with *Sagitta*. *Nature*, **201**, 211–213.
Reeve, M. R. (1966): Observations on the biology of a chaetognath. pp. 613–630. In *Some Contemporary Studies in Marine Science*, ed. by H. Barnes, George Allen, London.
Reeve, M. R. (1970): Complete cycle of development of a pelagic chaetognath in culture. *Nature*, **227**, 381.
Reeve, M. R. (1980): Comparative experimental studies on the feeding of chaetognaths and ctenophores. *J. Plankton Res.*, **2**, 381–393.
Reeve, M. R. and T. C. Cosper (1975): Chaetognatha. pp. 157–184. In *Reproduction of Marine Invertebrates*, ed. by A. C. Giese and J. S. Pearse, Vol. 2, Entoprocts: lesser coelomates, Academic Press, New York.
Reeve, M. R. and M. A. Walter (1972): Conditions of culture, food-size selection, and the effects of temperature and salinity on growth rate and generation time in *Sagitta hispida* Conant. *J. Exp. Mar. Biol. Ecol.*, **9**, 191–200.
Reeve, M. R., T. C. Cosper and M. A. Walter (1975): Visual observations on the process of digestion and the production of faecal pellets in the chaetognath *Sagitta hispida* Conant. *J. Exp. Mar. Biol. Ecol.*, **17**, 39–46.
Ryther, J. H. (1969): Photosynthesis and fish production. The production of organic matter and its conversion to higher forms of life vary throughout the world ocean. *Science*, **166**, 72–76.
Sameoto, D. D. (1972): Vearly respiration rate and estimated energy budget for *Sagitta elegans*. *J. Fish. Res. Bd. Can.*, **29**, 987–996.
Scott, T. (1891): The food of *Sagitta*. *Ann. Scott. Nat. Hist.*, **1**, 142–143.
Scott, T. (1893): The food of *Sagitta*. *Ann. Scott. Nat. Hist.*, **3**, 120.
Sheader, M. and F. Evans (1975): Feeding and gut structure of *Parathemisto gaudichaudi* (Guerin) Amphipoda, Hyperiidea. *J. Mar. Biol. Assoc. U.K.*, **53**, 641–656.
Shipley, A. E. (1910): Chaetognatha. Chap. 7. pp. 186–194. In *The Cambridge Natural History, Vol.* 2, ed. by S. F. Harmer and A. E. Shipley, MacMillan and Co., Ltd., London.
Steuer, A. (1910): *Planktonkunde*. Leipzig and Berlin, 723 pp.
Suarez-Caabro, J. A. (1955): Quetognatos de los mares Cubanos. *Mem. Soc. Cubana Hist. Nat.*, **22**,

125–280.

Sullivan, B. K. (1977): Vertical distribution and feeding of two species of chaetognaths at weather Station P. Ph.D. Diss., Oregon State Univ., 147 pp.

Sullivan, B. K. (1980): *In situ* feeding behavior of *Sagitta elegans* and *Eukrohnia hamata* (Chaetognatha) in relation to the vertical distribution and abundance of prey at Ocean Station "P". *Limnol. Oceanogr.*, **25**, 317–326.

Szyper, J. P. (1978): Feeding rate of the chaetognath *Sagitta enflata* in nature. *Estuar. Coast. Mar. Sci.*, **7**, 567–575.

Takano, H. (1971): Breeding experiments of a marine littoral copepod *Tigriopus japonicus* Mori. *Bull. Tokai Reg. Fish. Res. Lab.*, **64**, 71–80.

Taniguchi, A. (1972): Geographical variation of primary production in the Western Pacific Ocean and adjacent seas with reference to the relations between various parameters of primary production. *Mem. Fac. Fish. Hokkaido Univ.*, **19**, 1–13.

Tchindonova, J. G. (1959): Feeding of some group of macrozooplankton in the North-western Pacific. *Trudy Inst. Okeanolog. ANSSSR*, **30**, 166–189.

Terazaki, M. (1991): 10. Deep-sea chaetognaths. pp. 117–121. In *The Biology of Chaetognaths*, ed. by Q. Bone, H. Kapp and A. C, Pierrot-Bults, Oxford Univ. Press, Oxford.

Terazaki, M. (1993): Deep-sea adaptation of the epipelagic chaetognath *Sagitta elegans* in the Japan Sea. *Mar. Ecol. Prog. Ser.*, **98**, 79–88.

Terazaki, M. (1995): The role of carnivorous zooplankton, particularly chaetognaths in ocean flux. pp. 319–330. In *Biogeochemical Processes and Ocean Flux in the Western Pacific*, ed. by H. Sakai and Y. Nozaki, Terra Sci. Publ. Co., Tokyo.

Terazaki, M. (1996): Vertical distribution of pelagic chaetognaths and feeding of *Sagitta enflata* in the Central Equatorial Pacific. *J. Plankton Res.*, **18**, 673–682.

Terazaki, M. and R. Marumo (1982): Feeding habits of meso- and bathypelagic chaetognatha, *Sagitta zetesios* Fowler. *Oceanol. Acta*, **5**, 461–464.

Terazaki, M., R. Marumo and Y. Fujita (1977): Pigments of meso- and bathypelagic chaetognaths. *Mar. Biol.*, **41**, 119–125.

Terazaki, M., D. Kitagawa and Y. Yamashita (1985): Vertical distribution and migration of pelagic chaetognaths in the vicinity of Otsuchi in the spring season. *Otsuchi Mar. Res. Cent. Rep.*, **11**, 1–7.

Thomson, J. M. (1947): The Chaetognatha of south-eastern Australia. *Bull. Counc. Sci. Ind. Res.*, **222**, 1–43.

Tokioka, T. (1939): Chaetognaths collected chiefly from the bays of Sagami and Suruga, with some notes on the shape and structure of the seminal vesicle. *Rec. Oceanogr. Wks. Jap.*, **10**, 123–150.

Tsuruta, A. (1963): Distribution of plankton and its characteristics in the oceanic fishing ground, with special reference to their relation to fishery. *J. Shimonoseki College of Fish.*, **12**, 13–214.

Wimpenny, R. S. (1936): The distribution, breeding, and feeding of some important plankton organisms of the southwest North Sea in 1934. *Min. Aqr. Fish. Gr. Britain, Fish. Invest.*, **15**, 1–53.

Dynamics and Characterization of Marine Organic Matter, Eds., N. Handa, E. Tanoue and T. Hama, pp. 277–298.

Dynamics of Organic Marine Aggregates: Nanometer-Colloids to Marine Snow

Seiko HARA[1] and Isao KOIKE[2]

[1]*Miyazaki International College, 1405 Kano, Kiyotake-cho, Miyazaki 889-1605, Japan*
[2]*Ocean Research Institute, The University of Tokyo, 1-15-1, Minamidai, Nakano-ku, Tokyo 164-0014, Japan*

(Review, Received 21 October 1996; accepted 13 February 1997)

HISTORICAL BACKGROUND

A distinctive characteristic of the marine ecosystem is the existence of non-living organic carbon, which is two orders of magnitude larger than the biomass-carbon. In the terrestrial ecosystem, however, the quantity of biomass-carbon is approximately the same as that of organic carbon in soil, the largest detrital pool within the terrestrial system (Cauwet, 1978; Bolin, 1983). This difference can be attributed to the types of organisms dominant in those systems. Long-lived plants with large amounts of supporting tissue are predominant in the terrestrial ecosystem, while marine organisms are primarily single-celled and short-lived (Mopper and Degens, 1979; Lalli and Parsons, 1993). Most non-living organic matter in the ocean falls within a size range of less than 0.5–1.0 μm and is traditionally referred to as dissolved organic material. With an average depth of 4000 m, the ocean is often regarded as a huge but quite diluted incubation medium for various living organisms; in terms of its volume, however, less than a few percent of its volume is suitable for net organic matter production, that is, primary production. Marine organisms and their coupled carbon cycling are well adapted to this aquatic environment. The predominant small-sized organisms obtain their nutrients from the water. Because all metabolic processes of marine organisms are mediated through water, we would expect a close interaction between marine organisms and the various organic material found in the surrounding water column.

Importance of the interaction between marine organisms and the various forms of non-living organic materials in the surrounding water has been recognized for many years. For example, extensive studies on excretion and defecation by zooplankton have been conducted (Conover, 1966; Conover and Corner, 1968; Ikeda, 1974). With regard to food webs or carbon cycling in the upper ocean, inorganic nutrient excretion by zooplankton supports in part the theory of regenerated primary production, while fecal pellet production may contribute to the export of primary production (Dugdale and Goering, 1967; Honjo and Roman, 1978). Although the formation of organic detritus of various sizes including dissolved forms associated with so-called sloppy feeding and other feeding mechanisms would occur simultaneously, these feeding processes have attracted less attention and fewer studies have been reported (Lampert, 1978). The same is also true for microzooplankton feeding on bacteria and other small organisms. Due to the difficulty of observing these feedings, few studies have characterized the detritus and exudates formed by microzooplankton during the feeding process even though mass balance studies have suggested the significance of their production (Nagata, 1997).

We believe that a significant portion of the efforts connecting the biological processes with the dynamics of organic materials in marine environments should be credited to work by marine microbiologists (Azam *et al.*, 1983). To understand the life of heterotrophic bacteria in a water column, it is crucial to know about the biochemical nature and the dynamics of various organic materials. The idea of "phycosphere" proposed by Azam and Smith (1991) suggests that the dynamics of organic detritus and nutrients connect bacterioplankton and phytoplankton, two important populations of the marine ecosystem.

Through the close collaboration among various disciplines of marine science over the past decade, our knowledge about the different forms of non-living organic materials in marine environments has expanded significantly. We now know the presence and some of the characteristics of marine organic colloids and colloidal aggregates ranging in size from submicron to nanometer (Koike *et al.*, 1990; Wells and Goldberg, 1991; Nagata and Koike, 1995). Although the presence of marine snow or large organic aggregates has been recognized for more than 40 years, information on its biological and biochemical properties has only been accumulated during the past 10 to 15 years (Alldredge and Silver, 1988; Alldredge and Jackson, 1995). More recent findings of non-living particles such as transparent exopolymer particles (TEP), protein-rich particles, and bacterial ghosts stimulate the biological and physical interaction of organic detritus in the water column (Alldredge *et al.*, 1993; Zweifel and Hagström, 1995; Long and Azam, 1996).

In this paper, we examine recent studies on marine non-living organic materials of different size classes, focusing specifically on their dynamics and interaction with physical and biological forces. Although the available information remains fragmented and it is premature to draw a concert picture concerning the dynamics of organic detritus, our studies strongly suggest that aggregation is a very common process that is involved in the formation of marine detrital particles ranging in size from nanometer to centimeter. We hypothesize, therefore, that aggregation and disaggregation mechanisms are key processes to understanding the dynamics of marine organic materials.

NOMENCLATURE OF ORGANIC MATERIALS IN THE OCEAN

Traditionally, organic material in marine systems has been organized into several groups according to size or function (Cauwet, 1978; Romankevich, 1984). Operational definitions of size classification have been determined using filtration techniques. Residue remaining on a filter of ca. 0.6–1.0 μm in pore size is classified as particulate organic matter (POM), while matter passing through the filter is classified as dissolved organic matter (DOM). A part of DOM contains colloidal form, small particles having a diameter 1–1000 nm by difinition (Koike *et al.*, 1990; Wells and Goldberg, 1992). Marine snow is POM visible to the naked eye. When considering material dynamics in a marine system, organic material is divided into two groups, sinking and non-sinking forms (Fowler and Knauer, 1986). DOM and small sized POM correspond to the non-sinking form of organic materials, while marine snow and feces of zooplankton are typical of sinking organic materials. Marine organic material may be further categorized as living or non-living; non-living particles are generally referred to as organic detritus.

Here we classify organic material of non-living nature in marine systems into four major size categories: truly dissolved material (smaller than 1 nm); colloidal organic material (1 nm to 1 μm); suspended particulate organic material (1 μm to several hundred μm); and marine snow (larger than several hundred μm) (Wells and Goldberg, 1994). Within this context, organic detritus is defined as non-living particles greater than 1 nm in size.

The definition of marine aggregates, however, is still not clear. Riley (1963) referred to flocculent particles and heterogeneous amorphous aggregates as organic aggregates and referred to clear-shaped non-living particles such as diatom shells as organic detritus. Mel'nikov (1977), however, used the term organic detritus particles to refer to suspended, whole, non-living particles. His organic aggregates are one type of morphologically identifiable detritus. In this study, we define marine organic aggregates as comprised primarily of one or more types of organic detritus that have joined together to form a mass.

COMPARISON OF MARINE SNOW, SUSPENDED ORGANIC MATTER, AND ORGANIC COLLOIDS

Morphological and Physical Characteristics

The smallest group of marine detritus was described by Wells and Goldberg (1991) as marine colloids at least 1 nm in size. Recognized for many years, marine snow is the largest marine organic aggregate, varying in origin and ranging in size from several hundred micrometers to several centimeters (Alldredge and Silver 1988). The smallest group of marine organisms includes small bacteria and the recently described marine viruses of ca. 30–100 nm, all of which distribute abundantly in the oceans (Berg *et al.*, 1989; Proctor and Fuhrman, 1990; Hara *et al.*, 1991).

The typical size of marine organic particles (aggregates) varies by a factor of 10^7, ranging from nanometers to centimeters, and the size of marine organisms varies by a factor of 10^8, ranging from hundreds of nanometers to tens of meters. As can be seen from these view points, the size range of marine organic particles overlaps that of living organisms, since the major source of marine particles are organisms themselves. Both groups overlap the two categories of dissolved and particulate forms. The rough size ranges and the abundance of both living organisms and non-living organic particles are summarized in Fig. 1.

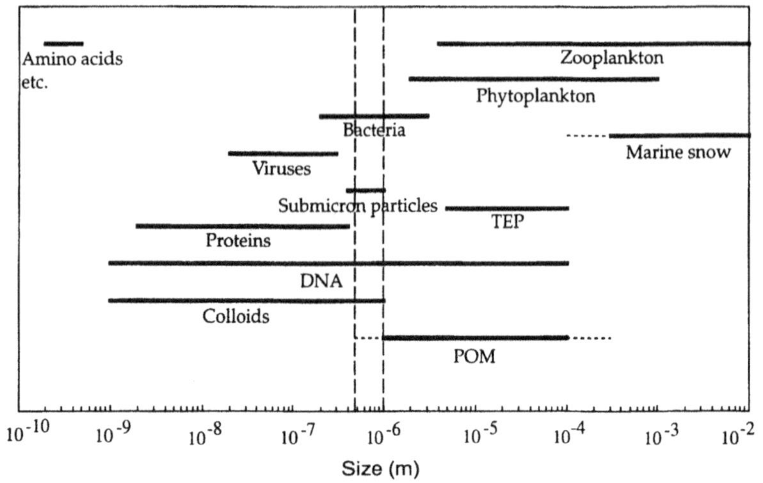

Fig. 1. Size distribution of organic materials in seawater. TEP: Transparent exopolymer particles. (Adapted from Zweifel, 1995.)

Marine snow is commonly a cluster-cluster aggregate visible to the naked eye. Simon *et al.* (1990) and Alldredge and Gotschalk (1990) classified marine snow into four groups based on its morphology or component particle composition: (1) larvacean houses centered, (2) diatom dominated flocs (at least >30%), (3) mostly unidentified aggregates, and (4) fecal aggregates (>50% occupied by fecal pellets). Although marine snow contains various phyto-and zooplankton (diatoms, protozoa and other picoplankton) as well as a bacterial population, estimates of its biomass-carbon (4–13%) seem to be a minor portion of the total organic carbon, irrespective of its composition (Caron *et al.*, 1982; Simon *et al.*, 1990). In addition to the dead bodies of phyto- and zooplankton, the major common components of marine snow include mucous and other organic aggregates of various shapes, sizes, and origins (Alldredge and Silver, 1988; Alldredge and Jackson, 1995). Because of its aggregate structure, marine snow is highly fractal and extremely fragile, well-known characteristics of this large detritus.

Suspended particles have been observed using light or electron microscopy after first concentrating the aggregates on a filter (Gordon, 1970; Romankevich, 1984). Mel'nikov (1972) and Romankevich (1984) identified three characteristic groups among suspended detrital particles (1–200 μm) in seawater: aggregates, flocs or flakes, and fragments. While aggregates appeared as heterogeneous masses comprised of separate particles bound together, flocs or flakes appeared to be fine, transparent, platy particles with irregular angular edges, possibly the result of organic colloid aggregation. Fragments consisted of the residue of plankter tissues such as the extremities of copepods (Mel'nikov, 1977; Romankevich, 1984). In surface waters, aggregates represented 60–80 % of all identified particles (Cauwet, 1981).

Loosely flocculated aggregates of various size including marine snow are easily broken up by routine collection methods and subsequent treatments (Alldredge and Silver, 1988; Lampitt *et al.*, 1993). A high porosity and the resulting high water content make detection of these aggregates particularly difficult, even after they are collected. Because of this, these loosely flocculated aggregates of various sizes (Alldredge and Silver, 1988; Wells and Goldberg, 1993) may not have been sufficiently observed and described in earlier works (Mel'nikov, 1977; Cauwet, 1981).

Using alcian blue, a stain specific for polysaccharides, Alldredge *et al.* (1993) observed an abundant presence of polysaccharide-rich, transparent, flat particles in the upper ocean. These transparent exopolymer particles (TEP) appeared as discrete sheets, films, discs, or strings, both with and without inclusions (Alldredge *et al.*, 1993; Passow and Alldredge, 1994). These plate-like particles may not be identical to flake detritus described by Mel'nikov (1977) and Romankevich (1984) because of their proteinaceous character (Cauwet, 1981). Gordon (1970) speculated that flakes, including those formed in the bubbling surface waters, originated from organic polymers of colloidal size. Disaggregation of TEP occurred by EDTA treatment (Alldredge *et al.*, 1993). They suggested that TEP were fractal aggregations of carbohydrate microfibrils and colloids exuded by phytoplankton and bacteria. TEP are easily collected on a filter using traditional filtration techniques, thereby suggesting a not so fragile character. The rather homogenous composition of TEP may explain the differences cited above. The length of TEP ranged from three to several hundred micrometers, and the average diameter was around 10–20 μm, except during flocculation of the diatom bloom when the diameter was greater than 100 μm. It was demonstrated that TEP are major agents in the aggregation of diatoms and in the formation of marine snow (Alldredge and Jackson, 1995; Dam and Drapeau, 1995). A considerable percentage of seemingly free-living bacteria (2–68%) were also attached to these TEP.

Colloids include three different particle structural types: primary colloids, ~2–5 nm; spherical colloids, <100 nm; and cluster colloids, >10 nm (Wells and Goldberg, 1993; Hara and Koike, in prep.). Primary colloids are the smallest particles observed. Relatively low molecular weight organic constituents are included in the primary colloids; for example, a spherical colloid of 2 nm in size with a density of 1.07 g ml^{-1} corresponds to an organic molecule of ~2,500 dalton (Dean 1948). Using electron microscopy, spherical colloids appear to be comprised of tightly bound primary colloids with almost no porosity (Wells and Goldberg, 1991). Secondary minima of potential energy curves are not deep enough to cause the flocculation of primary colloids less than 10 nm in size with reaction-limited colloidal aggregation (Shaw, 1992). Spherical colloid-size aggregations with fractal structures of diffusion-limited colloidal aggregations of primary colloids were rarely observed. The size and morphology of the spherical colloids indicate that they may be a monomeric type of molecule. Cluster colloids including both reaction-limited and diffusion-limited aggregates show the typical fractal structure due to coalescence of the spherical colloids (Fig. 2, Hara and

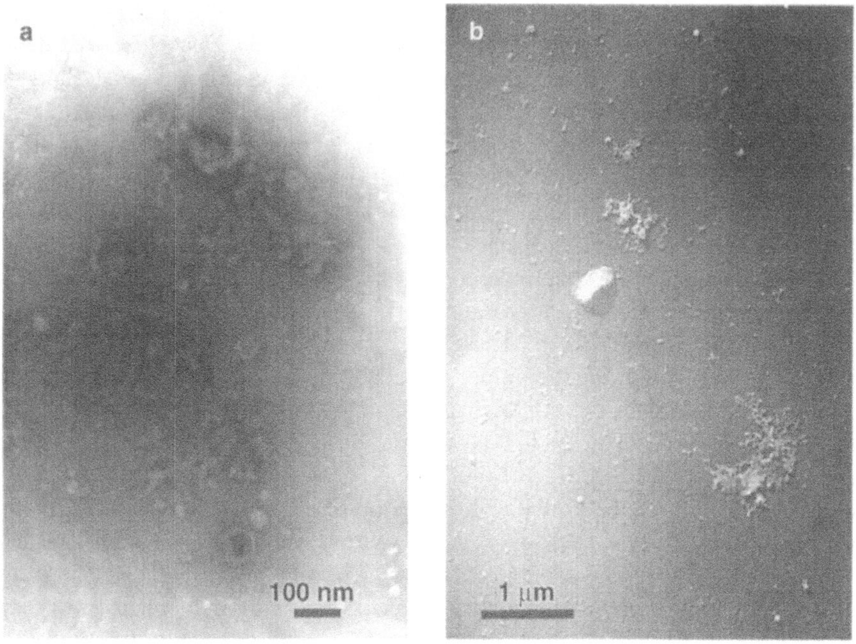

Fig. 2. TEM micrographs of submicron particles and other colloids: **a**. Negatively stained aggregates of smaller spherical colloids, virus-like particles, and a prymnesiophycean scale showing a typical fractal structure; **b**. Shadowed material of submicron sized fractal aggregates and of smaller colloid particles. A bacterium is at the center.

Koike, unpublished).

Until recently, the physical characteristics of detritus of colloidal size had not been examined, primarily because of the difficulty of separating this detritus from living organisms. Using a high resolution particle counter, Koike *et al.*, (1990) showed that the marine detritus ranging in size from 0.4–1.0 μm was much more fragile than living organisms or marine bacteria of a similar size. This finding strongly suggests the aggregated nature of marine detritus in this submicron size range. Further observation of submicron detritus using electron microscopy revealed the fractal and aggregated structure of these particles (Fig. 2). Further, the specific density of this submicron detritus is very close to that of the surrounding seawater as determined by ultra-centrifugation experiments, thereby indicating a high water content of more than 99 % or a composition of lipid-like compounds. The loosely aggregated nature of the submicron particles was further confirmed by filtration; more than two-thirds of the particles counted in the 0.4–1.0 μm range passed through a 0.2 μm Nuclepore filter (Koike *et al.*, 1990).

Zweifel and Hagström (1995) reported new types of detrital organic particles in the submicron-size range; they described the bacterial "ghosts" that were stainable by DAPI, but the DAPI fluorescence disappeared by Triton X-100 and

2-propanol washing. Zweifel and Hagström suggest that the ghosts are virus-lysed bacteria or cell wall material egested during protozoan grazing. However, because the preparation procedure for visualizing ghosts includes treatments that may damage the bacterial membrane, for example, immersing samples in Triton X-100, the possibility that ghosts are in fact bacterial cells with small amounts of nucleoid cannot be excluded. Actually, Choi *et al.* (1996) indicated that some bacteria without visible nucleoides were shown to become nucleoide visible cells with a change in nutritional condition. DNA in a nucleoid region could not be detected at least in some bacteria, using the method of Zweifel and Hagström (1995).

We may conclude that a significant portion of marine organic detritus is in the form of aggregates ranging in size from a few nanometers up to a few centimeters, and the physical characteristics of marine aggregates may be primarily defined by their structure or the condition of their aggregation. Aggregates are the most basic and stable form for most organic particles in the ocean.

Abundance and distribution

Suspended organic particulate material in the ocean represents only a small fraction (1–2%) of total organic carbon. Whereas, colloidal fraction may account for a significant fraction (8–16%) of total organic carbon (Sharp, 1973; Romankevich, 1984). Recently, high contribution of colloidal organic materials within the total organic carbon has been reported in the Gulf of Mexico (40–53%) and in the Kuroshio area of the Pacific Ocean (30–37%) (Guo *et al.*, 1994; Ogawa and Ogura, 1992). The abundance of non-living particles increases exponentially as their size decreases (Sheldon and Parsons, 1967; Sharp, 1973; Koike *et al.*, 1990; Wells and Goldberg, 1991). When considering the global carbon cycle, it is essential to determine the distribution and size range of non-living particles (Figs. 1, 3) as well as the in *situ* mechanisms for their production and removal.

Marine snow is large enough to be observed with the naked eye, but it is not easy to examine the abundance and the mass of thse particles in the sea. Underwater photography of marine snow has indicated ~1 to tens of marine snow flocks per liter in abundance (Asper, 1987; Tsuji and Sukizaki, 1991; Lampitt *et al.*, 1993). The abundance of these flocks increases as their size decreases (Asper, 1987; Tsuji and Sukizaki, 1991).

The abundance and size distribution of suspended organic matter of 1 to several hundred micrometers may be electrically enumerated using a particle counter (Sheldon *et al.*, 1972) The electric signal, however, cannot distinguish between non-living particles and living organisms, the majority of which in this size range is phyto- and microzooplankton. The microscopic observation of samples from euphotic layers indicated that less than half of the particles are commonly associated with living organisms, while suspended particles of this size range are mostly of organic origin (Sutcliffe *et al.* 1970) even though the contribution of the number of living particles among the total suspended particles in the euphotic layer is higher than that in the submicron size range (less than

10%) (Koike *et al.*, 1990). In the upper 30 m of the subarctic Pacific Ocean, the number of total particles in the size range of 1–10.2 μm was 7–8 × 10^3 ml^{-1}, while the number in the subtropical Pacific Ocean was 2 × $10^3 ml^{-1}$ (Seki *et al.*, 1972). Supplemental data of non-living suspended particles that was obtained using microscopic observation of filtered samples revealed that the abundance of TEP varied by four orders of magnitude (10^0 to 10^4 TEP ml^{-1}) in the euphotic layer (Alldredge *et al.*, 1993).

Colloids of submicron size, originally named sub-micrometer particles (Koike *et al.*, 1990), had an abundance of 10^7 ml^{-1} in surface seawater, about 10 times greater than that of bacteria (Koike *et al.*, 1990; Sieracki and Villes, 1992). Colloids of 1–380 nm can be enumerated and observed using electron microscopy after settling by ultracentrifugation (Wells and Goldberg, 1991). Spherical and cluster colloids (10–150 nm) had a distribution of ~10^9 ml^{-1} and ~10^{10} ml^{-1} in surface seawater, respectively (Wells and Goldberg, 1991; Hara and Koike, in prep.).

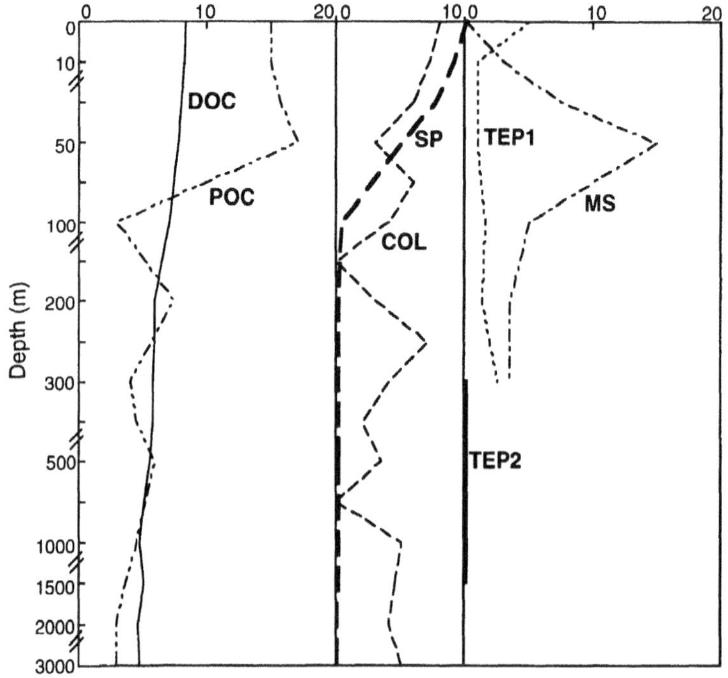

Fig. 3. Vertical distribution of organic materials in oceanic seawater. DOC (dissolved organic carbon), × 100 μg Cml^{-1}, after Carlson and Ducklow, 1995; POC (particulate organic carbon), × 10 μg Cml^{-1}, after Tanoue and Handa, 1979; COL (colloid particles), × 10^9 ml^{-1}, after Wells and Goldberg, 1994; SP (submicron particles), × 10^7 ml^{-1}, after Koike *et al.*, 1990; TEP1 (transparent exopolymer particles from coastal water), × 10^3 ml^{-1}, and TEP2 (transparent exopolymer particles from oceanic water), × 10^3 ml^{-1}, after Passow and Alldredge, 1994; MS (marine snow), × 10^{-3} ml^{-1}, after Lampitt *et al.*, 1993.

Vertical distributions of marine snow, suspended organic matter, and colloids reported in the open ocean are shown in Fig. 3. While living organisms, including bacteria and viruses, are found in high concentrations at the oceanic surface layer, each group of non-living particles shows characteristic patterns of distribution. Dissolved organic carbon decrease rapidly with depth in the upper 200 m, slowly decrease from 200 to 1000 m, and almost constant below 1000 m in the central equatorial Pacific Ocean (Carson and Ducklow, 1995). Particulate organic carbon is more concentrated in the euphotic layer and decrease gradually with depth in the middle and deep waters in the northern North Pacific Ocean (Tanoue and Handa, 1979). Concentrations of colloid aggregates smaller than 200 nm decrease with depth in the surface layer, and are high but erratic in the deep waters of the North Atlantic (Wells and Goldberg, 1994). Concentration of larger aggregates, larger than 1.0 μm in size, are also abundant but erratic throughout the water columns of the Atlantic and Pacific Oceans (Wells and Goldberg, 1993).

The numbers of submicron sized colloids are positively correlated with abundance of phytoplankton as well as suspended organic carbon in the upper ocean (Koike *et al.*, 1990). Also two to three orders of magnitude decrease in the number below the 500 m depth suggests biological origin of the submicron-sized colloids. As with phytoplankton, the distribution of TEP in the ocean is more abundant in coastal waters than in the open ocean and is concentrated in the euphotic layer (Passow and Alldredge, 1994). The non-homogenous distribution of colloid aggregates and TEP in correlation with depth indicates that aggregates are quite reactive and that colloidal dynamics vary widely in seawater.

The abundance of marine snow increases gradually with depth (in the Panama Basin, Asper, 1987; in Suruga Bay, Tsuji and Sukizaki, 1991) or shows a clear peak just below or at the base of the upper mixed layer (Lampitt *et al.*, 1993). A sharp peak of volume concentration of marine snow is usually encountered at the middle layer (Tsuji and Sukizaki, 1991) or just below or at the base of the upper mixed layer (Lampitt *et al.*, 1993).

Biochemical Characteristics

The primary source of organic aggregates in the open ocean is marine organisms belonging to various trophic levels. In coastal and estuarine waters, an input of organic detritus from terrestrial origins can occur (Sempére and Cauwet, 1995). These organic materials, however, occupy only a minor portion in the oceans except in some specific environments such as intertidal flats or the mouths of large rivers (Meyers-Schulte and Hedges, 1986). The breakdown of living organisms, including benthic macrophytes such as seaweed and seagrass, initiates the formation of organic aggregates in the ocean by means of biological and other forces (Alber and Valiela, 1994).

Because we have a good knowledge regarding the major biochemical composition of marine organisms, such as phytoplankton and zooplankton (Parsons *et al.*, 1984), the biochemical characterization of suspended marine aggregates should not be difficult if the biochemical modification of the detritus is not so extensive (Tanoue *et al.*, 1982b). Actually, up to 80 % of marine organic carbon collected on a glass fiber filter can be identified as typical biopolymers, i.e. carbohydrates, amino acids, lipids, and nucleic acids (Holm-Hansen *et al.*, 1968; Paul *et al.*, 1985; Tanoue *et al.*, 1986; Yanada and Maita, 1995). In the open ocean, some of the organic matter on the filter is living organisms, but more than half is usually organic detritus (Parsons, 1963; Romankevich, 1984).

The elemental composition of suspended particles collected on glass fiber filters has been studied extensively (Parsons and Strickland, 1962; Tanoue and Handa, 1979; Romankevich, 1984). While carbon/nitrogen ratios of suspended materials collected in the North Pacific Ocean increased from ca. 7–8 in the euphotic layer to 11–14 at a depth below the middle and deep layers, their organic contents remained almost the same (30–45%) in the whole water column (Tanoue and Handa, 1987). This trend has been observed in other regions of the Pacific, Atlantic and Indian Oceans (Romankevich, 1984). Among the total POC in the euphotic layers of the ocean, amino acid-carbon, carbohydrate-carbon, and lipid-carbon occupy 35–45, 11–19, and 13–17%, respectively. No large differences were observed in the biochemical composition when compared against the latitude (Tanoue *et al.*, 1982a).

Additional biochemical information regarding the suspended detritus in the ocean has been collected using microscopic observation after specific staining (Nemoto and Ishikawa, 1969; Gordon, 1970; Cauwet, 1981). Using a specific staining method, Alldredge *et al.* (1993) revealed TEP in the ocean to be polysaccharide-rich detritus. Protein-rich particles were also found in the upper oceans (Long and Azam, 1996). The microscopic identification of these suspended particles with different biochemical characteristics needs to be reviewed in conjunction with other chemical methods to evaluate the percentage of those particles among the total number of suspended particles.

Since there is a significant contribution of colloidal organic carbon within the total organic carbon in the ocean (30–50%), the importance of the biochemical characteristics of colloidal organic-carbon has been recognized (Ogawa and Ogura, 1992; Guo *et al.*, 1994). In the North Pacific Ocean, the bulk of the chemical composition of the total colloidal fraction (1 nm–0.2 μm) in surface water was compared with the composition of that at a depth of 2500 m (Benner *et al.*, 1992). The C/N ratio of colloids in the surface water was relatively low (15.3) when compared to that in deeper waters (19.6–22.5) although the relative concentration of polysaccharides in the surface water (ca. 50%) was significantly higher than that in the deeper water (25%). A minor portion of organic materials passing through the filter filtration can be identified using routine biochemical methods (15%: Williams and Druffel, 1988), suggesting that biological and other modifications occur after the detritus formation (Lee and Hendricks, 1993).

The chemical characterstics of marine snow apparently differ from those of suspended particles. The carbon and nitrogen ratios of marine snow in coastal waters range from 6:1 to 11:1, a range close that of the living organisms of their possible origin (phytoplankton and zooplankton) (Alldredge and Silver, 1988). Simultaneous determinations of the carbon content and the C/N ratio in marine snow (8% and C/N = 11–12, respectively) and suspended particles in the surrounding waters (16% and C/N = 7.7, respectively) obtained from the Gulf of California, demonstrated significant differences between the two (Alldredge, 1979). Marine snow samples obtained in the Santa Barbara channel, with its more eutrophic waters than the Gulf of California, revealed similar carbon contents when compared with that of the suspended particles. The highly variable biochemical nature of marine snow has also been confirmed by later studies (Alldredge and Silver, 1988; Wakeham and Canuel, 1988).

INTERACTION BETWEEN MARINE SNOW, SUSPENDED ORGANIC MATTER, AND ORGANIC CLLOIDS

Alldredge and Silver (1988) identified two major pathways of marine snow formation: (1) zooplankton-discarded materials such as larvacean houses and fecal pellets and (2) aggregates of smaller particles in the water column, especially in the size range of tens to hundreds of microns.

Formation of the primary organic detritus through biological processes

Although the size range of initial organic detritus originating from various sources of living organisms would cover the entire spectrum of organic matter in aquatic systems, the relative contribution for each size fraction varies depending on the source organism as well as the processes of formation. Bacteria, phytoplankton, and zooplankton produce various forms of exudates that range from simple metabolites such as gluconic acids to extracellular polymeric organic materials (Fogg, 1983; Decho, 1990; Decho and Lopez, 1993). Among the extracellular products produced by unicellular planktonic organisms, polysaccharides and mucopolysaccharides play a key role as binding agents in marine aggregation processes because of their sticky nature (Alldredge *et al.*, 1995; Mopper *et al.*, 1995). The production of these polymers by algae may be induced by nutrient stress, which is most active at the end of the blooms (Fogg, 1983). Some diatom species produce mucus sheaths around the cells. After the death of the cells, the mucus becomes an important component of marine snow through the aggregation process. In coastal and estuary areas, the dissolved material released as leachate from decomposing macrophyte is also a potential source of organic aggregates (Alber and Valiela, 1994). Appendicularians produce and discard numerous gelatinous houses every day (Taguchi, 1982). They use the houses as a sticky "net" to collect their food and, when discarded, these houses also become an identifiable source of marine snow (Alldredge and Silver, 1988).

The release of low molecular weight organic compounds such as dissolved free amino acids and organic acids are common in various marine organisms including bacteria and zooplankton (Webb and Johannes, 1967, 1969; Mague *et al.*, 1980). Because these chemically identifiable low molecular weight organic compounds are usually metabolized by the bacterial population rather quickly with a residence time in seawater ranging from minutes to hours, it would be difficult for them to react with other organic detritus in the water column (Fuhrman and Ferguson, 1986).

Cell-lysis resulting from various processes, including viral infection, would release a wide range of organic material including intercellular metabolites like amino acids and sugars, which are suitable substrates for heterotrophic bacteria. With the presence of large numbers of viral particles in the ocean and a significant portion of marine microorganisms, including cyanobacteria, being infected by viruses, formation of small size detritus resulting from the viral infection would be potentially important (Bergh *et al.*, 1989; Proctor and Fuhrman, 1990; Hara *et al.*, 1991; Heldal and Bratbak, 1991). Koike *et al.* (1993) have reported on the formation of submicron-sized colloids resulting from a viral burst of marine bacteria in laboratory cultures.

Food web interactions also produce primary organic detritus of varying sizes. For many years, fecal pellets and feces production through the feeding activities of zooplankton has been recognized as one of the major pathways of detritus formation in the ocean. The recognition of this pathway is important for the marine biogeochemist, primarily because of its role as a vertical transporter of organic material produced in the upper layers to the deep sea (Wiebe *et al.*, 1976; Honjo and Roman, 1978; Dunbar and Berger, 1981). In particular, fecal pellets produced by crustacean zooplankton were rigid and have often been considered as the most dominant form of sinking organic material (Gauld, 1957; Honjo and Roman, 1978; Sasaki and Nishizawa, 1981). With the accumulation of sediment trap data, however, this view has been modified somewhat because the contribution of pellets to the total settling material could be low (Pilskaln and Honjo, 1987). The rapid biological modification of fecal pellets produced by biological processes such as "coprorhexy" was proposed by Lampitt *et al.* (1990). Regardless of its mechanisms, the breakdown of intact fecal pellets and the formation of amorphous aggregates occur quickly in certain environments.

Some zooplankton and also microzooplankton such as heterotrophic flagellates produce amorphous feces (Stoecher, 1984; Tranvik, 1994). Koike *et al.* (1990) observed the production of microaggregates of submicron size through the feeding of bacteria by nanoflagellates. The aggregated nature of those "pellets" was confirmed using electron microscopy (Hara and Koike, unpublished). The "picopellet" production and its importance in material cycling are discussed in more detail in other reviews of this volume (Nagata, this volume).

Aggregation processes through physico-chemical and biological processes

In this section, we will focus on the process of aggregation or coagulation, beginning with the primary organic detritus of various sizes and its association with biological processes. The aggregation of marine detrital particles requires the two subsequent reactions, i.e., collision and sticking (Shanks and Edmondson, 1989; Kepkay, 1994). Several hydrodynamic factors including Brownian motion, fluid shear and the number of particles affect collision rates. Coagulation efficiency, however, is controlled by other physical and biological factors (Jackson, 1980). The size, structure, and surface condition of the particles including surface hydrophobicity are highly variable depending on the type of the detrital particles. Kepkay (1994) recently reviewed the colloid aggregation processes in marine environments.

Although the importance of marine colloids in coagulation process has been speculated for many years (Cauwet, 1978), interest in the study of colloids in the field of marine biogeochemistry increased recently with their direct enumeration and observation using a particle counter and electron microscopy, respectively (Koike *et al.*, 1990; Wells and Goldberg, 1991). Both abundance, 10^6–10^7 ml^{-1} for submicron-sized and 10^9–10^{10} ml^{-1} for nanometer-sized colloids, and a significant carbon content (10–50% of total organic carbon) in seawater suggest the importance of marine colloids within carbon cycling (Cauwet, 1978; Benner *et al.*, 1992; Ogawa and Ogura, 1992; Wells and Goldberg, 1992; Kepkay, 1994). Their substantial carbohydrate content and a rapid turnover rate in the upper ocean support this view (Benner *et al.*, 1992; Moran and Buesseler, 1992; Amon and Benner, 1994).

Colloids in marine waters are further subdivided into two major categories. The primary colloids in the size range from several nm to 100 nm, and the secondary colloid or colloid aggregate is larger than 10 nm (Wells and Goldberg, 1993; Hara and Koike, in preparation). Hydrophilic colloids are sufficiently stable in seawater. Hydrophobic colloids, however, are never stable in the thermodynamic sense and spontaneously form larger aggregates in seawater. Electron microscope observations have shown that there are many colloidal aggregates in seawater suggesting that "unit" colloidal particles are hydrophobic (Wells and Goldberg, 1993; Hara and Koike, in preparation). These unit colloids may include colloidal-sized molecules including proteins, polysaccharides, and nucleic acids or amphipathic compounds such as phospholipids. Relatively small hydrophilic molecules such as mono-and oligo-saccharides and dissolved free amino acids, however, do not form colloids in any great concentration. Not all materials released to the seawater form colloids.

Colloid aggregation is affected by the size of the colloid particles. For particles smaller than 10 nm, the secondary minimum of the potential energy that gives rise to flocculation is never deep enough. These particles will coagulate rapidly into the diffusion-limited colloid aggregation (DLCA), resulting in a condensed structure (Shaw, 1992). With larger particles, however, the energy barrier may cause observable effects, and the particles flocculate into a loose, easily reversible, reaction-limited colloid aggregation (RLCA). The materials composing the aggregates are another factor determing the mechanisms of aggregation (Logan and Wilkinson, 1990; Wells and Goldberg, 1993). Wells and Goldberg (1993) reported the presence of those two types of suspended aggregates at the different depths of the Sargasso Sea. The universal occurrence of RLCA-type aggregates suggests the involvement of surface-active biopolymers for this process. Logan and Wilkinson (1990) also discussed the difference of fractal structure between marine snow and inorganic colloidal aggragates.

Using microscopy, amorphous aggregates and flake-like aggregates can be identified in suspended organic particles (one to several hundred micrometers in size). Aggregates account for a major portion of the organic detritus in this size range (Mel'nikov, 1977; Cauwet, 1981; Romankevich, 1984). Marine snow is an aggregate comprised of detritus of different sizes and also of living particles that may possibly contain colloidal-sized particles (Alldredge and Silver, 1988). Larson and Shanks (1996) reported efficient incorporation of 1 μm beads into laboratory-made, diatom-based marine snow.

Although the three stages of organic detritus (colloid aggregates, suspended organic matter, and marine snow) have very large size ranges (from nanometer to centimeter), most show exactly the same fractal structure and are formed by the aggregation of smaller particles. It may be concluded that successive aggregation may be the major force in producing larger particles, ranging from colloid aggregates to marine snow.

Fate of various sized aggregates

Aggregate formation from colloidal size up to marine snow is particularly important for the removal of organic carbon from the upper ocean because colloidal detritus never settles by itself. This aggregation process of organic detritus is referred to as "aggregation-pump" and emphasizes the role of large-sized aggregates as a major vehicle for vertical carbon transport (Fowler and Knauer, 1986). As mentioned earlier, the fuel of the "aggregation-pump" is the primary detritus of varying size and morphology that originated directly from biological reactions (Fig. 4). Some marine snow consists mostly of small-sized amorphous detritus, while abandoned larvacean houses (1–20 mm) constitute one of the largest detritus components (Alldredge and Madin, 1982). Each size class of organic aggregates associated with this "aggregation-pump" is continuously fed by various-sized organic detritus. Furthermore, the pump begins to incorporate these aggregates as a part of the larger marine aggregates, thereby increasing their mass.

The settling of aggregates is primarily a non-biological process and complicated factors such as size, water content, and the structure of the aggregates may control the sinking rate for each aggregate. The mean settling velocity of marine snow (2.4–75 mm in maximum length) obtained by in situ measurements was 74 ± 39 m day^{-1} (Alldredge and Gotschalk, 1988). Sinking rates increases exponentially with the increase of aggregate size and dry weight. Based on various observations of the sinking behavior of marine aggregates, it has been suggested that organic aggregates larger than ca. 0.1 mm roughly correspond to the sinking aggregates (Asper, 1987; Alldredge and Gotschalk, 1988; Alldredge and Silver, 1988).

Aggregates may be consumed by feeding processes at different trophic levels and/or by microbial decomposition processes (Smith *et al.*, 1992; Lampitt *et al.*, 1993; Marchant and Scott, 1993; Larson and Shanks, 1996). The consumption of marine snow by zooplankton, micronekton, and fish has been reported

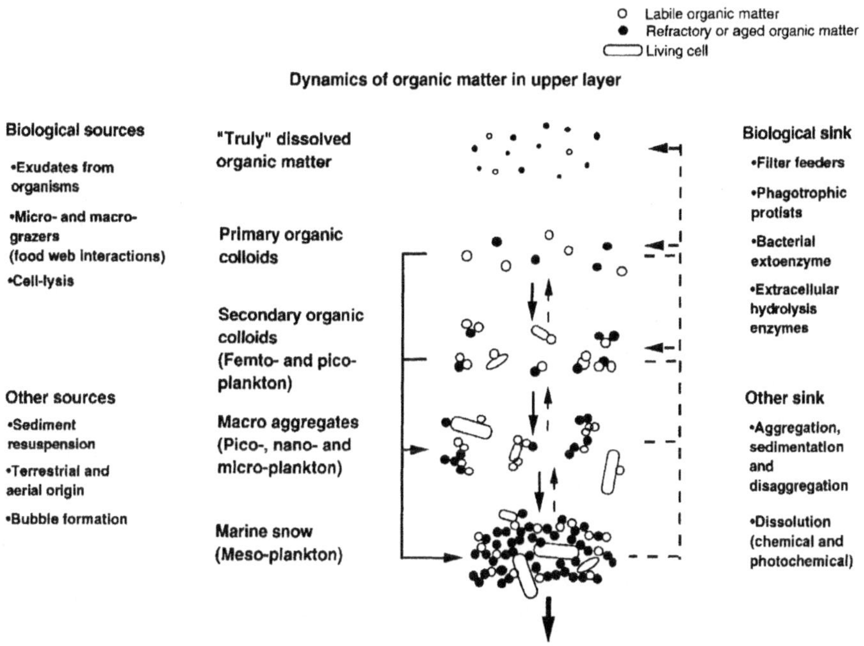

Fig. 4. Conceptual model of "Aggregation-pump" indicating the dynamics of DOM and aggregates in the upper layer of the ocean.

(Alldredge, 1972; Alldredge and Madin, 1982; Lampitt *et al.*, 1993; Larson and Shanks, 1996). Thus, the consumption of marine aggregates during the feeding process of animals and their subsequent production of fecal pellets may be a significant pathway of their removal. Gonzales and Suttle (1993), Marchant and Scott (1993) and Tranvik *et al.* (1993) have recently reported the uptake of collidal-sized particles by Antarctic choanoflagellates and heterotrophic flagellates, respectively. The "aggregation pump" may be one of the major supplier of foods in the detritus food chain.

Microbial activity appears to affect both the aggregation and the dissolution of marine aggregates (Alldredge and Silver, 1988; Smith *et al.*, 1995). Exopolymers exuded from microbes act as a medium for the coagulation of living and non-living particles (Alldredge *et al.*, 1993). Coagulation of colloidal organic materials into larger size aggregates may cause a rapid increase of microbial decomposition or respiration, especially for the "newer" colloids (Kepkay and Johnson, 1989; Kepkay *et al.*, 1990, 1990; Johnson and Kepkay, 1992). On the other hand, a reduced rate of microbial degradation of protein which is adsorbed to artificial colloid particles was reported by Nagata and Kirchman (1996). Bacteria and other microbes also decompose marine snow and other large organic aggregates (Alldredge and Silver, 1988; Biddanda, 1988; Simon *et al.*, 1990; Smith *et al.*, 1992, 1995; Karl and Tilbrock, 1994). Fragmentation of flocculent aggregates may occur with the rapid utilization of labile material within marine aggregates (Alldredge and Silver, 1988). In the degradation process of marine snow, it provides a nutrient rich semi-independent microenvironment surrounded by the nutrient-depleted surface waters (Shanks and Trent, 1979).

The "aggregation-pump" generally stimulates the consumption of organic detritus of any size at the euphotic layer through the formation of sinking aggregates. The production of larger aggregates provides some benefits for feeding and settling the microbial population. Because of this, organic detritus incorporated into the pump would have a shorter turnover rate when compared to that outside the pump with the exception of the organic compounds easily utilized by the microbial population.

Our "aggregation-pump" hypothesis may provide some insight into the accumulation of low molecular weight refractory organic materials (<ca. 1000 dalton) in the ocean (up to 70% of the total organic carbon: Amon and Benner, 1996). Traditionally, low molecular weight (LMW)-DOM in the ocean is regarded as being easily consumed by microbes, resulting in rapid turnover (Fuhrman and Ferguson, 1986). The recent dating of size-fractionated DOC samples, however, indicates that the residence time of LMW-DOM is >1000 years (Santschi *et al.*, 1995). Those refractory DOM may have been formed by abiotic reactions such as melanoidin reaction, becoming resistant to enzymatic attack by microbes (Hedges, 1988; Keil and Kirchman, 1994). Or they may comprise a highly heterogeneous mixture of organic compounds and the concentration of each molecule is not high enough for the substrate-specific hydrolytic enzyme. At present, the primary origin of this refractory organic materials, however, has not been understood. Because of the smaller and possibly hydrophilic character of this organic material, it may escape from the dynamic cycling of marine aggregates mediated by the aggregation-pump and, therefore, accumulate in the water column. We now know that the aggregation-pump is an efficient vehicle for removing organic materials of various sizes in the ocean. Therefore, the seemingly long residence time of smaller organic materials (<ca. 1000 dalton) in the ocean suggests that the aggregation-pump does not work for this particular organic material.

From the discussion presented above, we may distinguish two types of organic materials in the ocean as determined by its behavior with regard to the aggregation-pump. One type is associated with the inner workings of the aggregation-pump and the other is external to the pump. The truly soluble organic compounds may belong to the latter group and be comprized of minor fractions of metabolites from direct biological origin and also, possibly, be modified forms of soluble size. It is worthwhile to note that metabolites of biological origin such as free amino acids and free sugars, have an extremely short residence time (minute to an hour), while the potentially modified organic detritus of this category has the longest residence time within the total range of marine organic detritus (Fuhrman and Ferguson, 1986; Amon and Benner, 1996,).

CONCLUDING REMARKS

In this paper, we reviewed the characteristics of marine organic detritus including the colloids, suspended organic particles, and marine snow. Because the information on their morphology and dynamics is limited, we adopted the conventional size classification of organic detritus. According to the "aggregation-pump" hypothesis, however, which we discussed here as one of the primary mechanisms for the vertical transport of organic detritus, a major portion of marine detrital particles continuously interact with one other, increasing their size successively to form larger aggregates. Therefore, we postulate that no discontinuity exists among organic aggregates of variable sizes ranging from nanometer to centimeters. Information on the biochemical characteristics of marine organic detritus remains limited due to the difficulty in separating organic detritus from living organisms in any size range, especially in the upper ocean. This is also true for their morphological characterization because of the fragile nature associated with the wide size ranges. Given the importance of understanding the aggregation-disaggregation processes of marine detritus, chemical and morphological characteristics, residence time, and dynamics of detritus in the ocean should be investigated in greater detail in future studies.

The "aggregation-pump" may be an important mechanism for the vertical transport of organic carbon. To draw a general picture of carbon dynamics in the ocean, it is necessary to quantify its seasonal and regional contributions to the total flux together with information concerning other processes including biological transport such as fecal pellet and migration and physical transport such as the mixing of water masses.

The importance of collaboration between biological, chemical, and physical studies on marine aggregates was demonstrated in the recent multi-disciplinary studies on marine snow (Alldredge *et al.*, 1995). The same holds true for basin-scale studies on carbon cycling in the ocean (Ducklow and Harris, 1993; Murray, 1995). With research focused in this direction, information on marine aggregate dynamics would become less fragmented and be better understood in conjunction with other components of marine systems.

REFERENCES

Alber, M. and I. Vallela. 1994. Production of microbial organic aggregates from macrophyte-derived dissolved organic material. *Limnol. Oceanogr.*, **39**: 37–50.

Alldredge, A. L. 1972. Abandoned larvacean houses: A unique food source in the pelagic environment. *Science*, **117**: 885–887.

Alldredge, A. L. 1979. The chemical composition of macroscopic aggregates in two neritic seas. *Limnol. Oceanogr.* **24**: 855–866.

Alldredge, A. L. and C. Gotschalk. 1988. In situ setting behavior of marine snow. *Limnol. Oceanogr.*, **33**: 339–351.

Alldredge, A. L. and C. C. Gotschalk. 1990. The relative contribution of marine snow of different origins to biological processes in coastal waters. *Continental Shelf Res.*, **10**: 41–58.

Alldredge, A. L. and L. P. Madin. 1982. Pelagic tunicates: Unique herbivores in the marine plankton. *Bioscience*, **32**: 655–661.

Alldredge, A. L. and M. W. Silver. 1988. Characteristics, dynamics and significance of marine snow. *Prog. Oceanog.*, **20**: 41–82.

Alldredge, A. L. and G. A. Jackson. 1995. Aggregation in marine systems. *Deep-Sea Res.*, **42**: 1–7.

Alldredge, A. L., U. Passow and B. E. Logan. 1993. The abundance and significance of a class of large, transparent organic particles in the ocean. *Deep-Sea Res.*, **40**: 1131–1140.

Alldredge, A. L., C. Gotschalk, U. Passow and U. Riebesell. 1995. Mass aggregation of diatom blooms: Insights from a mesocosm study. *Deep-Sea Res.*, **42**: 9–27.

Amon, R. M. W. and R. Benner. 1994. Rapid cycling of high-molecular-weight dissolved organic matter in the ocean. *Nature*, 369: 549–552.

Amon, R. M. W. and R. Benner. 1996. Bacterial utilization of different size classes of dissolved organic matter. *Limnol. Oceanogr.*, **41**: 41–51.

Asper, V.L. 1987. Measuring the flux and sinking speed of marine snow aggregates. *Deep-Sea Res.*, **34**: 1–17.

Azam, F. and G. F. Smith. 1991. Bacterial influence on the variability in the ocean's biogeochemical state: A mechanistic view. p.213–236. In: *Particle Analysis in Oceanography*, ed. by Demers, S., Springer-Verlag., Heidelberg.

Azam, F.,T. Fenchel, J. G. Field, J. S. Gray, L. A. Meyer-Reil and F. Thingstad. 1983. The ecological role of water column microbes in the sea. *Mar. Ecol. Progr. Ser.* **10**: 257–263.

Benner, R., J. D. Pakulski, M. McCarthy, J. I. Hedges and P. G. Hatcher. 1992. Bulk chemical characteristics of dissolved organic matter in the ocean. *Science*, **255**: 1561–1564.

Bergh, O., K. Y. Borsheim, G. Bratbak and M. Heldal. 1989. High abundance of viruses found in aquatic environments. *Nature* (London), **340**: 467–468.

Biddanda, B. A. 1988. Microbial aggregation and degradation of phytoplankton-derived detritus in seawater. II. Microbial metabolism. *Mar. Ecol. Prog. Ser.*, **42**: 89–95.

Bolin, B. 1983. The carbon cycle. In: *The Major Biogeochemical Cycles and Their Interactions*, ed. by B. Bolin and R.B. Cook, SCOPE 21, John Wiley & Sons, Chichester.

Carlson, C. A. and H. W. Ducklow. 1995. Dissolved organic carbon in the upper ocean of the central equatorial Pacific Ocean, 1992: Daily and finescale vertical variations. *Deep-Sea Res.*, **42**: 639–656.

Caron, D. A., P. G. Davis, L. P. Madin and J. M. Sieburth. 1982. Heterotrophic bacteria and bacteriovorous protozoa in oceanic macroaggregates. *Science*, **218**: 795–797.

Cauwet, G. 1978. Organic chemistry of sea water particles: concepts and developments. *Oceanol. Acta*, **1**: 99–105.

Cauwet, G. 1981. Non-living particulate matter. p. 71–89. In: *arine organic chemistry-Evolution, composition, interactions and chemistry of organic matter in seawater*, ed. by E. K. Duursma and R. Dawson, Elsevier Scientific Publ. Co., New York.

Choi, J.,W., E. B. Sherr and B.F. Sherr. 1996. Relation between presence-absence of a visible nucleoid and metabolic activity in bacterioplanlton cells. *Limnol. Oceanogr.* **41**: 1161–1168.

Conover, R. J. 1966. Assimilation of organic matter by zooplankton. *Limnol. Oceanogr.* **11**: 338–345.

Conover, R. J. and Corner, E. D. S. 1968. Respiration and nitrogen excretion by some marine zooplankton in relation to their life cycles. *J. mar. biol. Ass. U.K.*, **48**: 49–75.

Dam, H. G. and D. T. Drapeau. 1995. Coagulation efficiency, organic-matter glues and the dynamics of particles during a phytoplankton bloom in a mesocosm study. *Deep-Sea Res.*, **42**: 111–123.

Dean, R. B. 1948. Modern colloids: *An introduction to the physical chemistry of large molecules and small particles.* Van Nostrand.

Decho, A. W. 1990. Microbial exopolymer secretions in ocean environments: Their role(s) in food webs and marine processes. *Oceanogr. Mar. Biol. Ann. Rev.*, **28**: 73–153.

Decho, A. W. and G. R. Lopez. 1993. Exopolymer microenvironments of microbial flora: Multiple and interactive effects on trophic relationships. *Limnol. Oceanogr.*, **38**: 1633–1645.

Ducklow, H. W. and R. P. Harris. (eds.) 1993. Topical Studies in Oceanography: JGOFS: The north Atlantic bloom experiment. *Deep-Sea Res.* **40**: 1–641.

Dugdale, R. C. and J. J. Goering. 1967. Uptake of new and regenerated forms of nitrogen in primary productivity. *Limol. Oceanogr.* **12**: 196–206.

Dunbar, R. B. and W. Berger. 1981. Fecal pellet flux to modern bottom sediments of Santa Barbara Basin (California) based on sediment trapping. *Bull. Geol. Soc. Am.*, **92**: 212–218.

Fogg, G. E. 1983. The ecological significance of extracellular products of phytoplankton photosynthesis. *Bot. Mar.*, **26**: 3–14.

Fowler, S. W. and G. A. Knauer. 1986. Role of large particles in transport of elements and organic compounds through the oceanic water column. *Prog. Oceanogr.*, **16**: 147–194.

Fuhrman, J. A., R. I. Ferguson. 1986. Nanomolar concentrations and rapid turnover of dissolved free amino acids in seawater: agreement between chemical and microbiological measurements. *Mar. Ecol. Prog. Ser.* **33**, 237–242.

Gauld, D. T. 1957. Aperitrophic membrane in calanoid copepods. *Nature* (London), **179:** 325–326.

Gonzalez, J. M. and C. A. Suttle. 1993. Grazing by marine nanoflagellates on viruses and virus-sized particles: ingestion and digestion. *Mar. Ecol. Prog. Ser.*, **94**: 1–10.

Gordon, Jr. and D. C. 1970. A microscopic study of organic particles in the North Atlantic Ocean. *Deep-Sea Res.* **17**: 175–185.

Guo, L., C. H. Coleman Jr. and P.H. Santschi. 1994. The distribution of colloidal and dissolved organic carbon in the Gulf of Mexico. *Mar. Chem.*, **45**: 105–119.

Hara, S., K. Terauchi and I. Koike. 1991. Abundance of viruses in marine waters: Assessment by epifluorescence and transmission microscopy. *Appl. Environ. Microbiol.*, **57**: 2731–2734.

Hedges, J. I. 1988. Polymerization of humic substances in natural environments, p. 45–58. In: H*umic substances and their role in the environment*, ed. by F.H. Frimmel and R.F. Christman, John Wiley & Sons, Chichester.

Heldal, M. and B. Bratbak. 1991. Production and decay of viruses in aquatic environments. *Mar. Ecol. Prog. Ser.* **72**: 205–212.

Holm-Hansen, O. W. H. Sutcliffe, Jr. and J. Sharp. 1968. Measurement of deoxyribonucleic acid in the ocean and its ecological significance. *Limnol. Oceanogr.* **13**, 507–514.

Honjo, S. 1978. Sedimentation of materials in the Sargasso Sea at a 5,367m deep station. *J. Mar. Res.*, **36**: 469–492.

Honjo, S., M. R. Roman. 1978. Marine copepod fecal pellets: production, preservation and sedimentation. *J. Mar. Res.*, **36**: 45–57.

Ikeda, T. 1974. Nutritional ecology of marine zooplankton. *Mem. Fac. Fish. Hokkaido Univ.*, **22**:1–97.

Jackson, G. A. 1990. A model of the formation of marine algal flocs by physical coagulation processes. *Deep Sea Res.*, **37**: 1197–1211.

Johnson, B. D. and P. E. Kepkay. 1992. Colloid transport and bacterial utilization on oceanic DOC. *Deep-Sea Res.*, **39**: 855–869.

Karl, D. M. and B. D. Tilbrook. 1994. Production and transport of methane in oceanic particulate organic matter. *Nature* (London), **368**: 732–734.

Keil, R. G. and Kirchman, D. L. 1994. Abiotic transformation of labile protein to refractory protein in seawater. *Mar. Chem.*, **45**: 187–196.

Kepkay, P. E. 1994. Particle aggregation and the biological reactivity of colloids. *Mar. Ecol. Prog. Ser.*, **109**: 293–304.

Kepkay, P. E. and B. D. Johnson. 1989. Coagulation on bubbles allows microbial respiration of oceanic dissolved organic carbon. *Nature* (London), **338**: 63–65.

Kepkay, P. E., W. G. Harrison and B. Irwin. 1990. Surface coagulation, microbial respiration and primary production in the Sargasso Sea. *Deep-Sea Res.*, **37**: 145–155.

Kepkay, P. E., D. K. Muschenheim and B. D. Johnson. 1990. Surface coagulation and microbial respiration at a tidal front on Georges Bank. *Continental Shelf Res.*, **10**: 573–588.

Koike, I, S. Hara, K. Terauchi and K. Kogure. 1990. Role of sub-micrometre particles in the ocean. *Nature* (London), **345**: 242–24.

Koike, I, S. Hara, K. Terauchi, A. Shibata and K. Kogure. 1993. Marine Viruses—Their role in upper ocean dissolved organic matter (DOM) dynamics. p. 311–314. In: *Trends in Microbial Ecology*, ed. by R. Guerrero and C. Pedros-Alio, Spanish Society for Microbiology.

Lalli C. and T. R. Parsons. 1993. *Biological oceanography—An introduction*. Pergamon Press, Oxford, pp. 115–150.

Lampert, W. 1978. Release of dissolved organic carbon by grazing zooplankton. *Limol. Oceanogr.* **23**: 831–834.

Lampitt, R. S., T. Noji and B. Bodungen. 1990. What happens to zooplankton faecal pellets? Implications for material flux. *Mar. Biol.*, **104**: 15–23.

Lampitt, R. S., K. F. Wishner, C. M. Turey and M. V. Angel. 1993. Marine snow studies in the Northern Atlantic Ocean: distribution, composition and role as a food source for migrating plankton. *Mar. Biol.*, **116**: 689–702.

Larson, E. T. and A. L. Shanks. 1996. Consumption of marine snow by two species of juvenile mullet and its contribution to their growth. *Mar. Ecol. Prog. Ser.*, **130**: 19–28.

Lee, S. and S. M. Hendricks. 1993. How the nature of dissolved organic matter might affect the analysis of dissolved organic carbon. *Mar. Chem.*, **41**: 105–120.

Logan, B. E, D. B. Wilkinson. 1990. Fractal geometry of marine snow and other biological aggregates. *Limnol. Oceanogr.* **35**: 130–136.

Long, R. A., F. Azam. 1996. Abundant protein-containing particles in the sea. Aquat. *Microb. Ecol.*, **10**: 213–221.

Mague, T. H., E. Friberg, D. J. Hughes and I. Morris. 1980. Extracellular release of carbon by marine phytoplankton; a physiological approach. *Limnol. Oceanogr.*, **25**: 262–279.

Marchant, H. J. and F. J. Scott. 1993. Uptake of sub-micrometre particles and dissolved organic material by Antarctic choanoflagellates. *Mar. Ecol. Prog. Ser.*, **92**: 59–64.

Mel'nikov, I. A. 1977. Morphological characteristics of organic detritus particles. *Oceanol. Acad. Sci. U.S.S.R.*, **16**: 401–403.

Meyers-Schulte, K. J. and J. H. Hedges. 1986. Molecular evidence for a terrestrial component of organic matter dissolved in ocean water. *Nature*, **321**: 61–63.

Mopper, K. and E. T. Degens. 1979. Organic carbon in the ocean: Nature and cycling, p. 293–316. In: *The global carbon cycle*, ed. by B. Bolin, E.T. Degens, S. Kempe and P. Ketner, John Wiley & Sons, Chichester.

Mopper, K., J. Zhou, K. Sri Ramana, U. Passow, H. G. Dam, D and T. Drapeau. 1995. The role of surface-active carbohydrates in the flocculation of a diatom bloom in a mesocosm. *Deep-Sea Res.*, **42**: 47–73.

Moran, S. B. and K. O. Buesseler. 1992. Short residence time of colloids in the upper ocean estimated from ^{238}U–^{234}Th disequilibria. *Nature* (London), **359**: 221–223.

Murray, J. W. (ed.) 1995. Topical Studies in Oceanography: A U.S. JGOFS Process Study in the Equatorial Pacific. *Deep-Sea Res.* **42**: 275–903.

Nagata, T. 1997. "Picopellets" produced by phagotrophic nanoflagellates: Role in the material cycling within marine environments. this volume.

Nagata, T. and D. L. Kirchman. 1996. Bacterial degradation of protein adsorbed to model submicron particles in seawater. *Mar. Ecol. Prog. Ser.*, **132**: 241–248.

Nagata, T. and I. Koike 1995. Marine colloids: Their roles in food webs and biogeochemical fluxes. p. 275–292. In: *Biogeochemical Processes and Ocean Flux in the Western Pacific*. ed. by H. Sakai and Y. Nozaki, Terra Scientific Publishing Co., Tokyo.

Nemoto, T. and K. Ishikawa. 1969. Organic particulate and aggregate matters stained by histochemical reagents in the East China Sea. *J. Oceanogr. Soc. Japan*, 4: 281–290.

Ogawa, H. and N. Ogura. 1992. Comparison of two methods for measuring dissolved organic carbon in sea water. *Nature* (London), **356**: 696–698.

Parsons, T. R. 1963. Suspended organic matter in sea water. pp.205. In: *Progress in Oceanography Vol. 1*, ed. by Sears, M., Pergamon, Oxford.

Parsons, T. R. and J. D. H. Strickland. 1962. Oceanic detritus. *Science*, **136**: 313–315.

Parsons, T. R., M. Takahashi and B. Hargrave. 1984. *Biological Oceanographic Processes*. 3rd. ed., Pergamon. Oxford, 330pp.

Passow, U. and A. L. Alldredge. 1994. Distribution, size and bacterial colonization of transparent exopolymer particles (TEP) in the ocean. *Mar. Ecol. Prog. Ser.*, **113**: 185–198.

Paul, J. H., W. H. Jefferey and M. DeFlaun. 1985. Particulate DNA in subtropical oceanic and estuarine planktonic environments. *Mar. Biol.*, **90**: 95–101.

Pilskaln, C. H. and S. Honjo. 1987. The fecal pellet fraction of biogeochemical particle fluxes to the deep sea. *Global Biogeochem. Cycles*, 1: 31–48.

Proctor, L. M. and J. A. Fuhrman. 1990. Viral mortality of marine bacteria and cyanobacteria. *Nature* (London), **343**: 60–62.

Riley, G. A. 1963. Organic aggregates in seawater and the dynamics of their formation and utilization. *Limnol. Oceanogr.*, **8**: 372–381.

Romankevich, E. A. 1984. *Geochemistry of organic matter in the ocean.* Springer-Verlag, Tokyo, pp. 89–95.

Santschi, P. H., L. Guo, M. Baskaran, S. Trumbore, J. Southon, T. S. Bianchi, B. Honeyman, and L. Cifuents. 1995. Isotope evidence for the contemporary origin of high-molecular weight organic matter in oceanic environments. Geochem. Cosmochem. *Acta*, **59**: 625–631.

Sasaki, H. and S. Nishizawa. 1981. Vertical flux profiles of particulate materials in the sea off Sanriku. *Mar. Ecol. Prog. Ser.*, **6**: 191–201.

Seki, F., I. Koike, E. Matsumoto, and A. Hattori. 1972. A study of the distribution of total bacteria, bacterial aggregates and heterotrophic bacteria in the sea. 1. I the subarctic Pacific region and the western north Pacific central region. *J. Oceanogr. Soc. Japan*, **28**, 103–108.

Sempéré, R. and G. Cauwet. 1995. Occurrence of organic colloids in the stratified estuary of the Krka River (Croatia). *Estuarine Coastal Shelf Sci.*, **40**: 105–114.

Shanks, A. L. and J. D. Trent. 1979. Marine snow: Microscale nutrient patches. *Limnol. Oceanogr.*, **24**: 850–854.

Shanks, A. L. and E. W. Edmondson. 1989. Laboratory-made artificial marine snow: a biological model of the real thing. *Mar. Biol.*, **101**: 463–470.

Sharp, J. H. 1973. Size classes of organic carbon in seawater. *Limnol. Oceanogr.*, **18**: 441–447.

Shaw, D. J. 1992. *Introduction to Colloid & Surface Chemistry.* 4th ed., Butterworth Heinemann, London.

Sheldon, R. W. and T. R. Parsons. 1967. A continuous size spectrum for particulate matter in the sea. *J. Fish. Res. Bd. Canada*, **24**: 909–915.

Sheldon, R. W., A. Prakash and W. H. Sutcliffe, Jr. 1972. The size distribution of particles in the ocean. *Limnol. Oceanogr.*, **17**: 327–340.

Sieracki, M. E. and C. L. Viles. 1992. Distributions and fluorochrome-staining properties of sub-micrometer particles and bacteria in the North Atlantic. *Deep-Sea Res.*, **39**: 1919–1929.

Simon, M., A. L. Alldredge and F. Azam. 1990. Bacterial carbon dynamics on marine snow. *Mar. Ecol. Prog. Ser.*, **65**: 205–211.

Smith, D. C., M. Simon, A. L. Alldredge and F. Azam. 1992. Intense hydrolytic enzyme activity on marine aggregates and implications for rapid particle dissolution. *Nature* (London), **359**: 139–142.

Smith, D. C., G. F. Steward, R. A. Long and F. Azam. 1995. Bacterial mediation of carbon fluxes during a diatom bloom in a mesocosm. *Deep-Sea Res.*, **42**: 75–97.

Stoecker, D. K. 1984. Particle production by planktonic ciliates. Limnol. Oceanogr., **29**: 930–940.

Sutcliffe, W. H., Jr., R. W. Sheldo and A. Prakash. 1970. Certain aspect of production and standing stock of particulate matter in the surface waters of the Northwest Atlantic ocean. *J. Fish. Res. Bd. Can.*, **27**: 1917–1926.

Taguchi, S. 1982. Seasonal study of fecal pellets and discarded houses of appendicularia in a subtropical inlet, Kaneoha Bay, Hawaii. *Estuar. Coast. Shelf Sci.* **14**: 545–555.

Tanoue, E. and N. Handa, H. 1979. Distribution of particulate organic carbon and nitrogen in the Bering Sea and northern North Pacific Ocean. *J. Oceanogr. Soc. Japan*, **35**: 47–62.

Tanoue, E., and N. Handa. 1987. Monosaccharide composition of phytoplankton, suspended particles, sinking particles and bottom sediments from the Bering Sea and northern North Pacific. *Oceanol. Acta*, **10**: 91–99.

Tanoue, E., N. Handa and M. Kato. 1982a. Horizontal and vertical distribution of particulate organic matter in the Pacific sector of the Antarctic Ocean. Trans. *Tokyo Univ. Fish.*, **5**: 65–83.

Tanoue, E., N. Handa and H. Sakugawa. 1982b. Difference of the chemical composition of organic matter between fecal pellet of *Euphausia superba* and its feed, Dunaliella tertiolecta. *Trans. Tokyo Univ. Fish.* **5**: 189–196.

Tanoue, E., M. Zenimoto, Y. Komaki and N. Handa. 1986. Distribution of particulate organic matter in the Pacific and Indian sector of the Antarctic Ocean in the austral summer. Mem. *Natl Inst. Polar Res., Spec. Issue*, **40**: 380–394.

Tranvik, L. 1994. Colloidal and dissolved organic matter excreted by a mixotrophic flagellate during bacteriovory and autotrophy. *Appl. Environ. Microbiol.*, **60**: 1884–1888.

Tranvik, L. J., E. B. Sherr and B.F. Sherr. 1993. Uptake and utilization of 'colloidal DOM' by heterotrophic flagellates in seawater. *Mar. Ecol. Prog. Ser.*, **92**: 301–309.

Tsuji, Y. and S. Sukizaki. 1991. Measurement of marine snow abundance using the submersible. *La mer*, **29**: 159–165.

Wakeham, S. G. and E. A. Canuel. 1988. Organic geochemistry of particulate matter in the eastern tropical North Pacific Ocean: Implications for particle dynamics. *J. Mar. Res.*, **46**: 183–213.

Webb, K. L. and R. E. Johannes. 1967. Studies of the release of dissolved free amino acids by marine zooplankton. *Limnol. Oceanogr.*, **12**: 376–382.

Webb, K. L. and R. E. Johannes. 1969. Do marine crustaceans release dissolved amino acids? *Comp. Biochem. Physiol.*, **29**: 875–878.

Wells, M. L. and E. D. Goldberg. 1991. Occurrence of small colloids in sea water. *Nature* (London), **353**: 342–344.

Wells, M. L. and E. D. Goldberg. 1992. Marine sub-micron particles. *Mar. Chem.*, **40**: 5–18.

Wells, M. L. and E. D. Goldberg. 1993. Colloid aggregation in seawater. *Mar. Chem.*, **41**: 353–358.

Wells, M. L. and E. D. Goldberg. 1994. The distribution of colloids in the North Atlantic and Southern Oceans. *Limnol. Oceanogr.*, **39**: 286–302.

Wiebe, P. H., S. H. Boyd and C. Winget. 1976. Particulate matter sinking to the deepsea floor at 2000 m in the tongue of the ocean, Bahamas, with a description of a new sedimentation trap. *J. Mar. Res.*, **34**: 341–354.

Williams, P. M. and E. R. M. Druffel. 1988. Dissolved organic matter in the ocean: comments on a controversy. *Oceanography*, **1**, 14–17.

Yanada, Y and Y. Maita. 1995. Regional and seasonal variations of biomass and bio-mediated materials in the North Pacific Ocean. p. 293–306. In: *Biogeochemical Processes and Ocean Flux in the Western Pacific*, ed. by H. Sakai and Y. Nozaki, Terra Scientific Publishing Co., Tokyo.

Zweifel, U. L. 1995. Degradation of dissolved organic carbon in the marine environment: Implication for bacterial growth. PhD thesis, University of Umea, Sweden. pp. 35.

Zweifel, U. L. and A. Hagström. 1995. Total counts of marine bacteria include a large fraction of non-nucleoid containing "ghosts". *Appl. Environ. Microbiol.*, **61**: 2180–2185.

Dynamics and Characterization of Marine Organic Matter, Eds., N. Handa, E. Tanoue and T. Hama, pp. 299–309.

Dissolved Organic Carbon in Sub-Arctic Northeast Pacific Ocean

C. S. WONG[1], Zhiming YU [2], Wm. K. JOHNSON[1], R. J. MATEAR[1] and F. A. WHITNEY[1]

[1]*Institute of Ocean Sciences, Department of Fisheries & Oceans, P.O. Box 6000, Sidney, B.C., V8L 4B2, Canada*
[2]*Institute of Oceanology, Academia Sinica, 7 Nan Hai Road, Qingdao, China*

(Original, Received 3 July 1996; accepted 15 November 1997)

Abstract—DOC measurements in the Northeast Pacific Ocean from May and September 1995 cruises were made by HTCO technique for Stations P04, P12, P16 and P26 on Line P between the west coast of Vancouver Island to Station P (50°N, 145°W). The DOC features together with oceanographic data of temperature, salinity, nitrate, and silicate were discussed.

INTRODUCTION

The sub-Arctic northeast Pacific Ocean is an oceanic regime characterized by high nitrate and low chlorophyll-*a* (HNLC) and by a phytoplankton assemblage dominated by pico-phytoplankton (Parsons and Lalli, 1988; Wong *et al.*, 1995). In this regime, total production is 140 gC $m^{-2}year^{-1}$ (Wong *et al.*, 1995). The new production accounts for 25%, and the regenerated production for 75% (104 gC $m^{-2}year^{-1}$). Regenerated production is a source of oceanic dissolved organic carbon (DOC). This paper describes a DOC data set in NE Pacific Ocean. DOC is one of the main themes of research in CJGOFS (Canadian Joint Global Ocean Flux Study), started in spring of 1995, in an attempt to understand the DOC distribution and cycling in both the coastal and open-ocean waters off the west coast of British Columbia, Canada.

Dissolved organic carbon (DOC) was measured by high-temperature catalytical oxidation (HTCO) in the western North and Subtropical Pacific waters and East China Sea (Sugimura and Suzuki, 1988; Suzuki and Tanoue, 1991), the western North Pacific (Tanoue, 1993), equatorial Atlantic Ocean (Thomas *et al.*, 1995). and central equatorial Pacific Ocean (Carlson and Ducklow, 1995). We report here recent measurements of DOC made in the NortheastPacific Ocean on Line P between the west coast of Vancouver Island and Station P (50°N, 145°W) using a similar HTCO technique and procedure as described in Suzuki and Tanoue, 1992.

METHODOLOGY

Sampling sites and methods

Samples were collected on the CSS *John P. Tully* in 1995 for cruises 9505

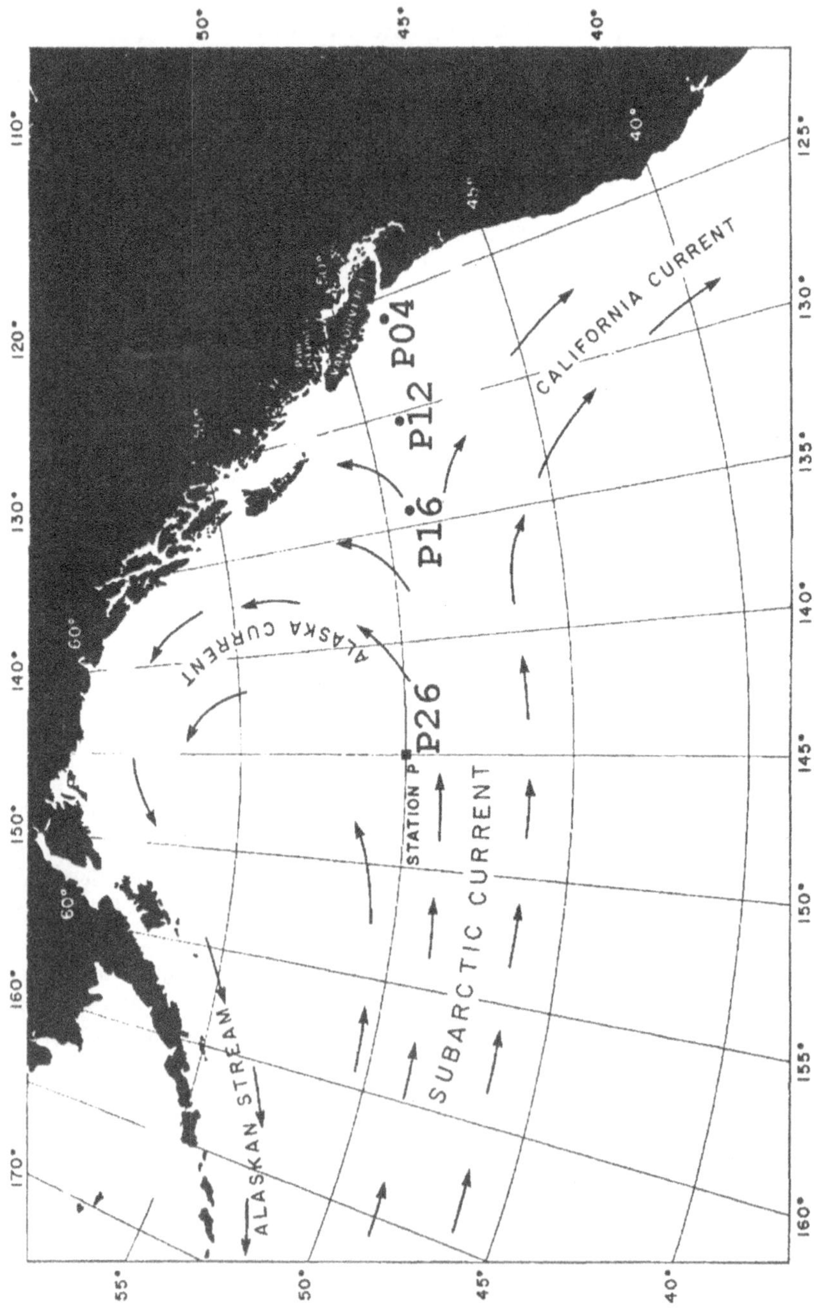

Fig. 1. Positions of sampling stations for DOC in Northeast Pacific Ocean.

(May) and 9512 (September) at oceanographic stations on Line P between the west coast of Vancouver Island and Station P (50°N, 145°W): Station P04 (48°39′ N, 126°40′ W), P12 (48°17′ N, 134°40′ W), P16 (48°58′ N. 130°40′ W) and P26, i.e. Station P itself (50°N, 145°N), as shown in Fig. 1, which also includes the pattern of surface circulation in Northeast Pacific waters. Seawater samples were collected using 101 Niskin PVC bottles and subsampled without using plastic hose, directly into pre-cleaned (by washing, then rinsing with distilled water, and then baking at 550°C for 1 hr) 125 ml or 250 ml Pyrex glass bottles with ground-glass stoppers. Sea water samples from deeper than 200 m were put into ampoules without filtration. Samples from ≤100 m depths were filtered through 0.7 μm acid GF/F filters (previously baked at 450°C) on PFA filter housings, which were rinsed with samples prior to filtration. Aliquots for DOC were taken by using a preset 10 ml Oxford pipette, rinsed twice with the seawater from the 125 ml bottle, into two precleaned 10 ml glass ampoules. The ampoules were acidified with 1 ml of 8–10% phosphoric acid, sparged for 10 min with UHP (ultra high purity) nitrogen at 20 psi at a flow rate of 150–175 ml min^{-1}, and sealed by propane torch while purging the mouth of the ampoule with UHP air flowing at 150–200 ml min^{-1}. The sealed ampoules were stored in the refrigerator at 4°C. The DOC seawater standard was prepared from Go-flo or Niskin samplers filled with water from 500 m to 2,000 m at Station P, and stored in a 15 l pre-baked glass bottle, capped with a Teflon-covered rubber stopper. Our tests showed that new PVC samplers should be avoided in DOC sampling. However, aged PVC samplers frequently used in seawater sampling gave consistent results irrespective of size of sampler (1.7 l to 10 l) and types (Go-flo or Niskin), with or without stainless steel springs.

DOC measurements

The DOC measurements were made by a high-temperature catalytic oxidation technique (HTCO) as described in Suzuki *et al.* (1992), using a Sumigraph TOC-analyzer model TOC-90 (Sumika Chemical Analysis Services Ltd.) of the original Suzuki design. Acidified samples (1 ml 9% phosphoric acid to 10 ml seawater) were injected (100 ml) into a gas stream of pure oxygen or TOC air/ultra zero air, with the gases passing through a quartz tube filled with a Pt. catalyst (3% Pt on Al_2O_3) at 680°C, and then into an infrared analyzer. The instrument was calibrated against a pure organic standard and blank assessment was also made, as described in the following.

DOC standards were prepared from a primary standard of glucose (SIGMA Co.) in water purifed by passing through a Millipore Organic-Q cartridge system. Working standards were prepared by adding glucose to open-ocean seawater to minimize matrix effects. There was no difference in slopes of the calibration curves using either seawater (aged seawater of low DOC, usually <40 μM), or Millipore Organic-Q filtered water. Typical standards were 0, 40, 80, 120, 180 and 200 μM carbon, and the slopes of the area/DOC plot for the May and September cruises are 10.2 and 9.92 × 10^3 units/μM carbon, respectively. Both standard and seawater samples were first diluted with 9% phosphoric acid (10 ml. seawater to 1 ml. of the acid), and sparged for 10 mins using UHP air flowing at 150–175 ml/min. The blank was estimated by injecting increasing amounts of organics Q water (50 to 225 ml). The intercept gave the amount of carbon that was not from the water added and was used for our system blank. Depending on the set of data points for each calibration curve, this blank ranged from –9 μM to +3 μM (May 1993). These values are in the same range as Suzuki *et al.* (1992). Since our analytical system is the same as Suzuki *et al.*, we are satisfied that the blank should not be a factor affecting the DOC values. The blank was low compared to the noise and considered to be insignificant. Peltzer and Brewer (1993) concluded that the blank was "zero" based on a similar procedure and catalyst but on a different type of instrument. However, since mid-1995, a significant but consistent change in the system blank has persisted. This was likely caused by an overflow of the scrubber, which allowed the phosphoric acid and silver nitrate solution to flow through the gas lines to the infrared analyzer. Since then, the blank has been much higher but quite consistent in magnitude at 31 ± 3 μMC. This blank adjustment has been applied to all our data collected after 5 June 1995, including the September 1995 cruise data used in this paper. This large blank adjustment to the September 1995 measurements gave DOC values below 600 m, the same as the measurements from the May 1995 cruise with a small blank adjustment.

RESULTS AND DISCUSSIONS

DOC, rather than TOC(total organic carbon), was measured in the upper 200 m since the sea-water samples were filtered. Samples deeper than 200 m were not filtered, and thus should be TOC as the sum of DOC and POC. POC at Station P (Fig. 1) was less than 3 μM below 50 m for September at the end of the productive summer. For May during the productive season, POC decreased to 3 μM at 200 m. Thus, below 200 m, POC was about 4% of the DOC at 70 μM. DOC should the same as TOC within the uncertainties of the measurements.

The changes in DOC, salinity and nitrate in the upper 600 m. for stations P04, P12, P16 and P26 from coastal to open-ocean sub-Arctic Pacific are shown in Figs. 2a to 2d, 3a to 3d and 4a to 4d, respectively. The changes in surface properties of temperature (T°C), salinity (psu), nitrate (μM) and silicate (μM), are shown in Fig. 5a for May 1995 and 5b for September 1995. The coastal station P04 near the west coast of Vancouver Island has the highest DOC concentration of 109 μM at 10 m depth with a low salinity of 31.56 psu Above this depth, the salinity was uniform but below it, there was a small increase in salinity. The higher DOC at the interface of the surface mixed layer and the more stable density structure below may be the result of an accumulation of suspended particles favorable to more intensive bacterial activities, thus producing more DOC. The nitrate in the upper 10 m was depleted, thus indicating the higher DOC was associated with high productivity. In May 1995, the surface DOC along Line P was at a level of about 90 μM at coastal stations P04, P12 and P16, then suddenly

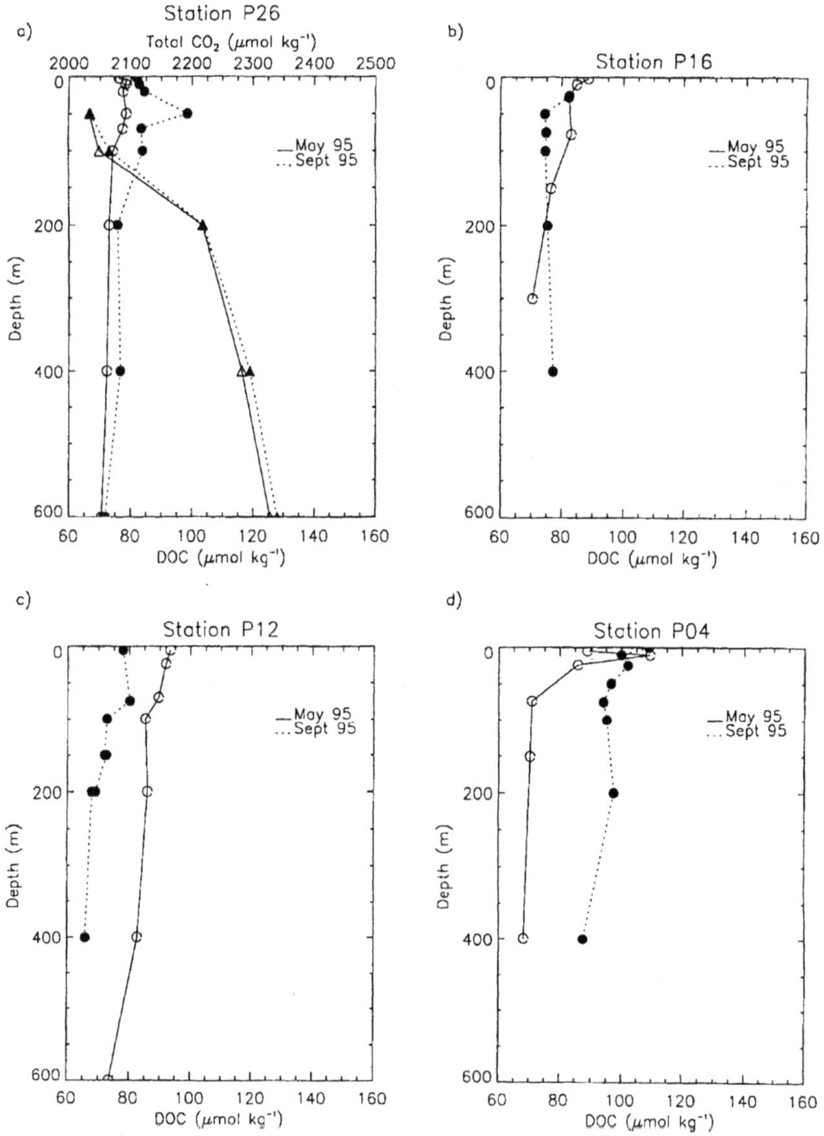

Fig. 2. Distribution of DOC for the depths 0–600 m at Stations (a) P26, (b) P16, (c) P12 and (d) P04 (open circle and solid line for May 1995 and filled circle and dotted line for September 1995). DIC is also shown in (a) P26.

dropped off to about 75 μM at Station P. The surface salinities of the Line P stations were all at about 32.5 psu, with the coastal station P04 at a slightly lower salinity of 31.5 psu, and the offshore Station P at the highest salinity of 32.7 psu. The surface nitrate was depleted at the coastal stations P04 and P12, and was

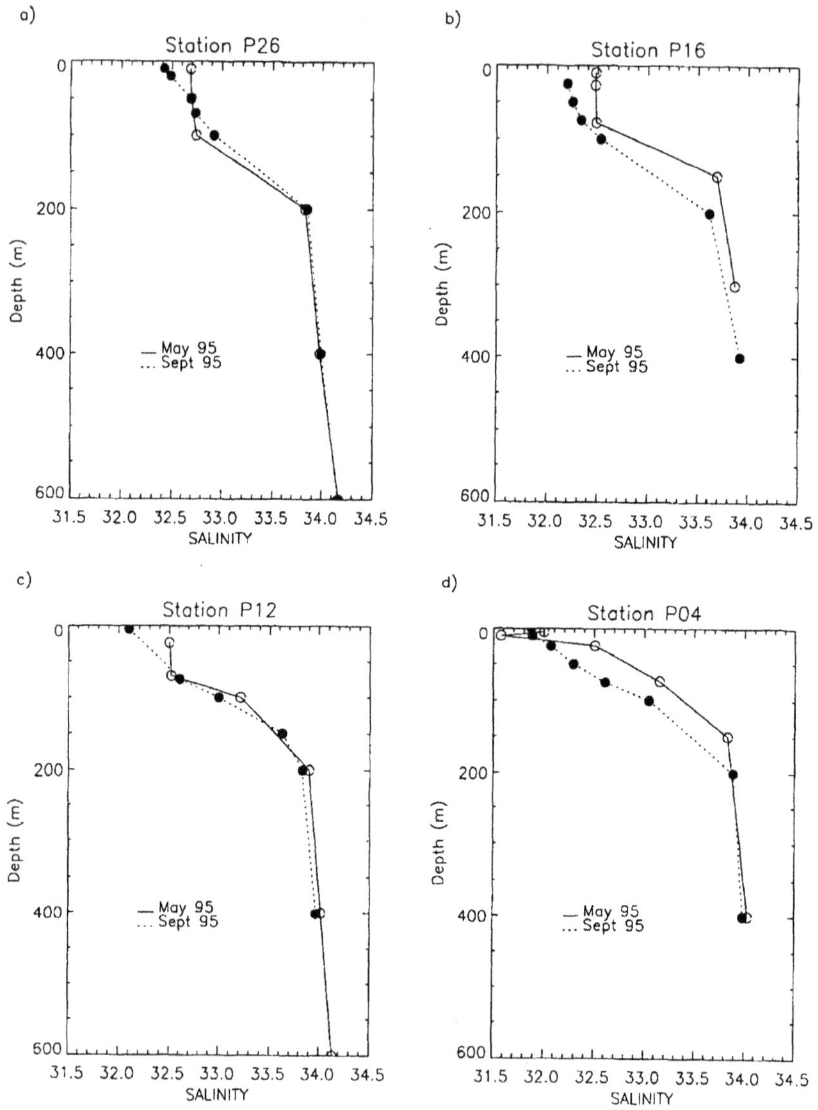

Fig. 3. Distribution of salinity for the depths 0–600 m at Stations (a) P26, (b) P16, (c) P12 and (d) P04 (open circle and solid line for May 1995 and filled circle and dotted line for September 1995).

increasing to about 9 and 12 μM in the offshore stations P16 and Station P, respectively. Higher surface DOC was associated with low nitrate, indicative of increased biological production and/or increased cycling rate in May.

Generally, in the upper ocean, DOC concentrations increased from May to September, reflecting increased biological production. At Station P, the DOC in

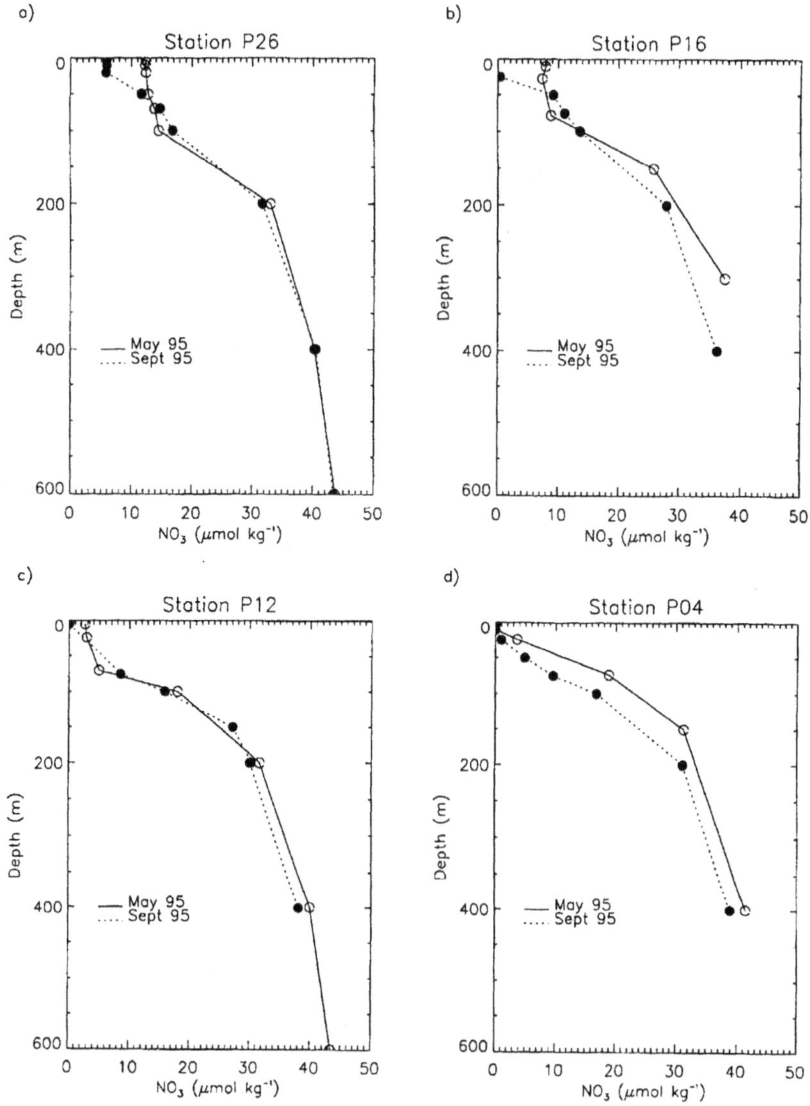

Fig. 4. Distribution of nitrate for the depths 0–600 m at Stations (a) P26, (b) P16, (c) P12 and (d) P04 (open circle and solid line for May 1995 and filled circle and dotted line for September 1995).

the upper 200 m showed an increase of about 10 μM between May and September 1995. DIC (Dissolved inorganic carbon) again showed an increase of about 5 ± 2 μM for the upper 600 m (Fig. 2a). An increase in both DIC and DOC in the upper water column indicated both the export and transformation of POC, as new production flux out of the surface mixed layer to deeper waters in the halocline.

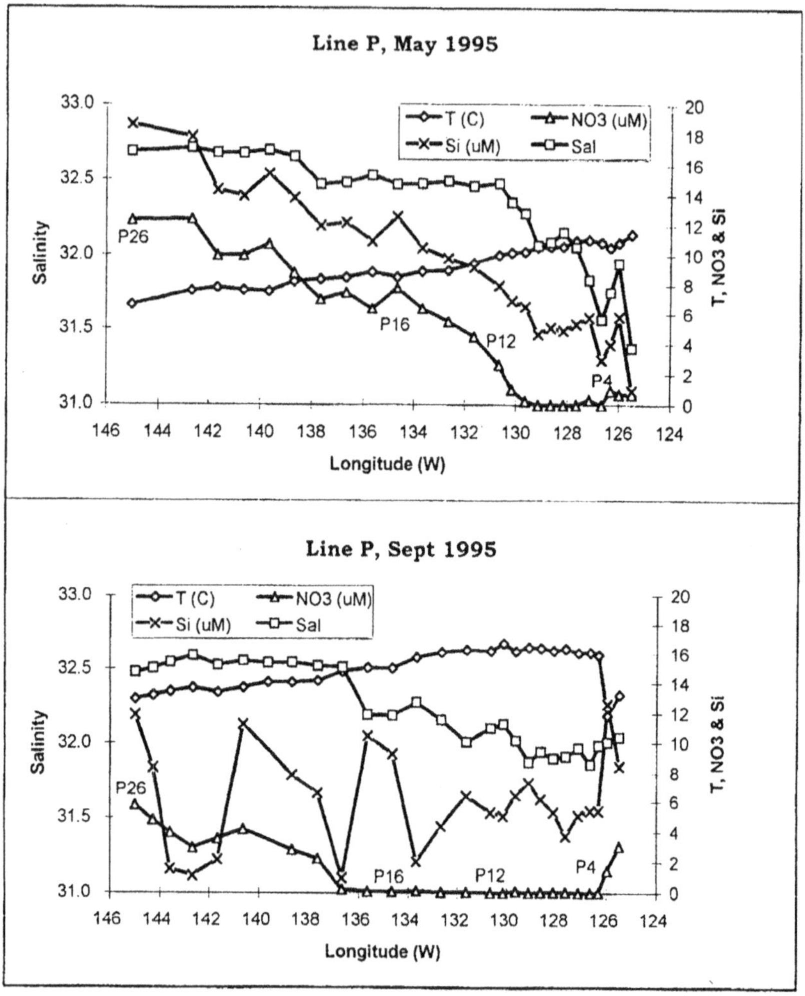

Fig. 5. Distribution of surface water properties (temperature in °C, salinity in psu, nitrate and silicate in μM), along Line P in (a) May 1995 and (b) September 1995.

The POC data in September 1995 from free-floating sediment traps indicated that the rate at 50 m (70 mgC $m^{-2}d^{-1}$) was almost double the rates at 100 and 200 m (38 and 41 mgC $m^{-2}d^{-1}$, respectively), and triple those at 800 and 1,000 m, (26 and 24 mgC $m^{-2}d^{-1}$ respectively). The POC flux in May 1993 was typically higher in the surface waters (90–150 mgC $m^{-2}d^{-1}$ for the upper 200 m) but similar for the bottom waters (22 and 24 mgC $m^{-2}d^{-1}$ at 800 and 1,000 m respectively). Subsequently, the exported POC was regenerated into both DOC and DIC assuming that the advective and diffusive transports were small. For example, assuming the diffusivity Kv to be 0.3 cm^2s^{-1} (Matear and Wong, 1997) as an upper bound, the upward diffusive flux of carbon as DIC at 100 m in 100 days is 1 μM, and assuming the upwelling velocity, w, to be 0.5 m yr^{-1} (Matear and Wong, 1997) as an upper bound, the upward advective flux equals 2 μM. These values were smaller than or equal to the uncertainty in DOC measurements. The new production of carbon exported from the mixed layer as POC would be regenerated as DIC and DOC in the water column below the mixed layer. The increased concentrations of DIC and DOC could total about 15 μM over the water column of 600 m, i.e. an inferred annual production of 133 gC $m^{-2}yr^{-1}$, assuming that the production in the other 8 months (October–April) amounts to one-third of the summer production of (May–September) as proposed by Parsons and Lalli (1988). This value is comparable to recent productivity measurements at about 140 gC $m^{-2}year^{-1}$ at Station P (Wong *et al.*, 1995) and 190 gC $m^{-2}year^{-1}$ from modelling (McClain *et al.*, 1996). However, this preliminary work is very speculative and needs improved DOC measurements to confirm its validity. The DOC in May of 80 μM and September of 85 μM in the upper 100 m, was near to 88 μM (Tanoue, 1993) at 45°N in the North Pacific. The average of the DOC at 400 m and 600 m is about 72 μM for the May and September values, slightly lower than 78 μM below 500 m of Tanoue (1993).

The coastal stations P04 and P12 have lower salinity and higher temperature. At the coastal station P04, the halocline was at about 175 m in May and deepened slightly to 200 m in September. Below 200 m, the salinity was constant for the May and September cruises. The stronger circulation of coastal waters with freshwaters from the Fraser River and drainage from the west coast produced low salinity waters in the shallower depths. For Station P04, DOC was about 20–30 μM higher at the end of the summer production period in September. At station P 12, higher DOC started to develop in May in the mixed layer (0–100 m), below which DOC was decreasing slowly to 75–85 μM. In September, the surface mixed layer became shallower and fresher, while the deeper water column structure remained virtually unchanged as shown by the salinity in Fig. 3c. However, DOC became lower by about 20 μM in September. Station P16, 640 km off the west coast of Vancouver Island, showed a decreasing trend of DOC with depth in the upper 300 m. The DOC was higher in May but lower by 15–20 μM in September. The circulation was very complex in September with fully developed eddies with typical zig-zag patterns in both salinity and silicate, as shown in Fig. 5. This may lead to large changes and different seasonal trends particularly for stations P12 and P04 in the coastal waters.

In the Northeast Pacific Ocean, oceanographic factors, such as upwelling and a strong but shallow halocline, favor the microbial loop and DOC production in summer. The microbial loop counteracts the action of the biological pump (Karasch *et al.*, 1996). The biology of the Northeast Pacific is dominated by small phytoplankton and small zooplankton grazers, which help to cycle the biomass carbon in the upper ocean. Bacteria further enhances the DOC pool in the upper ocean. On the other hand, episodic events such as blooms of larger diatoms possibly stimulated by aeolean iron input, will favor CO_2 removal by fast sedimentation into the deep ocean. DOC dynamics is important to the global carbon cycle and further work in the sub-Arctic waters is necessary to understand fully the roles of the biological pump and the counteraction by the microbial loop, which affect the uptake of CO_2 from the atmosphere.

CONCLUSIONS

(1) In the upper 400 m at Station P in Northeast Pacific waters, DOC increased from a nearly uniform value of about 70 μM in May (the beginning of the productive period), to a level 5–10 μM higher in September with a DOC maximum at 50 m, the top of the summer halocline. These values are similar to the range of 78–88 μM at 45°N (Tanoue, 1993). The integrated increase in DOC from May to September was about 24 g m^{-2}.

(2) For coastal stations, DOC was related to biological production as nitrate was being depleted. The highest DOC off the west coast of Vancouver Island was 109 μM at 10 m depth in May. The DOC distribution was affected by circulation, especially the fully developed eddies in September, as seen in the zig-zag patterns of silicate and salinity.

(3) At Station P, the summer increase in both DOC and DIC in the water column could be the result of the regeneration of carbon export from the surface mixed layer, assuming small diffusive and advective fluxes. However, this is very speculative due to uncertainties in the DOC measurements. There was a shift in the blank for the two cruises due to accidental flooding of the instrument for HTCO by reagents between these periods, although the blanks were consistent for measurements before and after flooding.

REFERENCES

Carlson, C. A. and H. W. Ducklow (1995): Dissolved organic carbon in the upper ocean of the central equatorial Pacific Ocean, 1992: daily and finescale vertical variations. *Deep-Sea Res. II*, **42**, 639–656.

Karrasch, B., H.-G. Hoppe, S. Ullrich and S. Podewski (1996): The role of mesoscale hydrography on microbial dynamics in the northeast Atlantic: results of a spring bloom experiment. *J. Mar. Res.*, **54**, 99–122.

Matear, R. J. and C. S. Wong (1997): Estimation of vertical mixing in the upper ocean at Station P from chloroflurocarbons. *J. Mar. Res.*, **55**, 507–521.

McClain, C.R., K.Arrigo, King-Shang Tai and D. Turk (1996): Observations and Simulations of physical and biological processes at ocean weather station P, 1951-1980. *J. Geophys, Res.*, **101**, (C2), 3697-3713.

Parsons, T. R. and C. M. Lalli (1988): Comparative oceanic ecology of the plankton communities of the subarctic Atlantic and Pacific Oceans. *Oceanogr. Mar. Biol. Ann. Rev.*, **26**, 317–359.

Peltzer, E. T. and P. G. Brewer (1993): Some practical aspects of measuring DOC-sampling artifacts and analytical problems with marine samples. *Mar. Chem.*, **41**, 243–252.

Sugimura, Y. and Y. Suzuki (1988): A high-temperature catalytic oxidation method for the determination of non-volatile dissolved organic carbon in seawater by direct injection of a liquid sample. *Mar. Chem.*, **24**, 105–131.

Suzuki, Y. and E. Tanoue (1991): Dissolved organic carbon enigma-implication for ocean margin. pp. 18–23. In *Ocean Margin Processes in Global Change*, ed. by R. F. C. Mantoura, J. M. Martin and R. Wollast, Dahlem Workshop, Berlin, John Wiley & Sons, Chichester.

Suzuki, Y., E. Tanoue and H. Ito (1992): A high-temperature catalytic oxidation method for the determination of dissolved organic carbon in seawater analysis and improvement, *Deep-Sea Res.*, **39**, 185–198.

Tanoue, E. (1993): Three vertical profiles of dissolved organic carbon in the North Pacific. *Mar. Chem.*, **41**, 261–264.

Thomas, C., G. Cauwet and J.-F. Minster (1995): Dissolved organic carbon in the equatorial Atlantic Ocean. *Mar. Chem.*, **49**, 155–169.

Wong, C. S., F. A. Whitney, K. Iseki, J. S. Page and Jiye Zeng (1995): Analysis of trends in primary productivity and chlorophyll-*a* over two decades at Ocean Station P (50°N, 145°W) in the subarctic Northeast Pacific Ocean. pp. 107–117. In *Climate Change and Northern Fish Populations*, ed. by R. J. Beamish, Can. Sp. Pub. of J. Fish. Aquat. Sci., 121.

Dynamics and Characterization of Marine Organic Matter, Eds., N. Handa, E. Tanoue and T. Hama, pp. 311–337.

Bulk Chemical Aspects of Dissolved Organic Matter in Seawater Review: The Recent Findings and Unsolved Problems

Hiroshi OGAWA

Ocean Research Institute, University of Tokyo, Minamidai, Nakano, Tokyo 164, Japan

(Review, Received 6 September 1996; accepted 28 October 1996)

INTRODUCTION

It is widely known that most dissolved organic matter (DOM) in seawater are not identified by traditional biochemical methods although their total standing stock in the seawater column is approximately 10 times larger than the particulate organic fraction including most marine organisms. Two features of seawater DOM, i.e. the size of the pool is very big but the nature is little known, have been repeatedly pointed out over the past four decades at least. We can learn the details of this matter from many previous syntheses based on past results (Duursma, 1965; Williams, 1971, 1975; Mopper and Degens, 1979; Dawson and Liebzeit, 1981; Thurman, 1984; Williams and Druffel, 1988; Hedges, 1992).

In the 1990's, this conclusion is essentially unchanged, although, it appears that the approach to the seawater DOM has been certainly changing. The form "dissolved" in seawater is operationally defined with a simple filtration, so does not directly mean the real form in water. It indicates only a set of various compounds passed through a glassfiber or membrane filter. Thus, it has been increasingly required that the previously defined DOM should be divided into further subsets on the basis of a more realistic and practical definition.

CLASSIFICATION OF DOM IN SEAWATER

In positively evaluating the role of DOM in marine material cycling, we need to extract some portions reactive to marine biogeochemical processes and to characterize their chemical nature because the large part of DOM in the ocean has been recognized as having very stable properties, as suggested by the measurements of ^{14}C age of the bulk DOM (1,500~6,000 yrs: Williams and Druffell, 1987; Bauer *et al.*, 1992). In particular, to extract the biologically reactive fraction from the bulk DOM has been consistently one of the most important approaches.

Carlson *et al.* (1994) indicated the existence of a "semi-labile" fraction of DOM that has a turnover time of months to a year through biological activities in the upper ocean water column, from findings that the vertical distribution of dissolved organic carbon (DOC) in the Sargasso Sea of the Atlantic Ocean seasonally changed. This type of classification of DOM based on its biological reactivity (labile, semi-labile and refractory) has also been implied by the traditional degradation experiment of seawater DOC (Ogura, 1972). However, the chemical characterization of each class of DOM has not yet been achieved .

The recent works on photochemical oxidation of DOM by Drs. Mopper and Kieber indicated the existence of a photochemically reactive fraction of DOM from UV irrdiation experiments and field observations. Carbonyl compounds and carbon monoxide were detected as phtooxidation products. However, the chemical identification of the source materials seems to be less determinable although they have been generally identified to be humic-like substances by fluorecent characterization (e.g. Mopper *et al.*, 1991).

On the other hand, as a classification based on the initiative of chemical definition, the size fractionation using the ultrafiltration technique has come to be widely used to separate the colloidal or high molecular weight (HMW)-DOM from the truly dissolved or low molecular weight (LMW) DOM fraction (Benner *et al.*, 1992; Ogawa and Ogura, 1992; Guo *et al.*, 1995). It seems that direct counting submicron or colloidal particles in seawater (Koike *et al.*, 1990; Wells and Goldberg, 1991) might partly promote the choice of this approach to extract the colloidal fraction from the previously defined DOM, although it is still ambiguous whether the observed fine particles would be consistent with the colloidal fraction separated by ultrafiltration.

The current ultrafiltration system of the tangential- or cross-flow type can separate and concentrate the colloidal or HMW-DOM from a large volume of seawater with a high efficiency. It has enabled the detailed chemical analyses of the DOM components in this size fraction, e.g. carbohydrates (Benner *et al.*, 1992), protein molecules (Tanoue, 1995), and natural radio and stable isotopes (^{234}Th: Honeyman and Santschi, 1989; ^{14}C and ^{13}C: Santschi *et al.*, 1995). By contrast, the chemical composition of the truly dissolved or LMW-DOM fraction has mostly remained unknown because the general ultrafiltration method could not completely separate this size class of DOM from sea salts, so that it could not be concentrated.

There are other chemical methods for classifying seawater DOM. One is separation based on the difference in the affinity of each constituent for a specific resin column. In particular, XAD resin has been traditionally used to extract aquatic humic substances. This combined method with varying the eluent pH and hydrophobicity can further separate the humic substance into several fractions. Characterization using the fluorescence of aquatic DOM is another method to determine the marine humic substances, which is not only a traditional tool but also one of the current, important methods used for studying humic constituents of aquatic DOM.

Thus, there seem roughly four types of viewpoint for the classification of marine DOM:

(1) Size distribution;
(2) Component identification (for biomolecules);
(3) Chemical properties (for humic substances);

(4) Biogeochemical reactivity.

Although the improvement of individual approach and further accumulation of individual information are necessary, their combined approach would be more important. We especially need the combined information about the biogeochemical reactivity and chemical aspects of the seawater DOM.

In the 1990's, this appears to have been elevated much higher by some important findings suggesting that some portion of DOM could be biogeochemically more reactive and play a more significant role in the marine carbon cycle than had been previously recognized. For instance, these include the fractions photochemically reactive (Mopper *et al.*, 1991), biologically semi-labile (Carlson *et al.*, 1994) and physicochemically transferable to the large particle (Moran and Busseler, 1992).

In the following sections, after reviewing of the current results and problems of individual approaches, some combined information will be presented.

SIZE DISTRIBUTION OF DOM IN SEAWATER

General features of size distribution of oceanic DOM

The ultrafiltration technique is a traditional means of separating DOM into several size (molecular weight) fractions. In an early method, the size distribution was displayed as differences in the DOC concentration of each filtrate passed through several ultrafiltration membranes (Ogura, 1974). Because this method essentially did not utilize the concentration process of ultrafiltration, it was dependent on how accurately the operator corrected the raw data of filtrate concentration for the procedural blank (mainly generated from the membrane). It was also difficult to estimate how a certain degree of loss of DOM occurred during the process through its adsorption onto the membrane surface. Thus, directly determining the concentrations of components in the retentate concentrated in the ultrafiltration process is a currently dominant method, which was established by Carlson *et al.* (1985).

The current (after 1990) results of the size fractionation of DOM are noted in Table 1. These works all confirmed that the carbon mass balance among the initial, filtrate and retentate during the ultrafiltration were approximately between 90 and 110%. Several types of ultrafiltration membrane or cartridge are available to investigate the size spectrum of DOM, which has a nominal molecular weight cut off (NMWCO) of generally between 1,000 (1 k) and 500,000 (500 k) Dalton (Da). However, in fact, the two types of NMWCO of 1 kDa and 10 kDa seem to be most widely used. For comparative purposes, the results from only these two size fractionations for various natural water samples are listed in Table 1.

If both data of the fractionations with 10 kDa and 1 kDa are available, the bulk of the DOM could be separated into the following three fractions: HMW (>10 kDa), middle molecular weight (MMW: 1–10 kDa) and LMW (<1 kDa). In the open ocean and coastal areas close to the ocean, their proportions to total DOC are typically 3–10%, 20–40% and 50–70%, respectively. Thus, the oceanic DOM

Table 1. Recent reports on the molecular weight distribution of DOC obtained by the ultrafiltration technique.

Study Area	Ultrafiltration Type	DOC (μM)	HMW % of DOC		Reference
			M.W.>10 kDa	M.W.>1 kDa	
Open Ocean & Coastal Area					
Pacific (off Hawaii)	Amicon: S10N1	38-82	-	22-33	Benner *et al*. (1992)
Western North Pacific	Amicon: YM10, YM2	70-88	3.8-4.9	30-37	Ogawa & Ogura (1992)
Middle Atlantic Bight	Amicon: S10N1, H10P10	45-102	2.5-10	31-51	Santschi *et al*. (1995); Guo *et al*.(1995)
Gulf of Mexico	Amicon: S10N1, H10P10	47-214	2.9-14	31-61	Guo *et al*. (1994); Santschi *et al*. (1995) Guo *et al*. (1995)
Eutrophic Coastal & Estuarine Area					
Tokyo Bay (Japan)	Amicon: YM10, YM2	95-291	7.7-16	32-43	Ogawa (1992)
Bedford Basin (Canada)	Millipore: PTGC	56-158	3.2-17	-	Kepkay *et al*.(1993); Niven *et al*.(1995)
Rhone delta (France)	Millipore: PTGC	82-148	8.2-22	-	Dai *et al*. (1995)
Venice Lagoon (Italy)	Millipore: PTGC	170-312	10-25	-	Martin *et al*. (1995)
Gulf of Mexico	Amicon: S10N1	152-451	-	30-62	Amon & Benner (1996)
River					
Amazon River	Amicon: S10N1	269-755	-	64-88	Benner & Hedges (1993)
	Amicon: S10N1	378-801	-	81-88	Amon & Benner (1996)
Trinity River	Amicon: S10N1, H10P10	495-575	11	68-70	Santschi *et al*. (1995); Guo *et al*.(1995)
Rhone delta (France)	Millipore: PTGC	144	30	-	Dai *et al*. (1995)
Venice Lagoon (Italy)	Millipore: PTGC	117	26	-	Martin *et al*. (1995)

is generally characterized by the large contribution of LMW (<1 kDa) and a small HMW (>10 kDa) fraction. These findings seem to be generally consistent with the results of the 1980's (Romankevich, 1984; Carlson *et al.*, 1985).

Guo *et al.* (1995) attempted to separate the MMW-DOM (1–10 kDa) further into 1–3 kDa and 3–10 kDa fractions, using a cartridge with a NMWCO of 3 kDa (Amicon Co.), for the oceanic samples in the Gulf of Mexico and the Middle Atlantic Bight. Then, these fractions accounted for 24% and 7–14% of the total DOC, respectively. It is interesting that the whole size distribution of the oceanic DOM could be characterized by the tendency that the smaller size fraction has the larger portion of DOM.

Although there seems to be little data on the vertical change in the size distribution of DOM in the oceanic environment, several studies have found that the contribution of the HMW fraction to the total DOC decreased with increasing depth (Benner *et al.*, 1992; Ogawa and Ogura, 1992; Guo *et al.*, 1995), suggesting a large size fraction might be more reactive.

Difference in the size distribution between marine and terrestrial DOM

The size distribution of the terrestrial DOM in river water is apparently different from that of marine DOM. It represents the high percentages of HMW fractions (>1 kDa, >10 kDa), with a relatively small contribution of the LMW (<1 kDa) fraction, compared with seawater DOM. On average, 70–85% of the total DOC could be recovered in the fraction of more than 1 kDa. For the eutrophic coastal and estuarine water samples in which both marine and terrestrial DOM would be included, the size distribution of DOM generally seemed to be intermediate between those of the two sources of DOM.

In Fig. 1, the all data sets from the literatures cited in Table 1 are plotted as a function of the carbon concentration in each size fraction (>10 kDa, >1 kD and <1 kDa) or its percentage of total DOC vs. total DOC concentrations. Since most of the concentrations of DOC in the open ocean water appear to be within the range of 35–90 μMC (Sharp *et al.*, 1995; Carlson and Ducklow, 1995; Thomas *et al.*, 1995; Guo *et al.*, 1995; Peltzer and Hayward, 1996), DOM in the coastal or estuarine environment with a high DOC concentration of more than 200 μMC is expected to contain a significant amount of the terrestrial source. Figures 1a and b indicate that the organic carbon concentrations in the HMW fractions increase with an increasing of the total DOC. Moreover, the percentages of the HMW fractions also show the similar tendency that they increase with the total DOC concentrations (Figs. 1d and e), suggesting the occurrence of a terrestrial source in the coastal and estuarine waters with high DOC concentrations and its apparent disappearance in the oceanic environments. By contrast, the relative contribution of the LMW fraction (<1 kDa) to the total DOC decrease with increasing total concentrations (Fig. 1f) although their concentrations increase with the total DOC (Fig. 1c). Also in individual cases, some works found that the contribution of the HMW (>10 kDa and/or >1 kDa) fraction to total DOC decreased from nearshore to offshore, while the predominance of the LMW fraction was relatively elevated (Ogawa, 1992; Dai *et al.*, 1995; Guo *et al.*, 1995; Amon and Benner, 1996).

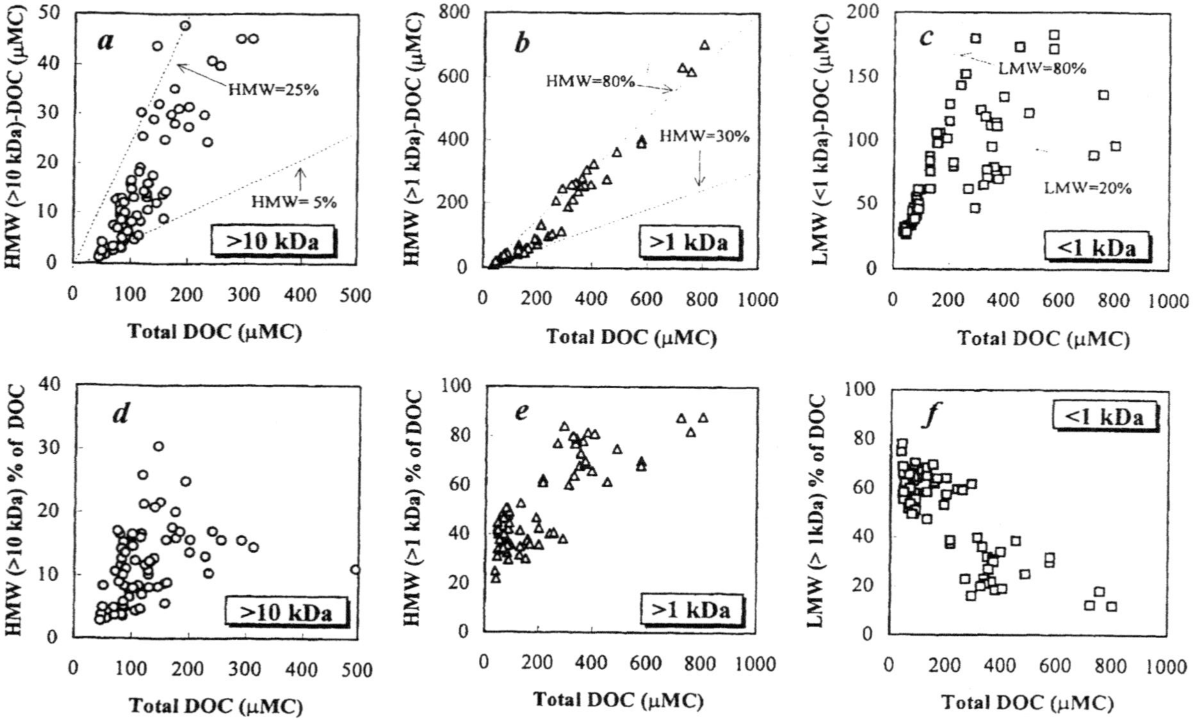

Fig. 1. The carbon concentration of each size fraction of DOM (a, b, c) and its proportion to the total DOC (d, e, f) as a function of the total DOC concentration. The results for the HMW fractions >10 kDa and >1 kDa are shown in a, d (symbols of circle) and b, e (triangle), respectively, while figures c, f (square) indicate those for the LMW-fraction <1 kDa. All the data from the literature cited in Table 1 were used. The closed and open symbols indicate the data of seawater (including oceanic, coastal and estuarine water) and river water, respectively.

Size fractionation of LMW-DOM

There is little information available on the detailed size distribution of the LMW (<1 kDa) fraction although this accounts for the largest portion (~60%) of the seawater DOM. Several previous and recent studies have used the ultrafiltration membrane with a NMWCO of 0.5 kDa (UM-3, YC-05, UAM-50, Amicon Co.: Barber, 1968; Ogura, 1974; Romakevich, 1984; Amador *et al.*, 1990). Ogura (1974) found the LMW fraction (<0.5 kDa) to be 22–43% (av. 30%) of the total DOC in the eutrophic coastal area of Tokyo Bay, while Romakevich (1984) and Amador *et al.* (1990) indicated that 60–80% of the total DOC was in the LMW (<0.5 kDa) fraction for the coastal and oceanic water samples. These suggest the possibility that a significant portion of what is generally called "LMW-DOM" might be considerably small in molecular size, probably several hundreds Da, although it would require that these previous data should be re-evaluated because the carbon mass balance during the ultrafiltration was not determined or unsatisfactory determined in most cases.

Recently, we have tried to apply the reverse osmotic (RO) filtration method to isolate the LMW-DOM in seawater. The employed system (Remolino; Millipore Co.) was usually equipped with a RO membrane filter with a NMWCO of 0.4 kDa. This process, under the high pressure of nitrogen gas (usually 30 kg cm^{-2}), excludes the very small size of DOM (<0.4 kDa) passed through a membrane, while the retentate fraction (>0.4 kDa) is concentrated. A part of the preliminary results of its application to a variety of marine environment samples is noted in Table 2 (Ogawa and Koike, unpublished data).

The results of normal ultrafiltration for separating the HMW-DOM (>10 kDa) using a NMWCO of 10 kDa membrane were shown together for comparative purposes. The HMW-DOM fractions (>10 kDa) recovered by this ultrafiltration were only a small portion of the bulk DOM, i.e. 9–14% (av. 12 ± 2%) of the total DOC, which are consistent with the previous results, while the RO filtration system with 0.4 kDa could recover a large portion of DOM, i.e. 68–97% (av. 78 ± 14%) of the total DOC. The tendency to indicate relatively higher recovery for the coastal samples (86–97%) compared to the oceanic ones (62–79%) also seems to be consistent with the general results obtained from the ultrafiltration described above.

Since the carbon mass balance during this process is generally good (93–109%) as well as that for the 10 kDa ultrafiltration (98–106%), these suggest that the very small size fraction (<0.4 kDa) might contribute less to the bulk DOM (3–38%, av. 22 ± 14% of DOC) than has been previously recognized (Romankevich, 1984). Although fractionation with 1 kDa was not conducted in this case, assuming that the general proportions of the LMW (<1 kDa) fraction to the oceanic DOM would be relatively constant (*ca.* 60–80%), it is possible that a considerable portion of the oceanic DOM might be in the 0.4–1 kDa fraction (20–60%).

To separate the seawater DOM into further small size fractions, the RO

Table 2. Low molecular weight fractionation of DOC using reverse osmotic filtration with a 0.4 kDa or 0.1 kDa membrane (Remolino system; Millipore Co.), as compared with the general ultrafiltration for high molecular weight fractionation (10 kDa) (Ogawa and Koike, unpublished).

	Western North Pacific[+]						Coast in Japan (surface water)				
Sampling sites	Stn.A (15m) [45° N,165° E]		Stn.B (100m) [24° N,170° E]		Stn.D2' (50m) [0° N,159° W]		Otsuchi (Jul.'94) [39° N,142° E]		Misaki (May '94) [35° N,140° E]		
NMWCO (Dalton)	10 k	0.4 k	10 k	0.4 k	10 k	0.4 k	10 k	0.4 k	10 k	0.4 k	0.1 k
[A] Initial DOC # (μMC)	-	59.5	65.6	65.6	61.7	61.7	79.4	79.4	94.9	94.9	94.9
[B] Filtrate (μMC)	-	16.0	57.8	15.9	57.0	25.6	68.1	10.1	81.0	21.1	18.3
[C] Retentate## (μMC)	-	46.8	5.7	44.8	7.5	38.3	11.5	76.7	12.2	78.0	81.4
* Recovery of Retentate (%)	-	79	9	68	12	62	14	97	13	82	86
** Mass Balance (%)	-	106	97	93	105	104	100	109	98	104	105

+The KH-93-4 cruise of R/V *Hakuho-maru* (Oct.–Nov., 1993).
#Seawater samples were all filtered through 0.1 μm Omnipore PTFE membrane filters (Millipore Co.).
##The data corrected for concentration factors (×2–12) were shown.
*[C]/[A] × 100%; **([B] + [C])/[A] × 100%.
The DOC concentrations were all determined by the HTCO method (Shimadzu TOC-5000).
The repeatable precision for triplicate measurements was typically ±0.5–1.5 μMC.

filtration using a membrane with a NMWCO of 0.1 kDa was preliminarily performed for a coastal sample (Misaki, the coast in Japan). The recovery of a fraction of more than 0.1 kDa corresponded to 86% of the total DOC, which was approximately comparable to the proportion for a fraction of more than 0.4 kDa (82%). As far as this one result is concerned, it may probably imply that most of materials in the small size fraction of less than 0.4 kDa are actually much smaller molecules of less than 0.1 kDa. However, further examination is necessary to confirm the resolution for this method to separate the LMW compounds in seawater. Moreover, the present result of carbon mass balance between before and after fractionation (105%) is not so satisfactory to accurately evaluate the more detailed size distribution of the LMW-DOM, of which the carbon concentrations are considerably low.

CHEMICAL IDENTIFICATION OF DOM COMPONENTS

As described at the beginning of this text, it has been normal to consider that the bulk chemical aspects of DOM are represented by the contribution of component molecules to the bulk amount (usually DOC), which could be detected by some traditional biochemical methods. Then, the general conclusion, however, has not been revised up to the present, i.e. the amount of identified components (mainly amino acids, carbohydrates, and lipids) could account for less than 30% of the bulk DOM in seawater (Williams and Druffel, 1988).

To resolve this problem, some approaches have been attempted to break the stagnation and progress in the elucidation, depending on varying viewpoints. In this section, they will be divided into the following three items.

(1) To improve the traditional methods for identifying thebiomolecules of DOM;

(2) To characterize DOM as humic-like substances;

(3) To identify biomolecules with size distribution.

Improvement of the traditional hydrolysis methods to determine combined biomolecules of DOM

It is one viewpoint that the real state of biomolecules dissolved in seawater would be somewhat different from its original condition in the living matter, and partly altered to be so complex in the structure and/or form of bond to other molecules that the normal biochemical method could not completely detect them. Then, it would be necessary to modify the previous analytical means in order to be more proper for the identification of the organic compounds in seawater.

Keil and Kirchman (1991) examined an application of the vapor-phase hydrolysis method (Tsugita *et al.*, 1987) to the HPLC-measurements of dissolved combined amino acids (DCAA) concentrations in seawater, instead of the traditional standard method using acid hydrolysis in the liquid phase (Parsons *et al.*, 1984). In consequence of the comparison between the two methods for a variety of marine samples, the concentrations of DCAA measured using the new methods were consistently higher than those measured by the conventional method. The ratio of vapor/liquid averaged 1.54 ± 0.43 ($n = 62$) with a maximum of 3.0. Further examination combined with molecular weight fractionation indicated that DCAA resistive to the liquid-phase acid hydrolysis were mainly in the HMW-DOM (>10 kDa) while there was no significant difference between the two results for the LMW-DOM fraction (<10 kDa).

Moreover, their results also showed the hydrolyzable DCAA only by the vapor-phase method were relatively abundant in the oceanic water (liquid/vapor = 2.00 ± 0.80 for five samples in the subarctic Pacific) rather than the estuarine and coastal samples. The concentrations of DCAA measured by the previous hydrolysis method have been estimated to account for ~15% of the total dissolved organic nitrogen (DON) in the ocean (Sharp, 1983). Thus, to re-evaluate the scale of the amino acids pool relative to the entire DOM in the sea, the further accumulation of the data of DCAA concentrations obtained by the vapor-phase method would be necessary.

On the other hand, an improved method for measuring the total amount of dissolved carbohydrate (DCHO) was presented by Pakulski and Benner (1992). The original method comprises a hydrolysis step with 0.1 N HCl followed by the colorimetric detection of formaldehyde produced from the periodic acid oxidation of monosaccharide-derived alditols with 3-methyl-2-benzothiazolinone hydrazone hydrochloride (MBTH; Johnson and Sieburth, 1977; Burney and Sieburth, 1977). In the new method, hydrolysis with 12 M H_2SO_4 was applied instead of HCl to more efficiently hydrolyze the structural polymeric carbohydrates. The original method using concentrated sulfuric acid has been employed to extract the cellulose carbohydrates in lake or coastal sediments, which would be mainly derived from terrestrial higher plants (Rogers, 1965; Cowie and Hedges, 1984).

Consequently, this method, improved for dissolved polymeric carbohydrates, gave 2–5 times higher concentrations of total DCHO than those by the previous hydrolysis method for various oceanic samples, which accounted for 10–46% (av. 21 ± 7%) of the total DOC (Pakulski and Benner, 1994).

According to a crude estimation from the results of the DCAA and DCHO concentrations obtained by the above two improved methods, it is possible that the amount of amino acids and carbohydrates would correspond to approximately half of the total DOM in the surface of the ocean, although their simultaneous measurements seem not to have been reported yet.

Characterization of DOM as a humic-like substance

It is a traditional point of view that a large portion of DOM, which could not be identified for any general biomolecules, might considerably change their original structure through some biogeochemical processes (humification). Several chemical assays for the characterization of humic substances have been traditionally applied to seawater DOM. In particular, the following two approaches have been consistently in general use.

One is solid phase extraction on macroreticular resins of the XAD type

(Mantoura and Riley, 1975), followed by some chemical analyses of the extracts, e.g. elemental composition (Stuermer and Harvey, 1974; Meyers-Schulte and Hedges, 1986; Hedges *et al.*, 1992), ^{1}H and ^{13}C NMR spectroscopy (Stuermer and Harvey, 1974; Harvey *et al.*, 1983; Hedges *et al.*, 1992), lignin-derived phenols composition (Meyers-Schulte and Hedges, 1986), amino acids composition (Gagosian and Stuemer, 1977; Hubberten *et al.*, 1994), natural abundance of stable ($\delta^{13}C$: Stuermer and Harvey, 1974; Gagosian and Stuermer, 1977; Meyers-Schulte and Hedges, 1986) and radio ($\Delta^{14}C$: Druffel *et al.*, 1992) carbon isotope ratio. The other is to estimate the contribution of the humic substances from the fluorescent characteristics of the bulk DOM (Traganza, 1969; Hayase *et al.*, 1988; Chen and Bada, 1992; Mopper and Schultz, 1993; Coble, 1996).

(a) XAD-extracted humic-like substance

The category of aquatic humic substances is operationally defined by fractionation using a macroreticular resin (Amberlite XAD-2 is currently in general use). It is based on the conceptual model that the aquatic humic substances would be essentially hydrophobic, but behave like weak organic acids that are ionized to be soluble in seawater under its ambient pH (~8). Thus, on lowering the pH in sample water (<2), they are completely adsorbed onto the XAD resin which has hydrophobic properties due to a lack of any functional group. Then, since they are to be desorbed from the resin under greater pH, they are recovered with alkaline and organic eluents (normally NH_4OH/CH_3OH). Extracted organic materials are generally called "humic substances", which could be operationally fractionated further into the soluble and insoluble fractions under acid conditions (called "fulvic acids" and "humic acids, respectively).

As mentioned above, the XAD method to isolate some parts of DOM from seawater has been successfully combined with advanced instrumental analyses to detail their chemical structure and origin. Studies using ^{1}H and ^{13}C NMR spectroscopy (Stuermer and Harvey, 1974; Harvey *et al.*, 1983; Hedges *et al.*, 1992) have suggested that the marine humic substances are characterized by highly branched aliphatic carbon skeletons and distinctly differs in the absence of aromaticity from the riverine dissolved humic substances. Meyers-Schulte and Hedges (1986) estimated that the contribution of terrestrial origin to the marine dissolved humic pool would be small (<10%) by determining the lignin-derived phenols and $\delta^{13}C$ in the XAD extracts from a large volume of seawater in the East Equatorial Pacific Ocean. The measurements of $\Delta^{14}C$ by Druffel *et al.* (1992) indicated that the ^{14}C-ages of dissolved humic fraction are about ~1500 years older than those of bulk DOC in the Sargasso Sea of the western North Atlantic Ocean.

In contrast with the relatively high achievement of chemical characterization, the XAD method has an essentially critical defect in evaluating the contribution of the humic substances obtained by this method to the bulk chemical composition of DOM. That is, this "humic substance" only means an operationally defined fraction, which is unlikely to include any distinct boundaries based on strong chemical evidence. As pointed out by Ishiwatari (1993), some part of identifiable biomolecules would be contained in this humic fraction, similarly, the fraction not extracted with the XAD resin would include a humic-like substance. In fact, Hubberten and Kattner (1994) indicated that on average 40–70% (almost 100% at the deep layer) of TDAA (total dissolved amino acids) was found in the dissolved humic fraction isolated with the XAD-2 resin in the Greenland Sea.

On the other hand, the XAD-2 fraction isolated from seawater in the East Equatorial Pacific accounted for only 5–15% of the concentrations of total DOC (Meyers-Schulte and Hedges, 1986) and, in general, also less than 20% (Ishiwatari, 1993). Thus, there seems no evidence that the XAD extractable humic substance would be representative of the unknown fraction that accounts for a large part of the bulk DOM in seawater.

The contribution of nitrogenous compounds to a dissolved humic substance is still controversial, including the pathways of humic formation (Laane, 1984; Harvey, 1984). The atomic C/N ratios of the XAD-2 humic substances in the Central and Equatorial Pacific were measured to be 32–47 with a solid CHN analyzer (Meyers-Schulte and Hedges, 1986; Hedges *et al.*, 1992). Although they are lower than those of the riverine humic substances from the Amazon River waters (44.6 ± 5.2: Ertel *et al.*, 1986), they are much higher when compared with that of the marine planktonic source (the Redfield Ratio = 6.6). Moreover, they are higher than the bulk DOC/DON ratios, e.g. ~25 in the Pacific Ocean using the UV oxidation method (Jackson and Williams, 1985), 16–38 in the Southern California Bight using the high temperature catalytic oxidation method (HTCO: Hansell *et al.*, 1993), and those of the HMW-DOM (>1 kDa) isolated from the deep or surface oceanic waters in the Gulf of Mexico and the Pacific Ocean (13–22 by the CHN analyzer: Benner *et al.*, 1992; McCarthy *et al.*, 1993). These suggest a nitrogen depletion in the marine humic substances relative to the bulk or the HMW-DOM.

However, Lara *et al.* (1993) indicated that, on average, 44% of the total DON was found in the XAD-2 humic fraction in the Greenland Sea and, furthermore, that amino acids accounted for 31% of this humic nitrogen (Hubberten and Kattner, 1994). This estimation is unlikely to be consistent with the previous findings for marine humic substances, that is, the low yield of DOC and the high C/N ratio (Meyers-Schulte and Hedges, 1986). Further examination would be necessary to elucidate the variety of the C/N ratio of marine humic substances, which could result from two possible sources of procedural and sample differences.

To increase the yield from the isolation of dissolved humic substances from seawater, the examination of a C-18(ODS)-hydrophobic solid-phase extraction was attempted instead of the XAD resin (Mills *et al.*, 1987; Amador, 1990). It appeared that the C-18 extraction gave slightly higher recoveries of DOM, but generally less than 40%. Fu and Pocklington (1983) indicated that the isolation method combining XAD-2 resin and activated carbon could recover the seawater DOC with high yields of ~90%, although little information on the application of this method seemed to be presented by other studies.

Lara and Thomas (1994) showed that the fractionation of DOM by the standard XAD method would miss about 15% of ^{14}C-labeled DOM from a marine diatom culture because of the hydrophobic fraction which was tightly bounded to the resin and could not be eluted. Then they found that this hydrophobic dissolved humic fraction consistently increased with degradation of the algal cells and accounted for up to 40–50% of the total labeled DOC after 7–8 months, suggesting that this operationally missing fraction would be a major component of recalcitrant humic substances in seawater (Lara and Thomas, 1995).

(b) Characterization of dissolved humic-like substances by the fluorescent properties

The fluorescent properties of DOM have been widely used to elucidate the distribution of dissolved humic substances in a variety of marine environments (Traganza, 1969; Hayase *et al.*, 1988; Chen and Bada, 1992; Mopper and Schultz, 1993; Coble, 1996). Since this method is very simple and without any inherent problems in extraction, hydrolysis and derivatization procedures in other wet-chemistry approaches, most of the findings appear to be very consistent. However, this method only exhibits some characteristics of DOM, but does not give any direct evidence for the abundance of specific components.

The vertical distribution of this fluorescence in the ocean is known to be unique. The typical profiles of the concentrations of the bulk and identifiable biogenic components of DOM indicate a tendency to decrease downward from the surface maximum, subsequently to be relatively constant in the deep layer below about 1,000 m in depth (e.g. Druffel *et al.*, 1992). This suggests a surface source for these components. By contrast, the distributions of humic-like substances determined by the fluorescent intensities (typically, Ex/Em = 320–360/420–460 nm) are characterized by the surface minimum followed by a downward increase with the maximum in the mid depth (~1000 m), subsequently gradually decreasing with increasing depth to the deep layer (e.g. Mopper and Zhou, 1990; Mopper *et al.*, 1991; Chen and Bada, 1992). Then it is recognized that the fluorescent humic substances would disappear in the surface layer by photochemical bleaching.

Recently, high-resolution fluorescence spectroscopy using the excitation-emission matrix (EEM) has been introduced into the studies on seawater DOM (Coble *et al.*, 1993; Mopper and Schultz, 1993; Coble, 1996). The EEM spectra of DOM in various natural waters showed the existence of some fluorescent components characterized by each excitation and emission maximum (Coble, 1996). Mopper and Schultz (1993) indicated that the seawater DOM fluorescence could be categorized into two groups, i.e., one is a previous humic-type fluorescence and the other is of a protein-type. The vertical distribution of the latter in the Sargasso Sea was markedly different from that of the former, which is similar to general biogenic components such as DCAA and DCHO.

Identification of biomolecules with size fractionation of DOM

DOM is composed of various size molecules. Assuming that the chemical composition of DOM would vary with its size classes, it might more or less reduce

Table 3. New approaches and findings of chemical components of the seawater DOM during 1985–1995.

Main Component	Method: Isolation	Method: Isolates % of the bulk	Method: Seawater Vol. (liters)	Method: Detection	Amount or Contribution	Study Area	Reference
Polysaccharide heteropolysaccharide β-Glucan	Dialysis/GPC Sephadex G-25 > 4 kDa	20-33 % of total DCHO	5-10	GC, GC/MS IR spectroscopy	39.5-133 μg l^{-1} 1.9-6.6 % of total DOC	Mikawa Bay (Japan)	Sakugawa & Handa (1985)
Lignin lignin-derived phenols	XAD-2 resin (Humic substance)	5-15 % of total DOC	400	GC, GC/MS	~0.2 % of humic-OC	Eastern Equatorial Pacific	Meyers-Schulte & Hedges (1986)
Carbohydrates C-O, O-C-O bond	Ultrafiltration Amicon: DC10L > 1 kDa	22-33 % of toal DOC	200	^{13}C-NMR	2-13 μMC 25-54 % of HMW-OC	North Pacific (ALOHA)	Benner *et al.* (1992)
Plant pigments chlorophyll *a, b* carotenoids	Ultrafiltration Amicon: DC10L > 1 kDa	31-53 % of total DOC	200	HPLC	~0.8 μg /mg HMW-OC	Gulf of Mexico	Bianchi *et al.* (1995)
Protein	Ultrafiltration Millipore:PELICON > 10 kDa	-	60	SDS-PAGE	2-12 % of total DON	North Pacific	Tanoue (1995)
^{14}C ages of HMW-DOM	Ultrafiltration Amicon: DC10,30L	of total DOC		TAMS		Gulf of Mexico Middle Atlantic	Santschi *et al.* (1995)
	> 1 kDa	31-63 %	~200		380-4500 y BP		
	>10 kDa	5-8%	~1000		Modern		

the value of pursuing bulk chemical characterization. Actually, most of the studies after 1990 on the chemical characterization of DOM have used the approach combining chemical identification and size fractionation.

From a technical point of view, the introduction of a high-performance (cross- or tangential-flow type) ultrafiltration system into the marine DOM study has been largely responsible for the recent advancement in the chemical characterization of the colloidal or HMW-DOM (Table 3). Most cases have succeeded in detecting a DOM component in seawater through sample preparation by this ultrafiltration technique, which enables high concentrations of the HMW-DOM to be obtained from a large volume of seawater (hundreds of liters scale).

Benner *et al.* (1992) isolated the HMW-DOM fraction (>1 kDa) with a cross-flow type ultrafiltration system (DC10L, Amicon Co.) from a large volume of oceanic water (200 L) in the North Central Pacific. The chemical structures of these isolates were determined by the ^{13}C-NMR technique. Consequently, they found that carbohydrates could account for approximately 50% of the HMW-DOM fraction in the surface and decrease to about 25% in deeper layers, suggesting that the HMW-DOM in the oceanic surface would be characterized by more abundant biomolecules and more reactive behavior rather than by little known and refractory components.

A similar conclusion was obtained from the determination of the ^{14}C ages with Tandem accelerator mass spectrometry (TAMS) and the plant pigments with a high performance liquid chromatography (HPLC) for the HMW-DOM fractions (>10 kDa or >1 kDa) isolated from the coastal and oceanic waters of the Gulf of Mexico and the Mid Atlantic (Santschi *et al.*, 1995; Bianchi *et al.*, 1995).

Tanoue (1995) first succeeded in detecting dissolved protein molecules in the oceanic waters of the North Pacific using the combining technique of the concentration of HMW-DOM (>10 kDa) by the tangential-flow ultrafiltration system (PELICON, Millipore Co.) and the separation and detection of protein molecules by sodium dodecylsulfate-polyacrylamide gel electrophoresis (SDS-PAGE). One of the major isolated protein molecules was further investigated by N-terminal sequence analysis, and its sequence revealed a homologue of a trans-outer-membrane channel protein (porin P) of Gram-negative bacteria, suggesting bacterial membranes would be a possible source of the HMW-DOM in seawater (Tanoue *et al.*, 1995).

Although there are some differences among these studies in the size range of HMW-DOM and the sampling sites, all cases seem to consistently suggest that the HMW-DOM components would be more biogenic rather than an unidentifiable fraction, as have been previously recognized.

Ogawa (1992) investigated the concentrations and compositions of DCAA (by the OPA-HPLC method after liquid-phase hydrolysis) and DCHO (by the acetylation-GC method after hydrochloric acid hydrolysis) with DOC in the two HMW-DOM fractions (>10 kDa and >1 kDa) isolated from the surface waters in the eutrophic coastal area of Tokyo Bay. The large size fraction of more than 10 kDa consistently had small portions of the total DOC (8–16%) rather than the

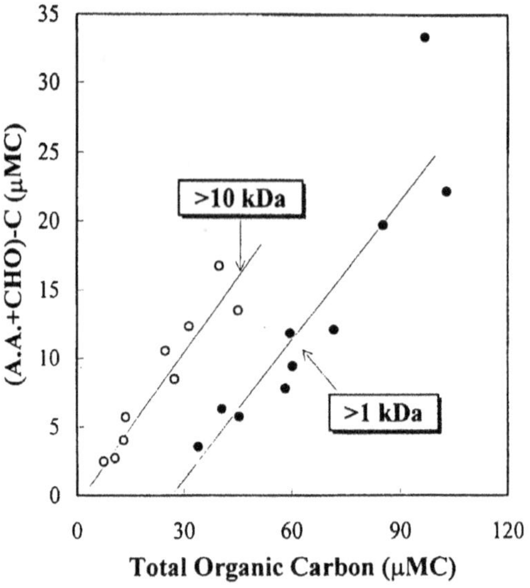

Fig. 2. The contribution of the biomolecular carbon (sum of combined amino acids and carbohydrates) to the total organic carbon for two HMW-DOM fractions (>10 kDa and >1 kDa) isolated by ultrafiltration (YM10 and YM2, respectively: Amicon Co.) from the surface waters at the same station of Tokyo Bay during March, 1990-January, 1991 (Ogawa, 1992). After samples were treated by liquid-phase hydrolysis (6N HCl, 100°C, 24 hrs), twenty liberated amino acids were determined by the fluorometric detection of their OPA-derivatives separated by a reversed-phase HPLC system. The measurements of carbohydrates were performed using the capillary GC-FID method for the acetylated alditols derived from eight neutral monosaccharides in hydrolyzate (1N HCl, 100°C, 7 hr). The total organic carbon was determined by the high temperature catalytic oxidation (HTCO) method (Shimadzu TOC-5000). The carbon concentrations of the amino acids and carbohydrates were calculated from the carbon contents of individual molecules. Regression equations were:

>10 kDa: $Y = (0.36 \pm 0.11)X - (0.1 \pm 2.9)$, $r = 0.949$ ($n = 9$,)

>1 kDa: $Y = (0.37 \pm 0.13)X - (10.4 \pm 9.1)$, $r = 0.914$ ($n = 10$,)

where X = total organic carbon, Y = amino acids plus carbohydrates carbon.

fraction of more than 1 kDa (32–43%), as shown in Table 1.

On the other hand, the contributions of carbon amounts in the DCAA and DCHO to the total carbon were significantly higher for the large size fraction >10 kDa (26–43%, av. 35 ± 6%) than the fraction >1 kDa (10–34%, av. 18 ± 7%). Assuming that amino acids and carbohydrates could account for the major part of biomolecules, this implies that the HMW-DOM (>10 kDa) would relatively abound in the biogenic molecules, while the smaller size fraction of between 1 and 10 kDa would be mainly composed of unknown materials.

As shown in Fig. 2, good correlations were found between the biomolecules (amino acids + carbohydrates) carbon and the total carbon concentrations for both fraction sizes. The two equations had similar slopes, but different constants as indicated by x-intercept. Then, DOM in each size fraction was estimated to be

Table 4. Three component categories of DOM applied to the two size fractions isolated from the coastal seawaters of Tokyo Bay*.

Category	> 10 kDa	>1 kDa
Intact Biomolecules	35	18
Associate Biomolecules	64	37
Background Substance	1	45
Total	100 %	100 %

*The mean values for the proportion of each class carbon to the total carbon in the two size fractions are shown, which were calculated from the data used in Fig. 2 as follows:

Intact Biomolecules: identified portion with combined amino acids and carbohydrates
$(B/T) \times 100\%$;

Associate Biomolecules: unidentified portion related with Intact Biomolecules
$((S - B) / T) \times 100\%$;

Background Substance: unidentified portion independent of Intact Biomolecules
$(I/ T) \times 100\%$,

where

B: amount of amino acids plus carbohydrates' carbon;
T: total organic carbon;
I: X-intercept in the correlation between T (X-axis) and B (Y-axis);
S: T – I (organic carbon dependent on the slope in the relation).

categorized into three possible classes (Table 4).

The first is the class of "Intact Biomolecules", which corresponds to the amount of DCAA and DCHO identified by standard methods. The second class, named "Associate Biomolecules", is the organic carbon fraction that is neither identifiable to amino acids or carbohydrates, but co-varying with them at a constant mass ratio, as indicated by the slope value. It is possible that this fraction might include other biomolecules such as lipids and nucleic acids. In this case, the contribution of the second class would be somewhat overestimated because these possible components should belong to the first category. Another possibility is that this fraction might be originally composed of biomolecules, but they would slightly change their original structure or bonding form to other molecules, so that they could not be extracted and/or detected by the standard methods. In particular, the tightly combined amino acids and carbohydrates possibly might belong this class because this study used the normal hydrolysis method for extracting DCAA and DCHO.

The last class, named "Background Substance", is another carbon fraction that could not be identified with any biomolecules and which has a constant concentration indicated by the x-intercept in Fig. 2. It is possible that if these components could be derived from biomolecules, their original structure would be considerably changed and nothing would remain but their unreactive portions to a variety of biological and/or physicochemical processes, so that they could not be identified with any biomolecules and their variation would be independently constant of intact biomolecules.

As shown in Table 4, the smaller size fraction includes relatively much more "Background Substance" rather than biomolecules, while in the large size fraction, there is little "Background Substance" and most of the unidentifiable organic carbon belongs to the "Associate Biomolecules" class. It is surprising that we can find similar results in a few previous studies in no-marine environments. Hama and Hnada (1980) separated DOM in Japanese lake waters into some size fractions by a gel-filtration technique (Sephadex G-25) and subsequently determined some biomolecules in each fraction. Their results showed that the HMW fractions were more abundant in the carbohydrates, proteins and amino acids (about 35–50% of the fraction >1.5 kDa) than the LMW-DOM fractions (about 3–10% of <0.2 kDa) which had the largest portion of the bulk DOM (about 35–70% of total DOC). Akiyama (1972) indicated that the HMW-DOM (>50 kDa) in degradation products of green algae consisted of high percentages of proteins and carbohydrates, while most of the LMW-DOM (<0.7 kDa) could not identified.

CHEMICAL ASPECTS AND BIOGEOCHEMICAL REACTIVITY OF DOM IN SEAWATER

There are, in general, two sources of DOM in the marine environment. One is the terrestrial origin transported through river input, and the other is the biological production in the euphotic layer of the marine water column. It is possible that we could find their behavior and reactivity in real marine environments through their qualitative and quantitative variations along the transect from nearshore to offshore in the coastal area or from the surface to deep layer in the ocean. Some recent works have found that the carbon concentrations of the HMW-DOM and its proportions to the total DOC were likely to decrease along the above two directional transects, while the contribution of the LMW-DOM was relatively elevated (Benner *et al.*, 1992; Ogawa and Ogura, 1992; Ogawa, 1992; Dai *et al.*, 1995; Guo *et al.*, 1995; Amon and Benner, 1996). It suggests that the HMW materials might be predominantly supplied in both source sites and relatively reactive to degradation or alteration in the marine environments. By contrast, the LMW-DOM pool appears to be more stable, like a background. Actually, in the deep ocean where DOM would be farthest from the source sites, *ca.* 60–80% of DOC is in the LMW fraction of less than 1 kDa (Benner *et al.*, 1992; Guo *et al.*, 1995).

Moreover, the recent study by Santschi *et al.* (1995) on ^{14}C-measurement of the HMW-DOM separated from Atlantic oceanic and coastal waters strongly supports that the HMW-DOM (especially the >10 kDa fraction) would be more reactive compared with the LMW or bulk DOM.

Some experimental and geochemical findings suggest possible processes to which HMW-DOM is more reactive compared with the LMW fraction. These possibly include three sink processes, i.e. biological degradation, photochemical degradation, and removal by coagulation to shift to sinking particles. Because each process is detailed by separate articles in this volume, this paper will only briefly deal with it.

Biological degradation

Amon and Benner (1994, 1996) indicated that the bacterial utilization rates were 2–4 times greater for the HMW-DOM (>1 kDa) than for the LMW-DOM (<1 kDa), from a direct comparison of bacterial growth and DOC degradation rates between these two fractions separated from the surface water in the Gulf of Mexico during a diatom bloom. From these findings, they proposed a size-reactivity continuum model. It implies both the bioreactivity and the size of aquatic organic matter including particles to LMW-DOM, would have a tendency to become smaller with proceeding degradation (Amon and Benner, 1996).

Their model has no chemical evidence but the size class of DOM. However, this concept could be partly supported by categorization combining the biomolecular identification and size classification of DOM, noted in Table 4, which suggests the smaller size fractions of DOM would be less identifiable (Ogawa, 1992).

Photochemical degradation

Much experimental and geochemical evidence has suggested that aquatic humic substance would be a major reactant to photochemical degradation, e.g., photochemical production rates were much higher in river and ocean deep water than oceanic surface water (Mopper and Zhou, 1990; Kieber *et al.*, 1990; Mopper *et al.*, 1991), the high photoreactivity of the XAD isolates from various natural waters (Kieber *et al.*, 1990), and the vertical distribution of humic-derived fluorescence in the ocean with a surface minimum (Mopper and Zhou, 1990; Mopper *et al.*, 1991; Chen and Bada, 1992; Mopper and Schultz, 1993; see review by Ohta (2000, in this volume)).

However, there seem to be only a small amount of information on the photodegradation of different size classes of DOM. Kieber *et al.* (1989) found no photochemical production in the ultrafiltered seawater through a membrane with NMWCO of 0.5 kDa (YC-05, Amicon Co.), which included only the LMW-DOM (<0.5 kDa), suggesting that the HMW-DOM (>0.5 kDa) would be a major photochemical reactant in seawater. It would correspond with the findings that riverine and estuarine DOM that were generally abundant in the HMW fraction (see Table 1), appeared to be highly photo-reactive. However, some previous studies indicated that the major part of marine humic substance was in the LMW-fulvic acids fraction (<0.7 kDa: Stuermer and Harvey, 1974; <2 kDa: Thurman, 1984) and the HMW-humic acids fraction was generally the minor portion. Further studies would be necessary to elucidate the relationship between the size classes of DOM and the photo-reactivity.

Table 5. A conceptual simple classification of the marine DOM and biogeochemical reactivity of each type of DOM.

		HMW	LMW
General Biomolecules		I	II
	Biological Reactivity	Reactive semi?-labile to microbial degradation	Reactive highly labile to microbial degradation
	Physicochemical Reactivity	Reactive adsorption and/or coagulation to particles	?
Denatured Molecules		III	IV
	Biological Reactivity	Stable ?	Stable ?
	Physicochemical Reactivity	Reactive reactive to photochemical degradation	Stable ?

Coagulation removal

Recently, many studies have used ^{234}Th as a chemical tracer for the in-situ measurement of colloidal-particle interaction, which is generally estimated from ^{238}U-^{234}Th disequilibria between the dissolved and colloidal phase (usually >10 kDa or >1 kDa) (Honeyman and Santschi, 1989; Moran and Busseler, 1992; Santschi *et al.*, 1995, Niven *et al.*, 1995). They suggest that the residence time of the colloidal (HMW-DOM) fraction in the upper ocean water column would be very short (days to weeks) because of its rapid coagulation to shift to sinking particles.

Combination of chemical aspects and biogeochemical reactivity of DOM in seawater

A conceptual classification model of marine DOM is proposed in Table 5, which is based on the relation between the chemical aspects and biogeochemical (biological and physicochemical) reactivities. The seawater DOM is roughly categorized into four groups by the combination of two size classes and two component identities. Although it would be difficult to give a clear border of size between the LMW and HMW, this case assumes the size boundary would be around 1–10 kDa. Similarly, the definition of two component categories is ideal, however, "General Biomolecules" mostly indicates the group of "Intact Biomolecules" presented in Table 4, while "Denatured Biomolecules" indicates "Background Substance". Although the group of "Associate Biomolecules" in Table 4 is conceptually intermediate in this case, their reactivities would possibly resemble the former rather than the latter.

Types I and III of the HMW fraction are recognized to be biologically and/or physicochemically reactive, according to our recent knowledge. Photochemical degradation is not adapted to the HMW-general biomolecules group (type I), because a few recent studies indicated that such materials, e.g. protein, were likely to be altered rather than degraded by the photochemical reaction (Keil and Kirchman, 1993; Naganuma *et al.*, 1996). On the other hand, the coagulation process characterizes type I rather than type III because the carbohydrate fraction seems to be closely concerned with this process (Mopper *et al.*, 1995).

In the estuarine system, the HMW protein-like materials dissolved in river water were found to be flocculated to shift to large particles during the mixing of fresh water with seawater (Ogura *et al.*, 1985). There were a few previous studies indicating some portion of the riverine humic substance also was selectively removed from the estuarine water column through the flocculation (Sholkovits, 1976; Fox, 1983). However, the bulk of riverine DOC has been generally found to be conservative during the estuarine mixing process (Mantoura and Woodward, 1983; Ogawa and Ogura, 1990; Ogawa, 1992).

Type II (LMW-general biomolecules) indicates a typical biogenic monomer or oligomer such as free amino acids, oligopeptides, mono- or oligo-saccharides and organic acids. The LMW-carbonyl compounds produced from photochemical degradation of marine humic substance (type III), such as formaldehyde, acetaldehyde, glyoxal, glyoxlate and pyruvate (Mopper and Stahovec, 1986; Mopper *et al.*, 1991), belong to this group. They are well known to have very short turn-over times (minutes to hours) through marine biological processes. Therefore, they are unlikely to accumulate in seawater and show their quantitatively small contribution to the bulk DOM. On the other hand, there is a little information on the physicochemical process of LMW-biomolecules in seawater. But, even if any pathway would be possible, the microbial process could be generally predominant (e.g. Carlson *et al.*, 1985).

The rest of the LMW fraction, which could not be chemically identified, belongs to the category type IV. According to some recent information, this fraction appeared to be characterized by a stable nature. Because the majority of the bulk DOM in the ocean would be in the LMW fraction and the biomolecular components would be minor in it; the largest portion of the DOM could be in this fraction (type IV).

The quantitative contribution of each type of DOM to the bulk could be estimated to be ordered as IV >> III > (<) I >> II. The degree of each difference is partly ambiguous, such as between types I and III. However, the relation between the scale of each class pool and its reactivity would be generally consistent in principle so that the more reactive fraction would have apparently the smaller portion of DOM. It also appears to be consistent with the general finding that the nature of bulk DOM in the deep oceanic water would be quite stable, which is mainly composed of type IV-DOM.

CONCLUSIONS AND FUTURE CHALLENGE

Some significant differences in nature between the HMW and LMW-DOM would be the most important finding obtained from our recent studies on the bulk DOM in seawater. Simultaneously, this would suggest a big problem for the future.

The bulk of DOM would be mainly composed of relatively smaller size molecules rather than HMW-materials. The HMW-DOM would be constituently biogenic and biogeochemically reactive, while the LMW-DOM would be chemically unknown and unreactive. This constitution of the bulk DOM could be compared to a pyramidal structure including the particulate living matter at its top. Materials at the upper stage have a larger size and more reactivity, but less standing stock. It is apparently different from the previous general concepts that the unknown (probably humic-like) and HMW fraction might be responsible for the stable nature of the bulk DOM in seawater.

The characterization of the HMW-DOM is just beginning. Some recent studies that have achieved detailed chemical identification of the HMW-DOM (Table 3) have suggested some possible sources of DOM. Further investigations have a chance of making evident what kind of process is predominant for the DOM production in the marine biogeochemical cycle.

However, the biggest target for our future challenge should be now quite clear. It is the LMW-DOM that is classified into type IV in Table 5. What kind of materials would compose this LMW-DOM fraction? What kind of process would be concerned with the production of this type of DOM? What kind of mechanism would make it stable in the marine environment, particularly to the microbial degradation? In fact, it is just a concept that the unknown-LMW-DOM might be denatured materials. It would be another possibility even that specific and refractory LMW-biomolecules could be selectively preserved in seawater, although such a type of DOM is not included in the present classification (Table 5). Vigorous investigation focussed on the LMW-DOM in seawater would be necessary.

Acknowledgements—The original data were all obtained from a joint study with Dr. N. Ogura of Tokyo University of Agriculture and Technology or Dr. I. Koike of the Ocean Research Institute, University of Tokyo. The author thanks both professors.

REFERENCES

Akiyama, T. (1972): Chemical composition and molecular weight distribution of dissolved organic matter produced by bacterial degradation of green algae. *Geochem. J. Jap.*, **6**, 93–104.

Amador, J. A., P. J. Milne, C. A. Moore and R. G. Zika (1990): Extraction of chromophoric humic substances from seawater. *Mar. Chem.*, **29**, 1–17.

Amon, R. M. W. and R. Benner (1994): Rapid cycling of high-molecular weight dissolved organic matter in the ocean. *Nature*, **369**, 549–552.

Amon, R. M. W. and R. Benner (1996): Bacterial utilization of different size classes of dissolved organic matter. *Limnol. Oceanogr.*, **41**, 41–51.

Barber, R. T. (1968): Dissolved organic carbon from deep water resists microbial oxidation. *Nature*, **220**, 274–275.

Bauer, J. E., P. M. Williams and E. R. M. Druffel (1992): ^{14}C activity of dissolved organic carbon fractions in the North Central Pacific and Sargasso Sea. *Nature*, **357**, 667–670.

Benner, R. and J. I. Hedges (1993): A test of the accuracy of freshwater DOC measurements by high-temperature catalytic oxidation and UV-promoted persulfate oxidation. *Mar. Chem.*, **41**, 161–165.

Benner, R., J. D. Pakulski, M. McCarthy, J. I. Hedges and P. G. Hatcher (1992): Bulk chemical characteristics of dissolved organic matter in the ocean. *Science*, **255**, 1561–1564.

Bianchi, T. S., C. Lambert, P. H. Santschi, M. Baskaran and L. Guo (1995): Plant pigments as biomarkers of high-molecular-weight dissolved organic carbon. *Limnol. Oceanogr.*, **40**, 422–428.

Burney, C. M. and J. McN. Sieburth (1977): Dissolved carbohydrates in seawater. 2. A spectrophotometric procedure for total carbohydrate analysis and polysaccharide determination. *Mar. Chem.*, **5**, 15–28.

Carlson, C. A., H. W. Ducklow and A. F. Michaels (1994): Annual flux of dissolved organic carbon from the euphotic zone in the northwestern Sargasso Sea. *Nature*, **371**, 405–408.

Carlson, C. A. and H. W. Ducklow (1995): Dissolved organic carbon in the upper ocean of the central equatorial Pacific Ocean, 1992: Daily and finescale vertical variations. *Deep-Sea Res.*, **42**, 639–656.

Carlson, D. J., M. L. Brann, T. H. Mague and L. M. Mayer (1985a): Molecular weight distribution of dissolved organic materials in seawater determined by ultrafiltration: a re-examination. *Mar. Chem.*, **16**, 155–171.

Carlson, D. J., L. M. Mayer, M. L. Brann and T. H. Mague (1985b): Binding of monomeric organic compounds to macromolecular dissolved organic matter in seawater. *Mar. Chem.*, **16**, 141–153.

Chen, R. F. and J. L. Bada (1992): The fluorescence of dissolved organic matter in seawater. *Mar. Chem.*, **37**, 191–221.

Coble, P. G. (1996): Characterization of marine and terrestrial DOM in seawater using excitation-emission matrix spectroscopy. *Mar. Chem.*, **51**, 325–346.

Coble, P. G., C. A. Schultz and K. Mopper (1993): Fluorescence contouring analysis of DOC intercalibration experiment samples: a comparison of techniques. *Mar. Chem.*, **41**, 173–178.

Cowie, G. L. and J. I. Hedges (1984): Carbohydrate sources in a coastal marine environment. *Geochim. Cosmochim. Acta*, **48**, 2075–2087.

Dai, M., J.-M. Martin and G. Cauwet (1995): The significant role of colloids in the transformation of organic carbon and associated trace metals (Cd, Cu and Ni) in the Rhone delta (France). *Mar. Chem.*, **51**, 159–175.

Dawson, R. and G. Liebezeit (1981): The analytical methods for the characterization of organics in seawater. pp. 445–496. In *Marine Organic Chemistry*, ed. by E. K. Duursma and R. Dawson, Elsevier, Amsterdam.

Druffel, E. R. M., P. M. Williams, J. E. Bauer and J. R. Ertel (1992): Cycling of dissolved and particulate organic matter in the open ocean. *J. Geophys. Res.*, **97**, 15639–15659.

Duursma, E. K. (1965): The dissolved organic constituents of sea water. pp. 433–475. In *Chemical Oceanography, Vol. 1*, ed. by J. P. Riley and G. Skirrow, Academic Press, New York.

Ertel, J. R., J. I. Hedges, A. H. Devol, J. E. Richey and M. Ribeiro (1986): Dissolved humic substances in the Amazon river system. *Limnol. Oceanogr.*, **31**, 739–754.

Fox, L. E. (1983): The removal of dissolved humic acid during estuarine mixing. *Estua. Coast. Shelf. Sci.*, **16**, 431–440.

Fu, T. and R. Pocklington (1983): Quantitative adsorption of organic matter from seawater on solid matrices. *Mar. Chem.*, **13**, 255–264.

Gagosian, R. B. and D. H. Stuermer (1977): The cycling of biogenic compounds and their diagenetically transformed products in seawater. *Mar. Chem.*, **5**, 605–632.

Guo, L., P. H. Santschi and K. W. Warnken (1995): Dynamics of dissolved organic carbon (DOC) in oceanic environments. *Limnol. Oceanogr.*, **40**, 1392–1403.

Guo, L., Jr. C. H. Coleman and P. H. Santschi (1994): The distribution of colloidal and dissolved organic carbon in the Gulf of Mexico. *Mar. Chem.*, **45**, 105–119.

Hama, T. and N. Handa (1980): Molecular weight distribution and characterization of dissolved organic matter from lake waters. *Arch. Hydrobiol.*, **90**, 106–120.

Hansell, D. A., P. M. Williams and B. B. Ward (1993): Measurements of DOC and DON in the Southern California Bight using oxidation by high temperature combustion. *Deep-Sea Res.*, **40**, 219–234.

Harvey, G. R. (1984): Comment on the structure of marine fluvic and humic acids. *Mar. Chem.*, **15**, 85–87.

Harvey, G. R., D. A. Boran, L. A. Chesal and J. M. Tokar (1983): The structure of marine fluvic and humic acids. *Mar. Chem.*, **12**, 119–132.

Hayase, K., H. Tsubota, I. Sunada, S. Goda and H. Yamazaki (1988): Vertical distribution of fluorescent organic matter in the North Pacific. *Mar. Chem.*, **25**, 373–381.

Hedges, J. I. (1992): Global Biogeochemical cycles: progress and problems. *Mar. Chem.*, **39**, 67–93.

Hedges, J. I., P. G. Hatcher, J. R. Ertel and K. J. Meyers-Schulte (1992): A comparison of dissolved humic substances from seawater with Amazon River counterparts by ^{13}C-NMR spectrometry. *Geochim. Cosmochim. Acta*, **56**, 1753–1757.

Honeyman, B. D. and P. H. Santschi (1989): A brownian-pumping model for trace metal scavenging: evidence from Th isotopes. *J. Mar. Res.*, **47**, 950–995.

Hubberten, R. J. and L. G. Kattner (1994): Amino acid composition of seawater and dissolved humic substances in the Greenland Sea. *Mar. Chem.*, **45**, 121–128.

Ishiwatari, R. (1992): Macromolecular material (humic substance) in the water column and sediments. *Mar. Chem.*, **39**, 151–166.

Jackson, G. A. and P. M. Williams (1985): Importance of dissolved organic nitrogen and phosphorus to biological nutrient cycling. *Deep-Sea Res.*, **32**, 223–235.

Johnson, K. M. and J. McN Sieburth (1977): Dissolved carbohydrates in seawater. 1. A precise spectrophotometric analysis for monosaccharides. *Mar. Chem.*, **5**, 1–13.

Keil, R. G. and D. L. Kirchman (1991): Dissolved combined amino acids in marine waters as

determined by a vapor-phase hydrolysis method. *Mar. Chem.*, **33**, 243–259.
Keil, R. G. and D. L. Kirchman (1994): Abiotic transformation of labile protein to refractory protein in sea water. *Mar. Chem.*, **45**, 187–196.
Kepkay, P. E., S. E. H. Niven and T. G. Milligan (1993): Low molecular weight and colloidal DOC production during phytoplankton bloom. *Mar. Ecol. Progr. Ser.*, **100**, 233–244.
Kieber, D. J., J. McDaniel and K. Mopper (1989): Photochemical source of biological substrates in sea water: implications fro carbon cycling. *Nature*, **341**, 637–639.
Kieber, R. J., X. Zhou and K. Mopper (1990): Formation of carbonyl compounds from UV-induced photodegradation of humic substances in natural waters: fate of riverine carbon in the sea. *Limnol. Oceanogr.*, **35**, 1503–1515.
Koike, I., S. Hara, K. Terauchi and K. Kogure (1990): Role of sub-micron particles in the ocean. *Nature*, **345**, 242–244.
Laane, R. W. P. M. (1984): Comment on the structure of marine fluvic and humic acids. *Mar. Chem.*, **15**, 85–87.
Lara, R. J. and D. N. Thomas (1994): XAD-fractionation of "new" dissolved organic matter: is the hydrophobic fraction seriously underestimated? *Mar. Chem.*, **47**, 93–96.
Lara, R. J. and D. N. Thomas (1995): Formation of recalcitrant organic matter: humification dynamics of algal derived dissolved organic carbon and hydrophobic fractions. *Mar. Chem.*, **51**, 193–199.
Lara, R. J., U. Hubberten and G. Kattner (1993): Contribution of humic substances to the dissolved nitrogen pool in the Greenland Sea. *Mar. Chem.*, **41**, 327–336.
Mantoura, R. F. C. and J. P. Riley (1975): The analytical concentration of humic substances from natural waters. *Anal. Chim. Acta*, **76**, 97–106.
Mantoura, R. F. C. and E. M. S. Woodward (1983): Conservative behaviour of riverine dissolved organic carbon in the Severn Estuary: chemical and geochemical implications. *Geochim. Cosmochim. Acta*, **47**, 1293–1309.
Martin, J. M., M. H. Dai and G. Cauwet (1995): Significance of colloids in the biogeochemical cycling of organic carbon and trace metals in a coastal environment-example of the Venice Lagoon (Italy). *Limnol. Oceanogr.*, **40**, 119–131.
McCarthy, M. D., J. I. Hedges and R. Benner (1993): The chemical composition of dissolved organic matter in seawater. *Chem. Geol.*, **107**, 503–507.
Meyers-Schulte, K. H. and J. I. Hedges (1986): Molecular evidence for a terrestrial component of organic matter dissolved in ocean water. *Nature*, **321**, 61–63.
Mills, G. L., E. McFadden and J. G. Quinn (1987): Chromatographic studies of dissolved organic matter and copper-organic complexes isolated from estuarine waters. *Mar. Chem.*, **20**, 313–325.
Mopper, K. and E. T. Degens (1979): Organic carbon in the ocean: nature and cycling. pp. 293–316. In *The Global Carbon Cycle*, ed. by B. Bolin, E. T. Degens, S. Kempe and P. Ketner, John Wiley and Sons, New York.
Mopper, K. and C. A. Schultz (1993): Fluorescence as a possible tool for studying the natural and water column distribution of DOC components. *Mar. Chem.*, **41**, 229–238.
Mopper, K. and W. L. Stahovec (1986): Sources and sinks of low molecular weight organic carbonyl compounds in seawater. *Mar. Chem.*, **19**, 305–321.
Mopper, K. and X. Zhou (1990): Hydroxyl radical photoproduction in the sea and its potential impact on marine processes. *Science*, **250**, 661–663.
Mopper, K., X. Zhou, R. J. Kieber, D. J. Kieber, R. J. Sikorski and R. D. Jones (1991): Photochemical degradation of dissolved organic carbon and its impact on the oceanic carbon cycle. *Nature*, **353**, 60–62.
Mopper, K., J. Zhou, K. Ramana, U. Passow, H. G. Dam and D. T. Drapeau (1995): The role of surface-active carbohydrates in the flocculation of a diatom bloom in a mesocosm. *Deep-Sea Res.*, **42**, 47–73.
Moran, S. B. and K. O. Busseler (1992) Short residence time of colloids in the upper ocean estimated from ^{238}U-^{234}Th disequilibria. *Nature*, **359**, 221–223.
Naganuma, T., S. Konishi, T. Inoue, T. Nakane and S. Sukizaki (1996): Photodegradation or photoalteration? Microbial assay of the effect of UV-B on dissolved organic matter. *Mar. Ecol.*

Prog. Ser., **135**, 309–310.
Niven, S. E. H., P. E. Kepkay and A. Boraie (1995): Colloidal organic carbon and colloidal ^{234}Th dynamics during a coastal phytoplankton bloom. *Deep-Sea Res.*, **42**, 257–273.
Ogawa, H. (1992): Biogeochemical study on dissolved organic matter in the estuarine and coastal seawater of Tokyo Bay: the chemical characteristics and the methodology. Ph.D. Thesis, Tokyo Univ. Agric. Tech., 227 pp.
Ogawa, H. and N. Ogura (1990): Source and behaviour of organic carbon of seawater in Tokyo Bay. *Chikyukagaku* (*Geochemistry*), **24**, 27–41 (in Japanese with English abstract).
Ogawa, H. and N. Ogura (1992): Comparison of two methods for measuring dissolved organic carbon in sea water. *Nature*, **356**, 696–698.
Ogura, N. (1972): Rate and extent of dissolved organic matter in surface seawater. *Mar. Biol.*, **13**, 89–93.
Ogura, N. (1974): Molecular weight fractionation of dissolved organic matter in coastal seawater by ultrafiltration. *Mar. Biol.*, **24**, 305–312.
Ogura, N., Y. Tanaka, J. Ito and K. Takahashi (1984): Behaviour of proteinaceous substances in the estuary of the Tama River. *J. Oceanogr. Soc. Jpn.*, **40**, 184–191.
Ohta, K. (1997): Review. p. 000–000. In *xxx*, ed. by N. Handa, Terra Pub., Tokyo (in this volume).
Pakulski, J. D. and R. Benner (1992): An improved method for the hydrolysis and MBTH analysis of dissolved and particulate carbohydrates in seawater. *Mar. Chem.*, **40**, 143–160.
Pakulski, J. D. and R. Benner (1994): Abundance and distribution of carbohydrates in the ocean. *Limnol. Oceanogr.*, **39**, 930–940.
Parsons, T. R., Y. Maita and C. M. Lalli (1984): pp. 44–45. In *A Manual of Chemical and Biological Methods for Seawater Analysis*, ed. by $. xxx, Pergamon, Oxford.
Peltzer, E. T. and N. A. Hayward (1996): Spatial distribution and temporal variability of total organic carbon along 140°W in the equatorial Pacific Ocean, 1992. *Deep-Sea Res.* (in press).
Rogers, M. A. (1965): Carbohydrates in aquatic plants and associated sediments from two Minnesota lakes. *Geochim. Cosmochim. Acta*, **29**, 183–200.
Romankevich, E. A. (1984): *Geochemistry of Organic Matter in the Ocean*. Springer-Verlag, Berlin, 334 pp.
Sakugawa, H. and N. Handa (1985): Isolation and chemical characterization of dissolved and particulate polysaccharides in Mikawa Bay. *Geochim. Cosmochim. Acta*, **49**, 1185–1193.
Santschi, P. H., L. Guo, M. Baskaran, S. Trumbore, J. Southon, T. S. Bianchi, B. D. Honeyman and L. Cifuentes (1995): Isotopic evidence for the contemporary origin of high-molecular weight organic matter in oceanic environments. *Geochim. Cosmochim. Acta*, **59**, 625–631.
Sharp, J. H. (1983): The distribution of inorganic and organic nitrogen and dissolved and particulate nitrogen in the sea. pp. 1–33. In *Nitrogen in the Marine Environments*, ed. by J. Carpenter and D. Capone, Academic, New York.
Sharp, J. H., R. Benner, L. Bennet, C. A. Carlson, S. E. Fitzwater, E. T. Peltzer and L. M. Tupas (1995): Analyses of dissolved organic carbon in seawater: the JGOFS EqPac methods comparison. *Mar. Chem.*, **48**, 91–108.
Sholkovitz, E. R. (1976): Flocculation of dissolved organic and inorganic matter during the mixing of river water and seawater. *Geochim. Cosmochim. Acta*, **40**, 831–845.
Stuermar, D. H. and G. R. Harvey (1974): Humic substance from seawater. *Nature*, **250**, 480–481.
Tanoue, E. (1995): Detection of dissolved protein molecules in oceanic waters. *Mar. Chem.*, **51**, 239–252.
Tanoue, E., S. Nishiyama, M. Kamo and A. Tsugita (1995): Bacterial membranes: Possible source of a major dissolved protein in seawater. *Geochim. Cosmochim. Acta*, **59**, 2643–2648.
Thomas, C., C. Cauwet and J.-F. Minster (1995): Dissolved organic carbon in the equatorial Atlantic Ocean. *Mar. Chem.*, **49**, 155–169.
Thurman, E. M. (1985): *Organic Geochemistry of Natural Waters*. Martinus Nijhoff/Dr W. Junk Publishers, Dordrecht, 497 pp.
Traganza, E. D. (1969): Fluorescence excitation and emission spectra of dissolved organic matter in sea water. *Bull. Mar. Sci.*, **19**, 897–904.
Tsugita, A., T. Uchida, H. W. Mewes and T. Atake (1987): A rapid vapor-phase acid (hydrochloric

acid and trifluoroacetic acid) hydrolysis of peptide and protein. *J. Biochem.*, **102**, 1593–1597.

Wells, M. L. and E. D. Goldberg (1991): Occurrence of small colloids in sea water. *Nature*, **353**, 342–344.

Williams, P. J. LeB. (1975): Biological and chemical aspects of dissolved organic material in sea-water. pp. 301–363. In *Chemical Oceanography, Vol. 2*, 2nd edn. by J. P. Riley and G. Skirrow, Academic Press, New York.

Williams, P. M. (1971): The distribution and cycling of organic matter in the ocean. pp. 145–163. In *Organic Compounds in Aquatic Environments*, ed. by S. J. Faust and J. V. Hunter, Marcel Dekker, New York.

Williams, P. M. and E. R. M. Druffel (1987): Radiocarbon in dissolved organic matter in the Central North Pacific Ocean. *Nature*, **330**, 246–248.

Williams, P. M. and E. R. M. Druffel (1988): Dissolved organic matter in the ocean: Comments on a controversy. *Oceanography*, **1**, 14–17.

Dynamics and Characterization of Marine Organic Matter, Eds., N. Handa, E. Tanoue and T. Hama, pp. 339–382.

Strong Organic Ligands in Seawater: Peculiar Functional Groups in Oceanic Organic Matter-Synthesis

Katsumi HIROSE

Geochemical Department, Meteorological Research Institute, Nagamine 1-1, Tsukuba, Ibaraki 305, Japan

(Synthesis, Received 3 August 1996; accepted 16 October 1996)

INTRODUCTION

Marine chemists have been gradually paying increasing attention to strong ligands of low levels involved in dissolved organic matter (DOM) and particulate organic matter (POM) (Dawson and Duursma, 1981; Farrington, 1992). However, studies on the strong organic ligands in seawater have been focused upon only the field of metal speciation. Knowledge about the natural strong organic ligands has been scanty until now. In this report, the strong organic ligands are defined as organics with some functional groups to form complexes with metals under the conditions of seawater. The strong organic ligands, therefore, may be classified as chelators (Buffle, 1988), which can sterically form a claw-like structure with coordination bonds; EDTA (ethlenediaminetetra-acetic acid)-chelation is the commonest example given. Such strong organic ligands may be closely related to the biogeochemical behavior of chemical constituents in the marine environments: first, strong ligands in DOM and POM play a significant role in controling chemical forms and levels of trace metals in seawater; second, strong ligands link the interaction between particulate matter and trace metals including radionuclides, which is related to removal processes of "so-called" particle-reactive metals (Hirose *et al.*, 1992); third, strong ligands are related to the bioavailability of trace metals from marine organisms (Barber and Ryther, 1969; Hirose and Sugimura, 1985). On the other hand, if strong ligands are well characterized chemically, they may become an indicator for understanding the behavior of organic matter in marine environments.

Chemical speciation of trace constituents, including inorganic and organic characterization, is one of the most important problems in the field of marine chemistry (Sillen, 1961; Buffle, 1988; Stumm and Morgan, 1996), where the major concern has been whether organic complexes of metals exist under the conditions of seawater or not. Existence of organic trace metal complexes had been postulated as long as 60 years ago (Harvey, 1928; Cooper, 1937), and subsequent studies were made to demonstrate the presence of these complexes and to characterize their nature. However, the presence of organic complexes of trace metals has been recognized by works since 1980, although the trace metals specified were limited, e.g., Cu, Zn and Fe. A recent common understanding of chemical forms is that significant parts of Cu, Zn and Fe(III) dissolved in seawater are present as organic complexes. This situation is in contrast to knowledge about the interaction between humic substances and trace metals before 1980, which was reviewed by Mantoura (1981). The findings on recent metal speciation studies have revealed that the existence of strong organic ligands, which are not "so-called" humic substance, is real in seawater.

Thorium isotopes have been extensively studied to understand scavenging processes in the water column (e.g., Bacon and Anderson, 1982; Cochran *et al.*, 1987; Clegg and Whitfield, 1990, 1991). Thorium isotopes can be powerful tracers in understanding the biogeochemical processes, because they are closely associated to the hypothesis that the removal process of thorium in the water column may directly reflect the sinking process of biogenic particulate matter (Eppley, 1989). Based on this hypothesis, the export flux of particles has been estimated from ^{234}Th-^{234}U disequilibria in the water column (Buesseler *et al.*, 1992). In these studies, it is important to clarify what is the functional groups in biogenic particulate matter connected with thorium isotopes. However, studies on the interaction between thorium isotopes and particulate matter are limited (Moore and Hunter, 1985; Moore and Millward, 1988), and most of the experimental studies have not taken account of the organic complexation between thorium isotopes and biogenic particulate matter. The speciation of metals in particulate matter (Chester *et al.*, 1988; Hirose and Sugimura, 1993) revealed that metal ions, including thorium isotopes, are associated with organic binding sites in particulate matter by complexation. A strong ligand with the properties of chelators in POM connects between thorium isotopes and POM (Hirose, 1995).

Trace metals such as Fe, Cu, Zn and others, which are nanonutrients, are essential for phytoplankton growth (Sunda, 1989). On the other hand, higher concentrations of some trace metals, which occur in high upwelling region in the ocean, show toxicity to marine organisms (Anderson and Morel, 1978). Metal toxicity for phytoplankton growth is reduced by the addition of artificial chelators such as EDTA. In natural systems, the presence of natural chelators has been postulated to detoxify higher concentrations of trace metals. Chelators, which are produced by micro organisms, play a significant role for the uptake processes of trace metals by marine organisms; such chelators are identified as siderophores (e.g., Lewis *et al.*, 1995), whose molecular structures are shown in Fig. 1, and phytochelatins ((γ-glu-cys)$_n$-gly: $n = 2$–11) (Grill *et al.*, 1985; Ahner *et al.*, 1995; Ahner and Morel, 1995). The biological production of siderophores is regulated by ambient iron levels, such that siderophores are synthesized and released in times of low-iron stress and repressed in high-iron environments. These findings support the presence of natural organic strong ligands in the marine environment.

A final goal on the study of strong organic ligands in seawater is to isolate and identify strong organic ligands in particulate matter and dissolved organic matter and to determine their molecular structure. However, actual characterization of strong organic ligands has not been enough until now. It is very difficult to isolate and identify the organic ligands by the present analytical method of organics because there is little possibility that organic ligands, comprising of

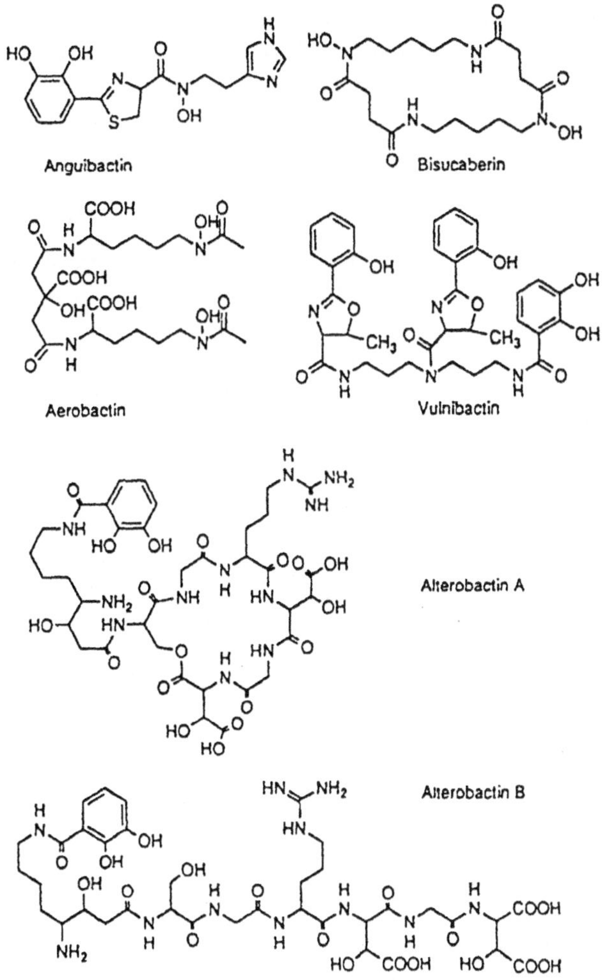

Fig. 1. Structures of marine siderophores: anguibactin form *Vibrio anguillarum* (Jalal *et al.*, 1989); bisucaberin from *Alteromonas haloplankis* (Takahashi *et al.*, 1987); vulnibactin from *Vibrio vulnificus* (Okujo *et al.*, 1994); aerobactin from an unknown *Vibrio* sp. (Haygood *et al.*, 1993); alterobactin A and alterobactin B from *Alteromonas luteoviolacea* (Reid *et al.*, 1993).

functional groups in organic macromolecules as a binding site, are present as only a molecule with a definite form. It is more likely to consider that strong organic ligands occupy only a residue of various and complex organic molecules. In order to characterize organic ligands in particulate matter and dissolved organic matter, therefore, it is effective to use metal ions, especially chemically well-characterized metals such as thorium, instead of the general organic analysis. The organic characterization using metals can introduce a new window to realize properties of POM and DOM.

The purpose of this paper is to review our knowledge of strong organic ligands in seawater and to elucidate the chemical and biogeochemical properties of strong organic ligands in particulate matter and dissolved organic matter. Finally, we discuss oceanic cycling of organic matter using the strong organic ligand as a probe.

STRONG ORGANIC-METAL COMPLEXES IN SEAWATER

The possibility of the existence of strong organic-metal complexes in seawater had been recognized qualitatively until 1980 (Williams, 1965; Barber and Ryther, 1969; Batley and Florence, 1976; Duinker and Kramer, 1977; Sugimura *et al.*, 1978). The development of bioinorganic chemistry has revealed the presence of strong organic-metal complexes in biological systems (Stryer, 1981); some metals (Fe, Cu, Zn, etc.) occupying the reactive center in enzymes, and highly stable complex, such as Fe(III) complexes of siderophores (Fig. 1), were also discovered.

The most important parameter to characterize the metal complexes with organic ligands is the stability constant, which is thermodynamically unique for a complex between a metal and an organic ligand (Stumm and Morgan, 1996). The stability constant contains information about chemical properties of both organic ligands and metals. If we can determine a rigorous stability constant of the complex of an organic ligand with an unique metal, we can obtain more information about the molecular structure of the organic ligand and the corresponding metal complex by comparison with the values determined for metal complexes with well-known ligands. Unfortunately, we have only conditional stability constants of metal complexes in natural systems through the lack of side reaction coefficients of both metals and ligands. However, the conditional stability constants are also useful for characterization of organic ligands if the method to determine the conditional stability constants of metal complexes is well understood chemically.

The concentrations of the strong organic ligands in seawater can be determined when accompanied by the measurements of the conditional stability constants of natural metal complexes. The level of the strong organic ligand in seawater is another significant parameter to characterize the organic ligand because it appears to closely depend on the magnitude of the conditional stability constant; the lower the concentration of the strong organic ligand, the stronger the stability of the corresponding metal complex. Therefore, the concentration of the organic ligand in seawater provides a powerful constraint to specify the strong organic ligand as well as the conditional stability constant.

Conditional stability constants of metal complexes

Since 1980, complexing characteristics of seawater have been widely studied for several transition metals (e.g., Cu, Zn) using two distinct concepts, i.e., metal titration and ligand titration, although most of the studies are focused on copper because of its strong complexation and useful analytical tools. The metal titration methods have been extensively employed for the chemical speciation of trace metals in natural waters, with MnO_2 adsorption (van den Berg, 1982a, b), electrochemical (van den Berg *et al.*, 1984; Kramer and Duinker, 1984; Nelson, 1985; Kramer, 1986; Buckley and van den Berg, 1986; Hering *et al.*, 1987; Coale and Bruland, 1988, 1990; Apte *et al.*, 1990; Midorikawa *et al.*, 1990), chromatographic (Sunda and Hanson, 1987; Miles *et al.*, 1987), bioassay (Sunda and Ferguson, 1983), solvent extraction (Moffett and Zika, 1987; Moffett *et al.*, 1990), chemi-luminescence techniques (Sunda and Huntsman, 1991) being used. However, recent studies have focused on the electrochemical technique with ligand competition (van den Berg, 1984; Donat and van den Berg, 1992; Yokoi and van den Berg, 1992; Rue and Bruland, 1995; van den Berg, 1995). On the other hand, ligand titration, which contains the competitive reactions between natural and artificial organic ligands, has been hardly developed although initial work to use EDTA was carried out in 1982 (Hirose *et al.*, 1982).

Recent common understanding based on the experimental studies, most of which has been focused on Cu(II), is that strong complexing ligands exist in seawater; more than 99% of copper dissolved in seawater being present as organic complexes whose conditional stability constants are in the range of between 10^9 and 10^{15}. The conditional stability constants of organic copper complexes in seawater, which have mainly been determined since 1982, are summarized in Table 1. It must be noted that strong organic ligands able to support quantitative complexation of Cu(II), are commonly present from coastal to open ocean. Hirose (1994a) carried out the classification of strong organic ligands dissolved in seawater in terms of the conditional stability constants of metal complexes and the corresponding ligand concentration.

Based on the conditional stability constants of organic Cu(II) complexes, organic ligands in seawater are conveniently divided into three classes: the strongest, L_1, ($\log K^c_{CuL} > 13$), the strong, L_2, ($\log K^c_{CuL} \sim 12$) and the weak ligand, L_3, ($\log K^c_{CuL} < 10$). It must be noted that there is no case for having simultaneously detected all three kinds of organic ligands in the same experiment. As a general trend, however, the concentrations of the organic ligands complexed with Cu(II) are of the following order: $L_1 < L_2 << L_3$. Taking into account the total concentration of Cu(II) dissolved in seawater, which is usually less than those of organic ligands, L_2, L_1, or $L_1 + L_2$, it is clear that most of the organic ligand L_3 is not bonded with Cu(II) under the conditions of seawater; in other words, most of the dissolved Cu(II) is present as complexes with the organic ligands L_1 and/or L_2. This situation is consistent with the fact that the organic ligand L_3 cannot be detected by the ligand titration method (Hirose *et al.*, 1982).

The chemical speciation of Zn(II) in seawater has been studied by several workers (van den Berg and Dharmvanji, 1984; van den Berg, 1985; Bruland, 1989; Donat and Bruland, 1990; Muller and Kester, 1991), although inorganic speciation of Zn(II) in seawater has been carried out from initial stage (Zirino and Yamamoto, 1972; Dyrssen and Wedborg, 1974, Stumm and Brauner, 1975). The conditional stability constants of organic Zn(II) complexes, which were determined by the ligand and metal titration methods, are summarized in Table 2. The

Table 1. Conditional stability constants of organic Cu complexes in seawater.

Sampling location	pH	pCu	log K^c_1	logK^c_2	log K^c_3	L_1 (nmol/L)	L_2 (nmol/L)	L_3 (nmol/L)	Ref.
western North Pacific	8.1	11.1	13.8	11.8	-	7	21	-	[1]
Irish Sea	8.2	-	-	-	10-10.4	-	-	60-150	[2]
Southeastern Gulf of Mexico	8.2	-	-	>12	9.8	-	5	15	[3]
Cape San Blas,Florida	8.2	11.5	-	11.2	9.0	-	13	80	[3]
Mississippi Riv.Plume	8.1	11.3	-	11.1	8.9	-	20	130	[3]
South Atlantic	7.7	-	-	12.2	10.2	-	11	33	[4]
Atlantic	-	-	-	-	9.9,9.0	-	-	31,87	[5]
Atlantic	-	-	-	-	9.7,8.6	-	-	60,120	[5]
North Sea	7.9-8.2	-	-	-	8.9-9.1	-	-	80-103	[6]
Tamor Estuary	-	-	-	-	8.6-9.1	-	-	390	[7]
North Atlantic	8.0-8.3	-	-	-	7.8-8.2	-	-	50-82	[8]
North Atlantic	-	12.2-12.7	-	9.8-12	7.4-8.9	-	4-144	2-440	[9]
Biscayne Bay	-	-	-	12	10.5	-	5.1	110	[10]
Naragansett Bay	8.0	12.5	-	12.4	10	-	50	100	[11]
Naragansett Bay	8.0	12.1	-	12	10	-	20	100	[11]
Coastal Peru	8.2	11.4	-	12.3	9.2	-	4.5	70	[11]
Chistiansen Basin	8.0	11.8	-	11.7	9.1	-	50	68	[12]
Montauk Point	8.2	12.2	-	11.7	9.1	-	20	50	[12]
Northeast Pacific									
Surface	-	13.9	-	11.9	9.5-10.6	-	1.8	7.6	[13]
Deep	-	11	-	-	8.1-9.1	-	-	5-10	[13]
North Pacific	-	-	13.0	-	10.0	1-3	-	5	[14]
Southwestern Sargasso Sea	-	-	13.2	-	9.7	2.0	-	80	[3]
Shelf water off North Carolina	8.1	12.5	13.2	-	10.0	3.3	-	26	[15]
Severn Estuary	-	11.1-12.8	-	11.4-12.8	-	-	13-196	-	[16]
Indian Ocean	-	-	-	12.6	-	-	4.13	-	[17]
North Sea	-	-	-	12.4	-	-	16.2	-	[17]
NE Atlantic	-	-	13.0-14.8	-	-	3-14	-	-	[18]
Mediterranean	-	-	12.2-13.1	-	-	3-8	-	-	[18]
Central Pacific	-	-	12.7-13.4	-	-	2.2-3	-	-	[19]
South San Francisco Bay	-	-	>13.5	-	9.0-9.6	13	-	60-80	[20]

[1] Hirose *et al.*, 1982, [2] van den Berg, 1982, [3] Sunda and Ferguson, 1983, [4] van den Berg, 1984, [5] van den Berg *et al.*, 1984, [6] Kramer and Duinker, 1984, [7] Nelson, 1985, [8] Kramer, 1986, [9] Buckley and van den Berg, 1986, [10] Moffett and Zika, 1987, [11] Sunda and Hanson, 1987, [12] Hering *et al.*, 1987, [13] Coale and Bruland, 1988, [14] Coale and Bruland, 1990, [15] Sunda and Huntsman, 1991, [16] Apte *et al.*, 1990, [17] Donat and van den Berg, 1992, [18] Campos and van den Berg, 1994, [19] Miller and Bruland, 1994, [20] Donat *et al.*, 1994.

conditional stability constants are divided into three classes as well as Cu(II). The stability of organic Zn(II) complexes is systematically lower than those of Cu(II). This tendency is reasonable, according to the Irving-Williams series for transition metals (Stumm and Morgan, 1996), although there is no discussion about whether the organic ligands associated with Zn(II) are the same as those of Cu(II) or not.

To summarize studies regarding metal speciation, it is obvious that reactions between natural organic ligands and metal ions are reversible, where reversibility is attributed to methodologies involving both metal exchange and ligand exchange reactions, and that the existence of strong organic ligands is common in marine environments; in other words, present strong organic ligands are not

Table 2. Conditional stability constants of organic Zn complexes in seawater.

Sampling location	pH	pZn	log K^c_1	logK^c_2	log K^c_3	L_1	L_2	L_3	Ref.
							nmol/L		
western North Pacific	8.1	8.7	10.7	9.3	-	5	8	-	[1]
South Atlantic	-	-	-	-	7.4	-	-	30	[2]
Irish Sea	-	-	-	8.4	7.5	-	26	64	[3]
central North Pacific Gyre	-	-	11.0	-	-	1.2	-	-	[4]
Northeast Pacific	8.2	-	10.2-11.3	-	-	1.6-2.3	-	-	[5]
Atlantic,etc.	-	-	-	6.3-8.1	5.8-6.6	-	-	-	[6]
Narragansett Bay	-	-	-	8.3-9.4	7.4-7.7	-	4-46	32-104	[7]

[1] Hirose *et al.*, 1982, [2] van den Berg and Dharmvanij, 1984, [3] van den Berg, 1985, [4] Bruland, 1989, [5] Donat and Bruland, 1990, [6] Stanley, Jr. and Byrne, 1990, [7] Muller and Kester, 1991.

species produced by marine organisms under special conditions such as exposure of highly toxic metals. These findings lead to the idea that it is significant to characterize the chemical properties of stronger organic ligands L_1 and/or L_2 in seawater. Now we propose here some problems about strong organic ligands dissolved in seawater.

1. How many discrete strong organic ligands exist in seawater?
2. Are the strong organic ligands elucidated by each metal ion peculiar to the corresponding metal? In the other words, do the organic ligands detected by Cu(II) completely differ from those by Zn(II) or not.
3. Are strong organic ligands deduced as a result of the artifacts used in the methods? This contains the problem of whether organic ligands determined by the titration methods should be analyzed by discrete or continuous models.

Why cannot the feature of the strong organic ligands in seawater still not be clarified? The major reason is present in the organic geochemistry pointed out by Mantoura (1981). A concern of marine chemists has been focused on the speciation of trace metals related to their bioavailability and not the organic characterization of strong ligands.

Classification of strong organic ligands

Organic matter in seawater are considered to be composed of many, diverse organic molecules. It is likely that many kinds of strong organic ligands are also present in seawater. We postulate that there are only one or two strong organic ligands related to the biogeochemical processes of trace metals, such as complex formation and bioavailability in the marine environments. Knowledge of chemical thermodynamics for metal complexation in solution, especially the linear-free energy relationship (LFER), will become a tool to characterize natural strong organic ligands. For example, the logarithmic stability constants of hydoxo complexes of metal ions are linearly related to those of DTPA (diethlenetrinitrilopenta-acetic acid) complexes as shown Fig. 2. By using this tool, the window of possible strong organic ligands complexed with metals in

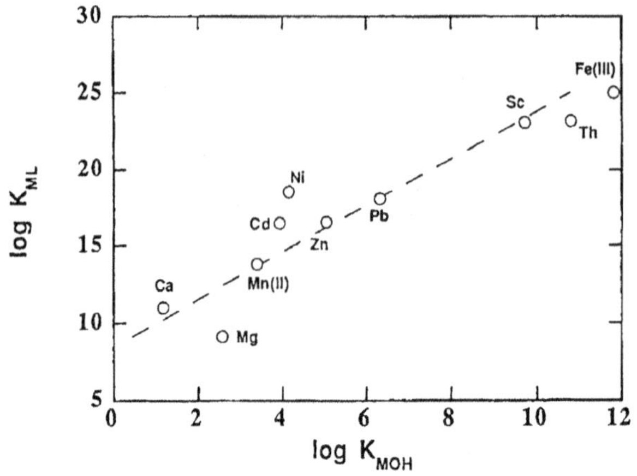

Fig. 2. Linear free energy relationship between hydroxo and DTPA complexes with a series of metal ions. Since chemical properties between the hydroxyl group (nonodentate ligand) and DTPA (chelator) differed from each other significantly, the slope of LFER is not unity. If the properties of a ligand are chemically the same as the reference ligand, the slope of LFER is nearly equal to unity.

natural systems can be narrowed by comparison with the well-known stability constants of complexes of artificial ligands with metal ions. Linear-free energy techniques have recently been applied for the estimation of missing values of the metal complex formation constant in several areas of chemical oceanography (Dyrssen, 1985; Elliott, 1988; Stanley and Byrne, 1990; Muller and Kester, 1991; Hirose, 1994a).

Until now we have little information on the properties of organic ligands, except that they may be chelators involved in macromolecules like humic compounds (Mantoura, 1981). Hirose (1994a) has applied the LFER technique to natural organic ligands, assuming that those in seawater are less specific and less selective chelators regarding the complexation of metal ions, similar to aminopolycarboxylic acids. In this case, it is expected that there is an artificial ligand with which logarithmic stability constants of metal complexes are linearly related to logarithmic conditional stability constants of metal complexes with a natural organic ligand with a slope of unity, when the chemical properties of binding the sites of a natural ligand, e.g., number of carboxylic and/or amino groups, are similar to those of an artificial ligand. Then, the following equation is established between the natural strong organic ligand and an artificial ligand as a reference:

$$\log K_{ML} = \log K_{MY} + A, \qquad (1)$$

where K_{ML} and K_{MY} are the stability constants of the metal complex with the natural organic ligand and an artificial ligand, respectively, and A the constant.

The conditional stability constants of natural organic metal complexes have been extensively measured using various techniques. However, the data set applied to the LFER technique must satisfy the following conditions: the conditional stability constants for more than one metal ion being determined using the same method and/or the same quality, with no effect of competitive reactions between metals. There are possibilities that the conditional stability constants obtained by metal titration methods contain side reactions related to metal exchange. Another important aspect is that, if the same water samples are used to determine the conditional stability constant of natural complexes, the concentration of the organic ligand determined by a metal should be of the same value as those of other metals. To apply a LFER technique, actually, a ligand concentration is required of the same order of magnitude between different metals because of different samples and experimental conditions. In other words, the LFER technique cannot be applied between the conditional stability constants corresponding to the different order of magnitude of the ligand concentration. For example, the concentration of the stronger ligand for Cu(II) and Zn(II) complexes was an order of nmol/L as shown in Tables 1 and 2, whereas it was less than nmol/L for organic Co and Ni complexes in seawater (van den Berg and Nimmo, 1987; Nimmo *et al.*, 1989; Zhang *et al.*, 1990). Furthermore, there is suspicion that kinetically inert complexes of Ni(II) and Co(III), in which Co(II) is easily oxidized to form complexes under the oxic conditions, are related to measurements of higher stability constants. In this connection, the conditional stability constants of organic Ni complexes in seawater estimated from the LFER technique were significantly lower than those determined by electrochemical methods (Table 3). In other words, currently there is no guarantee that the stronger

Table 3. Predicted stability constants of metal complexes with DTPA and EDTA-type organic ligands in seawater.

Metal ion	DTPA type ligand		EDTA type ligand	
	log K_{MY1}	log K_{ML1}	log K_{MY2}	log K_{ML2}
Th^{4+}	28.8	21.1	23.2	16.1
Fe^{3+}	28.0	20.1	25.1	18.0
Hg^{2+}	26.7	19.0	21.7	14.6
Cu^{2+}	21.6	13.8	18.8	11.7
Zn^{2+}	18.4	10.7	16.5	9.4
Ca^{2+}	10.8	3.1	11.0	3.9
Mg^{2+}	9.3	1.6	9.1	2.0
Ni^{+2}	20.3	12.2	18.6	11.5
UO^{2+}	-	14.5[a)]	(19.7)[b)]	(12.9)

K_{MY1} and K_{MY2} are the stability constants of metal complexes with DTPA and EDTA, respectively.

[a)]This value was determined from the relationship between the ThCC and particulate U (Hirose, 1995).

[b)]Estimated value. The stability constants of uranyl DTPA and EDTA complexes in aqueous solutions have not been determined directly due to the complexity of reactions.

ligand detected for Cu(II) and Zn(II) in seawater is chemically the same as that for Co and Ni. The data sets given at the same quality, therefore, are limited to the results of Cu(II) and Zn(II) obtained by ligand titration (Hirose *et al.*, 1982); two kinds of organic ligands, L_1 and L_2, coexist in seawater, in which the stability constants are of the same orders as those determined by recent studies (Tables 1 and 2). It must be noted that there is a smaller contribution from the side reactions of metal exchange on the conditional stability constant for metal titration methods, including ligand competition (Midorikawa and Tanoue, 1996).

Some artificial ligands, whose stability constants for Cu(II), Zn(II) and others, were determined under the same conditions (the same ionic strength and so on), were examined to satisfy the LFER for the natural organic ligands L_1 and L_2, respectively. We found that LFERs with a slope of unity are established between the organic ligand L_1 and DTPA, and between the organic ligand L_2 and EDTA: the constants, A, in Eq. (1) being 7.7 and 7.1 for the organic ligands L_1 and L_2, respectively. Now to avoid confusion between ligand terms for the ligand and metal titration methods, we use terms of DTPA and EDTA-type organic ligands to the strong organic ligands L_1 and L_2, respectively, determined by the ligand titration. The LFER technique permits us to estimate the stability constants of natural organic complexes in seawater for metals with known stability constants of EDTA and DTPA complexes. The calculated results are summarized in Table 3, together with the corresponding stability constants of EDTA and DTPA complexes (Anderegg, 1977; Kotrly and Sucha, 1985).

The predicted stability constants for alkaline earth-metals are relatively low ($\log K_{CaL} = 3.9$ and $\log K_{MgL} = 2.0$ for the EDTA-type ligand), which is consistent with the fact that the major species of Ca and Mg are free ion. Since specific interactions between the alkali and alkaline earth-metal ions and organic ligands in natural waters have, in general, escaped the attention of aquatic chemists, there has been only limited study to determine the conditional stability constants of Ca and Mg complexes with marine organic matter. Mantoura *et al.* (1978) measured the conditional stability constants of Ca and Mg with marine humic compounds: $\log K^c_{CaL} = 3.87$ and $\log K^c_{MgL} = 3.4$. It is surprising that the conditional stability constant of organic Ca complex in seawater is in fair agreement with that of the EDTA-type organic ligand predicted by the LFER technique. This finding does not simply imply that one of the natural organic ligands is completely characterized by the LFER technique, because the conditional stability constants for several transition metals with humic compounds determined by Mantoura *et al.* (1978) were much lower than the current data. However, these low conditional stability constants can be explained by the competitive reactions between metals as discussed by Hirose (1994a) and by lower sensitivity of measurements until 1980. Therefore, the result permits us to present the hypothesis that the chemical properties of the EDTA-type organic ligand are similar to the humic compounds specified by Mantoura *et al.* (1978). As discussed later, this hypothesis may be supported by the fact that the EDTA-type ligand has not been found in particulate matter samples, although the EDTA-type ligand concentrations in seawater are about one order of magnitude greater than those of the DTPA-type.

It is interesting to estimate the conditional stability constant of organic Fe complex because the chemical speciation of Fe is highlighted regarding the Martin hypothesis of oceanic Fe (Martin and Fitzwater, 1988; Martin and Gordon, 1988; Martin *et al.*, 1994). The conditional stability constants predicted for organic Fe^{3+} complexes by the LFER technique, both of which are higher than the lower limits of possible complex formation (Hirose, 1994a), suggest the possibility that Fe^{3+} in seawater forms organic complexes. Hudson *et al.* (1992) suggest that organic complexation of Fe(III) is possible in seawater from the results with model and natural systems. Hudson and Morel (1990) have studied Fe transport in marine phytoplankton, in which the half-saturation constant for the Fe cell surface complex of phytoplankton was measured kinetically. The stability constant of the Fe^{3+} cell surface complex can be evaluated to be 20.2 (= $\log K^{c}_{FeL}$) from the result of Hudson and Morel (1990), which is in fair agreement with the value ($\log K_{FeL} = 20.1$) predicted for the DTPA-type organic ligand. Studies of the chemical speciation of Fe(III) (Rue and Bruland, 1995; van den Berg, 1995; Wu and Luther, 1995) revealed that the stability of Fe(III) complexes dissolved in seawater is very high ($20.5 < \log K_{FeL}' < 22.5$) and virtually all iron occurs organically complexed. The values of the conditional stability constant of the Fe complex determined experimentally are in agreement with an estimate by the LFER technique when we take account of some uncertainties accompanied by both methodologies. It should be noted that the concentration of the strong organic ligand determined for Fe complexation (Rue and Bruland, 1995; van den Berg, 1995; Wu and Luther, 1995) was an order of nmol/L, which is the same order of magnitude as that of the DTPA-type ligand. This finding strongly supports that the stronger organic ligand, which is not specific to Fe(III) such as siderophores, classified as DTPA-type, exists at the order of nmol/L in seawater. It is noteworthy that the present conditional stability constant of the organic Fe complex contains relatively widespread uncertainty, which seriously affects the chemical form of dissolved Fe in seawater; e.g., for $\log K_{FeL} = 20.2$, about half of the total dissolved Fe(III) is complexed with organic ligand (Hirose, 1994a), whereas, for $\log K_{FeL} = 21.6$, Fe(III) occurs 99% complexed with organic ligands (van den Berg, 1995). Furthermore, studies about Fe speciation lead to the hypothesis that the DTPA-type organic ligand dissolved in seawater is characterized as fresh organic matter originating from the phytoplankton cell surface (Hudson and Morel, 1990). This hypothesis will be discussed in detail later.

Thorium in seawater has been extensively studied in the field of chemical oceanography regarding scavenging processes (Bacon and Anderson, 1982). A typical feature for Th in seawater is that the recent level (0.02–2 pmol/L) of ^{232}Th (Anderson *et al.*, 1983; Nozaki and Horibe, 1983; Huh and Bacon, 1985, Nozaki *et al.*, 1987; Nozaki and Yamada, 1987; Huh *et al.*, 1989) is supersaturated regarding the dissolution of inorganic salts (the solubility of thorianite: about 0.01 pmol/L) (Langmuir and Herman, 1980). In order to understand the mechanism of the scavenging and dissolution phenomena of Th, it is also significant to clarify the chemical form of both particulate and dissolved Th. A recent leaching study (Hirose and Sugimura, 1992) suggests that Th in particulate matter is present as organic complexes. Dissolved Th is also considered to be complexed with humic substances (Nash and Choppin, 1980; Cacheries and Choppin, 1987). Calculation using the conditional stability constant predicted by the LFER technique, possible free ligand and free thorium concentrations (Hirose, 1994a) suggests that only DTPA-type organic ligand, evolving in relatively fresh biogenic matter, can form a complex with Th. The EDTA-type organic ligand, which may be a major one of the humic compound type, cannot be complexed with Th under the conditions of seawater. Our model compounds for natural organic ligands such as DTPA and EDTA contain a large number of carboxylic acid groups. Santschi *et al.* (1995) suggest that carboxylic acid groups in colloidal organic matter in seawater are responsible for the adsorption of thorium.

Chemical forms of organic ligands

We have little information about the chemical properties and molecular structure of organic ligands, as described in the previous section. It, however, can be discussed whether a major form of organic ligands is free or not, although this problem has received little attention until now.

The ratio of the complexed ligand concentration to the free ligand concentration for the individual metal ion R_L, is easily deduced from the definition of the stability constant as follows:

$$R_L = [ML]/[L] = K_{ML}[M^{n+}] \tag{2}$$

where $[M^{n+}]$ shows the free ion concentration of a metal.

The free ion concentrations of metals are the most significant factor for understanding biological uptake processes and the chemical speciation of metals. For trace metals, one of the current problems is how high are the levels of free ions in seawater and how do these factors control free ion levels. However, the free ion concentration levels of major metals in seawater, i.e., Ca and Mg, can be estimated from an inorganic equilibrium system, although there may be some discussion on the estimation of rigorous concentrations of these free metal ions. On the other hand, for some trace metals such as Th(IV) and Fe(III), which are

Table 4. Free metal concentrations and the ratios of the complexed ligand concentration to the free ligand one under seawater conditions.

Metal ion	Free ion log $[M^{n+}]$	DTPA-type ligand log R_{L1}	EDTA type-ligand log R_{L2}
Th^{4+}	-23.0	-1.9	-6.9
Fe^{3+}	-19.3	0.8	-1.3
Cu^{2+}	-12.0	1.8	-0.2
Zn^{2+}	-11.0	-0.3	-1.6
Ca^{2+}	-2.07	1.0	1.8
Mg^{2+}	-1.49	0.1	0.5
Ni^{2+}	-11.5	0.7	0

supersaturated regarding the dissolution of inorganic salts, the free ion concentrations are controlled by the dissolution equilibrium of inorganic salts, which are calculated from the solubility products (Sillen and Martell, 1965; Biedermann and Chow, 1966). For metals undersaturated regarding the dissolution of inorganic salts in seawater, i.e., Cu(II), Zn(II), Ni(II) and so on, it is difficult to thermodynamically estimate the free ion concentrations. Then, the free metal concentrations of Cu(II) and Zn(II) in seawater were tentative values, taking the direct measurements of free metal ion concentrations into account (Table 4). It must be noted that these free ion concentrations of Cu(II) and Zn(II) are at optimal levels regarding phytoplankton growth (Sunda and Guillard, 1976; Jackson and Morgan, 1978; Anderson and Morel, 1978; Anderson *et al.*, 1978; Brand *et al.*, 1983). For Ni(II), the free ion concentration was estimated from conditional stability constants and free ligand concentrations, taking account of about 2 nmol/

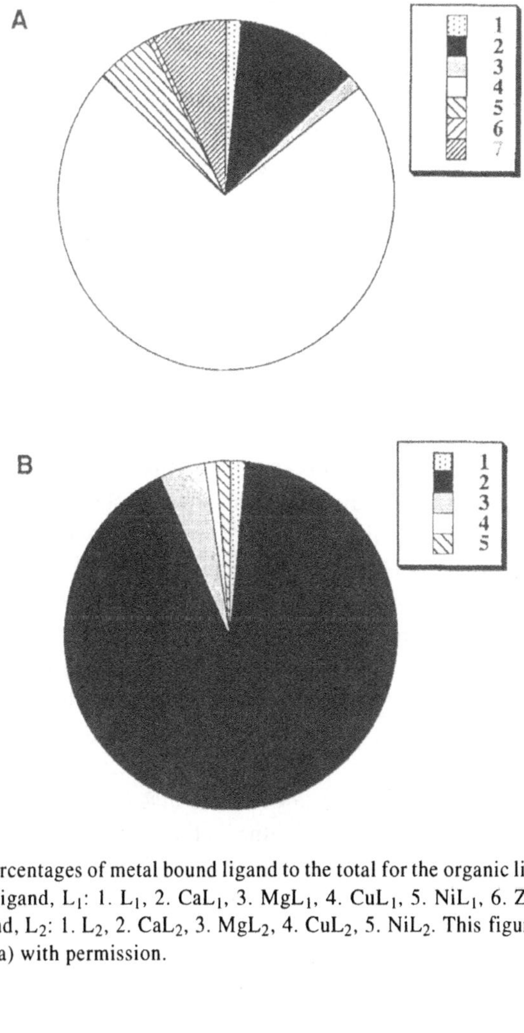

Fig. 3 The percentages of metal bound ligand to the total for the organic ligands, L_1 and L_2. Organic ligand, L_1: 1. L_1, 2. CaL_1, 3. MgL_1, 4. CuL_1, 5. NiL_1, 6. ZnL_1, 7. FeL_1. Organic ligand, L_2: 1. L_2, 2. CaL_2, 3. MgL_2, 4. CuL_2, 5. NiL_2. This figure is based on Hirose (1994a) with permission.

L of the total Ni concentration. The free metal levels of some metals in seawater are summarized in Table 4.

The ratios R_L for some metals are calculated form Eq. (2) using the corresponding stability constants predicted by the LFER technique. The results are shown in Table 4. A typical feature concerning the chemical form of the organic ligands, i.e., EDTA and DTPA-type ligands, is that most of the organic ligands are present as complexes with metals. Figure 3 shows the percentage of the metal bound ligands to the total for the DTPA and EDTA-type organic ligands, assuming that there is a smaller contribution from other trace metal ions (Co and others) because of their extremely low concentrations (less than 0.1 nmol/L). The percentages of the free ligands to the total are only less than 1 and 2% for the DTPA and EDTA-type organic ligands, respectively. Most of the binding site for the major EDTA-type ligand is occupied by calcium ion, whose possibility was discussed by Reuter and Predue (1977). There is no change for this situation as calculated from the data by Mantoura *et al.* (1978), even if estimates from the LFER technique cannot be recognized. On the other hand, one of the major forms of the DTPA-type organic ligand is estimated to be a Cu complex in the case when the total concentrations of Cu(II) and the DTPA-type ligand in seawater are 4.5 and 4 nmol/L, respectively. It is noteworthy that the portion of metal complexes of the total ligand for the DTPA-type ligand is largely dependent of concentrations of trace metals (especially, Cu(II)) and the organic ligand. These calculation supports the hypothesis that, mainly, two strong organic ligands, i.e., EDTA and DTPA-type ligands, are related to complexation of Cu, Zn, Fe, Th and others.

Strong ligands in particulate matter

Dynamics of oceanic particulate matter, which consists of sinking and suspended particles, is a major concern on for chemical oceanographers because particulate matter is related to the key process of carbon and metal cycling (Fowler and Knauer, 1986; Wangersky, 1986; Hurd and Spencer, 1991). Strong ligands in particulate matter may play a significant role in trace metal cycling in the ocean if the strong ligand connects directly between particulate organic matter and metal ions in the marine environment. On the other hand, the strong ligands in particulate matter are considered to be related to the origin of dissolved strong organic ligands. There is, however, little direct information about the strong organic ligands in particulate matter. When measured concentrations of certain metals, e.g., U(VI), Cu(II) and others, in seawater are compared with concentrations predicted from thermodynamic solubility calculation, these metals are found to be undersaturated (Goldberg, 1954; Krauskopf, 1956). Nevertheless, most of the trace metals in seawater are also found in suspended matter. These observation have led to the hypothesis that sorption of metals by particles is an important and efficient process in marine environments. In other words, the existence of the strong ligands in particulate matter has been only postulated by using metal ions as a probe.

Hemipelagic, organic rich sediments contain authigenic uranium derived from seawater (Veeh, 1967; Veeh *et al.*, 1974), and in some cases, uranium concentrations correlate with the organic matter content of sediment. Sediment trap experiments (Anderson, 1982) indicated that biogenic uranium dominated the total uranium measured in sediment trap samples, which has $^{234}U/^{238}U$ activity ratios of about 1.14, consistent with a seawater source. Anderson (1982) has found a positive correlation between bio-authigenic uranium and the organic matter contents in sediment trap samples. Taking account of the fact that uranium dissolved in seawater is present as a stable carbonate complex (Djogic *et al.*, 1986) and undersaturated to the dissolution of uranium minerals (Langmuir, 1978), these findings suggest that uranium is directly associated with organic matter in sediments and sinking particles; in other words, it indirectly suggests the presence of strong ligands in sediments and suspended particles. It is, therefore, important to obtain direct evidence about the existence of strong organic ligands in particulate matter and sediment by laboratory experiments.

Sequential leaching experiments (Chester *et al.*, 1988; Hirose and Sugimura, 1991, 1993) have been developed for the chemical speciation of particulate metals. The results revealed that trace metals, including Cu, U and Th, in oceanic suspended matter are associated with organic binding sites of particulate matter by complexation. For thorium isotopes in suspended matter, most of the thorium isotopes are preserved onto particulate matter at higher acidic media (pH = 1), which clearly implies that they are bound with the strong organic ligand of particulate matter, because possible inorganic adsorbers of trace metals, which comprise of metal hydroxides and oxides, are dissolved in such an acidic solution and most of the binding sites in inorganic adsorbers are masked by the hydrogen ion.

Hirose (1990) suggests that mass balance analysis can be applied to the seawater system including particulate metals if they are associated with the strong organic ligand in particulate matter by complexation. Mass balance considerations indicate that the conditional stability constant defined for organically bound metals in suspended particles has the same value as that determined for the corresponding dissolved metal-organic complex if the binding site in particulate organic matter is chemically similar that of dissolved organic matter. Hirose (1994b) applied the mass balance technique to the stepwise leaching experiment of particulate uranium in seawater. The results suggest that the interaction between uranium and the binding site in particulate matter is a reversible process by complexation and that the strong organic ligand in particulate matter is a kind of polydentate ligand, which is corresponding to the chelator.

Th-complexing capacity of particulate matter

According to thermodynamic considerations (Langmuir and Herman, 1980), the affinity of Th(IV) to organic functional groups is enhanced in more acidic solutions compared to other metals because the effect of side reactions of hydrolysis can be minimized, i.e., Th exists as a free 4+ ion, although the reactivity of organic ligands for metal ions is suppressed largely due to the protonation on acidic sites of organic ligands. These findings suggest that the experiment under the acidic conditions is an effective approach for studying chemical interactions between Th and organic ligands in particulate matter. Acidic experimental conditions have other merits; the system can be simplified to a singular reaction between Th and organic ligands without other complicated and complexing side-reactions where dissociation of inorganic salts is accomplished completely; i.e., effects of endogeneously occurring metals in particulate matter as well as contaminated metals are eliminated during the sampling and experimental processes.

Taking account of the benefit of using thorium isotopes for organic characterization, Hirose and Tanoue (1994) introduced a new parameter for the estimation of the complexation abilities of particulate organic matter (POM), namely, the Th-complexing capacity of POM. This concept corresponds to that of the copper-complexing capacity of DOM, which has been developed since *ca.* 1970 (see Mantoura, 1981). Measurements of the metal complexing capacity of natural waters have been widely adopted as the methodology for chemical speciation. The complexing capacity of a sample, defined as the moles of added metal (usually Cu^{2+}) which are complexed per liter of sample, is a measure of the metal-buffering capacity and is of fundamental importance for a quantitative assessment for the fate of polluting metals in natural waters (Mantoura, 1981). The complexing capacity is also a measure of the abundance of the organic ligand in natural waters.

The adsorption experiment to particulate matter in a 0.1 mol/L solution of HCl (Hirose and Tanoue, 1994) revealed that a strong organic ligand complexed with Th is present in particulate matter under experimental conditions. The Th-complexing capacity (ThCC) is defined as the total concentration of a strong organic ligand in particulate matter reacted with Th in acidic media. Since there is only a single reaction process for the Th adsorption onto particulate matter under the conditions of the 0.1 mol/L HCl solution, the Th complexing capacity of particulate matter is expressed as

$$\mathrm{ThCC} = [\mathrm{ThL_p}]\{1 + (K_{\mathrm{ThL}}[\mathrm{Th}^{4+}])^{-1}\}, \tag{3}$$

where K_{ThL} shows the conditional stability constant of an organic Th complex onto particulate matter, and $[ThL_p]$ and $[Th^{4+}]$ denote concentrations of adsorbed Th and free Th ion, respectively. The conditional stability constant of the organic Th complex in particulate matter was evaluated to be $10^{6.6}$ L/mol in 0.1 mol/L HCl solution. We examined the reproducibility of the Th-complexing capacity for the same surface water sample. The Th-complexing capacities were reproducible to about 5% for three repeated runs. It must be noted that the ThCC, the concentration of the strong organic ligand, is a powerful tool as a new quantitative indicator to characterize organic matter in seawater.

In order to understand the chemical properties of the strong organic ligand identified by Th, it is of importance to examine the relationship of the conditional stability constants between different metals, determined under the same experimental conditions as discussed in the previous section. Hirose and Tanoue (1994) examined the metal exchange reaction between Th and Fe(III). The metal exchange equilibrium constant between Fe(III) and Th for the binding site in POM can be directly compared with the corresponding constant for the natural organic ligand in seawater because the side reaction coefficients of proton and other metals of the organic ligand show no difference between Th and Fe(III), and the side reaction coefficients of hydrolysis for Th and Fe(III) are negligible in 0.1M HCl solution. On the other hand, the conditional stability constant of the metal complex dissolved in seawater should be equal to the corresponding constant for the binding site in POM (Hirose, 1990), if the organic ligand in DOM is chemically the same as that in POM. Hirose and Tanoue (1994) experimentally obtained the logarithmic value of −1.7 for the metal exchange equilibrium constant between Fe(III) and Th. A coordination chemistry model developed by Hirose (1994a) provides estimates of the conditional stability constants of organic Fe(III) and Th dissolved in seawater, although the value for Th has not been experimentally determined until now. The logarithmic metal exchange equilibrium constants between Fe(III) and Th for the DTPA-type and EDTA-type are calculated to be −1.0 and 1.9 from Table 3, respectively. Taking account of some uncertainties due to the difference of the experimental conditions and in the model, the chemical properties of the binding site in POM identified by Th are more similar to that for the DTPA-type organic ligand dissolved in seawater. This type of organic ligand corresponds to the strongest ones for copper and zinc which were determined using the solution chemistry techniques (Hirose *et al.*, 1982; Coale and Bruland, 1990; Donat and van den Berg, 1992).

The experimental conditions to determine the strong organic ligand identified by Th differ from those in seawater. Therefore, it is important to clarify whether the strong ligand in particulate matter corresponding to the ThCC is the same as the binding site related to particulate metals in seawater or not. Hirose (1995) examined the relationship between the ThCC and particulate U in seawater, in which uranium in seawater is one of the most adequate elements for applying the mass balance technique due to reasons to be discussed later. Since most of the particulate U is present as an organic complex in seawater (Hirose and Sugimura, 1991; Hirose, 1994b), the conditional stability constant of the organic U complex in particulate matter provides significant information for understanding the chemical properties of the strong organic ligand in seawater.

According to the mass balance, the following equilibrium is expected to be established between the U(VI) ion and the binding site in particulate matter under the conditions of seawater

$$UO_2^{2+} + L_p = UO_2L_p, \qquad (4)$$

where UO_2^{2+} is uranyl ion and L_p the organic binding site in particulate matter. The conditional stability constant, K_{UL} corresponding to Eq. (4) is defined as

$$K_{UL} = [U(VI)L_p][UO_2^{2+}]^{-1}[L_p']^{-1}, \tag{5}$$

where $[L_p']$ is the concentration of the binding site in particulate matter, which includes only free and protonated forms of the binding site. It must be noted that this conditional stability constant corresponds to that obtained by the ligand titration method (Hirose, 1994a).

The total concentration of the organic binding site, C_L, can be represented as follows:

$$C_L = [L_p'] + [M_1L_p] + \ldots + [M_nL_p] + [U(VI)L_p], \tag{6}$$

where M_1, ..., M_n are the metals other than U(VI), which react with the binding site in particulate matter under the conditions of seawater. The C_L value is written using the side reaction coefficient, $\alpha_{L(Mi)}$: $C_L = [L_p']\alpha_{L(Mi)}$.

From Eqs. (5) and (6), the concentration of the organic U complex in particulate matter can be expressed as follows:

$$[U(VI)L_p] = K_{UL}[UO_2^{2+}]\alpha_{L(Mi)}^{-1}C_L. \tag{7}$$

The physicochemical characterization of U(VI) (Langmuir, 1978; Djogic *et al.*, 1986) indicated that U(VI) dissolved in seawater is present predominantly as a tricarbonate complex and to a small extent as a trihydroxy complex under seawater conditions. Free UO_2^{2+} ion, therefore, is controlled by the buffer action of carbonate in seawater. Furthermore, dissolved U(VI) is conservative in marine environments (Ku *et al.*, 1977; Sugimura and Mayeda, 1980). These findings indicate that the concentration of free U(VI) ion is maintained at a constant level in surface waters, which is calculated to be $10^{-16.1}$ mol/L by using the total U(VI) concentration (Sugimura and Mayeda, 1980), the stability constant of U(VI) tricarbonate complex and carbonate ion concentration in seawater (Djogic *et al.*, 1986). Since the side reaction coefficient of the binding sites in particulate matter, $\alpha_{L(Mi)}$, can be approximated as a constant value in surface waters, the particulate U is expected to be linearly related to the total concentration of the binding site in particulate matter (Hirose and Sugimura, 1991).

Now we introduce a hypothesis that the ThCC of particulate matter, classified as the DTPA-type ligand, corresponds to the total concentration of the binding site for particulate metals. It then can be expected from Eq. (7) that the ThCC of particulate matter should be linearly related to the concentration of particulate U. The plot of the ThCC vs. particulate U is shown in Fig. 4. We have a good correlation between the ThCC in particulate matter and particulate U in seawater, whose slope corresponds to a value of $K_{UL}[UO_2^{2+}]\alpha_{L(Mi)}^{-1}$. The result indicates that the binding site associated with U(VI) in particulate matter is the strong organic ligand identified by Th; in other words, the binding site related to the complexation of U(VI) in particulate matter is the DTPA-type organic ligand. This finding supports that the DTPA-type organic ligand is responsible to metal complexation in both POM and DOM. The conditional stability constant of U(VI)

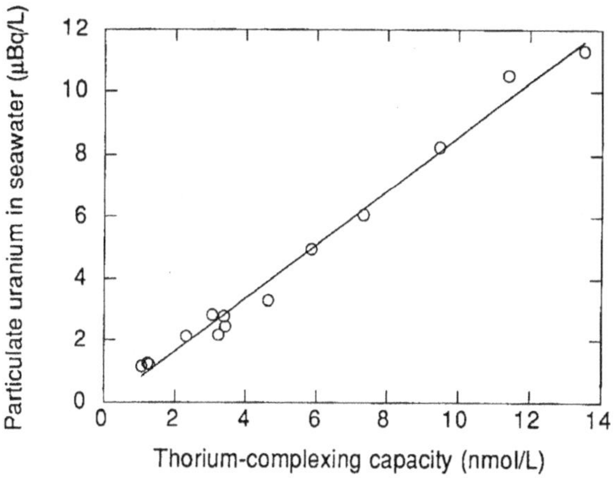

Fig. 4. The relationship between the strong organic ligand specified by Th (ThCC) in particulate matter and particulate uranium. The correlation coefficient was 0.994. (Hirose, 1995, with permission.)

complex in particulate matter is estimated to use the free U(IV) ion concentration and the side reaction coefficient of the organic ligand based on Eq. (7). As shown in Fig. 3, the concentration of the DTPA-type organic ligand not associated with metals was estimated to be about 1% of the total under seawater conditions, which means that the side reaction coefficient of the ligand is around 10^2. Therefore, the conditional stability constant of U(VI) complex in particulate matter is calculated to be about $10^{14.5}$ from Eq. (7).

In a previous paper (Hirose, 1994b), we evaluated a range for the conditional stability constant of the organic U(VI) complex in particulate matter under the conditions of seawater, which is calculated to be $10^{12} < K_{UL} < 10^{16}$, based on the application of the mass action technique to the sequential leaching experiments of particulate U(VI) in seawater. It is noteworthy that the conditional stability constant of the organic U(VI) complex determined by using the ThCC as the binding site in particulate matter exists within the previous estimation. This finding supports the hypothesis that the organic ligand corresponding to the ThCC is directly related to the complexation of U(VI) in particulate matter.

According to mass action consideration (Hirose, 1990), the conditional stability constant defined for organically binding metals in particulate matter should be of the same value as that determined for the corresponding dissolved organic metal complex if the binding site in POM is chemically similar to that of dissolved organic matter. Since the DTPA-type organic ligand is responsible for the complexation of metal ions with both POM and DOM, the conditional stability constant determined for the organic U(VI) complex in particulate matter can be directly compared with that for the dissolved metal complexes (Hirose *et al.*, 1982). A marked feature of the comparison between the conditional stability constants of trace metals (U(VI), Cu(II), Zn(II)) determined under the conditions of seawater is that the value for the organic U(VI) complex is larger than that of Cu(II) (the order of the conditional stability constants: U(VI) > Cu(II) > Zn(II)). This finding implies that the organic ligand in particulate matter shows a higher affinity to U(VI) than Cu(II) in non-saline weak acidic solution, because side reactions due to the formation of carbonates and hydroxides of these metals are negligible. The order of relative affinity to particulate matter between trace metals coincides with the results of adsorption experiments using microorganisms (Nakajima and Sakaguchi, 1993).

The total amount of trace metals adsorbed onto the microorganisms, which corresponds to the total concentration of the organic ligand in biota, was in the range from 10^{-5} to 10^{-4} mol/g (dry weight basis of organisms) (Nakajima *et al.*, 1979; Nakajima and Sakaguchi, 1993), whose values are of the same order of magnitude as the ThCC in particulate matter (10^{-5} to 10^{-4} mol/g (dry weight basis of suspended matter)). These findings suggest that the strong organic ligand corresponding to the ThCC in particulate matter, which is responsible for the complexation of trace metals under the conditions of seawater, is essentially biogenic. The DTPA-type ligand may be commonly found in marine microalgaes and bacteria.

It is well known that microorganisms, including microalgae, bacteria and basidiomycetes, show high affinity to trace metals (Pribil and Marvan, 1976; Nakajima *et al.*, 1979; DiSpirito *et al.*, 1983; Nakajima and Sakaguchi, 1993). Sakaguchi *et al.* (1978) found that some microalgae such as *Chlorella regularis* can accumulate large amounts of uranium from non-saline waters. The adsorption reaction of uranium onto microorganisms is reversible regarding adsorption-desorption treatments (Nakajima and Sakaguchi, 1993). "Non-living" microorganisms show the accumulation of uranium as well as "living" organisms, which implies that the binding site in the microorganisms reacts chemically with metal ions. The binding site in bacteria, which is present in the cell membrane, can strongly associate with thorium higher than uranium in an acidic medium (pH 3.5) (Sakaguchi and Nakajima, 1991). These findings suggest that marine microorganisms may support the amount of strong organic ligand in particulate matter and dissolved organic matter.

We concluded that the strong organic ligand (DTPA-type) identified by Th isotopes as a probe, which may be commonly found in marine microorganisms, is closely related to the complexation of trace metals in particulate matter under the conditions of seawater. In other words, a strong organic ligand can be extracted from numerous natural organic ligands by use of Th with high chemical reactivity and high acidic media for the adsorption. The ThCC can be measured with high sensitive because thorium isotopes with different half lives can be applied to determine the ThCC. Therefore, it will be expected that the chemical properties and the biological role of the strong ligand in microorganisms may be studied at a scale of individual cells by use of Th isotopes as a probe, although a problem is how organisms in the ocean possess the strong ligand.

Th-complexing capacity of dissolved organic matter

The strong organic ligand detected in particulate matter by using Th isotopes was classified as the DTPA-type ligand (Hirose and Tanoue, 1994). Studies of metal speciation in seawater (Hirose.1994a) revealed that the DTPA-type ligand exists in dissolved organic matter. These findings predict that the strong organic ligand corresponding to the Th-complexing capacity in particulate matter should be present in dissolved organic matter. Hirose (1996) developed a new method determining the Th-complexing capacity in dissolved organic matter using XAD-2 adsorption.

The experimental study of Hirose (1996) revealed that Th reacts quantitatively with a binding site of dissolved organic matter (DOM) in 0.16 mol/L H^+ solution by complexation, when Th(IV) is equilibrated with DOM over 24 hr. Mass balance analysis on Th complexation indicates that Th forms a 1:1 complex with the binding site in DOM, whose the conditional stability constant is $10^{6.7}$ L/mol. The conditional stability constant of the Th complex with the strong ligand in DOM is in good agreement with that determined in particulate matter under similar experimental conditions (Hirose and Tanoue 1994); that is, the strong organic ligand in DOM identified by Th is chemically the same as that in particulate matter.

The Th complexation experiment using the XAD-2 adsorption in acidic media provides a parameter concerning complexation properties of DOM. Since a single reaction occurs for Th complexation under the conditions of the 0.16 mol/L H^+ solution, the Th-complexing capacity of DOM, as well as the Th-complexing capacity of particulate matter, is defined by Eq. (3), where the value of [ThL] is the concentration of Th adsorbed onto XAD-2 resin. We examined the reproducibility of the Th-complexing capacity for the same deep water sample. The Th-complexing capacities were reproducible to less than 5% for three repeated runs (Hirose, 1996). When we adopt three sigma of counting for the detection limit and 200 mL of seawater as a sample volume, the method has a detection limit of approximately 0.05 nmol/L for the Th-complexing capacity of DOM by using ^{230}Th. There is a serious problem whether the strong organic ligand in DOM reacted with Th is quantitatively adsorbed onto the XAD-2 resin or not. It is very difficult to solve this problem because there is little information about dissolved organic matter involving the strong organic ligand. However, there is a great possibility that most of the strong organic ligand dissolved in seawater is adsorbed onto the XAD-2 resin, taking account of the level (a order of nmol/L) and reproducibility of the Th-complexing capacity in DOM. On the other hand, the fact that DOM involving the strong organic ligand is adsorbed onto XAD resin implies that a significant part of this DOM has hydrophobic properties. It is noteworthy that a significant part of the cell membrane in marine organismsis constructed by hydrophobic residues.

A preliminary result for the Th-complexing capacity of DOM was reported (Hirose 1996); in surface waters of the Japan Sea, the Th-complexing capacity in DOM was several nmol/L, which is of the same order of magnitude as the concentration of the DTPA-type organic ligand dissolved in seawater (Tables 1 and 2), although sampling sites determined the concentrations of the strong organic ligand dissolved in seawater differed from each other.

The Th-complexing capacity of DOM as well as particulate matter, which can be easily determined without the effect of interference by other metals, is a highly sensitive method of organic speciation (detection limit: approximately 0.05 nmol/L). This method requires relatively small volumes of samples (several 100 mL of waters). The simplification provided by the methodology yields some useful merits and versatility, although unsolved problems still remain.

Acid-base properties of strong ligands

In order to understand the chemical properties of organic ligands, one of the most important factors is their acid-base characteristics. In the case when organic ligands form chelate with metal ions, organic ligands are needed to be polydentate with several numbers of functional groups of acids and bases such as aminopolycarboxylic acids. Therefore, obtaining evidence on acid-base properties of natural strong organic ligands is a significant pathway to clarifying their chemical feature.

According to the result of sequential leaching experiments (Hirose, 1994b), the strong organic ligand in particulate matter has more than two acidic functional groups in the pH range between 2 and 8.2. This fact suggests that the strong organic ligand in particulate matter is a kind of chelator. The Th adsorption experiment onto particulate matter (Hirose and Tanoue, 1994) revealed that the strong ligand has a functional group of a strong acid with a logarithmic protonation constant of –1.1.

Experiments of Th adsorption in acidic media provide information about acid-base properties of the strong ligand comparing with complexation of Th under the conditions of seawater if the strong ligand reacting with Th in acidic media is the same as that in seawater. As discussed in earlier, the strong ligand identified by Th in acidic media is characterized as the DTPA-type estimated in seawater. On comparison of the conditional stability constant of the Th complex with the DTPA-type organic ligand under the conditions of seawater, the decrease of the apparent conditional stability constant of Th complex in a 0.1 mol/L HCl solution is attributable to protonation reactions on acidic functional groups of the strong ligand; in other words, a side reaction coefficient of the strong organic ligand due to the protonation corresponds to the difference between the conditional stability constants in 0.1 mol/L HCl solution and seawater. We make the pH of seawater a reference, where a side reaction coefficient of the strong ligand due to the protonation is unity. Since the protonation to acidic groups occurs in the range between pH^a (=1) and pH^s (=8.2), the following relationship is obtained

$$n(pH^s - pH^a) \geq \log\alpha_{L(H)}{}^a = \log K_{ThL}{}^s - \log K_{ThL}{}^a + \alpha_{L(M)}{}^s \quad (8)$$

where n is a number of acidic groups, and superscripts a and s show acid solution and seawater, respectively. When this relationship is applied to complexation of Th with the DTPA-type ligand, we obtained $n \geq 3$, which is consistent with an estimate from sequential leaching experiments (Hirose, 1994b). As a result, the DTPA-type organic ligand has, at least, more than four acidic functional groups (more than three in the pH range between 1 and 8.2, and more than one in pH of lower than 1). Therefore, we conclude that the DTPA-type organic ligand is polydentate, i.e., a kind of chelator.

Present feature of strong organic ligands

The strong organic ligands, which can quantitatively form complexes with some trace metals, such as Cu, Zn, Fe and Th, under the conditions of seawater, exist in DOM and POM. The most important feature is that the reactions between metal ions and the strong organic ligands are reversible. As a result, the strong organic ligands in seawater show the properties of chelators, which is also supported by consideration of the acid-base properties of the organic ligands. According to the present understanding of the strong organic ligand in marine environments, the kind (or number) of strong organic ligands existing in seawater is limited. The stability of metal complexed with the strong organic ligand, which characterizes the chemical properties of the strong organic ligand, is closely related to the level of the corresponding ligand concentration; e.g., the organic ligand concentrations for EDTA and DTPA-type dissolved in seawater are several tens to one hundred nmol/L and an order of nmol/L, respectively. The EDTA-type organic ligand, which has been found at the early stage of speciation studies and major species in dissolved organic matter, have not been detected in particulate matter until now. This fact supports a hypothesis that the EDTA-type organic ligand dissolved in seawater is involved in aged organics. There is, however, little information about the biogeochemical cycling of the EDTA-type ligand. On the other hand, the DTPA-type organic ligand, which is the strongest among the organic ligands detected in seawater, is found in both particulate and dissolved organic matter. The oceanic flow of the DTPA-type organic ligand, which will be discussed later, is postulated as follows:

$$\text{Marine organisms} \Rightarrow \text{Particulate matter} \Rightarrow \text{Dissolved organic matter}$$

or

$$\text{Particulate matter} \Leftarrow \text{Marine organisms} \Rightarrow \text{Dissolved organic matter},$$

where the scheme seriously depends on the implication and/or definition of particulate matter. The DTPA-type ligand, which is derived from marine micro-organisms, reflects a fresh part of organic matter. Although specific organic ligands, such as siderophores and phytochelatins (Ahner and Morel, 1995), may be present in seawater, such organic ligands with lower molecular weights may be only minor species with concentrations of less than nM in open ocean.

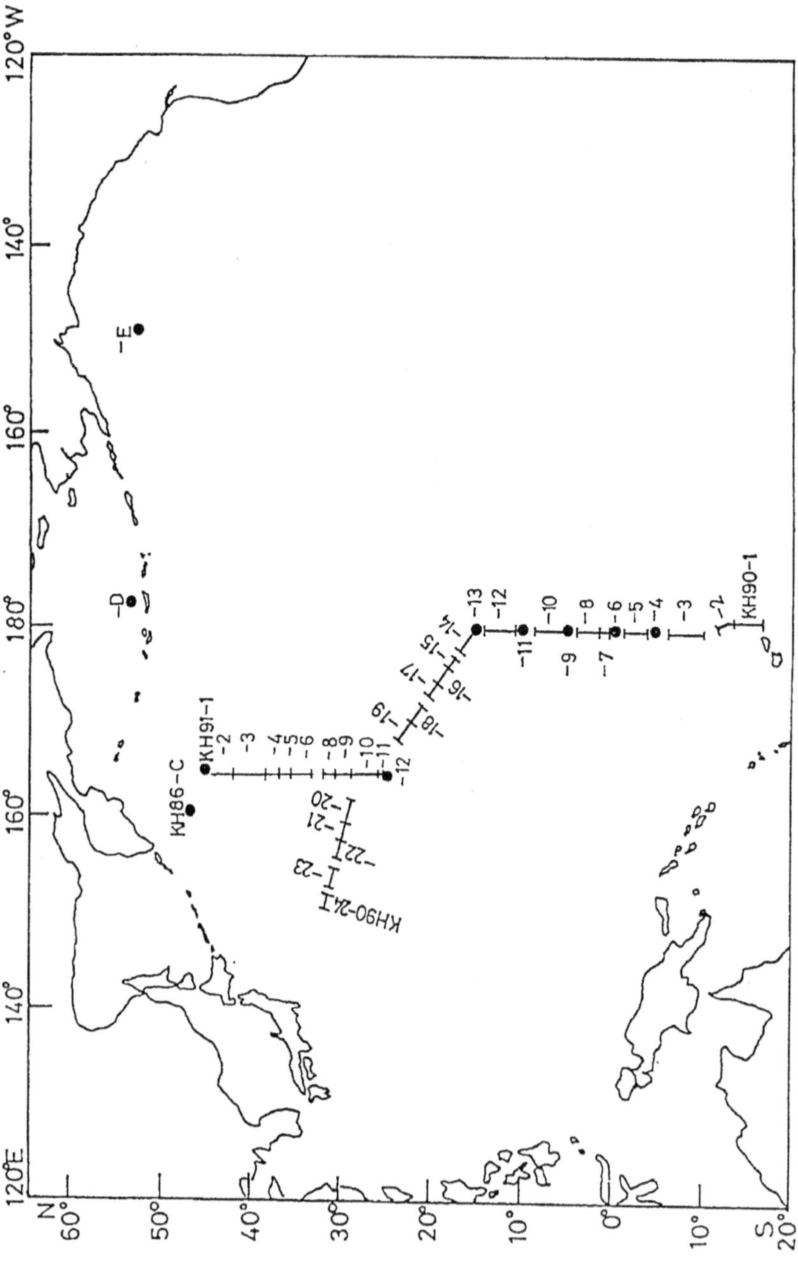

Fig. 5. Sampling locations of particulate matter in surface and vertical waters in the Pacific (Hirose and Tanoue, 1999).

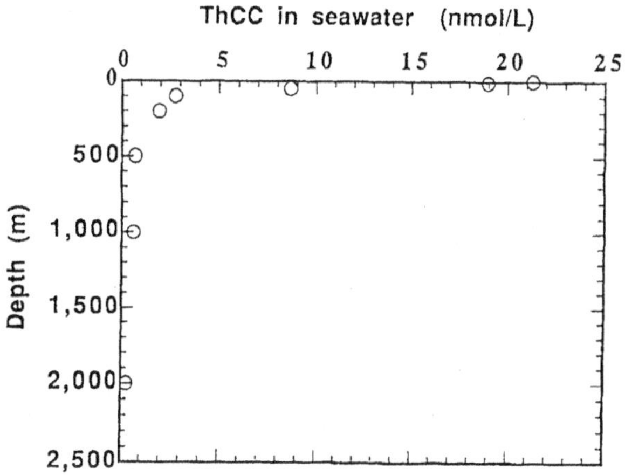

Fig. 6. Vertical profile of the strong organic ligand (ThCC) in particulate matter. The sampling station was 46°39.3′ N, 162°01.9′ E. (Hirose and Tanoue, 1998).

OCEANIC DISTRIBUTION OF THE STRONG ORGANIC LIGAND

In general, there is a possibility that the newly introduced parameter throws light on understanding the biogeochemical behavior of chemicals in oceanic environments. The oceanic distribution of the strong organic ligand, specified by thorium, may provide new information on the biogeochemical behavior and cycling of organic matter. In this section, we indicate the vertical and horizontal distributions of the strong organic ligand (ThCC) in particulate matter and dissolved organic matter collected in the Pacific and marginal seas (Hirose, 1996; Hirose and Tanoue, 1998, 1999) and compare the ThCC with organic components such as POC, PN, combined amino acids and others (see synthesis by Tanoue (2000, in this volume)).

Distribution of the strong organic ligand in particulate matter

a) Vertical

The vertical profiles of the strong organic ligand in particulate matter were determined at four stations, three of which (Stns. C, D and E) were located in relatively highly productive areas in the northern North Pacific and Bering Sea and one site (Stn. F) was located in the oligotrophic area of the central North Pacific (Hirose and Tanoue, 1997). Sampling sites are shown in Fig. 5. Values for POC, PN, proteins and combined amino acids at the same sampling site were determined.

The typical example of vertical profiles of the strong organic ligand in particulate matter for a North Pacific station (Stn C) is shown in Fig. 6. The concentration of the strong organic ligand (ThCC) in particulate matter in the surface layers decreased rapidly with depth at all stations examined. Relative decreases in concentrations of POC, PN, total amino acids and proteins in the identical filter samples were almost the same as those found in the ThCC (Tanoue, 1992a). Below the surface water layers, the ThCC, as well as levels of organic constituents, tended to decrease with depth toward the deep water layers. The levels of the ThCC in particulate matter in the intermediate and deep water layers were similar to each other, independent of the station, although biogeochemical conditions, such as vertical flux of particles, differed to some extent between the sampling stations (Martin *et al.*, 1987; Tsunogai and Noriki, 1991). The range of ThCC in particulate matter, from 3.99 to 21.3 nmol/L in surface waters of four stations, decreased to a range of 1.88 to 2.78 nmol/L at near 100 m depth, and from 0.22 to 0.35 nmol/L at the depth of 2,000 m at four stations. It is noteworthy that the strong organic ligand specified by thorium is present in deep particulate matter samples, although the strong organic ligand is a biologically fresh component. The differences between the two extremes of the ThCC values at different stations, apart from the surface layer, were not large. Similar phenomena were observed for organic constituents examined in the present study. The concentration range of POC, from 2.09 to 10.2 mmol/L in surface waters, decreased to the ranges from 1.57 to 2.02 mmol/L at a depth of around 100 m and to 0.51 to 1.37 μmol/L at a depth of 2,000 m. The vertical profiles of the organic ligand detected in the present study indicated that this organic ligand was a non-conservative organic constituent of particulate matter, as discussed in the previous section, and production, transport and decomposition were controlled by similar biogeochemical processes to those that governed other biochemical constituents of particulate matter. The ratio of ThCC to POC in particulate matter from deep waters, which ranged between 0.2×10^{-3} and 0.57×10^{-3}, decreased systematically with increasing depth. This finding suggests that the strong

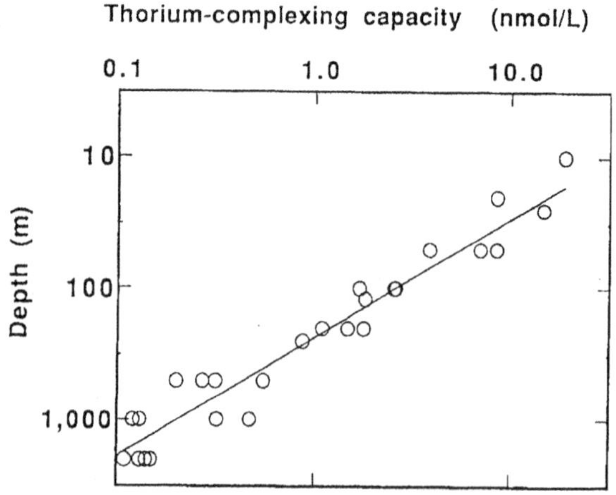

Fig. 7. Log-log relationship between corrected ThCC and depth. The data contained four vertical profiles of ThCC. The corrected ThCC was calculated from the assumption of a constant refractory fraction with 50% of deep ThCC. The correlation factor is 0.97.

organic ligand in particulate matter is more rapidly decomposed by biogeochemical reactions than the bulk organic carbon (POC).

The vertical distribution of the strong organic ligand in particulate matter was expressed by a power function, whose analysis was carried out for POC (Gordon and Cranford, 1985). Hirose and Tanoue (1998) found that, if each organic component in particulate matter is constructed by two parts, i.e., fresh and refractory fractions. The refractory fraction is independent upon depth, whereas the fresh fraction is a function of depth. The log-log plots of "fresh" ThCC to depth for the composite data of four stations are shown in Fig. 7. All data put were on the same line, which implies that profiles of "fresh" ThCC appear to be independent of sampling sites. This relationship is the same as an empirical equation of the particle export flux introduced by Martin *et al.* (1987) from sediment trap experiments.

It is important to discuss the implication of the particle fluxes measured by sediment traps, including a problem regarding sampling sites for the particle export fluxes and particulate matter, because depth-dependency of suspended matter and its constituents in the water column resembles that of the particle export flux determined by sediment trap experiments. Buesseler (1991) and his group (Buesseler *et al.*, 1994) suspected the measurement of the absolute fluxes by sediment traps. There are a number of potential problems with shallow traps: (1) a significant fraction of the organic material collected is often attributed to swimmers, which are extremely difficult to separate or even elucidate (Lee *et al.*, 1988; Michaels *et al.*, 1990), (2) in high shear environments, traps may serve as selective particles, preferentially over or under sampling particles of different size classes (Gardner, 1980a, b; Butman *et al.*, 1986), thus possible biasing composition as well as fluxes, (3) decomposition and dissolution of material may alter the composition (Knauer *et al.*, 1984) and (4) the vertical migration of zooplankton can cause traps to underestimate export fluxes if the migrators feed above the trap and defecate below (Angel, 1989).

The most important aspect regarding sediment trap studies (Martin *et al.*, 1987) is that the export fluxes of biogenic particles may be expressed as a function of depth with a power of b as a spatial and temporal average, which has been extensively used in modeling of carbon cycling in the ocean (Bacastow and Maier-Reimer, 1990; Najjar *et al.*, 1992; Maier-Reimer, 1993; Anderson and Sarmiento, 1995; Yamanaka and Tajika, 1996). The export flux of carbon, F_C, is expressed as follows:

$$F_C = F_{C,100}(z/100)^b, \tag{9}$$

where $F_{C,100}$ and z show the carbon export flux at 100 m depth and depth, respectively. This equation leads the rapid decrease of the particle export flux just below the euphotic layer. It must be noted that a serious problem regarding this phenomenon was discussed by Karl *et al.*, (1988). Therefore, if the empirical equation of the particle export flux from sediment trap experiments reflects the real feature of particle sinking processes, it is concluded that the vertical

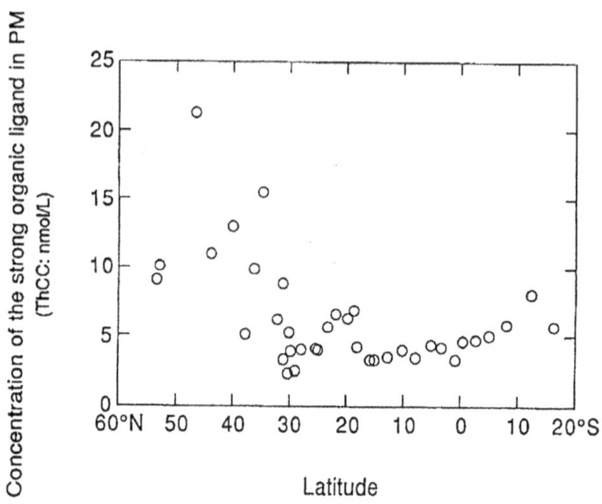

Fig. 8. Meridional distribution of the strong organic ligand (ThCC) in surface particulate matter.

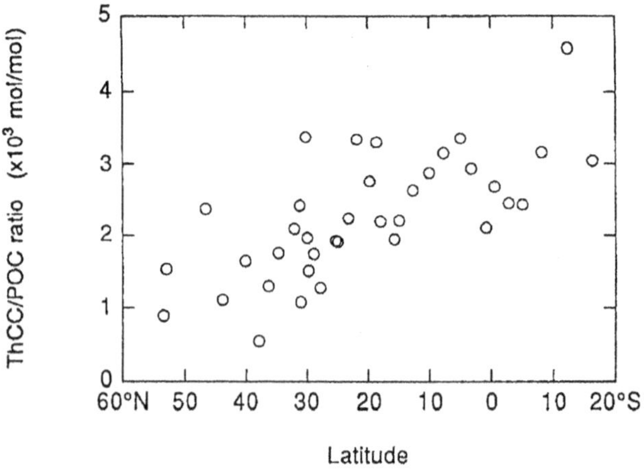

Fig. 9. Meridional distribution of the ratio of the strong organic ligand (ThCC) to organic carbon (POC) in surface particulate matter.

distribution of suspended matter is primarily controlled by the particle export flux (Hirose and Tanoue, 1998). On the other hand, as another aspect, there is the possibility that the empirical equation given to the particle export flux obtained from sediment traps reflects only the amount of particulate matter at a corresponding depth instead of the "real" particle export flux.

The b value of the power function may be related to the degradation rate of the corresponding chemical component of the POM (Hirose and Tanoue, 1998).

The decomposition rates of organic constituents in particulate matter can be ranked in the following order: proteins > PN > strong organic ligand > combined amino acids > POC. The result supports that the strong organic ligand is a biologically fresh component in particulate matter.

b) Meridional

The concentrations of the strong organic ligand (ThCC) in particulate matter were determined in surface water samples in the Pacific (see Fig. 5), together with values for POC, PN, and combined amino acids (CAA) (Hirose and Tanoue, 1999).

The strong organic ligand specified by Th generally exists in particulate matter of the Pacific surface at a basin scale. The concentration of the strong organic ligand (ThCC) in particulate matter of surface waters, which ranged from 2.32 to 21.3 nmol/L, showed a typical meridional distribution, as shown in Fig. 8. The higher surface ThCC, which showed larger variation, was observed in highly productivity areas (north of 35°N), whereas the lowest value corresponded to the oligotrophic area (near 30°N) in the central North Pacific. Relatively high ThCC was observed in the subtropical surface waters near 20°N. Another feature is that there is no peak of the ThCC in the equatorial Pacific, although the equatorial Pacific corresponds to a high productivity area.

The latitudinal distribution of the strong organic ligand in particulate matter was similar to those of POC, PN and proteins. However, detail distributions of chemical components in particulate matter differed from each other; e.g., a lower concentration of proteins in surface particulate matter was observed in the subtropical region near 20°N (Tanoue, 1996), whereas a peak of the strong ligand concentration occurred in the same region. We examine the mutual relationship between the ThCC and other chemicals in particulate matter. The ThCC/POC ratios in surface waters, which ranged from 0.55×10^{-3} to 3.9×10^{-3} (mol/mol C), were not geographically constant as seen in Fig. 9, although the order of magnitude of the ThCC/POC ratios was in agreement with that between the binding site and organic carbon in microorganisms (around 10^{-3} (mol/mol C organisms)) (Nakajima *et al.*, 1979; Nakajima and Sakaguchi, 1993). The finding is consistent with the hypothesis that the strong organic ligand in particulate matter is biogenic (Hirose, 1995; Hirose and Tanoue, 1998). The ThCC/POC ratios in surface waters show systematic variation in the Pacific; i.e., the ThCC/POC ratios decreased from the subtropical to the subarctic region. The relatively high ThCC/POC ratios occurred in the subtropical gyre near 20°N and in the region of the North Equatorial Counter Current (NECC), which is characterized by low salinity and extended from about 4°N to about 10°N (Philander, 1990). Tanoue (1996) found that the area of NECC appeared to coincide with the relatively low values of the protein/CAA ratio and the area of the North Equatorial Current (NEC) and the South Equatorial Current (SEC), which are located to the north and south of the NECC, respectively, corresponded to the higher protein/CAA ratios. The ThCC/POC ratios appeared to be relatively low in the NEC and SEC, which is the inverse of the spatial variation of the protein/CAA ratios. The result implies that the rather low ThCC/POC ratio occurred in a highly productive area. A similar tendency was observed in the ThCC/PN and ThCC/CAA ratios in the Pacific surface waters. The meridional distribution patterns of the ThCC/PN and ThCC/CAA ratios are completely different from the protein/POC and protein/CAA ratios in corresponding particulate matter samples (Tanoue, 1996). Although both of the proteins and the strong organic ligand in particulate matter, which are indicators of fresh organic matter from their vertical profiles (Hirose and Tanoue, 1998), are derived from marine organisms, the oceanic behavior of the strong ligand in particulate matter differs significantly from that of the proteins .

c) Sources of the strong organic ligand in particulate matter

The vertical profiles of the organic ligand in particulate matter (Hirose and Tanoue, 1998) indicated that the strong organic ligand, as well as the proteins, is a non-conservative organic constituent of particulate matter, and its production, transport and decomposition are controlled by similar biogeochemical processes to those that governed other biochemical constituents of particulate matter. However, the behavior of the strong ligand in surface particulate matter significantly differs from that of proteins. The particulate matter onto the GF/F filters consists both of living organisms and detritus matter. The result suggests that the strong organic ligand in particulate matter is closely related to living organisms (Hirose, 1995). Organisms retained on the GF/F filter were mainly a mixture of autotrophic phytoplankton and hetertrophic bacteria. Now it is important to evaluate the individual contributions of autotrophic phytoplankton and hetertrophic bacteria about the strong organic ligand. The number of bacteria in surface waters at the same locations where the POM samples designated KH91-1 and -12 had been collected, were about $1.2–1.4 \times 10^6$ cells/ml (depths of 0–11 m), and about $0.2–0.7 \times 10^6$ cells/ml (depths of 0–10 m), respectively (Hara *et al.*, 1996). On the other hand, concentrations of *Chl.a*, considered to be an indicator of the amount of phytoplankton, were 0.7 mg/L (depths of 0–10 m) and 0.05 mg/L (depths of 0–10 m) at sampling stations of KH91-1 and -12, respectively. Concentrations of the

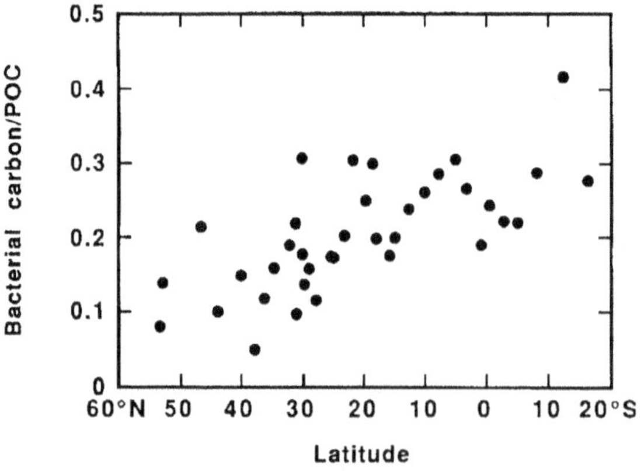

Fig. 10. Meridional distribution of the ratio of bacterial biomass-C to POC.

strong organic ligand in particulate matter at the corresponding stations were evaluated to be *ca.* 12 and 3.99 nmol/L, respectively. When the concentrations of the strong organic ligand were compared with amounts of bacteria and/or *Chl.a*, they corresponded to bacteria rather than *Chl.a*. Therefore, the major part of the strong organic ligand in particulate matter is attributable to bacteria. In this connection, Moffett *et al.* (1990) provided direct laboratory evidence for the production of a strong Cu-binding ligand by a marine photoautotroph. This ligand has a Cu-complexing strength identical to that of the stronger ligand observed in the Sargasso Sea and in the central northeast Pacific (Coale and Bruland, 1990), which corresponds to the DTPA-type organic ligand, which are to be able to form complexes with Cu, Fe, Th and others under the conditions of seawater, as discussed earlier. The laboratory results from a preliminary survey of four marine phytoplankton species (three species of eukaryotes and one prokaryote) (Moffett *et al.*, 1990) indicated that only the cyanobacterial species *Synechococcus* produced a chelator forming Cu complexes of the same strength as those observed in the Sargasso Sea. Furthermore, Cho and Azam (1988) suggested that bacteria could be responsible for radionuclide adsorption in the unchanging pool of fine, non-sinking particles. A recent study by Tortell *et al.*, (1996) revealed that oceanic bacteria contain more iron per biomass than phytoplankton.

The strong ligand in bacteria is calculated to be about 18 amol/cell on the assumption that 50% of the bacteria were retained on the GF/F filter (Lee and Fuhrman, 1987). By using the result that the amount of organic carbon in a bacterial cell is 20 fg C/cell (Lee and Fuhrman, 1987), a content of the strong organic ligand in bacteria was calculated to be about 1.1×10^{-2} mol/mol C, whose value corresponds to a conversion factor to estimate the bacterial biomass from the concentration of strong ligands in particulate matter. The portion of bacterial biomass to POC can be calculated from the ThCC using a conversion factor. The result is shown in Fig. 10. The bacterial biomass occupied 20 to 30% to POC in 30°N to 20°S, whereas north of 35°N, it ranged between 5 and 20%. Caron *et al.* (1995) estimated the contribution of microorganisms to particulate carbon in surface waters of the Sargasso Sea. The bacterial assemblage constituted a significant fraction of the total particulate carbon (=10–20%). Although sampling stations and periods differed from each other, the portion of bacterial biomass to POC in the Pacific estimated from the strong ligand, is nearly the same range as those in the Sargasso Sea. Kirchman *et al.* (1993) determined the biomass and production of heterotrophic bacterioplankton in the subarctic Pacific. The bacterial biomass nitrogen occupied 15% of particulate nitrogen. Taking account of the observation that bacteria concentrated nitrogen in comparison with phytoplankton and detritus matter, the portion of bacterial biomass to POC in the subarctic Pacific coincides with that estimated from the strong ligand.

Tanoue (1996) indicated that the proteins in particulate matter originate from both phytoplankton and bacteria. The major part of the proteins in particulate matter was derived from phytoplankton north of 35°N whereas the proteins from bacteria occupied a significant part in the subtropical Pacific. We conclude that the proteins in particulate matter reflect both phytoplankton and bacteria, whereas most of the strong organic ligand in particulate matter is derived from bacteria, although both the strong organic ligand and the proteins are a measure of "living" POM (Tanoue, 1996). The strong organic ligand identified by Th in particulate matter becomes a tracer of bacterial biomass.

d) The relationship between POC and the binding sites responsible for thorium isotopes

Eppley (1989) proposed that, if the particles to which thorium adsorbs are biogenic, then ^{234}Th and particulate carbon data could be used to estimate new production. Based on this hypothesis, the export flux of carbon has been estimated from ^{234}Th-^{238}U disequilibria (Coale and Bruland, 1985; Buesseler *et al.*, 1992). However, there is a little concern about the interaction between the binding site in sinking particles and thorium isotopes. If the organic ligand corresponding to the ThCC can be linked to biogenic POM by use of Th isotopes, we would provide additional information about the use of ^{234}Th for the estimation of the carbon export flux.

The present characterization of the strong organic ligand in particulate matter, as discussed in the previous section, suggests that the sites bound to ^{234}Th in particulate matter, which is a DTPA-type organic ligand originating from bacteria, is not directly related to the total amount of POM but to the amount of bacteria. In fact, the meridional distributions of POC and the strong organic ligand in particulate matter imply that the ratios of the strong organic ligand to POC, as shown in Fig. 9, which varied near to one order of magnitude, are not constant in the Pacific ocean surface waters; the lower ThCC/POC ratios occurred in the high productivity area north of 35°N. This suggests that the ratio of ^{234}Th to biogenic organic matter is not constant temporally and spatially. Therefore, there is the possibility that the carbon export flux determined from ^{234}Th-^{234}U

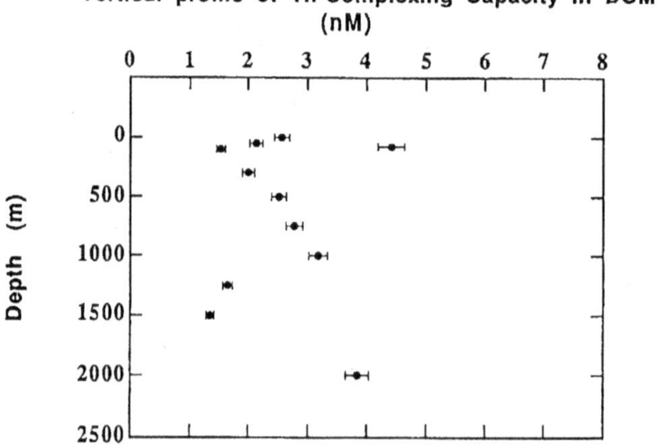

Fig. 11. Vertical profile of the strong organic ligand (ThCC) dissolved in seawater.

disequilibria is an underestimation for the subarctic area and the high flux season corresponding to phytoplankton blooming,where the increase of bacterial biomass cannot immediately follow the blooming of phytoplankton, and an overestimation for the oligotrophic area. It is concluded that the carbon export flux cannot be determined with higher accuracy from ^{234}Th-^{234}U disequilibria methods if the ratio of POC to ^{234}Th in the ocean is assumed to be constant.

Distribution of the strong ligand in dissolved organic matter

There is little information about the distribution of the strong organic ligand because the concentration of it dissolved in seawater cannot be determined easily. Another problem is that the strong organic ligand is considered to be biologically active; there is a possibility that the strong organic ligand is produced and decomposed in sample handling and/or experimental processes due to the contamination of bacteria. Although some difficult problems are present in methodologies, the vertical profiles of given the strong organic ligand significantly differ from those of well-characterized constituents in seawater such as DOC and others.

A preliminary vertical profile of the strong organic ligand (ThCC) dissolved in the Japan Sea waters (Hirose, 1996) is shown in Fig. 11. The concentration of the strong organic ligand in DOM in surface waters is of the same order of magnitude as the concentration of the stronger ligand of the DTPA-type (Hirose, 1994a; Campos and van den Berg, 1994). The vertical profile of Th-complexing capacities in the Japan Sea showed a subsurface maximum and increased gradually from 100 m to 1,000 m depth. Campos and van den Berg (1994) determined the vertical profiles of a strong organic ligand, being able to complex with Cu, in the Northeast (NE) Atlantic, which is specified as of the DTPA-type from the value of the conditional stability constant (*ca.* 10^{13}) (Table 1). The vertical distribution of the strong ligand dissolved in seawater of the NE Atlantic showed a subsurface maximum (78 m) and a minimum at 100 m depth (Campos and van den Berg, 1994), whose pattern is similar with that of the Th-complexing capacity of DOM, although sampling sites differed from each other. It must be noted that the vertical profile of the Th-complexing capacity in particulate matter, which showed a surface maximum and decreased rapidly in the subsurface layer, completely differs from that for DOM. The results suggest that the strong organic ligand dissolved in seawater specified with Th is chemically the same as the DTPA-type related to metal complexation in seawater. These findings lead to the hypothesis that dissolved organic matter in deep waters contains biologically fresh components because the DTPA-type ligand reflects a fresh part of organic matter. Another feature is the possibility that the vertical profiles of the strong organic ligand dissolved in seawater differ from that of DOC (Tanoue, 1992b, 1993), although both of the concentration of strong organic ligand and DOC has not been measured in the same samples.

The recent studies by Hirose and Tanoue (1994), and Hirose (1995, 1996), suggest that the Th-complexing capacity in both DOM and particulate matter will become a useful parameter for the fate and role of the DTPA-type strong ligand in the biogeochemical processes in the ocean, which is directly related to the complexation of trace metals in seawater. Further studies are necessary to clarify the chemical properties and oceanographic meaning of the Th-complexing capacity of DOM and particulate matter.

OCEANIC CYCLING OF ORGANIC MATTER

The oceanographic roles of the strong organic ligand in seawater have been discussed from the aspect of bioavailability of trace metals, i.e., regulation of free ion concentration of trace metals in seawater, which is related to detoxification, storage of metals as a soluble form, and direct uptake of metal complexes by marine organisms (Gavis, 1983; Hirose and Sugimura, 1985; Sunda, 1989). In other words, the strong organic ligands play an ecological role in oceanic environments. However, there are too few studies on the biogeochemical behavior of organic matter in the ocean to be able use the strong organic ligand as an indicator. It is important to understand the biogeochemical behavior of organic matter, including the production and degradation of particulate matter and dissolved organic matter, aggregation and disaggregation of fine organic particles, sinking of large organic particles, and vertical transport of DOM by physical processes, whose processes are directly related to carbon cycling.

Enigma and serious problems are present concerning oceanic behavior of organic matter. Although the particle export flux is one of the most important biogeochemical processes, as discussed in previous sections, we have no technique to determine the particle export flux in the water column with any great accuracy; studies indicated that in order to determine the export flux with great accuracy, it is very difficult to measure directly the export flux of carbon by using sediment traps and ^{234}Th methods. Dissolved organic carbon (DOC), whose measurements have some problems (Tanoue, 1992b, 1993), is the largest organic carbon pool in the ocean. The apparent age of DOC is very old ($1–6 \times 10^3$ years) due to an estimation from ^{14}C (Druffel and Williams, 1990). Nevertheless, a major part of DOC is constructed by organics with low molecular weights (Guo *et al.*, 1995). DOM with high molecular-weights was more rapidly decomposed than one with lower molecular weights (Santschi *et al.*, 1995). Recent studies (see Santschi *et al.*, 1995) indicate that a major fraction of DOM with high molecular weights have turnover times of weeks to months. Furthermore, DOM through whole water columns contains the strong organic ligand and proteins (Tanoue *et al.*, 1995; Hirose, 1996) as a signal of biological fresh matter. These findings strongly suggest that DOM contains biologically fresh components through whole water columns.

There are at least two solutions to solve the DOC enigma in oceanic environments (see the review by Ogawa (2000, in this volume)): one is the existence of very refractory fractions of DOC with low molecular weights (less than 1 kD), the discovery of which is being carried out by marine organic geochemists, and another is that biological processes including DOM, i.e., uptake and production of DOM, occur apart from other oceanic processes in corporating fresh DOM such as the supply of DOM from sinking particulate matter and out flow from DOM pools. For the latter case, it is not necessary to introduce very refractory organics with low molecular weights because DOM is biologically renewed through the uptake of aged DOM by bacteria and the reproduction of aged DOM by the attack of bacteriophages (this part is based on a discussion with Tanoue, see synthesis by Tanoue (2000, in this volume)). This implies that there is little supply of DOM from the major biogeochemical process, involving primary production and export flux, to the DOM pool; as a result, the DOC pool is not a major path in the oceanic carbon cycle, even if DOC is the major organic pool in the ocean. Unfortunately, we have no direct evidence to verify the hypothesis regarding the biogeochemical behavior of oceanic DOM. It must be noted, however, that knowledge on specific functional groups such as the strong organic ligand and macromolecules, e.g., proteins, in oceanic organic matter is useful for understanding the biogeochemical processes of organic matter in oceanic environments.

CONCLUDING REMARKS

Current understanding of the strong organic ligands in seawater is summarized as follows:

1. Methodologies for studying natural strong organic ligands, most of which have been based on mass balance techniques, revealed that reactions between natural strong organic ligands and metal ions, which occur in the ocean, are reversible; the strong organic ligands in particulate matter and dissolved organic matter have the properties of chelators.

2. The strong organic ligands in seawater can be classified by using the conditional stability constant of metal complexes and concentration of the corresponding organic ligand. The LFER technique is useful in characterizing the strong organic ligands. Two types of strong organic ligands coexist in seawater: one the DTPA type, whose concentration is relatively low, the other is the EDTA type, which is a major ligand.

3. The strong organic ligands are commonly found in marine environments. Especially, the DTPA-type ligand, which is responsible for complexation of trace metals such as Cu, Zn, Fe, Th, and others, exists in both particulate matter and dissolved organic matter.

4. Most of the strong organic ligands are present as metal complexes; alkaline-metals, which are major inorganic components in seawater, play a significant role for the complexation of strong organic ligands.

5. Acid-based properties of strong organic ligands are one of the most important factors in chemically characterizing strong organic ligands. The DTPA-type ligand shows properties of multi-acids, such as aminopolycarboxylic acids, which have more than four acidic sites. (More than three acidic groups are present in the pH range between 1 and 8.2, and one site exists in the lower pH region.) This fact strongly suggests that natural strong organic ligands in the ocean include the carboxylic acid group, which is a major functional group in DOM.

6. The DTPA-type ligand in both DOM and POM, which can be identified by Th in acidic media, is derived from marine bacteria, which may be located in cell membranes. This hypothesis is supported by the fact that a significant part of organic matter involving the DTPA-type ligand has hydrophobic properties. The bacterial biomass may be estimated from the concentration of DTPA-type ligands in particulate matter and the conversion factor (1.1×10^{-2} mol ThCC/mol bacteria C).

7. Metals become a good tool to characterize the chemical properties of the natural organic ligands. The organic ligands specified by metals are useful as a probe of the oceanic cycling of organic matter.

The most serious problem for studies of strong organic ligands in seawater is that both strong DTPA and EDTA-type organic ligands cannot be easily isolated and chemically identified. We, therefore, cannot provide direct information on the molecular properties of organic ligands in seawater. However, we can accumulate evidence on the presence of strong ligands and knowledge about their possible chemical properties and sources. Since the strong organic ligands have a chemical function to form metal complexes in seawater, directly related to bioavailability and removal processes of traces metals, their temporal and spatial distributions and oceanic behavior are important to understand the key processes of oceanic cycling of organic matter. Especially, the strong organic ligand in particulate matter and dissolved organic matter, identified by Th, which is an indicator of bacterial biomass, becomes a new parameter to understand the biogeochemical behavior of organic matter, especially microbial loop.

Acknowledgements—The author would like to thank E. Tanoue for giving him the chance to publish this work and for his helpful discussion, and T. Midorikawa and the anonymous reviewer for constructive comments.

REFERENCES

Ahner, B. A. and F. M. M. Morel (1995): Phytochelatin production in marine algae. 2. Induction by various metals. *Limnol. Oceanogr.*, **40**, 658–665.

Ahner, B. A., S. Kong and F. M. M. Morel (1995): Phytochelatin production in marine algae. 1. An interspecies comparison. *Limnol. Oceanogr.*, **40**, 649–657.

Anderegg, G. (1977): Critical survey of stability constants of EDTA complexes, IUPAC Chem. Data Ser. No. 14, Pergamon, London.

Anderson, D. M. and F. M. M. Morel (1978): Copper sensitivity of *Gonyaulax tamarensis*. *Limnol. Oceanogr.*, **23**, 268–295.

Anderson, M. A., F. M. M. Morel and R. R. L. Guillard (1978): Growth limitation of a coastal diatom by low zinc ion activity. *Nature* (London), **276**, 70–71.

Anderson, R. F. (1982): Concentration, vertical flux, and reminerarization of particulate uranium in seawater. *Geochim. Cosmochim. Acta*, **46**, 1293–1299.

Anderson, R. F., M. P. Bacon and P. G. Brewer (1983): Removal of ^{230}Th and ^{231}Pa from the open ocean. *Earth Planet. Sci. Lett.*, **65**, 7–23.

Angel, M. V. (1989): Dose mesopelagic biology affect the vertical flux? pp. 155–173. In *Productivity of the Ocean: Present and Past*, ed. by W. H. Berger, V. S. Smetacek and G. Wefer, Wiley, New York.

Apte, S. C., M. J. Gardner and J. R. Ravenscroft (1990): An investigation of copper complexation in the Severn Estuary using differential pulse cathodic stripping voltamrnetry. *Mar. Chem.*, **29**, 63–75.

Bacastow, R. and E. Maier-Reimer (1990): Ocean-circulation model of the carbon cycle. *Clim. Dyn.*, **4**, 95–125.

Bacon, M. P. and R. F. Anderson (1982): Distribution of thorium isotopes between dissolved and particulate forms in the deep sea. *J. Geophy. Res.*, **87**, 2045–2056.

Barber, R. T. and J. H. Ryther (1969): Organic chelators; factors affecting primary production in the Cromwell Current upwelling. *J. Exp. Mar. Biol. Ecol.*, **3**, 191–199.

Batley, G. E. and T. M. Florence (1976): Determination of the chemical forms of dissolved cadmium, lead and copper in sea water. *Mar. Chem.*, **4**, 347–363.

Biedermann, G. and J. T. Chow (1966): Studies on the hydrolysis of metal ions. Part 57. The hydrolysis of the iron(III) ion and the solubility product of $Fe(OH)_{2.7}Cl_{0.3}$. *Acta Chem. Scand.*, **20**, 1376–1388.

Billet, D. S. M., R. S. Lampitt, A. L. Rice and R. F. C. Mantoura (1983): Seasonal sedimentation of phytoplankton to the deep-sea benthos. *Nature*, **302**, 520–522.

Bishop, J. K. B., D. R. Ketten and J. M. Edmond (1978): The chemistry, biology and vertical flux of particulate matter from the upper 400 m of the Cape Basin in the southeast Atlantic Ocean. *Deep-Sea Res.*, **25**, 1121–1161.

Brand, L. E., W. G. Sunda and R. R. L. Guillard (1983): Limitation of marine phytoplankton reproductive rates by zinc, manganese, and iron. *Limnol. Oceanogr.*, **28**, 1182–1195.

Bruland, K. W. (1980): Oceanographic distribution of cadmium, zinc, nickel and copper in the North Pacific. *Earth Planet. Sci. Lett.*, **47**, 176–198.

Bruland, K. W. (1989): Complexation of zinc by natural organic ligands in the central North Pacific. *Limnol. Oceanogr.*, **34**, 267–283.

Buckley, P. J. M. and C. M. G. van den Berg (1986): Copper complexation profiles in the Atlantic Ocean. A comparative study using electro-chemical and ion exchange techniques. *Mar. Chem.*, **19**, 281–296.

Buesseler, K. O. (1991): Do upper-ocean sediment traps provide an accurate record of particle flux? *Nature*, **353**, 420–423.

Buesseler, K. O., M. P. Bacon, J. K. Cochan and H. D. Livingston (1992): Carbon and nitrogen export during the JGOFS North Atlantic Bloom Experiment estimated from ^{234}Th:^{238}U disequilibria. *Deep-Sea Res.*, **39**, 1115–1137.

Buesseler, K. O., A. F. Michaels, D. A. Siegel and A. H. Knap (1994): A three-dimensional time-dependent approach to calibrating sediment trap fluxes. *Global Biogeochem. Cycles*, **8**, 179–193.

Buffle, J. (1988): *Complexation Reactions in Aquatic Systems*. Ellis Horwood Limited, Chichester.

Butman, C. A., W. D. Grant and K. D. Stolzenback (1986): Predictions of sediment trap biases in turbulent flows: a theoretical analysis based on observations from literature. *J. Mar. Res.*, **44**, 601–644.

Byrne, R. H. and D. R. Kester (1976): Solubility of hydous ferric oxide and iron speciation in seawater. *Mar. Chem.*, **4**, 255–276.

Cacheris, W. P. and G. P. Choppin (1987): Dissociation kinetics of thorium-humate complex. *Radiochim. Acta*, **42**, 185–190.

Campos, M. L. A. M. and C. M. G. van den Berg (1994): Determination of copper complexation in sea water by cathodic stripping voltametry and ligand competition with salicylaldoxime. *Anal. Chim. Acta*, **284**, 481–496.

Caron, D. A., H. D. Dam, P. Kremer, E. J. Lessard, L. P. Madin, T. C. Malone, J. M. Napp, E. R. Peele, M. R. Roman and M. J. Youngbluth (1995): The contribution of microorganisms to particulate carbon and nitrogen in surface waters of the Sargasso Sea near Bermuda. *Deep-Sea Res.*, **42**, 943–972.

Chester, R., A. Thomas, F. J. Lin, A. S. Basaham and G. Jacinto (1988): The solid state speciation of copper in surface water particulates and oceanic sediments. *Mar. Chem.*, **24**, 261–292.

Cho, B. C. and F. Azam (1988): Major role of bacteria in biogeochemical fluxes in the ocean's interior. *Nature*, **332**, 441–443.

Clegg, S. L. and M. Whitfield (1990): A generalized model for the scavenging of trace metals in the open ocean—I. Particle cycling. *Deep-Sea Res.*, **37**, 809–832.

Clegg, S. L. and M. Whitfield (1991): A generalized model for the scavenging of trace metals in the open ocean—II. Thorium scavenging. *Deep-Sea Res.*, **38**, 91–120.

Coale, K. H. and K. W. Bruland (1985): ^{234}Th:^{238}U disequilibria within the California current. *Limnol. Oceanogr.*, **30**, 22–33.

Coale, K. H. and K. W. Bruland (1988): Copper complexation in the North-west Pacific. *Limnol. Oceanogr.*, **33**, 1084–1101.

Coale, K. H. and K. W. Bruland (1990): Spatial and temporal variability in copper complexation in the North Pacific. *Deep-Sea Res.*, **34**, 317–336.

Cochran, J. K., H. D. Livingston, D. J. Hirschberg and L. D. Surprenant (1987): Natural and anthropogenic radionuclide distributions in the Northwest Atlantic Ocean. *Earth Planet. Sci. Lett.*, **84**, 135–152.

Cooper, L.H.N.(1937):Some conditions govening to solubility of iron, *Proc. R.Soc. London*, **B129**, 299-305.

Dawson, R. and E. K. Duursma (1982): State of the art. pp. 497–511. In *Marine Organic Chemistry*, ed. by E. K. Duursma and R. Dawson, Elsevier, Amsterdam.

DiSpirito, A. A., J. W. Talnagi and O. H. Touvinen (1983): Accumulation and cellular distribution of uranium in Thiobacillus ferrooxidans. *Arch. Microbiol.*, **135**, 250–253.

Djogic, R., L. Sipos and M. Branica (1986): Characterization of uranium(VI) in seawater. *Limnol. Oceanogr.*, **31**, 1122–1131.

Donat, J. R. and K. W. Bruland (1990): A comparison of two voltamrnetric techniques for determining zinc speciation in Northeast Pacific Ocean waters. *Mar. Chem.*, **28**, 301–323.

Donat, J. R. and C. M. G. van den Berg (1992): A new cathodic stripping voltametric method for determining organic copper complexation in seawater. *Mar. Chem.*, **38**, 69–90.

Donat, J. R., K. A. Lao and K. W. Bruland (1986a): Speciation of dissolved copper and nickel in South San Francisco Bay: a multi-method approach. *Anal. Chim. Acta*, **284**, 547–572.

Donat, J. R., P. J. Statham and K. W. Bruland (1986b): An evaluation of a C-18 solid phase extraction technique for isolating metal-organic complexes from central North Pacific Ocean waters. *Mar. Chem.*, **18**, 85–99.

Druffel, E. R. M. and P. M. Williams (1990): Identification of a deep marine source of particulate organic carbon using bomb ^{14}C. *Nature*, **347**, 172–174.

Duinker, J. C. and C. J. M. Kramer (1977): An experimental study on the speciation of dissolved zinc, cadmium, lead, and copper in River Rhine and North Sea water by differential pulsed anodic stripping voltammetry. *Mar. Chem.*, **5**, 203–228.

Dyrssen, D. (1985): Metal complex formation in sulphidic seawater. *Mar. Chem.*, **15**, 285–293.

Dyrssen, D. and M. Wedborg (1974): Equilibrium calculations of the speciation of elements in seawater. pp. 181–196. In *The Sea, Vol. 5*, ed. by E. D. Goldberg, Wiley-Interscience, New York.

Elliott, S. (1988): Linear free energy techniques for estimation of metal sulfide complexation constants. *Mar. Chem.*, **24**, 203–213.

Eppley, R. W. (1989): New production: history, methods, problems. pp. 85–97. In *Productivity of the Ocean: Present and Past*, ed. by W. H. Berger, V. S. Smetacek and G. Wefer, Wiley, New York.

Farrington, J. W. (1992): Macromolecular organic matter working group report. In *Marine Organic Geochemistry: Review and Challenges for the Future*, ed. by J. W. Farrington, *Mar. Chem.*, **39**, 39–50.

Fowler, S. W. and G. A. Knauer (1986): Role of large particles in the transport of elements and organic compounds through the oceanic water column. *Prog. Oceanogr.*, **16**, 147–194.

Gardner, W. D. (1980a): Sediment trap dynamics and calibration: a laboratory evaluation. *J. Mar. Res.*, **38**, 17–39.

Gardner, W. D. (1980b): Field assessment of sediment traps. *J. Mar. Res.*, **38**, 41–52.

Gavis, J. (1983): Toxic binding of cupric ion by marine phytoplankton. *J. Mar. Res.*, **41**, 53–63.

Grill, E., E.-L. Winnacker and M. H. Zenk (1985): Phytochelations: The principal heavy-metal complexing peptides of higher plants. *Science*, **230**, 674–676.

Goldberg, E. D. (1954): Marine geochemistry. I. Chemical scavengers of the sea. *J. Geol*, **62**, 249–265.

Gordon, D. C., Jr. and P. J. Cranford (1985): Detailed distribution of dissolved and particulate organic matter in the Arctic Ocean and comparison with other oceanic regions. *Deep-Sea Res.*, **32**, 1221–1232.

Guo, L., P. H. Santschi and K. W. Warnken (1995): Dynamics of dissolved organic carbon (DOC) in oceanic environments. *Limnol. Oceanogr.*, **40**, 1392–1403.

Hanson, A. K., Jr., C. M. Sakamoto-Arnold, D. L. Huizenga and D. R. Kester (1988): Copper complexation in Sargasso Sea and Gulf Stream warm-core ring waters. *Mar. Chem.*, **23**, 181–203.

Hara, S., I. Koike, K. Terauchi, H. Kamiya and E. Tanoue (1996): Abundance of viruses in deep oceanic water. *Mar. Ecol. Prog. Ser.*, **145**, 269–277.

Harley,H.W.(1928): *Biological Chemistry and Physics of Sea Water*, Cambridge University Press, Cambridge.

Haygood, M. G., P. D. Holt and A. Butler (1993): Aerobactin production by a planktonic *Vibrio* sp. *Limnol. Oceanogr.*, **38**, 1091–1097.

Hering, J. G., W. G. Sunda, R. L. Ferguson and F. M. M. Morel (1987): A field comparison of two methods for the determination of copper complexation: bacterial bioassay and fixed-potential amperometry. *Mar. Chem.*, **20**, 299–312.

Hirose, K. (1990): Chemical speciation of trace metals in seawater: implication of particulate trace metals. *Mar. Chem.*, **28**, 267–274.

Hirose, K. (1994a): Conditional stability constants of organic metal complexes in seawater: past, present and a simple coordination chemistry model. *Anal. Chim. Acta*, **284**, 621–634.

Hirose, K. (1994b): Speciation of particulate uranium in seawater: mass balance analysis of sequential leaching experiments. *J. Radioanal. Nucl. Chem., Articles*, **181**, 11–24.

Hirose, K. (1995): The relationship between particulate uranium and thorium-complexing capacity of oceanic particulate matter. *Sci. Total Environ.*, **173/174**, 195–201.

Hirose, K. (1996): Determination of a strong organic ligand dissolved in seawater; thorium-complexing capacity of oceanic dissolved organic matter. *J. Radioanal. Nucl. Chem., Articles*, **204**, 193–204.

Hirose, K. and Y. Sugimura (1985): Role of metal-organic complexes in the marine environment. A comparison of the copper and ligand titration methods. *Mar. Chem.*, **16**, 239–247.

Hirose, K. and Y. Sugimura (1990): Chemical speciation of particulate uranium in seawater. *J. Radioanal. Nucl. Chem., Articles*, **149**, 83–96.

Hirose, K. and Y. Sugimura (1993): Chemical speciation of particulate ^{238}U, $^{239,240}Pu$ and thorium isotopes in seawater. *Sci. Total Environ.*, **130/131**, 517–524.

Hirose, K. and E. Tanoue (1994): Thorium-particulate matter interaction. Thorium complexing capacity of oceanic particulate matter: theory. *Geochim. Cosmochim. Acta*, **58**, 1–7.

Hirose, K. and E. Tanoue (1998): The vertical distribution of thorium-complexing capacity in particulate matter: implications with respect to the dynamics of particulate matter in oceanic water columns. *Mar. Chem.*, **59**, 235–252.

Hirose, K. and E. Tanoue (1999): Spatial distribution of the strong organic ligand in particulate matter of Pacific surface waters. *Mar. Chem.* (to be submitted).

Hirose, K., Y. Dokiya and Y. Sugimura (1982): Determination of conditional stability constants of organic copper and zinc complexes dissolved in seawater using ligand exchange method with EDTA. *Mar. Chem.*, **11**, 343–354.

Hirose, K., Y. Sugimura and M. Aoyama (1992): Plutonium and ^{137}Cs in the western North Pacific: estimation of residence time of plutonium in surface waters. *Appl. Radiat. Isot.*, **43**, 349–359.

Hudson, R. J. M. and F. M. M. Morel (1990): Iron transport in marine phytoplankton: kinetics of cellular and medium coordination reactions. *Limnol. Oceanogr.*, **35**, 1002–1020.

Hudson, R. J. M., D. T. Covault and F. M. M. Morel (1992): Investigations of iron coordination and redox reactions in seawater using ^{59}Fe radiometry and ion-pair solvent extraction of amphiphilic iron complexes. *Mar. Chem.*, **38**, 209–235.

Huh, C.-A. and M. P. Bacon (1985): Th-232 in the eastern Caribbean Sea. *Nature*, **316**, 718–721.

Huh, C.-A., W. S. Moore and D. Kadko (1985): Oceanic ^{232}Th: a reconnaissance and implications of global distribution from manganese nodules. *Geochim. Cosmochim. Acta*, **53**, 1357–1366.

Hurd, D. C. and D. W. Spencer (1991): Marine particles: analysis and characterization. Geophysical Monograph 63, AGU.

Jackson, G. A. and J. Morgan (1978): The metal-chelator interactions and phytoplankton growth in sea water media: theoretical analysis and comparison with reported observations. *Limnol. Oceanogr.*, **23**, 268–282.

Jalal, M. A. F., M. B. Hossain, D. van der Helm, J. Sanders-Loehr, L. A. Actis and J. H. Crosa (1989): Structure of anguibactin, a unique plasmid-related bacterial siderophore from the fish pathogen *Vibrio anguillarum*. *J. Am. Chem. Soc.*, **111**, 292–296.

Karl, D. M., G. A. Knauer and J. H. Martin (1988): Downward flux of particulate organic matter in the ocean: a particle decomposition paradox. *Nature*, **332**, 438–441.

Kirchman, D. L., R. G. Keil, M. Simon and N. A. Welschmeyer (1993): Biomass and production of heterotrophic bacterioplankton in the oceanic subarctic Pacific. *Deep-Sea Res.*, **40**, 967–988.

Knauer, G. A., D. M. Karl, J. H. Martin and C. N. Hunter (1984): *In situ* effects of selected preservatives on total carbon, nitrogen and metals collected in sediment traps. *J. Mar. Res.*, **42**, 445–462.

Kotrly, S. and L. Sucha (1985): *Handbook of Chemical Equilibria in Analytical Chemistry.* Ellis Horwood Limited, Chichester.

Kramer, C. J. M. (1986): Apparent copper complexation capacity and conditional stability constants in North Atlantic waters. *Mar. Chem.*, **18**, 335–349.

Kramer, C. J. M. and J. C. Duinker (1984): Complexation capacity and conditional stability constants for copper sea and estuarine waters, sediment extracts and colloids. pp. 1–15. In *Complexation of Trace Metals in Natural Waters*, ed. by J. M. Kramer and J. C. Duinker, Nijhoff/ Junk, The Hague.

Krauskopf, K. B. (1956): Factors controlling the concentration of thirteen trace metals in seawater. *Geochim. Cosmochim. Acta*, **12**, 331–344.

Ku, T. L., K. G. Knauss and G. G. Mathieu (1977): Uranium in open ocean: concentration and isotopic composition. *Deep-Sea Res.*, **24**, 1005–1017.

Landing, W. M. and K. W. Bruland (1989): The contrasting biogeochemistry of iron and manganese in the Pacific Ocean. *Geochim. Cosmochim. Acta*, **51**, 29–43.

Lampitt, R. S. (1985): Evidence for the seasonal deposition of detritus to the deep-sea floor and its subsequent resuspension. *Deep-Sea Res.*, **32**, 885–897.

Langmuir, D. (1978): Uranium solution-mineral equilibria at low temperatures with applications to sedimentary ore deposits. *Geochim. Cosmochim. Acta*, **42**, 547–569.

Langmuir, D. and J. S. Herman (1980): The mobility of thorium in natural waters at low temperatures, *Geochim. Cosmochim. Acta*, **44**, 1753–1766.

Lee, C., S. G. Wakeham and J. I. Hedges (1988): The measurement of oceanic particle flux—are "swimmers" a problem? *Oceanogr.*, **1**, 24–36.

Lee, S. and J. A. Fuhrman (1987): Relationships between biovolume and biomass of naturally derived marine bacterioplankton. *Appl. Environ. Microbiol.*, **53**, 1298–1303.

Lewis, B. L., P. D. Holt, S. W. Taylor, S. W. Wilhelm, C. G. Trick, A. Butler and G. W. Luther, III (1995): Voltammetric estimation of iron(III) thermodynamic stability constants for catechole siderophores isolated from marine bacteria and cyanobacteria. *Mar. Chem.*, **50**, 179–188.

Mackey, D. J. (1983): The strongcomplexing capacity of seawater—an investigaion of south-eastern Australian coastal waters. *Mar. Chem.*, **14**, 73–87.

Maier-Reimer, E. (1993): Geochemical cycles in an ocean general circulation model. Preindustrial tracer distributions. *Global Biogeochem. Cycles*, **7**, 645–677.

Mantoura, R. F. C. (1981): Organo-metallic interactions in natural waters. pp. 179–223. In *Marine Organic Chemistry*, ed. by E. K. Duursma and R. Dawson, Elsevier, Amsterdam.

Mantoura, R. F. C., A. Dickson and J. P. Riley (1978): The complexation of metals with humic materials in natural waters. *Estuarine Coastal Mar. Sci.*, **6**, 387–408.

Martell, A. E. and R. M. Smith (1974): *Critical Stability Constants. Vol. 1.* Plenum Press, New York.

Martell, A. E. and R. M. Smith (1982): *Critical Stability Constants. Vol. 5 First Supplement.* Plenum Press, New York.

Martin, J. H. and S. E. Fitzwater (1988): Iron deficiency limits phytoplankton growth in the N.E.

Pacific Subarctic. *Nature*, **331**, 341–343.
Martin, J. H. and R. M. Gordon (1988): North Pacific iron distribution in relation to phytoplankton productivity. *Deep-Sea Res.*, **35**, 177–196.
Martin, J. H., G. A. Knauer, D. M. Karl and W. W. Broenkow (1987): VERTEX: carbon cycling in the northeast Pacific. *Deep-Sea Res.*, **34**, 267–285.
Martin, J. H., R. M. Gordon, S. E. Fitzwater and W. W. Broenkow (1989): VERTEX: phytoplankton/ iron studies in the Gulf of Alaska. *Deep-Sea Res.*, **36**, 649–680.
Martin, J. H., K. H. Coale and others (1994): Testing the iron hypothesis in ecosystems of the equatorial Pacific Ocean. *Nature*, **371**, 123–129.
McKnight, D. M. and F. M. M. Morel (1980): Copper complexation by siderophores from filamentous blue-green algae. *Limnol. Oceanogr.*, **25**, 62–71.
Michaels, A. F., M. N. Silver, M. M. Gowing and G. A. Knauer (1990): Cryptic zooplankton "swimmers" in upper ocean sediment traps. *Deep-Sea Res.*, **37**, 1285–1296.
Midorikawa, T. and E. Tanoue (1996): Effects of ligand speciation on determinations of the complexing abilities of strong ligands in natural waters Detection of a strong ligand. *J. Oceanogr.*, **52**, 421–439.
Midorikawa, T., E. Tanoue and Y. Sugimura (1990): Determination of complexation ability of natural ligands in seawater for various metal ions using ion selective electrodes. *Anal. Chem.*, **62**, 1737–1746.
Miller, L. A. and K. W. Bruland (1994): Determination of copper speciation in marine waters by competitive ligand equilibration/liquid-liquid extraction: an evaluation of the technique. *Anal. Chim. Acta*, **284**, 573–586.
Mills, G. R. and J. G. Quinn (1984): Dissolved copper and copper-organic complexes in the Narrgansett Bay Estuarine. *Mar. Chem.*, **15**, 151–172.
Mills, G. R., E. McFadden and J. G. Quinn (1987): Chromatographic studies of dissolved organic matter and copper-organic complexes isolated from estuarine waters. *Mar. Chem.*, **20**, 313–325.
Moffett, J. W. and R. G. Zika (1987): Solvent extraction of copper acetyl-acetonate in studies of copper(II) speciation in seawater. *Mar. Chem.*, **21**, 301–313.
Moffett, J. W., L. E. Brand and R. G. Zika (1990): Distribution and potential sources and sinks of copper chelators in the Sargasso Sea. *Deep-Sea Res.*, **37**, 27–36.
Moore, R. M. and K. A. Hunter (1985): Thorium adsorption in the ocean: reversibility and distribution amongst particle sizes. *Geochim. Cosmochim. Acta*, **49**, 2253–2257.
Moore, R. M. and G. E. Millward (1988): The kinetics of reversible Th reactions with marine particles. *Geochim. Cosmochim. Acta*, **52**, 113–118.
Muller, F. L. L. and D. R. Kester (1991): Voltammetric determination of the complexation parameters of zinc in marine and estuarine waters. *Mar. Chem.*, **33**, 71–90.
Najjar, R., J. L. Sarmiento J. L. and J. Toggweiler (1992): Downward transport and fate of organic matter in the ocean: Simulations with a general circulation model. *Global Biogeochem. Cycles*, **6**, 45–76.
Nash, K. L. and G. R. Choppin (1980): Interaction of humic and fluvic acids with Th(IV). *J. Inorg. Nucl. Chem.*, **42**, 1045–1050.
Nelson, A. (1985): Voltammetric measurement of copper(II)/organic interactions in estuarine waters. *Anal. Chim. Acta*, **169**, 287–298.
Nakajima, A. and T. Sakaguchi (1993): Accumulation of uranium by basidiomycetes. *Appl. Microbiol. Biotechnol.*, **38**, 574–578.
Nakajima, A., T. Horikoshi and T. Sakaguchi (1979): Uptake of copper ion by green microalgae. *Agric. Biol. Chem.*, **43**, 1455–1460.
Nimmo, M., C. M. G. van den Berg and J. Brown (1989): The chemical speciation of dissolved nickel, copper, vanadium and iron in Liverpool Bay, Irish Sea. *Estuarine Coastal Shelf Sci.*, ***29***, 57–74.
Nozaki, Y. and Y. Horibe (1983): Alpha-emitting thorium isotopes in the northwest Pacific deep waters. *Earth Planet. Sci. Lett.*, **65**, 39–50.
Nozaki, Y. and M. Yamada (1987): Thorium and protoactinium isotope distributions in waters of the Japan Sea. *Deep-Sea Res.*, **34**, 1417–1430.

Nozaki, Y., H.-S. Yang and M. Yamada (1987): Scavenging of thorium in the ocean. *J. Geophy. Res.*, **87**, 2045–2056.

Okujo, N., M. Saito, S. Yamamoto, T. Yoshida, S. Miyoshi and S. Shinoda (1994): Structure of vulnibactin, a new polyamine-containing siderophore from *Vibrio vulnificus*. *Biometals*, **7**, 109–116.

Philander, S. G. (1990): *Oceanic Variability in the Tropics. in El Niño, La Niña, and the Southern Oscillation*. Academic Press, San Diego, pp. 58–102.

Pribil, S. and Marvan, P. (1976): Accumulation of uranium by the chlorococcal algae *Scenedesmum quadricauda*. *Arch. Hydrobiol. Suppl. 49, Algological Studies*, **15**, 214–225.

Reid, R. T., D. H. Live, D. J. Faulkner and A. Butler (1993): A siderophore from a marine bactrium with an exceptional ferric ion affinity constant. *Nature*, *366*, 455–458.

Reuter, J. M. and E. M. Perdue (1977): Importance of heavy metal organic matter interactions in natural waters. *Geochim. Cosmochim. Acta*, **41**, 325–334.

Rich, H. R. and F. M. M. Morel (1990): Availability of well-defined iron colloids to the marine diatom *Thalassiosira weissflogii*. *Limnol. Oceanogr.*, **35**, 652–662.

Rue, E. L. and K. W. Bruland (1995): Complexation of iron(III) by natural organic ligands in the Central North Pacific as determined by a new competitive ligand equilibration/adsorptive cathodic stripping voltammetric method. *Mar. Chem.*, **50**, 117–138.

Ryther, J. H. and R. R. Guillard (1959): Enrichment experiments as a means of studying nutrients limiting to phytoplankton production. *Deep-Sea Res.*, **6**, 65–69.

Sakaguchi, T. and A. Nakajima (1991): pp. 309–322. In *Mineral Bioprocessing*, ed. by R. W. Smith and M. Misra, The Minerals, Metals & Materials Society.

Sakaguchi, T., T. Horikoshi and A. Nakajima (1978): Uptake of uranium from sea water by microalgae. *J. Ferment. Technol.*, **56**, 561–565.

Santschi, P. H., L. Guo, M. Baskaran, S. Trumbore, J. Southon, T. S. Bianchi, B. Honeyman and L. Cifuentes (1995): Isotopic evidence for the contemporary origin of high-molecular weight organic matter in oceanic environments. *Geochim. Cosmochim. Acta*, **59**, 625–631.

Sillen, L. G. (1961): The physical chemistry of seawater. pp. 549–582. In *Oceanography*, ed. by M. Sears, Amer. Assoc. Adv. Sci., Washington, D.C.

Sillen, L. G. and A. E. Martell (1964): *Stability Constants of Metal-Ion Complexes*. Chem. Soc., London.

Stanley, J. K., Jr. and R. H. Byne (1990): Inorganic complexation of zinc(II) in seawater. *Geochim. Cosmochim. Acta*, **54**, 753–760.

Stumm, W. and P. A. Brauner (1975): Chemical speciation. pp. 175–234. In *Chemical Oceanography, Vol. 1*, ed. by J. P. Riley and G. Skirrow, 2nd edn., Academic Press, London.

Stumm, W. and J. J. Morgan (1996): *Aquatic Chemistry*. 3rd edn., Wiley, New York.

Stryer, L. (1981): *Biochemistry*. W. H. Freeman, San Francisco.

Sugimura, Y. and M. Mayeda (1980): The uranium content and activity ratio $^{234}U/^{238}U$ in sea water in the Pacific Ocean. pp. 211–246. In *Isotope Marine Chemistry*, ed. by E. D. Goldberg, Y. Horibe and K. Saruhashi, Uchida Rokakuho, Tokyo.

Sugimura, Y., Y. Suzuki and Y. Miyake (1978): Chemical forms of minor metallic elements in the ocean. *J. Oceanogr. Soc. Jpn.*, **34**, 93–96.

Sunda, W. G. (1989): Trace metal interactions with marine phytoplankton. *Biol. Oceanogr.*, **6**, 411–442.

Sunda, W. G. and R. L. Ferguson (1983): Sensitivity of natural bacterial communities to additions of copper and to cupric ion activity: a bioassay of copper complexation in sea water. pp. 871–891. In *Trace Metals in Sea Water*, ed. by C. S. Wong, E. Boyle, K. Bruland, J. Burton and E. D. Goldberg, Plenum, New York.

Sunda, W. G. and R. R. L. Guillard (1976): The relationship between cupric ion activity and the toxicity of copper to phytoplankton. *J. Mar. Res.*, **34**, 511–529.

Sunda, W. G. and A. K. Hanson (1987): Measurement of free cupric ion concentration in seawater by a ligand competition technique involving copper sorption onto C18 SEP-PAK cartridges. *Limnol. Oceanogr.*, **32**, 537–551.

Sunda, W. G. and S. A. Huntsman (1991): The use of chemiluminescence and ligand competition with EDTA to measure copper concentration and speciation in seawater. *Mar. Chem.*, **36**, 136–163.

Symes, J. L. and D. R. Kester (1985): Copper(II) interaction with carbonate species based on malachite solubility in perchloorate medium at the ionic strength of seawater. *Mar. Chem.*, **16**, 189–211.

Takahashi, A., H. Nakamura, T. Kameyama, S. Kurasawa, H. Naganawa, Y. Okami, T. Takeuchi, and H. Umezawa (1978): Bisucaberin, a new siderophore, sensitizing tumor cells to macrophage-mediated cytolysis II. Physico-chemical properties and structure determination. *J. Antibiot.*, **40**, 1671–1676.

Tanoue, E. (1992a): Occurrence and characterization of particulate proteins in the Pacific Ocean. *Deep-Sea Res.*, **39**, 743–746.

Tanoue, E. (1992b): Vertical distribution of dissolved organic carbon in the North Pacific determined by the high-temperature catalytic oxidation method. *Earth Planet. Sci. Lett.*, **111**, 201–216.

Tanoue, E. (1993): Distributional characteristics of DOC in the central equatorial Pacific. *J. Oceanogr.*, **49**, 625–639.

Tanoue, E. (1996): Characterization of the particulate protein in Pacific surface waters. *J. Mar. Res.* (in press).

Tanoue, E. (2000): Protein in the sea.In *Dynamics and Characterization of Marine Organic Matter*, ed. by N. Handa *et al.*, Terra Pub., Tokyo (in this volume).

Tanoue, E., S. Nishiyama, M. Kamo and A. Tsugita (1995): Bacterial membranes: possible source of a major dissolved protein in seawater. *Geochim. Cosmochim. Acta*, **59**, 2643–2648.

Tortell, P. D., M. T. Maldonado and N. M. Price (1996): The role of heterotrophic bacteria in iron-limited ocean ecosystems. *Nature*, **383**, 330–332.

Tsunogai, S. and S. Noriki (1991): Particulate fluxes of carbonate and organic carbon in the oceans. Is the marine biological activity working as a sink of the atmospheric carbon? *Tellus*, **43B**, 256–266.

Van den Berg, C. M. G. (1982a): Determination of copper complexation with natural organic ligands in seawater by equilibration with MnO_2. I. Theory. *Mar. Chem.*, **11**, 307–322.

Van den Berg, C. M. G. (1982b): Determination of copper complexation with natural organic ligands in seawater by equilibration with MnO_2. II. Experimental procedures and application to surface seawater. *Mar. Chem.*, **11**, 323–342.

Van den Berg, C. M. G. (1984a): Organic and inorganic speciation of copper in the Irish Sea. *Mar. Chem.*, **14**, 201–212.

Van den Berg, C. M. G. (1984b): Determination of the complexing capacity and conditional stability of complexes of copper(II) with natural organic ligands in seawater by cathodic stripping voltammetry of copper catechol complex ions. *Mar. Chem.*, **15**, 1–18.

Van den Berg, C. M. G. (1985): Determination of the zinc complexing capacity in seawater by cathodic stripping voltammetry of zinc-APCD complex ions. *Mar. Chem.*, **16**, 121–130.

Van den Berg, C. M. G. (1995): Evidence for organic complexation of iron in seawater. *Mar. Chem.*, **50**, 139–157.

Van den Berg, C. M. G. and S. Dharmvanij (1984): Determination of complexing capacities of zinc using equilibration of MnO_2 with ^{65}Zn. *Limnol. Oceanogr.*, **29**, 1025–1036.

Van den Berg, C. M. G. and M. Nimmo (1987): Determination of interactions of nickel with dissolved organic material in seawater using cathodic stripping voltammetry. *Sci. Total Environ.*, **60**, 185–195.

Van den Berg, C. M. G., P. M. Buckley and S. Dharmavanij (1984): Determination of ligands concentration and conditional stability constants in seawater. Comparison of the DPASV and MnO_2 adsorption techniques. pp. 213–216. In *Complexation of Trace Metals in Natural Waters*, ed. by J. M. Kramer and J. C. Duinker, Nijhoff/Junk, The Hague.

Veeh, H. H. (1967): Deposition of uranium from the ocean. *Earth Planet. Sci. Lett.*, **3**, 145–150.

Veeh, H. H., S. E. Calvert and N. B. Price (1974): Accumulation of uranium in sediments and phosphorites on the South West African shelf. *Mar. Chem.*, **2**, 189–202.

Wangersky, P. J. (1986): Biological control of trace metal residence time and speciation: a review and synthesis. *Mar. Chem.*, **18**, 269–297.

Wefer, G., E. Suss, W. Balzer, G. Liebezeit, P. J. Muller, C. A. Ungerer and W. Zenk (1982): Fluxes of biogenic components from sediment trap deployment in circumpolar waters of the Drake Passage. *Nature*, **299**, 145–147.

Williams, P. M. (1969): The association of copper with dissolved organic matter. *Limnol. Oceanogr.*, **14**, 156–158.

Wu, J. and G. W. Luther, III (1995): Complexation of Fe(III) by natural organic ligands in the Northwest Atlantic Ocean by a competitive ligand equilibration method and a kinetic approach. *Mar. Chem.*, **50**, 159–177.

Yamanaka, Y. and E. Tajika (1996): The role of the vertical fluxes of particulate organic matter and calcite in the oceanic carbon cycles: studies using an ocean biogeochemical general circulation model. *Global Biogeochem. Cycles*, **10**, 361–382.

Yokoi, K. and C. M. G. van den Berg (1992): The determination of iron in seawater using catalytic cathodic stripping voltammetry. *Electroanalysis*, **4**, 65–69.

Zeuhlke, R. W. and D. R. Kester (1983): Copper speciation in marine waters. pp. 773–788. In *Trace Metals in Sea Water*, ed. by C. S. Wong, E. Boyle, K. W. Bruland, J. D. Burton and E. D. Goldberg, Plenum Press, New York.

Zhang, H., C. M. G. van den Berg and R. Wollast (1990): The determination of interactions of cobalt(II) with organic compounds in seawater using cathodic stripping voltammetry. *Mar. Chem.*, **28**, 285–300.

Zirino, A. and S. Yamamoto (1972): Speciation of zinc, lead and cadmium in seawater. *Limnol. Oceanogr.*, **17**, 661–671.

Dynamics and Characterization of Marine Organic Matter, Eds., N. Handa, E. Tanoue and T. Hama, pp. 383–463.

Proteins in the Sea — Synthesis

Eiichiro TANOUE*

Geochemical Research Department, Meteorological Research Institute, Nagamine 1-1, Tsukuba 305, Japan

(Synthesis, Received 23 August 1996; accepted 22 October 1996)

INTRODUCTION

Background

Primary production by phytoplankton is the largest source of organic matter in the sea. Photosynthetically produced organic matter in the photic layer is transferred to higher and lower trophic levels through marine food webs. Organic matter in the surface water is also transported to the intermediate and deep waters, as well as to the sediment. As a consequence, organic matter plays a role in many biogeochemical processes in the water column, for example, the fueling of marine food webs and the control of the dynamics of other chemicals (by, for example, the regeneration of nutrients, the consumption of dissolved oxygen, and the scavenging of metals, radionuclides, pollutants, etc.), on the one hand, while, on the other hand, organic matter itself is transformed to detrital particulate organic matter (POM) and dissolved organic matter (DOM). The resultant detrital POM and DOM are also tightly coupled to the biogeochemical cycle in the marine system.

Despite the fact that the organic carbon cycle in the marine system starts from almost a single process (photosynthesis) in the photic layer, the relative abundance of phytoplankton is the result of complex processes, one order of magnitude lower than that of detrital POM and two orders of magnitude lower than that of DOM. Thus, processes by which the organic matter produced by phytoplankton is transformed to detrital POM and DOM are of central interest in studies of the organic geochemistry of the water column in the sea. More than 30 years ago, Duursma (1961) stated that organic matter can only be formed by living organisms, and it seems reasonable to assume a relationship exists between the living organic matter in suspension and the dissolved organic material. Indeed, from the beginning of this century, investigators have attempted to characterize this relationship, and they continue to do so. Lee and Wakeham (1989) emphasized that, in studies of the production, transformation and fate of organic matter in the sea, it is best to concentrate on the processes themselves, rather than to survey the hundreds of thousands of organic compounds in the sea, because detrital POM and DOM might contain numerous compounds that origi-

*Present address: Institute for Hydrospheric-Atmospheric Sciences, Nagoya University, Furo-cho, Chikusa-ku, Nagoya, 464-8601, Japan.

nated in living organisms and their derivatives, and analyses of every compound would be too costly in terms of time and money. This approach contrasts with the development of the inorganic chemistry of the ocean, where efforts have focused on the concentrations and distributions of elements. The elements of which the oceanic distributions have been clarified almost fill the entire Periodic Table (Broecker and Peng, 1982; Bruland, 1983; Quinby-Hunt and Turekian, 1983; Nozaki, 1992).

Marine organic geochemists have been trying to characterize the relationship between living organic matter in suspension and dissolved organic matter *via* two major approaches, namely, the use of biomarkers and bulk chemical characterization. Much excellent work has been done to clarify the quantitative and qualitative importance of the processes of transformation of the living biomass to inanimate organic pools (e.g., see the review by Lee and Wakeham, 1989, and references therein).

Biomarkers

Structurally well-defined biomarkers have been used as indicators of sources and processes, for example, lignin-derived vanillyl phenols have been used as an indicator of the input of allochthonous organic matter (e.g., Hedges and Parker, 1976; Meyer-Schulte and Hedges, 1986). In studies of autochthonous sources, the biochemical specificity of many molecular structures in marine organisms can be used to assign contributions from particular classes or families of biota (e.g., Farrington, 1992). For examples, dinosterol and related 4-methyl sterols are markers of dinoflagellates (Boon *et al.*, 1979) and alkenones reflect the contributions of prymnesiophytes (Marlowe *et al.*, 1990). It has been demonstrated in sinking particles that relative increases in levels of branched fatty acids or non-protein amino acids are signatures of bacterial degradation (Lee *et al.*, 1983; Lee and Cronin, 1984). The occurrence of hydrolyzed carotenoid esters (fucoxanthinol and peridininol) or esterified derivatives of chlorophyll *a* (steryl chlorin esters) is an indication that such pigments in phytoplankton were transformed as a result of ingestion by zooplankton (e.g., Repeta and Gagosian, 1982, 1984; Harradine *et al.*, 1996). Thus, the detection of such biomarkers provides clear evidence that a specified origin or process influences the chemical composition of the POM and DOM.

Major and commonly detected biochemicals, such as amino acids, fatty acids and monosaccharides, are universal chemical constituents of marine organisms, and the internal compositions of these components vary among taxa and among the physiological conditions of each taxon (e.g., Volkman, 1986; Wakeham, 1995). In general, biomarkers are minor constituents of marine organisms and they are selected as compounds that retain enough of their original structure to allow identification of the source organism. Therefore, the dynamics of biomarkers can differ from those of major organic constituents of marine organisms. For example, marine-derived lipids dominate in particles in the surface water, but terrestrial components increase in relative abundance in the ocean's interior, as well as in the deep-sea sediment, although the bulk of organic carbon in particles and deep sediments appears to be largely of marine origin (Emerson *et al.*, 1987;

Lee and Wakeham, 1989; Prahl, 1992). It has proved difficult to correlate the presence of trace levels of specific markers with other common and major organic components.

Bulk chemical characterization

Bulk chemical analyses have continued to provide an important framework for our ideas about the cycle of organic matter in the sea. Since 1960, technical developments in the elementary analysis of organic matter have made it possible to assess the basin-scale distribution of particulate organic carbon (POC) and dissolved organic carbon (DOC; e.g., Menzel and Vaccaro, 1964; Armstrong *et al.*, 1966). Menzel and Ryther (1970) stated that the concentration of POC is, as expected, highly variable at the surface and that it is, in most instances, directly related to the level of organic production. However, all attempts to correlate POC with the rate of primary production have yielded considerable scatter because detrital POC can also vary widely and its concentration is superimposed on the amount of living matter (Menzel, 1974). The influence of regional and spatial variations in concentrations of POC and DOC in the surface water was not evident in earlier studies of those in the intermediate and deep waters (e.g., Menzel and Ryther, 1970; Menzel, 1974). About a decade after the cited earlier studies, Gordon and Cranford (1985) found that a general relationship exists between primary production in surface water and the concentration of POM throughout the water column on a broad scale. They commented that their general conclusion was very similar to those obtained from sediment trap studies. To date, however, relationships between the POM (including sinking particles), as well as the DOM, in the surface water and in the deep water are not well understood.

The characteristics of the distribution of DOM, as well as of POM, indicate relative homogenity in the intermediate and deep waters and have led to the hypothesis that DOM and detrital POM comprise of stable compounds that might be refractory to biological degradation and might be inactive in the biogeochemical cycle in the sea. This hypothesis is supported by the very low levels of radiocarbon (^{14}C) of the DOM (Williams and Druffel, 1987; Bauer *et al.*, 1992) and of the POM (Druffel and Williams, 1990) and by the results of a degradation experiment (Barber, 1968). Bulk chemical analyses, such as the analysis of a stable carbon isotope (^{13}C; Williams and Gordon, 1970; Williams and Druffel, 1987; see also Williams *et al.*, 1992 and Druffel and Williams, 1992), ^{1}H and ^{13}C nuclear magnetic resonance (NMR) and infrared (IR) spectra of the isolated DOM fraction (Stuermer and Harvey, 1974, 1977; Stuermer and Payne, 1976; and also see, Rashid, 1985), indicate that the majority of DOM is marine in origin, although it is still a matter of discussion whether the source of DOM is land-derived or marine (Meybeck, 1982; Mantoura and Woodward, 1983). The bulk organic matter of the inanimate organic pools has been designated collectively "water humus" (Skopintsev, 1972) or "gelbstoff" (Kalle, 1966).

The humic substances are formed by condensation reactions that depend on intermolecular collisions, and the formation of the humic substances has been studied by soil organic chemists. Oceanic environments, with their extremely low levels of organic matter and low temperatures are different from soil environments. Gagosian and Lee (1981) mentioned the possibility that condensation reactions might occur in the sea through the many processes that result in high local concentrations of organic matter, such as those found in decaying organisms, the organic films on the sea surface, the organic matter in particles, the gut of filter-feeding organisms, and aggregates of surface-active organic material. Mechanistic models have been proposed by which humic substances are formed in aquatic environments (see, for example, Ishiwatari, 1992). However, it appears that there is no real evidence that the degraded oligomeric products are transformed to "humic" substances in the water column in open sea environments.

Introduction of the importance of bacterial production and microbial food-web processes in the marine ecosystem (e.g., Pomeroy, 1974; Azam *et al.*, 1983) has given new insight into the dynamics of DOM. About 50% (Ducklow and Carlson, 1992), with a range from 20% to >100%, of the organic matter (Kirchman *et al.*, 1993, 1995), generated by primary production is transferred to the DOM and then utilized by heterotrophic bacteria. Moreover, Amon and Benner (1996) demonstrated that DOM with molecular masses of more than 1,000 daltons (1 kDa) was readily remineralized by bacteria. The bulk chemical characterizations performed in the 1990's have also revealed that part of the DOM is more dynamic than previously believed (see the review by Ogawa, 1998, in this volume). Analyses by ^{14}C dating indicate that DOM with a molecular mass of more than 10 kDa is very young (a few decades), while DOM of low molecular mass is old (380–4,500 yr; Santschi *et al.*, 1995). The high-molecular-mass DOM consists of carbohydrate-like materials (Benner *et al.*, 1992) and has undergone extensive degradation (Amon and Benner, 1994; Guo *et al.*, 1994, 1995).

Chemical characterization of biochemical compounds at the macromolecular level

The use of biomarkers and bulk chemical characterization are important for specifying sources and processes, and for delineating the dynamics of the organic matter in the sea. However, each approach has intrinsic advantages and weaknesses. Despite the fact that amino acids, fatty acids and monosaccharides are the major detectable organic constituents of the POM and DOM, relationships between such biochemical compounds in the inanimate organic pools, and those of living organisms are not well defined by such approaches.

About 97% of the cellular material of living organisms consists of macromolecules, and three classes can be distinguished: lipids; periodic molecules, such as peptidoglycans and polysaccharides; and informational macromolecules, such as nucleic acids and proteins (Gottschalk, 1986; Chróst, 1991). Most chemical analyses of detrital POM and DOM have been made by the determination of elementary components (organic C, N, P and S) and by colorimetric bulk analysis, or by analysis at the monomer level of, for example, amino acids, fatty acids and monosaccharides, after hydrolysis of macromolecules. Since there are gaps in our understanding of the organic matter in organisms and in inanimate organic pools, in terms of chemical forms, and since it is difficult to deduce source macromolecules from monomer-level information (see below), it is hard to specify the source of detected biochemical compounds in detrital POM and DOM, even if these components are marine in origin. The chemical characterization of such biochemical compounds at the macromolecular level will provide further clues to the identification of sources and processes by which detrital POM and DOM are transferred from living organisms to inanimate organic pools.

Many organic compounds produced in the surface water are decomposed, for the most part, in the surface water, and the relative stability of individual compounds might be related to their chemical structure and the matrix in which they are formed (e.g., Lee and Wakeham, 1989). During the transformation of organic constituents of the living biomass to those of non-living organic pools, biomacromolecules are subjected to the activities of the autolytic enzymes of the organisms themselves, with successive degradation through higher and lower trophic levels. Since the first stage of the degradation of macromolecules is dependent on enzyme-substrate relationships, such alterations must be strongly controlled by the nature of the macromolecules themselves (Chróst, 1991; Billen, 1991). Thus, a knowledge of these macromolecules is clearly essential for an understanding of the relatively short time-scale of the dynamics of the organic matter.

With the exception of studies on polysaccharides in POM and DOM by Handa and his co-workers (Handa, 1970; Handa *et al.*, 1972; Sakugawa and Handa, 1985a, b), accumulation of biochemical inventories at the macromolecular level has attracted little attention, and there seem to be problems that block the continued progress of such an approach. One problem is that the majority of detrital POM and DOM is chemically amorphous "humic" material, as mentioned above. This issue has hindered efforts to identify marine organic matter at the macromolecular level because humic substances are known to be too complex for complete chemical characterization. Another problem involves methodological difficulties. It is difficult to collect and extract sufficient amounts of a macromolecule for chemical analysis. Our techniques of chemical analysis are also inadequate because analyses at the molecular level typically involve degradation (e.g., hydrolysis, oxidation or pyrolysis) for the release of components that can be quantified (Farrington, 1992).

Improvements in biochemical and molecular biological techniques have made it possible to separate, detect and examine the properties of biomacromolecules in a complex mixture with minimal damage to the molecules. Lists of biochemical compounds at the macromolecular level in non-living organic pools will provide new insight into sources of organic matter from correlation with the compounds in living organisms. The determination of biomacromolecules in the detrital POM and DOM presents a greater challenge as we try to widen the window through which we can see what really happens in the sea. This approach also has weaknesses. Biochemical compounds such as amino acids, carbohydrates and lipids comprise less than 30% of the DOM (Williams and Druffel, 1988) and approximately 50~80% of the POM (Tanoue, 1985). An approach for characterization of biochemical compounds at the macromolecular level needs to be developed in parallel with progress in the use of biomarkers and bulk chemical characterization.

Surveys of proteins in the water column

For the initial approach to a survey of the marine organic matter at the biomacromolecular level and to examine the traceability of such macromolecules from the organic constituents of living organisms to inanimate organic pools, an analysis of proteins was selected for the following reasons.

First, both particulate combined amino acids (PCAA) and dissolved combined amino acids (DCAA) represent the largest well-defined molecules in the POM and DOM. The majority of amino acids in living organisms exist as proteins, and proteins account for more than 50% of organic constituents (Romankevitch, 1984) and about 85% of the nitrogen of living organisms (Billen, 1984). Thus, proteins are the most probable source of PCAA and DCAA. However, amino acid analysis of PCAA, DCAA and sedimentary amino acids has not yielded source-related information (Lee and Bada, 1975, 1977; Henrich and Williams, 1985; Coffin, 1989; Cowie and Hedges, 1992; Hubberten, 1994; McCarthy *et al.*, 1995) because most proteins in marine organisms have similar amino acid compositions (Cowey and Corner, 1963; Strickland, 1965; Chau *et al.*, 1967; Degens, 1970; Cowie and Hedges, 1992). Direct estimations of protein in the POM have been undertaken by fluorometric and colorimetric methods, such as Lowry's method (Parsons and Strickland, 1962; Clayton, 1988), the biuret reaction (Danilenko, 1977), the ninhydrin method (Handa *et al.*, 1972), the fluorescamine assay (Packard and Dortch, 1975; Garfield *et al.*, 1979), the Coomassie blue assay (Setchell, 1981) and the bicinchoninic acid assay (Nguyen and Harvey, 1994). Such methods have verified the occurrence of protein in the POM but have provided limited information about the nature and molecular characteristics of the proteins. Such approaches do not give any clues that allow us to trace the processes of transformation of cellular proteins to detrital proteins in the POM. Thus, the chemical form of PCAA and DCAA and the molecular characteristics of detrital proteins are critically important for an understanding of the source and dynamics of proteins in the sea.

The versatile and well established technique of polyacrylamide gel electrophoresis (PAGE) can be applied to the separation and detection of protein molecules. PAGE in the presence of anionic detergent sodium dodecyl sulfate (SDS), has been one of the most widely used techniques for separation of proteins according to their molecular masses. SDS is a strong denaturing detergent that destroys noncovalent interactions, such as ionic and hydrogen bonds and apolar (hydrophobic) interactions, among proteins or between proteins and nonproteinaceous organic matter. Upon treatment with SDS, covalently linked polypeptide subunits are effectively separated from other organic material. Polypeptides bind large amounts of SDS: 1.4 gram of SDS per gram of protein is a typical ratio (Pitt-Rivers and Impiombato, 1968; Fish *et al.*, 1970; Reynold and Tanford, 1970a, b). Thus, the number of SDS molecules that bind to a protein is of the order of half the number of amino acid residues in the polypeptide chain. As a result, the intrinsic charge of each polypeptide is cancelled out and the net negative charge per unit mass becomes approximately constant (Andrew, 1986). Electrophoretic migration in a gel is then proportional to the effective molecular radius or is approximately proportional to the molecular mass of the polypeptide chain (Shapiro *et al.*, 1967; Neville, 1971). Proteins and/or their subunits (polypeptides), as characterized by SDS-PAGE, are defined as proteins in the present discussion.

Samples for analysis of particulate and dissolved proteins were collected during six cruises, between 1985 and 1993, from the northern North Pacific Ocean and the Bering Sea, through the equatorial Pacific and Indian Oceans, to the Antarctic Ocean. The next section provides a summary of mainly published data (Tanoue, 1991, 1992, 1995, 1996; Tanoue *et al.*, 1995, 1996; Suzuki *et al.*, 1997), with a discussion of the sources of proteins and the transfer of proteins from organisms to non-living particulate and dissolved proteins.

PARTICULATE PROTEINS

POM is a mixture of living organisms and the detritus, which is a fraction, that escape from short time-scale remineralization. To distinguish the living and detrital carbon in the POM, labile markers such as chlorophyll, ATP and DNA, are measured as indicatives of phytoplankton or living biomass, assume a carbon/marker ratio, and total carbon (POC) was interiorly divided. The relative abundances of the living and detrital carbon vary spatially and temporally (e.g., Holm-Hansen, 1969; Eppley *et al.*, 1977; Dortch and Packard, 1989; Cho and Azam, 1990; Kirchman *et al.*, 1993; Yanada and Maita, 1995). PCAA is the most prominent component of POM. Carbon and nitrogen in PCAA from acid-hydrolyzed POM account for approximately 30 of the POC and 50% or more of particulate nitrogen (PN) (Handa, 1970; Handa *et al.*, 1972; Siezen and Mague, 1978; Tanoue and Handa, 1979; Tanoue, 1985; Liebezeit and Bölter, 1986). Thus, PCAA might be a mixture of living and detrital PCAA. However, the temporal and spatial variations of the relative abundances of two types of PCAA have not been well documented, and their respective impacts on the biological and chemical processes are not understood.

The detritus is thought to be formed through complex processes, namely, fragments of living organisms including fecal pellets, aggregates of DOM and sorption of DOM onto mineral particles (Cauwet, 1978; see the review by Hara and Koike, 2000, in this volume). The non-biological formation of POM from DOM has long been discussed (e.g., Riley, 1970; Cauwet, 1978; Mopper and Degens, 1979; Romankevitch, 1984). Morphological and histochemical observations have revealed that there are non-living and discrete proteinaceous particles throughout the water column (Nemoto and Ishikawa, 1969; Gordon, 1970; Mel'nikov, 1974; Long and Azam, 1996). However, to date, their sources and chemical nature are unknown. Chemical characterization of PCAA in intermediate and deep waters will provide clues for an understanding of the formation of detrital POM where only bacteria are potential contributors of the living biomass to POM.

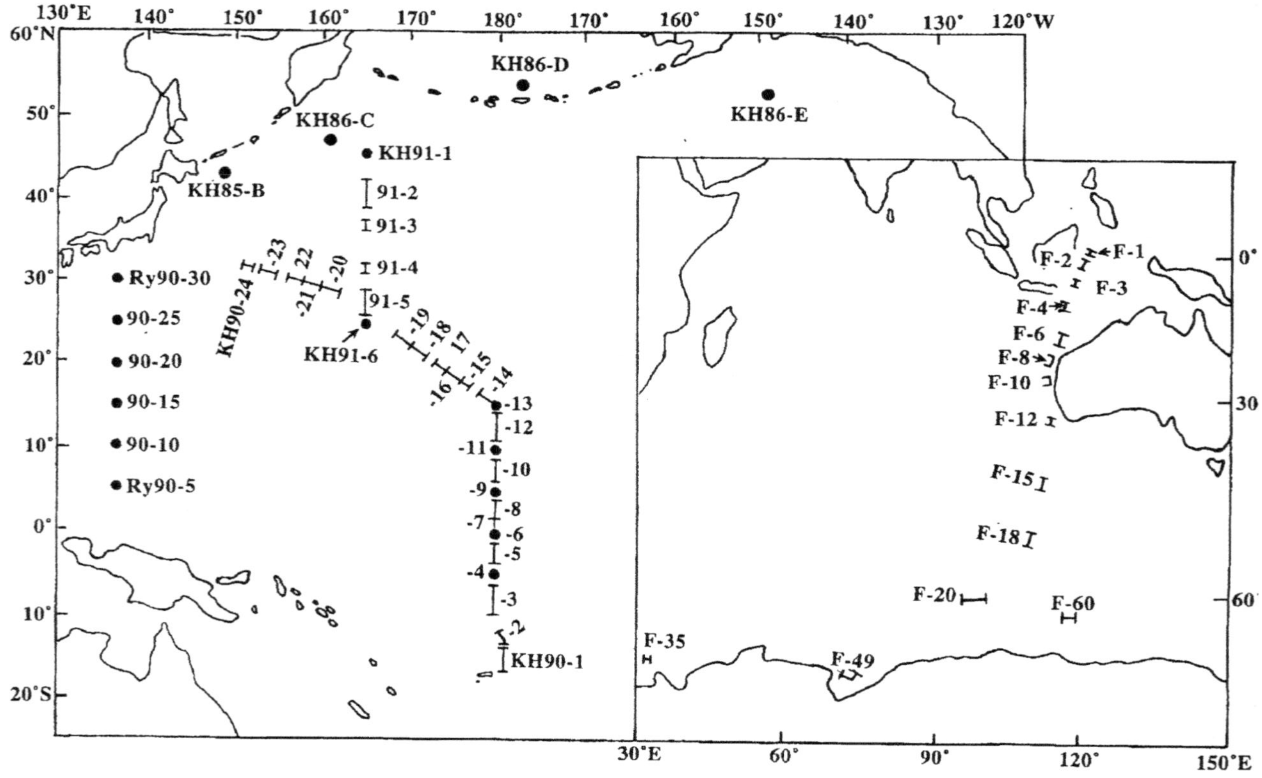

Fig. 1. The distribution of the sampling locations of particulate matter during the cruises Ry90-10 on January 1990 (sample Nos. Ry90-5 to Ry90-30), KH90-2 on September 1990 (KH90-1 to KH90-24), KH91-3 on May 1991 (KH91-1 to KH91-6), KH85-2 on April 1985 (KH85-B), KH86-3 from June to August 1986 (KH86-C, -D and -E) and JARE 34 from November 1992 to March 1993 (F-1 to F-60).

Materials

POM samples were collected by the filtration of seawater onto GF/F glass fiber filter during the six cruises (Fig. 1). Detailed procedures have been previously reported (Tanoue, 1992, 1996; Tanoue *et al.*, 1996) and are reported here only briefly. Samples of surface seawater were obtained by an underway pumping system during the cruises of the JARE 34 (sample Nos. F-1 ~ F-60 in Fig. 1), KH90-2 (sample Nos. KH90-1 ~ KH90-24) and KH91-3 (sample Nos. KH91-1 ~ KH91-6) and by a submergible pump during the Ry90-01 cruise (sample Nos. Ry90-30 ~ Ry90-5). Samples of surface through to deep waters were collected with a Van Dorn type, large-volume water sampler (200 l) during the KH85-3 and KH86-3 cruises (KH85-B and KH86-C, -D and -E). During the KH85-3 and KH86-3 cruises, plankton samples were collected with a Norpac net made of NXX 13 bolting cloth with a 0.095 mm mesh. Vertical tows from a depth of 100 m to the surface were made.

The particulate matter collected on the GF/F filter, as well as the plankton samples, were kept frozen at –30°C onboard ship and lyophilized ashore as soon as possible. The lyophilized samples were again stored in a deep freezer (–60°C) until analysis. Special attention was paid to keeping the samples frozen during the processes.

Methods

Analytical procedures of POC, PN, PCAA and particulate protein have been reported previously (Tanoue *et al.*, 1982; Tanoue, 1991, 1992) and are described here only briefly.

Prior to the analyses of POC and PN, each sample was allowed to stand in a desiccator filled with HCl vapor and then dried. POC and PN were determined with a CHN-analyzer.

The fluorescamine method (Udenfriend *et al.*, 1972) was used for the determination of PCAA after hydrolysis with hydrochloric acid (6 M) at 105°C for 24 hr. After the hydrolysate was dried *in vacuo*, the residue was redissolved in borate buffer (pH 9.0) and the solution was centrifuged. Fluorescamine reagent was added to the supematant and the reaction mixture was stirred. The intensity of fluorescence (excitation, 390 nm; emission, 480 nm) was calibrated with a solution of glycine. Reaction with fluorescamine results in products with different fluorescene yields with different amino acids and the reagent reacts with primary amines but not with proline or hydroxyproline, which are not primary amines (Udenfriend *et al.*, 1972). Therefore, amounts of PCAA are relative values (Tanoue, 1996).

Particulate protein was extracted with a sample buffer solution of Tris-HCl (pH 6.8. 62.5 mM), sodium dodecylsulfate (SDS, 2%, w/v), 2-mercaptoethanol (5%, v/v) and heated (100°C) for 3 min. After cooling, urea (8 M, final conc.) was added for fascilitating the solubilization of particulate protein. After centrifugation at 2,000 G for 5 min at room temperature, the supematant was adjusted to pH

6.8 and analyzed by the SDS-PAGE. In the present study, the simplest possible procedures have been employed, unless otherwise noted. Namely, no preconcentration, desalting or purification of particulate protein in the electrophoretic sample took place for avoiding biased molecular distribution of particulate protein due to any pretreatment. The GF/F filter was treated in the same manner as the POM sample to serve a control at the beginning and end of the sample preparation of each cruise and no protein was detected in the control samples.

The SDS-PAGE, and staining and destaining procedures (Coomassie brilliant blue-R250, CBB-R staining method) were performed by the method of Laemmli (1970), using the ready - made continuous-gradient gel (5–20%, PAGEL, NPG-520 type, Atto, Tokyo) or hand - made discontinuous gel (Tanoue, 1992). The following standard proteins (Sigma, St. Louis, MO), with their molecular masses given in kilo-daltons (kDa), were used: bovine serum albumin (66 kDa), egg albumin (45 kDa), glyceraldehyde-3-phosphate dehydrogenase (36 kDa), trypsinogen (24 kDa), β-lactoglobulin (18.4 kDa) and lysozyme (14.3 kDa).

Particulate proteins on the gel were quantified by the dye-elution method

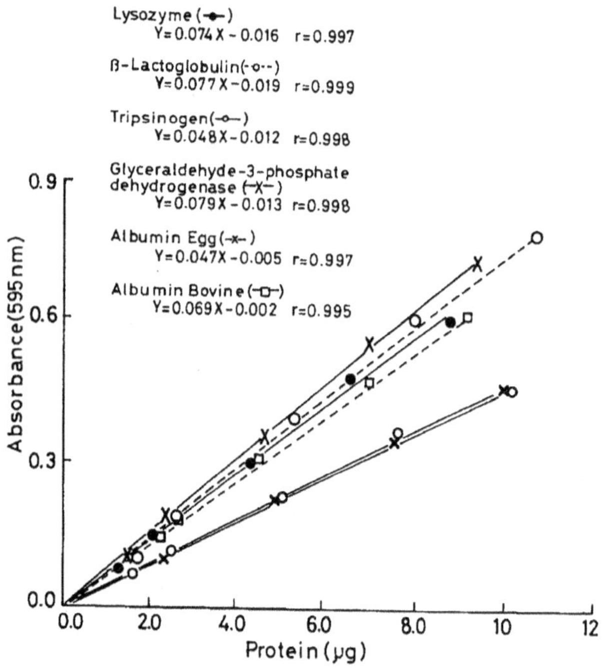

Fig. 2. Plot of absorbance (595 nm) of dye eluted from dye-protein complexes in the gel vs. amounts of six standard proteins in the gel. (Reprinted with permission from *Geophysical Monograph*, 63, E. Tanoue, Electrophoretic separation of particulate proteins in seawater. In *Marine Particles: Analysis and Characterization*, 1991, American Geophysical Union.)

(Ball, 1986; Tanoue, 1991, 1992). After SDS-PAGE, staining and destaining, 5-mm sections of the gel were cut out. Dye (CBB-R) was extracted with an organic solvent from a dye-protein complex in the matrix of each gel piece and absorbance of the dye was measured at 595 nm using the extraction solution as a blank. The total amount of protein in the gel was quantified by summing the amounts of protein in the individual gel piece with bovine serum albumin as the standard. The assay was applied to six marker proteins for the construction of standard curves (Fig. 2). All six proteins gave straight lines when absorbance at 595 nm was plotted against the amount of protein, but the slopes of the lines were different. CBB-R forms electrostatic bonds with NH_3^+ groups and non-covalent bonds with non-polar regions in the proteins (Andrew, 1986). There are, therefore, significant protein-to-protein variations in the efficiency of the formation of the dye-protein complexes. Bovin serum albumin (BSA), commonly used as a standard for the estimation of protein by the Lowry method, was found to be of intermediate activity, and thus used as the standard protein for quantification. It should be emphasized that the quantities of protein cited are relative values in the present study.

For amino-terminal amino acid sequencing, a protein band interest was extracted by passive diffusion from the gel after SDS-PAGE of the sample without pretreatment. The protein bands with the same electrophoretic mobility on the multiple gels of multiple samples collected from various areas, were cut out with a razor blade from the gels. The resulting gel pieces that contained a protein band interest were combined. For extraction of the protein from the gel matrix, the combined gel pieces in the diluted solution (1/10) of the sample buffer were allowed to stand for 3 months in a refrigerator (4°C) and then at room temperature for the next 3 months, with occasional agitation. The protein in the diluted sample buffer was concentrated by a pressure-driven disposable ultrafiltration unit (Molcut-II) with a 10,000 nominal molecular mass cut-off filter (filter material; low protein binding regenerated cellulose). The concentrated sample was subjected to re-electrophoresis and the protein in the gel was electroblotted onto the polyvinylidene difluoride membrane (PVDF membrane, Immobilon, 0.45 μm pore size, Millipore, Bedford, MA) as described by Matsudaira (1987). A protein band on the PVDF membrane was sequenced with a sequencer (model 477A; Applied Biosystems, Foster City, CA, U.S.A.) equipped with on-line reverse-phase HPLC for phenylthiohydantoin (PTH) amino acids that were generated by automated Edman degradation. Detailed procedures of amino-terminal amino acid sequence were reported by Tanoue *et al.* (1995).

Vertical distributions of particulate proteins in the North Pacific Ocean and the Bering Sea

The electrophoretograms of particulate proteins from the surface through to deep waters at four stations in the northern North Pacific Ocean and the Bering Sea (Fig. 1), as well as samples of plankton collected at respective stations, are shown in Figs. 3 and 4. The electrophoretic patterns of the surface samples were

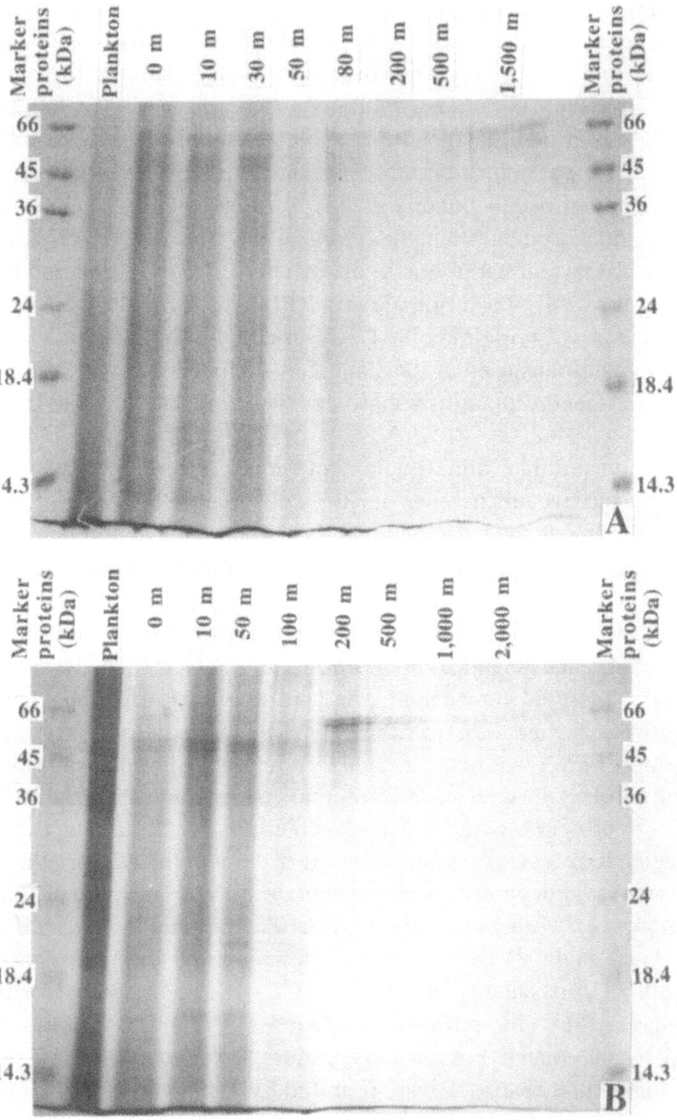

Fig. 3. Electrophoretograms of particulate proteins and plankton at sites KH85-B (A, 42°55.1′ N, 150°04.8′ E) and KH86-C (B, 46°39.9′ N, 162°01.9′ E). SDS-PAGE was performed with a hand-made discontinuous gel system, with 12.5% acrylamide for the separation gel and 2.5% acrylamide for the spacer gel. Proteins on the gels were visualized by staining with CBB-R.

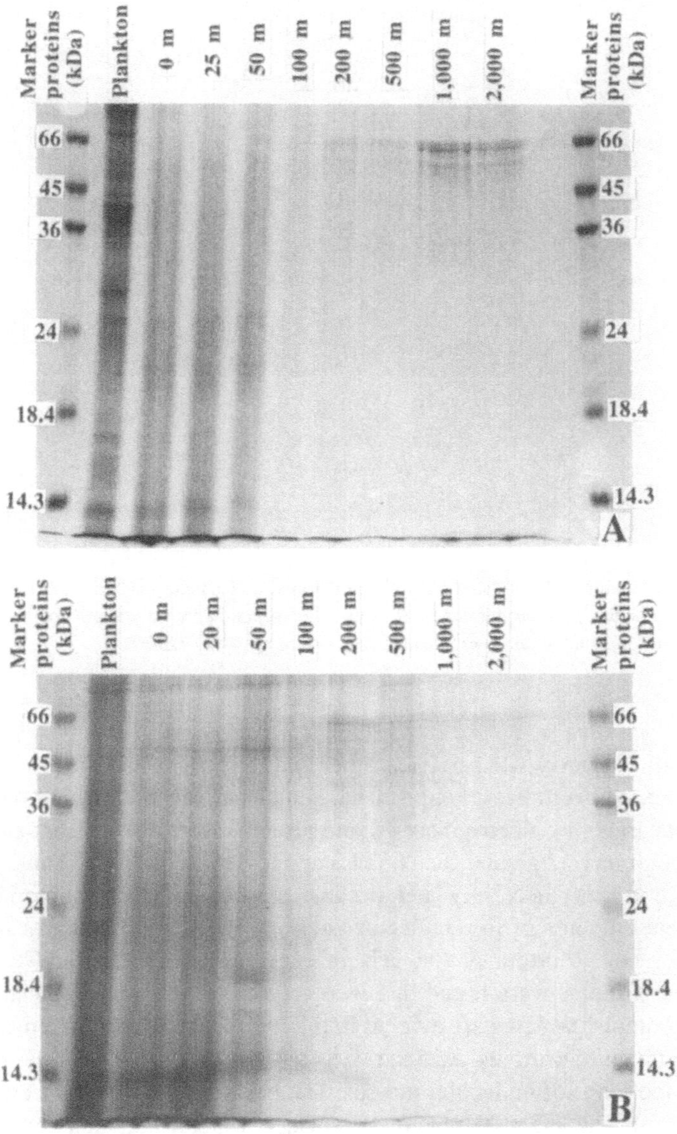

Fig. 4. Electrophoretograms of particulate proteins and plankton at sites KH86-D (A, 52°58.0′ N, 177°30.2′ E) and KH86-E (B, 52°59.4′ N, 144°57.1′ W). SDS-PAGE was performed with a hand-made discontinuous gel system, with 12.5% acrylamide for the separation gel and 2.5% acrylamide for the spacer gel. Proteins on the gels were visualized by staining with CBB-R. (Reprinted with permission from *Deep-Sea Research*, **39**, E. Tanoue, Occurrence and characterization of particulate proteins in the Pacific Ocean, 1992, Pergamon Press Ltd.)

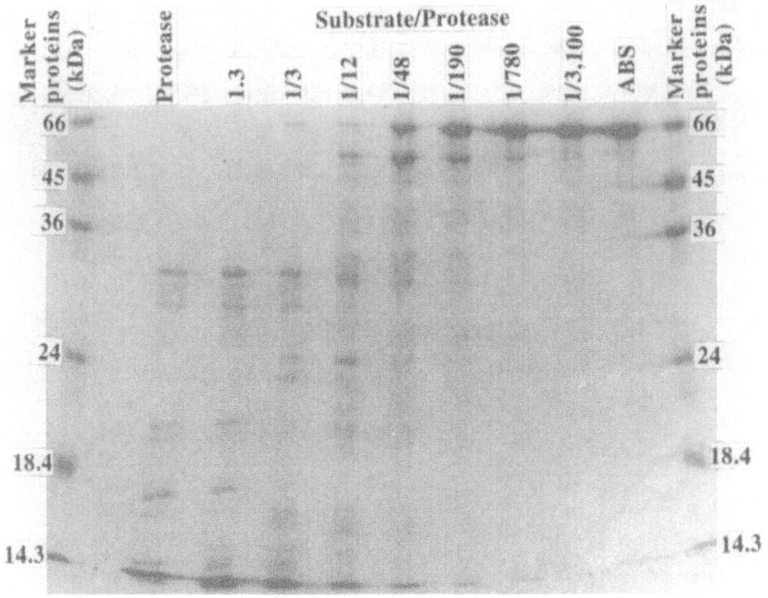

Fig. 5. Electrophoretograms of proteolytic digestion of the authentic protein, bovine serum albumin (BSA). SDS-PAGE was performed with a hand-made discontinuous gel system, with 12.5% acrylamide for the separation gel and 2.5% acrylamide for the spacer gel. Proteins on the gels were visualized by staining with CBB-R.

very similar in many ways to each other and to those of the plankton. Proteins were observed over a wide range of molecular masses, both from 14 kDa to 66 kDa as separated by electrophoresis, and proteins were also present as fractions of molecular masses greater than 66 kDa and less than 14.3 kDa. Thus, gels were stained relatively uniformly and became a smeared electrophoretic pattern. Significant amounts of low-molecular-mass proteins were concentrated at the migration front (bottom of the gels in Figs. 3 and 4). Distinctive bands of individual proteins were found that were superimposed on the smeared "background proteins" which were evident in the samples of plankton, indicating that a large number of proteins, each at relatively low levels, overlapped each other over a wide range of molecular masses. The background proteins were found in the plankton and the POM from the surface waters, but their levels rapidly decreased with depth, and the background proteins were barely evident in POM collected from waters deeper than 200 m at all the stations cited.

The most striking feature of the molecular mass distribution of POM was that the discrete bands in the molecular-mass range between 45 kDa and 66 kDa were clearly visible in the intermediate and deep waters. The relative abundance of these discrete bands varied with depth and locations. The protein band with an apparent molecular mass of 66 kDa was most evident and observed in all the samples of POM from the intermediate and deep waters. The band was found in

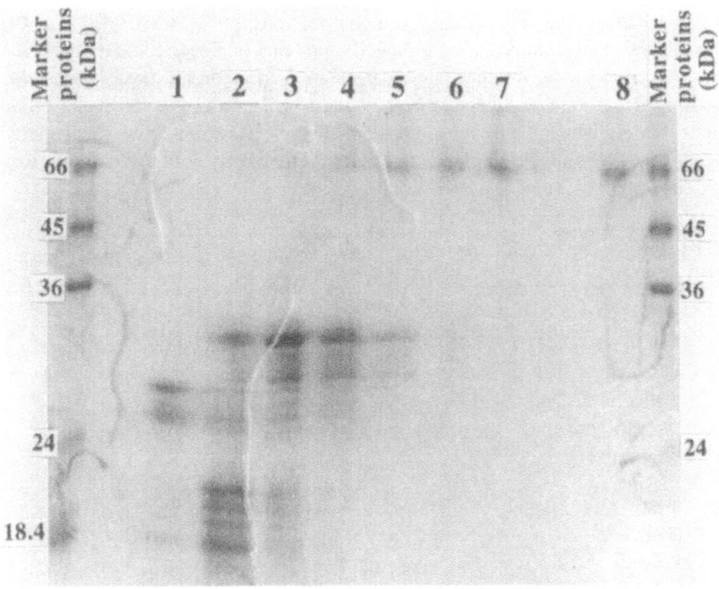

Fig. 6. Electrophoretograms of proteolytic digestion of the CBB-R stained band of molecular mass of 66 kDa. SDS-PAGE was performed with a hand-made discontinuous gel system, with 12.5% acrylamide for the separation gel and 2.5% acrylamide for the spacer gel. Proteins on the gels were visualized by staining with CBB-R. Twenty microliters of various concentrations of protease were loaded per sample well. The protease concentration was 24, 12, 2.4, 0.48, 0.096 and 0.019 μg per each 20 μl for lanes 2, 3, 4, 5, 6 and 7, respectively. Protease absence of the 66-kDa band and the 66-kDa band without protease were run for lanes 1 and 8, respectively. (Reprinted with permission from *Deep-Sea Research*, **39**, E. Tanoue, Occurrence and characterization of particulate proteins in the Pacific Ocean, 1992, Pergamon Press Ltd.)

the POM from the surface waters at sites KH85-B and KH86-C (Fig. 3), but it was not visible in the surface waters at sites KH86-D and -E (Fig. 4). The band was also not recognizable in all the samples of plankton. Thus, the vertical change in the composition of the particulate proteins was quite clear. The POM in the surface waters comprised many proteins with different molecular masses, each at a low level, while the POM in the intermediate and deep waters comprised a relatively small number of proteins as major constituents.

To verify the band of a particulate protein that was separated by the SDS-PAGE was true protein, the technique of partial enzymatic proteolysis (Cleveland *et al.*, 1977) was applied. Figure 5 shows an example of the application of this technique to the authentic standard, bovine serum albumin (BSA). Following the first SDS-PAGE of the BSA, staining and destaining, the band of the BSA was cut out from the gel. The gel slice was placed in the sample well of the second gel of the SDS-PAGE, and a given amount of protease of *Staphylococcus aureus* strain (strain V8) was overlaid into each sample well to allow proteolytic digestion to

Table 1. Particulate protein concentration and contributions to POC (Pr-C/POC) and PN (Pr-N/PN) and the relative contribution of specific proteins (molecular masses of 45–66 kDa) to the total protein in the intermediate and deep waters. Carbon and nitrogen of bovine serum albumin (BSA) were calculated to be 45.7 and 15.8%, respectively, from its amino acid composition. Carbon and nitrogen contents of the particulate protein were calculated by multiplying the protein values by 0.457 and 0.157, respectively. Protein concentrations are expressed as μg BSA equivalent per liter.

Depth (m)	0	10	25	50	100	200	500	1000	2000
KH86-C									
Protein	31.6	26.2		21.2	3.64	3.42	1.01	0.70	0.23
POC(μg/l)	108	98.4		44.8	23.5	15.5	9.27	6.63	7.48
PN(μg/l)	21.2	20.4		6.68	2.00	1.08	--*	--*	--*
Pr-C/POC (%)	13.4	12.2		21.6	7.1	10.1	5.0	4.8	1.4
Pr-N/PN (%)	24.8	20.3		50.1	28.8	50.0			
Specific/Total (%)							30	27	46
KH86-D									
Protein	4.1		41.6	27.7	2.56	1.82	0.87	1.07	0.72
POC(μg/l)	122		115	52.1	24.2	23.1	12.2	13.0	16.5
PN(μg/l)	21.6		22.9	7.99	2.57	2.20	0.71	--*	--*
Pr-C/POC (%)	16.5		16.5	24.3	4.8	3.6	3.3	3.8	2.0
Pr-N/PN (%)	32.2		28.7	54.8	15.7	13.1	19.0		
Specific/Total (%)							32	50	42
KH86-E									
Protein	13.7		14.6**	8.50	1.97	1.47	0.99	0.26	0.26
POC(μg/l)	78.7			44.7	18.8	19.9	6.07	6.73	7.27
PN(μg/l)	18.0			8.23	2.08	1.63	0.47	--*	--*
Pr-C/POC (%)	8.0			8.7	4.8	3.5	7.5	4.8	1.6
Pr-N/PN (%)	12.0			16.3	15.0	14.2	31.9		
Specific/Total (%)							27	32	43

*PN could not be quantified because the amount of PN in the deep water samples (20 l) was too low to detect, given the sensitivity of the measuring system.
**Datum at a depth of 20 m.

proceed, and then, subsequently, electrophoresis followed. The protein in the gel was digested using increasing concentrations of protease. The digestion products of the BSA were separated according to their molecular masses (Fig. 5).

The band with an apparent molecular mass of 66 kDa was subjected to the proteolytic digestion (Fig. 6). The protease only cleaves at the COOH-terminal side of aspartic and glutamic acid residues (Houmard and Drapeau, 1972; Drapeau *et al.*, 1972), consequently, digestion products contain appropriate-sized peptide fragments for the SDS-PAGE. Re-electrophoresis of the 66 kDa-band without protease gives again the visible band on the anticipated position on

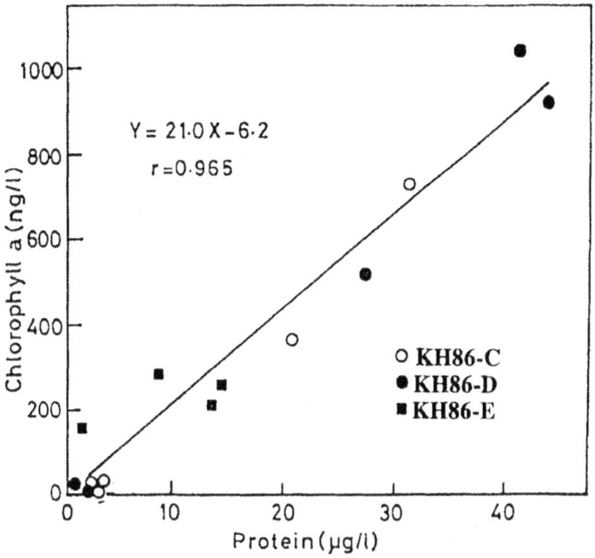

Fig. 7. Correlations between particulate protein (relative to a BSA standard) and Chl *a* at sites KH86-C, -D and -E. (Reprinted with permission from *Deep-Sea Research*, **39**, E. Tanoue, Occurrence and characterization of particulate proteins in the Pacific Ocean, 1992, Pergamon Press Ltd.)

the second SDS-PAGE gel (lane 8). A progressive disappearance of the band is seen with increasing levels of protease from lane 7 to lane 2. The bands were not digested significantly at lower concentrations of protease but disappeared at high concentrations (lanes 6 and 7). Depending on the concentration of the protease, cleavaged fragments of the band can be observed in the molecular mass range of 24–36 kDa (lanes 4 and 5). Digestion of the cleavaged fragments proceeded at higher concentrations of protease, and the cleavaged fragments of molecular mass of less than 24 kDa appeared in the gel with the highest levels of protease (lanes 2 and 3). These results indicate that the protease intrinsically worked upon the band in the gel; thus the 66-kDa band was true protein.

Vertical changes in the concentrations of particulate protein were examined at sites KH86-C, -D and -E (Table 1). The concentrations of the protein are high in the euphotic layer and decrease rapidly with depth. Higher concentrations of particulate protein were found at the sites which were associated with higher concentrations of POC, PN and Chl *a*. The coefficient of the correlation for protein and Chl *a* was highest (r = 0.965, n = 14, 0–200 m; Fig. 7), and the correlation coefficients for protein and POC, and for protein and PN were 0.928 (n = 22, 0–2000 m) and 0.872 (n = 14, 0–200 m), respectively. A steep gradient in the concentrations of the particulate protein below the euphotic layer (Table 1) and the highest correlation coefficient for protein and Chl *a* (Fig. 7) indicated that a turnover rate of the particulate protein in the surface water appeared to be almost the same as that of Chl *a* in the northern North Pacific Ocean and the

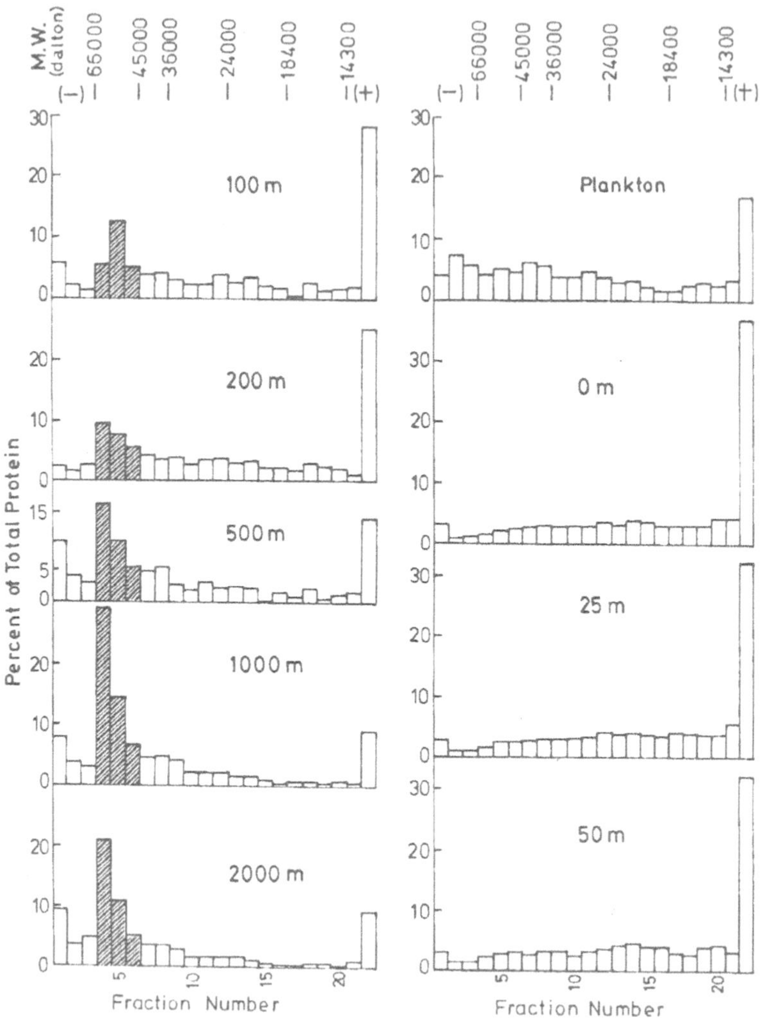

Fig. 8. Relative abundance of the proteins in the S-mm-long pieces of gel for samples, determined by the dye-elution method, collected at the site KH86-D. Shaded fraction indicates that specific proteins with molecular masses from 45 to 66 kDa were clearly visible. (Reprinted with permission from *Geophysical Monograph*, 63, E. Tanoue, Electrophoretic separation of particulate proteins in seawater. In *Marine Particles: Analysis and Characterization*, 1991, American Geophysical Union.)

Bering Sea. The SDS-PAGE of the plankton yielded smeared electrophoretograms, and the background proteins represented quantitatively almost 100% of the particulate proteins in the surface waters, and the background proteins also gave smeared electrophoretograms (Figs. 3 and 4). Thus, the majority of the particulate protein in the surface water are cellular proteins of living organisms, mostly phytoplankton, and they are susceptible to biological degradation.

A group of proteins with molecular masses between 45 kDa and 66 kDa was prominent in the POM from the intermediate and deep waters. Very clear bands of proteins were visible in these samples, indicating that the level of the background proteins were negligible over this range of molecular masses. Each sample lane on the gel was cut at 5 mm intervals and the relative abundance of proteins in each 5-mm fraction is shown in Fig. 8. With the exception of the fraction at the migration front (bottom of the gel), each protein fraction was uniformly stained and did not exceed 10% of the total protein in the samples of plankton and of the POM in depths shallower than 50 m. Specific proteins in the molecular mass range of 45 kDa to 66 kDa (shaded fractions) were the major proteins in the POM from depths greater than 500 m. Contributions of these specific proteins to the total particulate proteins are between 27 and 55% in the POM below 500 m (Table 1).

Organisms produce a variety of proteins. O'Farrell (1975) separated 1,100 different protein components from a single species of bacterium, *Escherichia coli*, by two-dimensional electrophoresis and commented that the resolution of any one-dimensional technique was inadequate for the total separation of proteins in biological systems more complex than a bacteriophage. Thus, the group of specific proteins found in the POM from the intermediate and deep waters is unlikely to represent the proteins in living organisms, but they represent the inanimate proteins. Since they were not recognizable in the electrophoretograms of samples of the natural plankton, they are minor components of proteins in living organism but survive selectively to be prominent components in the particulate proteins in the intermediate and deep waters. Thus, they are likely to be refractory to biological degradation.

Contributions of PCAA to POC and PN were found in the ranges of 30.6~31.3% and 55.8~62.0%, respectively, of the surface water (0–100 m depth) and decreased in the ranges of 17~22% and 50~55% below a depth of 500 m at two stations of the northern North Pacific Ocean and the Bering Sea (Tanoue and Handa, 1979). Contributions of protein to POC and PN at Stations C and D were determined with the ranges of 12~24% and 20~55%, respectively, in the surface water (0–50 m depth), and percentages of protein carbon to POC decreased in the ranges of 1.4~5.0% below the depth of 500 m. Although, it is difficult to discuss the relationship between levels of particulate proteins in the present study and those of PCAA in the literature, it appears that vertical decreases in contributions of protein to POC are more significant that those of PCAA to POC. It is suggested that the major proportion of the PCAA is in the form of protein in the euphotic layer, while the protein contributes a small portion of the PCAA, and non-proteinaceous amino acids are major constituents of the PCAA in the intermediate and deep waters.

In summary, two characteristic groups of proteins were identified from the vertical changes in their molecular distribution. The first group, derived directly from cellular proteins of living organisms, was made up of a large number of proteins, each present at relatively low levels, and with widely ranging molecular masses. The proteins of this group are very susceptible to biological degradation. The second group included specific proteins with a limited range of molecular masses, and were the most abundant protein constituents in the POM from the intermediate and deep waters. They might be resistant to biological attack. Thus, sources and the dynamics of proteins, as well as PCAA, are quite different in the POM from the surface water and from the deep water, even if they have similar amino acid compositions.

Relative abundances of the particulate protein and particulate combined amino acids (PCAA), and their meridional characteristics in the Pacific surface waters

Since PCAA is the most prominent component of POM, and since amino acids in organisms are in the form of proteins, amino acids in POM have been cited as "proteinaceous compounds". In the previous section, it was suggested that the amino acids in the POM after hydrolysis, did not necessarily represent amino acids that were liberated from protein molecules at the three depth stations located in the northern North Pacific Ocean and the Bering Sea. In this section, the relationship between the concentrations of the particulate protein and PCAA, and their meridional changes are examined in the POM from the surface waters along transects from 45°N to 25°S in the central Pacific Ocean (Fig. 1).

The meridional distributions of organic compounds in the POM from the surface waters in the Pacific Ocean are summarized in Figs. 9 and 10. The concentrations of POC, PN, PCAA and particulate protein were high in the subarctic waters (north of 41°N; Anma *et al.*, 1990). The concentrations of these materials and the intensity of *in situ* fluorescence decreased rapidly toward the south (Fig. 9). The steep gradients in such concentrations in the area north of 34°N correspond to the region of the Kuroshio extension (North Pacific Current; Kawai, 1972). This region is thought to be a transitional area between the subarctic and the subtropical regions (Masuzawa, 1972). The present results correspond to previous observations, in the North Pacific Ocean, of steep meridional gradients in the composition and biomass of phytoplankton and zooplankton species (Johnson and Brinton, 1963; McGowan and Williams, 1973; Ishizaka *et al.*, 1994), the *in situ* fluorescence of Chl *a* (Pak *et al.*, 1988) and the surface concentration of Chl *a* (Dickson and Wheeler, 1993).

In areas south of 34°N, which correspond to the subtropical gyre, the concentrations of the various materials did not show significant meridional variation, with the exception of the values of C/N and the non-proteinaceous amino acids (non-proteinaceous AA, Fig. 10). The C/N values were low, being less than 6 in the areas north 34°N and they increased to around 7 between approximately 30°N to 18°N. They appeared to decrease again to around 6 toward the south. In areas associated with high values of C/N, amounts of non-proteinaceous AA were significantly, high and the particulate protein made a lower contribution to the PCAA (protein/PCAA).

In the equatorial region, contributions of the particulate protein to the PCAA varied significantly even though the concentrations of POC, PN, PCAA and protein did not show clear meridional differences. The Equatorial Pacific area, encompassed by a relatively oligotrophic environment, is the site of upwelling and, as a consequence, of highly productive areas (Betzer *et al.*, 1984; Chavez and Barber, 1987; Bender and McPhaden, 1990). The productivity of this area is meridional and zonal (e.g., Thomas, 1979; Wyrtki and Kilonsky, 1984; Bender and McPhaden, 1990). The area of the North Equatorial Counter Current (NECC), extending from about 4°N to about 10°N (Philander, 1990), appeared to coincide with the relatively low values of protein/PCAA (Fig. 10). The protein/PCAA values appear to be high in areas of the North Equatorial Current (NEC) and the South Equatorial Current (SEC) that are located to the north and south of the NECC, respectively, at sites along the path of the westward-flowing surface current which is due to the upwelling in the east. The values of protein/PCAA decreased again in the area south of the SEC, namely, the subtropical region of the South Pacific Ocean. The resolution of the sampling in the equatorial region was inadequate. However, similar patterns of levels of dissolved organic carbon (DOC) and partial pressure of carbon dioxide in seawater (pCO_2) were observed during the same cruise (Tanoue, 1993; Ishii and Inoue, 1995), with both parameters being lower in the area of the NECC than those in the NEC and in the SEC, indicating relatively low biological productivity in areas of the NECC. The contributions of the particulate protein to the PCAA are positively correlated with the primary productivity of the respective water masses.

The meridional differences were evident in concentrations of POC, PN, PCAA and particulate protein along the transects. The correspondence between the meridional changes in the concentrations of POC, PN, PCAA and the *in situ* fluorescence indicates that the amount of particulate protein in the surface waters is closely associated with biological productivity. Contributions of the particulate protein to the PCAA appeared to be high in the northern North Pacific and Equatorial regions, namely, in productive areas, while relatively low values of protein/PCAA were found in the subtropical regions and the NECC region, namely, in the oligotrophic areas of the Pacific Ocean. A chemical form of PCAA is thus meridionally different. The majority of PCAA was in the form of protein molecules in the surface waters in productive areas, while a small fraction of PCAA was in the form of protein and PCAA was mainly present as non-proteinaceous AA in an oligotrophic area.

Horizontal characteristics of molecular mass distribution of particulate proteins in the surface waters from the Pacific and Southern Oceans

All electrophoretograms of particulate protein along the transects in the central Pacific are compiled in Fig. 11. In the northern three samples along 165°E (Fig. 11A; KH91-1, 2 and 3 in Fig. 1), the gel was heavily and uniformly stained for the fraction between 14 kDa and 66 kDa. Significant amounts of low molecular mass proteins were concentrated at the migration front, but no clear protein band was visible. The electrophoretograms were similar to those of the samples in the surface waters and of the plankton (Figs. 3 and 4) collected from the more northern areas (KH86-C, -D and -E in Fig. 1). In the electrophoretograms of the southern three samples along 165°E (KH91-4, 5 and 6 in Fig. 11A),

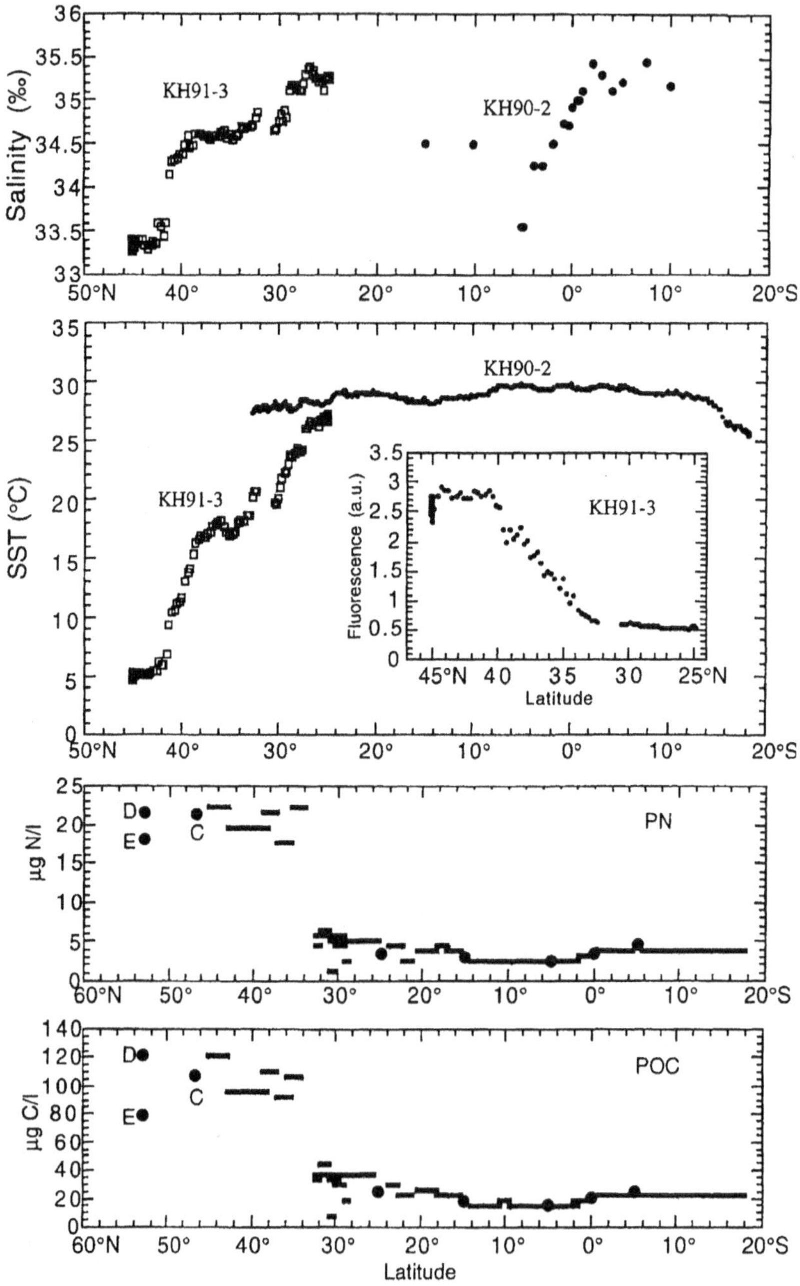

Fig. 9. Meridional distributions of POC (μgC/l), PN (μgN/l), salinity (‰) and SST (°C) along the cruise tracks of KH90-2 and KH91-3, and *in situ* fluorescence intensity (arbitrary unit. a.u.) along 165°E. Closed circles represent samples collected from fixed locations and the bars represent the range of locations of the collected sample (see Fig. 1).

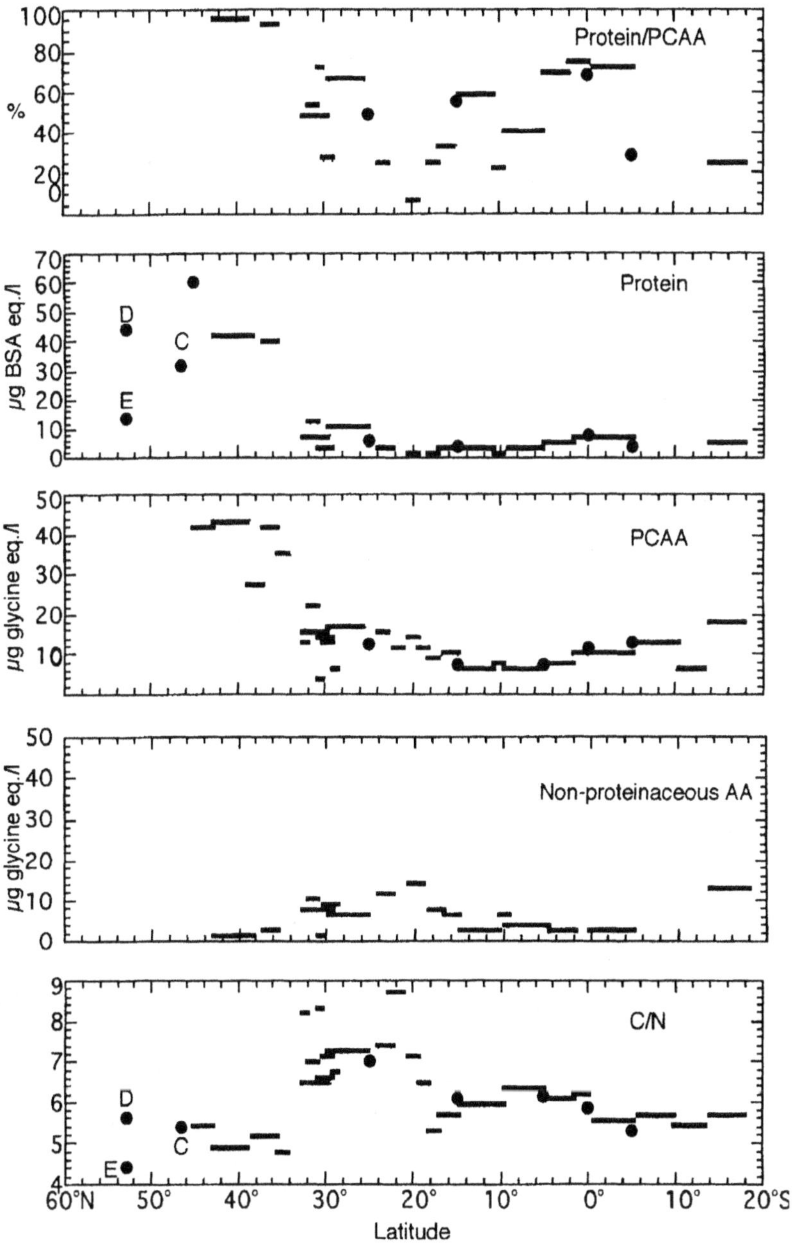

Fig. 10. Meridional distributions of C/N (w/w), non-proteinaceous AA (μg relative to a glycine standard/l), PCAA (μg relative to a glycine standard/l), protein (μg relative to a BSA standard/l) and protein/PCAA (%) along the cruise tracks of KH90-2 and KH91-3. Data from the northern North Pacific Ocean and the Bering Sea (Table 1) are included.

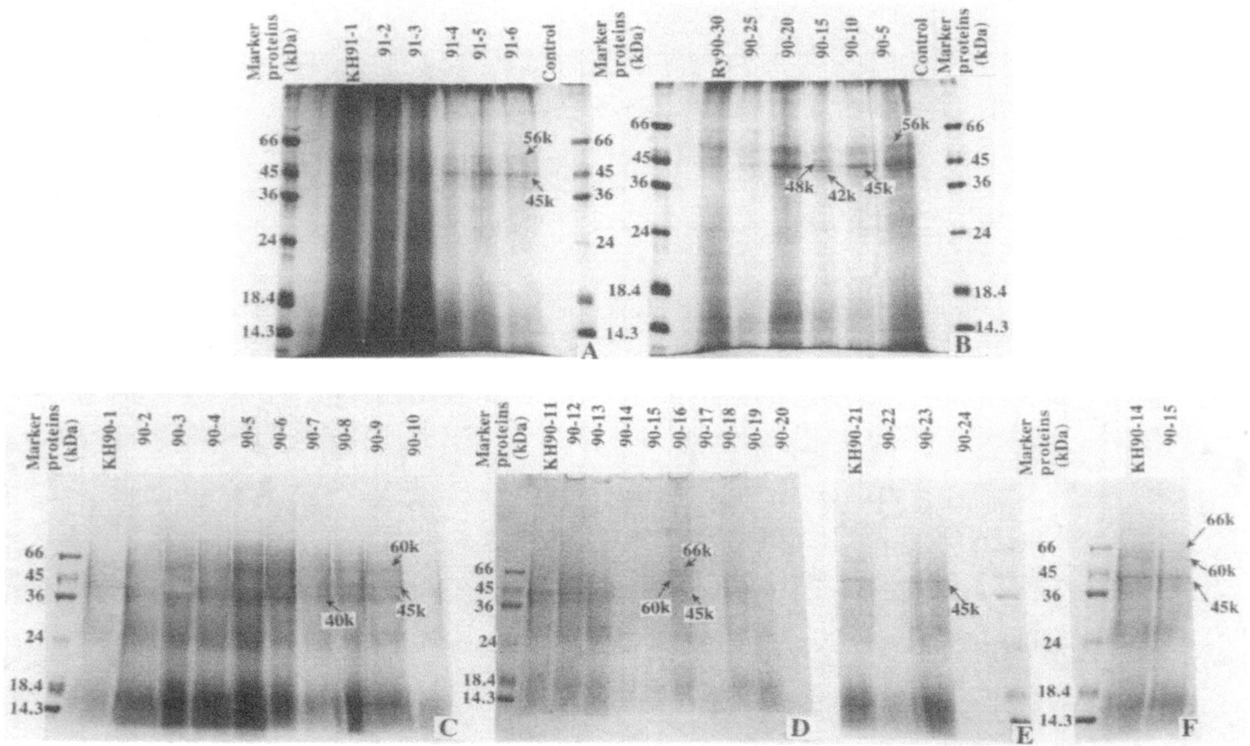

Fig. 11. Electrophoretograms of particulate protein in surface waters along 165°E (A), 137°E (B) and 180° (C, D, E and F). SDS-PAGE was performed using a hand-made discontinuous gel system, with 12.5% acrylamide for the separation gel and 2.5% acrylamide for the spacer gel for A and B. A ready-made continuous gel was used for C, D, E and F. Electrophoretograms obtained from samples that had been concentrated with a disposable ultrafiltration unit (Molcut-II) are shown in F. Proteins on the gels were visualized by staining with CBB-R.

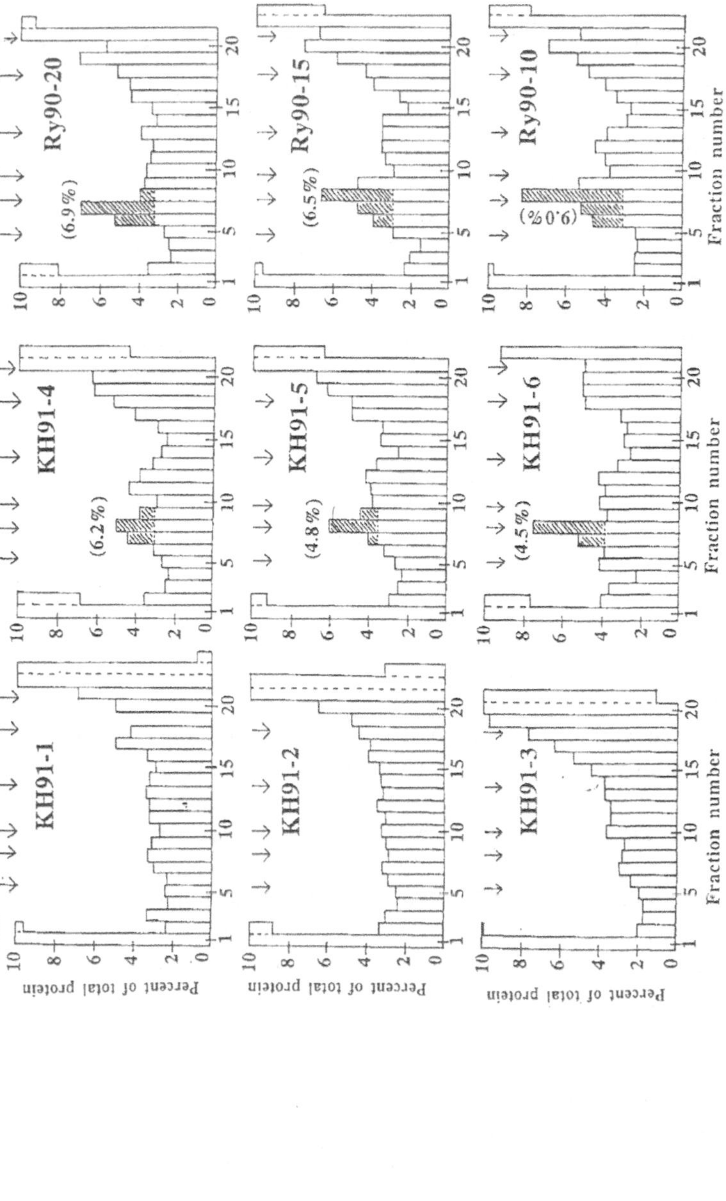

Fig. 12. Relative abundance of the proteins in 5-min pieces of gel for samples collected along 165°E and 137°E. Shaded fractions represent the contribution (%) of the proteins in the distinct bands to the total protein (from Tanoue, 1996).

however, distinctive bands of individual proteins that were found over a limited range of molecular masses were superimposed on the smeared background proteins. The boundary between the waters that yielded the two types of electrophoretograms was between the 30°~35°N along 165°E (Figs. 1 and 11A), corresponding to the transitional area between the subarctic and the subtropical regions.

Several clear protein bands are also visible over a limited range of molecular masses between 36 kDa and 66 kDa in samples collected from 30°N to 5°N along 137°E (Fig. 11B). In POM collected from wide areas from about 18°S to 32°N along 180° in the central Pacific (Fig. 1), electrophoretic patterns (Figs. 11C, D and E) are essentially the same as those along 137°E and the southern three samples along 165°E. Distinctive bands of individual proteins were found over a limited range of molecular masses between 36 kDa and 66 kDa, and were superimposed on the smeared background proteins. Relative staining intensities of individual protein bands as well as background proteins were different among the samples examined. For example, staining intensities were low and individual protein bands were unclear on the electrophoretograms of KH90-14, -15, -22 and -24 (Figs. 11D and E) collected from the subtropical region. However, distinctive bands occurred in such samples after the concentration of the electrophoretic samples of KH90-14 and KH90-15 using a pressure-driven disposable ultrafiltration unit (Molcut-II) (Fig. 11F).

Figure 12 shows the relative abundance of proteins determined in each 5 mm gel interval from the samples along 165°E and 137°E. Proteins concentrated at the migration front in each case in a range of 9.3~22.7% of the total protein, but relative abundances of proteins at the migration front decreased from north to south. In the separate fractions between 14.3 kDa and 66 kDa, each 5-mm interval was relatively uniformly stained or stained heavier towards low molecular mass areas, but a clear peak was not found in samples from the northern three samples (KH91-1, -2 and -3) so the electrophoretograms were smeared and distinct bands were unrecognizable (Fig. 11A). In the southern three samples (KH91-4, -5 and -6) where distinct bands were visible, the relative abundance of molecular mass fractions between 45 kDa and 66 kDa were a little higher than the higher and lower molecular mass fractions in each separate fraction. The peaks in the separate fraction corresponded to the molecular mass area that distinct bands were visible. The relative abundance of proteins determined in each 5-mm gel interval from the samples along 137°E (Fig. 12) also show clear peaks in the separate fractions between 45 kDa and 66 kDa (Fig. 11B).

To estimate the contributions of these distinctive bands towards the total protein, correction of the background proteins was made because this fraction also contained the background proteins. We assume that the background proteins are distributed uniformly and that the amounts of distinctive bands are equivalent to the excess portions compared with higher and lower molecular fractions. The contributions of the distinct bands to total proteins (shaded fractions in Fig. 12) were calculated to be in the range of 4.5~6.2% for the southern three samples along 165°E and in the range of 6.5~9.0% along 137°E, respectively. Since the samples along 137°E that exhibited clearer distinctive bands were selected, the values were maximal, thus, the specific proteins may not exceed 10% of the total particulate protein in the surface water, even if the distinctive bands were clearly visible.

Resolution of electrophoretograms of POM was inadequate due to the lack of a purification step of particulate protein during the preparation of electrophoretic samples. Inadequate purification of electrophoretic sample leads to error in estimations of molecular masses (e.g., Tanoue, 1995). Thus, the molecular masses of the specific proteins were tentatively determined by simple comparison of their electrophoretic mobilities between locations of specific proteins and those of marker proteins on the same gel. Proteins with apparent molecular masses of approximately 56 kDa and 45 kDa were found to be major specific proteins in the samples along the transects of 137°E and 165°E (Figs. 11A and B). Proteins with apparent molecular masses of 42 kDa and 48 kDa were also found in samples along 137°E (Fig. 11B). Among the series of 24 samples along 180° transect (Figs. 11C, D, E and F), 45-kDa protein was commonly found as one of the major specific proteins. Protein with apparent molecular masses of approximately 60 kDa was also found in the majority of samples and 66-kDa protein was found in the sample of 90-16.

Particulate proteins in the surface sample along the cruise track of JARE 34 (Fig. 1) is shown in Fig. 13. Figures 13A and B show the electrophoretograms of the particulate proteins that were prepared without any concentration step. The gel was less heavily stained in the samples from F-1 ~ F-12 that corresponded to the areas from the equatorial to subtropical regions (Fig. 1). The staining intensities increased in samples of F-15 ~ F-60 collected from the Southern Ocean (Fig. 13B). Proteins with apparent molecular masses of approximately 45 kDa and 66 kDa were detected as faint but discrete bands. In the case of particulate proteins, a large amount of glass-fiber filter necessitates the preparation of a relatively large volume of samples for electrophoresis, and the silver-staining method could not be used because of a high background. As a result, it was difficult for prepare a sample that was concentrated enough for the observation of molecular distribution after SDS-PAGE.

Particulate proteins in the same electrophoretic samples were concentrated by precipitation with TCA (Figs. 13C and D). The staining intensity of the background proteins increased and distinct bands of individual proteins were also found in the samples after precipitation with TCA. Bands of protein with apparent molecular masses of 66 kDa and 45 kDa were clearly visible in samples collected from the equatorial to subtropical regions (Fig. 13C). They were superimposed on the smeared background proteins and were unclear due to the high levels of background proteins in the samples from the Southern Ocean (Fig. 13D). To examine the influence of the precipitation with TCA, we compared electrophoretograms of cultured marine bacterium, *Marinomonas communis*, with and without the precipitation with TCA (Fig. 13E). Bulk proteins of *M. communis* were distributed over a wide range of molecular masses and gave smeared electrophoretograms. The molecular mass distributions and staining intensities of the

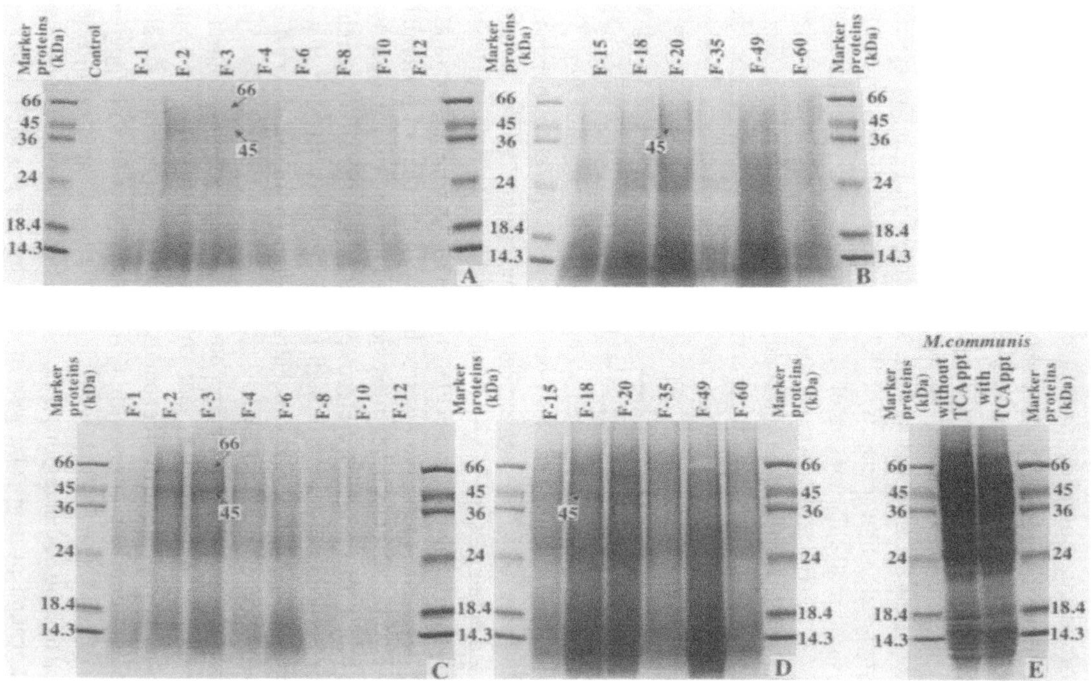

Fig. 13. Electrophoretograms of particulate proteins along the cruise of the JARE 34. A ready-made continuous gel was used and proteins were stained with CBB-R. Each marker protein in the left- and right-hand lanes of the sample lanes was loaded at 1 and 2 μg. A. Amount of samples of F-1, F-2, F-3, F-4, F-6, F-8, F-10 and F-12 were equivalent to 2.0, 1.9, 1.5, 2.0, 1.0, 1.8 and 2.0 liters of the original seawater, respectively. C, D. The amount of samples were equivalent to 3.0 liters of the original seawater in each case. E. The amount of samples of the bulk proteins of *Marinomonas communis* with and without precipitation with TCA were equal. (Reprinted with permission from *Limnology and Oceanography*, **41**, E. Tanoue *et al.*, Discrete dissolved and particulate proteins in oceanic waters, 1996, American Soc. Limnol. Oceanogr. Inc.)

two electrophoretograms did not differ. Differences were also not observed in electrophoretograms between particulate proteins with and without precipitation of TCA.

It was clearly observed that 45-kDa protein commonly occurred in the samples collected from the oligotrophic area of the north Pacific Ocean, through the equatorial and Indian Oceans, to the Southern Ocean. Fifteen electrophoretic samples of KH90-1, -2, -6, -7, -8, -9, -10, -11, -12, -13, -14, -16, -18, -20 and -22 (Figs. 11C, D and E) collected from 15°S to 30°N in the central Pacific Ocean were subjected to electrophoresis three times. The each band that represented the 45-kDa protein was cut out from the gel and the resulting 45 gel pieces were composited into one sample and the 45-kDa protein was extracted. Since the efficiency of extraction was extremely low in the preliminary experiments (data not shown), a pooled sample from the composited gel pieces was prepared for a repeated electrophoresis. Re-electrophoresis of the combined extract gave a clear, sharp, single and reproducible band at the anticipated position after the second SDS-PAGE. Thus, it was verified that the 45-kDa protein is a single protein species in terms of SDS-PAGE. In this case, the background proteins were insignificant, and this protein was subjected to N-terminal amino acid sequence analysis (Table 2). The first 12 residues from the amino-terminus of the protein were identified. A single amino acid was detected at every step of Edman degradation indicating that the 45-kDa protein visualized from the samples collected from a wide area of the central Pacific Ocean were single species. However, no similar amino-terminal amino acid sequence was retrieved from a homology search of the Protein Identification Resource International database (PIR-International, version 39.0) through the Japan International Protein Information Database (JIPID).

Particulate proteins and PCAA

The majority of the PCAA were in the form of protein molecules in the surface waters in the northern North Pacific and equatorial regions, namely, in productive areas, while a small fraction of the PCAA was in the form of protein molecules and the PCAA was mainly present as non-proteinaceous AA in the subtropical gyre, namely, in an oligotrophic area (Fig. 10). Amounts of non-proteinaceous AA, obtained by subtracting the amount of protein in a gel from

Table 2. N-terminal amino acid sequence of the 45-kDa protein obtained by automated Edman degradation (from Tanoue, 1996).

Cycle	1	2	3	4	5	6	7	8	9	10	11	12
45 kDa protein	Gly	Thr	Gln	Pro	Asn	Pro	Ser	Pro	Ala	Ser	Pro	Val

that of PCAA on a filter, were relatively high in the subtropical region.

SDS-PAGE resolves proteins with large molecular masses. Small peptides are lost to some extent and amino acids are lost completely because these small compounds diffuse from the gel matrix during staining and destaining. Dye (CBB-R) is not sensitive to small peptides and does not stain amino acids. Non-proteinaceous AA did not, therefore, consist of large proteins but might have consisted of peptides and/or amino acids. Since oligopeptides and amino acids are minor components of living organisms, as mentioned above, and since the levels of non-proteinaceous AA were not correlated with those of POC, PN, *in situ* fluorescence and particulate protein, it seems unlikely that non-proteinaceous AA was derived directly from living organisms. Such small compounds were retained on the GF/F filter probably *via* their association with macromolecules that were large enough to be also retained on the filter. Thus, it seems likely that the majority of the PCAA in oligotrophic areas might exist as oligopeptides and/or amino acids that have been incorporated into the polymeric detritus.

It has been demonstrated that monomeric organic compounds, including amino acids, bind abiologically to dissolved macromolecules or are transformed biologically into polymeric DOM (Carlson *et al.*, 1985; Brophy and Carlson, 1989). Authentic proteins have also been demonstrated to be transformed abiotically into refractory DOM in seawater (Keil and Kirchman, 1994). Laboratory experiments have focused on the processes of formation of refractory DOM from labile amino acids and protein. Although no direct evidence that such processes occur in the POM has yet been reported, it is likely that such reactions could occur to transform cellular proteins, peptides and amino acids from living organisms into detrital combined amino acids. Peptides and amino acids that have been incorporated into macromolecules are known to be quite refractory to degradation (Carlson *et al.*, 1985; Brophy and Carlson, 1989; Keil and Kirchman, 1994), while the majority of cellular proteins in organisms are readily remineralized within hours or days (Hollibaugh and Azam, 1983; Coffin, 1989; Tupas and Koike, 1990; Keil and Kirchman, 1993). Thus, the dynamics of the PCAA in the surface water might vary meridionally. Namely, the PCAA in subarctic waters might be readily remineralized because PCAA is mainly in the form of protein, while the PCAA in subtropical waters might be inactive because the majority of the PCAA is in the form of inanimate non-proteinaceous compounds.

Background proteins

As consequences of the vertical changes in the distribution of proteins, two groups of proteins, namely, background proteins and specific proteins, were characterized by samples at four depth stations in the northern North Pacific Ocean and the Bering Sea (Figs. 3 and 4). The background proteins and specific proteins were also identified from the samples collected from the surface waters of the North Pacific Ocean, through the equatorial Pacific and the Indian Oceans, to the Antarctic Ocean (Figs. 11 and 13). The background proteins were distributed over a wide range of molecular masses and were evident in the POM from the subarctic surface waters; they yielded smeared electrophoretograms. SDS-PAGE of the entire organisms also yielded smeared electrophoretograms (Figs. 3, 4 and 13). The background proteins quantitatively represented almost 100% of the particulate protein in the subarctic surface waters. Even in the samples from the subtropical gyre, in which the specific proteins were clearly detectable, the background proteins accounted for more than 90% of the particulate protein (Fig. 12). Vertical changes in their concentrations were closely correlated with those of Chl *a*. The levels of the background proteins were also correlated with those of POC, PN, PCAA, *in situ* fluorescence intensity and particulate protein in the surface waters (Figs. 9 and 10). Thus, the background proteins represent cellular proteins of living organisms, and they are susceptible to biological degradation.

Organisms retained on the GF/F filter were mainly a mixture of autotrophic phytoplankton and heterotrophic bacteria. Numbers of bacteria in the surface waters at the same locations as those at which the POM samples designated KH91-1 and KH91-6 had been collected, were about 1.2–1.4 × 10^6 cells/ml (depths of 0–11 m) and about 0.2–0.7 × 10^6 cells/ml (depths of 0–10 m), respectively (Hara, personal communication). The amounts of protein from the bacterial population were calculated to be 12–14 μg/l and 2–7 μg/l for samples from KH91-1 and KH91-6, respectively, on the assumption that the amount of organic matter in a bacterial cell is twice that of carbon (20 μg C/cell; Lee and Fuhrman, 1987), that 50% of the organic matter is protein (Romankevich, 1984), and that 50% of bacteria were retained on the GF/F filter (Lee and Fuhrman, 1987). Concentrations of Chl *a* were 0.7 μg/l (depths of 0–10 m) and 0.05 μg/l (depths of 0–10 m) at the sites of samples KH91-1 and KH91-6, respectively. The amounts of protein from the phytoplankton population were calculated to be 21 μg/l and 1.5 μg/l for samples designated KH91-1 and KH91-6, respectively, on the assumption that the amount of organic matter in phytoplankton is twice that of carbon while that of carbon is 50 times of that of Chl *a* (Cho and Azam, 1990), that 30% of organic matter is protein (Mayzard and Martin, 1975), and that 100% of phytoplankton is retained on the GF/F filter regardless of the latitude (Chavez *et al.*, 1995). The contribution of proteins from the phytoplankton population to the background proteins appeared to be equal to or larger than that from the bacterial population in the sample designated KH91-1, from the subarctic region. In sample KH91-6, from the subtropical region, bacterial proteins made a greater contribution than those from phytoplankton.

Such rough estimations are affected by the choice of conversion factors (see reviewed by Ducklow and Carlson, 1992). The relationship between particulate protein and Chl *a* obtained from their vertical changes in concentration, indicated that the main part of particulate protein was derived from that of phytoplankton at the more northerly stations, KH86-C, D and E (Fig. 7). It appears that phytoplankton biomass decreases more dramatically than that of bacterial biomass from eutrophic to oligotrophic environments (Bird and Kaliff, 1984; Cole *et al.*, 1988; Cho and Azam, 1990) as seen in the samples of KH91-1 and KH91-6. Consequently, the relative abundance of bacteria to phytoplankton biomasses is low in mesotrophic to eutrophic waters, while bacterial biomass is commonly greater than phytoplanktom biomass in oligotrophic water (Cho and Azam, 1990; Kirchman *et al.*, 1993; Caron *et al.*, 1995). Similar results were also obtained from the study of strong organic ligands for thorium in particulate matter (see the synthesis by Hirose, 2000 in this volume).

Hirose and Tanoue (1994) developed a method for detecting the strong organic ligands in POM and applied it to the same samples from which the particulate proteins were examined (Hirose and Tanoue, 1998a, c). They reported that the strong organic ligand in POM is directly derived from bacteria, and the relative abundance of strong organic ligand to POC is higher in the samples from the subtropical regions than from the subtropical regions. The ratio of bacterial carbon, estimated from the concentration of the strong organic ligand, to POC is around 0.1 in the subarctic area and increases toward the south to a range of between 0.3 and 0.4 in the subtropical area along the transect of the central Pacific Ocean (Fig. 1). Although, the electrophoretograms could not differentiate the source of the background proteins at present, contributions of cellular proteins of bacteria and phytoplanton to the background proteins may be different in the waters of oligotrophic and productive regions.

Specific proteins

Specific proteins were distributed over a limited range of molecular masses between approximately 40 kDa and 66 kDa and were prominent components of particulate proteins collected from the intermediate and deep waters in the northern North Pacific Ocean and the Bering Sea (Figs. 3 and 4 and Table 1). The specific proteins were not significant in the surface samples collected from the productive oceanic areas, but they were clearly visible and widely distributed in the samples collected from the oligotrophic surface waters (Figs. 11 and 13). The specific protein was superimposed on relatively low levels of background proteins in the samples from the low-latitude areas where low levels of POC, PN, PCAA, *in situ* fluorescence intensity and particulate protein were recorded. Only six specific proteins were recognizable as major specific proteins, as shown in Figs. 11 and 13. Among the specific proteins, a protein with an apparent molecular mass of 45 kDa was evident in the samples from the surface waters. The 66-kDa protein was also found but was insignificant in the samples from the surface waters. This protein was prevalent in the samples from the intermediate and deep waters. The 45-kDa protein was commonly visualized as one of the major recognizable proteins in the POM form the surface waters on the basin scale, while this protein was not observed in the samples from the intermediate and deep waters. Distributional differences between two proteins might imply differences in sources of these specific proteins.

The fact that multiple samples gave a single amino-terminal amino acid sequence (Table 2) indicates that the 45-kDa protein in multiple samples was derived from an identical source and was in an identical chemical form, from the amino-terminus to, probably, the carboxyl-terminus and that the protein survives without modification in the surface POM. Its sequence was characteristic (Table 2). Proline was found at the 4th, 6th, 8th and 11th positions among 12 amino acid residues from the amino-terminus. Several proline-rich proteins that are associated with the actin cytoskeleton of eukaryotic cells have been reported (Donnelly *et al.*, 1993). However, no homologies were found between proteins of this type and the 45-kDa protein isolated in the present study.

In summary, two characteristic groups of proteins, namely background proteins and specific proteins, were also identified from the horizontal distribution of the particulate proteins. The background proteins contributed greatly to the total particulate protein, and the concentrations of background proteins were positively related to those of Chl *a*. It was suggested that the relative significance of cellular proteins in phytoplankton and in bacteria might be meridionally different. Specific proteins are always minor components in POM from the surface waters, but they are universally visible in the oligotrophic areas, an indication that proteins from specific sources survive and accumulate as a consequence of their resistance to degradation. The 45-kDa protein, a member of the specific proteins, appears to be very widespread in the POM of the surface waters from the North Pacific Ocean, through the equatorial Pacific and the Indian Oceans, to the Southern Ocean.

PCAA were distinguishable into three main types, namely, the background proteins, the specific proteins and the non-proteinaceous AA, in terms of their dynamics. The background proteins are readily remineralizable, while the specific proteins survive and accumulate in the POM without modification. The non-proteinaceous compounds in POM were significant in the oligotrophic regions, indicating that they have already undergone degradation, and resultant oligopeptides and/or amino acids might be transformed to polymeric detrital organic matter.

DISSOLVED PROTEINS

Dissolved organic matter (DOM) in sea water is one of the three active reservoirs on the Earth, the other two are living plants and soil organic matter on land, and holds approximately as much carbon as occurs in atmospheric carbon dioxide (Hedges, 1992). DOM has important influences on oceanic biogeochemical processes and, through biogeochemical processes, it plays an active role in the global carbon cycle (Hedges, 1987). Amino acids in seawater have been cited in terms of dissolved free amino acids (DFAA) and dissolved combined amino acids (DCAA) after hydrolysis of macromolecules. Dissolved combined amino acids (DCAA) represent the largest well-defined molecules of DOM (Coffin, 1989; Keil and Kirchman, 1991) and indirect evidence suggested that protein is the dominant form of DCAA in eutrophic waters (Billen, 1991). However, the hypothesis that DOM is a recalcitrant organic pool that is composed of a complex mixture of diagenetically altered organic materials, as well as inadequate technical improvements, have hindered us trying to define the chemical form of DCAA (see Introduction). Thus, sources of DCAA is unclear and chemical forms of

DCAA are unknown. However, the occurrence of inanimate protein molecules in the particulate phase strongly suggested that there could be protein molecules in the dissolved phase and encouraged our attempts to extract dissolved proteins from seawater. In this section, the detection and molecular characteristics of dissolved proteins are described, and sources of the detected proteins are discussed.

Materials

Samples of seawater of the subarctic, subtropical and equatorial Pacific Ocean were taken from various depths at five stations, namely, A (45°10.3′ N, 165°34.4′ E), B (24°35.0′ N, 170°0.1′ E), C (22°47.1′ N, 158°04.6′ W), D_2' (00°0.2′ S, 158°59.1′ W), and D_3 (08°0.01′ S, 160°00.0′ W) during a cruise of the RV *Hakuho-Maru* (cruise KH-93-4) from 19 October to 17 November 1993 (Fig. 14). Samples of seawater were collected with diaphragm-type air-driven pump above a depth of 200 m at Stations A, C, and D_2'. Samples of seawater were also

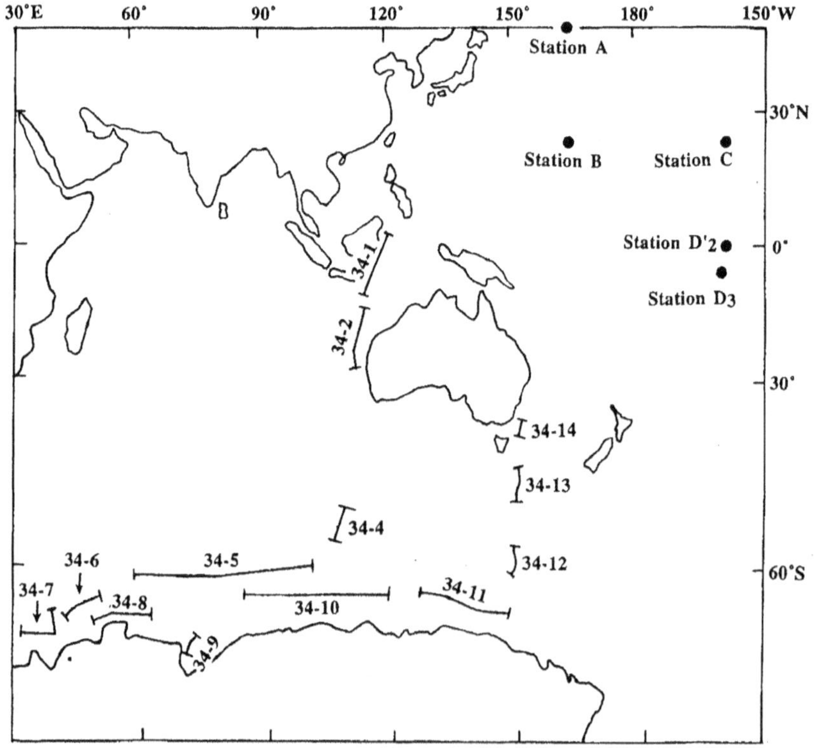

Fig. 14. The distribution of the sampling locations of the composite samples of surface seawaters for extraction of dissolved proteins along the cruise track of the JARE 34 (sample Nos. 34-1 to 34-14), and locations of depth Stations A, B, C, D_2' and D_3 for extraction dissolved proteins in the Pacific Ocean during the cruise of KH93-4.

collected with a Niskin-type bottle, mounted on a CTDO-Rosette Multiple sampler below a depth of 200 m at Stations C and D_2' and throughout the water column at Station D_3. Each sample (18.2–69.3 liters) was filtered through a GF/F glass-fiber filter (Whatman), immediate after sampling.

Samples of surface water of the Antarctic Ocean were obtained by an underway pumping system from 20 November 1992 to 19 March 1993, during the cruise of the 34th Japan Antarctic Research Expedition (JARE 34, sample Nos. 34-1 ~ 34-14, Fig. 14). Seawater was pumped up from the bottom of the ship (about 5 m below the surface) and introduced into a tank (capacity, 20 liters) with a continuous overflow. Seawater from the tank was filtered continuously through a GF/F filter *via* a specially designed filter apparatus with a filtration area of 285 $\times$ 420 mm^2 under a hydraulic pressure of ca 75 mm Hg. The GF/F filtrate (5 liters) was collected from time to time, usually twice a day. The filtrate was kept frozen at –30°C on board ship and at –20°C on land prior to analysis. Samples of the JARE 34 were thawed ashore and 3–10 individual samples were pooled to yield one composite sample. Composite samples, ranging in volume from 17.4 to 60.4 liters, were subjected to extraction of dissolved proteins.

Methods

The entire processes of extraction and detection of dissolved proteins are schematically shown in Fig. 15. The procedure involves three separate steps: crude concentration of dissolved protein from seawater using tangential flow ultrafiltration systems (Minitan and Pelicon; Millipore) with a filter with a nominal molecular mass cut-off of 10 kDa (the filter was low protein-binding regenerated cellulose; type, PLGC 000 05, Millipore); further concentration and purification of the dissolved protein in the crude concentrate by precipitation with trichloroacetic acid (TCA); and separation and detection of individual dissolved proteins by the SDS-PAGE.

After the addition of sodium dodecyl sulfate (SDS; 0.01%, final conc) to the GF/F filtrate, the filtrate was concentrated using a Pelicon system and then a Minitan system. The concentrate was desalted with 35 mM ammonium bicarbonate buffer (pH 7.8) that contained 0.01% SDS. The desalted concentrate was dried in a Speed Vac concentrator (model SC210A; Savant). The dried sample was redissolved as a crude concentrate in a solution of 35 mM ammonium bicarbonate buffer plus 0.01% SDS. The volume of each crude concentrate was less than 10 ml, and this material was stored frozen at –30°C until use.

A subsample (1–5 ml) of the crude concentrate was allowed to stand at 4°C for at least 12 h after the addition of TCA to a final concentration of 5% (vol/vol). The solution was centrifuged at 14,000 G, for 30 min at 4°C to remove TCA-soluble material. The TCA-insoluble fraction was subjected to sequential washing with TCA (5%), ethanol (100%) and diethyl ether (100%) to remove the inorganic residue, excess SDS, non-proteinaceous DOM, and residual TCA. The final pellet after centrifugation was air-dried. The dried sample was redissolved in a sample buffer solution that consisted of Tris-HCl (62.5 mM, pH 6.8), SDS (2%, wt/vol), 2-mercaptoethanol (5%, vol/vol) and glycerol (10%, vol/vol) and

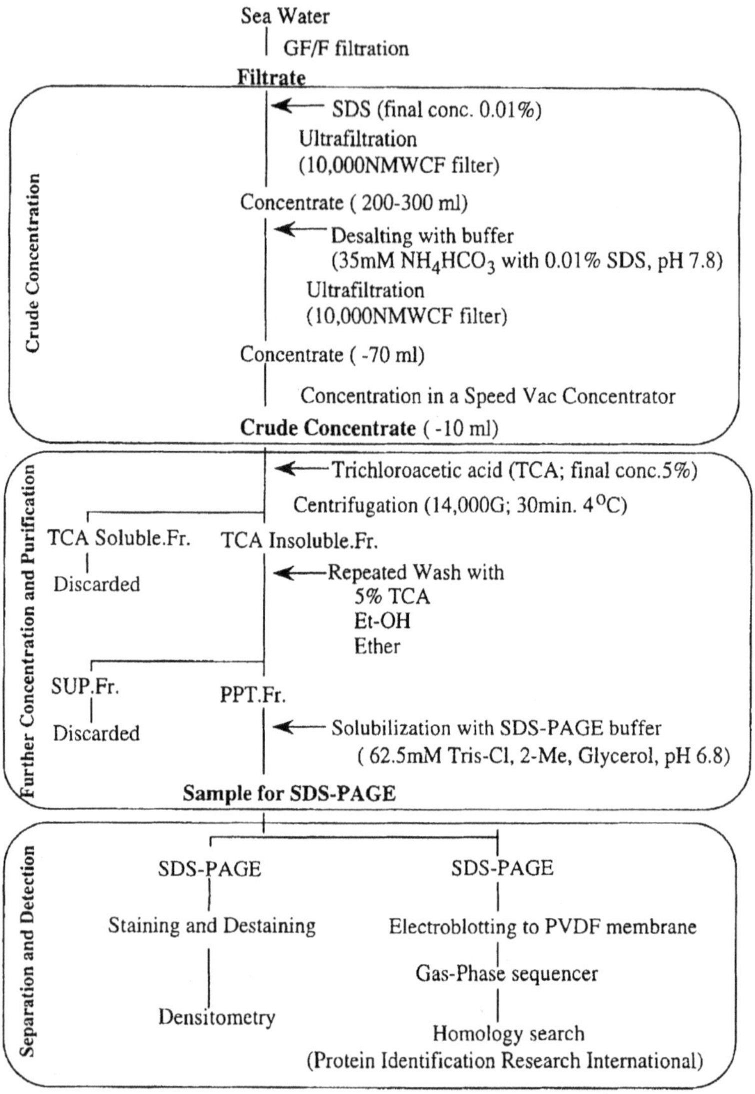

Fig. 15. Schematic procedures for extraction, detection and characterization of dissolved protein.

heated at 100°C for 3 min. Then it was subjected to SDS-PAGE. To examine the effect of the reduction of disulfide bonds, we redissolved some samples in a sample buffer with and without 2-mercaptoethanol and heating. In this case, samples with reduction were heated at 100°C for 6 min. Heating times of 3 or 6 min did not effect the electrophoretogram (data not shown).

The CBB-R staining method was used for detecting the protein bands on the gel. Removal of non-protein, dissolved organic materials in the crude concentrate by TCA precipitation produced a low background on the gel after electrophoresis and allowed the use of the highly sensitive silver-staining method. Silver staining was performed according the manufacturer's instructions using a silver-staining kit (2D-silver stain-II; Daiichi) that was based on the method of Oakley *et al.* (1980).

For the sequence of amino-terminal amino acids of dissolved proteins, proteins that had been separated by SDS-PAGE were electroblotted onto a polyvinylidene difluoride membrane (PVDF membrane; Immobilon, 0.45 μm pore size; Millipore) as described by Matsudaira (1980). The band of the protein of interest on the PVDF membrane was cut out with a clean razor and the protein on the membrane strip was sequenced with an automated sequencer (model 477A; Applied Biosystems), as mentioned above.

Methodological studies were presented elsewhere (Tanoue, 1995) and are not shown in detail. However, it was confirmed that the proteins detected as dissolved proteins were not derived from bacteria that were originally present in the filtrate, nor from bacteria that grew during the concentration steps. The proteins were also not due to contamination from filters and reagents or during handling.

Vertical distributions of dissolved proteins in the Pacific Ocean

Electrophoretograms of dissolved proteins in seawater at Station A are shown in Fig. 16. At Station A, located in the subarctic region, proteins were observed with a wide range of molecular masses. Proteins with molecular masses from 14 kDa to 66 kDa were separated by electrophoresis, and proteins were also present in fractions of molecular masses greater than 66 kDa and less than 14.3 kDa. Fewer than 30 bands were clearly visible as major bands of proteins in the samples from depths of 8 m to 200 m. The patterns from each depth were very similar in many ways. For example, the higher and lower limits of molecular masses detected in each lane on the gel were approximately the same. Protein species detected at each depth were quite similar to each other throughout the samples of the surface water examined. Bands of proteins of 48 kDa, 40 kDa, 37 kDa and 34 kDa, estimated from a Ferguson plot of Rf versus $\log_{10}$ molecular mass (Neville, 1971), were commonly observed as the major bands. A protein with an apparent molecular mass of approximately 48 kDa was the most prominent on the gel.

The apparent molecular masses of dissolved proteins estimated from the Ferguson plot showed that the major dissolved proteins tended to migrate more rapidly than the marker proteins. Among samples of dissolved proteins, proteins in the sample from a depth of 45 m appeared to migrate more rapidly. The electrophoretic mobility of a protein is dependent upon a number of experimental variables (e.g., polymerization of the gel, the buffer, the temperature during PAGE, the pH and ionic strength of the sample, etc.; e.g., Andrew, 1986). In the present case, the samples of dissolved proteins might vary as a result of the extent of purification of the dissolved proteins. The purification of dissolved proteins from the crude concentrate should be improved for a description of the molecular

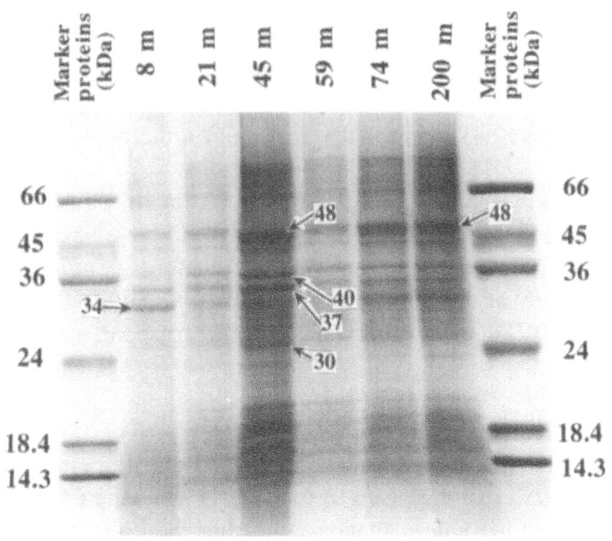

Fig. 16. Depth profiles of dissolved proteins at Station A. A ready-made continuous gel was used and proteins were stained with CBB-R. Amounts of samples loaded on the gel were equivalent to 1 liter of the original seawater in each case. Each marker protein in the left- and right-hand lanes was loaded on the gel at 1 and 2 μg, respectively. Arrows represent bands that have been confirmed as single protein bands by the analysis of the amino-terminal amino acid sequence (Table 3). (Reprinted with permission from *Marine Chemistry*, **51**, E. Tanoue, Detection of dissolved protein molecules in oceanic waters, 1995, Elsevier Science Ltd.)

inventory of dissolved proteins in the water column.

At Station B, located in the subtropical area, the patterns after the PAGE (Fig. 17) were a little different from those at Station A (Fig. 16). A relatively limited number of bands were visualized by staining with CBB-R 250 above a depth of 53 m (Fig. 17A). The intensities of staining were low at the surface and increased with depth. Between depths of 75 m and 211 m, the number of visualized bands increased. Proteins with apparent molecular masses of approximately 48 kDa and 37 kDa were clearly visible as the major proteins. Proteins of low molecular mass were also concentrated at the electrophoretic front in the gel. For samples below a depth of 211 m (Fig. 17B), the gel was heavily stained, indicating that high levels of protein in the dissolved phase were present in the intermediate waters. However, the electrophoretograms of intermediate samples were smeared. Inadequate resolution of electrophoretograms leads to errors in estimations of molecular mass. Thus, no Ferguson plot was constructed and molecular masses of proteins were tentatively determined by simple comparison of locations of dissolved proteins and of marker proteins on the same gel for Station B. Proteins of approximately 66 kDa and 34 kDa were predominant at depths from 462 m to 2,000 m. The 66-kDa protein was not evident in the surface waters (Fig. 17A) but was found in the intermediate and deep waters. The 48-kDa

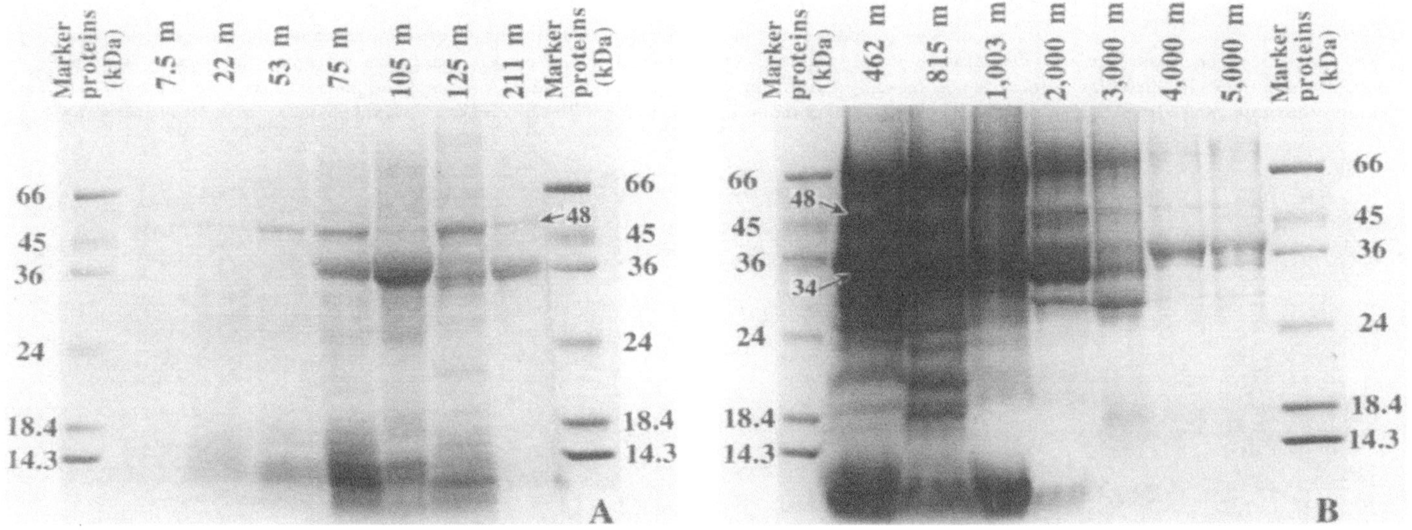

Fig. 17. Depth profiles of dissolved proteins from the surface through the deep waters at Station B. A ready-made continuous gel was used and proteins were stained with CBB-R. A. Amounts of samples from depths of 7.5, 22, 53, 75, 105, 125 and 211 m were equivalent to 1.0, 1.6, 1.5, 0.5, 0.8, 0.8 and 0.7 liter of the original seawater, respectively. B. Amounts of samples loaded on the gel were equivalent to 0.25 liter of the original seawater. Each marker protein was loaded at 1 and 2 μg in the left- and right-hand lanes, respectively, in A and B. Arrows represent bands that have been confirmed as single protein bands by the analysis of the amino-terminal amino acid sequence (Table 3). (Reprinted with permission from *Marine Chemistry*, **51**, E. Tanoue, Detection of dissolved protein molecules in oceanic waters, 1995, Elsevier Science Ltd.)

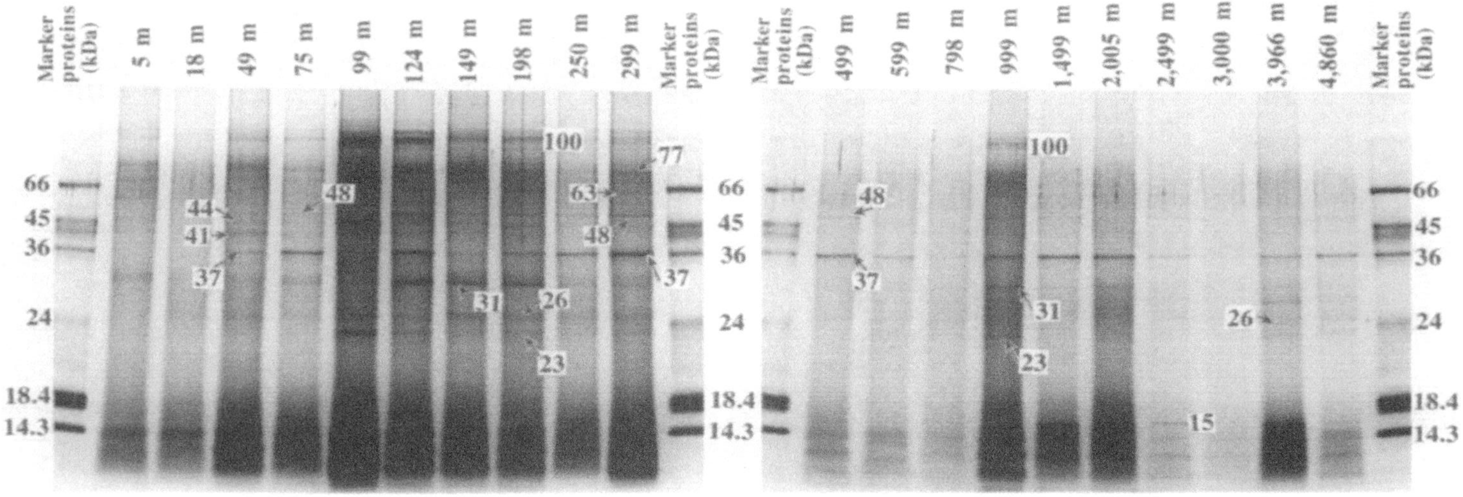

Fig. 18. Depth profiles of dissolved proteins at Station D$_2'$. A ready-made continuous gel was used and proteins were visualized by the silver-staining method. Amounts of samples loaded on the gel were equivalent to 100 ml of the original seawater in each case. Each marker protein was loaded at 25 and 50 ng in each left- and right-hand lanes of each gel. (Reprinted with permission from *Limnology and Oceanography*, **41**, E. Tanoue *et al.*, Discrete dissolved and particulate proteins in oceanic waters, 1996, American Soc. Limnol Oceanogr. Inc.)

protein, found as the major component in the surface water, was also found in the intermediate waters. The two samples from the greater depths of 4,000 m and 5,000 m, had lower levels of protein in the dissolved phase and relatively few proteins were visualized as bands on the gel. The 37 kDa-protein was the dominant protein below 4,000 m, and the 48-kDa protein was also found in the deep waters. High levels of dissolved protein were evident in the intermediate and deep waters at Station B.

Depth profiles of dissolved proteins at Stations D_2' and D_3, located in the South Equatorial Current, in the equatorial Pacific Ocean are shown in Figs. 18 and 19. The dissolved proteins had a wide range of molecular masses throughout the water columns. At Station D_2', fewer than 10 major bands were visible in the samples of the surface water from depths of 5 m to 75 m but the staining of each band was less intense. Between depths of 99 m and 299 m, the number of visualized bands increased and the gel was heavily stained. Below a depth of 499 m, the gel was less stained and the number of visualized bands also decreased. At Station D_3, the gel was less stained above a depth of 20 m and the number of visualized bands increased below a depth of 49 m. The staining intensities of individual bands in the surface waters were relatively low as compared with those in the intermediate water at Station D_3.

Although the number of visualized bands was different between the stations

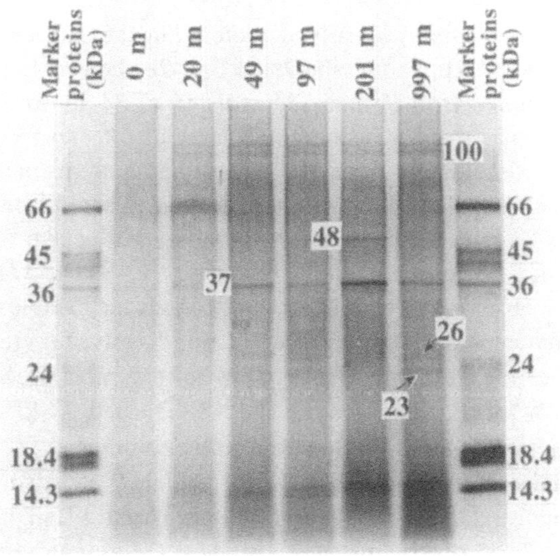

Fig. 19. Depth profiles of dissolved proteins at Station D_3. A ready-made continuous gel was used and proteins were visualized by the silver-staining method. Amounts of samples loaded on the gel were equivalent to 100 ml of the original seawater in each case. Each marker protein was loaded at 25 and 50 ng in the left- and right-hand lanes. (Reprinted with permission from *Limnology and Oceanography*, **41**, E. Tanoue *et al.*, Discrete dissolved and particulate proteins in oceanic waters, 1996, American Soc. Limnol. Oceanogr. Inc.)

and between the depths at each station, the electrophoretic patterns were similar to each other. Proteins with apparent molecular masses of approximately 63 kDa, 48 kDa, 44 kDa, 41 kDa, 37 kDa, 31 kDa, 26 kDa, 23 kDa and 15 kDa, were clearly visible at Station D_2' (Fig. 18). Particularly, proteins with apparent molecular masses of 48 kDa and 37 kDa were clearly visible as the major proteins throughout the water columns at both stations (Figs. 18 and 19). In the unresolved fraction, some bands with apparent molecular masses of more than 66 kDa were also found. For example, bands with apparent molecular masses of around 100 kDa, were found at Stations D_2' and D_3. Because their electrophoretic mobilities were close and because each band was not differentiated, they were designated collectively as 100-kDa protein. Protein with an apparent molecular mass of around 77 kDa, designated 77-kDa protein, was also found in the surface and intermediate waters at Station D_2'. Proteins with apparent molecular masses of less than 14.3 kDa were also found throughout the water columns at both stations.

Chl *a* maxima at three depth Stations B, D_2' and D_3 were found at depths of around 100 m, 50 m and 75 m, respectively. Concentrations of Chl *a* at each station decreased rapidly toward a depth of 300m, reaching zero below that depth, while relative high levels of dissolved proteins in the intermediate and deep waters were evident. Vertical distributions of Chl *a* and the dissolved proteins indicated that the levels of dissolved proteins did not correlate to those of Chl *a*.

Horizontal distribution of dissolved proteins in the surface waters from the Southern Ocean

Electrophoretograms of dissolved proteins in the surface water along the cruise track of the JARE 34 are shown in Fig. 20. Fewer than 30 bands were visualized in samples from cruise segments from 34-11 to 34-14, while the number of bands visualized was small in the case of samples from 34-1 to 34-10. The distribution of molecular mass of the dissolved proteins in Fig. 20 was quite similar to each other, even though samples were collected from quite different oceanic regions (Fig. 14). The major proteins did not appear to vary spatially. Bands of proteins with apparent molecular masses of 48 kDa, 37 kDa, and 15 kDa were commonly observed as the major bands in all the samples examined. The most striking feature was that the detected major dissolved protein species in the surface waters of the Southern Ocean were quite similar to those observed at depth stations in the Pacific Ocean. The 48-kDa and 37-kDa proteins were detected as major proteins throughout the water columns at Stations D_2' and D_3 of the equatorial region (Figs. 18 and 19), and they are also observed at the subarctic and subtropical Stations A and B (Figs. 16 and 17) in the Pacific Ocean. The 48-kDa and 37-kDa proteins were also detected as major proteins in all the samples of the surface waters along the cruise track. Other proteins detected at both stations (i.e., 100 kDa, 26 kDa, 23 kDa, and 15 kDa) in the Pacific Ocean, were also found in the samples from 34-11 to 34-14 (Fig. 20) in the Southern Ocean.

Figure 21 shows the concentrations of Chl *a* at the various sampling areas

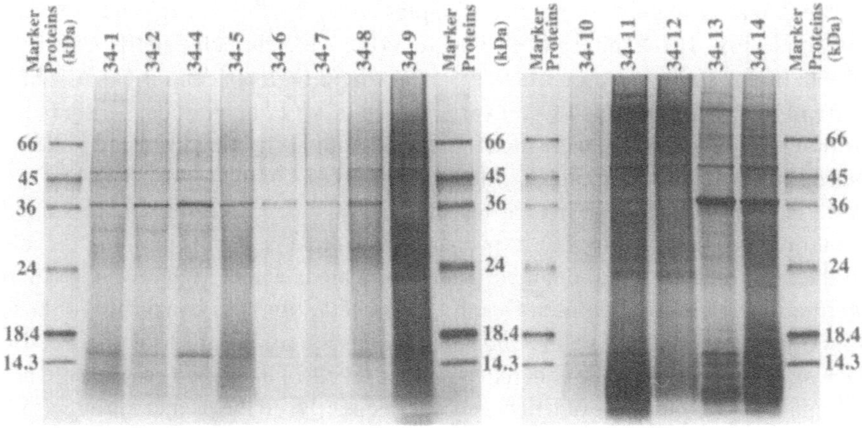

Fig. 20. Electrophoretograms of dissolved proteins in the surface water along the cruise track of the JARE 34. A ready-made continuous gel was used and proteins were visualized by the silver-staining method. Amounts of samples loaded on the gel were equivalent to 100 ml of the original seawater. Each marker protein was loaded at 25 and 50 ng in each left- and right-hand lanes of each gel. (Reprinted with permission from *Limnology and Oceanography*, **41**, E. Tanoue *et al.*, Discrete dissolved and particulate proteins in oceanic waters, 1996, American Soc. Limnol. Oceanogr. Inc.)

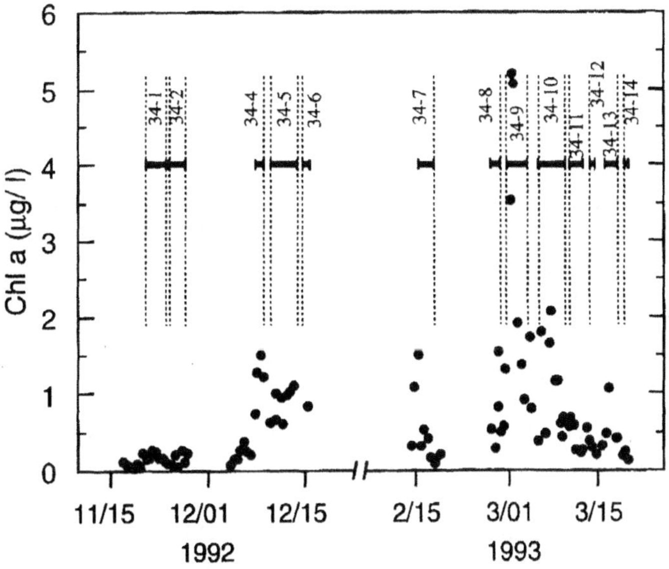

Fig. 21. Concentrations of Chl *a* along the cruise track of the JARE 34 and the range of locations from which composite samples of seawater for extraction of dissolved proteins.

along the cruise track of the JARE 34. Chl *a* concentrations were low in regions north of approximately 50°S, where samples 34-1, -2, -11, -12, -13 and -14 were collected (Fig. 14). Samples from 34-5 to 34-12 were collected in the Antarctic Ocean (areas south of approximately 55°S) where high concentrations of Chl *a* were observed but levels of Chl *a* varied by an order of magnitude. The gels for samples 34-13 and 34-14 were heavily stained, indicating that dissolved proteins were abundant in these samples, whereas the gels of samples 34-6, -7, -8 and -10 were less heavily stained, indicating that only low concentrations of dissolved protein were present (Fig. 20). It is clear that sample segments with high levels of the dissolved protein did not coincided with areas where high levels of Chl *a* were observed in the Southern Ocean. The observation was consistent to those found at depth stations in the subtropical and equatorial areas of the Pacific Ocean. Therefore, it can be concluded that the levels of dissolved proteins did not correlate directly to those of Chl *a*, and, consequently, to the primary production.

Partial amino-terminal amino acid sequence analysis

Among the protein molecules detected in the present study, some proteins were observed as common constituents of dissolved proteins in samples from a vast area that extended from the subarctic, subtropical and equatorial areas in the Pacific Ocean, through the Indian Ocean, to the Antarctic Ocean, regardless of locations and depths. Particularly, a protein with an apparent molecular mass of approximately 48 kDa was commonly observed as one of the major proteins in all samples examined. This protein was subjected to an analysis of its amino-terminal amino acid sequence, and the first amino-terminal 14 or 15 residues were identified (Table 3). The amino-terminal sequences of the 48-kDa protein from various depths at different sites were identical to each other. Table 3 indicates that the 48-kDa proteins visualized on the gels as bands were true proteins and that the protein of sample from each area was the same, even though samples were collected from quite different oceanographic locations. A homology search in the Protein Identification Resource International database (PIR-International, version 39.0), through the Japan International Protein Information Database (JIPID), indicated that this protein is a homolog of porin P from the Gram-negative bacterium *Pseudomonas aeruginosa* (Worobec *et al.*, 1988; Siehnel *et al.*, 1990; Siehnel *et al.*, 1992).

The other major proteins from each sample were also subjected to amino-terminal amino acid sequencing (Figs. 16 and 17, and Table 3). Successful sequencing analyses indicated that the various proteins visualized as major bands on the gels, were true proteins and single species. However, no similar amino-terminal amino acid sequences were identified in the PIR database. The resolution of one-dimensional electrophoresis is inadequate for the total separation of protein in a biological system more complex than a bacteriophage (O'Farrell, 1975). It is consistent that electrophoretograms of the particulate protein in the surface water as well as plankton and cultured bacterium, were smeared as a result of thousands of the background proteins. However, successful sequencing analysis after one-dimensional SDS-PAGE indicates that a limited number of dis-

Table 3. N-terminal amino acid sequences of the proteins in the dissolved phase. "Xaa" indicates that an identifiable phenylhydantoin derivative was not recovered. (Reprinted with permission from *Geochim. Cosmochim. Acta*, **59**, E. Tanoue *et al.*, Bacteria membrane: Possible source of a major dissolved protein in seawater, 1995, Elsevier Science Ltd.)

Cycle	mol.mass of protein	1	2	3	4	5	6	7	8	9	10	11	12	13	14	15
Station A (45°10.3'N, 165°34.4'E; depth, 5,934 m)																
45 m	48 kDa	Gly	Thr	Val	Thr	Thr	Asp	Gly	Ala	Asp	Ile	Val	Ile	Lys	Thr	
		Gln*														
	40 kDa	Thr	Val	Thr	Val	Thr	Pro	Leu	Met	Leu	Gly	Tyr	Thr	Phe	Gln	Leu
		Gly*														
	37 kDa	Ala	Asp	Val	Lys	Ile	Tyr	Gly	Arg	Ala	His	Val	Ser	Leu	Asp	Tyr
	30 kDa	Thr	Gln	Ala	Glu	Val	Gly	Ala	Ser	Ala	Gly	Leu	Ile	Asp	Pro	Asp
200 m	48 kDa	Gly	Thr	Val	Thr	Thr	Asp	Gly	Thr**	Asp	Ile	Val	Ile	Lys	Thr	Lys
		Ala*														
Station B (24°35.0'N, 170°00.1'E; depth, 5,966 m)																
211 m	48 kDa	Gly	Thr	Val	Thr	Thr	Asp	Gly	Ala	Asp	Leu**	Val	Ile	Lys	Thr	Lys
462 m	48 kDa#	Gly	Thr	Val	Thr	Thr	Asp	Gly	Ala	Asp	Xaa	Val	Ile	Lys	Thr	Lys
		Met	Lys	Asp	Gly	Leu	Val	Glu	Arg	Thr	Gln	Gly	Ser	Glu	Val	Asp
	34 kDa	Val	Thr	Gly	Gly	Tyr	Ala	Arg	Leu	Pro	Val	Glu	Leu	Tyr	Lys	
Station C (22°47.1'N, 158°04.6'W ; depth, 4744 m)																
49 m	48 kDa	Gly	Thr	Val	Thr	Thr	Asp	Gly	Xaa	Asp	Ile	Val	Ile	Lys	Thr	Lys
	39 kDa	Ala	Val	Val	Gly	Gly	Gly	Ala	Thr	Leu	Pro	Gln	Asn	Leu	Tyr	Asn
200 m	48 kDa	Gly	Thr	Val	Thr	Thr	Asp	Gly	Xaa	Asp						
	Porin P##	Gly	Thr	Val	Thr	Thr	Asp	Gly	Ala	Asp	Ile	Val	Ile	Lys	Thr	Lys

* This amino acid was also detected.

** The amino acid was detected but not at a significant level.

\# The protein in this sample with a molecular mass of 48 kDa was a mixture of two proteins, and two amino acids were detected at every step. One protein was identified as a homolog of porin P and no sequence resembling that of the other protein was found in the PIR database.

\## After Worobec *et al.* (1988) and Siehnel *et al.* (1990).

solved proteins present a relatively pure and high level, and they have not been reported previously with the exception of the homolog of porin P.

Sources of the dissolved proteins

The homolog of porin P

The Pseudomonadaceae are a very large and important family of Gram-negative bacteria. They comprise a substantial proportion of the microflora in soils, fresh water, marine environments, and many other natural habitats (Palleroni, 1992). *Pseudomonas* is one of the principal genera of bacteria represented in the ocean (Sieburth, 1979). Although strains of *Ps. aeruginosa* have been isolated from freshwater environments (e.g., Ogunseitan *et al.*, 1990), relatively little is known about the species that make up the community of bacterioplankton in the sea (Giovannoni *et al.*, 1990). Bacterial fauna and the occurrence of *Ps. aeruginosa* at the sampling sites are unknown. Schmidt *et al.* (1991) examined the phylogenetic diversity of the picoplankton community by analyzing the sequences of cloned ribosomal genes. They isolated bulk genomic DNA from picoplankton collected at station ALOHA, which is at almost the same location as Station C in the present study (Fig. 14), and they commented that some of the isolated rDNA clones were closely related to members of the r proteobacteria, such as *Alteromonas*, *Vibrio* and *Pseudomonas* spp. It is not necessary to postulate that the homolog of porin P is directly derived from a single pathogenic species of *Ps. aeruginosa*. However, strong homology between the amino-terminal amino acid sequences indicates that Gram-negative bacteria with genetic characteristics resembling those of *Ps. aeruginosa*, in terms of the biosynthesis of porin P, might be responsible for the production of the homolog of porin P found in the present study.

Porin P is a member of the class of proteins known collectively as porins. Porins from 32 species of bacteria have been identified and characterized. Porins are found in 10 separate families of Gram-negative bacteria and in the outer membranes of mitochondria and chloroplasts of eukaryotic cells (Zalman *et al.*, 1980; Lugtenberg and Alphen, 1983; Benz, 1985; Hancock, 1987). Porins form trans-outer-membrane and water-filled channels, and serve as a molecular sieve of the outer membrane of Gram-negative bacteria, permitting the uptake of small substrate molecules but excluding potentially harmful enzymes and other large hydrophilic molecules (Hancock, 1987; Nikaido, 1994). Porins fall into two classes. One consists of non-specific porins that form the channels that permit general diffusion of hydrophilic molecules below a certain size and whose dimensions determine the exclusion limit (usually less than 600 Daltons). The second consists of substrate-specific porins that act as channels for facilitated diffusion of specific substrates by virtue of specific substrate-binding sites in their channels.

Many non-specific porins are known but only seven substrate-specific porins have been characterized (Siehnel *et al.*, 1992; Nikaido, 1994) as follows: the maltose and maltodextrin-specific porin Tsx (Maier *et al.*, 1988) and the plasmid-encoded sucrose-specific ScrY (Schulein *et al.*, 1991) of *Escherichia coli*; and the imipenem/basic amino acid-specific porin OprD (Trias and Nikaido, 1990), the phosphate-specific porin OprP (Hancock *et al.*, 1982; Hancock and Benz, 1986), the glucose-specific porin OprB (Trias *et al.*, 1988) and the polyphosphate-specific OprO (Hancock *et al.*, 1992) of *Pseudomonas aeruginosa*. The amino-terminal amino acid sequence and molecular mass of the homologue of porin P found in the present study are identical to those of the phosphate-specific porin Opr P, known as porin P that is called Opr P after the gene designation (Siehnel *et al.*, 1990), and the polyphosphate-specific porin OprO of *Pseudomonas aeruginosa*, out of the seven substrate-specific porins (Hancock *et al.*, 1982; Hancock and Benz, 1986; Hancock *et al.*, 1992). Thus, the homolog of porin P extracted from seawater represents a unique type of porin.

In laboratory experiments, the synthesis of porin P is enhanced in wild-type cells of *Ps. aeruginosa* in response to phosphate limitation, and wild-type cells of *Ps. aeruginosa* do not produce porin P *in vitro* when they are grown in a medium that contains more than 0.62 mM phosphate, while they do produce porin P as a major outer-membrane protein in a medium that contains 0.2 mM phosphate (Hancock *et al.*, 1982). Porin P mediates the uptake of phosphate and forms the pores that are most effective in facilitating the diffusion of phosphate and small phosphorus-containing compounds (Poole *et al.*, 1987). The accumulation of polyphosphate is widespread among bacteria. Polyphosphate functions as a phosphorus-storage material and is utilized for the synthesis of nucleic acids and phospholipids under conditions of phosphorus starvation. Some microorganisms do not form any phosphorus-reserve material; one of these is *Ps. aeruginosa* (Gottschalk, 1986).

Since the homolog of porin P was found in seawater where the concentration of phosphate is always less than 5 μM (Redfield *et al.*, 1963), it is postulated that certain bacteria that form no phosphorus-storage material are likely to develop a phosphate-starvation-inducible, high-affinity, phosphate-uptake system for growing under the phosphate-limited conditions in the sea. The fact that synthesis of porin P is coregulated with that of periplasmic phosphate-scavenging enzymes, namely, alkaline phosphatase and phospholipase C (Hancock *et al.*, 1982), and the fact that periplasmic alkaline phosphatase activity is detected in estuarine and oceanic waters (Rath *et al.*, 1993) and in bacteria isolated from seawater (Martinez and Azam, 1993), seem to support the above prediction. It has not been well documented that bacteria in oceanic environments grow under phosphate-starved conditions. Recently, Pomeroy *et al.* (1995) showed phosphate to be the primary limiting factor for bacterial production and microbial community respiration in the Gulf of Mexico. It was also suggested that phosphorus is the principal limiting factor for bacterial production in the Sargasso Sea during summer (Ammerman *et al.*, 1994; Cotner *et al.*, 1994).

The apparent molecular mass of authentic porin P determined by SDS-PAGE and staining with CBB-R was 48 kDa (Poole and Hancock, 1986) and the apparent molecular masses of the homolog of porin P in the present study was approximately 48 kDa. The agreement of the amino-terminal sequences and the apparent molecular masses between the protein in the dissolved phase and porin P (Table

3) indicates that an almost complete homolog of porin P, from the amino-terminus to, probably, the carboxyl-terminus survives without modification in the water column. These results suggest that the homolog of porin P extracted from seawater very likely maintains its tertiary structure, because the conformation of a protein is governed by its amino acid sequence (Anfinsen, 1973). It is known that the trimeric form of porin P is resistant to proteases and no proteolysis is observed after intact bacteria are subjected to proteolysis (Worobec *et al.*, 1988; Siehnel *et al.*, 1990). Although no oligomeric form of the homolog of porin P has been recognized to date, the resistance of porin P could be one of the reasons why this protein is one of the major proteins in the water column (see below).

Proteins other than the homologue of porin P

No amino-terminal amino acid sequences similar to those of dissolved proteins with molecular masses of 30 kDa, 34 kDa, 37 kDa, 39 kDa and 40 kDa were retrieved from the PIR database. The amino-terminal amino acid sequences of dissolved proteins other than the homolog of porin P have not provided any source-related information to date. From the results of electrophoretic patterns and depth profiles of dissolved proteins, it was proposed that the bulk of the dissolved protein pool is not related directly to primary production. The majority of bacteria found in the sea are Gram-negative (e.g., Sieburth, 1979). Although, no porin in marine bacterioplankton has been studied, the Gram-negative bacteria examined in the laboratory have porins in their outer membranes as channel protein as mentioned above. The physical properties of the 44 porins examined to date are quite similars due to their similar functions. They have monomer molecular masses in the range of 28 kDa to 48 kDa and are present in membranes as oligomers, usually trimers (Hancock, 1986).

If we considered the general functions of porins, many porins with different molecular masses are very likely produced in the sea. However, previous studies of porins represent only a small percentage of the known species, and only 25% of the families of Gram-negative bacteria is examined. Such studies have also been strongly biased towards medically important pathogens (Hancock, 1987). The amino-terminal amino acid sequences of the porins of bacteria in the PIR database correspond to only nine species, namely, *Escherichia coli, Salmonella typhimurium, Neisseria spp., Pseudomonas aeruginosa, Brucella abortus, Comamonas acidovorans, Haemophilis influenzae, Bordetella pertussis* and *Klebsiella pneumoniae*. Therefore, failure to retrieve similar amino-terminal amino acid sequences from the database does not mean that dissolved proteins other than the homolog of porin P are not porins. In general, all porins are extremely resistant to proteases (Lugtenberg and Alphen, 1983; Cowan *et al.*, 1992) and no proteolysis of the porin P is observed when intact bacteria are subjected to proteolysis (Worobec *et al.*, 1988; Siehnel *et al.*, 1990). The resistance and similar physical properties of porins suggests that they may be preserved in seawater, thus other dissolved proteins observed in the present study could also be porins, but do not appear in the database. Immunochemical detection of porins and more detailed discussion are presented below.

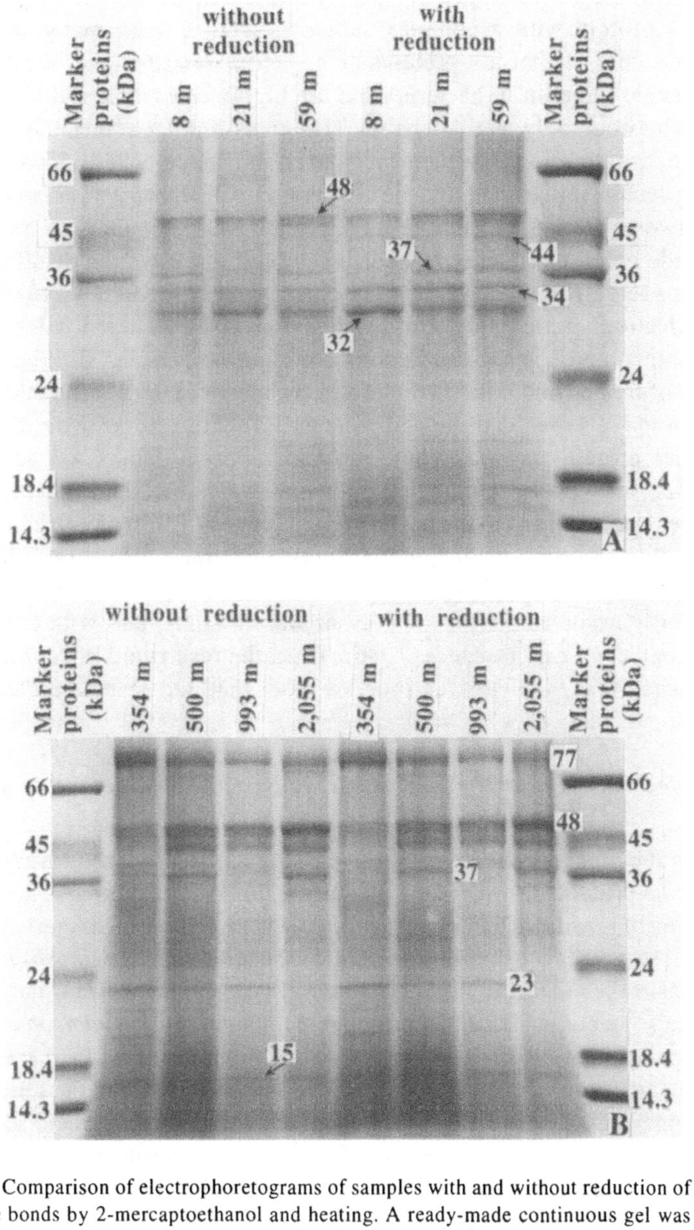

Fig. 22. Comparison of electrophoretograms of samples with and without reduction of disulfide bonds by 2-mercaptoethanol and heating. A ready-made continuous gel was used and proteins on the gels were stained with CBB-R. A. Samples of seawater were collected at subarctic Station A. The 48-kDa protein at different depths (45 and 200 m) was identified as a homolog of porin P (see Table 3). B. Samples of seawater were collected at subtropical Station C. The 48-kDa protein at different depths (49 and 200 m) was identified as a homolog of porin P (see Table 3). (Reprinted with permission from *Limnology and Oceanography*, **41**, E. Tanoue *et al.*, Discrete dissolved and particulate proteins in oceanic waters, 1996, American Soc. Limnol. Oceanogr. Inc.)

Disulfide bonds contribute greatly to the stabilization of the tertiary structure of protein with a complex subunit structure (e.g., many enzymes). An extreme case is fibrous proteins (e.g., actin, myosin, and keratin in many organisms, flagellin in bacteria) that are highly cross-linked through disulfide bonds between polypeptide chains. This cross-linking contributes to the resistance to denaturation (Haschemeyer and Haschemeyer, 1973). SDS and TCA are strong denaturing reagents that destroy non-covalent interactions among proteins or between protein and non-proteinaceous DOM, but they do not cleave the disulfide bond; a reducing reagent is required to cleave the disulfide bonds of proteins (e.g., Haschemeyer and Haschemeyer 1973; Andrews 1986).

Electrophoretic patterns of dissolved proteins with and without reduction with 2-mercaptoethanol and heating are compared in Fig. 22. Electrophoretograms of treated and untreated samples are identical, indicating that the 48-kDa protein and other major dissolved proteins did not interact with the disulfide bond to other proteins *in situ*. Thus, proteins whose subunits are aggregated by disulfide bridges are eliminated as being a possible source of the dissolved protein. Native porins in the bacterial membrane are present as trimers with non-covalent interaction; the major contribution to trimer formation is hydrophobic interaction (Cowan *et al.*, 1992). It is also known that porins are strongly but non-covalently associated with lipopolysaccharide (LPS) and with the underlying peptidoglycan (e.g., Hancock, 1986). Thus, the results did not conflict with the conclusion that the dissolved proteins other than the 48-kDa protein might be porins.

Immunochemical characterization of dissolved proteins and their source organisms

In the preceding section, two hypotheses were proposed for the sources of dissolved proteins. It was proposed that: 1) the 48-kDa protein is derived from a homolog of porin P of a marine bacterium whose genetic characteristics, in terms of the biosynthesis of porin P, resemble those of *Ps. aeruginosa*; and 2) porin proteins, other than the 48-kDa protein, that originated from marine bacteria could also occur as dissolved proteins. Although, the *Pseudomonaceae* are the principal genera of marine bacteria, as mentioned above, *Ps. aeruginosa* is a pathogenic bacteria and the occurrence of *Ps. aeruginosa* in the sea is unknown and somewhat questionable. Since no amino-terminal amino acid sequence of a dissolved protein other than the 48-kDa protein was retrieved from the database, sequence analysis has provided no source-related information about these proteins. Immunochemical approaches have been introduced for the detection of porin among dissolved proteins and to identify the source organism(s) of such porin in the sea.

The procedures for the immunochemical study have been reported elsewhere (Suzuki *et al.*, 1997) and are described here only briefly. A survey of porins among dissolved proteins was performed by Western blotting (Towbin *et al.*, 1979; Morita *et al.*, 1996). Two antisera were prepared for this purpose. One was an antiserum against an amino-terminal 14-mer of the 48-kDa protein. An attempt to produce an antiserum against the 48-kDa protein using the extracted and purified 48-kDa protein failed because of the difficulty in preparing sufficient 48-kDa protein from seawater. Since the amino-terminal amino acid sequence was known, a polypeptide that was identical from the amino terminus to the fourteenth amino acid residue of the 48 kDa-protein was synthesized and used as the antigen. The resulting antiserum (α-48DP N-14) react with the 14-meric fragment of the amino terminus of the 48-kDa protein. Antiserum against the porin protein (Omp35La) of *Listonella* (*Vibrio*) *anguillarum* was also prepared as described previously (Suzuki *et al.*, 1993). *L. anguillarum* is a well-known marine bacterium that is a fish pathogen. One of the outer membrane proteins, Omp35La, of *L. anguillarum* is a constitutively expressed porin and it has been characterized in terms of its biochemical and serological properties (Suzuki *et al.*, 1993, 1996). The antiserum (α-Omp35La) had a broad spectrum of immunoreactivity and it reacted with Omp35La of *L. anguillarum* and similar porins from bacteria that belong to the genus Vibrio and Photobacterium (Suzuki *et al.*, 1996).

Immunological cross-reactivities of the two antisera were examined with cultured marine bacteria. Proteins after SDS-PAGE of outer-membrane fractions of bacteria (Fig. 23A) were transferred electrophoretically to a nitrocellulose membrane and probed with α-48DP N-14 (Fig. 23B) and α-Omp35La (Fig. 23C). The antiserum α-48DP N-14 specifically detected porin P, with a molecular mass of 48 kDa, from *Ps. aeruginosa* (lane 1 in Fig. 23B). This antiserum reacted with proteins with molecular masses of 18 kDa from *Alteromonas macleodii* (lane 6) and of 78 kDa from *Aeromonas caviae* (lane 10). In the case of other tested bacteria, no positive bands were detected. In the case of the antiserum α-Omp35La, proteins with molecular masses of 33 kDa from *Vibrio parahaemolyticus* and of 40 kDa from *Photobacterium phosphoreum* were strongly positive (lanes 8 and 9 in Fig. 23C). These results are consistent with previous observations that α-Omp35La reacted with homologs of Omp35La from *Vibrio* and *Photobacterium* (Suzuki *et al.*, 1996). Several positive bands were also detected in analyses of *Ps. aeruginosa* (lane 1), *D. marina* (lane 3), *Al. macleodii* (lane 6) and *A. caviae* (lane 10). The immunoreactive proteins must have similar primary structure(s) in their molecules to that of Omp35La.

The cross-reactivities of the two antisera with dissolved proteins collected from the subarctic depth Station A (original electrophoretogram; Fig. 16) and from the surface waters in the Southern Ocean (original electrophoretogram; Fig. 20) are shown in Figs. 24 and 25, respectively. α-48DP N-14 reacted specifically with proteins of 48 kDa among the dissolved proteins collected from quite separate areas, the subarctic area of the Pacific Ocean and the Southern Ocean (Figs. 24B and 25B). Although a band of 48-kDa protein was visible on the gel after SDS-PAGE (Fig. 24A), no positive signal was detected with the sample from a depth of 5 m (Fig. 24B). The result might be due to an insufficient amount of protein in the assay. α-Omp35La reacted with a 34-kDa protein on samples from depths of 50 m and 200 m (Fig. 25C). This antiserum also reacted with the 48-kDa proteins. However, the signals were weak. An examination of the cross-reactivities of the two antisera with the outer-membrane proteins of cultured

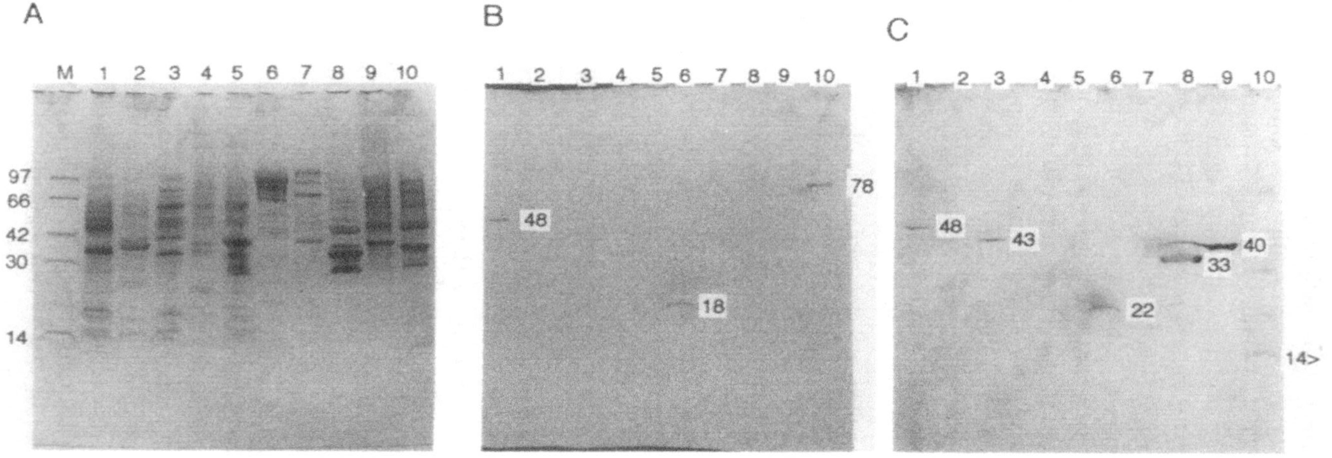

Fig. 23. Electrophoretograms (A) and Western blots with the α-48DP N-14 antiserum (B) and the α-Omp35La antiserum (C) of outer membrane proteins from 10 strains of cultured bacteria. Lanes: 1, *Pseudomonas aeruginosa*; 2, *Ps. nautica*; 3, *Deleya marina*; 4, *Marinomonas communis*; 5, *M. vega*; 6, *Alteromonas macleodii*; 7, *Al. haloplanktis*; 8, *Vibrio parahaemolyticus*; 9, *Photobacterium phosphroeum* and 10, *Aeromonas caviae*. A ready-made continuous gel was used and membrane proteins on the gel were stained with CBB-R (A). Proteins were immunostained with antisera and horseradish peroxidase-linked second antibodies after Western blotting (B and C). Molecular masses of positive proteins on Western blots are indicated in kDa. M in the left-hand lane represents the molecular markers. (Reprinted with permission from *Marine Ecology Progress Series, 158*, Suzuki *et al.*, Immunochemical detection of dissolved proteins and their source bacteria in marine environments, 1997, Inter-Research.)

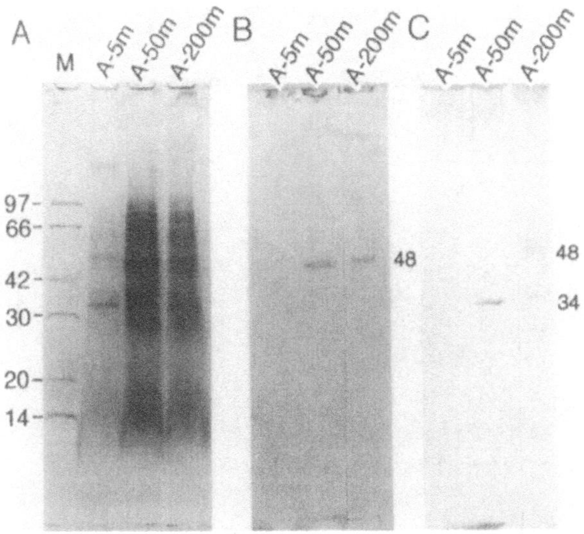

Fig. 24. Electrophoretograms (A) and Western blots with the α-48DP N-14 antiserum (B) and the α-Omp35La antiserum (C) of dissolved proteins at Station A. A ready-made continuous gel was used and dissolved proteins on the gel were stained with CBB-R (A). Proteins were immunostained with antisera and horseradish peroxidase-linked second antibodies after Western blotting (B and C). Molecular masses of positive proteins on Western blots are indicated in kDa. M in the left-hand lane represents the molecular markers. (Reprinted with permission from *Marine Ecology Progress Series*, *158*, Suzuki *et al.*, Immunochemical detection of dissolved proteins and their source bacteria in marine environments, 1997, Inter-Research.)

bacteria and with dissolved proteins extracted from seawater, revealed that the antiserum raised against amino-terminal amino acid sequences of the 48-kDa protein interacted with porin P of *Ps. aeruginosa* and the 48-kDa protein in the dissolved phase, regardless of the oceanic region and depth. The interaction between α-Omp35La and dissolved proteins strongly suggests that there are some porins other than the 48-kDa protein among the dissolved proteins, which might be derived from porins of marine bacteria, in particular members of the genera *Vibrio* and/or *Photobacterium*.

An attempt was made to identify the source bacteria (bacterium) of the 48-kDa protein by indirect fluorescent-antibody (FA) staining with α-48DP N-14. Since bacteria living in open-sea habitats are too small to be recognized by their cellular shape, a direct viable count method (DVC, Kogure *et al.*, 1979) was used to enumerate the antiserum-reactive bacteria. Samples of seawater were incubated at 20°C for 6 hr in darkness after the addition of yeast extract and nalidixic acid which inhibit DNA synthesis leaving other biosynthetic pathways to operate. The DVC affords an increase of bacterial cellular volume to easily detect minute bacterial particles. Then the bacteria of which cellular margins had been stained with fluorescent dye-conjugated secondary antibodies in α-48DP N-14, were

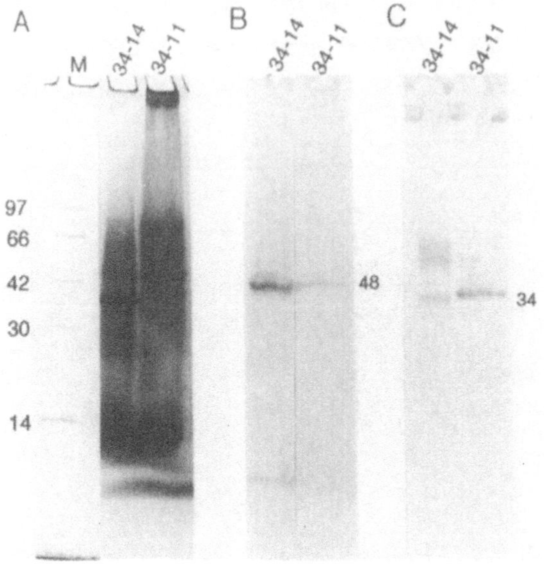

Fig. 25. Electrophoretograms (A) and Western blots with the (α-48DP N-14 antiserum (B) and the α-Omp35La antiserum (C) of dissolved proteins in surface waters at JARE 34. A ready-made continuous gel was used and dissolved proteins on the gel were stained with CBB-R (A). Proteins were immunostained with the antisera and horseradish peroxidase-linked second antibodies after Western blotting (B and C). Molecular masses of positive proteins on Western blots are indicated in kDa. M in the left-hand lane represents the molecular markers. (Reprinted with permission from *Marine Ecology Progress Series, 158*, Suzuki *et al.*, Immunochemical detection of dissolved proteins and their source bacteria in marine environments, 1997, Inter-Research.)

counted under a fluorescence microscope. Both total bacteria and DVC were also enumerated by epifluorescence microscopy after staining with a DNA-specific fluorochrome (DAPI, Porter and Feig, 1980).

The results are summarized in Fig. 26 for four stations, namely, Station 3 (34°42.1′ N, 137°07.0′ E), Station 4 (34°34.0′ N, 137°09.5′ E), Station 7 (35°01.6′N, 138°37.5′E) and Station 8 (34°49.9′N, 138°26.0′E). Bacterial cells that reacted with the α-48DP N-14 antiserum were detected at all stations and at all depths examined. Their abundance was approximately two orders of magnitude lower than that of DVC and approximately four orders of magnitude lower than the total bacterial counts, which were estimated by the routinely employed DAPI method. Nalidixic acid is effective only against susceptible Gram-negative bacteria and some Gram-negative bacteria are resistant to the antibiotic (Roszak and Colwell, 1987). Nalidixic acid-resistant mutants of *Ps. aeruginosa* strains are also isolated from the freshwater environments (Ogunseitan *et al.*, 1990), potentially causing underestimations of the numbers of DVC and of antiserum-reactive bacteria. Vertical profiles of antiserum-reactive bacteria were close to those of the DVC or the total number of bacteria. However, it is of interest that subsurface maxima of the antiserum-reactive bacteria were found in the lower euphotic layer

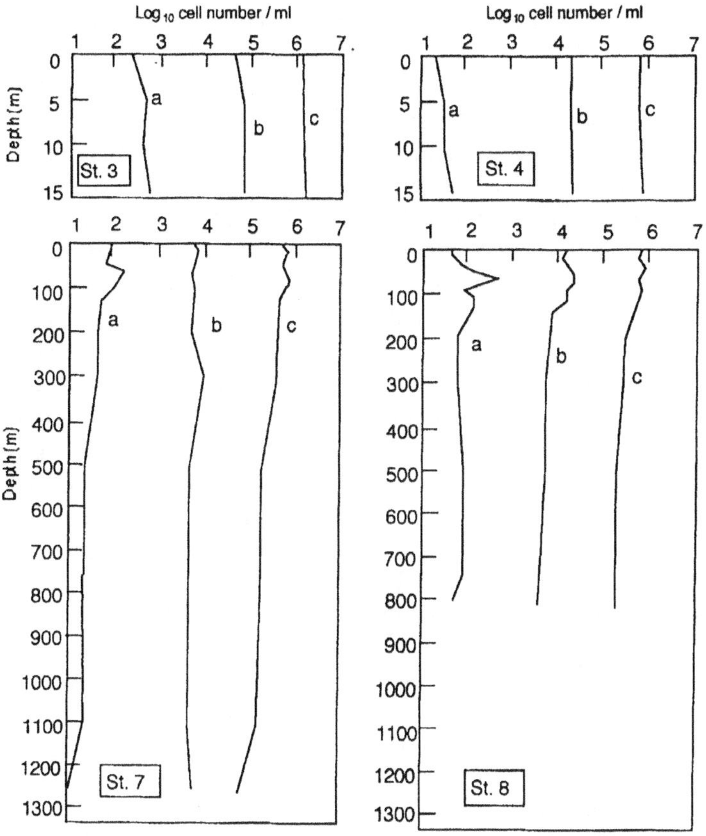

Fig. 26. Vertical distributions of numbers of the α-48DP N-14-reactive bacteria (a), of the direct viable count method (DVC, b) and of total bacteria (c) at Stations 3, 4, 7 and 8 during the KT 96-2 cruise of R/V *Tansei-Maru* (Ocean Research Institute) from January to February 1996. (Reprinted with permission from *Marine Ecology Progress Series*, *158*, Suzuki *et al.*, Immunochemical detection of dissolved proteins and their source bacteria in marine environments, 1997, Inter-Research.)

(75 m depth) at Stations 7 and 8. It is now clear that the bacteria (bacterium) that produce the outer-membrane protein that reacts with the antiserum raised against the 48-kDa dissolved protein, occur in the natural bacterial assemblage in marine environments.

Abundance of dissolved proteins

In preliminary recovery experiments, standard proteins added to seawater were not recovered or were poorly recovered in the absence of desalting or after desalting with deionized water at the end of the crude concentration step (Fig. 15). Only when a solution of ammonium bicarbonate that contained 0.01% SDS was used as the desalting buffer solution were the standard proteins recovered. The dissolved proteins were obtained only when the crude concentrate was prepared with the addition of SDS and no protein was clearly detectable when the sample was prepared without SDS. Therefore, SDS appears to be essential for the extraction of dissolved proteins from seawater.

The anionic long-chain amphiphile SDS the binds avidly to lipids, as well as to proteins (Helenius and Simons, 1975). Upon the addition of SDS to seawater, SDS binds to proteins, lipids and other polar and apolar dissolved organic materials in the seawater. SDS might form various mixed micelles, such as micelles of protein-SDS-lipid or of protein-SDS-other dissolved constituents, and so on, during preparation of the crude concentrate (Fig. 15). Since the micellar mass of SDS in a solution of 0.5 M NaCl is 36 kDa (Helenius and Simons, 1975), it is possible that mixed micelles of SDS and not only proteins but also non-protein organic constituents of the DOM are concentrated during the crude concentration step. Monomers and various mixed micelles of SDS might prevent the selective adsorption of dissolved proteins to the ultrafiltration membrane and to the inside of the tubing of the ultrafiltration system during the crude concentration. Thus, dissolved proteins in the samples with SDS would be recovered while those in samples without SDS would not.

While the formation of mixed micelles after the addition of SDS is taking place, the crude concentrate become a cloudy slurry and its sometimes separated into three layers after centrifugation during precipitation with trichloroacetic acid (TCA, Fig. 15). A pellet was formed at the bottom of the centrifuge tube, there was an intermediate layer, and there was a layer of low-density material at the surface of the supernatant. An appreciable amount of protein was eventually found to be associated with the low-density materials. Since mixed micelles with incorporated proteins could not be completely separated by centrifugation, the precipitation of dissolved proteins by treatment with TCA was incomplete and led to a significant loss of dissolved proteins. Recoveries of six standard proteins by the procedure that was used for the detection of dissolved proteins in seawater ranged only from 21 to 56% (average, 33%). The yield of dissolved proteins is not known. It is clear that methodological improvements are required for future quantitation of dissolved protein and, at present, quantitative estimation of the concentration of dissolved protein is impossible. A first-order approximation of the abundance of dissolved protein is, however, possible at this time. The lower limit of detection of protein by the Coomassie Brilliant Blue R250 (CBB-R) staining method corresponds, in general, to about 1 μg of protein per band (e.g., Andrew, 1976) and the intensity of staining is linear over a range from 0 to 10 μg of protein per band on a gel (Fig. 2). Since, in our experiments, any given protein whose level on the gel was below approximately 0.1 μg was not visualized as a single band, the concentrations of major proteins might exceed 0.1 μg per liter of the original seawater, if we compare the intensities of staining of dissolved proteins and marker proteins on the same gel in Figs. 16 and 17. Levels of total dissolved protein in the surface water at Station A were tentatively estimated by densitometry to be within a range of 12~38 μg protein/l (equivalent to 0.14~0.43 μg at.N/l) from the relationship between the total intensity of staining and total amount of standard marker proteins on the same gel. This value accounts for roughly 30% of DCAA, if we assume that the nitrogen content of dissolved protein is 16% and the level of DCAA ranges from 0.5 to 1.5 μg at.N/l in oceanic waters (Sharp, 1983), even though DCAA has not been determined by the most modern methodology (e.g., Keil and Kirchman, 1991). The value also accounts for 2 to 12% of the total dissolved organic nitrogen (DON) at Station A, where levels of DON ranged from 2.7 to 9.1 μg at.N/l (M. Yanada, personal communication).

Our first-order approximation suggests that the dissolved protein molecules make a major contribution to the oceanic pool of organic nitrogen. From the separation of dissolved proteins by SDS-PAGE, a first-order approximation is also possible for levels of specific proteins. The 48-kDa protein was found to be one of the major dissolved proteins in the surface waters from vast oceanic areas stretching from the North Pacific Ocean, through the equatorial Pacific and the Indian Oceans, to the Antarctic Ocean (Figs. 16, 17, 18, 19 and 20). The 48-kDa protein was also a major dissolved protein species throughout all water columns examined (Figs. 17, 18 and 22B), and the level of this protein was more than 0.1 μg/l in each case. If this protein occurs throughout the entire ocean, the total mass of the protein can be calculated to be more than 10^{14} g. This remarkable result suggests that the amount of this single protein molecule is equivalent to or greater than the total living biomass of zooplankton or living bacteria in the sea (Cauwet, 1978).

The importance of bacteria not only as consumers but also as producers of organic matter in the sea has emerged from first-order approximations of the levels of dissolved proteins. The bacterial biomass has been reported to be greater than the phytoplankton biomass (Fuhrman *et al.*, 1989; Cho and Azam, 1990; Caron *et al.*, 1995) or approximately equal to it (Li *et al.*, 1992; Garrison *et al.*, 1993; Kirchman *et al.*, 1993). In general, the ratio of bacterial to phytoplankton biomass increases with decreasing phytoplankton biomass, namely, the value is low in eutrophic regions and increases considerably in oligotrophic environments (Simon *et al.*, 1992). The large bacterial biomass is potentially important as a source of non-living organic material. However, the primary stage of production of organic matter is photosynthesis. About 50% of primary production is transferred to the secondary production by bacteria (see the Introduction) and is transformed into the bacterial biomass, with bacterial respiratory losses along the way. Therefore, the flux of organic matter from the primary production to the sea must be greater than the role of bacterial secondary production. The bacterial biomass itself is greater in eutrophic regions than in oligotrophic regions, and it is also high in the surface water and decreases with depth. Thus, a large biomass of bacteria is not necessarily the key factor which controls the abundance of dissolved proteins in the water column. A knowledge of the nature of the organic constituents of bacteria and their transfer processes is critically important for an understanding of the molecular distribution of the proteins characterized in the present study.

SOURCES AND TRANSFER PROCESSES OF PARTICULATE AND DISSOLVED PROTEINS FROM MARINE ORGANISMS

Marine organisms produce a huge variety of proteins. Indeed, patterns after SDS-PAGE of particulate proteins in surface waters and in entire organisms reveal an array of smears. These proteins are the background proteins and they are the cellular constituents of living organisms. They can be expected to be transferred to non-living organic pools through various processes. However, the electrophoretograms of dissolved proteins in the entire water column and of particulate proteins in the intermediate and deep waters are quite different from those of living organisms. The background proteins are not evident in such samples and only a limited number of proteins are evident. It can be concluded, therefore, that the majority of proteins produced by organisms are readily remineralized and do not accumulate, while very specific proteins are selectively transferred to and accumulate in the non-living organic pools.

It has been proposed that the majority of DOM exists as "humic" substances (see Introduction) although only 5–15% of the total DOM has been isolated and characterized as humic substances (Thurman, 1985). The humic substances are thought to be a complex mixture of "heteropolycondensate" molecules that have been formed by the decomposition and copolymerization of organic constituents of organisms by biological and non-biological processes. The present observations suggest another pathway for the production of detrital POM and DOM, wherein selected protein from particular marine organisms are transferred to detrital POM and DOM and accumulate by virtue of their intrinsic properties.

Largeau and co-workers demonstrated, from comparative chemical and morphological studies of fossil remains, that biomacromolecules of extant organisms are preserved without any or with only minor transformation during the processes of sedimentation and diagenesis (e.g., Largeau *et al.*, 1986, 1991; Tegelaar *et al.*, 1989). Such biomacromolecules are selectively preserved because of their nonhydrolyzable and refractory nature, which is based on a network of long, saturated hydrocarbon chains, in the cell walls or protective envelopes of organisms (Largeau, 1995). Proteins are a different type of biomacromolecule altogether. The specific proteins in both the particulate phase and the dissolved phase were denaturated by SDS. Moreover, these proteins, once denaturated and transferred to a membrane after SDS-PAGE, released amino-terminal amino acids upon Edman degradation. The experimental results indicated that the specific proteins were not recalcitrant biopolymers and, thus, the persistence of these proteins might not be explicable simply by the "selective preservation" hypothesis.

Specific proteins were observed in both the particulate and dissolved phases, which, by definition, are separated by filtration through a GF/F glass-fiber filter. However, the electrophoretograms of dissolved and particulate proteins were quite different. The background proteins were not evident among the dissolved protein, even in the productive surface waters (e.g., Figs. 16 and 20). Moreover, in terms of the specific proteins themselves, there were differences between the non-living particulate and dissolved protein pools. For example, the POM sample designated KH91-6 (Fig. 1) was collected at the same location, namely Station B, as that at which the sample used for extraction and analysis of dissolved proteins was collected (Fig. 14). At this location, the 45-kDa protein was clearly detected in the surface POM (Fig. 11A, sample No. KH91-6), while the 48-kDa and 37-kDa proteins were the major dissolved protein components in the surface water (Fig. 17A).

The 48-kDa protein, one of the most common protein species in the dissolved phase, had an electrophoretic mobility close to that of the 45-kDa protein in the particulate phase. However, the amino-terminal amino acid sequence of the 45-kDa protein (Table 2) differed from that of the 48-kDa protein in the dissolved phase (Table 3). It should be noted that a homology search is effective for any part of a polypeptide chain. No amino acid sequence similar to the amino-terminal amino acid sequence of the 45-kDa protein was found in any part of the peptide chain of the 48-kDa protein in the dissolved phase and *vice versa*. Therefore, the 45-kDa protein in the surface POM and the 48-kDa protein in the dissolved phase must be different proteins. Six major dissolved protein species, other than the 48-kDa protein, were also distinct from the 45-kDa protein in the particulate phase. The specific proteins in the POM appear to be different from those in the DOM, although the respective specific proteins in both the POM and the DOM were ubiquitous.

The clear difference in the occurrence of the specific proteins in the POM and DOM led to a further hypothesis that the sources and/or transfer processes of specific proteins differ between proteins in the POM and the DOM. In this section, the sources and transfer processes of specific proteins, in particular, the 45-kDa protein in the particulate phase and the 48-kDa protein in the dissolved phase, are discussed.

Sources and transfer processes of specific proteins in the POM

The question now arises as to why the 45-kDa protein in the POM was not detected among the dissolved proteins. The levels of specific proteins in the particulate phase might be at least one order of magnitude lower than those of the major dissolved proteins. Amounts of samples KH91-4, -5 and -6 that were used for the detection of specific proteins were equivalent to 8.7–12 liters of original seawater, while the specific dissolved proteins in surface waters were detected from the equivalent of 1 liter of original seawater (Figs. 16 and 17) at the same locations as those at which samples of POM designated KH91-1 and -6 were collected. Since specific proteins in the particulate phase are minor components in the water column, it is possible that the 45-kDa protein might have been present in the dissolved phase but that it was present below the limit of detection when we examined the dissolved proteins. The question also arises as to why the 48-kDa protein was not detected in the particulate phase. Differences in sources and in the *in situ* physical states of specific proteins between the dissolved and particulate phases provide possible explanations for the differences in the molecular profiles of particulate and dissolved proteins. In the present study, complete denaturation and reduction of any disulfide bonds in the particulate proteins were accomplished by heating with SDS and 2-mercaptoethanol. Therefore, covalently linked polypeptide subunits were effectively separated and any disulfide bonds or other noncovalent bonds, such as ionic and hydrogen bonds, and apolar interactions were disrupted. Although we have no information about the possible oligomeric form of the 45-kDa protein in the particulate phase, the size (diameter) of the native form of the 45-kDa protein cannot exceed more than several dozen nanometers and must be less than 0.1 μm, since the length of a fibrous protein such as myosin (molecular mass, 500 kDa) is less than 140 nm (Haschemeyer and Haschemeyer, 1973). Since the 45-kDa protein was retained on the GF/F filter (nominal pore diameter, 0.7 μm), the "native" 45-kDa protein *in situ* might interact noncovalently with other macromolecules that are large enough to be retained on a GF/F filter. By contrast, the major dissolved proteins might not interact with such macromolecules *in situ* and, thus, these proteins would pass through the GF/F filter during filtration.

Laursen *et al.* (1996) demonstrated, using a biomimetic approach with proteolytic enzymes, that detrital POM in estuarine waters might be selectively enriched for membrane fragments of cells with associated proteins, since lower rate constants for proteolysis were obtained for the membrane fraction than for the cytoplasmic fraction after the physical separation of subcellular fractions of phytoplankton cells. There are also indications that proteins in membranes or cell walls are incompletely assimilated by protozoa, zooplankton and deposit feeders (Nagata and Kirchman, 1992; Mayer *et al.*, 1995) because such proteins are protected from proteolytic enzymes (Walker and Fearnley, 1986). Membrane proteins of organisms in surface waters seem to be the probable sources of the specific proteins in the detrital POM.

The 48-kDa protein in the dissolved phase was identified as a porin, a membrane protein formed in Gram-negative bacteria. If the specific proteins, including the 45-kDa protein, in the particulate phase are similar to proteins derived from bacteria, we might have expected their behavior in the water column, as well as during the process of filtration, to be similar. Since the 45-kDa protein was only observed in the surface POM but not in POM from deep waters and, moreover, since the physical state of the 45-kDa protein might be different *in situ* from that of dissolved proteins, the source of the 45-kDa protein in POM appears not to be similar to that of dissolved proteins. A membrane protein of an organism that is not a bacterioplankton but commonly lives in surface waters seems to be the likely source of the 45-kDa protein in the POM. This suggestion does not, however, exclude other possibilities. For example, a process might occur whereby cytoplasmic proteins that have been incorporated into the detrital macromolecules become refractory to degradation. Such processes were proposed by Keil and Kirchman (1994) for the formation of refractory DOM, as mentioned above, and by Mayer and co-workers for the sedimentary pool of large oligopeptides (>7–15 amino acid residues) that are protected by some kind of matrix (Mayer *et al.*, 1986, 1995).

We should bear in mind that sources and transfer processes seem to be dissimilar, even among specific proteins in the POM. The 66-kDa protein was observed, albeit not at significant levels in the surface POM. This protein was prevalent in the POM from intermediate and deep waters where only bacteria are the essential biota. However, the 66-kDa protein was not found as a major component of the dissolved protein and, as mentioned in the discussion on the source of the 45-kDa protein, bacteria might not be a source of the 66-kDa protein in the intermediate and deep waters. The possible source and transfer processes, namely, formation *in situ* by the coagulation of dissolved protein, production *in situ* by animals at a given depth, or derivation from remnants of biological tissues that have migrated from the surface, can be proposed but they seem to provide an unsatisfactory explanation for the occurrence of the 66-kDa protein. At present, the distributional characteristics of the 66-kDa protein in the POM remain an enigma.

All the processes discussed here do not give us an answer to the question of why only a small number of proteins appears to be immune to biological attack and to survive in the POM without obvious modification. Even if we restrict possible candidates for the source of the specific proteins to membrane proteins, we know that a variety of membrane proteins are produced in surface waters. The nature of the proteins themselves, as well as unknown particular processes, might be involved in the distribution of specific proteins in the POM. A similar question also arises from the molecular distribution of the dissolved proteins (see below).

Sources and transfer processes of dissolved proteins

Three factors might control the levels and specific identity of dissolved proteins: 1) the abundance of source organisms; 2) the transfer processes from the source organisms to the DOM; and 3) the resistance of transferred proteins to biological degradation.

The 48-kDa protein was identified as a homolog of porin P in a specific species of Gram-negative bacterium, *Pseudomanas aeruginosa*. The results of indirect fluorescent-antibody staining indicated that a very minor component of the natural bacterial assemblage (1% to 0.1% of the total number of bacteria; Fig. 26) might be the source of the 48-kDa protein, one of the major dissolved proteins. Thus, a large flux of the 48-kDa protein to the DOM due to the high abundance of the source bacteria seems unlikely to control the abundance of the dissolved protein.

The resistance of porin proteins to enzymatic hydrolysis might be important for the survival and accumulation of the 48-kDa protein in the sea. Immunochemical studies indicated that some dissolved proteins, other than the 48-kDa protein, were also porins derived from bacteria, probably, *Vibrio* and/or *Photobacterium*. Gram-negative bacteria are the major components of marine bacterioplankton and every Gram-negative marine bacterium might have its own porin(s) whose physical properties might be quite similar to those of other porins. Thus, we might have expected many kinds of porin with different molecular masses to occur in seawater. However, the fact that we found fewer than 30 detectable bands of proteins indicates that the majority of porins that are produced

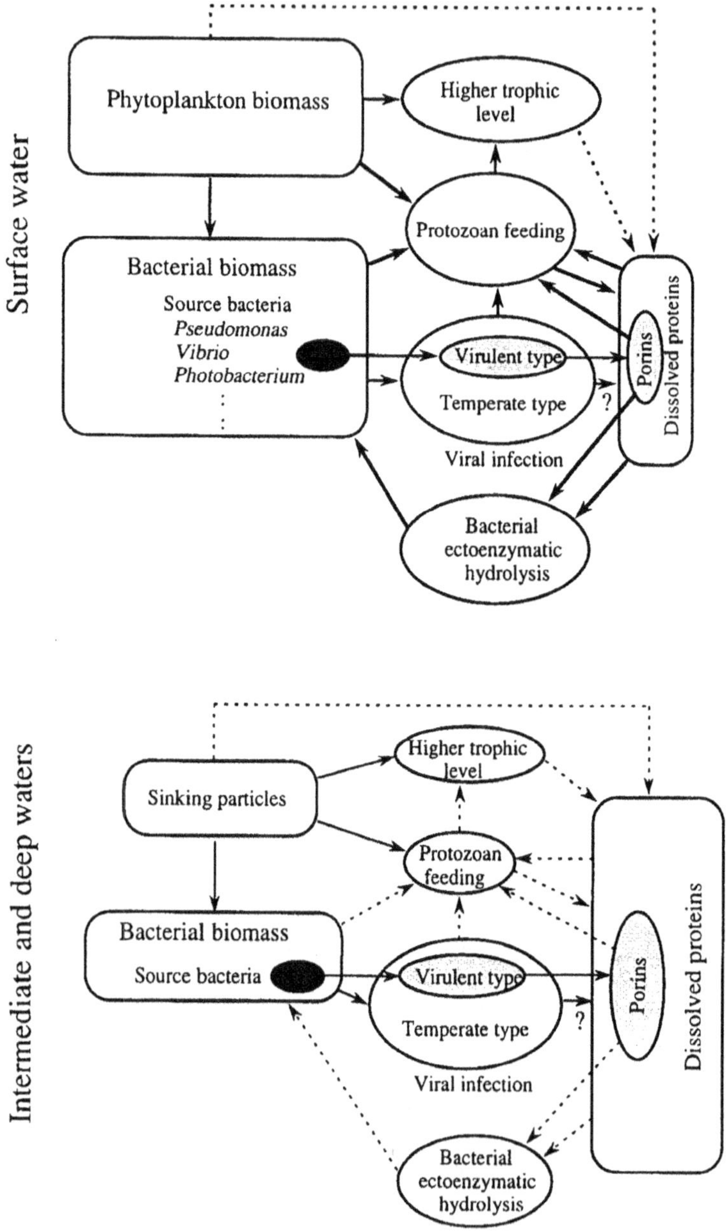

Fig. 27. Possible processes for transfer of dissolved proteins from bacteria to the pool of dissolved proteins. Broken arrows represent less significant processes (see text for details).

in the sea are degraded and do not accumulate. Resistance of porins to proteases or microbial attack is insufficient to explain the present observations. Thus, this factor cannot be the only one that controls the distribution of dissolved proteins. The process by which a particular porin of a particular bacterium is transferred to the dissolved protein pool might be the key to the control of the abundance of particular porin proteins in seawater.

Possible processes controlling the abundance of dissolved proteins are summarized schematically in Fig. 27. The grazing by protista has been proposed to be the major cause of loss of bacteria in aquatic environments and it might function as a direct link in the flow of energy from the DOM to higher trophic levels (i.e., organic matter flows from the DOM through bacteria to protista to higher trophic levels; e.g., Haas and Webb, 1979; Porter *et al.*, 1985; Tanoue and Hara, 1986; Sherr, 1988; Flood *et al.*, 1992; Tranvik *et al.*, 1993; Marchant and Scott, 1993). This process also releases bacterial components to the DOM (e.g., McManus and Fuhrman, 1988; Pace, 1988; Jumars *et al.*, 1989; Koike *et al.*, 1990; Nagata and Kirchman, 1992; Tranvik, 1994, see the review by Nagata, 1998, in this volume). Nagata and Kirchman (1992) proposed the hypothesis that heterotrophic flagellates release incompletely digested prey materials as colloidal "fecal pellets", in which bacterial membranes and other cellular materials are included.

Viruses have also been recognized as important agents in the mortality of bacteria (see, for example, the review by Fuhrman and Suttle, 1993, and references therein). Individual infected bacterial cells lyse and release cell debris that inevitably includes membrane proteins. Fuhrman and Noble (1995) estimated that the relative significance of viruses and protists in bacterial mortality is similar in coastal seawater from mesocosm experiments. Thus, viral lysis and grazing by protista are the most probable processes that mediate the transfer of porins from bacteria to the non-living dissolved phase. In terms of the vertical distribution, levels of organic matter are, in general, high in surface water and they decrease with depth because the site of the primary input of organic matter in the sea is the surface water. The relatively high levels of dissolved proteins in the intermediate and deep waters indicate that processes that control the levels of dissolved proteins could differ between the surface and deep waters (see below). For convenience, possible processes controlling the dynamics of dissolved proteins are discussed separately for surface waters and deep waters.

Dynamics of dissolved proteins in surface waters

The rate of production of porin proteins by bacteria is expected to be high in the surface water and two processes, namely, grazing by protista and viral lysis, for the transfer of bacterial proteins to dissolved proteins are also likely to be active in the surface water for the following reasons.

1) In general, there are significant correlations between the relative abundance of bacteria and chlorophyll in the open ocean, and the bacterial biomass is high in the surface water and decreases with depth (e.g., Bird and Kalff, 1984; Ducklow and Carlson, 1992).

2) The abundance of viruses parallels that of bacteria, with the abundance decreasing from the surface to the deep waters (e.g., Fuhrman and Suttle, 1993; Maranger and Bird, 1995).

3) The relative abundance of protista is also high in surface water and decreases with depth (Patterson *et al.*, 1993), although the vertical distribution of protista in open ocean habitats remains to be studied in detail (e.g., Fournier, 1966).

The proposed high rate of production and active transfer processes of dissolved proteins in surface waters are also supported by observations at the subarctic and subtropical Stations A and B, the sources of sample from which dissolved proteins were extracted (Fig. 14). The numbers of bacteria and viruses were high in the surface waters and decreased by one order of magnitude in the deep water at the two stations (Hara *et al.*, 1996). Total bacterial counts, DVC counts and the number of bacteria that reacted with the (α-48DPN-14 antiserum raised against the 48-kDa proteins were also higher in the surface water and decreased with depth (Fig. 26). Since the rate of production and the flux of the 48-kDa protein from bacteria to dissolved proteins were relatively high in the surface water, we might expect that concentrations of dissolved proteins would be high in surface waters. As mentioned above, however, depth profiles of dissolved proteins at Stations B and D_2' indicated that levels of individual proteins in the surface waters were relatively low as compared with those in the intermediate and deep waters (Figs. 17 and 18). Horizontal distributions of dissolved proteins also indicated that the abundance of dissolved proteins was not directly correlated with primary production at a given site (Figs. 20 and 21). Porins transferred from bacteria to the dissolved phase should, therefore, be degraded and should not accumulate significantly in surface waters.

The porin proteins are not truly dissolved proteins because of their rigid structure. The general shape of a porin is assumed to be a cylinder or a barrel (Benz and Hancock, 1981; Cowan *et al.*, 1992) with a length of 7.5 nm, which corresponds to the thickness of the outer membrane of Gram-negative bacteria (Rosenbusch, 1974). X-ray crystallographic analysis demonstrated that porins of *Escherichia coli* have an elliptical cross-section of 2.7 to 3.8 nm between main-chain atoms in the barrel walls (Cowan *et al.*, 1992). The shape and dimensions of the *in situ* physical state of porins extracted from seawater, has been assumed to be a cylinder or barrel with a length of 5–10 nm and a diameter of 2–5 nm (Tanoue *et al.*, 1995).

Heterotrophic flagellates utilize the colloidal DOM (for review, see Nagata and Koike, 1995). For example, protozoan grazers incorporate colloidal particles as small as 50 nm in diameter (Gonzalez and Suttle, 1993), dextran with a molecular mass of 4 kDa (Marchant and Scott, 1993) or dextrans and proteins of 55 to 2,000 kDa (Sherr, 1988; Tranvik *et al.*, 1993). Therefore, the grazing by protista might function not only for the transfer of bacterial proteins to the dissolved protein but also for the consumption of dissolved proteins (Fig. 27) which proteins might also be subjected to bacterial ectoenzymatic hydrolysis. This process has been theorized as a bacterial-viral loop, with organic matter cycling in this loop (i.e., organic matter flows repeatedly from bacteria to viruses to DOM to bacteria, *via* ectoenzymatic hydrolysis, Bratbak *et al.*, 1990; Fuhrman, 1992).

The dissolved proteins, once transferred from bacteria to the DOM, are subject to degradation by bacterial ectoenzymatic hydrolysis or to phagotrophic ingestion by protozoa. Relative intensities, in terms of flux from bacteria to DOM and the rate of degradation via the two processes of consumption, might reduce the relative abundance of particular dissolved proteins in surface waters.

Dynamics of dissolved proteins in intermediate and deep waters

Low rate of production and less active transfer processes are expected in the intermediate and deep waters because of the limited abundance of biota. The relative significance of the two above-mentioned transfer processes, namely, viral lysis and grazing by protista, could also differ between the surface and deep waters (Fig. 27).

Hara *et al.* (1996) examined the vertical distribution of bacteria and viruses at subarctic and subtropical Stations A and B. They stated that bacteriophagous activity by heterotrophic protista is unlikely to take place in the intermediate and deep waters because the bacterial density decreased to less than 10^5 cells/ml at depths greater than 200 m, and this level is less than the minimum density for bacteriophagous protista. Vertical distributions of protista and their activity have not been documented in oceanic environments, but it can be assumed that the abundance and activity of protista in the intermediate and deep waters are insignificant. For example, it was reported that the number of heterotrophic flagellates decreased by approximately two orders of magnitude from the surface to the deep waters, while the decrease in numbers of bacteria was about one order of magnitude in a water column in the North Atlantic (Patterson *et al.*, 1993).

While, Kokjohn *et al.* (1991) examined the effects of the density of host-cell on phage replication and found that there was no limiting host-cell density for phage replication under laboratory experiments. They reported that the infection of three *Ps. aeruginosa* bacteriophages was not impaired at the density of *Ps. aeruginosa* as low as 10 cells/ml. The results might rule out the possibility that bacteriophages detected in natural environments are dominated by inactive species (e.g., Wiggins and Alexander, 1985), although the dynamics of becteriophage-bacteria in the intermediate and deep waters of oceanic environments where the abundances of becteria and viruses are extremely low, are unknown.

At Stations A and B, the vertical distributions of viruses changed in parallel with those of bacteria, with some variation, and the ratio of viruses to bacteria ranged from 1.1 to 8.7 in the entire water column from the surface to a depth of 5,000 m (Hara *et al.*, 1996). In general, the number of viruses exceeds that of bacteria in marine environments (e.g., Fuhrman and Suttle, 1993; Maranger and Bird, 1995). Hara *et al.* (1996) discussed the source of viruses in deep waters and concluded that they might be produced *in situ* by the active viral infection. It seems that viral lysis might take place throughout the water column, while grazing by protista might be insignificant in the intermediate and deep waters.

The above possibility appears to address the question of why a very limited number of protein species was detected universally as major dissolved proteins. Immunochemical examination revealed that very minor components of the natural bacterial assemblage serve as the source of the universally observed 48-kDa protein, a major component of dissolved proteins. The host-virus relationship is very specific. In many cases, a virus infects a single species or sometimes a subspecies, or even an entire genus (Fuhrman and Suttle, 1993). Porins are known as receptors of viruses (Enderman *et al.*, 1978; Nakae, 1979; Lucky and Nikaido, 1980; Lugtenberg and Alphen, 1983; Nikaido and Vaara, 1985). It can be assumed, therefore, that the porin(s) in every marine bacterium has a direct relationship to a specific phage. Bacteriophages are of two types, namely, virulent and temperate phages. Virulent phages always lyse the host bacteria and bacteria that have been infected by a virulent phage release cell membranes without modification. Bacteria infected by temperate phages continue normal division unless induction occurs, and more than 90% of all known bacteriophages are of the temperate type (Freifelder, 1987). If we assume that the porins detected among the dissolved proteins are receptors of the virulent phages (Fig. 27), we can propose the following scenario. A specific and virulent phage recognizes and infects a specific bacterium *via* a particular porin protein. Consequently, very specific porin proteins from a minor component of the bacterial assemblage are released as dissolved proteins without modification.

The relative rates of consumption of dissolved proteins might be insignificant in the intermediate and deep waters (Fig. 27). Phagotrophic digestion, as a process of consumption, might be insignificant, as mentioned above. Nagata (1995) found that the bacterial activity in deep waters was approximately four orders of magnitude lower than that in surface waters at Stations A and B, while numbers of bacteria decreased by one to two orders of magnitude from surface to deep waters, an indication that rates of bacterial degradation, another aspect of the consumption of dissolved proteins, are also lower in deep waters than in surface waters. Such differences might control the vertical distributions of dissolved proteins. Possible explanations are as follows:

1) The production and transfer processes are active in the surface water, but transferred proteins are also actively remineralized by bacteria or protista. As a result, dissolved proteins are maintained at relatively low levels.

2) In the intermediate and deep waters, dissolved proteins are produced less actively but mainly by viral lysis, in particular by infection by virulent-type phages. However, transferred proteins survive and accumulate because of lower rates of phagotrophic digestion or bacterial ectoenzymatic hydrolysis. As a result, relatively high levels of dissolved proteins are maintained in the intermediate and deep waters.

The species diversity of the bacterial assemblage and the corresponding viral diversity are unknown. Discussions about consumption and transfer processes are almost entirely speculative at this time and it is too soon to draw any conclusions about the processes by which bacterial proteins are transferred to the pool of inanimate dissolved protein. In freshwater environments, *Ps. aeruginosa* bacteriophages are abundant and these phages exist in a dynamic equilibrium with the host population (Ogunseitan *et al.*, 1990). The relationships between viruses and bacteria appear to be important for an understanding of the persistence of specific proteins in the DOM.

Very recently, Yoshihara *et al.* (1996) discovered protein D2, a different porin from porin P, bore the protease activity. They reported that the protein D2 of the outer membrane of *Ps. aeruginosa* possessed a serine protease activity in addition to the channel function. The unexpected finding that a channel protein bears a protease activity raises another possible process by which selective porin proteins survive and accumulate in dissolved phase. Even the porin proteins are non-selectively transferred to the inanimate dissolved pool, porins that bear the protease activity might hydrolyze themselves, as a result, such porins might become easily degradable, while porins that do not possess such activity might not have undergone such processes. The porin P appears to have no protease activity because very little degradation of the purified porin P trimer occurred, even after sitting at room temperature for 4–6 months (Worobec *et al.*, 1988). The function and ubiquity of such protease activity among the porin family remain to be clarified.

Our discussions seem to suggest that the dissolved proteins detected in the intermediate and deep waters are produced *in situ* (Fig. 27). In other words, organic matter derived from the surface water as sinking particles is transferred to the DOM (Smith *et al.*, 1992) at greater depths and fuels the microbial loop. As a result, dissolved proteins are produced at greater depths, but organic matter of sinking particles might not serve as a direct source of DOM. A similar vertical profile is also obtained from the study of strong organic ligands for thorium in the dissolved phase in the Japan Sea (Hirose, 1996 and see the synthesis by Hirose, 1998 in this volume). Hirose (1996) examined a vertical profile of the strong organic ligand in the water column, and found that the concentration of the strong organic ligand was low in the surface water and increased gradually from depths of 100 m to 1,000 m and then decreased toward the depth. Since this strong organic ligand is thought to be a fresh organic compound that is derived from bacteria (Hirose and Tanoue, 1998a, b, c), the vertical profile observed by Hirose (1996) implies that such an organic compound is also produced *in situ*.

Mechanistic processes, such as sloppy feeding and excretion by zooplankton (e.g., Jumars *et al.*, 1989), leakage of organic matter from plankton cells through lysis or excretion (e.g., Cole *et al.*, 1982; Brock and Clyne, 1984; Malinsky-Rushansky and Legrand, 1996), by which organic matter generated by primary production is transferred to the DOM, have been proposed. Such processes are undoubtedly active and they might be important in a rapidly cycling fraction of the DOM in surface water. For example, short-lived plant pigments have been detected at extremely low levels in a high-molecular-mass fraction (>1 kDa) of DOM (Bianchi *et al.*, 1995).

In terms of proteins, however, quite a different picture is emerging with respect to the source of a major fraction of the DOM. Proteins produced by phytoplankton appear to be readily remineralized and do not accumulate in the dissolved phase. Proteins generated by secondary production are the direct source of dissolved proteins. Although the potential importance of the production of high-molecular-weight DOM or secretion of exopolymers (polymeric carbohydrates) by bacteria have been noted (Decho, 1990; Bratbak *et al.*, 1992; Heissenberger and Hemdl, 1994), little is known about the release of DOM from bacteria. In the "traditional" concept of the microbial loop, bacteria are thought to transform DOM to a living biomass. As shown in Fig. 27, it is proposed that the abundance and composition of dissolved protein result from coupling of the relative abundance, as well as the activities of three groups of organisms, namely, viruses, bacteria and protists. Thus, clarification of the processes proposed in Fig. 27 will provide important information about the ecology and structural dynamics of such biota in the microbial loop in oceanic environments. The hypotheses also include the possibility that the DOM cycles repeatedly *via* the microbial loop, even in the intermediate and deep waters.

SUMMARY AND CONCLUSIONS

The molecular characteristics of proteins have given us new insights into the chemical forms and dynamics of amino acids in inanimate forms, as well as into the sources and transfer processes of this major detectable biochemical material in the POM and the DOM, as follows.

1. Two characteristic groups of particulate proteins, background proteins and specific proteins, were identified from the meridional and vertical differences in their molecular distributions. The background proteins, derived directly from cellular proteins in living organisms, consist of a large number of proteins of widely ranging molecular masses, each present at a relatively low level. The background proteins contribute considerably to the total protein, as well as to particulate combined amino acids in surface waters, and they appear to be readily remineralized.

2. The specific proteins in the POM are distributed over a limited range of molecular masses, from approximately 45 kDa to 66 kDa. They are always minor components but they are prevalent in the POM from oligotrophic surface waters, and they are the major protein constituents in the POM from the intermediate and deep waters, an indication that proteins from specific sources survive and accumulate as a consequence of their resistance to degradation.

3. A protein with an apparent molecular mass of 45 kDa, one of the specific proteins, was commonly found at low latitudes from the North Pacific Ocean, through the equatorial Pacific and the Indian Oceans, to the Southern Ocean. The amino-terminal amino acid sequence indicated that the 45-kDa protein was a single protein species. Thus, a single and identifiable protein molecule appears to be very widespread in the surface POM at low latitudes.

4. The molecular characteristics of dissolved proteins were found to be quite different from those of particulate proteins. The background proteins were not evident among the dissolved proteins, even in the surface waters. Most of the dissolved proteins were distributed over a wide range of molecular masses, but

fewer than 30 individual proteins were detected.

5. The molecular distribution of dissolved proteins changed both horizontally and vertically, but some proteins were universally observed in all the samples examined from the surface to deep waters at the stations in the subarctic, subtropical and equatorial Pacific Ocean, and in the surface waters of the Indian and the Southern Oceans.

6. The amino-terminal amino acid sequence of the 48-kDa protein, an universal and major constituent of dissolved proteins, was found to be a homolog of porin P, an outer-membrane protein of the Gram-negative bacterium *Pseudomonas aeruginosa.* The correspondence of amino-terminal amino acid sequences and apparent molecular masses between this dissolved protein and authentic porin P indicates that almost the complete homolog of porin P, from the amino-terminus to (probably) the carboxyl-terminus, survives without modification in the water column.

7. The successful (without exception) amino-terminal amino acid sequencing of the proteins in bands on gels after SDS-PAGE demonstrated that six major dissolved proteins other than the 48-kDa protein were also discrete single species and were present at relative high levels, although they has not been reported previously.

8. Immunochemical examination revealed that some of the dissolved proteins were also derived from porins from bacterioplankton that belong to particular genera, namely, *Vibrio* and *Photobacterium.* Fluorescent-antibody staining suggested that a very minor component of the natural bacterial assemblage might be the source of the universally observed 48-kDa protein.

9. A first-order approximation of the levels of dissolved protein suggested that the dissolved protein accounts for roughly 30% of the dissolved combined amino acids. The abundance of dissolved protein in the entire ocean was calculated to be equivalent to (or greater than) the total living biomass of zooplankton or living bacteria. Thus, bacteria emerge not only as important consumers but also as important producers of organic matter in the sea.

10. The universal occurrence of the specific proteins in the POM and in the DOM suggests that very similar processes, by which specific proteins of particular organisms are transferred to the detrital POM and to the DOM, operate in every ocean.

11. DOM has been thought to be the pool of recalcitrant organic material, being a mixture of heteropolycondensate compounds, collectively called "humic substances". However, the present observations led to another hypothesis for the formation of DOM, whereby selected biopolymers survive and accumulate in seawater due to their intrinsic nature.

12. The distributional characteristics of dissolved proteins led to the hypothesis that the dissolved proteins below the surface water are produced *in situ.* The suggestion implies that DOM cycles repeatedly, *via* the microbial loop, even in the intermediate and deep waters.

From our observations of the molecular distribution of proteins, we have learned a great deal about the relationships between proteins in source organisms and proteins in the inanimate organic pools. Such relationships might not be recognizable from approaches that involve biomarkers or bulk chemical characterization (see the Introduction). However, it is quite clear that we have only just begun to understand the occurrence of discrete protein molecules in the sea. There is so much more to learn. For example, if we can trace the source of the 45-kDa protein in the POM and the 48-kDa protein in the DOM, we should gain information not only about the dynamics of particular proteins but also about the processes involved in the transfer of biochemical materials from cellular components of organisms to inanimate pools in the sea. Further information on the specific proteins observed in the present study is required to trace the processes of their production, transportation and decomposition, as well as to characterize the relevant dynamics of biopolymers in the sea. Attempts to identify biochemical materials at the macromolecular level should help us to reveal the sources and to trace the transfer processes of organic matter from the source organisms to the inanimate organic pools.

Acknowledgements—This work has been made in collaboration with A. Tsugita, M. Kamo, S. Suzuki, K. Kogure, S. Nishiyama, M. Ishii and T. Midorikawa. The author is grateful to C. Lee and I. Koike for their continual helps during the investigation. I also thank T. Nagata, K. Fukami and A. Tomaru for information about protista. The manuscript benefited from the comments of T. Hama, Y. Maita and S. Suzuki.

REFERENCES

Ammerman, J. W., E. Bentzen, E. R. Peele and J. B. Cotner (1994): Phosphorus cycling at the Bermuda time-series station: an overview. *EOS*, **75**, 100.

Amon, R. M. and R. Benner (1994): Rapid cycling of high-molecular-weight dissolved organic matter in the ocean. *Nature*, **369**, 549–552.

Amon, R. M. and R. Benner (1996): Bacterial utilization of different size classes of dissolved organic matter. *Limnol. Oceanogr.*, **41**, 41–51.

Andrew, A. T. (1986): *Electrophoresis: Theory, Techniques, and Biochemical and Clinical Applications.* 2nd ed. Clarendon, Oxford, 452 pp.

Anfinsen, C. B. (1973): Principles that govern the folding of protein chains. *Science*, **181**, 223–230.

Anma, G., K. Masuda, G. Kobayashi, H. Yamaguchi, T. Meguro, S. Sasaki and K. Ohtani (1990): Oceanographic structures and changes around the transition domain along 180° longitude, during June 1979–1988. *Bull. Fac. Fish. Hokkaido Univ.*, **41**, 73–88.

Armstrong, F. A. J., P. M. Williams and J. D. H. Strickland (1966): Photo-oxidation of organic matter in seawater by ultra-violet radiation, analytical and other applications. *Nature*, **211**, 481–483.

Azam, F., T. Fenchel, J. G. Field, J. S. Gray, L. A. Meyer-Reil and F. Thingstad (1983): The ecological role of water-column microbes in the sea. *Mar. Ecol. Prog. Ser.*, **10**, 257–263.

Ball, E. H. (1986): Quantitation of proteins by elution of Coomassie brilliant blue R from stained bands after sodium dodecyl sulfate-polyacrylamide gel electrophoresis. *Anal. Biochem.*, **155**, 23–27.

Barber, R. T. (1968): Dissolved organic carbon from deep waters resists microbial oxidation. *Nature*, **220**, 274–275.

Bauer, J. E., P. M. Williams and E. R. M. Druffel (1992): ^{14}C activity of dissolved organic carbon fractions in the north-central Pacific and Sargasso Sea. *Nature*, **357**, 667–670.

Bender, M. L. and M. J. McPharden (1990): Anomalous nutrient distribution in the equatorial Pacific in April 1988: evidence for rapid biological uptake. *Deep-Sea Res.*, **37**, 1075–1084.

Benner, R. J., J. D. Pakulski, M. D. McCarthy, J. I. Hedges and P. G. Hatcher (1992): Bulk chemical characteristics of dissolved organic matter in the ocean. *Science*, **255**, 1561–1564.

Benz, R. (1985): Porin from bacterial and mitochondrial outer membranes. *CRC Crit. Rev. Biochem.*,

19, 145–190.

Benz, R. and Hancock, R. E. W. (1981): Properties of the large ion-permeable pores formed from protein F of *Pseudomonas aeruginosa* in lipid bilayer membranes. *Bioch. Biophys. Acta*, **646**, 298–308.

Betzer, P. R., W. J. Showers, E. A. Laws, C. D. Winn, G. R. diTullio and P. M. Kroopnick (1984): Primary productivity and particule fluxes on a transect of the equator at 153°W in the Pacific Ocean. *Deep-Sea Res.*, **31**, 1–11.

Bianchi, T. S., C. Lambert, P. H. Santschi, M. Baskaran and L. Guo (1995): Plant pigments as biomarkers of high-molecular-weight dissolved organic carbon. *Limnol. Oceanogr.*, **40**, 422–428.

Billen, G. (1984): Heterotrophic utilization and regeneration of nitrogen. pp. 313–355. In *Heterotrophic Activity in the Sea*, ed. by J. E. Hobbie and P. J. leB. Williams, Plenum Press, New York.

Billen, G. (1991): Protein degradation in aquatic environments. pp. 123–143. In *Microbial Enzymes in Aquatic Environments*, ed. by R. J. Chróst, Springer-Verlag, New York.

Bird, D. F. and F. Kalff (1984): Empirical relationships between bacterial abundance and chlorophyll concentration in fresh and marine waters. *Can. J. Fish. Aquat. Sci.*, **41**, 1015–1023.

Boon, J. J., W. I. C. Rijpstra, F. de Lange, J. W. de Leeuw, M. Yoshioka and Y. Shimizu (1979): black Sea sterol—a molecular fossil of dinoflagellate blooms. *Nature*, **277**, 125–127.

Bratbak, G., M. Heldal, S. Norland and T. F. Thingstad (1990): Viruses as partners in spring bloom microbial trophodynamics. *Appl. Environ. Microbiol.*, **56**, 1400–1405.

Bratbak, G., M. Heldal, T. F. Thingstad, B. Riemann and O. H. Haslund (1992): Incorporation of viruses into the budget of microbial C-transfer. A first approach. *Mar. Ecol. Prog. Ser.*, **83**, 273–280.

Brock, T. D. and J. Clyne (1984): Significance of algal excretory products for growth of epilimnetic bacteria. *Appl. Environ. Microbiol.*, **47**, 731–734.

Broecker, W. S. and T.-H. Peng (1982): *Tracers in the Sea*. Eldigo Press, New York, 690 pp.

Brophy, J. E. and D. J. Carlson (1989): Production of biologically refractory dissolved organic carbon by natural seawater microbial populations. *Deep-Sea Res.*, **36**, 497–507.

Bruland, K. W. (1983): Trace elements in seawater. pp. 157–220. In *Chemical Oceanography*, Vol. 8, ed. by J. P. Riley and R. Chester, Academic Press, London.

Carlson, D. J., L. M. Mayer, M. L. Brann and T. H. Mague (1985): Binding of monomeric organic compounds to macromolecular dissolved organic matter in seawater. *Mar. Chem.*, **16**, 141–153.

Caron, D. A., H. G. Dam, P. Kremer, E. J. Lessard, L. P. Madin, T. C. Malone, J. M. Napp, E. R. Peele, M. R. Roman and M. J. Youngbluth (1995): The contribution of microorganisms to particulate carbon and nitrogen in surface waters of the Sargasso Sea near Bermuda. *Deep-Sea Res.*, **42**, 943–972.

Cauwet, G. (1978): Organic chemistry of seawater particulates. Concepts and developments. *Oceanol. Acta*, **1**, 99–105.

Chau, Y. K., R. Chuecas and J. P. Riley (1967): The component combined amino acids of some marine phytoplankton species. *J. Mar. Biol. Ass. U.K.*, **47**, 543–554.

Chavez, F. P. and R. T. Barber (1987): An estimate of new production in the equatorial Pacific. *Deep-Sea Res.*, **34**, 1229–1243.

Chavez, F. P., K. R. Buck, R. R. Bidigare, D. M. Karl, D. Hebel, M. Latasa, L. Campbell and J. Newton (1995): On the chlorophyll *a* retention properties of glass-fiber GF/F filters. *Limnol. Oceanogr.*, **40**, 428–433.

Cho, B. C. and F. Azam (1990): Biogeochemical significance of bacteria biomass in the ocean's euphotic zone. *Mar. Ecol. Prog. Ser.*, **63**, 253–259.

Chróst, R. J. (1991) (editor): Environmental control of the synthesis and activity of aquatic microbial ectoenzymes. pp. 29–59. In *Microbial Enzymes in Aquatic Environments*, Springer-Verlag, New York.

Clayton, J. R., Jr., Q. Dortch, S. S. Thoresen and S. I. Ahmed (1988): Evaluation of methods for the separation and analysis of proteins and free amino acids in phytoplankton samples. *J. Plank. Res.*, **10**, 341–358.

Cleveland, D. W., S. G. Fischer, M. W. Kirschner and U. K. Laemmli (1977): Peptide mapping by limited proteolysis in sodium dodecyl sulfate and analysis by gel electrophoreisi. *J. Biol. Chem.*, **252**, 1102–1106.

Coffin, R. G. (1989): Bacterial uptake of dissolved free and combined amino acids in estuarine waters. *Limnol. Oceanogr.*, **34**, 531–542.

Cole, J. J., G. E. Likens and D. L. Strayer (1982): Phtosynthetically produced dissolved organic carbon: an important source for planktonic bacteria. *Limnol. Oceanogr.*, **27**, 1080–1090.

Cole, J. J., S. Findlay and M. S. Pace (1988): Bacterial production in fresh and saltwater ecosystems: a cross-system overview. *Mar. Ecol. Prog. Ser.*, **43**, 1–10.

Cotner, J. B., E. R. Peele, J. W. Ammerman and E. Bentzen (1994): Phosphorus-limited plankton growth in the Sargasso Sea (BATS) and implication for the pelagic food web. *EOS*, **75**, 75.

Cowan, S. W., T. Schirmer, G. Rummel, M. Steiert, R. Ghosh, R. A. Pauptit, J. N. Jansonius and J. P. Rosenbusch (1992): Crystal structure explain functional properties of two *E. coli* porins. *Nature*, **358**, 727–733.

Cowey, C. B. and E. D. S. Corner (1963): On the nutrition and metabolism of zooplankton. 2. The relationship between the marine copepod Calanus helgolandicus and particulate material in Plymouth sea water. *J. Mar. Biol. Ass. U.K.*, **43**, 495–511.

Cowie, G. L. and J. I. Hedges (1992): Sources and reactivities of amino acids in a coastal marine environment. *Limnol. Oceanogr.*, **37**, 703–724.

Danilenko, A. F. (1977): Some results of protein analysis in Black Sea suspended matter. *Oceanol.*, **17**, 529–531.

Decho, A. W. (1990): Microbial exopolymer secretions in ocean environments: their role(s) in food webs and marine processes. *Oceanogr. Mar. Biol. Ann. Rev.*, **28**, 73–153.

Degens, E. T. (1970): Molecular nature of nitrogenous compounds in seawater and recent marine sediments. pp. 77–106. In *Organic Matter in Natural Waters*, ed. by D. W. Hood, Univ. Alaska, Inst. Mar. Sci., Occasional Pub. No. 1.

Dickson, M. L. and P. A. Wheeler (1993): Chlorophyll *a* concentrations in the North Pacific: Does a latitudinal gradient exist? *Limnol. Oceanogr.*, **38**, 1813–1818.

Donnelly, S. F. H., M. J. Pockongton, D. Pallotta and E. Orr (1993): A proline-rich protein, verprolin, involved in cytoskeletal organization and cellular growth in the yeast Saccharomyces Cerevisiae. *Mol. Microbiol.*, **10**, 585–596.

Dortch, Q. and T. T. Packard (1989): Differences in biomass structure between oligotrophic and eutrophic marine ecosystems. *Deep-Sea Res.*, **36**, 223–240.

Drapeau, G. R., Y. Boily and J. Houmard (1972): Purification and properties of an extracellular protease of Staphylococcus aureus. *J. Biol. Chem.*, **247**, 6720–6726.

Druffel, E. R. M. and P. M. Williams (1990): Identification of a deep marine course of particulate organic carbon using bomb ^{14}C. *Nature*, **347**, 172–174.

Druffel, E. R. M. and P. M. Williams (1992): Importance of isotope measurements in marine organic geochemistry. *Mar. Chem.*, **39**, 209–215.

Ducklow, H. W. and C. A. Carlson (1992): Oceanic bacterial production. *Adv. Microb. Ecol.*, **12**, 113–181.

Duursma, E. K. (1961): Dissolved organic carbon, nitrogen and phophorus in the sea. *Nethl. J. Sea Res.*, **1**, 1–148.

Endermann, P., I. Hindennach and U. Henning (1978): Major proteins of the *Escherichia coli* outer cell envelope membrane. Preliminary characterization of the phage λ receptor protein. *FEBS Lett.*, **88**, 71–74.

Emerson, S., C. Stump, P. M. Grootes, M. Stuiver, G. W. Farwell and F. H. Schmidt (1987): Organic carbon in surface deep-sea sediments: C14 concentration. *Nature*, **329**, 51–54.

Eppley, R. W., W. G. Harrison, S. W. Chisholm and E. Stewart (1977): Particulate organic matter in surface waters off Southern California and its relationship to phytoplankton. *J. Mar. Res.*, **35**, 771–696.

Farrington, J. W. (1992): Marine organic geochemistry: Review and challenges for the future. *Mar. Chem.*, **39**, 242 pp.

Fish, W. W., J. A. Reynolds and C. Tanford (1970): Gel chromatography of protein in denaturing solvents. Comparison between sodium dodecyl sulfate and guanidine hydrochloride denaturants. *J. Biol. Chem.*, **245**, 5166–5168.

Flood, P. R., D. Deibel and C. C. Morris (1992): Filtration of colloidal melanin from seawater by planktonic tunicates. *Nature*, **355**, 630–632.

Fournier, R. O. (1966): North Atlantic deep-sea fertility. *Science*, **153**, 1250–1252.

Freifelder, G (1987): *Microbial Genetics*. Jones and Bartlett Publishers, Boston.

Fuhrman, J. A. (1992): Bacterioplankton roles in cycling of organic matter: the microbial food web. pp. 361–383. In *Primary Production and Biogeochemical Cycles in the Sea*, ed. by P. G. Falkowski and A. D. Woodhead, Pleunum Press, New York.

Fuhrman, J. A. and R. T. Noble (1995): Viruses and protists cause similar bacterial mortality in coastal seawater. *Limnol. Oceanogr.*, **40**, 1236–1242.

Fuhrman, J. A. and C. A. Suttle (1993): Viruses and marine planktonic systems. *Oceanography*, **6**, 51–63.

Fuhrman, J. A., T. D. Sleeter, C. A. Carlson and L. M. Proctor (1989): Dominance of bacterial biomass in the Sargasso Sea and its ecological implications. *Mar. Ecol. Prog. Ser.*, **57**, 207–217.

Gagosian, R. B. and C. Lee (1981): Processes controlling the distribution of biogenic organic compounds in seawater. pp. 91–123. In *Marine Organic Chemistry*, ed. by E. K. Duursma and R. Dawson, Elsevier Sci. Publ., Amsterdam.

Garfield, P. C., T. T. Packard and L. A. Codispoti. (1979): Particulate protein in the Peru upwelling system. *Deep-Sea Res.*, **26**, 623–639.

Garrison, D. L., K. R. Buck and M. M. Gowing (1993): Winter plankton assemblage in the ice edge zone of the Weddell and Scotia Seas: composition, biomass and spatial distribution. *Deep-Sea Res.*, **40**, 311–338.

Giovannoni, S. J., T. B. Britschgi, C. L. Moyer and K. G. Field (1990): Genetic diversity in Sargasso Sea bacterioplankton. *Nature*, **345**, 60–63.

Gonzalez, J. M. and C. A. Suttle (1993): Grazing by marine nanaflagellates on viruses and virus-sized particles: ingestion and digestion. *Mar. Ecol. Prog. Ser.*, **94**, 1–10.

Gordon, D. C., Jr. (1970): A microscopic study of organic particles in the North Atlantic Ocean. *Deep-Sea Res.*, **17**, 175–185.

Gordon, D. C., Jr. and P. J. Cranford (1985): Detailed distribution of dissolved and particulate organic matter in the Arctic Ocean and comparison with other oceanic regions. *Deep-Sea Res.*, **32**, 1221–1232.

Gottschalk, G. (1986): *Bacterial Metabolism*. 2nd ed. Springer-Verlag, New York, 359 pp.

Guo, L., C. H. Coleman, Jr. and P. H. Santschi (1994): The distribution of colloidal and dissolved organic carbon in the Gulf of Mexico. *Mar. Chem.*, **45**, 105–119.

Guo, L., P. H. Santschi and K. W. Warnken (1995): Dynamics of dissolved organic carbon (DOC) in oceanic environments. *Limnol. Oceanogr.*, **40**, 1392–1403.

Haas, L. W. and K. L. Webb (1979): Nutritional mode of several non-pigmented microflagellates from the York River estuary, Virginia. *J. Exp. Mar. Biol. Ecol.*, **39**, 125–134.

Hancock, R. E. W. (1986): Model membrane studies of porin function. pp. 187–225. In *Bacterial Outer Membranes as Model Systems*, ed. by M. Inoue, John Willey & Sons, New York.

Hancock, R. E. W. (1987): Role of porins in outer membrane permeability. *J. Bacteriol.*, **169**, 929–933.

Hancock, R. E. W. and R. Benz (1986): Demonstration and chemical modification of a specific phosphate binding site in the phosphate-starvation-inducible outer membrane porin protein P of *Pseudomonas aeruginosa*. *Biochim. Biophys. Acta*, **860**, 699–707.

Hancock, R. E. W., K. Poole and R. Benz (1982): Outer membrane protein P of *Pseudomonas aeruginosa*.: regulation by phosphate deficiency and formation of small anion-specific channels in lipid bilayer membrane. *J. Bacteriol.*, **150**, 730–738.

Hancock, R. E. W., C. Egli, R. Benz and R. J. Siehnel (1992): Overexpression in *Escherichia coli* and functional analysis of a novel PPi selective porin, OprO, from *Pseudomonas aeruginosa*. *J. Bacteriol.*, **174**, 471–476.

Handa, N. (1970): Dissolved and particulate carbohydrates. pp. 128–152. In *Organic Matter in Natural Waters*, ed. by D. W. Hood, Univ. Alaska, Inst. Mar. Sci., Occasional Publ. No. 1.

Handa, N., K. Yanagi and K. Matsunaga. (1972): Distribution of detrital materials in the Western Pacific Ocean and their biochemical nature. pp. 53–71. In *Detritus and Its Role in Aquatic Ecosystems*, ed. by U. Melchiorri-Santolini and J. W. Hopton, Memorie Dell Instituto Italiano Di Idrobiologia, 29 (Suppl.).

Hara, S. and I. Koike (1998): Dynamics of organic marine aggregates; Nanometer-colloids to marine snow. This volume.

Hara, S., I. Koike, K. Terauchi, H. Kamiya and E. Tanoue (1996): Abundant viruses in deep oceanic waters. *Mar. Ecol. Prog. Ser.*,145, 269–277.

Harradine, P. J., P. G. Harris, R. N. Head, R. P. Harris and J. R. Maxwell (1996): Steryl chlorin esters are formed by zooplankton herbivory. *Geochim. Cosmochim. Acta*, **60**, 2265–2270.

Haschemeyer, R. H. and E. V. Haschemeyer (1973): *Proteins: A Guide to Study by Physical and Chemical Methods*. John Willey & Sons, New York, 445 pp.

Hedges, J. I. (1987): Organic matter in sea water. *Nature*, **330**, 205–206.

Hedges, J. I. (1992): Global biogeochemical cycles: progress and problem. *Mar. Chem.*, **39**, 67–93.

Hedges, J. I. and P. L. Parker (1976): Land-derived organic matter in surface sediments from the Gulf of Mexico. *Geochim. Cosmochim. Acta*, **40**, 1019–1029.

Heissenberger, A. and G. J. Hemdl (1994): Formation of high molecular weight material by free-living marine bacteria. *Mar. Ecol. Prog. Ser.*, **111**, 129–135.

Helelnius, A. and K. Simon (1975): Solubilization of membranes by detergent. *Biochim. Biophys. Acta*, **415**, 29–79.

Henrichs, S. M. and P. M. Williams (1985): Dissolved and particulate amino acids and carbohydrates in the sea-surface microlayer. *Mar. Chem.*, **17**, 141–163.

Hirose, K. (1996): Determination of a strong organic ligand dissolved in seawater, thoriumcomplexing capacity of oceanic dissolved organic matter. *J. Radioanal. Nucl. Chem., Articles*, **204**, 193–204.

Hirose, K. (1998): Strong organic ligands in seawater: Peculiar functional group in oceanic organic matter. This volume.

Hirose, K. and E. Tanoue (1994): Thorium-particulate matter interaction. Thorium complexing capacity of oceanic particulate matter: thery. *Geochim. Cosmochim. Acta*, **58**, 1–7.

Hirose, K. and E. Tanoue (1998a): The vertical distribution of Thorium-complexing capacity in particulate matter: Implication with respect to the dynamics of particulate matter in oceanic water columns. *Mar. Chem.*, **59**, 235–252.

Hirose, K. and E. Tanoue (1998b): The strong organic ligand in marine organisms. *Fresenius J. Anal. Chem.* (in press).

Hirose, K. and E. Tanoue (1998c): Characteristics of the strong organic ligand in particulate matter of Pacific surface waters (submitted).

Hollibaugh, J. T. and F. Azam (1983): Microbial degradation of dissolved proteins in seawater. *Limnol. Oceanogr.*, **28**, 1104–1116.

Holm-Hansen, O. (1969): Determination of microbial biomass in ocean profiles. *Limnol. Oceanogr.*, **14**, 740–747.

Houmard, J. and G. R. Drapeau (1972): Staphylococcal protease: a proteolytic enzyme specific for glutamoyl bonds. *Proc. Natl. Acad. Sci. USA*, **69**, 3506–3509.

Hubberten, U., R. J. Lara and G. Kattner (1994): Amino acid composition of seawater and dissolved humic substances in the Greenland Sea. *Mar. Chem.*, **45**, 121–128.

Ishii, M. and H. Y. Inoue (1995): Air-sea exchange of CO_2 in the central and western equatorial Pacific in 1990. *Tellus*, **47B**, 447–460.

Ishiwatari, R. (1992): Macromolecular material (humic substance) in the water column and sediments. *Mar. Chem.*, **39**, 151–166.

Ishizaka, J., H. Kiyosawa, K. Ishida, K. Ishikawa and M. Takahashi (1994): Meridional distribution and carbon biomass of autotrophic picoplankton in the Central North Pacific Ocean during Late Norhtern Summer 1990. *Deep-Sea Res.*, **41**, 1745–1766.

Johnson, M. W. and E. Brinton (1963): Biological species, water-masses and currents. pp. 381–414. In *The Sea*, ed. by M. N. Hill, Vol. 2, John Wiley & Sons, New York.

Jumars, P. A., D. L. Penry, J. A. Baross, M. J. Perry and B. W. Frost (1989): Closing the microbial loop: dissolved carbon pathway to heterotrophic bacteria from incomplete ingestion, digestion and adsorption in animals. *Deep-Sea Res.*, **36**, 483–495.

Kalle, K. (1966): The problem of Gelbstoff in the sea. *Oceanogr. Mar. Biol. Annu. Res.*, **4**, 91–104.

Kawai, H. (1972): Hydrography of the Kuroshio extension. pp. 235–352. In *KUROSHIO—It Physical Aspects*, ed. by H. Stommel and K. Yoshida, Univ. Tokyo Press, Tokyo.

Keil, R. G. and D. L. Kirchman. (1991): Dissolved combined amino acids in marine waters as determined by a vapor-phase hydrolysis method. *Mar. Chem.*, **33**, 243–259.

Keil, R. G. and D. L. Kirchman (1993): Dissolved combined amino acids: chemical form and utilizationbymarinebacteria. *Limnol. Oceanogr.*, **38**, 1256–1270.

Keil, R. G. and D. L. Kirchman (1994): Abiotic transformation of labile protein to refractory protein in sea water. *Mar. Chem.*, **45**, 187–196.

Kirchman, D. L., R. G. Keil, M. Simon and N. A. Welschmeyer (1993): Biomass and production of heterotrophic bacterioplankton in the oceanic subarctic Pacific. *Deep-Sea Res.*, **40**, 967–988.

Kirchman, D. L., J. H. Rich and R. T. Barber (1995): Biomass and biomass production of heterotrophic bacteria along 140 in the equatorial Pacific: Effect of temperature on the microbialloop. *Deep-Sea Res.*, **42**, 603–619.

Kogure, K., U. Shimidu and N. Taga (1979): A tentative direct microscopic method for counting living marine bacteria. *Can. J. Microbiol.*, **25**, 415–420.

Koike, I., S. Hara, K. Terauchi and K. Kogure (1990): The role of submicrometre particles in the ocean. *Nature*, **345**, 242–244.

Kokjohn, T. A., G. S. Sayler and R. V. Miller (1991): Attachment and replication of *Pseudomonas aeruginosa* bacteriophages under conditions simulating aquatic environments. *J. Gen. Microbiol.*, **137**, 661–666.

Laemmli, U. K. (1970): Cleavage of structural protein during the assembly of the head of bacteriophage T4. *Nature*, **227**, 680–685.

Largeau, C. (1995): Formation of refractory organic matter from biological precursors. pp. 275–292. In *Role of Nonliving Organic Matter in the Earth's Carbon Cycle*, ed. by R. G. Zepp and C. H. Sonntag, John Wiley & Sons, Chichester.

Largeau, C., S. Derenne, E. Casadevali, A. Kadouri and N. Sellier (1986): Pyrolysis of immature Torbanite and of the resistant biopolymer (PRB A) isolated from extant alga *Botryococcus braunii*. Mechanism of formation and structure of Torbanite. *Org. Geochem.*, **10**, 1023–1032.

Largeau, C., S. Derenne, C. Clairay, E. Casadevali, J. F. Raynaud, B. Lugardon, C. Berkaloff, M. Corolleur and B. Rousseau (1991): Characterization of various kerogens by scanning electrom microscopy (SEM) and transmission electron microscopy (TEM), Morphological relationships with resistant outer walls in extant microorganisms. pp. 91–101. In *Proceedings Intl. Sympo. Organic Petrology, Zeist 1990*, ed. by W. J. J. Fermont and J. W. Weegink, Geological Survey Netherlands, Special Issue.

Laursen, A. K., L. M. Mayer and D. W. Townsend (1996): Lability of proteinaceous material in estuarine seston and subcellular fractions of phytoplankton. *Mar. Ecol. Prog. Ser.*, **136**, 227–234.

Lee, C. and J. Bada (1975): Arnino acids in Equatorial Pacific ocean water. *Earth Planet. Sci. Lett.*, **26**, 61–68.

Lee, C. and J. Bada (1977): Dissolved amino acids in the equatorial Pacific, the Sargasso Sea, and Biscayne Bay. *Limnol. Oceanogr.*, **22**, 502–510.

Lee, C. and C. Cronin (1984): Particulate amino acids in the sea: effects of primary productivity and biological decomposition. *J. Mar. Res.*, **42**, 1075–1097.

Lee, C. and S. G. Wakeham (1989): Organic matter in seawater: biogeochemical processes. pp. 2–51. In *Chemical Oceanography. Vol. 9*, ed. by J. P. Riley, Academic Press, New York.

Lee, C., S. G. Wakeham and J. W. Farrington (1983): Variations in the composition of particulate organic matter in a time-series sediment trap. *Mar. Chem.*, **13**, 181–194.

Lee, S. and J. A. Fuhrman (1987): Relationships between biovolume and biomass of naturally derived marine bacterioplankton. *App. Environm. Microbiol.*, **53**, 1298–1303.

Li, W. K. W., P. M. Dickie, B. D. Irwin and A. M. Wood (1992): Biomass of bacteria, cyanobacteria, prochlorophytes and photosynthetic eukaryotes in the Sargasso Sea. *Deep-Sea Res.*, **39**, 501–

519.
Liebezeit, G. and M. Bölter (1986): Distribution of particulate amino acids in the Bransfield Strait. *Polar Biol.*, **5**, 199–206.
Long, R. A. and F. Azam (1996): Abundance protein-containing particles in the sea. *Aquat. Microbial Ecol.*, **10**, 213–221.
Lucky, M. and H. Nikaido (1980): Specificity of diffusion channels produced by λ phage receptor protein of *Escherichia coli*. *Proc. Natl. Acad. Sci. USA*, **77**, 167–171.
Lugtenberg, B. and L. V. Alphen (1983): Molelcular architecture and functioning of the outer membrane of *Escherichia coli* and other Gram negative bacteria. *Biochim. Biophys. Acta*, **737**, 51–115.
Maier, C., E. Bremer, A. Schmidt and R. Benz (1988): Pore-forming activity of the Tsx protein from the outer membrane of *Escherichia coli*. *J. Biol. Chem.*, **263**, 2493–2499.
Malinsky-Rushansky, N. Z. and C. Legrand (1996): Excretion of dissolved organic carbon by phytoplankton of different sizes and subsequent bacterial uptake. *Mar. Ecol. Prog. Ser.*, **132**, 249–255.
Mantoura, R. F. C. and E. M. S. Woodward (1983): Concervative behavior of riverine dissolved organic carbon in the Severn Estuary: chemical and geochemical implications. *Geochim. Cosmochim. Acta*, **47**, 1293–1309.
Maranger, R. and D. F. Bird (1995): Viral abundance in aquatic systems: a comparison between marine and fresh waters. *Mar. Ecol. Prog. Ser.*, **121**, 217–226.
Marchant, H. J. and F. J. Scott (1993): Uptake of sub-micrometer particles and dissolved organic material by Antarctic choanoflagellates. *Mar. Ecol. Prog. Ser.*, **92**, 59–64.
Marlowe, I. T., S. C. Brassell, G. Eglinton and J. C. Green (1990): Long-chain alkenones and alkyl alkenoates and the fossil coccolith record in marine sediments. *Chem. Geol.*, **88**, 349–375.
Martinez, J. and Azam, F. (1993): Periplasmic aminopeptidase and alkaline phosphatase activities in a marine bacterium: implications for substrate processing in the sea. *Mar. Ecol. Prog. Ser.*, **92**, 89–97.
Masuzawa, J. (1972): Water characteristics of the North Pacific central region. pp. 95–127. In *KUROSHIO—Its Physical Aspects*, ed. by H. Stommel and K. Yoshida, Univ. Tokyo Press, Tokyo.
Matsudaira, P. (1987): Sequence from picomole quantities of proteins electroblotted onto polyvinylidene difluoride membranes. *J. Biol. Chem.*, **262**, 10035–10038.
Mayer, L. M., L. L. Schick and F. W. Setchel (1986): Measurement of protein in nearshore marine sediments. *Mar. Ecol. Prog. Ser.*, **30**, 159–165.
Mayer, L. M., L. L. Schick, T. Sawyer and C. J. Plante (1995): Bioavailable amino acids in sediments: A biomimetic, kinetics-based approach. *Limnol. Oceanogr.*, **40**, 511–520.
Mayzaud, P. and J.-L. M. Martin (1975): Some aspects of the biochemical and mineral composition of marine plankton. *J. Exp. Mar. Biol. Ecol.*, **17**, 297–310.
McCarthy, M., J. I. Hedges and R. Benner (1995): Geochemistry of high molecular weight dissolved organic matter in the sea as implied by amino acids and organic nitrogen. pp. 1150–1152. In *Organic Geochemistry: Developments and Applications to Energy, Climate, Environment and Human History*, selected papers from the 17th International Meeting on Organic Geochemistry, ed. by J. O. Grimalt and C. Doronsoro, A.I.G.O.A., The Basque Country, Spain.
McGowan, J. A. and P. M. Williams (1973): Oceanic habitat differences in the North Pacific. *J. Exp. Mar. Biol. Ecol.*, **12**, 187–217.
McManus, G. B. and J. A. Fuhrman (1988): Control of marine bacterioplankton population: measurement and significance of grazing. *Hydrobiologia*, **159**, 51–62.
Mel'nikov, I. A. (1974): Use of histochemical reagents to determine the biochemical composition of detritus. *Oceanol.*, **14**, 922–926.
Menzel, D. W. (1974): Primary productivity, dissolved and particulate organic matter, and the sites of oxidation of organic matter. pp. 659–678. In *The Sea, Vol. 5*, ed. by E. D. Goldberg, Wiley-Intersci. Publ., New York.
Menzel, D. W. and J. H. Ryther (1970): Distribution and cycling of organic in the oceans. pp. 31–54. In *Organic Matter in Natural Waters*, ed. by D. W. Hood, Inst. Mar. Sci., Univ. Alaska, Pub.

No. 1.

Menzel, D. W. and R. F. Vaccaro (1964): The measurement of dissolved organic and particulatecarboninseawater. *Limnol. Oceanogr.*, **9**, 138–142.

Meybeck, M. (1982): Carbon, nitrogen, and phosphorus transport by world rivers. *Am. J. Sci.*, **282**, 401–450.

Meyer-Schulte, K. and J. I. Hedges (1986): Molecular evidence for a terrestrial component of organic matter dissolved in ocean water. *Nature*, **321**, 61–63.

Mopper, K. and E. T. Degens (1979): Organic carbon in the ocean: nature and cycling. pp. 293–316. In *The Global Carbon Cycle*, ed. by B. Bolin, E. T. Degens, S. Kempe and P. Ketner, SCOPE report 13, John Wiley & Sons, Chichester.

Morita, J., S. Suzuki and R. Kusuda (1996): Metalloprotease produced by Listonella anguillarum shows similar activity to plasma activated protein C in rainbow trout coagulation cascade. *Fish Pathol.*, **31**, 9–17.

Nagata, T. (1995): Preliminary report of the Hakuho Maru Cruise KH95-3. *Ocean Res. Institute, Univ. Tokyo*, 84 pp.

Nagata, T. (1998): "Picopellets" produced by phagotrophic nanoflagellates: Role in the material cycling within marine environments. This volume.

Nagata, T. and D. L. Kirchman (1992): Release of macromolecular organic complexes by heterotrophic marine flagellates. *Mar. Ecol. Prog. Ser.*, **83**, 233–240.

Nagata, T. and I. Koike (1995): Marine colloids: Their roles in food webs and biogeochemical fluxes. pp. 275–292. In *Biogeochemical Processes and Ocean Flux in the Western Pacific*, ed. by H. Sakai and Y. Nozaki, Terra Sci. Publ., Tokyo.

Nakae, T. (1979): A porin activity of purified λ-receptor protein from *Escherichia coli* in reconstituted vesicle membranes. *Biochem. Biophys. Res. Commun.*, **88**, 774–781.

Nemoto, T. and K. Ishikawa (1969): Organic particulate and aggregate matters stained by histological reagents in the East China Sea. *J. Oceanogr. Soc. Japan*, **25**, 281–290.

Neville, D. W. (1971): Molecular weight determination of protein-dodecyl sulfate complexes by gel electrophoresis in a discontinuous buffer system. *J. Biol. Chem.*, **246**, 6328–6334.

Nguyen, R. and H. R. Harvey (1994): A rapid micro-scale method for the extraction and analysis of protein in marine samples. *Mar. Chem.*, **45**, 1–14.

Nikaido, H. (1994): Porins and specific diffusion channels in bacterial outer membranes. *J. Biol. Chem.*, **269**, 3905–3908.

Nikaido, H. and M. Vaara (1985): Molecular basis of bacterial outer membrane permeability. *Microbiol. Rev.*, **49**, 1–32.

Nozaki, Y. (1992): Trace elements in sea water: Their mean concentrations and North Pacific profiles. *Chikyukagaku* (*Geochemistry*), **26**, 25–39 (in Japanese with English abstract).

Oakley, B. R., D. R. Kirsch and N. R. Morris (1980): A simplified ultrasensitive silver stain for detecting proteins in polyacrylamide gels. *Anal. Biochem.*, **105**, 361–363.

O'Farrell, P. H. (1975): High-resolution two-dimensional electrophoresis of proteins. *J. Biol. Chem.*, **250**, 4007–4021.

Ogawa H. (1998): Bulk chemical aspects of dissolved organic matter in seawater. This volume.

Ogunseitan, O. A., G. S. Sayler and R. V. Miller (1990): Dynamic imteractions of *Pseudomonas aeruginosa* and bacteriophages in lake water. *Microb. Ecol.*, **19**, 171–185.

Pace, M. L. (1988): Bacterial mortality and the fate of bacterial production. *Hydrobiologia*, **159**, 41–49.

Packard, T. T. and Q. Dortch (1975): Particulate protein-nitrogen in North Atlantic surface water. *Mar. Biol.*, **33**, 347–354.

Palleroni, N. J. (1992): Introduction to the family Pseudomonadaceae. pp. 3071–3085. In *The ProKaryotes*, ed. by A. Balows, H. G. Truper, M. Dworkin, W. Harder and K. H. Schleifer, Springer-Verlag, Berlin.

Pak, H., D. A. Kiefer and J. C. Kitchen (1988): Meridional variations in the concentration of chlorophyll and microparticles in the North Pacific Ocean. *Deep-Sea Res.*, **35**, 1151–1171.

Parsons, T. R. and J. D. H. Strickland (1962): Oceanicdetritus. *Science*, **136**, 313–314.

Patterson, D. J., K. Nygaard, G. Steinberg and C. M. Turley (1993): Heterotrophic flagellates and other protists associated with oceanic detritus throughout the water column in the mid North Atlantic. *J. Mar. Biol. Ass. U.K.*, **73**, 67–95.

Philander, S. G. (1990): *El Nino, La Nina, and the Southern Oscillation*, Academic Press, San Diego, 293 pp.

Pitt-Rivers, R and F. S. A. Impiombato (1968): The binding of sodium dodecyl sulphate to various proteins. *Biochem. J.*, **109**, 825–830.

Pomeroy, L. R. (1974): The ocean's food web: A changing paradigm. *BioScience*, **24**, 499–504.

Pomeroy, L. R., J. E. Sheldon, W. M. Sheldon and F. Peters, Jr. (1995): Limits to growth and respiration of hacterioplankton in the Gulf of Mexico. *Mar. Ecol. Prog. Ser.*, **117**, 259–268.

Poole, K. and R. E. W. Hancock (1986): Phosphate-starvation-induced outer membrane proteins of members of the Families *Enterobacteriaceae* and *Pseudomonaceae*: demonstration of immunological cross-reactivity with an antiserum specific for porin protein P of *Pseudomonas aeruginosa*. *J. Bacteriol.*, **165**, 987–993.

Poole, K., T. R. Parr, Jr. and R. E. W. Hancock (1987): Phosphate-selective porins from the outer membranes of fluorescent *Pseudomonas* sp. *Can. J. Miocrobiol.*, **33**, 63–69.

Porter, K. G. and Y. S. Feig (1980): The use of DAPI for identifying and counting aquatic microflora. *Limnol. Oceanogr.*, **25**, 943–948.

Porter, K. G., E. B. Sherr, B. F. Sherr, M. Pace and R. W. Sanders (1985): Protozoa in planktonic food webs. *J. Protozool.*, **32**, 409–415.

Prahl, F. G. (1992): Prospective use of molecular paleontology to test for iron limitation on marine promary productivity. *Mar. Chem.*, **39**, 167–185.

Quinby-Hunt, M. S. and K. K. Turekian (1983): Distribution of elements in sea water. *EOS*, **64**(14), 130–132.

Rashid, M. A. (1985): *Geochemistry of Marine Humic Compounds*. Springer-Verlag, New York, 300 pp.

Rath, J., C. Schiller and G. J. Herndle (1993): Ectoenzymatic activity and bacterial dynamics along a trophic gradient in the Caribbean Sea. *Mar. Ecol. Progr. Ser.*, **102**, 89–96.

Redfield, A. C., B. H. Ketchum and F. A. Richard (1963): The influence of organisms of the composition of sea-water. pp. 26–87. In *The Sea*, ed. by M. N. Hill, Vol. 2, John Wiley & Sons, New York.

Repeta, D. J. and R. B. Gagosian (1982): Carotenoid transformations in coastal marine waters. *Nature*, **295**, 51–54.

Repeta, D. J. and R. B. Gagosian (1984): Transformation reactions and recycling of carotenoids and chlorins in the Peru upwelling region (15°S, 75°W). *Geochim. Cosmochim. Acta*, **48**, 1265–1277.

Reynolds, J. A. and C. Tanford (1970a): Binding of dodecyl sulphate to protein at high binding ratios. Possible implications for the state of proteins in biological membrane. *Proc. Natl. Acad. Sci. USA*, **66**, 1002–1007.

Reynolds, J. A. and C. Tanford (1970b): The gross conformation of protein-sodium dodecyl sulfate complexes. *J. Biol. Chem.*, **245**, 5161–5165.

Riley, G. A. (1970): Particulate organic matter in sea water. *Adv. Mar. Biol.*, **8**, 1–118.

Romankevich, E. A. (1984): *Geochemistry of Organic Matter in the Ocean*. Springer-Verlag, Berlin, 334 pp.

Rosenbusch, J. P. (1974): Characterization of the major envelope protein from *Escherichia coli*. Regular arrangement on the peptidoglycan and unusual dodecyl sulfate binding. *J. Biol. Chem.*, **249**, 8019–8029.

Roszak, D. B. and R. R. Colwell (1987): Survival strategies of bacteria in the natural environment. *Microbiol. Rev.*, **51**, 365–373.

Sakugawa, H. and N. Handa (1985a): Isolation and chemical characterization of dissolved and particulate polysaccharides in Mikawa Bay. *Geochim. Cosmochim. Acta*, **49**, 1185–1193.

Sakugawa, H. and N. Handa (1985b): Chemical studies on dissolved carbohydrates in the water samples collected from the North Pacific and Bering Sea. *Oceanol. Acta*, **8**, 185–196.

Santschi, P. H., L. Guo, M. Baskaran, S. Trumbore, J. Southon, T. S. Bianchi, B. Honeyman and L. Cifuentes (1995): Isotopic evidence for the contemporary origin of high-molecular-weight organic matter in oceanic environments. *Geochim. Cosmochim. Acta*, **59**, 625–631.

Schulein, K., A. Schmidt and R. Benz (1991): The sugar-specific outer membrane channel ScrY constains functional characteristics of general diffusion pores and substrate-specific porins. *Mol. Microbiol.*, **5**, 2233–2241.

Schmidt, T. M., E. F. DeLong and N. R. Pace (1991): Analysis of a marine picoplankton community by 16S rRNA gene cloning and sequencing. *J. Bacteriol.*, **173**, 4371–4378.

Setchell, F. W. (1981): Particulate protein measurement in oceanographic samples by dye binding. *Mar. Chem.*, **10**, 301–313.

Shapiro, A. L., E. Vinuela and J. V. Maizel, Jr. (1967): Molecular weight estimation of polypeptide chain by electrophoresis in SDS-polyacrylamide gels. *Biochem. Biophys. Res. Commun.*, **28**, 815–820.

Sharp, J. H. (1983): The distributions of inorganic nitrogen and dissolved and particulate organic nitrogen in the sea. pp. 1–35. In *Nitrogen in the Marine Environments*, ed. by E. J. Carpenter and D. G. Capone, Academic Press, New York,

Sherr, E. B. (1988): Direct use of high molecular weight polysaccharide by heterotrophic flagellates. *Nature*, **335**, 348–351.

Sieburth, J. McN. (1979): *Sea Microbes*. Oxford Univ. Press, Oxford, 491 pp.

Siehnel, R., N. L. Martin and R. E. W. Hancock (1990): Sequence and relatedness in other bacteria of the *Pseudomonas aeruginosa* oprP gene coding for the phosphate-specific porin P. *Mol. Microbiol.*, **4**, 831–838.

Siehnel, R., C. Egli and R. E. W., Hancock (1992): Polyphosphate-selective porin OpO of *Pseudomonas aeruginosa*: expression, purification and sequence. *Mol. Microbiol.*, **6**, 2319–2326.

Siezen, R. J. and T. H. Mague (1978): Arnino acids in suspended particulate matter from oceanic and coastal waters of the Pacific. *Mar. Chem.*, **6**, 215–231.

Simon, M., B. C. Cho and F. Azam (1992): Significance of bacterial biomass in lakes and the ocean: comparison to phytoplankton biomass and biogeochemical implications. *Mar. Ecol. Prog. Ser.*, **86**, 103–110.

Skopintsev, B. (1972): On the age of stable organic matter-aquatic humus in oceanic waters. pp. 205–207. In *The Changing Chemistry of the Oceans*, ed. by D. Dryssen and D. Jagner, Proceedings of the 12th Nobel Symposium, 1971, Goteborg, Sweden, John Wiley & Sons, New York.

Smith, D. C., M. Simon, A. L. Alldredge and F. Azam (1992): Intense hydrolytic enzyme activity on marine aggregates and implications for rapid particle dissolution. *Nature*, **359**, 139–142.

Strickland, J. D. H. (1965): Production of organic matter at the primary stages of the marine food chain. pp. 477–610. In *Chemical Oceanography, Vol. 1*, ed. by J. P. Riley and G, Skirrow, Academic Press, New York.

Stuermer, D. H. and G. R. Harvey (1974): Humic substances from seawater. *Nature*, **250**, 480–481.

Stuermer, D. H. and G. R. Harvey (1977): The isolation of humic substances and alcoholsoluble organic matter from seawater. *Deep-Sea Res.*, **24**, 303–309.

Stuermer, D. H. and J. R. Payne (1976): Investigation of seawater and terrestrial humic substances with carbon-13 and proton nuclear magnetic resonance. *Geochim. Cosmochim. Acta*, **40**, 1109–1114.

Suzuki, S., K. Kuroe and R. Kusuda (1993): Characteristics of porin-like major outer membrane proteins of Listonella anguillara serotypes J-O-1, -2 and -3. *Biochem. Mol. Int.*, **32**, 605–613.

Suzuki, S., K. Kuroe, K. Yasue and R. Kusuda (1996): Antigenicity and N-terminal amino acid sequence of a 35 kDa porin-like protein of Listonella (Vibrio) anguillarum: Comparison among different serotypes and other bacterial species. *Lett. Appl. Microbiol.*, **23**, 257–260.

Suzuki, S., K. Kogure and E. Tanoue (1997): Immunochemical detection of dissolved proteins and their source organisms. *Mar. Ecol. Prog. Ser.*, 158, 1–9.

Tanoue, E. (1985): Distribution and chemical composition of particulate organic matter in the Pacific sector of the Antarctic Ocean. *Trans. Tokyo Univ. Fish.*, **6**, 43–57.

Tanoue, E. (1991): Electrophoretic separation of particulate proteins in seawater. pp. 163–169. In *Marine Particles: Analysis and Characterization*, ed. by D. C. Hurd and D. W. Spencer, American Geophysical Union, Geophys. Monogr., 63, Washington, D.C.

Tanoue, E. (1992): Occurrence and characterization of particulate proteins in the Pacific Ocean. *Deep-Sea Res.*, **39**, 743–761.

Tanoue, E. (1993): Distributional characteristics of DOC in the Central Equatorial Pacific. *J. Oceanogr.*, **49**, 625–639.

Tanoue, E. (1995): Detection of dissolved protein molecules in oceanic waters. *Mar. Chem.*, **51**, 239–252.

Tanoue, E. (1996): Characterization of the particulate protein in Pacific surface waters. *J. Mar. Res.*, **54**, 967–990.

Tanoue, E. and N. Handa (1979): Distribution of particulate organic carbon and nitrogen in the Bering Sea and northern North Pacific Ocean. *J. Oceanogr. Soc. Japan*, **35**, 47–69.

Tanoue, E. and S. Hara (1986): Ecological implications of faecal pellets produced by the Antarctic krill Euphausia superba in the Antarctic Ocean. *Mar. Biol.*, **91**, 359–369.

Tanoue, E., N. Handa and M. Kato (1982): Horizontal and vertical distributions of particulate organic matter in the Pacific sector of the Antarctic Ocean. *Trans. Tokyo Univ. Fish.*, **5**, 65–83.

Tanoue, E., S. Nishiyama, M. Kamo and A. Tsugita (1995): Bacterial membranes: Possible source of a major dissolved protein in seawater. *Geochim. Cosmochim. Acta*, **59**, 2643–2648.

Tanoue, E., M. Ishii and T. Midorikawa (1996): Discrete dissolved and particulate proteins in oceanic waters. *Limnol. Oceanogr.*, **41**, 1334–1343.

Tegelaar, E. W., J. W. deLeeuw, S. Derenne and C. Largeau (1989): A reappraisal of kerogen formation. *Geochim. Cosmochim. Acta*, **53**, 3103–3106.

Thomas, W. H. (1979): Anomalous nutrient-chlorophyll interrelationships in the offshore eastern tropical Pacific Ocean. *J. Mar. Res.*, **37**, 327–335.

Thurman, E. M. (1985): *Organic Geochemistry of Natural Waters*. W. Junk, Boston, 497 pp.

Towbin, H., T. Staehelin and J. Gordon (1979): Electrophoretic transfer of proteins from polyacrylamide gels to nitrocellulose sheets: procedure and some applications. *Proc. Natl. Acad. Sci. USA*, **76**, 4350–4354.

Tranvik, L. J. (1994): Colloidal and dissolved organic matter excreted by a mixotrophic flagellate during bacterivory and autotrophy. *Appl. Environ. Microbiol.*, **60**, 1884–1888.

Tranvik, L. J., E. B. Sherr and B. F. Sherr (1993): Uptake and utilization of "colloidal DOM" by heterotrophic flagellates in seawater. *Mar. Ecol. Prog. Ser.*, **92**, 301–309.

Trias, J. and H. Nikaido (1990): Protein D2 channel of the *Pseudomonas aeruginosa* outer membrane has a binding site for basic amino acids and peptides. *J. Biol. Chem.*, **265**, 15680–15684.

Trias, J., E. Y. Rosenberg and H. Nikaido (1988): Specificity of the glucose channel formed by protein D1 of *Pseudomonas aeruginosa*. *Bioch. Biophys. Acta*, **938**, 493–496.

Tupas, L. and I. Koike (1990): Amino acid and ammonium utilization by heterotrophic marine bacteria grown in enriched seawater. *Limnol. Oceanogr.*, **35**, 1145–1155.

Udenfriend, S., S. Stein, P. Bohlen, W. Dairman, W. Leimgraber and M. Weigle (1972): Fluorescamine: a reagent for assay of amino acids, peptides, protein and primary amines in the picomole range. *Science*, **178**, 871–872.

Volkman, J. K. (1986): A review of sterol markers for marine and terrigenous organic matter. *Org. Geochem.*, **9**, 83–99.

Wakeham, S. G. (1995): Lipid biomarkers for heterotrophic alteration of suspended particulate organic matter in oxygenated and anoxic water column of the ocean. *Deep-Sea Res.*, **42**, 1749–1771.

Walker, J. E. and I. M. Fearnley (1986): Sequence analysis of membrane proteins. pp. 243. In *Techniques for the Analysis of Membrane Proteins*, ed. by C. I. Ragan and R. J. Cherry, Chapman and Hall, London.

Wiggins, B. A. and M. Alexander (1985): Minimum bacterial density for bacteriophage replication: implications for significance of bacteriophages in natural ecosystems. *Appl. Environ. Microbiol.*, **49**, 19–23.

Williams, P. M. and E. R. M. Druffel (1987): Radiocarbon in dissolved organic matter in the central North Pacific Ocean. *Nature*, **330**, 247–248.

Williams, P. M. and E. R. M. Druffel. (1988): Dissolved organic matter in the ocean: comments on a controversy. *Oceanography*, **1**, 14–17.

Williams, P. M. and L. I. Gordon (1970): Carbon-13: carbon-12 ratios in dissolved and particulate organic matter in the sea. *Deep-Sea Res.*, **17**, 19–27.

Williams, P. M., K. J. Robertson, A. Soutar, S. M. Griffin and E. R. M. Druffel (1992): Isotopic signatures (^{14}C, ^{13}C, ^{15}N) as tracers of sources and cycling of soluble and particulate organic matter in the Santa Monica Basin, California. *Prog. Oceanogr.*, **30**, 253–290.

Worobec, E. A., N. L. Martin, W. D. McCubbin, C. M. Kay, G. D. Brayer and R. E. W. Hancock (1988): Large-scale purification and biochemical characterization of crystallization-grade porin protein P from *Pseudomonas aeruginosa*. *Biochim. Biophys. Acta*, **939**, 366–374.

Wyrtki, K. and B. Kilonsky (1984): Mean water and current structure during the Hawaii-to-Tahiti shuttle experiment. *J. Phys. Oceanogr.*, **14**, 242–254.

Yanada, M. and Y. Maita (1995): Regional and seasonal variations of biomass and biomediated materials in the North Pacific Ocean. pp. 293–306. In *Biogeochemical Processes and Ocean Flux in the Western Pacific*, ed. by H. Sakai and Y. Nozaki, Terra Sci. Publ., Tokyo.

Yoshihara, E., N. Gotoh, T. Nishino and T. Nakae (1996): Protein D2 porin of the *Pseudomonas aeruginosa* outer membrane bears the protease activity. *FEBS Lett.*, **394**, 179–182.

Zalman, L. S., H. Nikaido and Y. Kagawa (1980): Mitochondrial outer membrane contains a protein producing nonspecific diffusion channels. *J. Biol. Chem.*, **255**, 1771–1774.

Dynamics and Characterization of Marine Organic Matter, Eds., N. Handa, E. Tanoue and T. Hama, pp. 465–484.

Total Carbon and Nitrogen Contents and Molecular Composition of Water Soluble Organic Matter in the Marine Aerosols from Western North to Tropical Central Pacific

Kimitaka KAWAMURA

Institute of Low Temperature Science, Hokkaido University, N19 W8, Kita-ku, Sapporo 060-0819, Japan

(Original, Received 5 August 1996; accepted 28 October 1996)

Abstract—Marine aerosol samples collected from the western North to equatorial Pacific Ocean (16°S-35°N, 140°E-150°W), were studied for total carbon (TC: 0.069–5.27 μgm^{-3}, av. 0.39 μgm^{-3}) and nitrogen (TN: 0.026–1.44 μgm^{-3}, av. 0.12 μgm^{-3}) contents employing an elemental analyzer. Homologous series of dicarboxylic acids (C_2–C_{10}) and ω-oxocarboxylic acids (C_2–C_9) in the aerosols were determined by capillary gas chromatography (GC) and GC/mass spectrometer (GC/MS). Their molecular distributions indicated a predominance of oxalic acid (C_2) and glyoxylic acid (C_2). The concentrations of total diacids and ω-oxoacids were 10–250 ngm^{-3} (av. 63 ngm^{-3}) and 0.3–33.3 ngm^{-3} (av. 3.4 ngm^{-3}), respectively. The dicarboxylic acids showed a spatial distribution with higher concentrations in the western North Pacific region and lower concentrations in the central tropical Pacific. Relative abundances of total diacid carbon in TC, however, were highest (up to 16%) in the central equatorial Pacific, suggesting that small diacids are secondarily produced in the remote marine atmosphere. Further, ω-oxoacids were also abundant in the western North and tropical Pacific. In the remote marine atmosphere, oxalic acid and glyoxylic acid are likely produced by oxidative chain reactions occurring in the gaseous and particulate organic compounds which are transported long distances from the Asian continent.

INTRODUCTION

Recently, aerosols have received much attention because they potentially influence the global radiation budget by acting as negative climate forcings (Li *et al.*, 1996; Tegen *et al.*, 1996). In the marine atmosphere, aerosol concentrations are generally in a range of a few tens to several hundreds of micrograms per cubic meter (Warneck, 1988). They are mainly comprised of inorganic materials such as sea salt spray and mineral dust particles. In contrast, organic compounds are less abundant and account for less than 1 to 10% of the total aerosol masses. In spite of relatively small mass concentrations, organic aerosols are important because organic carbon exists in much smaller size fractions and has more number concentrations compared to sulfate aerosols (Novakov and Penner, 1993). Recently, Saxena *et al.* (1995) proposed that water-soluble organics alter the hygroscopic behavior of atmospheric particles and reported that in nonurban locations, organics enhance water absorption by inorganics and that organics account for 25–40% of the total water uptake in the relative humidity range of 80–88%. Thus, water soluble organic compounds potentially play an important role as cloud condensation nuclei in controlling cloud albedo (Kawamura and Usukura, 1993; Saxena and Hildemann, 1996).

Molecular studies of organic aerosols are important in marine geochemistry and atmospheric chemistry because their chemical structures and distributions provide unique information on the origin and long-range transport of terrestrially derived materials (e.g., Simoneit *et al.*, 1977; Gagosian, 1986; Kawamura, 1995) as well as photochemical reaction pathways of both marine and continental organic compounds (e.g., Kawamura and Gagosian, 1987). For example, lipid class compounds such as n-alkanes, fatty alcohols and fatty acids in marine aerosols have been used as source indicators of epicuticular waxes of terrestrial higher plants (Simoneit *et al.*, 1977; Gagosian, 1986; Kawamura and Gagosian, 1990; Kawamura, 1995). In contrast, the water soluble fraction has recently been found to contain a number of polar organic compounds such as low molecular weight dicarboxylic acids and other bifunctional carboxylic acids (Kawamura and Ikushima, 1993; Kawamura, 1993; Saxena and Hildemann, 1996). They are most likely produced in the atmosphere by photochemical reactions (Grosjean *et al.*, 1978; Hatakeyama *et al.*, 1987; Satsumabayashi *et al.*, 1990; Kawamura and Ikushima, 1993) and, thus, can be used as tracers to understand photochemical oxidation mechanisms and their precursor compounds (e.g., Kawamura and Gagosian, 1987).

As a homologous series or a single compound, dicarboxylic acids (C_2–C_{10}) have been reported in the mid-latitudinal continental aerosols (Grosjean *et al.*, 1978; Norton *et al.*, 1983; Yokouchi and Ambe, 1986; Kawamura and Kaplan, 1987; Satsumabayashi *et al.*, 1990; Kawamura and Ikushima, 1993, Sempéré and Kawamura, 1994), and Arctic and Antarctic atmosphere (Li and Winchester, 1993; Kawamura *et al.*, 1995a; Kawamura *et al.*, 1996a). However, water soluble organic compounds have rarely been studied in the marine atmosphere. In a previous study of marine aerosols from North Pacific, we detected a homologous series of dicarboxylic acids in the range of C_2–C_{10} (Kawamura and Usukura, 1993). These organic acids were found to account for up to 1.6% of the total aerosol mass and suggested that they significantly contribute to the aerosol carbon in the remote marine atmosphere.

In this study, we extended the aerosol sample collection in the central Pacific Ocean, including tropical regions (16°S-35°N, 140°E-150°W), for the molecular distributions of water soluble organic compounds. Aerosol samples are also analyzed for total carbon and nitrogen contents. Here, we report spatial distributions of C_2–C_{10} dicarboxylic acids, C_2–C_4 ω-oxocarboxylic acids, pyruvic acid and glyoxal as well as total carbon and nitrogen in the western North Pacific to Central Tropical Pacific. We also discuss the possible photochemical transformation and production of low molecular weight dicarboxylic acids in the remote marine atmosphere.

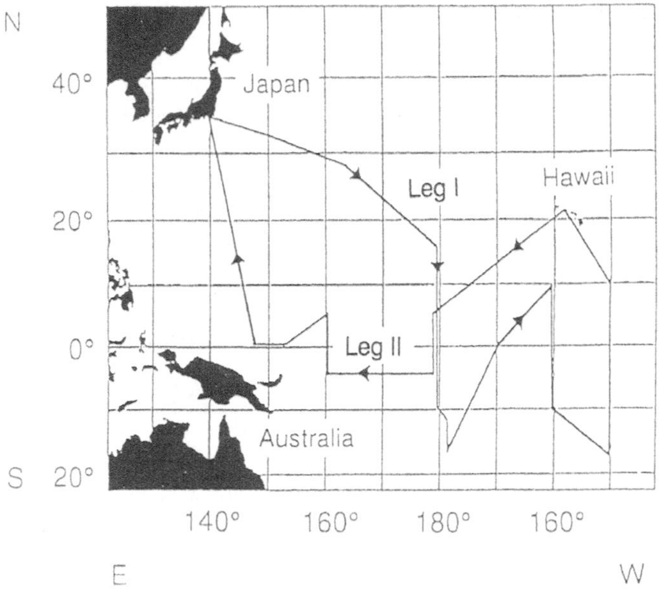

Fig. 1. Cruise track of R/V *Hakuho-Maru* (KH-90-2 and 3) for collecting marine aerosol samples in the Pacific Ocean. Leg I means a cruise from Japan to Hawaii whereas Leg II means a cruise from Hawaii to Japan.

SAMPLES AND METHODS

Thirty-three marine aerosol samples were collected on a precombusted (500°C) Pallflex quartz fiber filter (20 × 25 cm) by means of a high-volume air sampler during a cruise of R/V *Hakuho Maru* (KH90-2 and 3) (Leg I: September 3 to October 25, 1990 and Leg II: October 31 to December 14, 1990) in the western North Pacific to tropical Central Pacific. The cruise track is shown in Fig. 1. The air sampler was set up on the upper deck of the ship (*ca.* 14 m above the sea surface). The air sample operation during the cruise was automatically controlled with a wind sector (±45°) and wind speed (≥5 m/sec) system to avoid contamination from the ship's exhausts. The collected samples were stored in a precleaned glass jar with a Teflon-lined cap at –20°C prior to analysis.

One-eighth of the sample filters was analyzed for dicarboxylic acids by the procedure previously reported (Kawamura and Ikushima, 1993). Briefly, the filter was cut into small pieces and the carboxylic acids were extracted with pure water. The water extracts were concentrated by a rotary evaporator under vacuum and then by a nitrogen blow-down system almost to complete dryness. The carboxylic acids in the extracts were esterified with 14% BF_3/n-butanol to their dibutyl esters. The dicarboxylic acid dibutyl esters were determined with a Hewlett Packard HP 5890 capillary gas chromatograph (GC) equipped with an FID detector, and GC/MS (Finnigan-MAT ITS-40). The blank filters were always analyzed in parallel with the aerosol samples. However, the blank levels were negligible with small GC peaks of oxalic and phthalic acid butyl esters. All the data reported here were corrected for blanks. The recovery of oxalic acid was 70% and those of other diacids were better than 90%. Duplicate analyses showed an error of less than 13% for major diacids.

A part of the filter samples (*ca.* 10 cm^2) was cut off and subjected to a CHN elemental analyzer (Yanagimoto MT-3) for the determination of total carbon (TC) and total nitrogen (TN). Analytical errors for TC and TN based on duplicate analyses were within 10% (Sempéré and Kawamura, 1994).

RESULTS AND DISCUSSION

Concentrations of aerosol mass, total carbon and total nitrogen in the marine atmosphere

Table 1 presents aerosol mass concentrations and total carbon and nitrogen contents in the Pacific atmosphere. Total aerosol mass concentrations ranged from 19.2 to 403.5 μgm^{-3} with an average of 47.9 μgm^{-3}. The aerosol mass concentrations in the Pacific atmosphere are comparable to those (54–220 μgm^{-3}, av. 107 μgm^{-3}) reported in the urban atmosphere from Tokyo (Kawamura *et al.*, 1995b). The aerosol mass concentrations are on a similar level and no systematic trend is observed in the spatial distributions, except for one aerosol sample (QFF281) collected from the western North Pacific which showed a high aerosol mass concentration. This sample was collected during strong winds, suggesting a contribution from sea-salt spray. However, the total carbon concentration was not high in the aerosol sample (QFF 281, see Table 1). This suggests that seawater-derived carbonate is not a major source of the total carbon in the marine aerosols.

Concentration ranges of TC and TN were 0.069–5.27 μgm^{-3} (av. 0.38 μgm^{-3}) and 0.026–1.440 μgm^{-3} (av. 0.12 μgm^{-3}), respectively. These values are similar to TC (0.39–11, av. 1.7 ± 2.2) and TN (0.10–1.8, av. 0.37 ± 0.34) reported for the marine aerosols in the North Pacific (Uematsu *et al.*, 1995). However, these TC concentrations are about 50 times lower than those of the urban aerosol carbon (av. 19 μgm^{-3}) reported in Tokyo (Kawamura *et al.*, 1995b). Higher TC values are obtained in the western North Pacific near Japanese Islands (Fig. 2(a)), suggesting a significant contribution of continent-derived organic materials over the Pacific Ocean. The abundances of TC in aerosol mass (0.14–4.8%, av. 0.75%), showed higher values in the western North Pacific and western tropical Pacific (Fig. 2(b)). These results suggest a large contribution of terrestrial organic matter over the western Pacific Ocean. However, the TC/aerosol mass ratios in the marine atmosphere are much lower than those (8–39%, av. 20%) reported for urban aerosols from Tokyo (Kawamura and Ikushima, 1993), indicating that marine organic aerosols are diluted in the marine atmosphere by inorganic materials.

Similar spatial distributions are obtained for total nitrogen contents. As seen in Fig. 3(a), the highest TN value was obtained in the western North Pacific near

Table 1. Concentrations of aerosol mass, total carbon (TC) and nitrogen (TN) in the marine aerosols from the Pacific Ocean.

Sample ID	Latitude	Longitude	Ambi. Temp.	Aero. Mass	TC	TN	C/N ratios	TC/Aero	TN/Aero
(Unit)	N	E	C	ug/m3	ng/m3	ng/m3		%	%
QFF 249	33.75	145.01	27.2	55.0	215	28	7.6	0.39	0.05
QFF 250	31.38	153.97	27.2	26.3	279	103	2.7	1.06	0.39
QFF 251	27.60	162.35	26.8	19.5	286	98	2.9	1.47	0.50
QFF 252	23.03	169.64	28.0	30.6	122	31	3.9	0.40	0.10
QFF 253	18.05	176.01	28.2	37.9	106	57	1.9	0.28	0.15
QFF 254	9.75	179.99	28.4	29.6	113	70	1.6	0.38	0.24
QFF 255	3.94	179.99	28.6	30.9	82	90	0.9	0.27	0.29
QFF 256	-1.96	179.98	29.1	33.7	147	85	1.7	0.44	0.25
QFF 257	-7.63	180.02	28.7	32.4	227	135	1.7	0.70	0.42
QFF 258	-13.27	180.50	27.3	35.0	121	40	3.0	0.35	0.11
QFF 259	-12.49	183.11	28.1	29.7	159	59	2.7	0.54	0.20
QFF 260	-5.12	187.17	28.4	20.5	117	47	2.5	0.57	0.23
QFF 261	4.15	194.17	27.9	50.2	70	36	1.9	0.14	0.07
QFF 262	2.66	199.99	26.9	39.7	150	82	1.8	0.38	0.21
QFF 263	-2.46	200.02	27.7	31.3	194	98	2.0	0.62	0.31
QFF 264	-7.51	199.94	27.8	28.2	78	104	0.7	0.28	0.37
QFF 265	-12.84	203.97	26.9	49.8	69	36	1.9	0.14	0.07
QFF 266	-4.51	210.05	27.3	32.0	91	85	1.1	0.28	0.27
QFF 267	5.34	210.00	27.1	48.3	106	68	1.6	0.22	0.14
QFF 268	15.16	206.25	26.9	41.4	94	77	1.2	0.23	0.19
QFF 269	15.97	194.50	27.1	42.2	293	63	4.6	0.69	0.15
QFF 270	7.92	183.08	28.3	28.6	202	87	2.3	0.71	0.31
QFF 271	-0.09	179.02	28.8	19.2	106	80	1.3	0.55	0.41
QFF 272	-5.01	174.05	29.4	20.9	203	63	3.2	0.97	0.30
QFF 273	-4.99	164.40	28.1	43.7	79	26	3.1	0.18	0.06
QFF 274	0.02	160.02	28.5	21.7	418	113	3.7	1.93	0.52
QFF 276	2.39	154.81	27.9	37.8	1027	124	8.3	2.72	0.33
QFF 277	1.60	148.99	28.3	33.4	576	111	5.2	1.72	0.33
QFF 278	9.04	146.76	27.6	34.6	164	64	2.6	0.47	0.18
QFF 279	16.54	143.99	27.0	50.2	142	64	2.2	0.28	0.13
QFF 280	22.54	142.56	25.7	33.2	143	74	1.9	0.43	0.22
QFF 281	29.48	140.86	19.0	403.5	1299	289	4.5	0.32	0.07
QFF 282	34.33	139.94	12.3	109.4	5270	1437	3.7	4.82	1.31

Aerosol samples were collected during KH-92-2 and 3 cruise in September 3 to December 13, 1990.

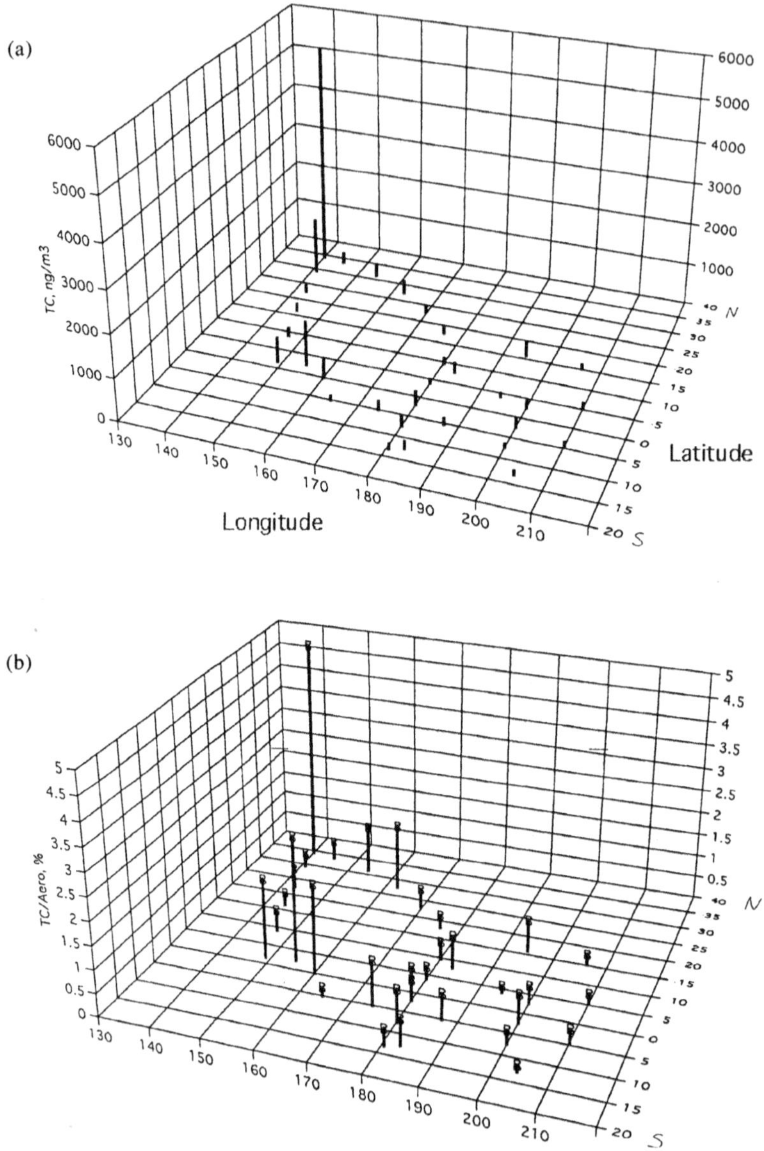

Fig. 2. Distributions of (a) total carbon (TC)(ngm^{-3}) and (b) TC/aerosol mass ratios(%) in the marine atmosphere from Pacific Ocean.

the Japanese Islands. Except for higher values in the western Pacific atmosphere, aerosol TN contents are rather constant over the Pacific Ocean. TN comprise 0.05–1.3% (av. 0.27%) of aerosol mass in the marine atmosphere. The TN values for the marine aerosols (av. 0.12 μgm^{-3}) are about 40 times lower than those (av. 5.3 μgm^{-3}) reported for the Tokyo atmosphere (Kawamura *et al.*, 1995b), being

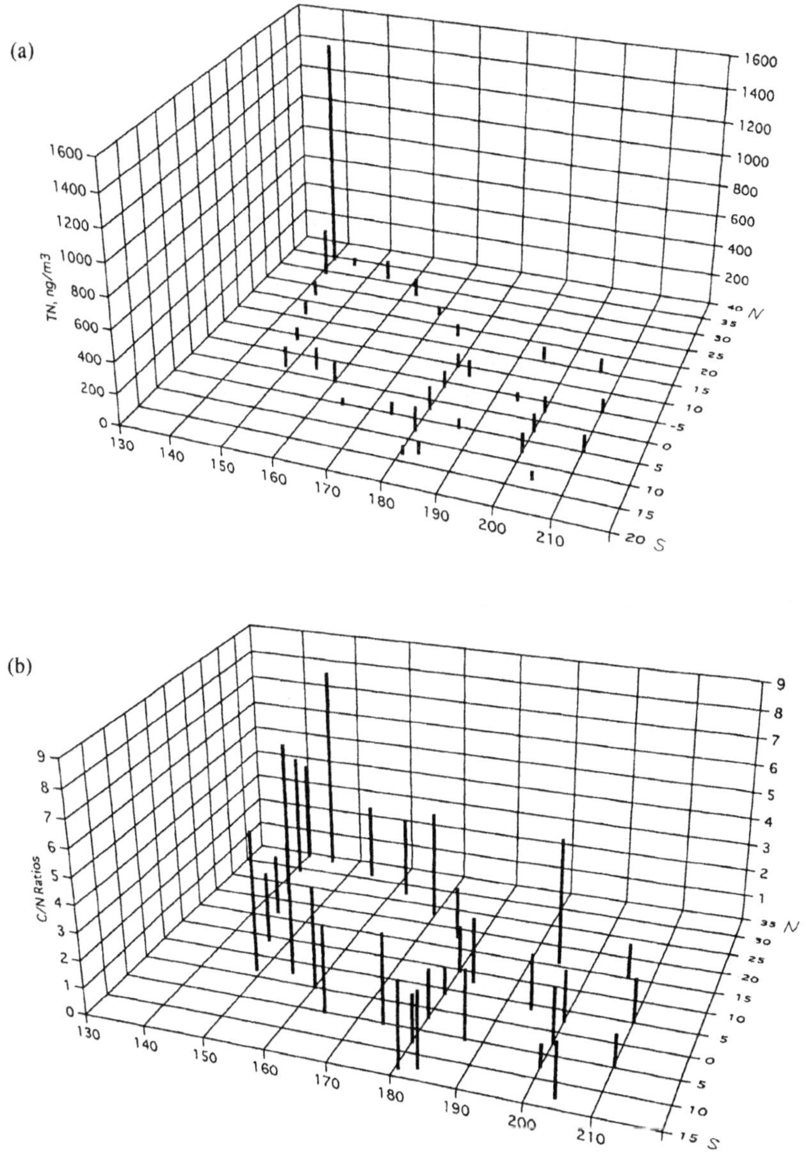

Fig. 3. Distributions of (a) total nitrogen (TN) contents (ngm^{-3}) and (b) C/N weight ratios in the marine atmosphere from Pacific Ocean.

similar to the case of TC. This suggests that particulate nitrogen is of continental origin and transported long distances to the remote marine atmosphere. However, C/N weight ratios fluctuate in the Pacific Ocean. As shown in Fig. 3(b), the marine aerosol C/N ratios in the Pacific are in a range of *ca.* 1–8 (av. 2.8). However, they are fairly low (values: 1–2) in the equatorial Central Pacific. Such low values are sometimes observed in the aerosols from urban Tokyo during a photochemical smog event in winter (Kawamura *et al.*, 1995b). These results may suggest that particulate nitrogen in the marine atmosphere is partly due to the formation of particulate nitrate through the oxidation of gaseous NO_x (Warneck, 1988), which originates from combustion sources in the Asian continent and is transported long distances over the Pacific. Alternatively, inorganic ammonium salt may contribute to the lower C/N ratios in the marine atmosphere (Uematsu *et al.*, 1995).

Molecular distributions of diacids, ketoacids, dicarbonyls

Figure 4 shows a representative capillary gas chromatogram of the derivatized water soluble fraction extracted from the marine aerosols. A homologous series of low molecular weight (LMW) dicarboxylic acids (C_2–C_{10}) and ω-oxocarboxylic acids (C_2–C_9) were detected. Pyruvic acid and glyoxal were also detected. Figure 5 gives the chemical structures of the water soluble organic acids and aldehydes detected in the marine aerosols. These polar organic compounds have been reported in aerosols and rainwater from Tokyo (Kawamura and Ikushima, 1993; Kawamura, 1993; Sempéré and Kawamura, 1994; Imai and Kawamura, 1996), from Los Angeles (Kawamura *et al.*, 1985; Kawamura and Kaplan, 1987; Kawamura *et al.*, 1996b) and remote marine atmospheres (Kawamura and

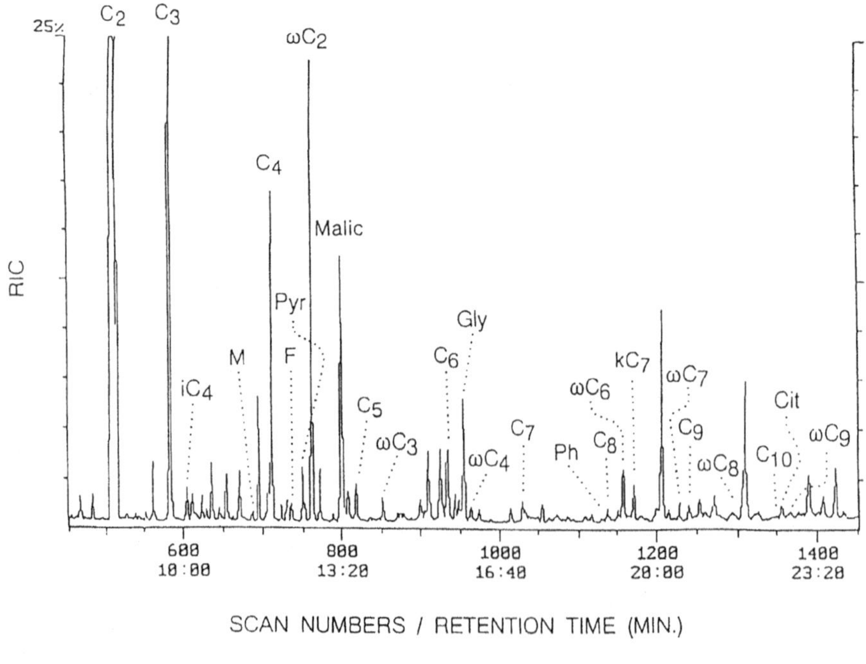

Fig. 4. A typical capillary gas chromatogram of the diacids fraction collected in the remote marine aerosol samples from the equatorial Pacific atmosphere. For abbreviation and compound structure, see Fig.5 and Table 2.

Usukura, 1993; Sempéré and Kawamura, 1996).

In all the aerosol samples, oxalic acid (C_2) is the most abundant species among the water soluble organic compounds detected. The second most abundant diacid is generally malonic acid (C_3). Few aerosol samples showed a predominance of C_4 over C_3 diacid. Adipic acid (C_6), fumaric acid (C_4 unsaturated diacid) and malic acid (C4 hydroxy diacid) were also detected as major diacid species. These diacid species have been reported as aerosol constituents in the continental atmosphere (Kawamura and Ikushima, 1993). 4-Oxoheptanedioic acid (C_7 keto diacid) which was identified in the marine atmosphere (Sakaguchi and Kawamura, 1994), was also detected in the Pacific aerosol samples studied, although its concentrations are low. As shown in Table 2, total concentrations of LMW diacids are in a range of 10–250 ngm^{-3} (av. 63 ngm^{-3}). These values are *ca.* 10

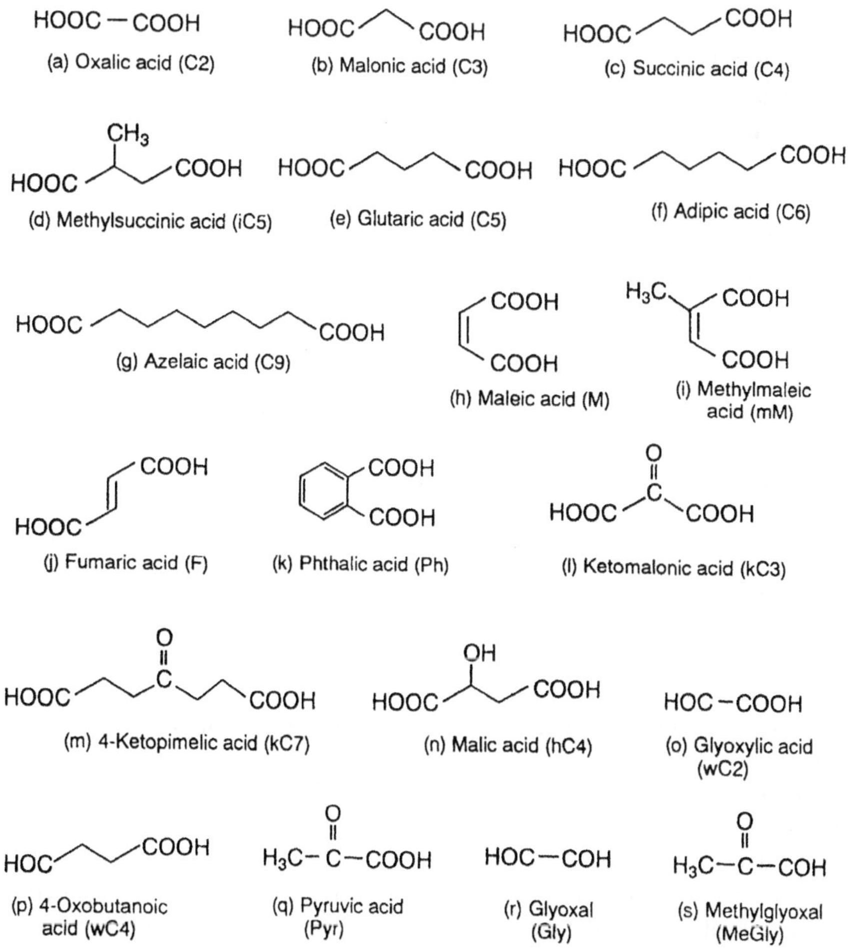

Fig. 5. Chemical structures of polar organic compounds detected in the marine atmosphere.

Table 2. Concentrations of dicarboxylic acids and related polar compounds in the remote marine aerosol samples collected from western North to tropical Central Pacific (33 samples).

Compounds	Carbon number	Abbr.	Concentrations, ngm-3 Range	Average ± stdev.
Dicarboxylic acids				
Oxalic	2	C2	6.5 - 161	40 ± 37
Malonic	3	C3	1.3 - 54	11 ± 11
Succinic	4	C4	0.29 - 16	2.8 ± 4.2
Glutaric	5	C5	<0.01 - 6.3	0.64 ± 1.2
Adipic	6	C6	0.60 - 13	2.1 ± 2.2
Pimelic	7	C7	0.03 - 3.3	0.34 ± 0.63
Suberic	8	C8	0.09 - 3.3	0.38 ± 0.56
Azelaic	9	C9	0.17 - 4.0	0.57 ± 0.63
Sebacic	10	C10	<0.01 - 3.9	0.23 ± 0.66
Methylmalonic	4	iC4	0.05 - 1.3	0.31 ± 0.30
Maleic	4	M	0.04 - 3.8	0.31 ± 0.70
Fumaric	4	F	0.16 - 4.5	1.0 ± 0.89
Phthalic	8	Ph	<0.01 - 7.6	0.66 ± 1.3
Malic	4	Malic	<0.01 - 17	2.2 ± 4.1
4-oxopimelic	7	kC7	<0.01 - 2.5	0.35 ± 0.66
Total			10 - 248	62 ± 60
w-Oxoacids				
Glyoxylic	2	wC2	0.13 - 18	2.0 ± 3.6
3-Oxopropanoic	3	wC3	0.02 - 1.7	0.19 ± 0.33
4-Oxobutanoic	4	wC4	0.01 - 2.3	0.31 ± 0.50
6-Oxohexanoic	6	wC6	0.04 - 3.4	0.39 ± 0.63
7-Oxoheptanoic	7	wC7	0.03 - 3.4	0.24 ± 0.61
8-Oxoocatnoic	8	wC8	0.03 - 2.2	0.38 ± 0.67
9-Oxononanoic	9	wC9	0.02 - 2.4	0.29 ± 0.55
Total			0.27 - 33	3.4 ± 6.4
a-Ketocarboxylic				
Pyruvic	3	Pyr	0.03 - 6.0	0.52 ± 1.2
Dicarbonyl				
Glyoxal	2	Gly	0.11 - 19	1.2 ± 3.3
Tricarboxylic				
Citric	6	Cit	0.03 - 2.3	0.4 ± 0.6

times lower than those of continental aerosols (Kawamura and Ikushima, 1993) and comparable to those of the previously reported marine aerosol samples from the northern North Pacific (27-32°N, 134°E-128°W, Kawamura and Usukura, 1993).

Relative abundances of C_2 plus C_3 diacids in the total diacids are $82 \pm 7\%$ in the Pacific aerosol samples and reach to a maximum (89%) in the tropical Pacific. This is in contrast to the continental aerosols collected in the mid-latitude of Japan which showed lower values (av. $64 \pm 9\%$) for the C_2 plus C_3 diacids (Kawamura and Ikushima, 1993). Accordingly, the abundances of C_4 diacid relative to total diacids in the marine atmosphere are lower (av. 3.6 ± 1.7%) than those (av. 7.9 ± 1.7%) of the continental aerosols (Kawamura and Ikushima, 1993). An increased relative abundance of C_2 and C_3 diacids and the depletion of C_4 diacid in the marine atmosphere suggest that these species are linked through photochemical chain reactions. It is likely that the smaller diacids are photochemically produced by the oxidation of C_4 diacid during the long-range atmospheric transport over the Pacific Ocean, a point to be discussed later.

Glyoxylic acid (ωC_2) is generally the most abundant in an homologue of ω-oxocarboxylic acids, followed by ωC_4 or ωC_6. C_9 ω-oxocarboxylic acid, which has been considered as a specific photooxidation product of unsaturated fatty acids (Kawamura and Gagosian, 1987) is relatively abundant. Concentration ranges of total ω-oxoacids are 0.3–33.3 ngm^{-3} (av. 3.4 ngm^{-3}). Although their concentrations are less abundant than those of diacids, they are possible precursors of dicarboxylic acids (Kawamura, 1993). Pyruvic acid and glyoxal are generally less abundant than oxalic acid. Their concentrations were 0.11–19.4 ngm^{-3} (av. 1.23 ngm^{-3}) and 0.03–5.98 ngm^{-3} (av. 0.52 ngm^{-3}), respectively.

Spatial distributions of diacids and related compounds

Figure 6(a) shows a spatial distribution of dicarboxylic acid concentrations in the Pacific atmosphere. Higher concentrations are found in the western Pacific. However, they are much lower than those (90–1400 ngm^{-3}, av. 480 ± 380 ngm^{-3}) of the continental aerosols from Tokyo (Kawamura and Ikushima, 1993). This suggests that the diacids in the marine atmosphere originate on the Asian continent and are scavenged by dry and wet processes during a west-to-east transport. Being different from the spatial distributions of total carbon which showed a maximum in the western North Pacific (Fig. 2(a)), dicarboxylic acids indicated higher concentrations in the equatorial western Pacific. Although this area is generally considered as a trade wind regime during the cruise in this region, winds predominantly came from the west or southwest. Thus, diacids and their precursors probably originate from southeastern islands such as New Guinea and Kalimantan. These results suggest that dicarboxylic acids and their precursors are emitted from tropical vegetation and transported over the western tropical Pacific.

Relative abundances of total diacid carbon content in the total aerosol carbon (diacid-C/TC, %) showed a spatial distribution, which is different from those of diacid concentrations and TC contents. In the western North Pacific, the diacid-C/TC are relatively low (2–5%). In contrast, they increased in the central equatorial Pacific, where the relative abundances showed high values of up to 16% (see Fig. 6(b)). The values (1–16%, av. 8.7%) obtained in the marine atmosphere are extensively high compared to those (0.18–1.8%, av. 0.95%) for the urban aerosols from Tokyo (Kawamura and Ikushima, 1993). This suggests that the diacids are produced in the marine atmosphere by a photochemical reaction and/or selective scavenging of the TC over the diacids in the marine atmosphere. However, the latter is unlikely. Thus, the enhanced diacid-C/TC values, coupled with an enhanced relative abundance of oxalic acid in the central

(a)

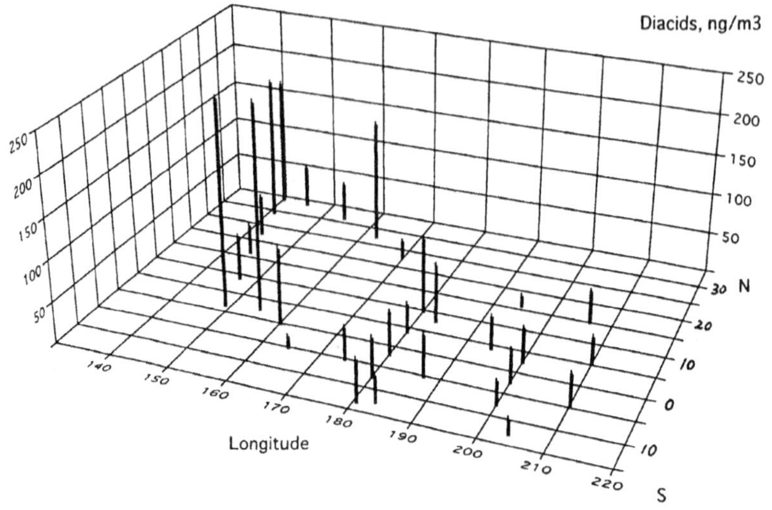

(b)

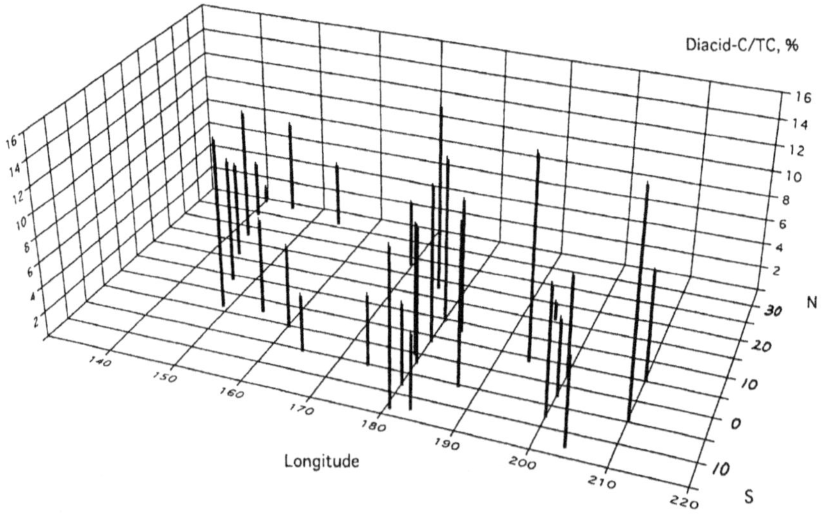

Fig. 6. Spatial distributions of (a) concentrations (ngm^{-3}) low molecular weight dicarboxylic acids (C_2–C_{10}), and (b) relative abundance(%) of the diacid-C in the total aerosol carbon in the remote marine aerosols from the Pacific Ocean.

equatorial Pacific, are most likely interpreted by a secondary production of small diacids in the marine atmosphere through a gas-to-particle conversion of volatile organic compounds, which are probably emitted from land vegetation and anthropogenic sources on the Asian continent, as well as heterogeneous oxidation of the organic compounds on aerosols.

ω-Oxoacids showed a spatial distribution similar to that of dicarboxylic

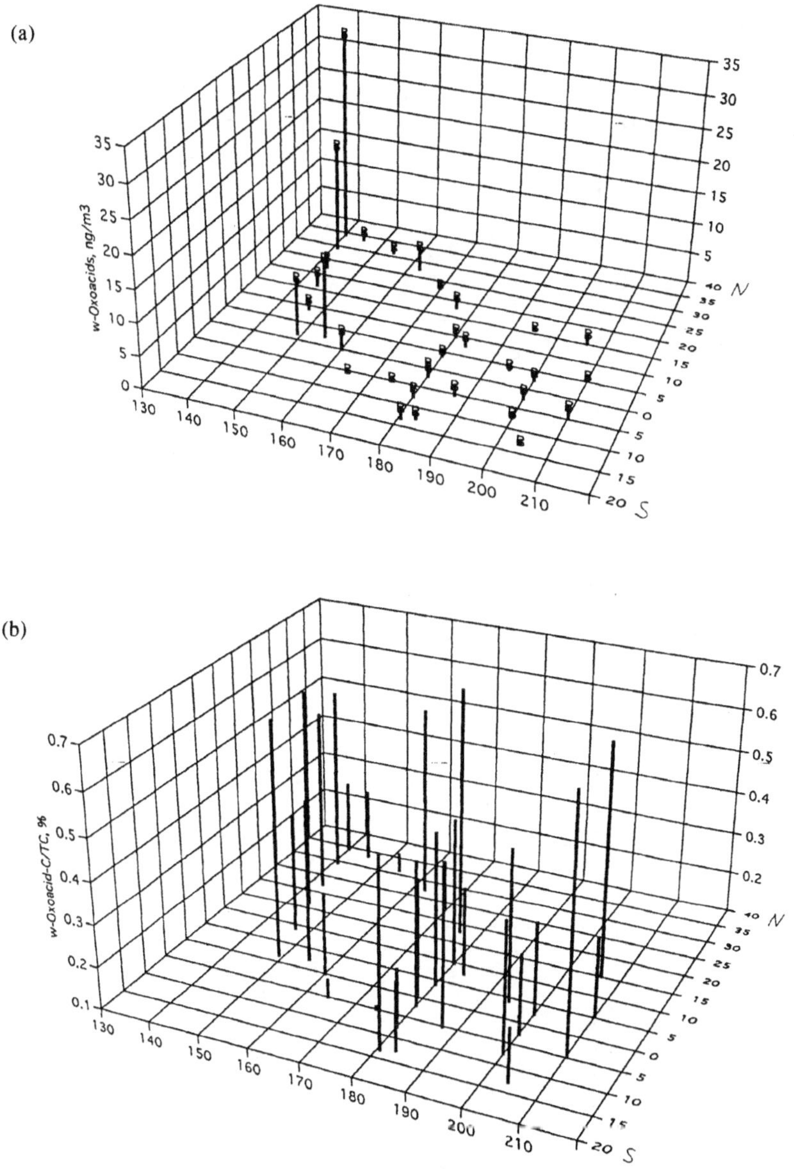

Fig. 7. Spatial distributions of (a) ω-oxocarboxylic acids (ngm^{-3}) and (b) relative abundance (%) of ω-oxocarboxylic acid-C in the aerosol carbon in the atmosphere of the Pacific Ocean.

acids, as shown in Fig. 7(a). Their concentrations are high in the western Pacific and low in the central Pacific. However, relative abundances of the ω-oxoacids in aerosol TC are higher in the equatorial regions and rather lower in the mid latitudes (Fig. 7(b)). These results are similar to diacids, as stated above, and

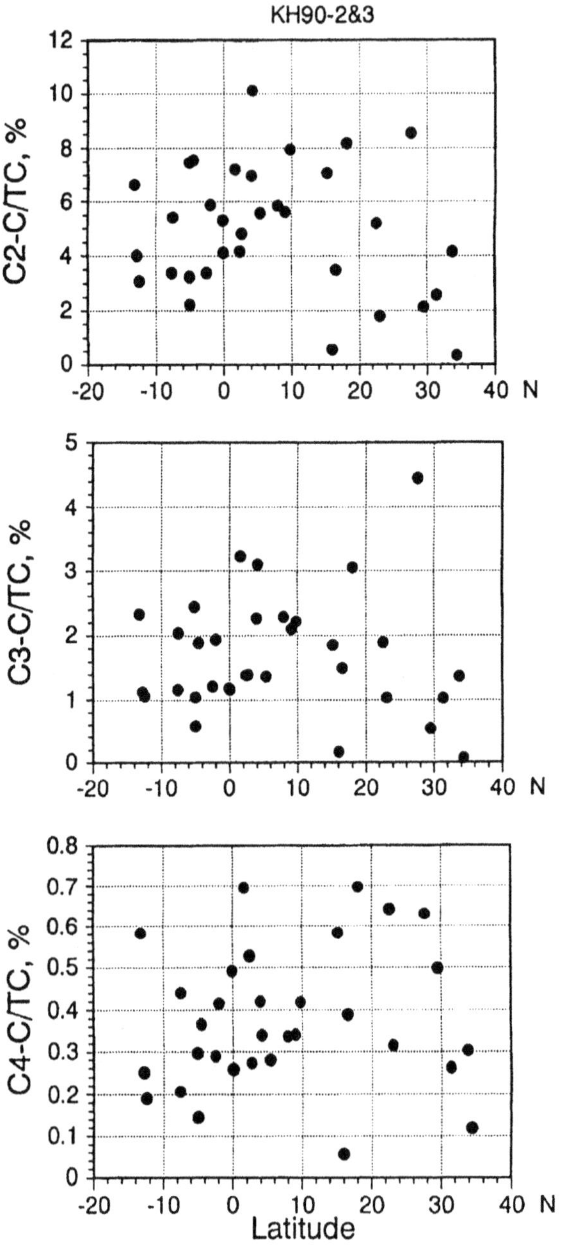

Fig. 8. Latitudinal distributions of small dicarboxylic acid-carbon in TC (%) in the Pacific atmosphere.

suggest that the oxoacids are also produced in the marine atmosphere during a long-range transport of terrestrially derived volatile and particulate organics. This is reasonable because ω-oxoacids have been considered as photochemical oxidation intermediates of cycloolefins and aromatic hydrocarbons (Kawamura, 1993) as well as biogenic unsaturated fatty acids (Kawamura and Gagosian, 1987).

Pyruvic acid and glyoxal showed spatial distributions similar to those of glyoxylic acid. Relative abundances of these compounds in TC are higher in the central tropical Pacific (the results are not shown here).

Photochemical production of diacids

Figure 8 shows a latitudinal distribution of relative abundance of dicarboxylic acid-C/TC for C_2–C_4 diacids. The oxalic acid-C/TC ratios fluctuate in a range of 0.1–10% (av. 5.0%). However, they seem to maximize in the tropical Pacific and decrease towards the mid-latitudes. Malonic acid and succinic acids also indicate a latitudinal distribution similar to oxalic acid. These latitudinal distributions demonstrate that oxalic acid, malonic and succinic acids may be preferentially produced in the tropical Pacific, probably by photochemical oxidation of both gaseous and particulate precursor organics. Based on seasonal

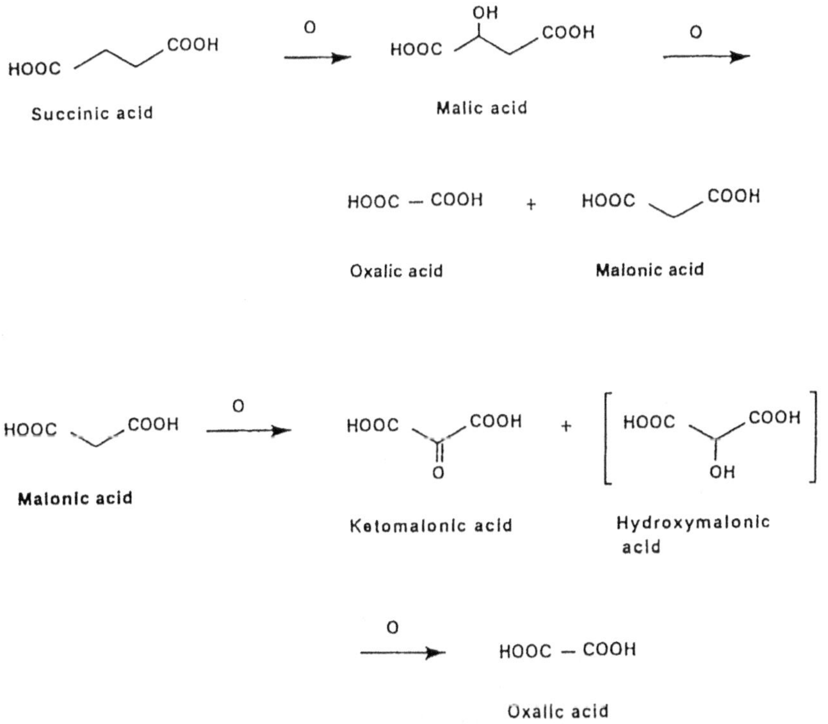

Fig. 9. Reaction mechanisms proposed for the photochemical production of oxalic (C_2) and malonic (C_3) and malic acids from succinic (C_4) acid in the marine atmosphere. Hydroxymalonic acid has not been detected in the marine aerosols.

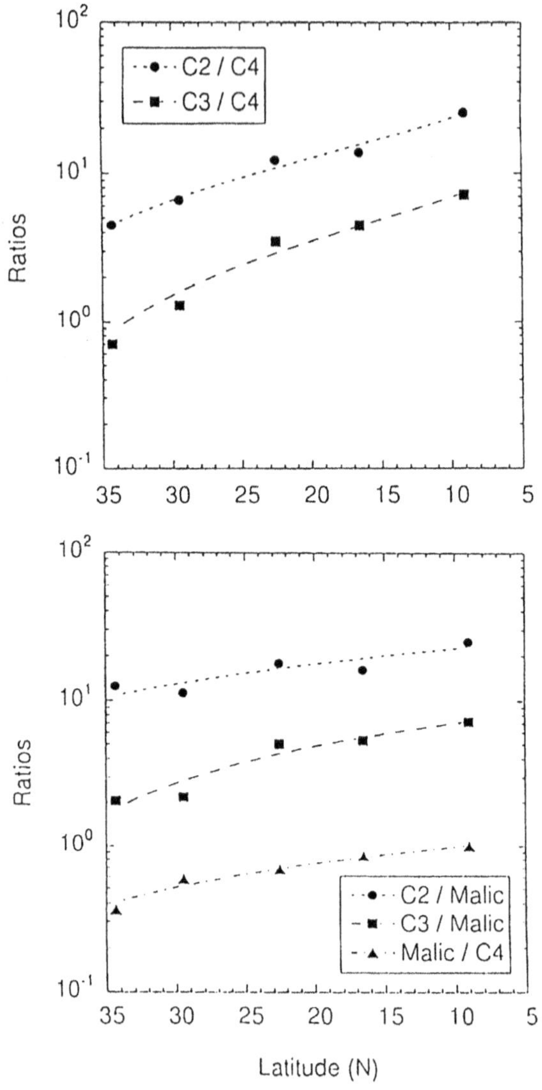

Fig. 10. Latitudinal distributions of the concentration ratios of C_2 and C_3 diacids to C_4 diacid and the ratios of C_2 and C_3 diacids to malic acid and malic to C_4 diacid in the western Pacific

observation of low molecular weight dicarboxylic acids in the Tokyo atmosphere, Kawamura and Ikushima (1993) proposed a photochemical reaction scheme for the production of oxalic acid and malonic acid which are enhanced in the summer time (see Fig. 9). With an analogy of the photochemical production of oxalic acid in summer (Kawamura and Ikushima, 1993), evidence to support such a process could be obtained in the marine atmosphere during the cruise at the latitudinal transect between low and mid-latitudes in the western North Pacific.

During Leg. II, aerosol sample collection was conducted in the latitudinal transect from the tropical Pacific (9°N, 147°E) to the western North Pacific near the Japanese Islands (34°N, 140°E). Figure 10 shows latitudinal changes in the abundances of C_2, C_3, C_4 diacids and malic acid relative to succinic and/or malic acid. The C_2/C_4 and C_3/C_4 concentration ratios largely increase from mid to low latitude in the western North Pacific. These results are consistent with the reaction scheme (Fig. 9) and support the idea that C_4 diacid is photooxidized to result in C_3 and C_2 diacids in the marine atmosphere, especially, the tropical Pacific. It is obvious that the precursor of the C_2 and the C_3 diacid is not only C_4 diacid but also other organic compound classes. However, this study indicates that longer chain diacids such as C_4 can serve as precursors and intermediates to result in C_2 and C_3 diacids in the low-latitude atmosphere where solar radiation is stronger and ambient temperature is higher. Interestingly, malic acid (hydroxy succinic acid) was found to be abundant in the tropical marine aerosols and the malic/succinic acid ratios also increase from mid latitudes (*ca.* 0.4) to lower latitudes (*ca.* 1.0), suggesting that the hydroxy diacid is an intermediate in the oxidative chain reactions where oxalic and malonic acids are produced (see Fig. 9).

Interestingly, oxalic acid-carbon/TC ratios (0.1–10%) in the marine aerosols are *ca.* 10 times greater than those (0.05–1%) of continental urban aerosol samples from Tokyo, which showed a summer maximum of C_2-diacid carbon/TC ratios (Kawamura and Ikushima, 1993), and also that (0.8%) of rural summer aerosols collected from Mt. Haruna, Gunma Prefecture, Japan (Kawamura and Ikushima, unpublished results). This probably indicates that photochemical transformations of both gas and aerosol phase organic compounds are much more significant in the marine atmosphere than in the continental atmosphere. 4-oxocarboxylic acids have been proposed as a likely source of C_4 diacid in the marine atmosphere (Kawamura and Gagosian, 1990; Nishikiori *et al.*, 1996). These mid-chain ketocarboxylic acids are produced in the marine atmosphere by a further oxidation of semi-volatile monocarboxylic acids which are photochemical reaction products of biogenic unsaturated monocarboxylic acid (Kawamura and Gagosian, 1990). 4-Oxononanoic and 4-oxoundecanoic acids have been reported as major mid-chain oxocarboxylic acids in the remote marine atmosphere from the North Pacific (Kawamura and Gagosian, 1990) and equatorial Pacific (Kawamura and Tanaka, unpublished results, 1994). These 4-oxocarboxylic acids may be oxidized in the marine atmosphere to result in succinic acid (C_4). Further, 4-oxopimelic acid was detected in the marine aerosols studied (Table 1). This compound can also serve as a precursor of C_4 diacid.

In the Pacific aerosols, both *cis* and *trans* configurations of C_4 unsaturated diacid were detected. However, their *cis/trans* composition is very different from those of continental aerosols in which maleic acid is much more abundant than fumaric acid. In the Pacific atmosphere, trans configuration (fumaric acid) is found to be much more abundant than *cis* configuration (maleic acid) except for one sample (QFF 281) which was collected near Japanese Islands. The maleic acid to fumaric acid ratios were very low (0.06 to 1.33, av. 0.26 ± 0.24) in comparison to those (0.8–3.9, av. 1.5 ± 1.2) reported in the Tokyo atmosphere (Kawamura and Ikushima, 1993) and that (0.90) of rural aerosols from Mt. Haruna (Kawamura and Ikushima, unpublished data). Photochemical oxidation of aromatic hydrocarbons such as benzene and toluene predominantly produce maleic acid and methylmaleic acid (Bandow and Washida, 1985: Bandow *et al.*, 1985; Kawamura and Ikushima, 1993), the former is further transformed to *trans* configuration (fumaric acid) in the atmosphere (Kawamura and Ikushima, 1993). Thus, the depletion of maleic acid suggests that photo-isomerization of maleic to fumaric acid is significant in the marine atmosphere or that maleic acid is selectively decomposed. These C_4 unsaturated diacids can also be photo-oxidized to result in C_2 diacid (oxalic acid).

SUMMARY AND CONCLUSIONS

(1) Total carbon (TC) and nitrogen (TN) concentrations were 0.069–5.27 μgm^{-3} (av. 0.38 μgm^{-3}) and 0.026–1.44 μgm^{-3} (av. 0.12 μgm^{-3}), respectively. TC and TN comprised 0.14–4.8% (av. 0.75%) and 0.05–1.3% (av. 0.27%) of total aerosol mass, respectively. The C/N weight ratios in the western Pacific are 4–8 (av. 2.8). However, they are fairly low (values: 1–2) in the equatorial central Pacific, suggesting a gas-to-particle conversion of NO_x during long-range transport over the tropical Central Pacific.

(2) This study demonstrates that homologous series of low molecular weight dicarboxylic acids are present in the remote marine aerosols from the western North Pacific to the tropical Pacific Ocean and indicates that oxalic acid is the most abundant species followed by malonic acid and then succinic acid. Another series of bifunctional carboxylic acids, i.e., ω-oxocarboxylic acids, are also present in the marine atmosphere with the smallest species (glyoxylic acid) being most abundant. These small dicarboxylic acids and ω-oxocarboxylic acids were found to contribute up to 16% and 0.7% of total marine aerosol carbon, respectively.

(3) The dicarboxylic acids and their precursors are transported long distances to the Pacific through the atmosphere from the Asian continent. However, spatial distributions of the diacids relative to total aerosol carbon suggested that small diacids are largely produced in the marine atmosphere in the equatorial Pacific where solar radiation is more extensive. Latitudinal distributions of C_2–C_4 diacids in the western North Pacific supported a selective production of oxalic acid by the photochemical oxidation of precursor compounds including longer chain diacids.

(4) Because water soluble organic compounds detected in the marine aerosols are highly oxygenated, their functional groups such as carboxyl and aldehyde should interact with water vapors in the marine atmosphere and act as a cloud condensation nuclei (CCN). They may potentially influence on the radiation balance in the marine atmospheric environments.

Acknowledgements—We thank the cruise members of R/V *Hakuho Maru* (KH90-2 and 3) for their collaboration in collecting the samples. We also acknowledge the help of Drs. H. Inoue, M. Ishii and Y. Suzuki in the aerosol sampling. Total carbon and nitrogen measurements were performed by T. Satoh. Dicarboxylic acids and related compounds were measured by F. Sakaguchi. Experimental work of this study has been done at the Department of Chemistry, Faculty of Science, Tokyo Metropolitan University. This study was supported in part by the Ministry of Education, Science and Culture through Grant-in-Aid 03248104.

REFERENCES

Bandow, H. and N. Washida (1985): Ring-cleavage reactions of aromatic hydrocarbons studied by FTIR spectroscopy II. Photooxidation of o-, m-, and p-xylenes in the NO_x-air system. *Bull. Chem. Soc. Jpn.*, **58**, 2541–2548.

Bandow, H., N. Washida and H. Akimoto (1985): Ring-cleavage reactions of aromatic hydrocarbons studied by FTIR spectroscopy. I. Photooxidation of toluene and benzene in the NO_x-air system. *Bull. Chem. Soc. Jpn.*, **58**, 2531–2540.

Gagosian, R. B. (1986): The air-sea exchange of particulate organic matter: The sources and long-range transport of lipids in aerosols. pp. 409–442. In *The Role of Air-Sea Exchange in Geochemical Cycling*, ed. by P. Buart-Menard, D.Reidel, Dordrecht, The Netherlands.

Grosjean, D., K. Van Cauwenberghe, J. P. Schmid, P. E. Kelley and J. N. Pitts, Jr. (1978): Identification of C3–C10 aliphatic dicarboxylic acids in airborne particulate matter. *Environ. Sci. Technol.*, **12**, 313–317.

Hatakeyama, S., M. Ohno, J. Weng, H. Takagi and H. Akimoto (1987): Mechanism for the formation of gaseous and particulate products from ozone-cycloalkene reactions in air. *Environ. Sci. Technol.*, **21**, 52–57.

Imai, Y. and K. Kawamura (1996): Distributions and seasonal variations of dicarboxylic acids, ketocarboxylic acids and dicarbonyls in rain, snow and aerosols in Tokyo. *Res. Org. Geochem.*, **11**, 61–66.

Kawamura, K. (1993): Identification of C2–C10 ω-oxocarboxylic acids, pyruvic acid and C2–C3 α-dicarbonyls in wet precipitation and aerosol samples by capillary GC and GC/MS. *Anal. Chem.*, **65**, 3505–3511.

Kawamura, K. (1995): Land-derived lipid class compounds in the deep-sea sediments and marine aerosols from North Pacific. pp. 31–51. In *Biogeochemical Processes and Ocean Flux in the Western Pacific*, ed. by H. Sakai and Y. Nozaki, Terrapub, Tokyo.

Kawamura, K. and R. B. Gagosian (1987): Implications of ω-oxocarboxylic acids in the remote marine atmosphere for photo-oxidation of unsaturated fatty acids. *Nature*, **325**, 330–332.

Kawamura, K. and R. B. Gagosian (1990): Mid-chain ketocarboxylic acids in the remote marine atmosphere: distribution patterns and possible formation mechanisms. *J. Atmos. Chem.*, **11**, 107–122.

Kawamura, K. and K. Ikushima (1993): Seasonal changes in the distribution of dicarboxylic acids in the urban atmosphere. *Environ. Sci. Technol.*, **27**, 2227–2235.

Kawamura, K. and I. R. Kaplan (1987): Motor exhaust emissions as a primary source for dicarboxylic acids in Los Angels ambient air. *Environ. Sci. Technol.*, **21**, 105–110.

Kawamura, K. and K. Usukura (1993): Distributions of low molecular weight dicarboxylic acids in the North Pacific aerosol samples. *J. Oceanogr.*, **49**, 271–283.

Kawamura, K., H. Kasukabe, O. Yasui and L. A. Barrie (1995a): Production of dicarboxylic acids in the arctic atmosphere at polar sunrise. *Geophys. Res. Lett.*, **22**, 1253–1256.

Kawamura, K., M. Kosaka and R. Sempéré (1995b): Distributions and seasonal changes of hydrocarbons in urban aerosols and rainwaters. *Chikyukagaku* (*Geochemistry*), **29**, 1–15.

Kawamura, K., R. Sempéré, Y. Imai, M. Hayashi and Y. Fujii (1996a), Water soluble dicarboxylic acids and related compounds in the Antarctic aerosols. *J. Geophys. Res.*, **D13**, 18,721–18,728.

Kawamura, K., S. Steinberg and I. R. Kaplan (1996b): Concentrations of monocarboxylic and dicarboxylic acids and aldehydes in southern California wet precipitations: Comparison of urban and nonurban samples and compositional changes during scavenging. *Atmos. Environ.*, **30**, 1035–1052.

Li, S. M. and J. W. Winchester (1993): Water soluble organic constituents in arctic aerosols and snow pack. *Geophys. Res. Lett.*, **20**, 45–48.

Li, X., H. Maring, D. Savoie, K. Voss and J. M. Prospero (1996): Dominance of mineral dust in aerosol light-scattering in the North Atlantic trade winds. *Nature*, **380**, 416–419.

Nishikiori, M., K. Kawamura, M. Hayashi and Y. Fujii (1996): Mono- and di-carboxylic acids and mid-chain oxocarboxylic acids in the Antarctic aerosols. *Chikyukagaku* (*Geochemistry*), **30**, 27–34.

Novakov, T. and J. E. Penner (1993): Large contribution of organic aerosols to cloud-condensation-nuclei concentrations. *Nature*, **365**, 823–826.

Norton, R. B., J. M. Roberts and B. J. Huebert (1983): Tropospheric oxalate. *Geophys. Res. Lett.*, **10**, 517–520.

Sakaguchi, F. and K. Kawamura (1994): Identification of 4-oxoheptanedioic acid in the marine atmosphere by capillary gas chromatography-mass spectrometry. *J. Chromatogr. A*, **687**, 315–321.

Satsumabayashi, H., H. Kurita, Y. Yokouchi and H. Ueda (1990): Photochemical formation of particulate dicarboxylic acids under long-range transport in central Japan. *Atmos. Environ.*, **24A**, 1443–1450.

Saxena, P. and L. M. Hildemann (1996): Water-soluble organics in atmospheric particles: A critical review of the literature and application of thermodynamics to identify candidate compounds. *J. Atmos. Chem.*, **24**, 57–109.

Saxena, P., L. M. Hildemann, P. H. McMurry and J. H. Seinfeld (1995): Organics alter hygroscopic behavior of atmospheric particles. *J. Geophys. Res.*, **100**, 18755–18770.

Sempéré, R. and K. Kawamura (1994): Comparative distributions of dicarboxylic acids and related polar compounds in snow, rain and aerosols from urban atmosphere. *Atmos. Environ.*, **28**, 449–459.

Sempéré, R. and K. Kawamura (1996): Low molecular weight dicarboxylic acids and related polar compounds in the remote marine rain samples collected from western Pacific. *Atmos. Environ.*, **30**, 1609–1619.

Simoneit, B. R. T., R. Chester and G. Eglinton (1977): Biogenic lipids in particulates from the lower atmosphere over the eastern Atlantic. *Nature*, **267**, 682–685.

Uematsu, M., K. Kawamura, T. Ibusuki and T. Kimoto (1995): Chemical composition of marine aerosols over the Central North Pacific. Results from the 1991 cruise of Hakurei Maru No. 2. pp. 3–14. In *Biogeochemical Processes and Ocean Flux in the Western Pacific*, ed. by H. Sakai and Y. Nozaki, Terrapub, Tokyo.

Warneck, P. (1988): *Chemistry of the Natural Atmosphere*. Academic Press, New York, 757 pp.

Tegen, I., A. A. Lacis and I. Fung (1996): The influence on climate forcing of mineral aerosols from disturbed soils. *Nature*, **380**, 419–422.

Yokouchi, Y. and Y. Ambe (1986): Characterization of polar organics in air borne particulate matter. *Atmos. Environ.*, **20**, 1727–1734.

Dynamics and Characterization of Marine Organic Matter, Eds., N. Handa, E. Tanoue and T. Hama, pp. 485–508.

Carbon Monoxide and Methane in Surface Seawater of the Tropical Pacific Ocean

Hidekazu MATSUEDA[1], Hisayuki Y. INOUE[1], Ichio ASANUMA[2], Michio AOYAMA[1] and Masao ISHII[1]

[1]*Geochemical Research Department, Meteorological Research Institute (MRI/GRD), 1-1, Nagamine, Tsukuba-shi, Ibaraki-ken 305, Japan*
[2]*Japan Marine Science and Technology Center (JAMSTEC), 2-15, Natsushima, Yokosuka 237, Japan*

(Original, Received 6 August 1996; accepted 19 September 1996)

Abstract—Methane and carbon monoxide in the surface seawater and the overlying atmosphere were measured in the equatorial Pacific from 160°E and 160°W. Both gases were supersaturated with respect to the atmosphere. The longitudinal distribution of oceanic methane showed an interannual variation between 1987 to 1990. This variation suggested that the production and/or transport of methane were largely changed by a strong upwelling during the 1988/89 La Niña event. The measurements of oceanic carbon monoxide in 1994 showed higher concentrations in the central and eastern regions than in the western region. This longitudinal distribution was mainly related to the solar radiation intensity. Carbon monoxide concentrations showed clear diurnal cycles that had a phase delay of 3–4 hours against the solar radiation cycle. However, the change rate of carbon monoxide concentration with time coincided well with the solar radiation pattern with no phase lag, suggesting that the rate of photochemical production quickly responded to the solar radiation intensity. A higher carbon monoxide concentration was observed in a water depth of 3 m rather than 5 m, suggesting a steep concentration gradient. From these results, it is concluded that longitudinal, interannual and vertical variations of the both gases are important for more accurate estimation on total oceanic fluxes in the tropical Pacific.

INTRODUCTION

Atmospheric methane (CH_4) is an important greenhouse gas that influences the radiative balance and climate of the Earth by warming the troposphere and cooling the stratosphere. Recent increases in atmospheric methane concentrations and interannual variations in these concentrations have been observed during the past two decades by several research groups (Blake and Rowland, 1988; Scheel *et al.*, 1990; Aoki *et al.*, 1992; Fraser *et al.*, 1992; Khalil *et al.*, 1993; Dlugokencky *et al.*, 1994; Lowe *et al.*, 1994; Sugawara *et al.*, 1994; Matsueda and Inoue, 1996; Matsueda *et al.*, 1996). Atmospheric carbon monoxide (CO) is not itself a greenhouse gas, but can influence the oxidative capacity in the atmosphere (Thompson, 1992). Recent changes in global atmospheric CO mixing ratios have been measured by several research groups (Scheel *et al.*, 1990; Fraser *et al.*, 1992; Brenninkmeijer, 1993; Derwent *et al.*, 1994; Khalil and Rasmussen, 1994; Novelli *et al.*, 1994). Such CO concentration changes could affect the future atmospheric CH_4 increase (Thompson and Cicerone, 1986), since the main sink of the both gases is an oxidative reaction with OH radicals in the atmosphere (Cicerone and Oremland, 1988). Thus, an understanding of both CH_4 and CO is needed for better prediction of a future climate changes due to the greenhouse gases.

The ocean acts on a source of atmospheric CH_4 and CO, since it is generally supersaturated with respect to the atmospheric concentrations. The oceanic CH_4 source was estimated to be less than 10% of the total sources of CH_4 to the atmosphere (Intergovernmental Panel on Climate Change, 1994). Oceanic CO is also a minor source to the atmosphere, but recent model sensitivity calculations suggest that the oceanic CO flux may significantly affect the CO pool in the marine boundary air (Erickson and Taylor, 1992; Springer-Young *et al.*, 1996). It is suggested that the oceanic source change could affect the chemical cycles in the marine boundary air. However, large uncertainties still remain on the estimation of global oceanic fluxes of both CH_4 and CO (Intergovernmental Panel on Climate Change, 1994). Thus, it is necessary for more measurements in the surface seawater to evaluate the impact of oceanic sources on the atmosphere.

Oceanic CH_4 and CO play an important role in the carbon cycle in seawater, because both gases are produced by the degradation of organic matter. CH_4 is mainly produced by biological degradation in anoxic marine environments such as the digestive track in zooplankton (Traganza *et al.*, 1979; de Angelis and Lee, 1994) and sinking particles (Karl and Tilbrook, 1994). Karl and Tilbrook (1994) reported that CH_4 associated with the sinking particles was vertically transported and released in the water column. On the other hand, the major production process of CO is a photochemical degradation of organic matter in the surface seawater. Several studies have shown that the photochemical production of CO is a large sink of bio-refractory organic matter in the deep water (Mopper *et al.*, 1991; Valentine and Zepp, 1993; Miller and Zepp, 1995). Both gases in the surface seawater are removed mainly by microbiological oxidation and air-sea exchange. The generated CO_2 from the CH_4 and CO oxidation is re-utilized by photosynthesis in the euphotic zone and converted to organic matter. Thus, the production and consumption of these trace gases are important processes in the understanding of the global carbon cycle in the ocean.

Since the 1960's, CH_4 distributions in the surface seawater have been observed in various oceanic environments (see the summaries by Lambert and Schmidt (1993) and Bange *et al.* (1994)). Surface CO concentrations have been measured in several oceanic environments such as the Pacific Ocean (Lamontagne, *et al.*, 1973; Bates *et al.*, 1993), the Atlantic Ocean (Conrad *et al.*, 1982), the Southern Ocean (Williams and Bainbridge, 1973) and the Arabian Sea (Owens *et al.*, 1991). Systematic measurements of CH_4 and CO in surface seawater were made between 1987 and 1994 to obtain consistent regional and temporal variations in the North and South Pacific (Bates *et al.*, 1995, 1996). These results showed higher supersaturations in the tropical Pacific than in the area outside the tropics. The high supersaturations are assumed to be enhanced by active biological and photochemical production due to the equatorial upwelling and stronger solar radiation in the tropics. It is also suggested that the turnover times of these gases are short due to active biological consumption in the upwelled equatorial region. In the present study, we focus on the CH_4 and CO cycles in the surface seawater of the equatorial Pacific. Surface CH_4 concentrations in the equatorial Pacific were observed between 1987 and 1990 to clarify interannual variations in the geographical CH_4 distribution. Surface CO measurements were made onboard ship in 1994 along the equator to observe short-term variations with a time scale of hours. Oceanographic and meteorological parameters relating to the CO variations were also measured during the 1994 cruise. On the basis of these observational results, we will discuss oceanic CH_4 and CO distributions as well as the major factors controlling their variations in the tropical Pacific. We will also estimate the air-sea fluxes of these gases in the tropical Pacific.

SAMPLING AND ANALYTICAL METHODS

Study region

The Ocean LIDAR Cruise (K94-06) aboard the R/V *Kaiyo* of JAMSTEC (Japan Marine Science and Technology Center) was performed from November 8, 1994 to December 15, 1994 to observe CO in the surface seawater and marine

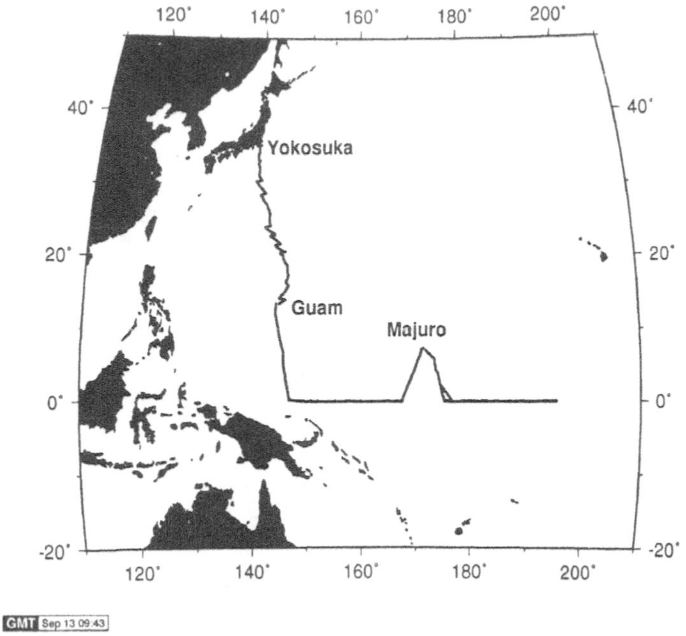

Fig. 1. Cruise track of the R/V *Kaiyo* during November–December 1994.

boundary air (Fig. 1). This cruise consisted of three legs: Leg P (Yokosuka in Japan to Guam), Leg I (Guam to Majuro) and Leg II (Majuro to Majuro). Leg P, the pre-K94-06 cruise, covered the subtropical regions in the western Pacific Ocean from 35°N to 14°N. Leg I and Leg II were mainly along the equator in the tropical Pacific Ocean between 145°E and 165°W. Leg I was performed in the western equatorial Pacific between 146°E and 170°E, while Leg II covered the central and eastern equatorial Pacific between 170°E and 165°W.

In 1987, 1989 and 1990, flask sampling was performed during cruises of the R/V *Natsushima* of JAMSTEC to measure CH_4 in the surface seawater and marine boundary air in the tropical Pacific. Samples for CH_4 measurement were mainly obtained along the equator between 5°N and 5°S to observe a longitudinal distribution between 160°E and 160°W. The details for these cruise tracks were previously described in Inoue *et al.* (1995) and Matsueda *et al.* (1996).

CO measurement

CO analytical system

Continuous CO measurements were performed during the cruise of 1994. Figure 2 shows a schematic diagram of the surface seawater measuring system for CO concentration and several oceanographic parameters. Seawater was pumped

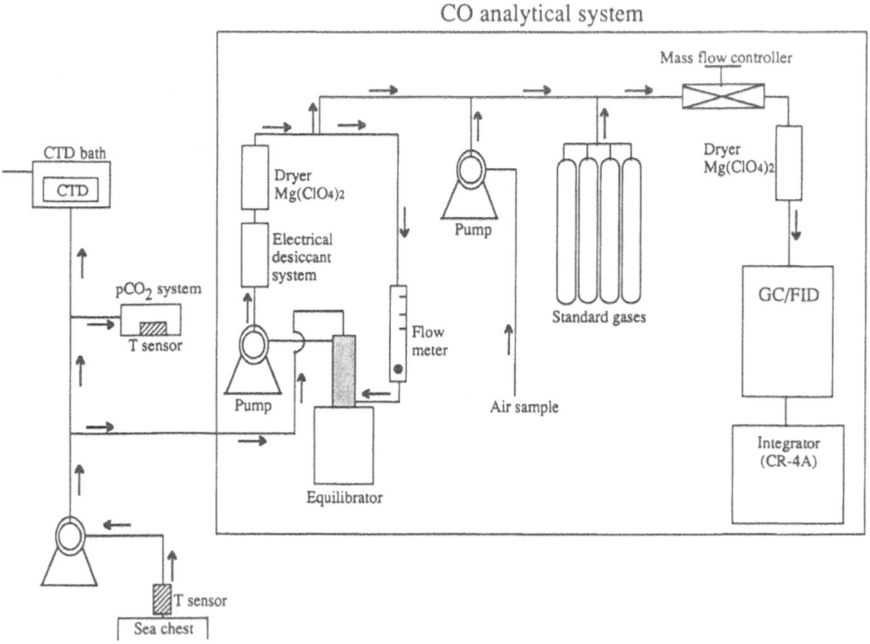

Fig. 2. Schematic diagram of the CO analytical system and seawater flow used by the R/V *Kaiyo* cruise in 1994.

up from a sea chest at the bottom of ship to the CO measuring system in a laboratory as well as to the CTD bath system on the upper deck. Surface seawater was also supplied to the MRI CO_2 analytical system in the laboratory in order to measure pCO_2 and DIC. Two temperature sensors (Pt-100Ω) were installed in the sea chest and a pCO_2 measuring system to monitor the rise of surface seawater temperature (SST) between the bottom of the ship and the laboratory. The average rise in SST was 0.4–0.5 °C. The sampling depth of the surface seawater was estimated to be about 5 m in comparison with the vertical profiles of the temperature and salinity data observed at several stations.

The equilibrator technique used in the pCO_2 measurement (Inoue *et al.*, 1995) was used for determining CO concentration in the surface seawater (Fig. 2). The seawater was continuously showered into the equilibrator at a flow rate of about 6 L/min. Air in the head space (about 1.5 L) of the equilibrator was circulated at a flow rate of more than 1 L/min using a diaphragm pump until the air in the head space was equilibrated with respect to the seawater CO. We confirmed through several experiments on board ship that equilibrium was reached between CO in the seawater and air in the equilibrator within about 30 min. Thus, the air was circulated for more than 30 min and then introduced into a gas chromatograph equipped with a flame ionization detector (GC/FID) for CO analysis. The circuit loop included an electrical desiccant system and chemical dryer tube ($Mg(ClO_4)_2$) to reduce the water vapor. In addition, another chemical dryer tube ($Mg(ClO_4)_2$) was connected just before the GC to completely dry the equilibrated air sample for the GC/FID analysis. This dryer was also used for drying the atmospheric sample in the marine boundary layer pulled through Teflon tubing using a diaphragm pump.

The GC/FID (GC-9A, Shimadzu Japan) was improved with a catalytic reduction converter to detect CO as CH_4, since CO is not sensitive for FID. The sample air was injected using a 5 ml sample loop into a pre-cut column (Silica Gel/ TritonX-305) and then the CO was separated from CH_4 using a main-column (Molecular Sieve-5A). The oven temperature was maintained at 90°C. The separated CO was catalytically converted to CH_4 by passing through a nickel furnace at 420°C with hydrogen and then quantified by the FID. A peak area of CH_4 converted from CO was integrated by a chromatographic integrator (CR-4A, Shimadzu Japan). One cycle of the GC analysis was 10 min., so that six measurements in one hour consisted of one air equilibrator, three atmospheric samples and two working standard gases. The working standard gas was used to monitor a stability of the FID sensitivity. All of the valves for switching the sample air and GC column were operated by a relay interface (PRG-102A; Shimadzu, Japan) coupled to the chromatographic integrator to automatically measure CO mixing ratios in both the sampled air and standard gases.

CO concentration in the sample air was calculated on the basis of measurements every 3 hours using three secondary standard gases with CO mixing ratios of about 0.1, 1 and 3 ppm in air. These standards were gravimetrically prepared by Nippon Sanso Co., Ltd. in Japan according to a similar procedure described in Matsueda (1993). No significant drift of the CO mixing ratio was found in all standard gas cylinders before and after the cruise in comparison with our primary standard gases. The precision of the CO analysis was about 3 ppb at our laboratory, but the shipboard measurements showed a larger precision of about 30 ppb due to the vibration of research vessel.

Oceanographic and meteorological observation system

The seawater pumped up was continuously overflowed into the CTD bath on the upper deck during Legs I and II. The CTD (SBE16 SEACAT Profiler), with a fluorometer (SEA TECH INC.) and a transmissometer (SEA TECH), was placed in the CTD bath to obtain temperature, salinity, chlorophyll fluorescence and transparency data every 2 mins. The CTD bath system was cleaned every 2 days to prevent a reduction in the sensitivity of sensors. The seawater in the bath was regularly sampled to calibrate the sensors from the CTD system.

An irradiance radiometer (MER-1010) was installed on the compass deck during Legs I and II to obtain the solar radiation information. Photosysnthetically available radiation of the downwelling irradiance (EdPAR) was measured at 12 wavelengths from 410 to 765 nm. Meteorological conditions during the whole cruise were observed every 3 hours to obtain wind speed, wind direction, atmospheric pressure and other meteorological parameters. These oceanographic and meteorological data were reported in the Ocean LIDAR Cruise Report (K94-06) of JAMSTEC.

CH_4 measurement

A similar equilibrium technique, as described above, was used for measuring the CH_4 concentration in the surface seawater during the cruises of 1987, 1989 and 1990. Air equilibrated with surface seawater (Inoue and Sugimura, 1992) was collected in a 200 ml cylindrical glass flask with high-vacuum greaseless stopcocks at both ends. Atmospheric CH_4 samples were simultaneously pulled from the bow of the vessel using a diaphragm pump and collected into the 200 ml glass flasks (Matsueda *et al.*, 1996).

The analytical method used to determine the CH_4 mixing ratio in the sampled air was previously reported in detail (Matsueda *et al.* 1992; 1993). The overall analytical precision was about 0.12%. All of the sample measurements were calibrated using our MRI/GRL standard gas scale that had been previously intercompared with the National Oceanic and Atmospheric Administration/Climate Monitoring and Diagnostics Laboratory (NOAA/CMDL) standard gas scale (Matsueda, 1993). This intercomparison showed that the MRI/GRL scale was 23 ppb higher than the NOAA/CMDL. All CH_4 measurements reported here were calculated on the basis of the MRI/GRL scale.

RESULTS AND DISCUSSION

CO concentration

Geographical distribution

Figure 3 shows the CO mixing ratios in air equilibrated with seawater and the overlying atmosphere obtained during the whole cruise from November 9, 1994

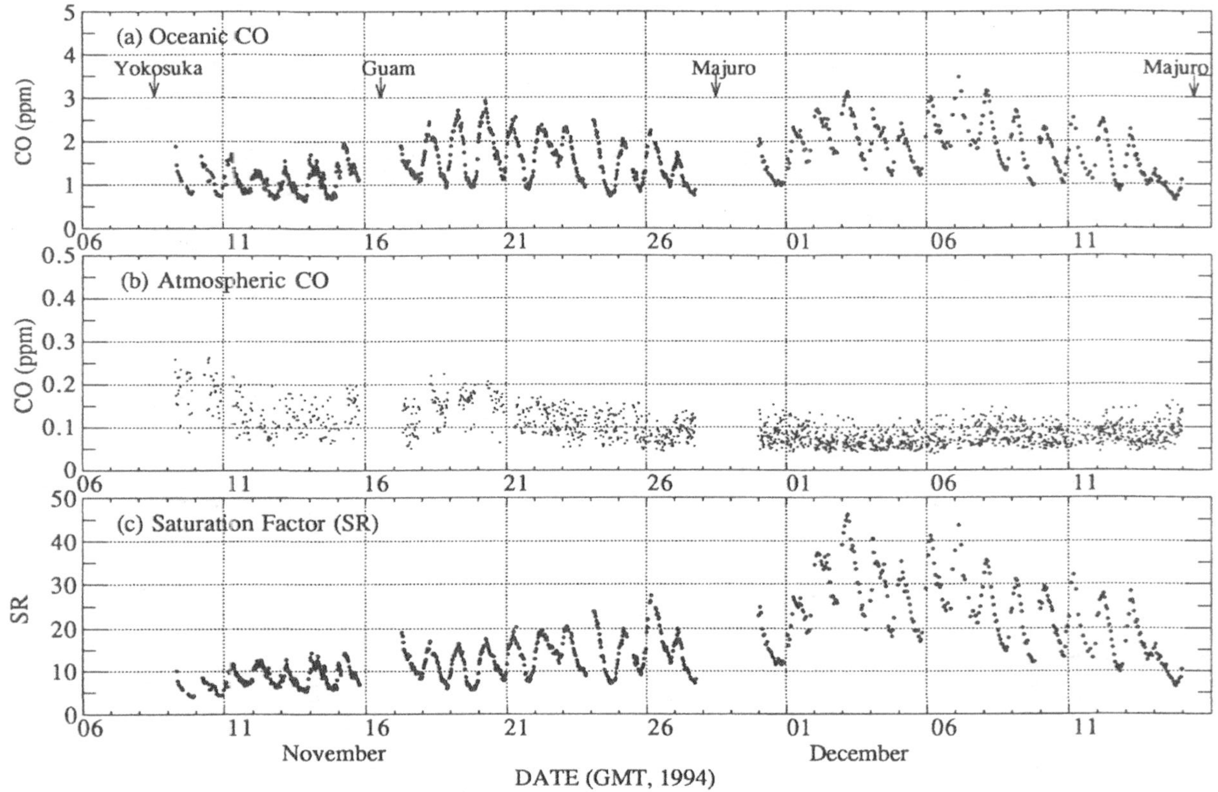

Fig. 3. CO mixing ratios in air equilibrated with surface seawater $[(X)_w]$ (a), mixing ratios in the overlying atmosphere $[(X)_a]$ (b), and saturation factor $[SR = (X)_w/(X)_a]$ (c), observed by the R/V *Kaiyo* cruise during November–December, 1994.

to December 14, 1994. All of the seawater CO measurements were corrected for the SST rise during pumping from the sea chest to the equilibrator, according to the gas solubilities (Wissenburg and Guinasso, 1979), although this correction reduced the measured value by only about 1%. The CO mixing ratio in the air equilibrated with the seawater $[(X)_w]$ varied from 0.6 to 3.5 ppm with a clear diurnal cycle. The oceanic CO level showed a markedly geographical difference among Legs P, I and II. The area outside the tropical Pacific during Leg P showed a lower oceanic CO level than in the tropical regions during Legs I and II. A similar increase in CO concentrations near the equator were observed on previous cruises in the same area of the Pacific Ocean (Swinnerton and Lamontagne, 1974; Swinnerton *et al.*, 1976; Bates *et al.*, 1995). In addition, CO values were higher in the central and eastern regions than in the western region.

Atmospheric CO mixing ratios $[(X)_a]$ showed one order of magnitude lower values from 0.05 to 0.25 ppm than the seawater measurements (Fig. 3). No diurnal cycle was found in the atmospheric measurements. The latitudinal distribution of the atmospheric CO in the western Pacific showed a gradual decrease toward the equator during Leg P. In the equatorial Pacific, slightly higher CO mixing ratios were found in the western region, but the CO in the central and eastern regions showed a constant level with a mean mixing ratio of about 80 ppb. The short-term variability of about 30 ppb was mainly due to the analytical error in shipboard measurement, as described above. Our measured values in the central equatorial Pacific were similar to atmospheric CO mixing ratios of about 75 ppb observed at Christmas Island (1°42′ N, 157°10′ W) in the same central Pacific during November–December in 1993 (Novelli *et al.*, 1994), although the long-term trend and difference of standard gas scale was not accounted for in this comparison.

Oceanic CO was supersaturated with respect to the atmosphere in all Pacific regions during this cruise. The Saturation factors (SR), which are the ratio of the measured mixing ratios in air equilibrated with seawater and the measured mixing ratios in the overlying atmosphere $[(X)_w/(X)_a]$, are shown in Fig. 3. The SR values clearly show a diurnal cycle due to the strong daily cycle of oceanic CO. The mean SR level showed a regional difference between Legs P, I and II. The SR distribution clearly showed an increase toward the equator as well as an increase from west to east in the equatorial Pacific. A similar geographical distribution was observed in the same region of the Pacific Ocean by Bates *et al.* (1995).

Figure 4(a) shows the CO concentration (nM) in the surface seawater calculated on the basis of the oceanic CO mixing ratios $[(X)_w]$, SST and salinity according to the equilibrium solubility equation of Wissenburg and Guinasso (1979). The data analysis of CO concentrations were made according to the following procedure to separate the mean regional variation from the large diurnal cycle. First, the measured data were fitted using a straight line to obtain an interpolated value every 10 min. Second, these equally-spaced data were smoothed by different low-pass filters with cut-off points of about 1.5 hr and about 2 days. One filter described the whole CO variation, and the other captured the regional variation of the mean CO level without daily cycle (Fig. 4(a)).

The mean CO level in the tropical Pacific clearly showed higher CO

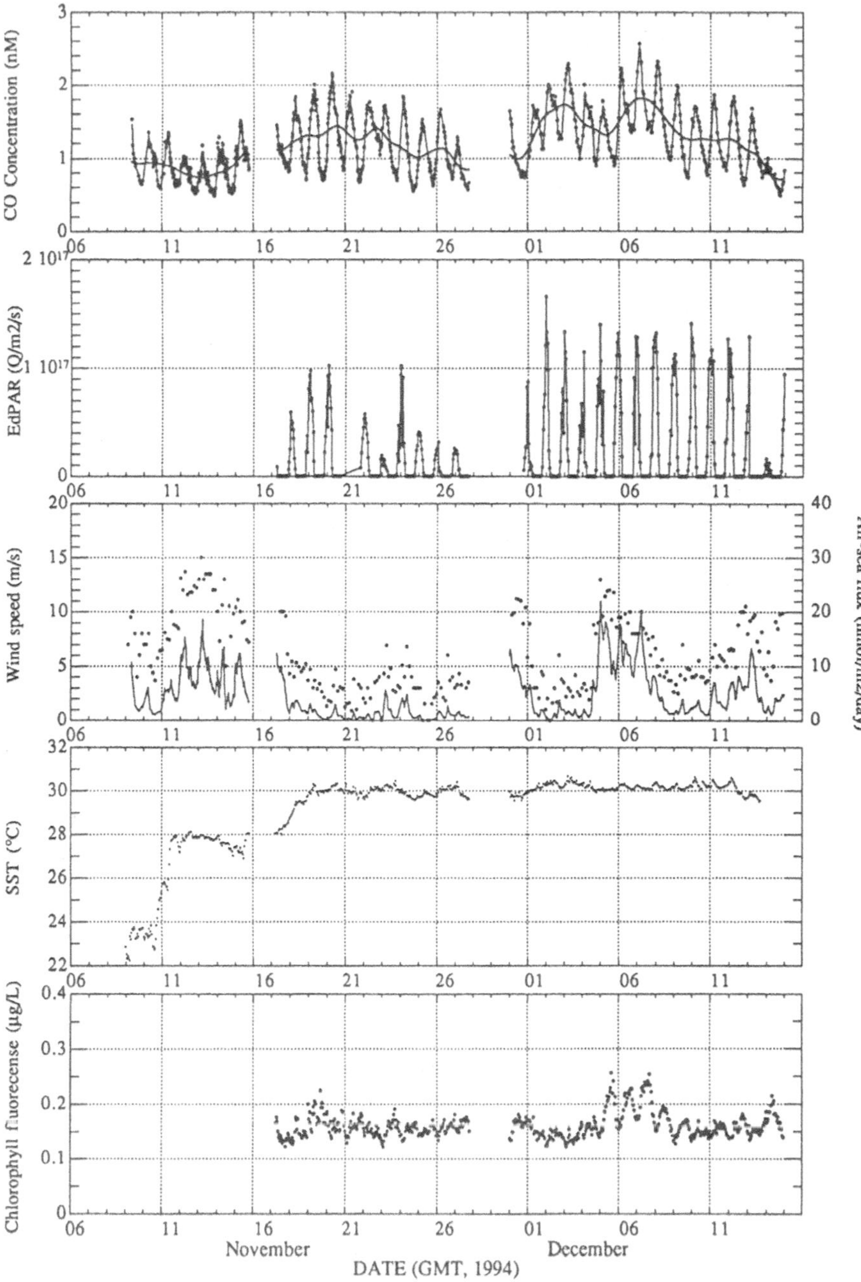

Fig. 4. CO concentrations in the surface seawater (a), photosynthetically available radiation of the downwelling irradiance (EdPAR) (b), wind speed (c), surface seawater temperature (SST) (d), and chlorophyll fluorescence (e), observed by the R/V *Kaiyo* cruise during November–December, 1994. The curves fitted to the CO concentration data in (a) are described in the text. Air-sea fluxes are calculated and shown by the solid line in (c). The flux calculation procedures are described in the text.

Table 1. Daily mean CO concentration in the surface seawater, integrated EdPAR (photosynthetically available radiation of the downwelling irradiance), wind speed, calculated CO flux, surface seawater temperature (SST) and chlorophyll fluorescence from the R/V *Kaiyo* cruise (K94-6) in 1994. Latitude and longitude were recorded at 00:00 for each day.

Date (GMT)	Latitude	Longitude	CO concentration (nM)	Integrated EdPAR (x10^{21}Q/m2/day)	Wind speed (m/s)	CO flux (μmol/m2/day)	SST (°C)	Chlorophyll fluorescence (μg/L)
Leg P								
1109	35°12'N	140°40'E	0.856		7.5	4.26	23.3	
1110	32°52'N	140°29'E	0.942		5.9	1.98	23.7	
1111	28°22'N	141°01'E	0.894		8.2	5.23	26.8	
1112	23°43'N	142°20'E	0.763		12.3	9.35	27.8	
1113	20°33'N	145°53'E	0.726		13.2	9.58	27.9	
1114	20°33'N	145°54'E	0.815		9.2	5.82	27.5	
1115	16°40'N	146°24'E	1.143		9.2	7.77	27.5	
Leg I								
1117	13°27'N	144°35'E		1.04	8.9		28.3	0.14
1118	9°59'N	145°03'E	1.241	1.42	4.8	2.35	29.3	0.15
1119	5°09'N	146°00'E	1.267	2.37	3.3	1.23	30.0	0.18
1120	0°46'N	146°57'E	1.477		2.7	1.01	30.1	0.16
1121	0°00'	150°59'E	1.248		2.0	0.44	29.9	0.16
1122	0°00'	155°18'E	1.424	1.15	2.5	0.81	30.0	0.15
1123	0°00'	159°49'E	1.178	0.83	3.6	1.76	30.2	0.16
1124	0°01'N	164°27'E	1.068	1.67	4.2	1.93	29.8	0.15
1125	0°05'N	167°26'E	1.062	0.91	1.4	0.31	29.8	0.15
1126	0°54'N	167°48'E	1.165	0.59	3.3	1.18	30.1	0.16
1127	5°16'N	170°01'E		0.55	3.4		29.9	0.16
Leg II								
1130	6°59'N	171°32'E	0.962		10.3	8.77	29.8	0.17
1201	3°59'N	173°45'E	1.395	1.83	4.6	2.58	30.1	0.15
1202	0°00'	175°13'E	1.661	2.92	2.4	1.09	30.3	0.14
1203	0°01'N	179°45'E	1.664	2.29	2.8	1.30	30.4	0.14
1204	0°01'N	175°53'W	1.426	2.62	5.0	3.98	30.2	0.16
1205	0°01'N	171°23'W	1.283	2.83	11.0	13.96	30.0	0.19
1206	0°02'N	167°12'W	1.712	2.34	8.9	12.53	30.2	0.20
1207	0°00'	165°04'W	1.812	3.07	8.0	9.98	30.1	0.21
1208	0°00'	166°32'W	1.545	2.83	4.7	3.13	30.2	0.17
1209	0°01'N	171°12'W	1.288	2.90	4.1	1.91	30.1	0.15
1210	0°01'N	175°41'W	1.251	2.69	4.1	2.04	30.3	0.15
1211	0°00'	179°60'W	1.249	2.92	5.6	3.72	30.2	0.15
1212	0°00'	179°60'W	1.207	2.41	8.0	6.34	30.1	0.16
1213	0°02'N	176°22'E	1.101		7.9	6.50	29.7	0.15
1214	3°19'N	173°17'E	0.705	1.09	7.1	3.10		0.18

concentrations in the central and eastern regions than in the western region. Such longitudinal differences in the tropical Pacific were also observed in the latitudinal transects across the tropical Pacific by Bates *et al.* (1995). To elucidate the causes of these longitudinal differences, the CO distribution was compared with the data of solar radiation of EdPAR, wind speed, SST and chlorophyll fluorescence observed during each cruise (Table 1). These parameters are important factors controlling the CO concentration in the surface seawater (Conrad *et al.*, 1982). The longitudinal difference in CO concentration in the equatorial Pacific was primarily related to an intensity of solar radiation, because the integrated EdPAR values in the central and eastern regions were approximately 2–3 times higher than that in the western regions (Fig. 4(b)). It is most likely that more photochemical production under stronger light conditions occurred in the central and eastern equatorial Pacific in 1994.

Early studies suggested that the higher CO levels in the eastern tropical Pacific were associated with the South Equatorial Current and equatorial upwelling (Swinnerton and Lamontagne, 1974; Swinnerton *et al.*, 1976; Bates *et al.*, 1995). Since CO precusors will be enhanced by an increase of dissolved organic matter (DOM) due to high biological activities (Wilson *et al.*, 1970; Setser *et al.*, 1982; Mopper *et al.*, 1991), Conrad *et al.* (1982) proposed parameters like chlorophyll as indicators for photochemical precursors in seawater. During our cruise in 1994, chlorophyll fluorescence showed no significant longitudinal difference along the equator (Fig. 4(e)). Slightly higher chlorophyll fluorescence appeared around 165°W, but the concentrations were lower than those in the same region during the non-El Niño event (Asanuma and Aoyama, 1994). The SST distribution observed in 1994 was quite similar to that observed during the 1986/88 El Niño event (Inoue and Sugimura, 1992). In addition, nutrients in the surface seawater was reported in the Ocean LIDAR Cruise Report (K94-06) to be near zero along the equator. These results suggested that biologically produced DOM was significantly limited due to the reduced upwelling in the equatorial Pacific during the 1994 cruise. These results suggest that solar radiation rather than biological activity is mainly responsible for the higher CO level in the central and eastern equatorial Pacific rather than in the western equatorial regions in 1994. It is possible that the CO distribution in this equatorial Pacific will be changed during different years due to the La Niña and/or non-El Niño events. The solar radiation pattern in the tropical Pacific is affected by the shift of convection toward the west during the La Niña event. During such events, biological activities increase due to the stronger upwelling. Thus, more long-term measurements are necessary to evaluate the interannual variation of oceanic CO distribution in the tropical Pacific.

CO flux between ocean and atmosphere

A remarkable change of surface wind speed was observed over the central and eastern Pacific during Leg II, but the solar radiation showed a relatively uniform distribution (Fig. 4(c)). Increased wind speed of around 10 m/s in this region almost coincided with a decrease in the CO level, suggesting a large loss of CO by air-sea exchange and vertical mixing. A similar decrease in CO level was also found in the subtropical region with a high wind speed up to 10 m/s. On the other hand, the western equatorial Pacific with a low wind speed of less than 5 m/s showed no clear relationship between the CO level and the wind speed. These results suggest that the CO level in the surface seawater was influenced by the enhanced air-sea exchange and vertical mixing due to the higher wind speeds.

To evaluate the air-sea exchange process, the ocean-atmosphere flux (F) was estimated according to the following equation as described by Bates *et al.* (1995),

$$F = K_l L \, \Delta pCO. \qquad (1)$$

Where K_l is the gas transfer velocity, L is the gas solubility at the ambient SST and salinity, and ΔpCO is the CO partial pressure difference between surface seawater and the overlying atmosphere. Although K_l largely depended on the wind speed/transfer velocity relationship, the equation for the spot measurement described by Wanninkhof (1992) was used in the present study. The Schmidt number as a parameter of K_l was calculated according to Bates *et al.* (1993). ΔpCO were calculated according to the following equation, as described by Bates *et al.* (1995),

$$\Delta pCO = (P_a - P_w)\,[(X)_w - (X)_a], \tag{2}$$

where P_a is the atmospheric pressure, and P_w is the saturation water vapor according to Weise and Price (1980).

Figure 4(c) shows the calculated CO flux from the surface seawater to the atmosphere. The estimated CO flux varied from 0.2 to 20 μmoles/m^2/day in the equatorial Pacific. This variability was mainly governed by a large variation of surface wind speed up to 10 m/s. The CO fluxes in the equatorial Pacific ranged from 0.3 to 4 μmoles/m^2/day under a low wind speed of less than 5 m/s. These values were similar to the CO fluxes of 0.8 to 6.5 μmoles/m^2/day in the equatorial Pacific calculated from the observed data of Bates *et al.* (1993), but they were lower than the model estimates of Erickson (1989). In the central and eastern equatorial Pacific, the CO flux increased by one order of magnitude with increased wind speed to about 10 m/s. In these higher wind speed regions, the CO flux clearly showed a large diurnal cycle through the enhancement of ΔpCO daily variation due to the high gas transfer velocity. These high fluxes are a major portion of the total fluxes in the central and eastern regions during the year. This result indicates that such a high CO flux under higher wind speeds should be taken into account for estimation of the total oceanic flux in the tropical Pacific. Bates *et al.* (1995) also pointed out that the wind fields for calculating the gas transfer velocity was one of the major uncertainties for CO flux estimation.

There are several major uncertainties in K_l and ΔpCO for estimation of oceanic CO flux, as pointed out by Bates *et al.* (1995). The other uncertainty proposed here is a practical problem concerning the water depth of the sample for measuring CO in the surface seawater. In general, an intake for surface seawater sampling is located at the bottom of the ship to continuously measure the CO concentration during the cruise. The sampling depth is usually around 5 m below the sea surface, but it is necessary for CO flux estimation to measure the CO concentration near the sea surface. Figure 5 shows a comparison of CO concentrations in the water samples at about 3 m and 5 m observed at two sites in the equatorial Pacific during the cruise in 1994. The water sample at 5 m was pumped up from the sea chest at the bottom of the ship, while a submersible pump was used for pumping up the water sample at about 3 m. These water samples at different depths were supplied into the equilibrator in turn to measure the mixing ratio of

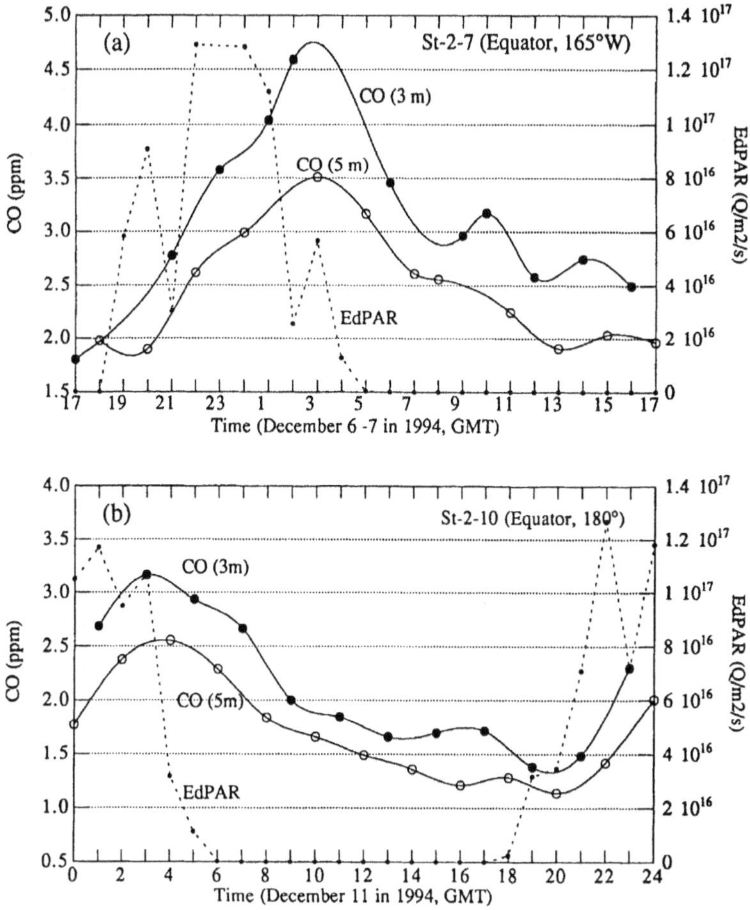

Fig. 5. Diurnal variations of CO mixing ratios in air equilibrated with seawater at 3 m (closed circles) and 5 m depths (open circles), observed along the equator at 165°W (a) and 180° (b). Photosynthetically available radiation of the downwelling irradiance (EdPAR) is shown in both plates by a dashed line.

CO equilibrated with the seawater.

A clear diurnal variation of CO was found at both depths, but the CO concentrations at 3 m were higher by about 30% than those at 5 m. This difference suggested that a steep gradient of CO concentration was formed in the water column above 5 m in the equatorial Pacific. Such a gradient may be caused by an active photochemical production of CO enhanced by stronger solar radiation near the sea surface. A similar vertical gradient in the surface seawater was also found in the vertical profiles of CO concentrations in the equatorial Pacific (Ohta, 1997). On the other hand, other measurements in the Pacific showed no vertical gradient of CO in the upper 5 m (Dr. T. S. Bates, personal communication). Both cases were found in the vertical distributions of CO in the Atlantic Ocean (Conrad

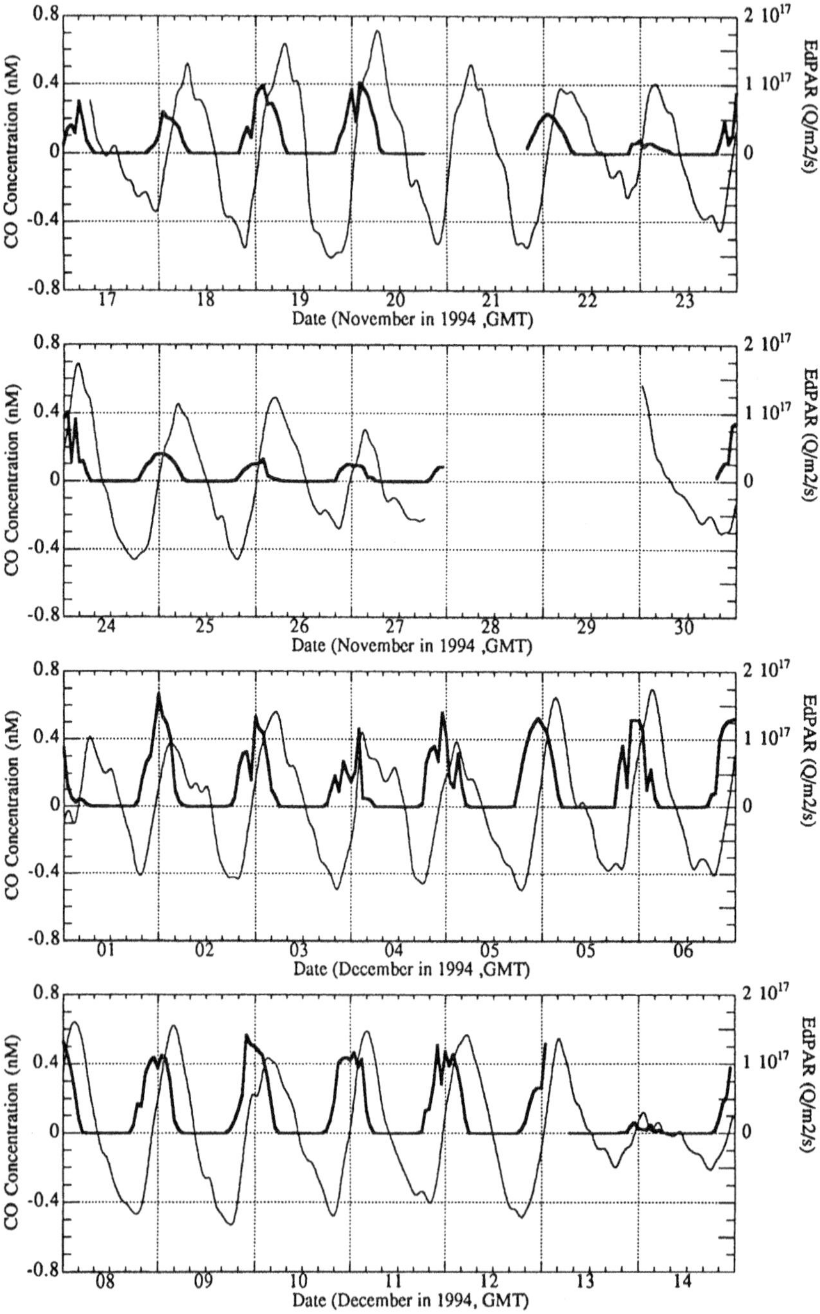

Fig. 6. Comparison of diurnal variations between CO concentration in the surface seawater (thin line) and photosynthetically available radiation of the downwelling irradiance (EdPAR) (bold line), observed in the tropical Pacific by the R/V *Kaiyo* cruise during November–December, 1994. The calculation procedure of the CO diurnal cycle is described in the text.

et al., 1982). Although it is difficult to explain a favorite oceanic condition for the formation of a CO vertical gradient near the surface seawater, our results suggest that the air-sea CO flux calculated using measurements at 5 m water depth is underestimated, at least in the tropical Pacific. In addition, Conrad and Seiler (1988a) pointed out a significance of surface microlayer for air-sea exchange of CO, because high biological activities occur due to the enrichment of organic matter in that layer. Thus, it is important for the estimation of CO flux to understand the production and consumption in the near surface in the tropical region.

Diurnal variation of CO

A diurnal variation in seawater CO concentration was clearly apparent during the entire cruise. Similar CO diurnal variations have been observed in various oceanic environments (e.g. Lamontagne *et al.*, 1971; Conrad *et al.*, 1982). To extract a diurnal cycle during Legs I and II, the CO data filtered with a cut-off point of about 1.5 hr were subtracted from the data filtered with a cut-off point of about 2 days (Fig. 6). The EdPAR data are also plotted together with the CO cycle to distinguish daytime from nighttime. A peak-to-peak amplitude of CO concentration ranged from 0.5 to 1.4 nM with a mean of about 1.2 nM.

The phase of the CO diurnal cycle clearly showed a time delay against the solar radiation pattern for all days. A CO maximum concentration was delayed for about 3–4 hours against the peak of the EdPAR diurnal cycle, while a minimum in the CO concentration appeared after 1–2 hours of the sunrise. A similar time lag of 2–3 hours between the surface CO concentration and solar radiation was also observed in the tropical Pacific Ocean (Johnson and Bates, 1996; Ohta, 1997) and the Atlantic Ocean (Conrad *et al.*, 1982). Conrad *et al.* (1982) pointed out that such a time lag could be caused by the delayed production by photosensitized reactions involving oxygen radicals. Several possible pathways of photooxidative CO production were proposed, but data are limited concerning the chemical structure of dissolved organic matter as well as radical formation in the seawater (Zafiriou *et al.*, 1984). Thus, it is difficult to quantitatively estimate the reaction time for CO photooxidative production in the seawater at the present time.

To qualitatively evaluate the production and/or loss rates of CO in the surface seawater, a time derivative of the diurnal cycle was calculated on the basis of a slope of the straight line between two data points. The calculated time derivatives (D), shown in Fig. 7, corresponded to increased or decreased rates of CO concentration (nM/day) due to an imbalance between production and loss in the surface seawater. It is of interest that the pattern of D during the daytime was quite similar to that of the solar radiation during all the days. This similarity suggests that the diurnal variation of D represents an apparent rate of photochemical CO production, because the variation of D also includes CO loss such as microbial consumption, air-sea exchange and vertical mixing. The relative contribution of CO loss to D cannot be quantitatively estimated in the present study, but the time variabilities are thought to be smaller than the time change in CO production rate. In particular, the maximum value of D coincided well with the solar radiation peak without a time lag, although the clear phase delay was

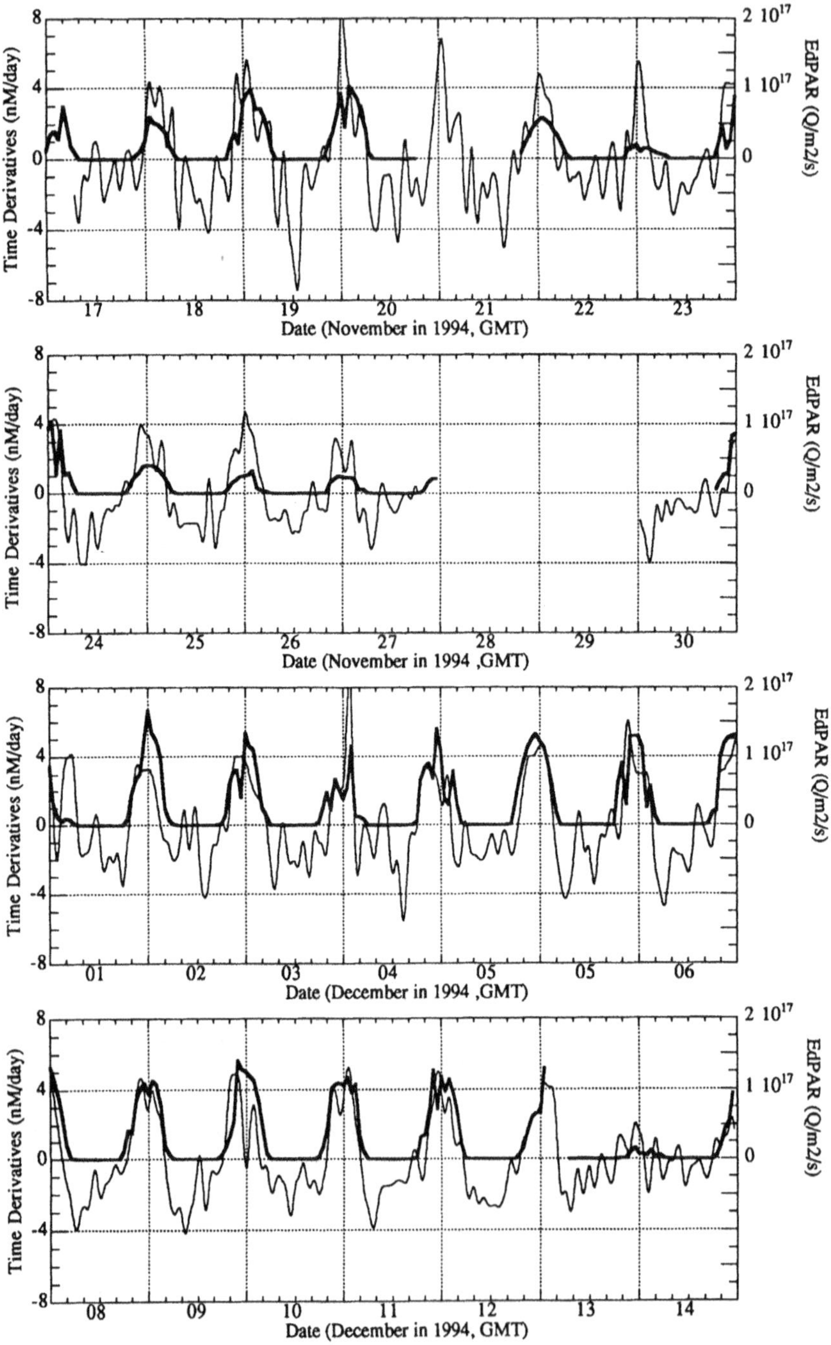

Fig. 7. Comparison of diurnal variation between calculated time derivatives of CO concentration (thin line) and photosynthetically available radiation of the downwelling irradiance (EdPAR) (bold line), observed in the tropical Pacific by the R/V *Kaiyo* cruise during November–December, 1994.

found in the CO concentration. This phase consistency suggests that the photochemical production rate of CO quickly responds to the change of the solar radiation within a time period of less than 1 hour. These results show that the CO production rate decreases after the solar radiation maximum, but the CO accumulation continues in the surface seawater until the production rate decreases to less than the total loss rate. A similar conclusion was obtained by an analysis of the CO budget using the vertical profiles of CO concentrations observed in the mixed layer (Johnson and Bates, 1996). The D values calculated in the present study are useful indicators for clarifying the relationship between the CO production and solar radiation, but it is necessary for more frequent data with a high time-resolution to clarify the mechanisms of photochemical CO production in the seawater.

During the night, D almost showed negative values with a mean of about –1.4 nM/day (Fig. 7) due to CO losses by microbial oxidation, air-sea exchange and vertical mixing. Jones (1991) reported biological CO oxidation rates from 0.001 to 0.09 nM/hr using a radioisotope technique in the highly oligotrophic southeastern Sargasso Sea. These oxidation rates are similar to the apparent loss rate of the D during the nighttime. This similarity suggest that microbial oxidation plays an important role in CO loss during the nighttime, although relative contributions of air-sea exchange and vertical mixing were not quantitatively estimated in the present study.

CH_4 CONCENTRATION

Figure 8 shows the longitudinal distributions of CH_4 mixing ratios in the air equilibrated with surface seawater between 160°E and 160°W in 1987, 1989 and 1990. All of the measurements were obtained in the equatorial Pacific between 5°N and 5°S each year. The measured mixing ratios were corrected for a 0.5°C SST rise during sampling using the gas solubilities (Wissenburg and Guinasso, 1979). The CH_4 mixing ratios ranged from 1.74 to 1.99 ppm. The CH_4 distributions in the tropical Pacific clearly showed an interannual variation (Table 2). A remarkable change in the longitudinal distribution was found in 1989. Higher mixing ratios of more than 1.9 ppm were observed in the eastern region between 175°W and 160°W, although the measurements in the western region west of 180° showed lower mixing ratios of about 1.77 ppm. Such a large longitudinal difference occurred during the 1988/89 La Niña event of low SST. In contrast, the measurements in 1987 showed no longitudinal difference in the CH_4 mixing ratio with a mean of about 1.82 ppm. In that year, high SST around 30°C was maintained in the equatorial Pacific due to the 1986/88 El Niño event. In 1990, a normal year, longitudinal distribution of the oceanic CH_4 showed a weak west-to-east gradient that was similar to the SST distribution. These results indicated that the interannual variation of the CH_4 level in the equatorial Pacific was largely affected by the La Niña and/or El Niño events. Such interannual variation of oceanic CH_4 in the tropical Pacific was also reported by Bates *et al.* (1996).

The CH_4 mixing ratio in the air was slightly lower than that in the surface

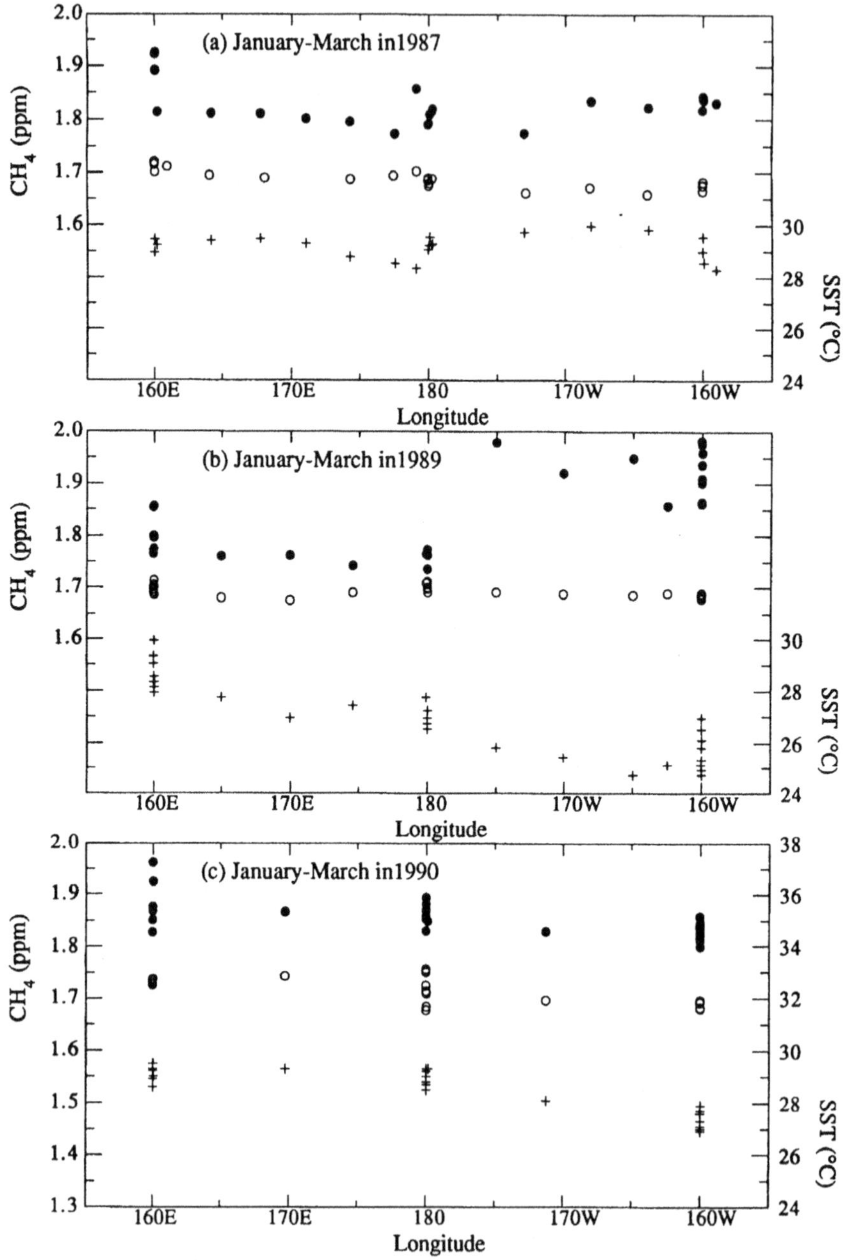

Fig. 8. Longitudinal distribution of CH_4 mixing ratios in air equilibrated with the surface seawater (closed circles), CH_4 mixing ratios in the overlying atmosphere (open circles) and surface seawater temperature (SST) (+), observed in the equatorial Pacific between 5°S and 5°N by the R/V *Natsushima* cruises during January-March in 1987 (a), 1989 (b) and 1990 (c).

Table 2. Mean methane mixing ratios in the air equilibrated with surface seawater (oceanic CH_4), saturation factor and surface seawater temperature (SST) in the western (160°E–180°E) and eastern region (175°W–160°W) of the equatorial Pacific from the R/V *Natsuhima* cruises in 1987, 1989 and 1990.

		1987		1989		1990	
		160E-180	175W-160W	160E-180	175W-160W	160E-180	175W-160W
Oceanic CH4	(ppm)	1.829	1.823	1.773	1.924	1.866	1.826
		(0.048)	(0.023)	(0.037)	(0.046)	(0.031)	(0.015)
Atmospheric CH4	(ppm)	1.694	1.668	1.696	1.686	1.728	1.690
		(0.014)	(0.009)	(0.011)	(0.004)	(0.024)	(0.006)
Saturation factor		1.081	1.096	1.044	1.142	1.080	1.080
		(0.023)	(0.010)	(0.020)	(0.027)	(0.019)	(0.015)
SST	(°C)	29.1	29.3	28.0	25.4	29.0	27.4
		(0.4)	(0.7)	(0.9)	(0.7)	(0.3)	(0.4)

*( ): Standard deviation.

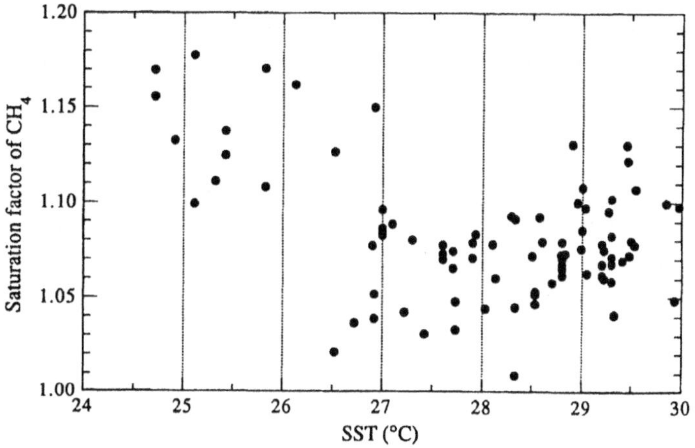

Fig. 9. Saturation factor (SR) of CH_4 observed in the equatorial Pacific during 1987–1990 plotted against the SST.

seawater, suggesting a flux of CH_4 from the ocean to the atmosphere in the equatorial Pacific. An interannual variation of atmospheric CH_4 mixing ratios in the tropical Pacific was previously reported in detail (Matsueda *et al.*, 1996), but a magnitude of atmospheric variation was smaller than the interannual variability of seawater CH_4. It is suggested that the interannual variation of oceanic CH_4 is driven not by the air-sea interaction but the changes in production and consumption rates in the water column.

Figure 9 shows the saturation factors (SR), which are the ratio of the measured mixing ratios in air equilibrated with surface seawater and the measured mixing ratios in the overlying atmosphere $[(X)_w/(X)_a]$, as a function of SST. The SR in higher SST of more than 26.5°C almost showed lower values of less than 1.10. These values measured here are quite similar to the average SR of 1.085 ± 0.035 measured in the same equatorial Pacific during 1987–1994 (Bates *et al.*, 1996). In contrast, higher SR values from 1.10 to 1.18 were found in a lower SST of less 26.5°C due to the La Niña event in 1989. A similar SR increase to more than 1.10 was also observed during the La Niña event in 1988 (Bates *et al.*, 1996), although the values of about 1.3 reported by Watanabe *et al.* (1995) were not observed by Bates *et al.* (1996) nor by our measurements. It is suggested that a strong upwelling of the cold water is responsible for the elevation of SR in the equatorial Pacific.

The strong upwelled region in the eastern Pacific in 1989 coincides well with higher nutrient concentrations (Inoue and Sugimura, 1992), suggesting active biological production. Since CH_4 is produced in the digestive tracks of zooplankton (Traganza *et al.*, 1979; Bianchi *et al.*, 1992; Marty, 1993; de Angelis and Lee, 1994) and released into the seawater through zooplankton fecal pellets (Karl and Tilbrook, 1994; Tilbrook and Karl, 1995), it is possible that such biological production could enhance CH_4 production (Conrad and Seiler, 1988b). On the other hand, Bates *et al.* (1996) pointed out that physical circulation process could also affect the SR, because the high CH_4 concentration associated with pycnocline was distributed in the eastern tropical Pacific (Burke *et al.*, 1983). It is suggested that the SR increase during the La Niña event could be a result of intrusion of CH_4 rich water into the surface layer with the strong upwelling. It is difficult to quantitatively evaluate the relative contribution of biological and physical processes to the SR increase, but both processes could be enhanced by a strong upwelling to elevate the surface CH_4 concentration. The strength of the upwelling will also affect the CH_4 consumption process of microbial oxidation, because the CH_4 biological oxidation rate was reported to increase through high microbial activities and high nutrient concentrations in the marine environment (Griffiths *et al.*, 1982; Jones and Amador, 1993). However, the in situ turnover time of oceanic CH_4 was estimated to be several years due to the slow rate of microbial oxidation in the ocean (Butler *et al.*, 1987; Ward *et al.*, 1987; Jones 1991). This turnover time seems to be long compared with the time scale of the SR variability due to the La Niña and/or El Niño events. Thus, biological production and/or transport are more likely to be responsible for the interannual variation of surface CH_4 in the equatorial Pacific.

CONCLUSION

Oceanic CH_4 and CO measured in surface seawater showed a supersaturation with respect to the atmosphere in the tropical Pacific. Saturation ratios of both gases along the equator varied longitudinally, interannually and vertically. These variations are important for estimating global oceanic fluxes to the atmosphere, because saturation ratios in the tropical Pacific are generally higher than those outside the tropical regions (Bates *et al.*, 1995; Bates *et al.*, 1996). The longitudinal distributions along the equator depended on the biological, physical, meteorological and photochemical parameters controlling the production and consumption rates of CH_4 and CO. These parameters changed interannually due to La Niña and/or El Niño events and appeared to affect the CH_4 longitudinal distribution in the equatorial Pacific. These events could also perturb the CO distribution, since they will affect important factors for oceanic CO variation such as the solar radiation and biological activities. Thus, more long-term measurements are necessary to clarify the longitudinal and interannual variations of both gases. In addition, our results suggest that more observations of vertical profiles in the mixed layer are necessary to more accurately estimate the oceanic CO flux in the tropical Pacific.

Acknowledgements—We would like to thank the captains, crew members and scientists of the R/V *Natsushima* and R/V *Kaiyo* cruises for their kind cooperation in the sampling of seawater and air. We thank Dr. T. S. Bates of NOAA/PMEL and Professor S. Tsunogai of Hokkaido University for their helpful and valuable reviews. This research was supported in part by the Science and Technology Agency (STA). The EdPAR data were results of the Special Coordination Funds of STA in 1994.

REFERENCES

Aoki, S., T. Nakazawa, S. Murayama and S. Kawaguchi (1992): Measurements of atmospheric methane at the Japanese Antarctic Station, Syowa. *Tellus*, **44B**, 273–281.

Asanuma, I. and M. Aoyama (1994): Monitoring of phytoplankton by shipboard ocean lidar. *Umi to Sora*, **70**, 109–121 (in Japanese).

Bange, H. W., U. H. Bartell, S. Rapsomanikis and M. O. Andreae (1994): Methane in the Baltic and North Seas and a reassessment of the marine emissions of methane. *Global Biogeochem. Cycles*, **8**, 465–480.

Bates, T. S., K. C. Kelly and J. E. Johnson (1993): Concentrations and fluxes of dissolved biogenic gases (DMS, CH_4, CO, CO_2) in the equatorial Pacific during the SAGE 3 experiment. *J. Geophys. Res.*, **98**, 16969–16977.

Bates, T. S., K. C. Kelly, J. E. Johnson and R. H. Gammon (1995): Regional and seasonal variations in the flux of oceanic carbon monoxide to the atmosphere. *J. Geophys. Res.*, **100**, 23093–23101.

Bates, T. S., K. C. Kelly, J. E. Johnson and R. H. Gammon (1996): A reevaluation of the open ocean source of methane to the atmosphere. *J. Geophys. Res.*, **101**, 6953–6961.

Bianchi, M., D. Marty, J.-L. Teyssié and S. W. Fowler (1992): Strictly aerobic and anaerobic bacteria associated with sinking particulate matter and zooplankton fecal pellets. *Mar. Ecol. Prog. Ser.*, **88**, 55–60.

Blake, D. R. and F. S. Rowland (1988): Continuing worldwide increase in tropospheric methane, 1978 to 1987. *Science*, **239**, 1129–1131.

Brenninkmeijer, C. A. M. (1993): Measurement of the abundance of ^{14}CO in the atmosphere and the $^{13}C/^{12}C$ ratio of atmospheric CO with application in New Zealand and Antarctica. *J. Geophys. Res.*, **98**, 10595–10614.

Bulter, J. H., R. D. Jones, J. H. Garber and L. I. Gordon (1987): Seasonal distribution and turnover of reduced trace gases and hydroxylamine in Yaquina Bay, Oregon. *Geochim. Cosmochim. Acta*, **51**, 697–706.

Burke, R. A., Jr., D. F. Reid, J. M. Brooks and D. M. Lavoie (1983): Upper water column methane geochemistry in the eastern tropical North Pacific. *Limnol. Oceanogr.*, **28**, 19–32.

Cicerone, R. J. and R. S. Oremland (1988): Biogeochemical aspects of atmospheric methane. *Global Biogeochem. Cycles*, **2**, 299–327.

Conrad, R. and W. Seiler (1988a): Influence of the surface microlayer on the flux of nonconservative trace gases (CO, H_2, CH_4, N_2O) across the ocean-atmosphere interface. *J. Atmos. Chem.*, **6**, 83–94.

Conrad, R., W. Seiler, G. Bunse and H. Giehl (1982): Carbon monoxide in seawater (Atlantic

Ocean). *J. Geophys. Res.*, **87**, 8839–8852.
Conrad, R. and W. Seiler (1988b): Methane and hydrogen in seawater (Atlantic Ocean). *Deep-Sea Res.*, **35**, 1903–1917.
de Angelis, M. A. and C. Lee (1994): Methane production during zooplankton grazing on marine phytoplankton. *Limnol. Oceanogr.*, **39**, 1298–1308.
Derwent, R. G., P. G. Simmonds and W. J. Collins (1994): Ozone and carbon monoxide measurements at a remote maritime location, Mace Head, Ireland, from 1990 to 1992. *Atmos. Environ.*, **28**, 2623–2637.
Dlugokencky E. J., L. P. Steele, P. M. Lang and K. A. Masarie (1994): The growth rate and distribution of atmospheric methane. *J. Geophys. Res.*, **99**, 17021–17043.
Erickson, D. J., III (1989): Ocean to atmosphere carbon monoxide flux: Global inventory and climate implications. *Global Biogeochem. Cycles*, **3**, 305–314.
Erickson, D. J., III and J. A. Taylor (1992): 3-D tropospheric CO modeling: The possible influence of the ocean. *Geophys. Res. Lett.*, **19**, 1955–1958.
Fraser, P. J., R. Harriss, S. A. Penkett, Y. Makide and E. Sanhueza (1992): Source gasese: Concentrations, emissions and trends. In *Scientific Assessment of Ozone Depletion: 1991*, WMO Global Ozone Research and Monitoring Project-Report No. 25.
Griffiths, R. P., B. A. Caldwell, J. D. Cline, W. A. Broich and R. Y. Morita (1982): Field observations of methane concentrations and oxidation rates in the southeastern Bering Sea. *Appl. Environ. Microbiol.*, **44**, 435–446.
Inoue, H. Y. and Y. Sugimura (1992): Variations and distributions of CO_2 in and over the equatorial Pacific during the period from the 1986/88 El Niño event to the 1988/89 La Niña event. *Tellus*, **44B**, 1–22.
Inoue, H. Y., H. Matsueda, M. Ishii, K. Fushimi, M. Hirota, I. Asanuma and Y. Takasugi (1995): Long-term trend of the partial pressure of carbon dioxide (pCO_2) in surface waters of the western North Pacific, 1984–1993. *Tellus*, **47B**, 391–413.
Intergovernmetal Panel on Climate Change (IPCC) (1994): *Radiative Forcing of Climate Change*, edited by J. T. Houghton *et al.*, 339 pp., Cambridge Univ. Press, New York.
Johnson, J. E. and T. S. Bates (1996): Sources and sinks of carbon monoxide in the mixed layer of the tropical South Pacific Ocean. *Global Biogeochem. Cycles*, **10**, 347–359.
Jones, R. D. (1991): Carbon monoxide and methane distribution and consumption in the photic zone of the Sargasso Sea. *Deep-Sea Res.*, **38**, 625–635.
Jones, R. D. and J. A. Amador (1993): Methane and carbon monoxide production, oxidation, and turnover times in the Caribbean Sea as influenced by the Orinoco River. *J. Geophys. Res.*, **98**, 2353–2359.
Karl, D. M. and B. D. Tilbrook (1994): Production and transport of methane in oceanic particulate organic matter. *Nature*, **368**, 732–734.
Khalil, M. A. K. and R. A. Rasmussen (1994): Global decrease in atmospheric carbon monoxide concentration. *Nature*, **370**, 639–641.
Khalil, M. A. K., R. A. Rasmussen and F. Moraes (1993): Atmospheric methane at Cape Meares: Analsis of a high-resolution data base and its environmental implications. *J. Geophys. Res.*, **98**, 14753–14770.
Lambert, G. and S. Schmidt (1993): Reevaluation of the oceanic flux of methane: uncertainties and long term variation. *Chemosphere*, **26**, 579–589.
Lamontagne, R. A., J. W. Swinnerton and V. J. Linnenbom (1971): Nonequilibrium of carbon monoxide and methane at the air-sea interface. *J. Geophys. Res.*, **76**, 5117–5121.
Lamontagne, R. A., J. W. Swinnerton, V. J. Linnenbom and W. D. Smith (1973): Methane concentrations in various marine environments. *J. Geophys. Res.*, **78**, 5317–5324.
Lowe, D. C., C. A. M. Brenninkmeijer, G. W. Brailsford, K. R. Lassey and A. J. Gomez (1994): Concentration and ^{13}C records of atmospheric methane in New Zealand and Antarctica: Evidence for changes in methane sources. *J. Geophys. Res.*, **99**, 16913–16925.
Marty, D. G. (1993): Methanogenic bacteria in seawater. *Limnol. Oceanogr.*, **38**, 452–456.
Matsueda, H. (1993): Intercalibration experiment of methane standard gas scale between NOAA/

CMDL and MRI/GRL. *Papers Meteorol. Geophys.*, **44**, 45–56.

Matsueda, H. and H. Y. Inoue (1996): Measurements of atmospheric CO_2 and CH_4 using a commercial airliner from 1993 to 1994. *Atmos. Environ.*, **30**, 1647–1655.

Matsueda, H., H. Inoue and Y. Sugimura (1992): Atmospheric methane over the western Pacific and the Antarctic Ocean from 1978 to 1986. *Geochem. J.*, **26**, 21–28.

Matsueda, H., H. Inoue and M. Ishii (1993): Latitudinal distributions of methane in the upper troposphere and marine boundary air over the Pacific in 1990. *Geophys. Res. Lett.*, **20**, 695–698.

Matsueda, H., H. Y. Inoue, M. Ishii and Y. Nogi (1996): Atmospheric methane over the North Pacific from 1987 to 1993. *Geochem. J.*, **30**, 1–15.

Miller, W. L. and R. G. Zepp (1995): Photochemical production of dissolved inorganic carbon from terrestrial organic matter: Significance to the oceanic organic carbon cycle. *Geophys. Res. Lett.*, **22**, 417–420.

Mopper, K., X. Zhou, R. J. Kieber, D. J. Kieber, R. J. Sikorski and R. D. Jones (1991): Photochemical degradation of dissolved organic carbon and its impact on the oceanic carbon cycle. *Nature*, **353**, 60–62.

Novelli, P. C., K. A. Masarie, P. P. Tans and P. M. Lang (1994): Recent changes in atmospheric carbon monoxide. *Science*, **263**, 1587–1590.

Ohta, K. (1997): Diurnal variations of carbon monoxide concentration in the equatorial Pacific upwelling region. *J. Oceanogr*,**53**,173-178 .

Owens, N. J. P., C. S. Law, R. F. C. Mantoura, P. H. Burkill and C. A. Llewellyn (1991): Methane flux to the atmosphere from the Arabian Sea. *Nature*, **354**, 293–296.

Scheel H. E., E.-G. Brunke and W. Seiler (1990): Trace gas measurements at the monitoring station Cape Point, South Africa, between 1978 and 1988. *J. Atmos. Chem.*, **11**, 197–210.

Setser, P. J., J. L. Bullister, E. C. Frank, N. L. Guinasso, Jr. and D. R. Schink (1982): Relationships between reduced gases, nutrients, and fluorenscence in surface waters off Baja, California. *Deep-Sea Res.*, **29**, 1203–1215.

Springer-Young, M., D. J. Erickson, III and T. P. Carsey (1996): Carbon monoxide gradients in the marine boundary layer of the North Atlantic Ocean. *J. Geophys. Res.*, **101**, 4479–4484.

Sugawara, S., T. Nakazawa, G. Hashida, C. Sakai and M. Tanaka (1994): Variations of atmospheric methane concentration over the western Pacific Ocean. *Proceedings of the International Symposium on Global Cycles of Atmospheric Greenhouse Gases*, 121–128.

Swinnerton, J. W. and R. A. Lamontagne (1974): Carbon monoxide in the South Pacific Ocean. *Tellus*, **26**, 136–142.

Swinnerton, J. W., R. A. Lamontagne and D. Smith (1976): Carbon monoxide concentrations in surface waters of the east tropical Pacific in 1974: Comparison with earlier values. *Mar. Chem.*, **4**, 57–65.

Tilbrook, B. D. and D. M. Karl (1995): Methane sources, distributions and sinks from California coastal waters to the oligotrophic North Pacific gyre. *Mar. Chem.*, **49**, 51–64.

Traganza, E. D., J. W. Swinnerton and C. H. Cheek (1979): Methane supersaturation and ATP-zooplankton blooms in near-surface waters of the Western Mediterranean and the subtropical North Atlantic Ocean. *Deep-Sea Res.*, **26A**, 1237–1245.

Thompson, A. M. (1992): The oxidizing capacity of the earth's atmosphere: Probable past and future changes. *Science*, **256**, 1157–1165.

Thompson, A. M. and R. J. Cicerone (1986): Possible perturbations to atmospheric CO, CH_4, and OH. *J. Geophys. Res.*, **91**, 10853–10864.

Valentine, R. L. and R. G. Zepp (1993): Formation of carbon monoxide from the photodegradation of terrestrial dissolved organic carbon in natural waters. *Environ. Sci. Technol.*, **27**, 409–412.

Wanninkhof, R. (1992): Relationship between wind speed and gas exchange over the ocean. *J. Geophys. Res.*, **97**, 7373–7382.

Watanabe, S., N. Higashitani, N. Tsurushima and S. Tsunogai (1995): Methane in the western North Pacific. *J. Oceanogr.*, **51**, 39–60.

Ward, B. B., K. A. Kilpatrick, P. C. Novelli and M. I. Scranton (1987): Methane oxidation and methane fluxes in the ocean surface layer and deep anoxic waters. *Nature*, **327**, 226–229.

Weiss, R. F. and B. A. Price (1980): Nitrous oxide solubility in water and seawater. *Mar. Chem.*, **8**,

347–359.

Williams, R. T. and A. E. Bainbridge (1973): Dissolved CO CH_4, and H2 in the Southern Ocean. *J. Geophys. Res.*, **78**, 2691–2694.

Wilson, D. F., J. W. Swinnerton and R. A. Lamontagne (1970): Production of carbon monoxide and gaseous hydrocarbons in seawater: Relation to dissolved organic carbon. *Science*, **168**, 1577–1579.

Wissenburg, D. A. and N. L. Guinasso, Jr. (1979): Equilibrium solubilities of methane, carbon monoxide, and hydrogen in water and sea water. *J. Chem. Engin. Data*, **24**, 356–360.

Zafiriou, O. C., J. Joussot-Dubien, R. G. Zepp and R. G. Zika (1984): Photochemistry of natural waters. *Environ. Sci. Technol.*, **18**, 358–371.

Dynamics and Characterization of Marine Organic Matter, Eds., N. Handa, E. Tanoue and T. Hama, pp. 509–531.

CO_2 Exchange between the Atmosphere and the Ocean: Carbon Cycle Studies of the Meteorological Research Institute Since 1968

Hisayuki YOSHIKAWA INOUE

Geochemical Research Department, Meteorological Research Institute, Nagamine 1-1, Tsukuba, Ibaraki 305 Japan

(Synthesis, Received 2 August 1996; accepted 27 December 1996)

Abstract—Since 1968, measurements of carbon dioxide in the atmosphere and in surface seawater have been made to clarify the role of the ocean in the global carbon cycle. Temporal and spatial variations in the oceanic carbonate system are summarized here, along with the techniques for atmospheric and oceanic CO_2 measurements developed by the Meteorological Research Institute, Japan Meteorological Agency.

INTRODUCTION

Atmospheric CO_2 is the most important greenhouse gas that has been increasing due to anthropogenic activities. Precise and direct measurements of atmospheric CO_2 using a non-dispersive infra-red gas analyzer (NDIR analyzer) were first made in 1957/58 by the Scripps Institution of Oceanography, USA at the South Pole and Mauna Loa in Hawaii (Keeling *et al.*, 1989). Since then, atmospheric CO_2 has been monitored at stations established in remote areas (see, for example, Trends '93, 1993; WMO WDCGG Data Report, 1995). The growth rate of atmospheric CO_2 during the1980s has been reported to be 1.5 ppm/yr (IPCC, 1994), which corresponds to about 58% of the total emissions from fossil-fuel burning and cement production. To predict future climate changes due to the increases of greenhouse gases on the basis of given CO_2 emission scenarios, it is essential to understand the current global carbon cycle that controls the atmospheric CO_2 level: the exchange fluxes between the atmosphere and surface oceans and between the atmosphere and the terrestrial biota. According to the carbon isotope measurements ($^{13}C/^{12}C$) of atmospheric CO_2, net fluxes among carbon reservoirs vary interannually (Francey *et al.*, 1995, Keeling *et al.*, 1995).

The net flux of CO_2 between the sea and the atmosphere (Fx) is given by the product of the difference in partial pressure of CO_2 (ΔpCO_2) between the sea (pCO_2^s) and the air (pCO_2^a) and the gas transfer coefficient (E).

$$\mathrm{Fx} = E(pCO_2^s - pCO_2^a) = E\Delta pCO_2. \quad (1)$$

The gas transfer coefficient E is expressed as a function of wind speed (Liss and Merlivat, 1986; Tans *et al.*, 1990; Wannikhof, 1993). Changes in ΔpCO_2 are mainly caused by pCO_2^s showing much larger spatial and temporal variability than pCO_2^a.

If there occurs a net exchange of CO_2 between the ocean and the atmosphere, the mole fraction of each of the dissolved inorganic carbon species in seawater varies from its original value because CO_2 acts as a weak acid. In seawater most of the dissolved inorganic carbon exists as bicarbonate ion (HCO_3^-) and carbonate ion (CO_3^{2-}), and only 1% of the total is the aqueous CO_2 that exchanges with the atmosphere. The relationship in changes between pCO_2^s and dissolved inorganic carbon concentration (TCO_2) is conventionally expressed by the homogeneous buffer factor (Revelle and Suess, 1957):

$$\beta = [(dpCO_2^s/pCO_2^s)/(dTCO_2/TCO_2)] \quad (2)$$

where $dpCO_2^s$ and $dTCO_2$ are small changes in pCO_2^s and TCO_2, respectively. Equation (2) gives the CO_2 uptake capacity of ocean water and reported values (Sundquist 1979; Wagner 1979) indicate that the relative change in pCO_2^s is about one order of magnitude larger than that of TCO_2.

Since the mid-1960s, the Meteorological Research Institute (MRI), in the Japan Meteorological Agency (JMA), began to measure the CO_2 mixing ratio in the atmosphere (xCO_2^a) and that in the air, which establishes equilibrium with a stream of flowing seawater (xCO_2^s). Miyake *et al.* (1974) developed a CO_2 measuring system consisting of a shower-head type equilibrator, chemical desiccants [$Mg(ClO_4)_2$], and a NDIR analyzer. From 1968 to 1972, they measured xCO_2^s in the Pacific on board the R/V *Hakuho-maru* (Ocean Research Institute, University of Tokyo) to find out whether the ocean acts as a sink or a source for atmospheric CO_2 (Miyake *et al.*, 1974).

After the cessation of the research program measuring xCO_2^s and xCO_2^a during the period from 1973 to 1980, the MRI restarted in January 1981 a new program to investigate temporal and spatial variations in the carbonate system of surface waters in the open ocean (Inoue *et al.*, 1987, 1995). Some of the results were reported earlier (Inoue and Sugimura, 1986, 1988a, 1988b, 1992; Inoue *et al.*, 1987, 1995, 1996; Fushimi, 1987, Ishii and Inoue 1995). In this report, we will summarize the results of research conducted by MRI researchers since the mid-1960s along with the techniques for measureing of xCO_2^a and xCO_2^s.

EXPERIMENT

Measurements of oceanic and atmospheric CO_2 prior to 1973

Measurements of xCO_2^s and xCO_2^a using a NDIR analyzer started in 1966 (Miyake and Sugimura, 1969). This system was designed for the measurements of xCO_2^s in discrete samples of seawater. Then, to investigate the role of the ocean for the global carbon cycle, Miyake *et al.* (1974) developed a CO_2 measuring system that could measure xCO_2^s and xCO_2^a quasi-continuously on board the ship (Fig. 1). Figure 1 represents a schematic diagram of the system that consisted of a NDIR analyzer (Beckman 315A), two columns of $Mg(ClO_4)_2$, two diaphragm

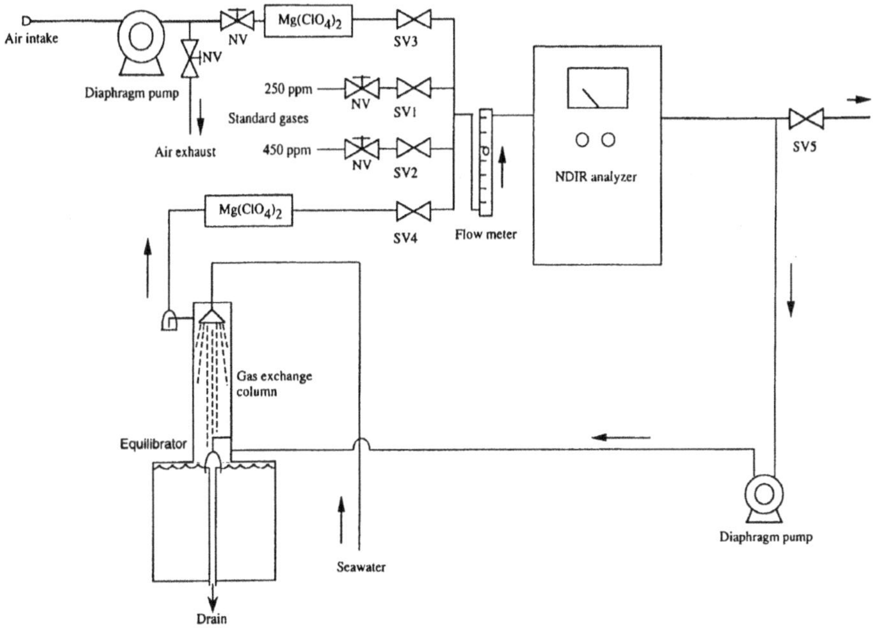

Fig. 1. A schematic diagram of the atmospheric and oceanic CO_2 measuring system used prior to 1973 (Miyake *et al.*, 1974). Arrows show the direction of air flow either for the measurement of xCO_2^a or xCO_2^s. The output voltages of the NDIR analyzer was recorded on a strip chart recorder, from which the CO_2 mixing ratio of sample airs was computed manually. SV meants the solenoid valve, and NV the flow control valve.

pumps, a flow meter, a series of electromagnetic valves and a showerhead-type equilibrator. The system was operated by an electromechanical timing device that repeated the same switching with an hourly cycle. For calibration, CO_2 standard gases (250 ppm and 450 ppm, CO_2 in N_2) were introduced into the sample cell of the NDIR analyzer at 0.5 l/min alternately for 5 minutes each. Following calibration, 25 minutes were allocated for each of the measurements of xCO_2^s and for xCO_2^a. Air was pumped at 10 l/min from an inlet installed at the bow of the ship to avoid contamination. Air were introduced (0.5 l/min) into the sample cell of the NDIR analyzer after drying. Sea water was pumped up continuously from 4 meters below the surface and introduced into the equilibrator. The fixed volume of air (*ca.* 2 l) was circulated at 0.5 l/min in a closed circuit consisting of the NDIR analyzer, the diaphragm pump, the equilibrator (countercurrentwise flow), and the column of $Mg(ClO_4)_2$. The output voltage of the NDIR analyzer was recorded on a strip chart.

As reported earlier (Inoue and Sugimura, 1988), however, there were a few problems in comparing data sets collected during this period and later. We have to take into account the pressure broadening effect of the NDIR analyzer due to the use of CO_2-in-N_2 standards, changes in curvature between the output voltage of the NDIR analyzer and the CO_2 mixing ratio, temperature increase between the seawater in equilibrator and at the sea surface, and the pressure difference in the sample cell of the NDIR analyzer during oceanic CO_2 measurements from those of standard gases and background air.

Pressure broadening effect of the NDIR analyzer

The CO_2-in-N_2 standards were used for the 1968/72 R/V *Hakuho-maru* cruises. Therefore, we have to determine the pressure broadening effect on the NDIR analyzer (Beckman 315A). Inoue and Sugimura (1988) estimated the pressure broadening effect using the standard gases (CO_2-in-N_2) prepared by Takachiho Kagaku Co., Ltd. The mole fraction of CO_2 in N_2 was determined by the manufacturer using the NDIR analyzer and CO_2-in-N_2 standards produced by the gravimetric method. A round number (3 digits) was reported as the CO_2 mixing ratio of standards. This means that the reproducibility of CO_2 measurement by the NDIR analyzer could have been as large as 1 ppm prior to 1973. We compared the values determined by CO_2-in-N_2 standards with those determined by CO_2-in-air standards (Inoue and Sugimura, 1988). The least-squares fit to the data yields:

$$xCO_2 = -18.54 + 1.14xCO_2(N_2) - 2.198 \times 10^{-4}xCO_2(N_2)^2, \quad (3)$$

where xCO_2 is the CO_2 mixing ratio determined by the CO_2-in-air standards and $xCO_2(N_2)$ is the apparent CO_2 mixing ratio determined by the CO_2-in-N_2 standards. The difference between xCO_2 and $xCO_2(N_2)$ is 2.7 ppm at 250 ppm, 3.7 ppm at 300 ppm, 3.5 ppm at 350 ppm, 2.3 ppm at 400 ppm and –0.1 ppm at 450 ppm.

The relationship between the output voltage of the NDIR analyzer and the CO_2 mixing ratio

Miyake *et al.* (1974) assumed the linearity between the output voltage of the NDIR analyzer and the CO_2 mixing ratio, in calculating the xCO_2 using only two working standard gases. However, the relationship between the output voltage and CO_2 mixing ratio varies with time. To estimate changes in this relationship, Inoue and Sugimura (1988) used the latitudinal distribution of atmospheric CO_2 reported by Bolin and Keeling (1963). Detrended seasonal variation in the atmospheric CO_2 over the ocean was assumed to be equal to that of Bolin and Keeling (1963). xCO_2^s was calculated from the reading of an analog reorder chart using Eq. (3) and the atmospheric CO_2 data.

Temperature increase between the temperature of seawater in the equilibrator and the sea surface temperature (SST)

Miyake *et al.* (1974) did not correct the temperature effect for xCO_2^s measurements. Assuming that total barometric pressure is 1 atm and salinity is 35 psu, we calculate the pCO_2^s in surface seawaters using the average temperature change (+0.3°C) during the 1983/84 BIOMASS cruise of R/V *Hakuho-maru*.

Pressure effect on xCO_2^s

Gas circulation by the diaphragm pump produced pressure changes in the closed loop. Changes in pressure in the sample cell were dependent on the flow rate and the order of connections among the parts in this loop. As discussed in

Inoue *et al.* (1995), a correction was needed for pressure change during the air circulation. However, there were no data to estimate the pressure effect on xCO_2^s. According to the archive in our laboratory, this problem was realized in earlier cruises but not checked out completely. A solution to this problem was not clear. In this study, we estimated the pressure effect by circulating the air in the closed circuit of the CO_2 measuring system used for the period from 1987 to 1993. The size of equilibrator was almost the same as that used prior to 1973, and the diameter of tubing (1/4 inch), and the orifice of the electromagnetic valves (3 mm) were the same as those used prior to 1973, although the system used in 1987/93 (Fig. 2) is more complicated. Because the diaphragm pump was installed upstream of the equilibrator (Fig. 1), the pressure in the sample cell of the NDIR analyzer decreased slightly during circulation of the equilibrating air, but increased during the measurements of standards and atmospheric CO_2. The pressure effect on xCO_2^s was examined on board the R/V *Natsushima* (Japan Marine Science and Technology Center) and M/S *Hokuto-maru* (Institute for Sea Training), during the cruises in the North Pacific where the xCO_2^s ranged from 300 to 400 ppm. The apparent decrease of the CO_2 mixing ratio during the circulation at 0.5 l/min is -1.8 ± 0.4 ppm. The pressure change in the sample cell of the NDIR analyzer between xCO_2^s measurements and standards, therefore, required a correction in the range from -2.2 to -1.4 ppm.

For the showerhead-type equilibrator developed by the MRI, we hardly observed changes in water level in the equilibrator (typically less than 3 cm change), which supported an assessment of the relatively small change in pressure for the pre-1973 CO_2 measuring system.

Measurements of oceanic and atmospheric CO_2, 1981–1986

xCO_2^a and xCO_2^s were measured basically using the same analytical system described above except for the introduction of an electric dehumidifier (Fushimi 1987; Inoue *et al.*, 1987), and the placing of the diaphragm pump between the electric dehumidifier and the equilibrator (Fig. 3 in Inoue *et al.* (1987)). For this reason, the pressure effect on the xCO_2^s has been corrected as reported (Inoue *et al.*, 1995).

The effect of the seawater temperature change between the equilibrator and surface seawater was corrected using the equation given by Gordon and Jones (1973):

$$\frac{\delta pCO_2}{\delta t} = 4.4 \times 10^{-2}(pCO_2) - 4.6 \times 10^{-6}(pCO_2)^2. \tag{4}$$

Equation (4) was integrated and water vapor pressure in the equilibrator and sea surface was calculated, taking into account the effect of salinity (assumed to be 35 psu). The temperature increase was estimated from measurements of water temperature at the equilibrator and sea surface twice a day using a Hg-thermometer which was calibrated at the JMA.

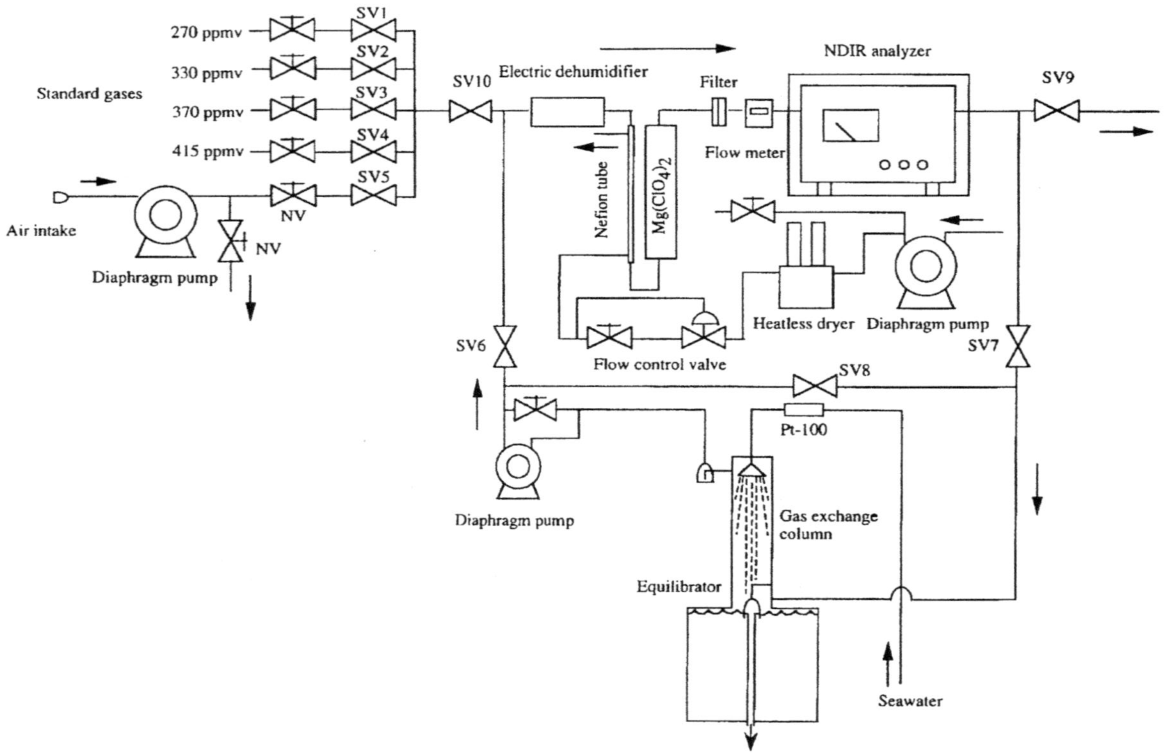

Fig. 2. A schematic diagram of the atmospheric and oceanic CO_2 measuring system used during the period from 1987 to 1993. Arrows shows the direction of air flow for either the measurement of xCO_2^a or xCO_2^s. The output voltage of the NDIR analyzer and water temperature were digitized and stored in floppy disks. SV means the solenoid valve, and NVthe flow control valve.

Other than two standard gases (250 ppm and 450 ppm CO_2 in N_2), the working standard gas (350 ppm) was usually used once a day to estimate the non-linearity between the CO_2 mixing ratio and the output voltage of NDIR analyzer.

At the end of February 1985, we improved our analytical system by using four standard gases to measure xCO_2^a and xCO_2^s instead of three (Fig. 3 in Inoue *et al.*, 1987). The four standard gases were introduced into the NDIR analyzer's cell in succession every hour. Each standard gas was introduced into the NDIR analyzer cell for 5 minutes, and the next 20 minutes was allocated for xCO_2^a. The remaining 20 minutes was used for measurements of xCO_2^s. We used this system for the period from May 1985 to July 1986.

Measurements of oceanic and atmospheric CO_2, 1987–1993

In April 1987, we changed our measurement methodology by introducing a system (Fig. 2) that was operated by a personal computer (HP 85) and a data acquisition unit (Inoue and Sugimura, 1988a, 1992). The output voltage of the NDIR analyzer and the seawater temperature at the inlet of the equilibrator were digitized and stored on floppy disks.

Changes in both the pressure and temperature in the NDIR analyzer's cell (Beckman model 864, model 865, and model 880) and water vapor in the sampled air were factors affecting the determination of xCO_2^a and xCO_2^s. The NDIR analyzer was installed in a plastic box in an air-conditioned room to keep the effects of room temperature change to within ±0.1°C. 20 seconds before the signal integration (A/D conversion), the flow of the respective sample gases was stopped and the outlet solenoid valve (No. 9) opened (open-end configuration) to equilibrate the temperature and pressure in the NDIR analyzer's cell and those of the standard gases.

The apparent time which was required to establish equilibrium between the equilibrator air in the closed circuit and seawater depends on the magnitude of disequilibrium, air flow rate, and water flow rate. We circulated air at 0.6 l/min for 10 minutes, which was enough to establish equilibrium for open ocean water. Immediately after stopping the air flow in the closed circuit, the solenoid valve No. 9 was also opened for the measurement of xCO_2^s. Because the pressure in the sample cell of the NDIR analyzer during circulation was slightly higher than the ambient air pressure, temperature and pressure equilibrium was readily attained without contamination. The length of tubing (>2.5 m, od. 0.635 mm) attached to solenoid valve No. 9 was long enough to prevent the ambient air from diffusing back into the NDIR analyzer sample cell.

To remove the water vapor from the sample gases, we used a Nefion tube (Perma Pure Ltd.) between the electric dehumidifier and the chemical desiccant column [$Mg(ClO_4)_2$]. Dry air was supplied to the Nefion tube via a diaphragm pump, a heatless-dryer (CKD) in which molecular sieves (13X) were used to remove water vapor, and a mass flow controller. The water vapor in the sample gas was removed by molecular diffusion through a film of ion exchange resin. The dew point of sample air flowing at 0.6 l/min was lower than 2°C at the outlet of the electric dehumidifier and –20°C of the Nefion tube.

Fluctuations of electric power frequency and voltage on the ship affected the determination xCO_2^a and xCO_2^s. The electric frequency and voltage supplied to, the chopper motor, detector, low-pass filter, etc., of the NDIR analyzer should be stable. But, we found that switching the heater in the NDIR analyzer disturbed them. For this reason, we isolated the electric lines that supplies the power to the chopper motor,low-pass filter, detector, etc., from that of the heating and the electric fan, etc. A voltage and frequency stabilizer (Takasago AF330) was used for the chpper motor, etc.

The precision to replicate analyses with this system in our laboratory on land is better than 0.05 ppm for Beckman models 865 and 880, and better than 0.1 ppm for Beckman model 864 (Inoue and Sugimura, 1992), and slightly worse during the shipboard analysis. It was affected mostly by the weather conditions and vibrations of the ship's engine. Taking into account the fluctuations of temperature increase of seawater in the tubing (<0.1°C) and pressure broadening effect due to the super/undersaturation of oxygen in seawater, the precision of pCO_2^s is estimated to be ± 2 μatm. Seawater temperatures at the inlet of the equilibrator and that in the surface water were measured at least twice a day by an Hg-thermometer to calibrate the temperature sensor at the inlet of the equilibrator and sea surface during respective cruises.

Measurements of oceanic and atmospheric CO_2, 1993–1996

The MRI system developed in 1986/87 was replaced by a new system in September 1993, because some of the spare parts could not be obtained. A satellite navigation unit (GPS) was also introduced into the system. To measure xCO_2^a and xCO_2^s, we also adopted the "open-ended configuration" by opening solenoid valve No. 9 when A/D integration commenced. Figure 3 shows the schematic diagram currently used for quasi-continuous measurements on board. The NDIR analyzer (BINOS 4.1 Rosemount) acted as an comparator of the CO_2 mixing ratio in ambient air and the air equilibrated with seawater relative to known CO_2 mixing ratio in standard gas cylinders. As described above the NDIR analyzer was also placed in a plastic box, and the electric power for the chopper motor, the low pass-filter, the detector, etc., was supplied via a voltage and frequency power stabilizer (PCR-500L, Kikusui; AA150F or AA330F, Takasago).

Duplicate analyses of the same sample air revealed that the standard deviation (1-sigma) was less than 0.04 ppm for BINOS 4.1 in the laboratory on land (Fig. 4).

The gas exchange column of the equilibrator was thermostated by the sampled seawater, and the temperature of the seawater was monitored at the inlet and water bath of the equilibrator (Pt-100 ohm). Before entering the gas exchange column of the equilibrator, air was passed through a jacket to bring its temperature close to that of the seawater. The temperature of the air at the outlet was also monitored (Pt-100 ohm). For the safe operation of the system, a water level sensor was attached to the gas exchange column of the equilibrator.

We set up the switching sequence of the standard and sample gases within a given time interval. Figure 5 shows the standard and sample gas sequencing

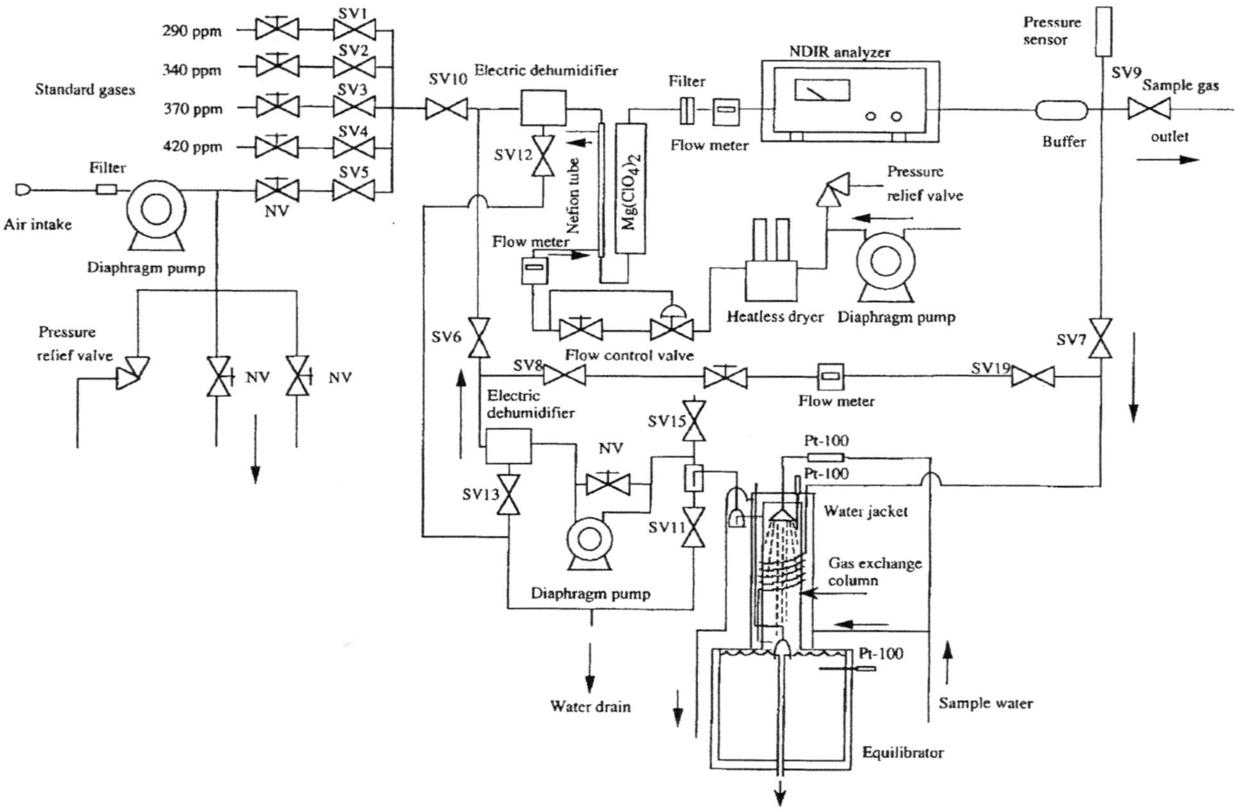

Fig. 3. A schematic diagram of the atmospheric and oceanic CO_2 measuring system that has been in use since 1993. Arrows shows the direction of air flow either for the measurement of xCO_2^a or xCO_2^s. The standard and sample gas sequence within a given time interval could be changed easily and the data was stored on memory cards. SV meants the solenoid valve, and NV the flow control valve.

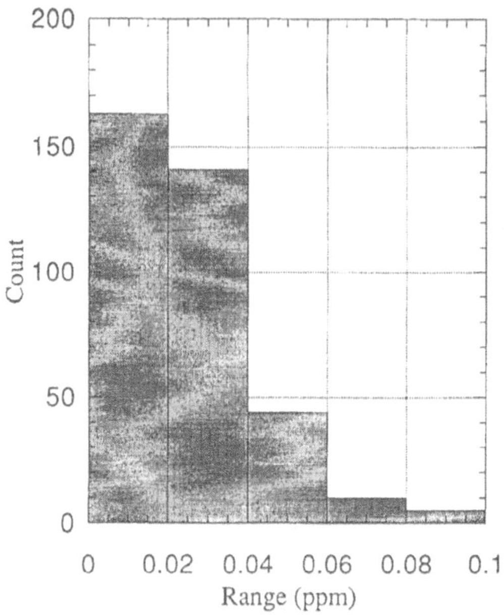

Fig. 4. Distribution of the standard deviation (1-s) of the CO_2 concentration for mean values as determined by the method of least-square fitting. Measurements were made by the NDIR analyzer (BINOS 4.1 and Beckman model 865) during the period from February 1996 to June 1996.

schemes employed for the international intercomparison of the underway pCO_2 system held on board the FS *Meteor* in the North Atlantic in June 1996. Some 22 minutes were allowed for the calibration by 4 working standard gases (250 ppm, 300 ppm, 350 ppm, 400 ppm), 16.5 minutes for measuring CO_2 in the ambient air (measured 3 times), and 50 minutes for measuring CO_2 in the air equilibrated with seawater (measured 4 times). The 1.5-hour cycle was repeated continuously. It was possible to change the 1.5-hour time intervals (max. 2-hour intervals), and order and number of measurements (max. 16). If we focused on the short-term or small-scale variations in oceanic (atmospheric) CO_2, we could measure only (xCO_2^a) xCO_2^s quasi-continuously. Raw data of output voltages of the NDIR analyzer, temperature sensors, pressure sensor, and geographical positions during each measurement were stored digitally (ASCII files) on memory cards (1 Mbytes) and transferred to a personal computer (PC 98 NEC) via an RS 232C interface.

CO_2 MIXING RATIO OF AIR SAMPLE

The output voltage for each working standard at the time of air sample measurements (V_i in Fig. 5) was obtained by linear interpolation derived from the output voltages between two successive measurements of the working standard

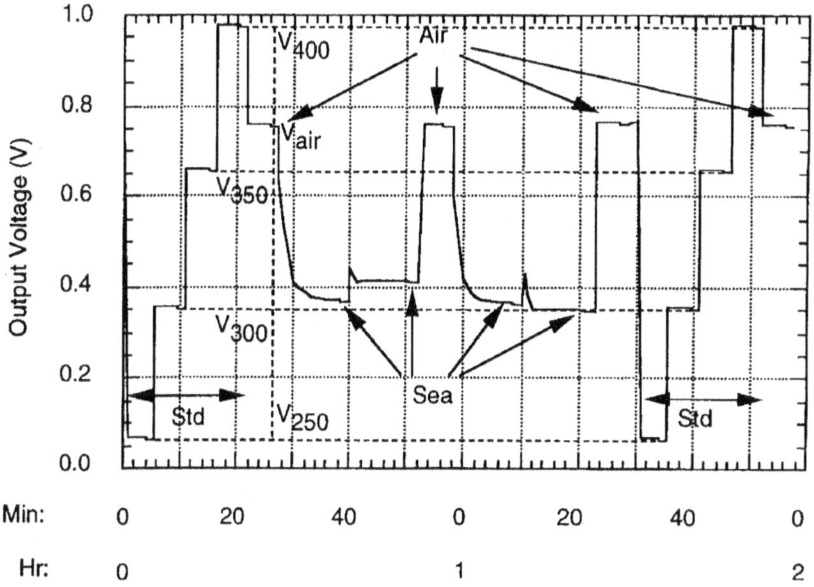

Fig. 5. Standard and sample gas sequencing schemes employed for international intercomparison of underway pCO_2 measurements on board the FS *Meteor* in the North Atlantic in June 1996. The output voltages of standard gases V_i at the time of sample gas measurements were obtained by linear interpolation between successive calibrations (see text). The subscript i meants the CO_2 mixing ratio of standard gas used on board FS *Meteor*.

gas. The coefficients of the second-degree polynomial (a, b, c) were determined by the method of least-squares fitting, and the mole fraction in the dry air was calculated using these coefficients and the output voltage of the NDIR analyzer:

$$xCO_2 = a + bV + cV^2. \tag{5}$$

The root mean squares of the residuals from the four-point fit were usually within 0.1 ppm.

The relationship between the partial pressure of CO^2 was expressed by:

$$pCO_2 = xCO_2 \, (P_{atm} - p_{wat}), \tag{6}$$

where P_{atm} is the atmospheric pressure at the sea surface, and p_{wat} the saturated vapor pressure of seawater. As can be readily seen in Eq. (6), the "partial pressure of CO_2" assumed an ideal gas system with no inter-molecule interaction. When we treat the "real" gas system based on thermodynamic theory, we should take into account the inter-molecule interaction which causes deviations from those of ideal gas behaviour. When we use the fugacity for the oceanic carbonate system, we need to calculate the fugacity of CO_2 in the moist air (DOE, 1994). However, this calculation could be the subject for future change, because it does not contain the inter-molecule interaction between CO_2 and H_2O. For this reason, we will use the partial pressure of CO_2 in wet air as reported historically.

DATA SELECTION OF xCO_2^a AND xCO_2^s

The objective of selecting atmospheric CO_2 data was to identify xCO_2^a values not affected by local sources and sinks. Since air CO_2 variabilities on a time-scale from a few minutes to an hour were clearly due to local contamination, we first rejected data showing the instability of the CO_2 mixing ratio, seen in an analog recording of the NDIR analyzer output on a strip chart recorder. If changes in the CO_2 mixing ratio measured within a few hours or a few tens of kilometers were large, we could reject the CO_2 values based on the assumption that variabilities of CO_2 mixing ratio in representative air should only vary by a small amount. The amount of CO_2 change is, however, dependent on time and geographical position. The CO_2 mixing ratio lying outside the 1-sigma (or 1.5-sigma) of an average at each latitude (or longitude) was flagged. Visual inspection of the remaining data sometimes showed that variabilities over hours at an oceanographic station were still large in comparison with those of adjacent latitudes measured by a ship under steam, probably due to local contamination from the stationary ship and station. It was dependent on the wind direction and speed. We rejected a high CO_2 mixing ratio until the standard deviation of the CO_2 mixing ratio decreased to the level of the adjacent latitudes. The standard deviation at each latitude was typically within the range of between 0.1 and 0.4 ppm for the period from 1987 to 1993.

We rejected xCO_2^s data that were contaminated by ambient air or were not obviously equilibrated due to a malfunction of the system. The xCO_2^s values falling the outside of 2-sigma of latitudinal (or longitudinal) mean were flagged.

Standard gases

During the period from January 1981 to April 1986, CO_2-in-N_2 mixtures were used as working standard gases. These gases were calibrated against the secondary standard gases (CO_2-in-air). CO_2 in N_2 gas mixtures were used because their CO_2 mixing ratio was stable with time. The secondary standard gases were calibrated at least twice a year against primary standards prepared gravimetrically by the Takachiho Co., Ltd. Both primary and secondary standard gases are composed of CO_2 in synthetic air without argon. The NDIR analyzer used for shipboard measurements (Beckman 315A) was employed to calibrate the working standard gases. Therefore, the observed value was based on a CO_2 mole fraction in synthetic air.

Since April 1987, standard gases (CO_2 in natural air) made by Nippon Sanso have been used as primary, secondary, and working standard gases. The primary standard gases were made gravimetrically following the same procedures reported by Tanaka *et al.* (1987). Within the range between 280 and 410 ppm, the relationship between the scale on the basis of primary standards made by Nippon Sanso (MRI87) and the Takachiho scale was given by

$$xCO_2O_2^{NS} = 0.247 + 1.028(xCO_2^{TK}) - 7.650 \times 10^{-5}(xCO_2^{TK})^2, \quad (7)$$

where xCO_2^{NS} is the mixing ratio in dry air based on the MRI87 scale and xCO_2^{TK} is that based on the Takachiho scale (Inoue and Sugimura, 1992). We calculated the CO_2 mixing ratios of the primary standards by a least-square fitting assuming a quadratic relationship between the output voltage of the NDIR analyzer and the mixing ratio (Table 1). This was adopted to minimize the uncertainties in calibrated values (Tanaka *et al.*, 1987) and to check the possible drift in CO_2 mixing ratio with time.

Table 1. Primary standard gases prepared by the gravimetric method in 1987. An example of least-square fitting assuming a quadratic relation between the output voltage of the NDIR analyzer (Beckman model 865) and the given CO_2 mixing ratio

Cylinder	xCO_2^{grv} (ppm)	xCO_2^{lsf} (ppm)	1-s (ppm)	No
DF4772	249.55	249.63	0.00	5
DF4764	275.45	275.38	0.01	5
DF4763	299.63	299.67	0.03	5
DF4762	325.31	325.37	0.01	5
DC9359	334.76	334.72	0.03	5
DC9357*	345.10	344.72	0.02	5
DF4776	349.90	349.90	0.03	5
DC9358	356.28	356.29	0.03	5
DF4775	375.79	375.69	0.02	5
DF4774	400.74	400.79	0.03	5
DF4773	425.07	425.10	0.03	5

xCO_2^{grv} means the CO_2 mole fraction determined by the gravimetrical method and xCO_2^{lsf} by the least-square fitting.

*CO_2 mixing ratio decreased with time. This standard gas was not used as the primary standard.

Table 2. Mean CO_2 mixing ratio determined by the MRI on the basis of MRI87 scale and those of USA-NOAA (Peterson 1993, private communication). A set of three aluminum cylinders (#11413, #13763, #11051) was circulated among laboratories inthe USA, Canada, Japan, New Zealand, and Australia during the period from April 1991 to March 1993. The other set of standards (#11429, #6272, #11062) was circulated among Germany, France, Spain, Italy, and Hungary.

Laboratory	Date	#11413	#13763	#11051
MRI	July 1992	341.60	347.48	375.15
NOAA/CMDL	Sept.1991	341.47	347.41	375.16
NOAA/CMDL	Dec. 1992	341.49	347.45	375.18

We participated in the CO_2 round robin intercomparison conducted in 1991 and 1992. The results of our analysis system show good agreement with the 1985 WMO mole fraction scale (determined by NOAA/CMDL) within the range from 330 to 370 ppm (Table 2) to within 0.12 ppm.

The JMA has been using four CO_2 working reference gases ranging from 290 ppm to 400 ppm to conduct operational observations from the R/V *Ryofu-maru*. These working standard gases were calibrated before and after each cruise against secondary standard gases, which were calibrated twice annually against the 1985 WMO standard gases. We sent our cylinders to the JMA annually to compare our MRI87 scale with the 1985 WMO scale. The relationship between the MRI87 scale and the 1985 WMO scale was given by

$$xCO_2^{WMO} = 5.410 + 0.975(xCO_2^{NS}) + 2.530 \times 10^{-5}(xCO_2^{NS})^2, \qquad (8)$$

where xCO_2^{WMO} is the mole fraction based on the 1985 WMO scale. However, this relationship is only tentative and may change considerably in the future. The

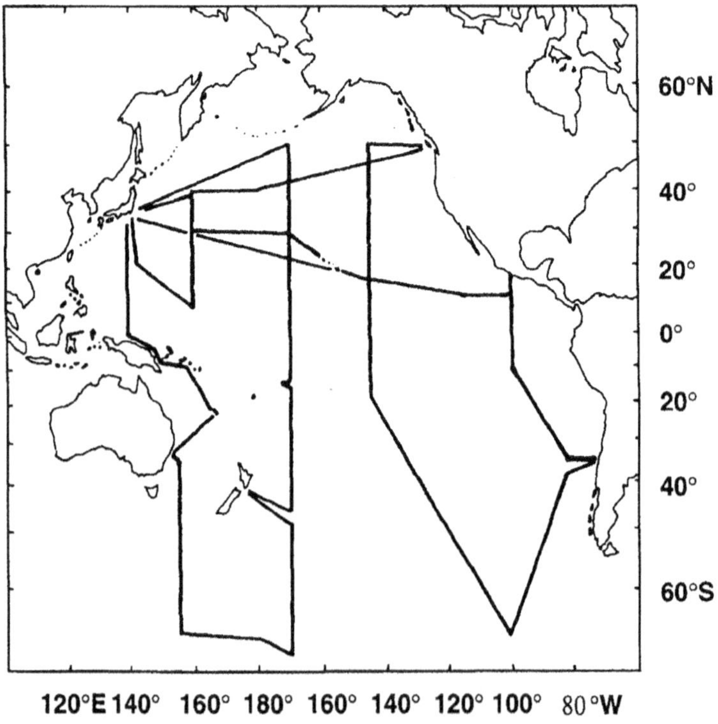

Fig. 6. The cruise tracks of R/V *Hakuho-maru* during which time measurements of pCO_2 were made throughout the period from 1968 to 1972. Fig. 1 in Miyake *et al.* (1974) was redrawn.

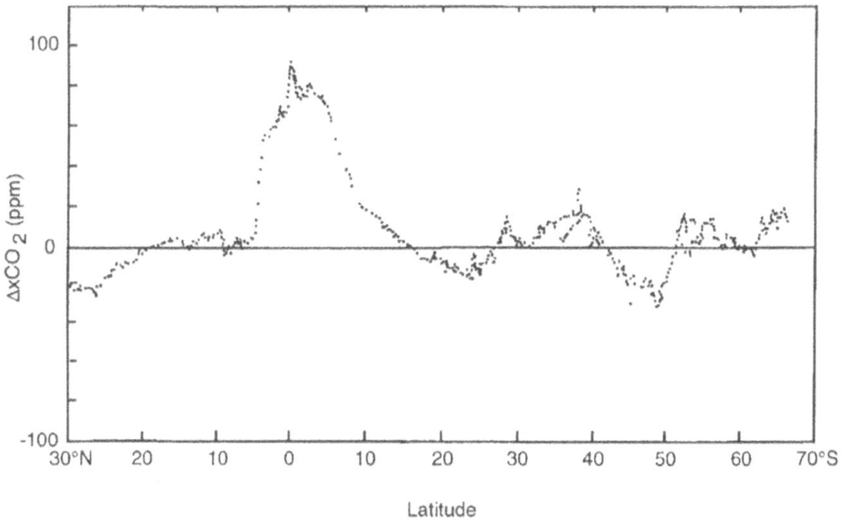

Fig. 7. Latitudinal distribution of xCO_2^a and xCO_2^s measured during the period from November 1968 to January 1969. Data south of 30°N were read from Fig. 4 in Miyake *et al.* (1974). The unit of vertical axis was changed from pCO_2 (ppm) to xCO_2 (ppm), and data north of 30°N were omitted because they were measured during a different period from April to June in 1970.

CO_2 mixing ratio sent to the JMA ranged from 290 to 400 ppm. In this paper we reported CO_2 mixing ratios based on the 1985 WMO scale as calculated by Eq. (8).

SUMMARY OF pCO_2 MEASUREMENTS

Results of pCO_2 measurements prior to 1973

Figure 6 shows the cruise tracks of R/V *Hakuho-maru* during which time quasi-continuous pCO_2 measurements were made using the CO_2 measuring system described in above. Figure 7 shows the latitudinal distribution of ΔxCO_2 along 170°W observed during the period from November 1968 to January 1969 redrawn from Fig. 4 of Miyake *et al.* (1974). The unit on vertical axis in Fig. 4 of Miyake *et al.* (1974) was expressed as ΔpCO_2 (ppm), but this should be ΔxCO_2 (ppm) after reviewing the archives in our laboratory. Because the ΔxCO_2 data north of 30°N in Fig. 4 of Miyake *et al.* (1974) were taken from another cruise conducted during the period from April to June in 1970, we did not plot the ΔxCO_2 data north of 30°N. Figure 7 shows similar features of ΔxCO_2 distribution as the current observations (Feely *et al.*, 1995; Ishii and Inoue, 1995), though corrections were needed to made both data sets compatible. In the area of equatorial upwelling, xCO_2^s was highly supersaturated with respect to xCO_2^a, and the xCO_2^s changed abruptly at the boundary (4°N) between the North Equatorial Countercurrent (NECC) and the South Equatorial Current (SEC). In contrast to the rapid change between the NECC and the SEC, xCO_2^s decreased gradually from the equator southward.

By compiling data from 1968 to 1972, Miyake *et al.* (1974) produced a ΔpCO_2 map of the Pacific Ocean (Fig. 5 in Miyake *et al.* (1974)), and concluded that the Pacific Ocean was a source for atmospheric CO_2. In their treatment, however, seasonal variations in pCO_2^s were not considered. Even if their conclusion was not correct, we are reluctant to criticize their pioneer ΔpCO_2 map and conclusions. The method that they adopted was so excellent that their xCO_2^s data were still valid. We believe that it is invaluable to retrieve xCO_2^s data measured prior to 1973 to infer the long-term changes in the oceanic carbonate system of the Pacific. Inoue and Sugimura (1988) used this approach to deduce the increase of pCO_2^s in the Pacific between 1984 and 1969.

Seasonal variation and long-term trend of pCO_2^s in the western North Pacific

Since 1981, measurements of atmospheric and oceanic pCO_2 have been made periodically along the same cruise tracks (Inoue *et al.*, 1995) aboard the JMA ship R/V *Ryofu-maru*. Every year the R/V *Ryofu-maru* leaves Tokyo in the middle of January and arrives at the equator (or 3°N) at the end of month, while conducting oceanographic and meteorological observations along 137°E. Prior to 1989, MRI observed pCO_2^s and pCO_2^a and in 1990 JMA took over the program of operational observations aboard the R/V *Ryofu-maru* using basically the same analytical system as that of MRI (Hirota *et al.*, 1993).

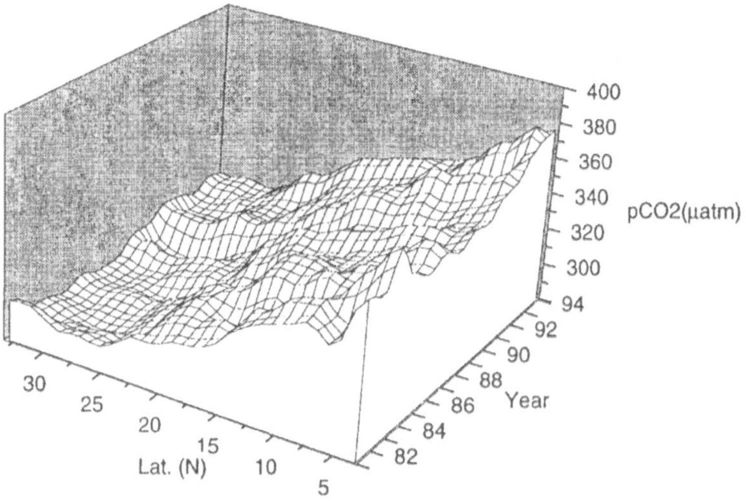

Fig. 8. Distribution of pCO_2^s in the western North Pacific (along 137°E) measured during every boreal winter since 1981. Due to the enhancement of vertical mixing, the pCO_2^s in the western equatorial Pacific increased during the El Niño event except in January and February 1987.

Inoue *et al.* (1995) reported a long-term increase of pCO_2^s. Figure 8 shows the latitudinal distribution of pCO_2^s along 137°E observed during every boreal winter since 1981. Every year, the pCO_2^s distribution in the subtropical area showed a similar pattern: off the coast of Japan the pCO_2^s was undersaturated with respect to pCO_2^a, and increased gradually toward the south. Near the equator, pCO_2^s was slightly larger than pCO_2^a. As reported earlier (Fushimi 1987; Inoue *et al.*, 1987), the pCO_2^s distribution near the equator is affected by the El Niño events.

Figure 8 clearly shows an increase of pCO_2^s in the whole area over periods from 1981 to 1996. By fitting a linear function to the averages at each latitude over periods from 1984 to 1993, Inoue *et al.* (1995) reported that the pCO_2^s in the subtropics has been increasing at a rate of 1.8 μatm/yr, equal to the atmospheric increase, while the rate of 0.5 μatm/yr in the western equatorial Pacific was less than that of the atmosphere. During the El Niño event, pCO_2^s data in the western equatorial Pacific were not used for this calculation. The difference in growth rate suggests temporal changes in ΔpCO_2 distribution (Volk and Bacastow, 1989). Inoue *et al.* (1995) examined thermodynamic factors controlling the oceanic carbonate system in the western North Pacific and concluded that the increase was caused by the oceanic CO_2 uptake. The oceanic TCO_2 increase was estimated to be 1 μmol/kg by using the homogeneous buffer factor.

Weiss *et al.* (1982) reported that the seasonal variations of fCO_2^s (fugacity) in the subtropics of the Pacific were mainly controlled by the thermodynamic temperature effect. In the western North Pacific, seasonal variations in pCO_2^s were found to correlate well with the sea surface temperature (Inoue *et al.*, 1995). The apparent relationship between pCO_2^s and SST is different from that of the thermodynamics. Details of the difference from the thermodynamic effects was discussed in Inoue *et al.* (1995). Seasonal variations in CO_2 flux between the sea and the air calculated from Eq. (1) showed an active CO_2 uptake during the winter season in the area of the Subtropical Mode Water formation, south of the Kuroshio and east of Japan (Inoue *et al.*, 1995).

Long-term variations in pCO_2^s have been confirmed by direct and precise measurements. However, there is a lack in data to show changes in the oceanic carbonate system. At least temporal variations in one of the other variables describing the oceanic carbonate system (TCO_2, pH and alkalinity) are needed. If the condition of the homogeneous buffer factor maintains, the seasonal variation in Fx and the long-term trend of pCO_2^s require rapid transfers of anthropogenic CO_2 to middle and deep waters (Inoue *et al.*, 1995). Phytoplankton in surface waters uses CO_2 and/or bicarbonate ion to conduct photosynthesis. If phytoplankton uses bicarbonate ion and releases the OH^- ion during the photosynthesis, as reported in fresh waters (Lucas, 1983), this may change the pCO_2-TCO_2 relationship. At the moment, we do not know the long-term trend of pH in seawaters. If the pH of surface seawater remains at the same level due to the biological activities, more TCO_2 can be stored in surface mixing layers with small changes in pCO_2 as compared with the pCO_2-TCO_2 relation defined by the homogeneous buffer factor. Accurate and continuous measurements of at least two variables of the carbonate system are needed to provide useful information about the fate of anthropogenic CO_2.

Interannual changes in CO_2 flux in the tropical ocean

In this section, we describe both the spatial and temporal variations in pCO_2^s in the central and western equatorial Pacific based on measurements between 1987 and 1994. The Equatorial Pacific is known as a strong natural oceanic source of CO_2 to the atmosphere. The CO_2 outflux from the equatorial Pacific were estimated to be 1–2 Gt-C/yr (Tans *et al.*, 1990) corresponding to 15–30% of the current CO_2 emission due to fossil-fuel combustion (Marland *et al.*, 1994). Compared with that of the subtropical regions in the western North Pacific (Inoue *et al.*, 1995) and in the Indian Ocean (Poisson *et al.*, 1993), the distribution of pCO_2^s values in the equatorial Pacific exhibit more variability (Feely *et al.*, 1987, 1995; Inoue and Sugimura, 1992; Wong *et al.*, 1993). During the 1982/83 El Niño event, Feely *et al.* (1987) reported that the pCO_2^s in the central and equatorial

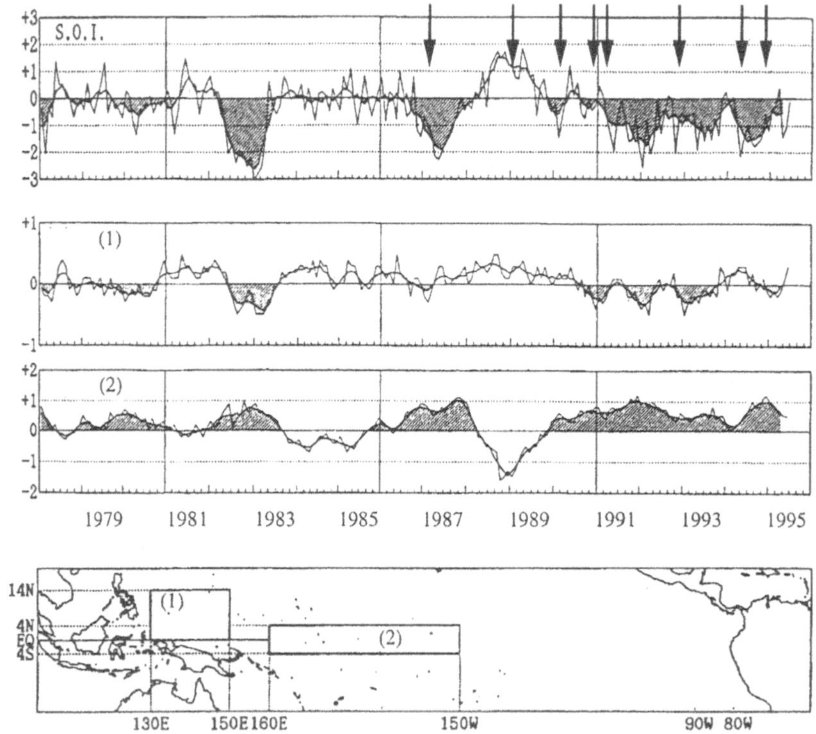

Fig. 9. Time series of SOI (solid line: running mean for 5 months) and SST anomalies (°C) in the central (160°E-150°W, 4°N-4°S) and western (130°E–150°E, 0°N–14°N) equatorial Pacific (Monthly Ocean Report, 1996). Positive anomalies in the central equatorial Pacific and negative anomalies in the western equatorial Pacific are shaded. Arrows in the top panel indicate the time for which pCO_2^s measurements were made.

Pacific decreased to a level almost equal to that of the pCO_2^a. While, during the 1988/89 La Niña event the pCO_2^s in the central and western equatorial Pacific increased considerably (Inoue and Sugimura, 1992). Figure 9 shows the anomalies of SST in the central and western equatorial Pacific (JMA, 1996). During the 1982/83 El Niño event, the pCO_2^s in the western equatorial Pacific increased with a SST decrease (Inoue *et al.*, 1987; Fushimi, 1987). In the equatorial Pacific, upwelling/vertical mixing play a major role in determining the pCO_2^s distributions.

In Fig. 10, the equatorial distributions of pCO_2^a and pCO_2^s for the period of January and February 1990 are shown. Inoue *et al.*, 1996 showed a "boundary" at which the pCO_2^s changes steeply. In the present study, the longitude where the steep change in pCO_2^s occurred is defined as P. The steep change in pCO_2^s

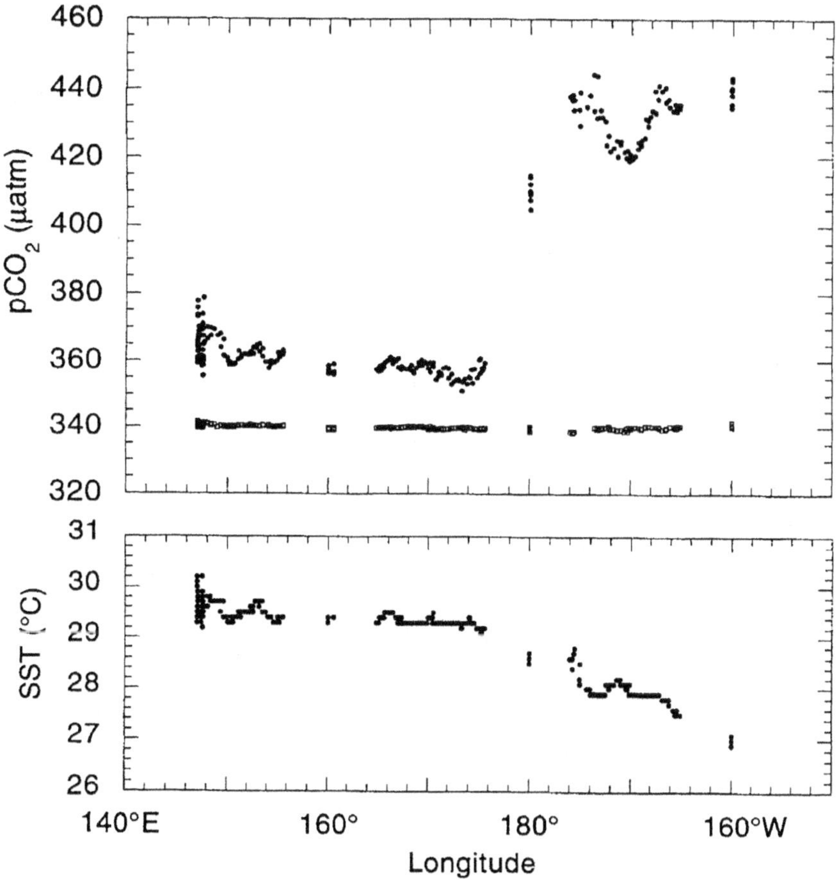

Fig. 10. The equatorial distribution of pCO_2^s and pCO_2^a (upper panel), and sea surface temperature (lower panel) for the period between January and February in 1990. A solid circle meants pCO_2^s and an open square pCO_2^a.

occurred at 180°, where the SST changed gradually. Over longitudes 176°W and 160°W, the seawater surface pCO_2^s (420–450 μatm) was highly supersaturated with atmospheric pCO_2 (340 μatm). West of 176°E the pCO_2 (350–380 μatm), was slightly supersaturated with respect to the pCO_2^a, and increased gradually toward the west. West of 176°E, the concentrations of $NO_3^- + NO_2^-$ were close to 0 μmol/kg, the sea surface salinity (SSS) was lower than 34.5 psu, and the SST around 29–30°C.

For the period November–December 1994, relatively low and rather constant pCO_2^s values were present west of 166°W (Fig. 3 in Inoue *et al.* (1996)). The pCO_2^s values began to change near 165°W as did the values of SSS. West of 166°W, the hydrographic property and nutrient concentrations exhibited patterns similar to those west of 176°E for January–February 1990.

Lower pCO_2 values always occurred in regions of high temperature (>29°C), low salinity (<34.5 psu), and low concentrations of nutrients observed west of 166°W in 1994. Conditions with low salinity and nutrient concentration and high temperature suggested a "warm water pool" in the western equatorial Pacific, closely connected with the El Niño/Southern Oscillation phenomenon. The correlation between the longitude where sharp change in pCO_2 occurs (P) and the El Niño/Southern Oscillation phenomenon was examined (Inoue *et al.*, 1996) by using the SOI (Monthly Ocean Report, 1996). A quadratic least-squares fit to the

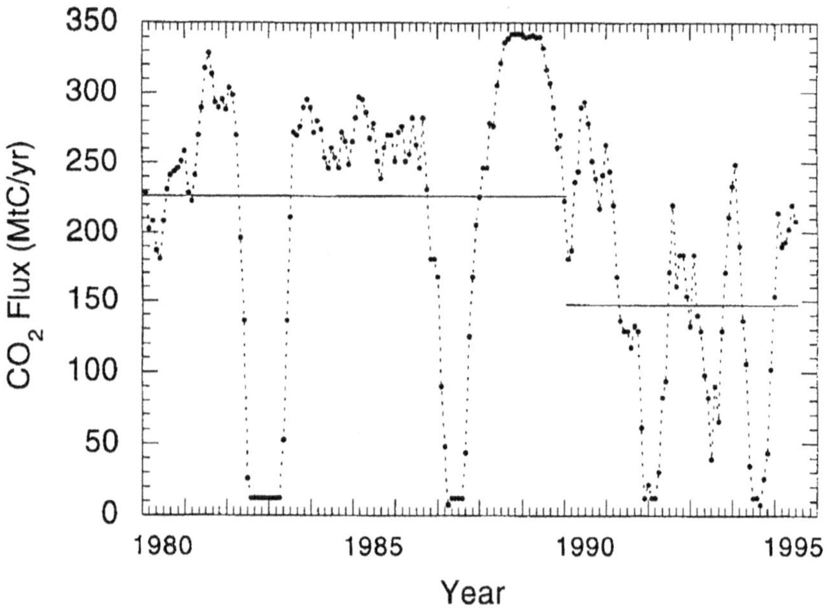

Fig. 11. Temporal changes in CO_2 outflux from the central and western equatorial Pacific (5.5°S-5.5°N, 130°E-160°W). The unit of vertical axis was Mt-C/yr (10^{12} g-C/yr). Horizontal bars indicate the average annual CO_2 outflux over 1980s and early 1990s. In this treatment, changes in ΔpCO_2 and wind field are not taken into account.

data yields $P = 163.85 - 20.33 \times (SOI) + 7.32 \times (SOI)^2$ ($N = 7$, $r = 0.97$). Clearly, the P moves eastward with a decrease in SOI, indicating that with strong CO_2 source region of the equatorial Pacific decreases with a decrease of the SOI.

In order to estimate the CO_2 outflux from the central and western equatorial Pacific, it is important to know the temporal changes in the areas of higher pCO_2^s. To know temporal changes in CO_2 outflux quickly and simply, we divided the central and western equatorial area (5.5°S-5.5°N, 130°E-160°W) into a higher pCO_2 region and a lower pCO_2 region. Then we estimated the CO_2 outflux for these areas. On the basis of CO_2 flux data by Ishii and Inoue (1995), the CO_2 outflux was estimated to be 9.6 mmol $m^{-2}day^{-1}$ for the higher pCO_2 region and 0.34 mmol $m^{-2}day^{-1}$ for lower pCO_2^s region. The average outflux for each region was simply taken from Ishii and Inoue (1995), where the relationship between gas exchange and wind speed was estimated by an equation given by Tans *et al.* (1990). The CO_2 outflux from the central and western equatorial Pacific decreases during the El Niño event, but increases during the La Niña event. This suggests significant intra- and interannual fluctuations of CO_2 outflux from the central and western equatorial Pacific (Fig. 11). This is not inconsistent with Francey *et al.* (1995) and Keeling *et al.* (1995), who suggested significant changes in ocean/plant uptake on time scales of a year to years. Over the 1980s, the average annual CO_2 outflux west of 160°W was estimated to be 230 MtC yr^{-1}, while during the period from January 1990 to June 1995 150 MtC yr^{-1} (Fig. 11). This preliminary result showed that the annual average CO_2 outflux from the central and western equatorial Pacific (5.5°S-5.5°N, 130°E-160°W) during the early 1990s decreased by one-third of the value in the 1980s.

REFERENCES

Boden, T. A., D. P. Kaiser, R. J. Sepanski and F. W. Stoss (ed.) (1994): *Trends '93: A Compendium of Data on Global Change.* ORNL/CDIAC-65, Carbon Dioxide Information Analysis Center, Oak Ridge National Laboratory, Oak Ridge, Tenn, U.S.A.

Bolin, B. and C. D. Keeling (1963): Large-scale atmospheric mixing as deduced from the seasonal and meridional variations of carbon dioxide. *J. Geophys. Res.*, **68**, 3899–3920.

Conway, T. J., P. P. Tans and L. S. Waterman (1994): Atmospheric CO_2 records from sites in the NOAA/CMDL air sampling network. pp. 41–119. In *Trends '93: A Compendium of Data on Global Change*, ed. by T. A. Boden, D. P. Kaiser, R. J. Sepanski and F. W. Stoss, ORNL/CDIAC-65, Carbon Dioxide Information Analysis Center, Oak Ridge National Laboratory, Oak Ridge, Tenn, U.S.A.

DOE (1994): *Handbook of Methods for the Analysis of the Various Parameters of the Carbon Dioxide in Sea Water*; version 2, ed. by A. G. Dickson and C. Goyet, ORNL/CDIAC-74, Carbon Dioxide Information Analysis Center, Oak Ridge National Laboratory, Oak Ridge, Tenn, U.S.A.

Feely, R. A., R. H. Gammon, B. A. Taft, P. E. Pullen, L. S. Waterman, T. J. Conway, J. F. Gendron and D. P. Wisegarver (1987): Distribution of chemical tracers in the eastern equatorial Pacific during and after the 1982–83 El Niño/Southern Oscillation Event, *J. Geophys. Res.*, **92**, 6545–6558.

Feely, R. A., R. Wanninkhof, C. A. Cosca, P. P. Murphy, M. F. Lamb and M. D. Steckiey (1995): CO_2 distributions in the equatorial Pacific during the 1991–1992 ENSO event. *Deep-Sea Res.*, **42**, 365–386.

Francey, R. J., P. P. Tans, C. E. Allison, I. G. Enting, J. W. C. White and M. Troller (1995): Changes in oceanic and terrestrial carbon uptake since 1982. *Nature*, **373**, 326–330.

Fushimi, K. (1987): Variation of carbon dioxide partial pressure in the western North Pacific surface water during the 1982/83 El Niño Event. *Tellus*, **39B**, 214–227.

Gordon, L. I. and L. B. Jones (1973): The effect of temperature on carbon dioxide partial pressure in seawater. *Mar. Chem.*, **1**, 317–322.

Hirota, M., K. Nemoto, A. Murata and K. Fushimi (1991): Observation of carbon dioxide in air and surface seawater in the western North Pacific Ocean. *Ocenographical Mag.*, **41**, 19–28.

Inoue H. and Y. Sugimura (1986): Distribution of pCO_2 and $\delta^{13}C$ in the air and surface sea water in the Southern Ocean, south of Australia. *Mem. Natl. Inst. Polar Res., Spec. Issue*, **40**, 454–461.

Inoue, H. and Y. Sugimura (1988a): Distribution and variations of oceanic carbon dioxide in the western North Pacific, eastern Indian, and Southern Ocean south of Australia. *Tellus*, **40B**, 308–320.

Inoue, H. and Y. Sugimura (1988b): Distribution of the pCO_2 in surface seawater of the western and central equatorial Pacific during the 1986/87 El Niño/Southern Oscillation Event. *Geophys. Res. Lett.*, **15**, 1499–1502.

Inoue, H. and Y. Sugimura (1992): Variations and distributions of CO_2 in and over the equatorial Pacific during the period from the 1986/88 El Niño Event to the 1988/89 La Niña Event. *Tellus*, **44B**, 1–22.

Inoue, H., Y. Sugimura and K. Fushimi (1987): pCO_2 and $\delta^{13}C$ in the air and surface sea water in the western North Pacific. *Tellus*, **39B**, 228–242.

Inoue, H., H. Matsueda, M. Ishii, K. Fushimi, M. Hirota, I. Asanuma and Y. Takasugi (1995): Long-term trend of the partial pressure of carbon dioxide (pCO_2) in surface waters of the western North Pacific, 1984–1993. *Tellus*, **47B**, 391–413.

Inoue, H., M. Ishii, H. Matsueda, M. Aoyama and I. Asanuma (1996): Changes in longitudinal distribution of the partial pressure of CO_2 (pCO_2) in the central and western equatorial Pacific, west of 160°W. *Geophys. Res. Lett.*, **23**, 1781–1784.

Ishii, M. and H. Inoue (1995): Air-sea exchange of CO_2 in the central and western equatorial Pacific in 1990. *Tellus.*, **47B**, 447–460.

Inter-Governmental Panel on Climate Change (1990): Greenhouse gases and aerosols. pp. 5–40. In *Climate Change. The IPCC Scientific Assessment*, ed. by J. T. Houghton, G. J. Jenkins and J. J. Ephraums, Cambridge University Press, Cambridge.

Japan Meteorological Agency (1989): Marine climatological tables of the North Pacific Ocean for 1971–1980. Marine Department, JMA. Tokyo.

Keeling, C. D., R. B. Bacastow, A. F. Carter, S. C. Piper, T. P. Whorf, T. P. Heimann, W. G. Mook and H. Roeloffozen (1989): A three-dimensional model of atmospheric CO_2 transport based on observed winds: 1 Analysis of observational data. pp. 165–236. In *Aspect of Climate Variability in the Pacific and Western America*, ed. by D. M. Peterson, AGU Geophysical Monograph, Washington, D.C.

Keeling, C. D., T. P. Whorf, M. Wahlen and J. van der Plicht (1995): Interannual extremes in the rate of rise of atmospheric carbon dioxide since 1980. *Nature*, **375**, 666–670.

Liss, P. S. and L. Merlivat (1986): Air-sea gas exchange rates: introduction and synthesis. pp. 113–127. In *The Role of Air-Sea Exchange in Geochemical Cycling*, ed. by P. Buat-Menard, D. Reidel, Dordrecht, The Netherlands.

Lucas, W. J. (1983): Photosynthetic assimilation of exogenous HCO_3^- by aquatic plants. *Ann. Rev. Plant Physiol.*, **34**, 71–104.

Marland, G., R. J. Andres and T. A. Boden (1994): Global, regional, and national CO_2 emissions. pp. 508–584. In *Trends '93: A Compendium of Data on Global Change, 1994, ORNL/CDIAC-65*, ed. by T. A. Boden, D. P. Kaiser, R. J. Sepanski and F. W. Stoss, Carbon Dioxide Information Analysis Center, Oak Ridge National Laboratory, Oak Ridge, Tenn., U.S.A.

Miyake, Y. and Y. Sugimura (1969): Carbon dioxide in the surface water and the atmosphere in the Pacific, the Indian and the Antarctic Ocean areas. *Rec. Oceanogr. Wks. Japan.*, **10**, 23–28.

Miyake, Y., Y. Sugimura and K. Saruhashi (1974): The carbon dioxide content in the surface water in the Pacific Ocean. *Rec. Oceanogr. Wks. Japan.*, **12**, 45–52.

Monthly Ocean Report (1996): El Niño Monitoring Center, Climate and Marine Department. Japan Meteorological Agency, Tokyo.

Poisson, A., N. Metzl, C. Brunet, B. Schauer, B. Bres, D. Ruiz-Pino and F. Louanchi (1993): Variability of sources and sinks of CO_2 in the western Indian and southern oceans during the year 1991. *J. Geophys. Res.*, **98**, 22,759–22,778.

Revelle, R. and H. E. Suess (1957): Carbon dioxide exchange between atmosphere and ocean, and the question of an increase of atmospheric CO_2 during the past decades. *Tellus*, **9**, 18–27.

Sundquist, E. T., L. N. Olummer and T. M. L. Wigley (1979): Carbon dioxide in the ocean surface: the homogeneous buffer factor. *Science* **204**, 1203–1205.

Tanaka, M., T. Nakazawa, and S. Aoki (1987): Time and space variations of tropospheric carbon dioxide over Japan. *Tellus*, **39B**, 3–12.

Tans, P. P., I. Fung and T. Takahashi (1990): Observational constrains on the global atmospheric CO_2 budget. *Science*, **247**, 1431–1438.

Volk, T. and R. Bacastow (1989): The changing patterns of ΔpCO_2 between ocean and atmosphere. *Global Biogeochem. Cycles*, **3**(2), 179–189.

Wagner, K. (1979): The carbonate system of the ocean. pp. 251–258. In *The Global Carbon Cycle*, ed. by B. Bolin, E. T. Degens, S. Kempe and P. Ketner, John Wiley & Sons, Chichester.

Wanninkhof, R. (1992): Relationship between wind speed and gas exchange over the ocean. *J. Geophys. Res.*, **97**, 7373–7382.

Weiss, R. F., R. A. Jahnke and C. D. Keeling (1982): Seasonal effects of temperature and salinity on the partial pressure of CO_2 in sea water. *Nature* **300**, 511–513.

WMO WDCGG Data Report (1995): *WDCGG No. 7. GAW DATA. Volume IV—Greenhouse Gases and Other Atmospheric Gases*. Japan Meteorological Agency, Tokyo.

Wong, C. S., Y.-H. Chan, J. S. Page, G. E. Smith and R. D. Bellegay (1993): Changes in equatorial CO_2 flux and new production estimated from CO_2 and nutrient levels in Pacific surface waters during the 1986/87 El Niño. *Tellus*, **45B**, 64–79.

Dynamics and Characterization of Marine Organic Matter, Eds., N. Handa, E. Tanoue and T. Hama, pp. 533–560.

North Pacific Water's Larger Potential Sink Capacity for Absorbing Anthropogenic CO_2 and the Processes Recovering It

Shizuo TSUNOGAI

Marine and Atmospheric Geochemistry Lab., Graduate School of Environmental Earth Science, Hokkaido University, Sapporo 060, Japan

(Synthesis, Received 27 December 1996; accepted 30 January 1997)

Abstract—The potential sink capacity of water for absorbing atmospheric CO_2 has been defined. The capacity is the amount of CO_2 absorbable when all the nutrients contained are transformed to organic matter. The Pacific Deep Water should have an excess potential sink capacity relative to the Atlantic Deep Water due to the gas exchange not consuming nutrients in the Antarctic Ocean in winter and the dissolution of $CaCO_3$ in the deep Pacific. Although the capacity has not yet been exactly estimated because of the lack of accurate and worldwide data of all the oceanic carbon components, the fact that the concentration of CO_2 equilibrated with the North Pacific Deep Water of null AOU is somewhat lower than that of the pre-industrial atmosphere may be caused by excess potential sink capacity. The total carbonate content corrected for phosphate and alkalinity clearly shows an invasion of anthropogenic CO_2 into the North Pacific Intermediate Water. The ^{13}C data depict the warm surface water having absorbed much anthropogenic CO_2, while its total CO_2 profile indicates inversely a small net loss by warming of the water. This means the warm water can absorb much CO_2 when it is cooled. Thus, the more the upwelling water in the eastern and equatorial Pacific, the greater the amount of CO_2 absorbed in the high-latitudes. Four processes are raised and discussed for realizing the large potential sink capacity and the absorption of much anthropogenic CO_2 in the North Pacific. These seem to be overlooked or underestimated by the present modelers giving smaller values for the oceanic uptake of CO_2. They are (1) the formation of the North Pacific Intermediate Water in the northwestern North Pacific occupying the layer down to about 1000 m with a residence time of a few tens to one hundred years, (2) the continental shelf system making the dense water which contains more total carbonate and supplying it to the subsurface layer of the open ocean, (3) the gas exchange process in the high-latitudes where the gas transfer velocity, being accelerated by bubbles, is larger than that accepted at present by the majority of scientists in this field, and (4) the abundant silica controlling the ecosystem by producing larger particles sinking into the abyss with higher speeds and larger organic-C/carbonate-C ratios in the western North Pacific. All these processes indicate the peculiar and important role of the North Pacific, especially its western part, in the oceanic uptake of anthropogenic CO_2.

INTRODUCTION

Tsunogai *et al.* (1993a) have determined the water column inventory of total CO_2 in the western North Pacific (19–40°N along 165°E) and estimated the increase

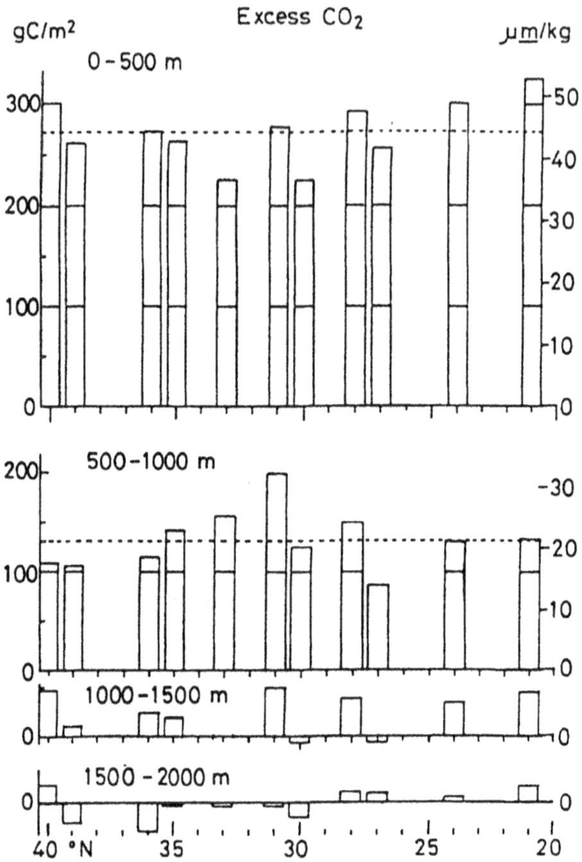

Fig. 1. Excess CO_2 calculated from the data obtained in the North Pacific along 165°E during the KH-91-5 cruise of the R/V *Hakuho Maru* in 1991 by applying the method of Chen (1982). Its inventories and mean concentrations are shown, respectively, on the left and right hand side axes. The details of the calculation have been given in Tsunogai *et al.* (1993a).

in total CO_2 to be 180±41 gC/m^2/yr during 18 years from 1973 to 1991 (Table 1). Figure 1 shows a considerable fraction of anthropogenic CO_2 has been penetrated into the intermediate water of 500–1000 m depth. The observed increase rate corresponds to the net absorption of 3.6±0.8 GtC/yr for the entire ocean area. On the other hand, a recent IPCC report (1996) has described that the oceanic uptake of anthropogenic CO_2 was 2.0±0.8 GtC/yr in the 1980's, citing results obtained from model calculations. The report has also given its terrestrial sinks, the uptake of CO_2 by the Northern Hemisphere forest regrowth of 0.5±0.5 GtC/yr and carbon uptake due to nitrogen and CO_2 fertilization and climate effects of 1.3±1.5 GtC/yr. The discrepancy between our observed results and the IPCC value for the oceanic uptake seems to be too large to be reasonably explained, even though the North Pacific may absorb more anthropogenic CO_2 than the mean ocean does.

Table 1. Mean integrated excess CO_2 in the water column in the western North Pacific (in gC/m^2). Their 95% confidence limits are also shown (after Tsunogai *et al.*, 1993a).

Year	1991 (KH-91-5)	1973 (GEOSECS)	Difference
Depth (m)			
0–500	272±21	150±24	122±32
500–1000	130±20	104±48	26±52
		(72±17)*	(58±26)*
1000–1500	24±14	26±26	–2±30
1500–2000	–1±11	1±16	–2±19
0–1000	402±29	222±29	180±41

*Calculated by assuming that the excess carbonate in the 500–1000 m layer increased at the same rate as that in the surface 500m, because the increase rate (the proportion not the absolute amount) should not be smaller than the shallower layer.

In this paper, first, we define a new term, the potential sink capacity of water for the absorption of atmospheric CO_2, and show the North Pacific water having a large potential sink capacity for anthropogenic CO_2. Next, we discuss the processes bringing the large potential sink capacity water to the surface and attaining relatively rapid equilibration for realizing the absorption of much anthropogenic CO_2 in the North Pacific. If these processes are adequately considered in a model calculation, the discrepancy will be reduced and the important role of the North Pacific in the absorption of anthropogenic CO_2 will be recognized.

POTENTIAL SINK CAPACITY OF WATER FOR THE ABSORPTION OF ATMOSPHERIC CO_2

Definition

We define the potential sink capacity for the absorption of atmospheric CO_2 (Θ) as the difference between the total CO_2 equilibrated with the present level atmospheric CO_2 at *in situ* temperature and salinity (TC_{STA}), and the expected total CO_2 content when all the nutrients contained are transformed into organic matter (TC_{P0});

$$\Theta = TC_{STA} - TC_{P0} \tag{1}$$

$$= TC_{STA} - (TC_{obs} - R_{C/P}P_{obs}) \tag{2}$$

$$= TC_{STA} - [(TC_{pre} + TC_{ox}) - R_{C/P}(P_{pre} + P_{ox})] \tag{3}$$

$$= TC_{STA} - (TC_{pre} - R_{C/P}P_{pre}) - (TC_{ox} + R_{C/P}P_{ox}) \tag{4}$$

$$= TC_{STA} - (TC_{pre} - R_{C/P}P_{pre}), \tag{5}$$

because

$$TC_{ox} = R_{C/P}P_{ox}, \tag{6}$$

where C and P are the concentrations of total CO_2 and nutrient (phosphate or nitrate), respectively, and $R_{C/P}$ is the so-called Redfield ratio of carbon to a chosen nutrient (Redfield *et al.*, 1963), the subscripts, obs, pre and ox, refer to the observed, preformed and oxidative quantities. For simplicity, a term relating to the formation or dissolution of $CaCO_3$ is omitted in the above definition and equation, which can be evaluated by computing the alkalinity change, as described later.

If the subsurface water system is completely closed after its formation by sinking from the surface, the decrease in its oxygen content is proportional to the amount of nutrients regenerated including carbonate and the potential sink capacity is not affected. Equation (5) means that the potential sink capacity of the respective subsurface water increases only by the amount corresponding to the atmospheric CO_2 increase, if CO_2 had been equilibrated at the air-water interface and contained no nutrient when it left the surface. The equilibration for CO_2 needs more time than that for O_2, but it is much faster than that for the isotopic equilibration of dissolved carbon, as explained later.

Otherwise, if O_2 reduced in the upwelling water is supplied at the surface but nutrients are not consumed in winter when the Pacific Deep Water (PDW) is formed, the amount of preformed nutrient content (P_{pre}) should be increased. However, the gas exchange should decrease TC_{pre} , because some CO_2 is conversely expelled at the same time of the O_2 dissolution. This depends on the difference in the degree of equilibration of gas exchange between O_2 and CO_2 at the surface. The potential sink capacity, therefore, is not necessarily proportional to the amount of preformed nutrients.

The Pacific Deep Water (PDW) contains more preformed nutrients than the Atlantic deep water (Redfield *et al.*, 1963). This is due to the fact that, when the PDW originated in the deep North Atlantic passes through the Antarctic Ocean, a part of the deep water is brought to the surface and exchanges dissolved gases without use of nutrients in winter. Although the preformed nutrients in the PDW are increased due to the oxygen exchange at the surface, the increase in its preformed carbonate should not be proportional to the increase in preformed nitrate or phosphate, because a part of the CO_2, at least, is lost from the surface water in the stormy winter Antarctic Ocean. The upwelling of the North Pacific Deep Water (NPDW) reduced in the dissolved oxygen content during the northward flow first expels CO_2 from the water to the atmosphere, especially in the warm region in the North Pacific. When all the nutrients are consumed and the water is cooled, however, the surface water should absorb much CO_2 in excess of the amount corresponding to the recent CO_2 increase in the atmosphere. This situation is shown in the next section.

Equations (1), (2) and (5) are rewritten as follows;

$$TC_{P0} = TC_{STA} - \Theta \tag{7}$$

$$= TC_{obs} - R_{C/P}P_{obs} \tag{8}$$

$$= TC_{pre} - R_{C/P}P_{pre}. \tag{9}$$

These equations mean that the uncertainty in the potential sink capacity, Θ, depends highly on uncertainties in the equilibrium constants used for obtaining TC_{STA} and systematic errors in the observed concentrations. However, TC_{P0}, which is different from Θ only by a definite amount, is simply calculated and the systematic error can be later corrected. In this study, we calculate only TC_{P0} and discuss it instead of Θ.

Although we have not given the detailed discussion, the dissolution of $CaCO_3$ in the abyssal water of the Pacific also makes water having a greater potential sink capacity.

Phosphate corrected total carbon dioxide concentration of the North Pacific water

The North Pacific water substantially originates in the Pacific Deep Water coming from the Antarctic Ocean, although its salinity and temperature are changed at the surface in the Pacific. The effect of salinity change can be corrected by using the quantities normalized to a salinity of, say, 35. Equations (1) and (2) are then rewritten as

$$N\Theta = NTC_{STA} - NTC_{P0} \tag{10}$$

$$= NTC_{STA} - (NTC_{obs} - 35/S \cdot R_{C/P}P_{obs}) \tag{11}$$

$$= (NTC_{STA} + 35/S \cdot R_{C/P}P_{obs}) - NTC_{obs}, \tag{12}$$

where N refers to the normalized quantity and S is the salinity, and we obtain

$$NTC_{P0} = NTC_{STA} - N\Theta = NTC_{obs} - 35/S \cdot R_{C/P}P_{obs}. \tag{13}$$

The alkalinity corrected NTC_{P0}, NTC_{AP0}, is calculated from the following equation:

$$NTC_{AP0} = NTC_{P0} - DNA/2 = NTC_{obs} - 35/S \cdot R_{C/P}P_{obs} - (NA_{obs} - NA_{std})/2. \tag{14}$$

The above equations indicate that we can discuss the matters on Θ almost similarly using NTC_{AP0}, because NTC_{STA} is nearly constant for the Pacific Deep Water. We have calculated the NTC_{AP0} values using the data obtained during the KH-93-2 cruise of the R/V *Hakuho Maru* in the western North Pacific and assuming a C/P ratio of 115 based on the results of Redfield *et al.* (1963) and Tsunogai (1972). The vertical profiles of observed total carbon dioxide concentration (NTC_{obs}) at a meridional section between 27°N and 45°N along 165°E are shown in Fig. 2. The maxima were found at depths between 1.0 and 1.5 km at every station, mainly due to the decomposition of organic matter. In the surface layer, however, there was a great difference between the stations south of 30°N and north of 40°N. This difference is not due to the decomposition of organic matter as shown in Fig. 3, where the NTC_{obs} values are plotted against AOU showing also a large difference between them in the near surface water. The small difference in the deep water is due to the production and dissolution of $CaCO_3$ (Tsunogai *et al.*, 1973; Tsunogai and Watanabe, 1981), which disappears in the plot of total carbonate corrected for alkalinity (Fig. 4). In the correction, we have

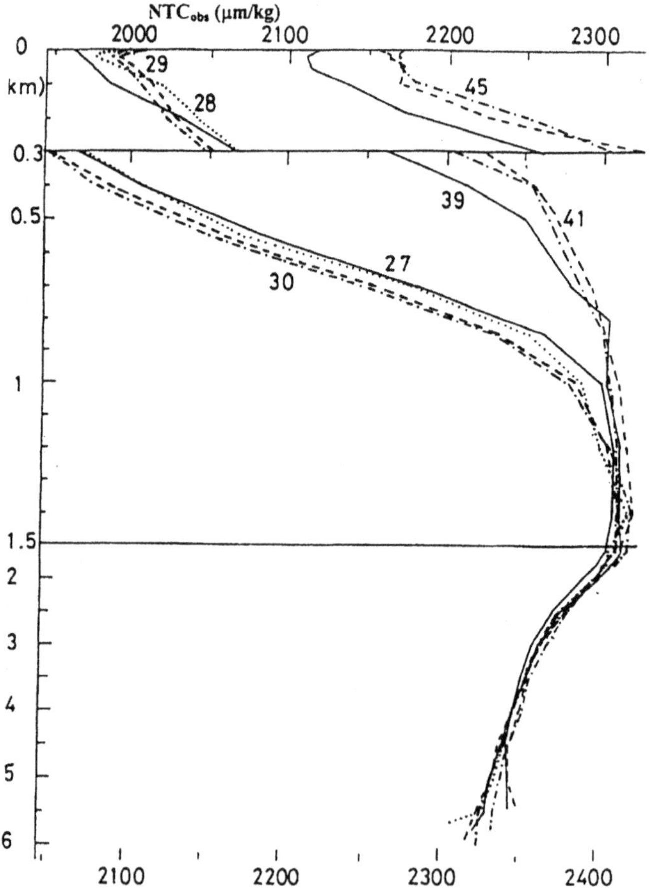

Fig. 2. Vertical profiles of total carbonate normalized to a salinity of 35psu (NTC_{obs} in mm/kg) obtained in the North Pacific along 165°E during the KH-93-2 cruise of the R/V *Hakuho Maru*. The numerals attached to the respective lines refer to the latitudes of the sampling stations.

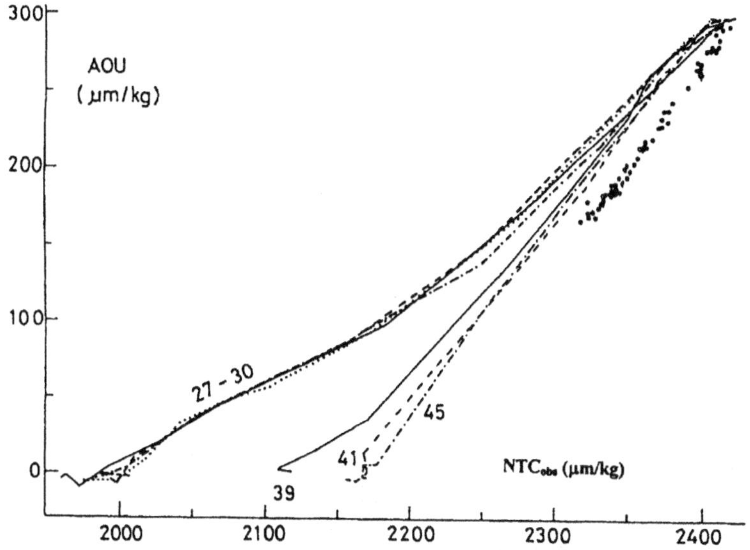

Fig. 3. Normalized total carbonate (NTC_{obs} in mm/kg) versus atomic oxygen utilization (AOU in mm/kg) for the data given in Fig. 2. The small open circles refer to those of water below the oxygen minimum layer.

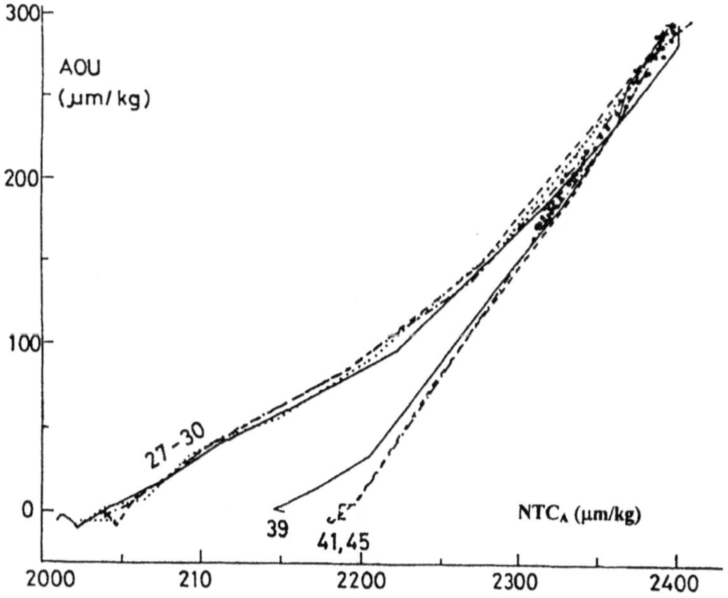

Fig. 4. Alkalinity corrected total carbonate (NTC_A in mm/kg) versus atomic oxygen utilization (AOU in mm/kg) for the data given in Fig. 2.

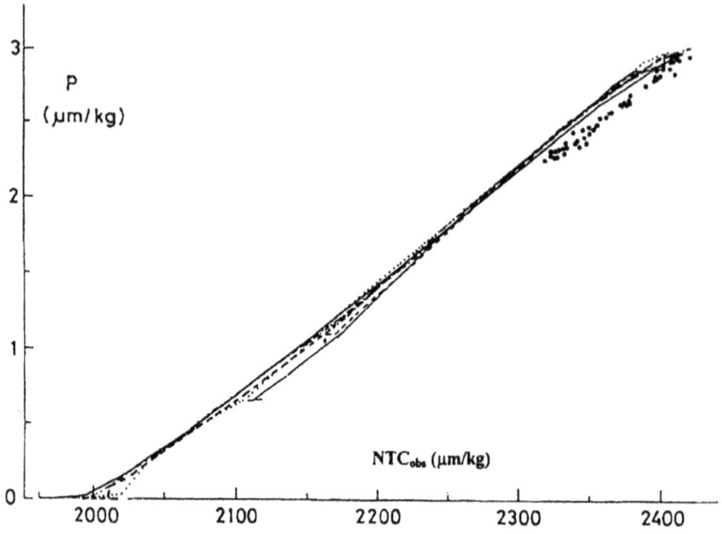

Fig. 5. Normalized total carbonate (NTC_{obs} in mm/kg) versus phosphate ((P_{obs} in mm/kg) for the data given in Fig. 2.

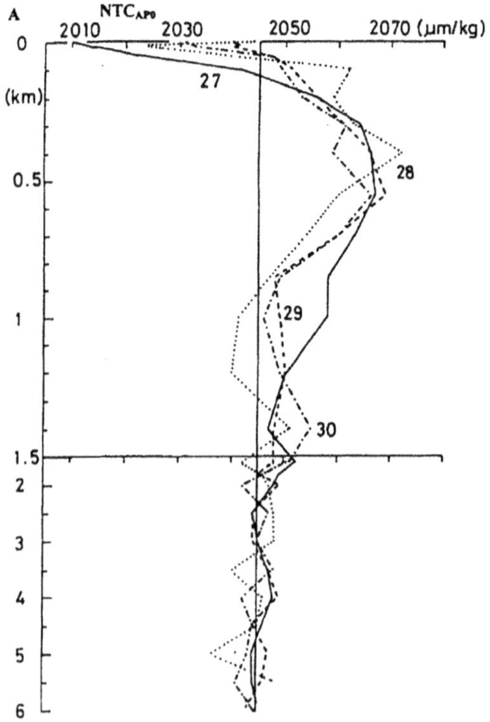

Fig. 6. Vertical profiles of phosphate and alkalinity corrected normalized carbonate (NTC_{AP0} in mm/kg) for the data obtained in the region south of 30°N (Fig. 6a) or north of 39flN (Fig. 6b) and given in Fig. 2.

chosen a standard value of 2.400 meq/kg for alkalinity. The large difference in the surface NTC_{obs} diminishes in the plot of NTC_{obs} against the phosphate concentration instead of AOU (Fig. 5). This means that the small NTC_{obs} in the surface layer is chiefly due to the biological assimilation of preformed nutrients, but the O_2 produced has been lost to the atmosphere. This also means that the AOU value cannot be used for the correction of biological activity and the consumption of preformed nutrients induces the extraordinary small NTC_{obs} in the warm water. The smaller effect due to the anthropogenic CO_2 is not appreciable in this rough plot (Fig. 5) and then we have calculated the NTC_{AP0} values.

The vertical profiles of NTC_{AP0} show interesting features (Fig. 6). At the stations south of 30°N (Fig. 6a), larger values are found at depths centering at 500 m, which are considered to be due to anthropogenic CO_2. In the warm surface 100 m layer exhausted in nutrients, the NTC_{AP0} values are inversely smaller than those of the Pacific Deep Water below 1.5 km depth even when the atmospheric CO_2 has been increased. This is due to a loss of CO_2 from the surface water by warming (the temperature effect). The 1°C increase in the water temperature reduces about 6 µmol/kg of total carbonate in seawater under an air-sea equilib-

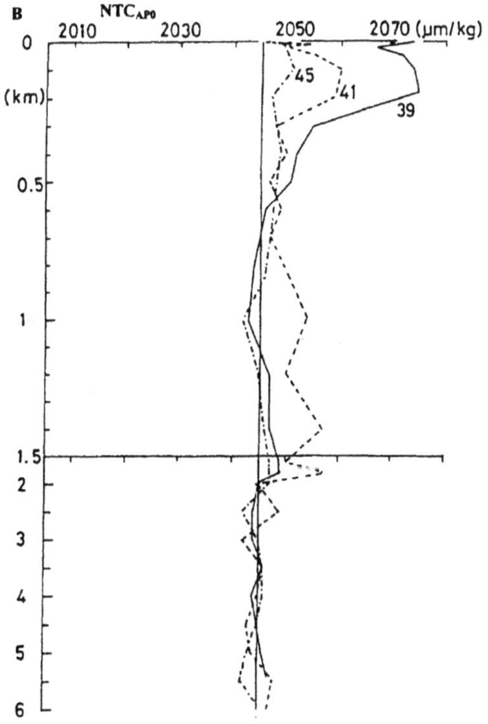

Fig. 6. (continued).

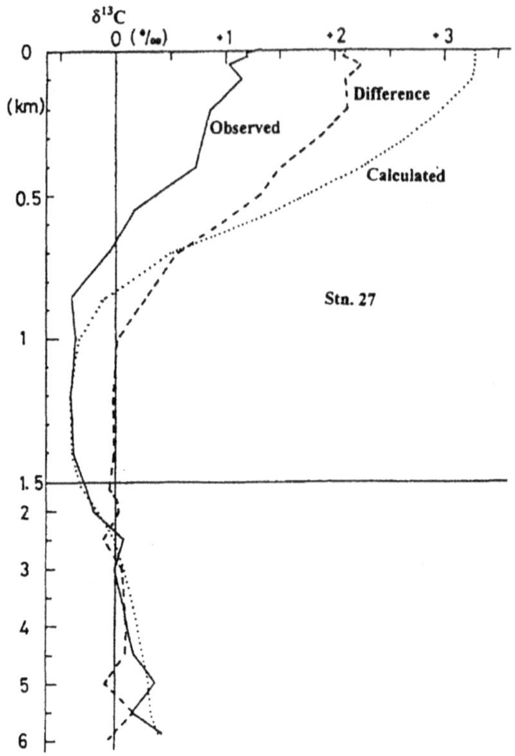

Fig. 7. Vertical profiles of observed ($\delta^{13}C_{obs}$) and calculated ($\delta^{13}C_{APO}$) stable carbon ratios at Stn. 27 (27°N, 165°E) in the western North Pacific. The difference is also shown as an example of those given in Fig. 8. The method of calculation, the correction method for phosphate and alkalinity, is given in the text.

rium at a constant atmospheric CO_2. The smaller values indicate that the surface water can absorb enormously much CO_2 when it is cooled. At the stations north of 39°N, the profiles are quite different (Fig. 6b). The excess of NTC_{AP0} in the surface over the deep water value, namely the absorbed anthropogenic CO_2, decreases with increasing latitude. This decrease is due to the nutrients still remaining, which are supplied from the upwelling deep water. Their penetration depths of the excess are shallower than those at stations around 30°N, due to the shallower penetration depth of the North Pacific Intermediate Water shown by Tsunogai *et al.* (1995).

Unfortunately, the NTC_{AP0} values calculated above have systematic errors and are applicable only to the data set standardized against our laboratory standards. Before the calculation of Θ and the amount of absorbed anthropogenic CO_2, it is necessary to correct the systematic errors between the various data sets and to correct exactly the alkalinity change and the temperature effect. The calculated results will be given in a later paper. Ono *et al.* (1996), however, have estimated the pre-industrial atmospheric CO_2 concentration from the deep water

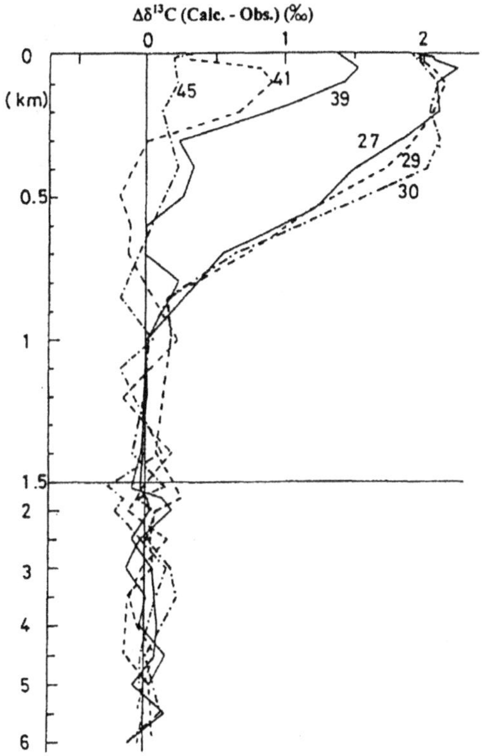

Fig. 8. Vertical profiles of the difference between observed $\delta^{13}C_{obs}$ and calculated $\delta^{13}C_{APO}$ at six stations. The positive values indicate the decrease in $\delta^{13}C$ due to the anthropogenic CO_2. The calculation has been done by introducing the $\delta^{13}C$ values of −21‰ for organic and +1‰ for carbonate particles into Eq. (14).

in the western North Pacific only considering the AOU to be about 270 ppm, which is somewhat lower than the usually accepted pre-industrial value. This may be due to the gas exchange in the Antarctic Ocean.

Information from the isotopic data is also interesting and valuable, because the temperature effect on the total carbonate can be ignored and it gives information on the gross gas exchange. We have similarly calculated the $\delta^{13}C_{APO}$ values from the data obtained by Okuda *et al.* (1995) for the same samples as for the total carbonate. The assumptions are that the organic matter produced in the pre-industrial era to be −21‰ and that the deep water below 3 km depth has not been contaminated with anthropogenic CO_2. An example obtained at 27°N is shown in Fig. 7. The observed $\delta^{13}C_{obs}$ in the near-surface layer are much smaller than the calculated $\delta^{13}C_{APO}$ values. This is due to the dissolution of light atmospheric carbon contaminated by the anthropogenic CO_2. We have then plotted only the differences between the two values, or the recent decrease due to the anthropogenic CO_2 for those stations in Fig. 8. The difference, the decrease in the observed value, is larger at the southern stations than at the northern stations, depending on the penetration depth of the Intermediate Water and the residence time of water in the surface mixed layer. We can estimate the amount of anthropogenic CO_2 from the profiles, which will be published in a forthcoming paper. It is clear, however, that the anthropogenic CO_2 is not significantly accumulated in the North Pacific water north of 45°N by letting the upwelled water flow southward, and the water constituting the subtropical gyre, having a longer residence time accumulates much anthropogenic CO_2. We can conclude that even the total carbonate reduced by the temperature effect in the warm surface water has been increased by reflecting the atmospheric CO_2 increase, probably proportionally to its increase in recent years.

PROCESSES DISSOLVING CO_2 IN THE NORTH PACIFIC

I will point out the following four processes as the oceanic sinks of anthropogenic CO_2 overlooked or underestimated by modelers. All these are based principally on our experimental results.

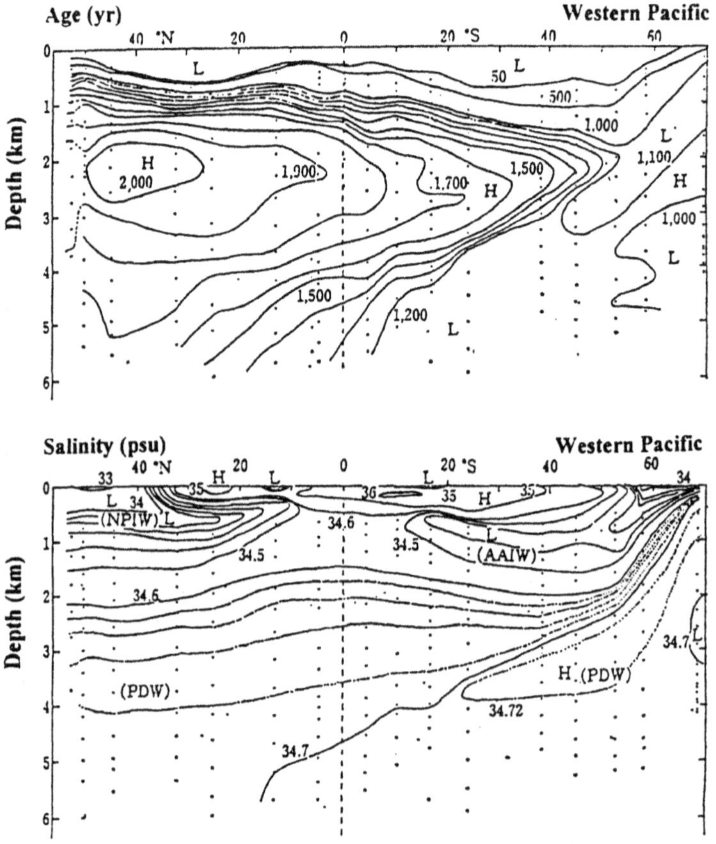

Fig. 9. North-south sections of age (A in years) and salinity (B in psu) in the western Pacific (after Tsunogai and Noriki, 1983). The ages are the lapse of time after leaving the surface.

North Pacific and Antarctic Intermediate Waters

The North Pacific Intermediate Water (NPIW) has been well studied by physical oceanographers (e.g., Tally and Nagata, 1991) and can be easily recognized by its less saline property (Fig. 9). While the outcrop of the lower NPIW has not been found at the surface, even in winter, its surface component should be derived from the surface water in the region north of the subarctic boundary (Fig. 10). In the South Pacific, the Antarctic Intermediate Water (AAIW) is considered to be similarly formed but in a somewhat higher latitude region than NPIW (Fig. 9).

Although the natural ^{14}C data cannot adequately depict the intermediate water (Tsunogai, 1981; Tsunogai and Noriki, 1983) because of its prevailing lateral flow (Fig. 9), the concentration of transient tracers clearly shows the intermediate water down to about 1000 m deep (Watanabe *et al.*, 1991; Tsunogai *et al.*, 1995; Tokieda *et al.*, 1996). The NPIW in 1973 was already contaminated by the artificial tritium and ^{14}C which were produced mainly by the H bomb tests in the early 1960's (Figs. 11 and 12). The distribution of artificial and stable CFCs (chlorofluorocarbons) released into the atmosphere mainly after the 1960's depicts more clearly the fraction of the surface component and its age, namely the lapse of time after leaving the surface (Figs. 13 and 14). These results suggest that the NPIW is formed in the northwestern North Pacific, but the region is centered at 40–45°N and 150–180°E.

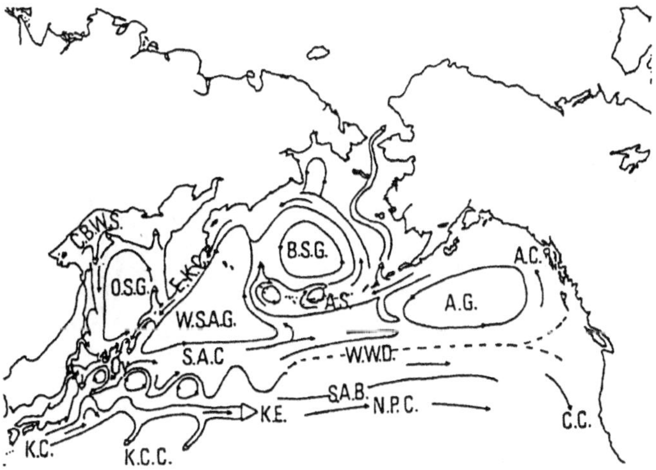

Fig. 10. Surface current system of the subarctic North Pacific (after Ohtani, 1991). The abbreviated names refer to Alaska Current (A.C.), Alaska Gyre (A.G.), Alaskan Stream (A.S.), Bering Sea Gyre (B.S.G.), Cold Saline Bottom Water (C.B.W.S.), California Current (C.C.), East Kamchatka Current (E.K.C.), Kuroshio (K.C.), Kuroshio Counter Current (K.C.C.), Kuroshio Extension (K.E.), North Pacific Current (N.P.C.), Okhotsk Sea Gyre (O.S.G.), Oyashio (Oy), Subarctic Boundary (S.A.B.), Subarctic Current (S.A.C.), Western Subarctic Gyre (W.S.A.G.) and West Wind Drift (W.W.D.).

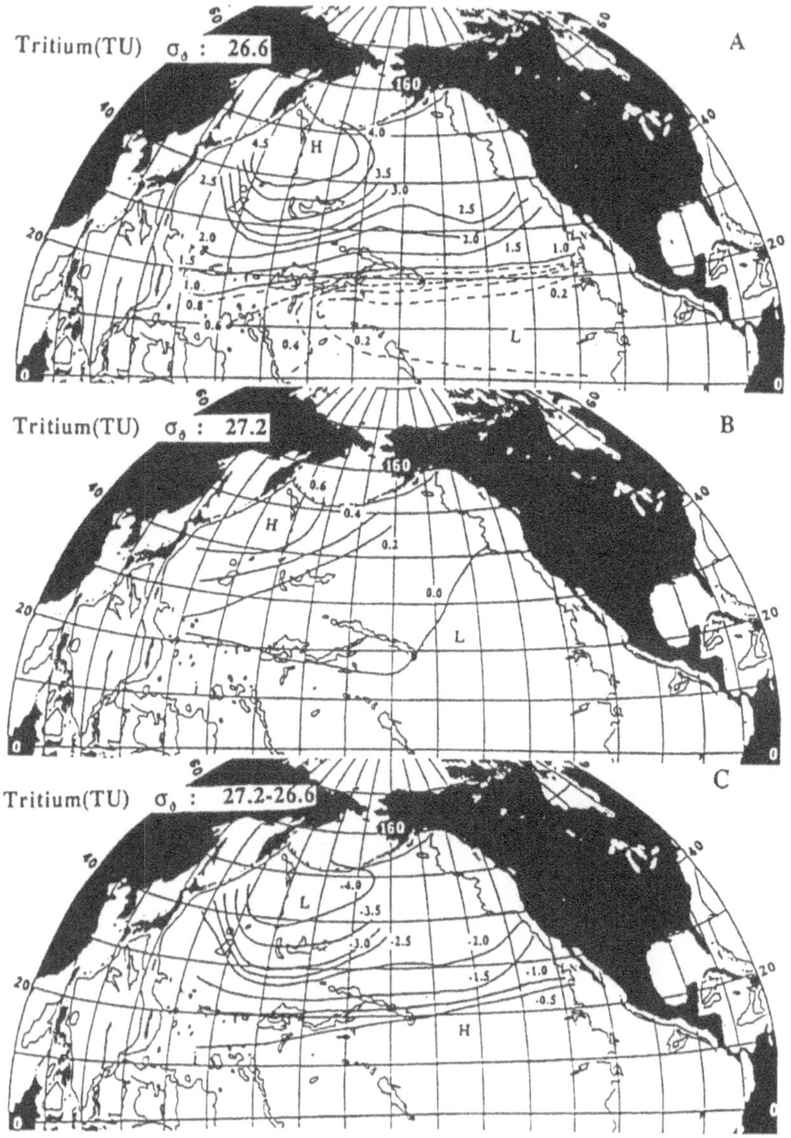

Fig. 11. Isopleths of tritium (in TU normalized to the 1980 level at the isopycnal surfaces of σ_θ 26.6 (A) and σ_θ 27.2 (B), and the difference between the two surfaces (C) (after Tsunogai *et al.*, 1995). The data were obtained during the GEOSECS expedition in 1973.

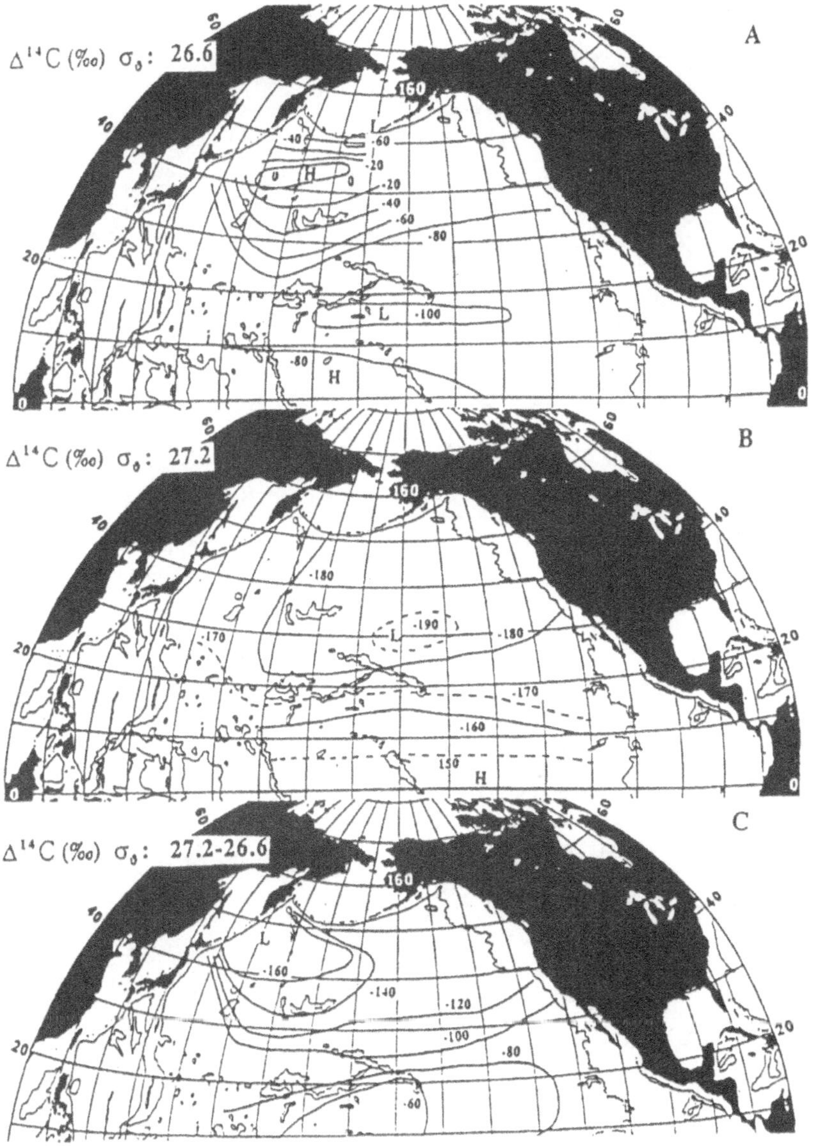

Fig. 12. Isopleths of $\Delta^{14}C$ (in ‰) at the isopycnal surface of σ_θ 26.6 (A) and σ_θ 27.2 (B), and the difference between the two surfaces (C) (after Tsunogai *et al.*, 1995). The data were obtained during the GEOSECS expedition in 1973.

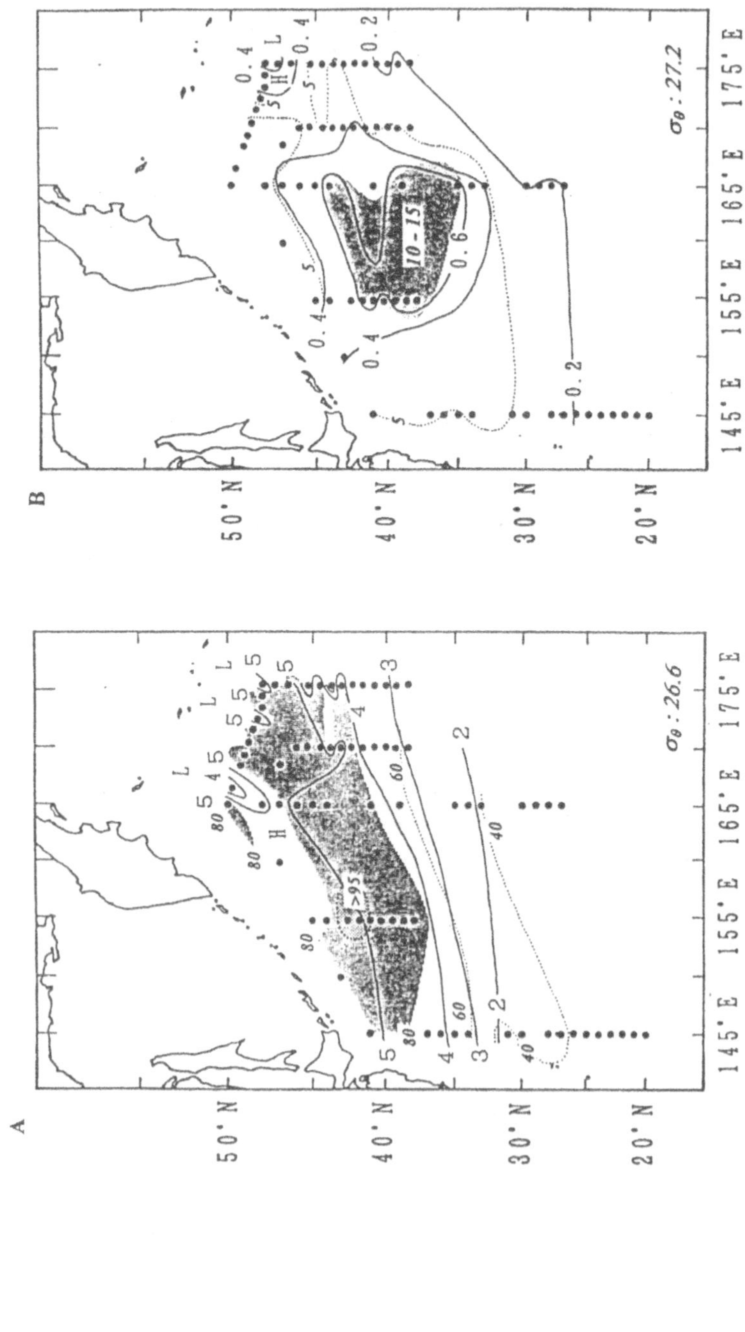

Fig. 13. Concentration of CFC-11 in 1993 (numerals of normal form and solid lines in pmol/kg) and its saturation degree relative to the atmospheric concentration in 1993 (numerals in italics and dotted lines) at the isopycnal surfaces of σ_θ 26.6 (A) and σ_θ 27.2 (B) (after Tokieda *et al.*, 1996).

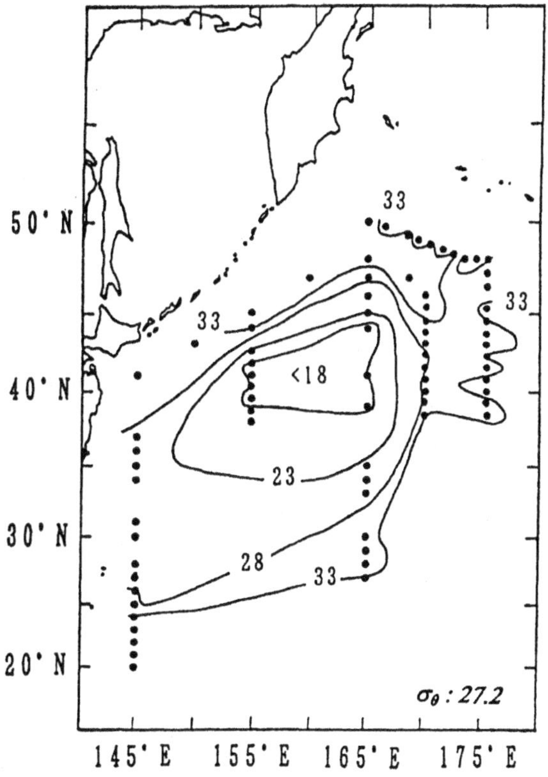

Fig. 14. Lapse of time after leaving the surface (in years) derived from CFC-11/CFC-12 ratios at the isopycnal surface of σ_θ 27.2. The plot at the σ_θ 26.6 surface is not shown, because all the ages are less than 18 years. The lower limit of 18 years is due to the fact that the CFC-11/CFC-12 ratio of the atmospheric CFCs is virtually constant after 1975.

The volume of NPIW is about 1/10 of the North Pacific Deep Water (NPDW), but its age, a few tens to one hundred years, is more than one order of magnitude less than that of NPDW in the North Pacific. This means that the production rate of NPIW is larger than that of NPDW and that NPIW formed in the North Pacific is more important as a sink for anthropogenic CO_2 than the NPDW formed in the Atlantic and the Southern Oceans, if the degree of equilibration of CO_2 at the surface is not significantly less for NPIW than for NPDW. It is necessary to estimate exactly the age of NPIW and the temporal changes in its amount formed in the North Pacific as well as AAIW in the Southern Ocean.

Continental shelf and hemipelagic zones absorbing CO_2

As a representative of the continental shelf zone in the temperate marine environment, the MASFLEX (Marginal Sea Flux Experiment in the western North Pacific) group studied the East China Sea and its preliminary result has been published in Tsunogai *et al.* (1997). In Fig. 15, the activity (concentration) of dissolved CO_2 in the surface water [f(CO_2)] is plotted against the longitude from the Chinese coast to the Kuroshio region. The activities seem to depend on temperature and salinity, because the activities were higher in the warmer summer, and lower in the region nearer the coast where the salinities were also lower. Using the least-square method, we have obtained the following regression equation,

$$f(CO_2) = 4.58T + 18.925S - 405 \quad \text{(in ppm or } 10^{-6}\text{)} \qquad (15)$$

where T and S are the temperature and salinity. Figure 16 shows the differences between the observed activities and the calculated ones from the above equation putting observed temperature and salinity. The standard deviation of the differences is only 7 ppm (2%) and the distribution of plus or minus signs is at random. Later we have found the f(CO_2) values also depend on the concentration of nutrients and obtained the following revised equation.

$$f(CO_2) = 6.85T + 16.10S + 99P - 372 \qquad \text{(in ppm)} \qquad (16)$$

where P is the phosphate concentration (in μmol/kg). The standard deviation of the difference between the observed and calculated values is 6.6 ppm, which is

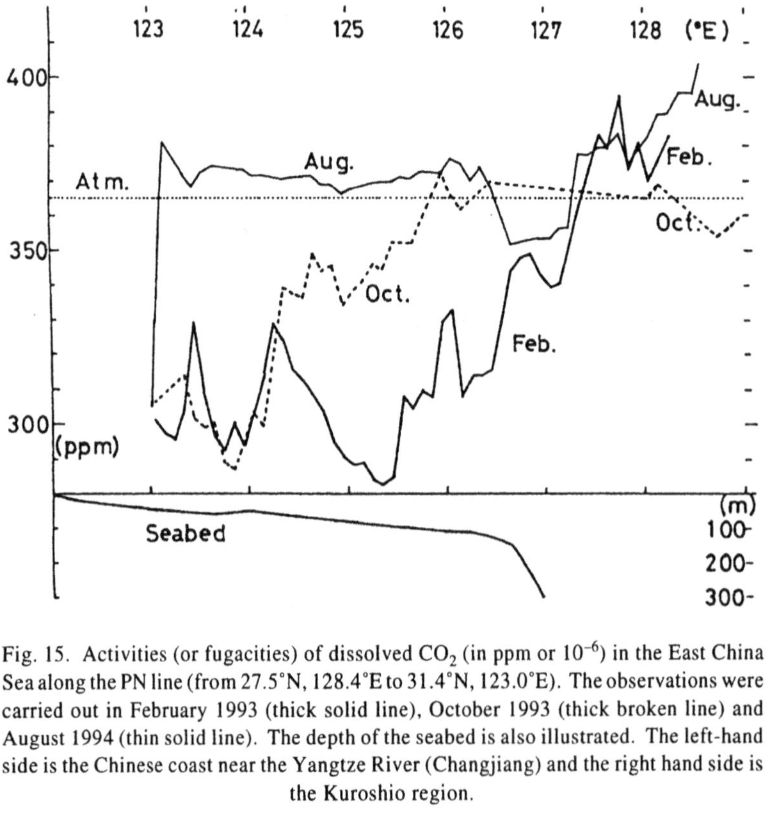

Fig. 15. Activities (or fugacities) of dissolved CO_2 (in ppm or 10^{-6}) in the East China Sea along the PN line (from 27.5°N, 128.4°E to 31.4°N, 123.0°E). The observations were carried out in February 1993 (thick solid line), October 1993 (thick broken line) and August 1994 (thin solid line). The depth of the seabed is also illustrated. The left-hand side is the Chinese coast near the Yangtze River (Changjiang) and the right hand side is the Kuroshio region.

also about 2% of the $f(CO_2)$ value. The coefficient of the above temperature dependent term corresponds to about a half of the value at which gas exchange would not occur, or the half equilibration value (14ppm/°C/2). The phosphate-dependent term is also approximately a half equilibration value, if usual biological uptake has occurred. We expect the phosphate concentration of river water to be 0.2 ppm, if phosphate in river water works similarly as the phosphate term stated above. These seem to be reasonable. Of course, this equation is peculiar to the East China Sea, a simple water system and rather fortuitously obtained, but it may be worthwhile to examine the relation in other seas and oceans.

The Earth is mild due to the transport of heat stored in water from the low latitudes and its release in the high latitudes. This release of heat lowers the water temperature of shallow seas more than that of the ocean at the same latitude in the temperate and high latitudes. The colder water dissolves more CO_2 and the dissolution of CO_2 in the shallow seas is accelerated by nutrients supplied in excess of carbonate from the continent, the bottom and possibly the open ocean by nektons including birds. The CO_2 fixed by organisms sinks toward the bottom

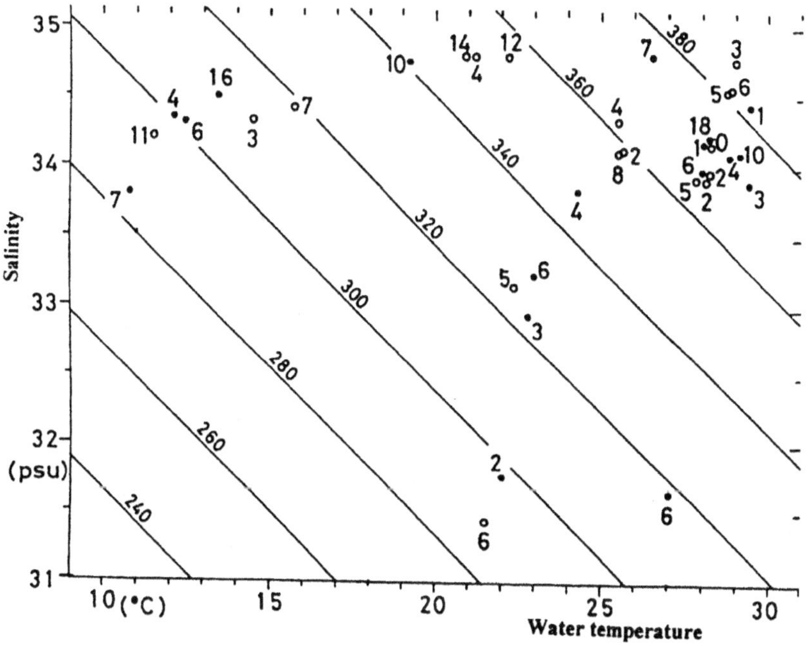

Fig. 16. Difference between the observed (given in Fig. 15) and calculated (from the equation 12 in the text) activities (fugacities) of dissolved CO_2 shown in the diagram of water temperature (T) and salinity (S). The open and solid circles refer to positive (namely, the observed values are larger than the calculated ones) and negative values, respectively, by the amount of numerals attached (in ppm or 10^{-6}). The straight lines are drawn according to Eq. (15) in the text, with the activity of CO_2 shown by the attached numerals.

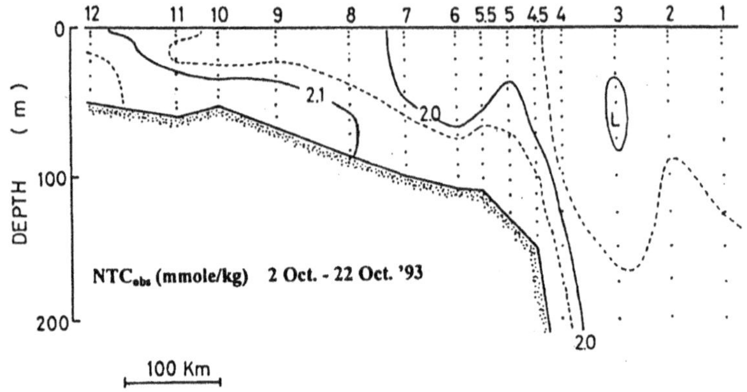

Fig. 17. Normalized total carbonate (in mmol/kg) observed in October 1993 along the PN line in the East China Sea (after Tsunogai *et al.*, 1997).

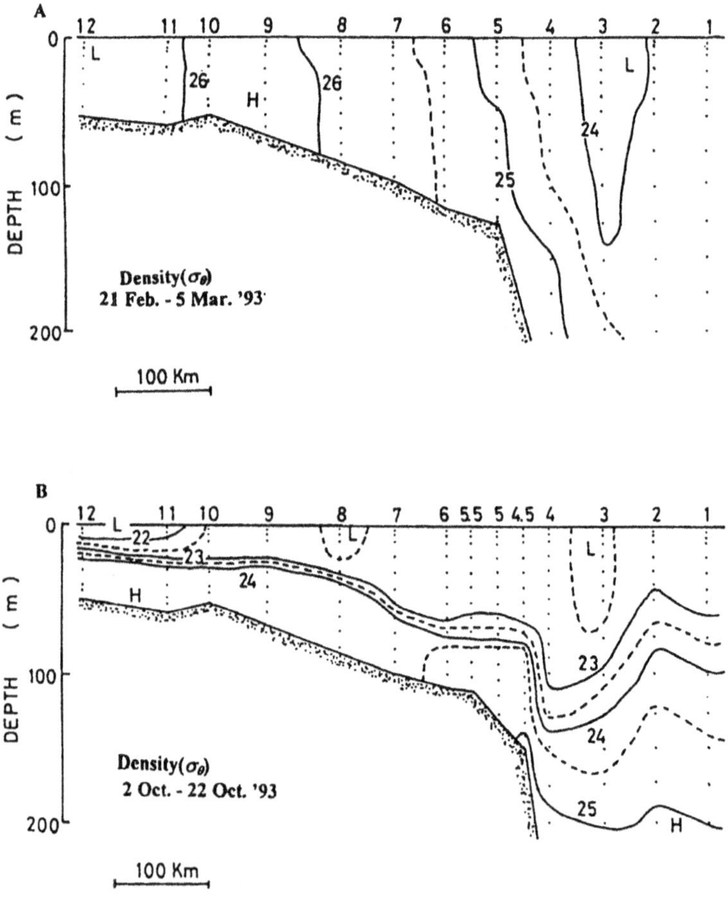

Fig. 18. Density of water (in the s values) observed in February (A) and October (B) 1993 along the PN line in the East China Sea (after Tsunogai *et al.*, 1997).

and is regenerated at the bottom. One example of the observation is given in Fig. 17. The water containing regenerated CO_2 in the subsurface layer of the shallow sea mixes isopycnally with water in the open sea. The colder water in the shallow sea is denser than the warmer water in the open sea (Fig. 18). Therefore, the continental shelf water, containing more total carbonate, is taken into the lower layer of the open sea water. This situation, the continental shelf acting as a sink of the atmospheric CO_2, is clearly depicted in Figs. 17 and 18.

Gas exchange in the high-latitudes

The mean gas transfer velocity of the world oceans estimated from ^{14}C (Fig. 19) is larger than 3 m/day which has been obtained from an equation expressed as a function of wind velocity (Liss and Merlivat, 1986) and the world mean wind velocity. At present, a majority of scientists in this field believe the value 3 m/day to be true as a most probable world mean value, although Tans *et al.* (1990) gave somewhat larger values. Liss and Merlivat (1986) equation was accom-

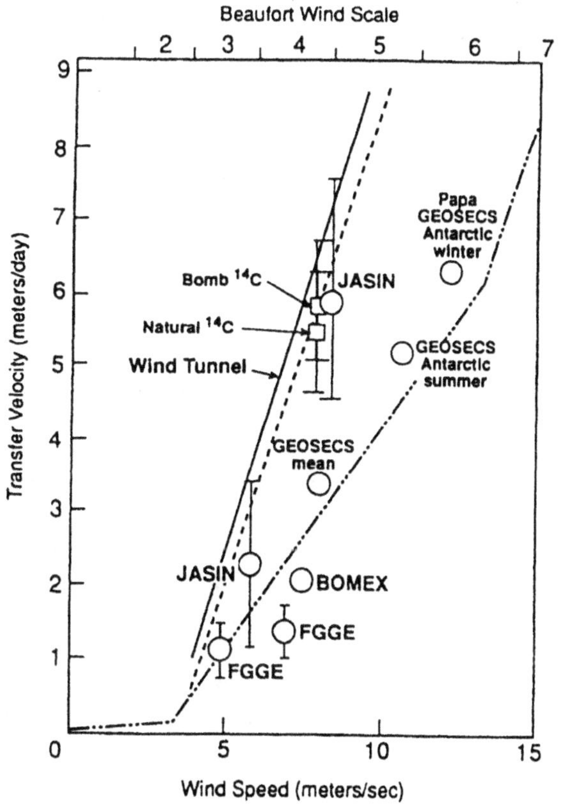

Fig. 19. Gas transfer velocity vs. wind velocity at the sea surface (after Murphy *et al.*, 1991). The lower dot-dashed and upper dashed lines represent the equations of Liss and Merlivat (1986) and Tans *et al.* (1990), respectively.

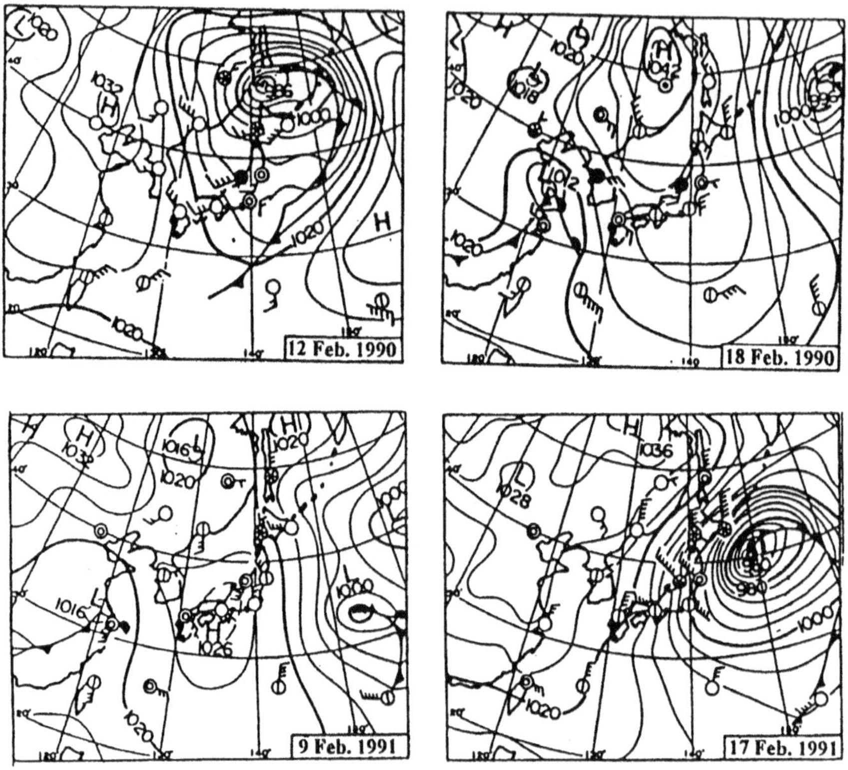

Fig. 20. Weather maps in the western North Pacific region. The observations were carried out a few days after the mapping dates.

plished with the data obtained chiefly from smaller-scale experiments in relatively calm conditions.

We (Tsunogai and Tanaka, 1980) have found that the gas transfer velocity is proportional to more than the 2nd power of the wind velocity and much larger in winter, while it was extremely small in the calm summer in the Japan Sea (Tsunogai *et al.*, 1993b). We (Tsunogai *et al.*, 1993c) have also obtained extraordinary large gas transfer velocities (7–18 m/day) under the stormy winter monsoon in the western North Pacific (Fig. 20) by the Rn method as given in Table 2. If we put the above transfer velocities into the Liss and Merlivat (1986) equation, the wind velocity in this period turns out to be 14–21 m/sec, which is unrealistically large as a mean velocity and it may not be the case. This large gas transfer velocity must be caused by some mechanisms other than the usual ones, such as the effect of the duration of storms amplifying the sea surface roughness and that of bubbles occurring during heavy storms (Keeling, 1993; Asher *et al.*, 1996; Loeman *et al.*, 1996).

The heavy storms are inclined to occur more frequently in the high latitudes where the deep and intermediate waters are formed. The effect of heavy storms should be greater for CO_2 than for O_2, because CO_2 dissociates and has a larger

Table 2. Observed ^{222}Rn deficiencies relative to its parent ^{226}Ra and calculated transfer velocities (k) for Rn and CO_2 in the western North Pacific (after Tsunogai *et al.*, 1993c). The wind velocities corresponding to the transfer velocities are computed according to the equation of Liss and Merlivat (1986) and also given (*).

Date	Location (dpm/m^2)	Rn deficiency (m/day)	k: Rn (m/day)	k : CO_2 (m/sec)	Wind*
14 Feb. 1990	41.5°N, 144.7°E	5060	10.3	12.0	16.8
22 Feb. 1990	40.0°N, 144.0°E	3640	6.5	7.6	13.7
12 Feb. 1991	41.5°N, 144.0°E	3330	7.2	8.4	14.3
20 Feb. 1991	40.0°N, 144.0°E	6440	15.3	17.8	20.9

absorption capacity, of which the gas exchange continues even after the O_2 gas exchange is virtually equilibrated. The equilibration for total carbonate in seawater with atmospheric CO_2, however, is not so slow as the isotopic equilibration of carbon. This is due to the fact that the most abundant carbonate species, HCO_{3^-}, is inert at the carbonate equilibration, but the ion participates the isotopic equilibration which has not been attained in most of the world's surface water. These indicate that the mean exchange rate of CO_2 for the whole ocean water is larger, not only than those for other sparingly soluble gases such as O_2, but also than that obtained from a simple function of only wind velocity due to the active gas exchange in the high latitudes.

Dissolved silica in the North Pacific Deep Water

In the temperate Northern Hemisphere, the Pacific Deep Water contains seven times more for dissolved silica (silicate) but only twice more for nitrate and phosphate than the Atlantic Deep Water (Tsunogai and Noriki, 1983). The extreme is the Bering Sea (Tsunogai *et al.*, 1979). The Si/N ratio of the water upwelling in the northern North Pacific, including the Okhotsk Sea and the Bering Sea, is higher than that of water upwelling in the eastern and equatorial Pacific, and more similar to the deep water (Tsunogai, 1987a; Tsunogai *et al.*, 1995). This is due to the fact that the water in the eastern and equatorial Pacific is not directly derived from the deep water beneath the surface but rather from the intermediate water formed in the western North Pacific (Tsunogai, 1987a; Tsunogai *et al.*, 1995). The intermediate water is strongly enriched with nitrate and phosphate but is weak in silicate due to the preferential regeneration of nitrate and phosphate from settling particles during the upwelling (Tsunogai and Noriki, 1987; Tsunogai *et al.*, 1990). Fine particles existing in the water column have negligibly small sinking velocities, but giant particles (marine snow) produced *in situ*, convey the fine particles like a train transports many passengers (Tsunogai *et al.*, 1974, 1980; Tsunogai, 1987b). Tsunogai and Noriki (1987) have concluded that the contribution of dissolved organic carbon flux into the deep sea is small and the isopycnal mixing transports water reduced in the O_2 content from the coastal and hemipelagic seas to the open oceans.

Tsunogai (1979) has placed emphasis on the role of silicate in the ecosystem of the ocean and proposed a hypothesis on the primary factor controlling the plankton community, which is silicate in seawater. The hypothesis is partly demonstrated in a coastal sea, Funka Bay (Tsunogai and Watanabe, 1983). In the western North Pacific, the water concentrated in silicate makes not only many

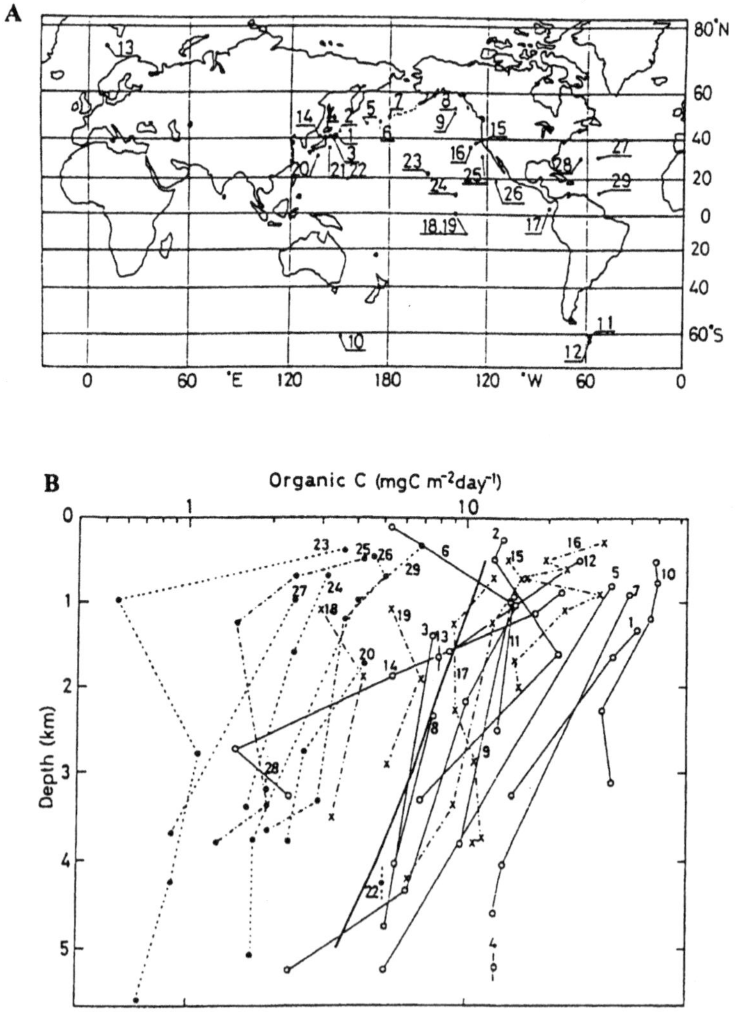

Fig. 21. Results obtained from the sediment trap experiments on the organic and carbonate carbon fluxes in the world oceans. Open circles are from the high-latitudes (>40°), crosses are from the transitional and equatorial zones, and solid circles are from the low-latitude zone (<35°). A thick line indicates the median values. The attached numerals refer to the station numbers (after Tsunogai and Noriki, 1991). A: Location of sampling stations. B: Organic carbon fluxes. C: carbonate carbon fluxes. D: Carbonate-C/organic-C flux ratios.

diatoms, but also makes particles of a high organic-C/carbonate-C ratio. Tsunogai and Noriki (1987, 1991) have summarized the results obtained by sediment traps (Fig. 21). The organic carbon flux varies widely and is high in the western North Pacific, while the carbonate C is less variable and higher in its eastern part. The more prominent feature is the high organic-C/carbonate-C ratio in the western

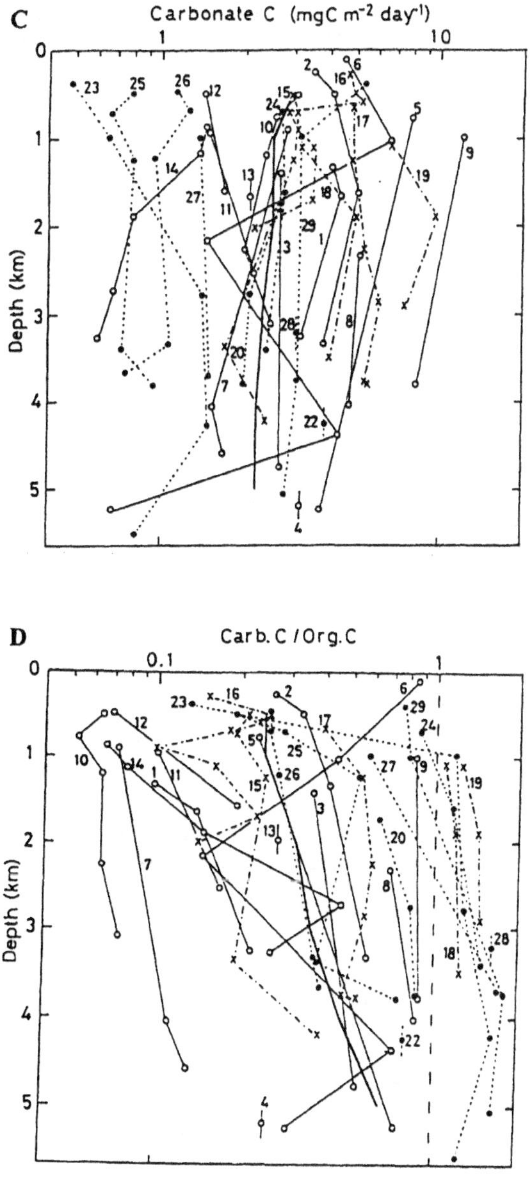

Fig. 21. (continued).

North Pacific. The ratio is a key factor for the biotic activity as a sink of atmospheric CO_2, because the formation of carbonate particles acts as a source for the atmospheric CO_2. Furthermore, the diatom bloom produces large settling particles penetrating into the deep sea (Tsunogai *et al.*, 1982, 1986; Kusakabe *et al.*, 1988) and the deep sea floor (Tsunogai, 1980) with a larger proportion before their decomposition.

The oceanic ecosytem also may be influenced by the atmospheric fallout (Tsunogai, 1991). Some mineral components such as P, Fe and Si are transported to the ocean via the atmosphere. Their deposition rates are extremely large in the western North Pacific nearer the Asian continent, as observed by the atmospheric components (Tsunogai and Kondo, 1982; Tsunogai, 1985; Suzuki and Tsunogai, 1988; Duce *et al.*, 1991) and by the sediment trap experiments (Noriki and Tsunogai, 1986; Saito *et al.*, 1992). Therefore, the biological pump for the CO_2 removal is more effective in the western part of the North Pacific than in its eastern part and in the Atlantic.

CONCLUSION

All the points described above indicate the importance of the North Pacific, especially its western part, as the sink of anthropogenic CO_2. The Pacific water excluding its marginal sea waters comprises 52% of the world's ocean water, but the information from the Pacific is poor as compared to that from the Atlantic. This may induce the overlooking or wrong estimation of processes occurring in the North Pacific. I am convinced that the role of the North Pacific in the future global warming due to CO_2 is more significant than that accepted by the present modelers.

Acknowledgments—I am extremely grateful to Drs. S. Watanabe, K. Kawabata, T. Ono, K. Okuda and T. Sato for providing me with the unpublished data. I am also indebted to Drs. S. Noriki, K. Harada, N. Tanaka, H. Narita and my staff and students for the valuable discussions with them. Finally, I would like to express my hearty thanks to Prof. N. Handa. I have known him since the 1960's, but our intimate relation began during the *Hakuho Maru* KH-70-2 cruise in 1970 when we promised to make new chemical oceanography in Japan. He also gave me support for my first sediment trap experiment carried out during the KH-78-3 cruise in the northern North Pacific in 1978.

REFERENCES

Asher, W. E., L. M. Karle, B. J. Higgins, P. J. Farley, E. C. Monahan and I. S. Leifer (1996): The influence of bubble plumes on air-seawater gas transfer velocities. *J. Geophys. Res.*, **101**, 12027–12041.

Chen, C. T. A. (1982): Oceanic penetration of excess CO_2 in a section between Alaska and Hawaii. *Geophys. Res. Lett.*, **9**, 117–119.

Duce, R., P. Liss, J. Merrill, E. Atlas, P. Buat-Menard, B. Hicks, J. Miller, J. Prospero, R. Arimoto, T. Church, E. Ellis, J. Galloway, L. Hansen, T. Jickells, A. Knap, K. Reinhardt, B. Schneider, A. Soudine, T. Tokos, S. Tsunogai, R. Wollast and M. Zhou (1991): The atmospheric input of trace species to the world ocean. *Global Biogeochem. Cycles*, **5**, 193–259.

IPCC (The International Panel on Climate Change) (1996): *Climate Change 1995, The Science of Climate Change*. ed. by J. T. Houghton, L. G. Meira Filho, B. A. Callander, N. Harris, A. Kattenberg and K. Maskell, Cambridge Univ. Press, Cambridge, UK, 572 pp.

Keeling, R. F. (1993): On the role of large bubbles in air-sea gas exchange and supersaturation in the ocean. *J. Mar. Res.*, **51**, 237–271.

Kusakabe, M., T.-L. Ku, K. Harada, K. Taguchi and S. Tsunogai (1988): Rapid transport of Chernobyl radioactivities into the North Pacific. *Geophys. Res. Lett.*, **15**, 44–47.

Liss, P. S. and L. Merlivat (1986): Air-sea gas exchange rates: introduction and synthesis. p. 113–127. In *The Role of Air-Sea Exchange in Geochemical Cycling*, ed. by P. Buat-Menard, D. Reidel, Dordrecht, The Netherlands.

Loeman, M. R., M. A. O'Dor and M. G. Skafel (1996): Bubbles entrained by mechanically generated breaking waves. *J. Geophys. Res.*, **101**, 20759–20769.

Murphy, P. P. R. A. Feely, R. H. Gammon, D. E. Harrison, K. C. Kelly and L. S. Waterman (1991): Assessment of the air-sea exchange of CO_2 in the South Pacific during austral autumn. *J. Geophys. Res.*, **96**, 20455–20465.

Noriki, S. and S. Tsunogai (1986): Particulate fluxes and major components of settling particles from sediment trap experiments in the Pacific Ocean. *Deep-Sea Res.*, **33**, 903–912.

Ohtani, K. (1991): To confirm again the characteristics of the Oyashio. *Bull. Hokkaido Natl. Fish. Inst.*, No. 55, 1–25.

Okuda, K., S. Tsunogai, S. Watanabe and M. Kusakabe (1995): Anthropogenic CO_2 dissolved in the western North Pacific estimated from $d^{13}C$ of dissolved inorganic carbon. pp. 400–405. In *Global Fluxes of Carbon and Its Related Substances in the Coastal Sea-Ocean-Atmosphere System*, ed. by S. Tsunogai *et al.*, M & J International, Yokohama.

Ono, T., S. Watanabe, T. Fukazawa and S. Tsunogai (1996): An improved calculation scheme of excess total carbonate dissolved in the North Pacific. *Caribbean J. Science*, **32**, 313–315.

Redfield, A. C., B. H. Ketchum and F. A. Richards (1963): The influence of organisms on the composition of seawater. pp. 26–77. In *The Sea*, Vol. 2.ed. by M. N. Hill, John Wiley, New York.

Saito, C., S. Noriki and S. Tsunogai (1992): Particulate flux of Al, a component of land origin, in the western North Pacific. *Deep-Sea Res.*, **39**, 1315–1327.

Suzuki, T. and S. Tsunogai (1988): Origin of calcium in aerosols over the western North Pacific. *J. Atmos. Chem.*, **6**, 363–374.

Tally, L. and Y. Nagata (1991): Oyashio and mixed water regions as a formation area of the North Pacific Intermediate Water. *Umi to Sora*, **4**, 65–74.

Tans, P. P., I. Y. Fung and T. Takahashi (1990): Observational constraints on the global atmospheric carbon dioxide budget. *Science*, **247**, 14321–1438.

Tokieda, T., S. Watanabe and S. Tsunogai (1996): Chlorofluorocarbons in the western North Pacific in 1993 and formation of North Pacific Intermediate Water. *J. Oceanogr.*, **52**, 475–490.

Tsunogai, S. (1972): An estimate of the rate of decomposition of organic matter in the deep water of the Pacific. pp. 517–533. In *Biological Oceanography of the Northern North Pacific Ocean*, dedicated to Dr. S. Motoda, ed. by A. Y. Takenouti (in chief), Idemitsu Shoten, Tokyo.

Tsunogai, S. (1979): Dissolved silica as the primary factor determining the composition of phytoplankton classes in the ocean. *Bull. Faculty Fisheries Hokkaido Univ.*, **30**, 314–322.

Tsunogai, S. (1980): Quantitative analysis of biological activity on the deep sea floor. *La mer* (*Bull. Soc. franco-japonaise d'oceanogr.*), **18**, 217–226.

Tsunogai, S. (1981): A method for chronology of the Pacific and Atlantic deep water and its application. *Chikyukagaku* (*Geochemistry*), **15**, 70–76.

Tsunogai, S. (1987a): Deep water circulation in the North Pacific deduced from Si-O diagrams. J. *Oceanogr. Soc. Japan*, **43**, 77–87.

Tsunogai, S. (1987b): "Train-passengers model" as an oceanic removal mechanism of chemical elements in seawater. *Chikyukagaku* (*Geochemistry*), **21**, 75–82.

Tsunogai, S. (1991): Contribution of Kosa to oceanic biological production. pp. 233–239. In *Kosa; Science of the Atmosphere and Hydrosphere*, ed. by Water Res. Inst., Nagoya Univ., Kokin Shoin, Tokyo.

Tsunogai, S. and T. Kondo (1982): Sporadic transport and deposition of continental aerosols to the

Pacific Ocean. *J. Geophys. Res.*, **87**, 8870–8874.
Tsunogai, S. and M. Minagawa (1978): Settling model for the removal of insoluble chemical elements in seawater. *Geochem. J.*, **12**, 47–56.
Tsunogai, S. and S. Noriki (1983): *Marine Chemistry - To Solve the Problems Occurring in the Sea with Chemistry*, Sangyo Tosho, Tokyo, 286 pp.
Tsunogai, S. and S. Noriki (1987): Organic matter fluxes and the sites of oxygen consumption in deep water. *Deep-Sea Res.*, **34**, 755–767.
Tsunogai, S. and S. Noriki (1991): Particulate fluxes of carbonate and organic carbon in the ocean: Is the marine biological activity working as a sink of the atmospheric carbon? *Tellus*, **43B**, 256–266.
Tsunogai, S. and N. Tanaka (1980): Flux of oxygen across the air-sea interface as determined by the analysis of dissolved components in sea water. *Geochem. J.*, **14**, 227–234.
Tsunogai, S. and Y. Watanabe (1981): Ca in the North Pacific water and the effect of organic matter on the Ca-alkalinity relation. *Geochem. J.*, **15**, 95–107.
Tsunogai, S. and Y. Watanabe (1983): Role of dissolved silicate in the occurrence of phytoplankton bloom. *J. Oceanogr. Soc. Japan*, **39**, 231–239.
Tsunogai, S., H. Yamahata, S. Kudo and O. Saito (1973): Calcium in the Pacific Ocean. *Deep-Sea Res.*, **20**,, 717–726.
Tsunogai, S., Y. Nozaki and M. Minagawa (1974): Behavior of heavy metals and particulate matters in seawater expected from that of radioactive nuclides. *J. Oceanogr. Soc. Japan*, **30**, 251–259.
Tsunogai, S., M. Kusakabe, H. Iizumi, I. Koike and A. Hattori (1979): Hydrographic features of the deep water of the Bering Sea -'The Sea of Silica'. *Deep-Sea Res.*, **26**, 641–659.
Tsunogai, S., M. Uematsu, N. Tanaka, K. Harada, E. Tanoue and N. Handa (1980): A sediment trap experiment in Funka Bay, Japan:"Upward flux" of particulate matter in seawater. *Mar. Chem.*, **9**, 321–334.
Tsunogai, S., M. Uematsu, S. Noriki, N. Tanaka and M. Yamada (1982): Sediment trap experiment in the northern North Pacific: Undulation of settling particles. *Geochem. J.*, **16**, 129–147.
Tsunogai, S., T. Suzuki, T. Kurata and M. Uematsu (1985): Seasonal and areal variation of continental aerosol in the surface air over the western North Pacific region. *J. Oceanogr. Soc. Japan*, **41**, 427–434.
Tsunogai, S., S. Noriki, K. Harada, T. Kurosaki, Y. Watanabe and M. Maeda (1986): Large but variable particulate flux in the Antarctic Ocean and its significance on the chemistry of the Antarctic water. *J. Oceanogr. Soc. Japan*, **42**, 83–90.
Tsunogai, S., Noriki, K. Harada and K. Tate (1990): Vertical change index for the particulate transport of chemical and isotopic components in the ocean. *Geochem. J.*, **24**, 229–243.
Tsunogai, S., T. Ono and S. Watanabe (1993a): Increase in the total carbonate in the western North Pacific water and a hypothesis on the missing sink of anthropogenic carbon. *J. Oceanogr.*, **49**, 305–315.
Tsunogai, S., Y. W. Watanabe, K. Harada, S. Watanabe, S. Saito and M. Nakajima (1993b): Dynamics of the Japan Sea deep water studied with chemical and radiochemical tracers. pp. 105–119. In *Deep Ocean Circulation, Physics and Chemical Aspects*, Elsevier Oceanogr. Ser., ed. by T. Teramoto, Elsevier Sci. Publ., Amsterdam.
Tsunogai, S., H. Kawabata and S. Watanabe (1993c): Directly observed gas transfer velocities at the ocean surface. pp. 139–141. In *Circulation and Control of Man-made Substances in Environment (1990–992)* ed. by N. Soga,"Man-Env. System", Res. Rep., CT083-N10B, Monbusho Kakenhi, Tokyo.
Tsunogai, S., S. Watanabe, M. Honda and T. Aramaki (1995): North Pacific Intermediate water studied chiefly with radiocarbon. *J. Oceanogr.*, **51**, 519–536.
Tsunogai, S., S. Watanabe, J. Nakamura, T. Ono and T. Sato (1997): A preliminary study of carbon system in the East China Sea. *J. Oceanogr.*, **53**, 9–17.
Watanabe, Y. W., S. Watanabe and S. Tsunogai (1991): Tritium in the northwestern North Pacific. *J. Oceanogr. Soc. Japan*, **47**, 80–93.

Ocean Sciences Research (OSR)

Vol. 1 Coastal Oceanography, Tetsuo YANAGI, 1999
ISBN 0-7923-5895-3

TERRAPUB, Tokyo / KLUWER, Dordrecht, London, Boston

GPSR Compliance

The European Union's (EU) General Product Safety Regulation (GPSR) is a set of rules that requires consumer products to be safe and our obligations to ensure this.

If you have any concerns about our products, you can contact us on ProductSafety@springernature.com

In case Publisher is established outside the EU, the EU authorized representative is:

Springer Nature Customer Service Center GmbH
Europaplatz 3
69115 Heidelberg, Germany

Batch number: 10809436

Printed by Printforce, the Netherlands